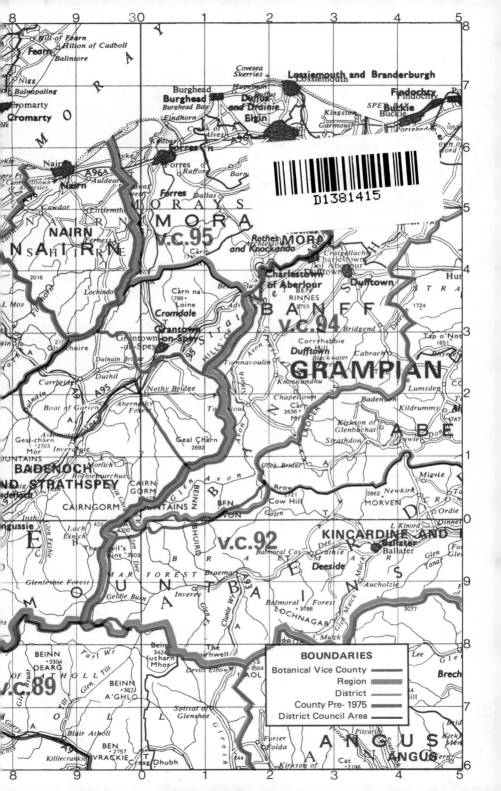

FLORA OF
MORAY, NAIRN &
EAST INVERNESS

Parnassia palustris   *Grass of Parnassus*

T. G. Collett

FLORA
OF
MORAY, NAIRN
&
EAST INVERNESS

Botanical vice-counties 95 Elgin and 96 Easterness

by

MARY McCALLUM WEBSTER

ILLUSTRATED
by
Olga M. Stewart
Edith M. Legge and Ronald Melville

Aberdeen University Press

First published 1978

© Mary McCallum Webster 1978

Printed in Great Britain
at Aberdeen University Press

ISBN 0 900015 42 X

CONTENTS

ILLUSTRATIONS

Photographs by the author unless otherwise stated.

Colour Plates

Black and White Plates

Note. The line drawings, originally life size, have been reduced by approximately 20 per cent. Scales where marked apply to the original size.

FOREWORD

I am pleased to be able to write a short note at the beginning of this new Flora. More than fifty years careful and devoted field study underlie this work. Mary McCallum Webster has examined personally practically every 10 km square in the vice-counties covered by the Flora with the exception of two very remote parts of Inverness, and anyone who has been involved in recording in the Highlands will realise what a remarkable achievement this is. The author has, in addition carried out much painstaking research in the relevant literature and Herbaria. The product is a really valuable Flora, in the best classical tradition, but with modern nomenclature and geared to the recently-developed system of grid mapping. The book represents a very satisfactory achievement of a life's work by one whose heart was always, and remains, 'in the Highlands'. There are really few better places for one's heart to be!

S. M. Walters
May 1977

PREFACE

At the end of the eighteenth century short lists of the plants of Moray, mostly trees, appeared on a parish basis in the *Statistical Accounts for Scotland* edited by Sir John Sinclair, followed in 1839 by Dr George Gordon's *Collectanea for a Flora of Moray*. Not until nearly a hundred years later was a complete *Flora of Moray* published by the Moray Field Club edited by James J. Burgess.

No flora has been published for Nairn or Inverness, but fairly comprehensive lists of local plants were compiled by Dr Thomas Aitken in *Transactions of the Inverness Scientific Society and Field Club* in 1875, by the Rev. Thomas Fraser of Croy in 1881 and by Dr Robert Thomson in *The Natural History of a Highland Parish* (*Ardclach*) published by the *Nairnshire Telegraph* in 1900.

This present work is the result of over fifty years personal observations in the area and includes much research in herbaria and literature. I have visited every 10 km square on the national grid except the remote areas of Inverness in grids 27/46; 27/56 and 27/77. Grids 27/77 and 38/23 have been excluded from the distribution maps owing to the very small areas they cover.

Unfortunately space and time has not permitted the inclusion of *Charophyta* or *Bryophyta* and much other useful information has of necessity been excluded.

The area covered is that of the Watsonian vice-counties of 95 Elgin (Moray) and 96 Easterness (East Inverness and Nairn) and in general follows the arrangement of most County Floras.

It is not a text book, but includes some useful *keys* to critical genera as well as many *pocket descriptions* of other species. National grid numbers have been excluded but a list of less well-known place names with 1 km grid references has been added in Appendix 2.

Not every plant has been illustrated but those that are somewhat similar in general appearance and difficult to describe are included. The distribution maps are also of selected species.

It would be impossible to thank individually everyone who gave assistance in the preparation of this work, but a full list of contributors is added in Appendix 1.

To the directors and staff of the following institutions I am grateful for permission to visit, on many occasions, the herbaria and libraries at the Universities of Aberdeen, Cambridge, Glasgow, Oxford, St Andrews and Stirling (the temporary home of the herbarium of the Smith Institute at Stirling). Also to the staff at the British Museum (Natural History), the Royal Botanic Gardens at Edinburgh and Kew, and the Curators of the museums at Forres, Elgin, Inverness, Nairn Literary Society at Nairn and Perth.

My thanks are also due to the many referees of critical genera in the Botanical Society of the British Isles who have examined my specimens and manuscript, especially to Dr J. Cullen (Anthyllis), J. E. Dandy (Potamogeton), E. S. Edees (Rubus), Dr R. M.

Harley (Mentha), Dr C. E. Hubbard (Gramineae), R. D. Meikle
(Salix), Dr R. Melville (Rosa), Dr A. J. Richards (Taraxacum),
P. D. Sell and Dr C. West (Hieracium and Pilosella), Dr S. M.
Walters (Alchemilla, Montia and Eleocharis) and Dr P. F. Yeo
(Euphrasia).

My gratitude is also given to Dr J. H. Gauld of the Macaulay
Institute for Soil Research for his account of the geology and soil,
a subject well outside my capability.

I would also like to thank the many landowners and the Forestry
Commission for access to their properties so readily given over many
years.

Most of the black and white figures have been drawn by Olga M.
Stewart with the exception of the Equisetaceae drawn by Edith M.
Legge and Rosa drawn by Ronald Melville. The silhouette compara-
tives have been photographed by kind permission of the Regius
Keeper of the Royal Botanic Gardens at Edinburgh. In addition
Dr M. C. F. Proctor has supplied some excellent black and white
photographs and T. G. Collett, Mary Collett, Stanley Heyward,
Alan Silverside and John Whitcombe some of the colour plates.
Joan Clark has cheerfully re-typed most of the manuscript, patiently
correcting grammatical mistakes and the spelling of Gaelic place
names. To all these friends I am more than grateful.

I am also greatly indebted to the Director of the Macaulay
Institute for Soil Research, to the Ordnance Survey for permission
to reproduce part of one of their maps and the Ray Society for
permission to print in full the intricate vice-county boundaries for
Elgin and Easterness. I would also like to thank the staff of the
Moray District Libraries, George Dixon for research on James
Straith and Isobel Rae for assistance in extracting information from
the Nairn Literary Society records and to Dr Ralph Kirkwood for
the list of specimens held in the herbarium at Strathclyde University.

With the rapidly rising cost of printing, the publication of this
flora would not have been possible without the considerable financial
support of two friends, Anne Berens and Anna Younger to whom
my thanks can hardly be adequate. Neither would it have been
possible without the continued encouragement of Peter Sell, who
not only undertook, with Cyril West, the mammoth task of writing
a key to Hieracia, but has corrected the manuscript, read the
proofs and given me much advice over the years as well as spurring
me on to set down on paper the results of many years work.

DISCOVERY OF THE FLORA

MORAY (Elgin)

There is little, or no evidence of the study of Botany in Moray until the end of the eighteenth century, although arboriculture was well advanced.

In *The History of the Province of Moray* by L. Shaw (1775) mention is made of the cultivation of flax, hemp and mustard and of the abundance of hazel, service trees (rowan), sloes and brambles to be found.

Later (1792–), in *The Statistical Accounts for Scotland* short lists of wild plants, consisting mainly of trees, were published on a parish basis, usually written by the Parish Minister or the Schoolmaster. Natural woods of oak, birch and allar (alder) were common and large enclosures of firs, ash and sycamore were planted in the policies of the larger houses, especially by the Earl of Fife, the Earl of Moray and Alexander Brodie of Brodie.

In the parish of Bellie the Rev. James Gordon observed that the only rare plant in the district was *Satyrium repens* (*Goodyera repens*) growing in plenty a mile from Fochabers.

In Birnie the Rev. Joseph Anderson records 'some oak, birch and hazel, ash and plane trees (sycamore) and that some large trunks of oak and fir are dug out of the mosses (lowland bogs). Shrubs of broom, furze, juniper, sloes, hips and brambles are innumerable, but the white water lily is the only herb peculiar to the parish and grows at Gedloch'.

In Cromdale the Rev. Lewis Grant mentions the cultivation of flax and in Dallas the Rev. David Milne remarks on 'the natural oak, birch and allar, geans and some large enclosures of planted firs and that a little flax is grown and a few fields laid down to pasture'.

In Duffus an 'anonymous friend' notes plantations of fir and in the west of the parish elms, ash and oaks and that by the sea at Spynie was a great deal of waste ground covered with a grass called 'bent' only fit for sheep-pasture.

In the parish of Duthil the Rev. Patrick Grant noted that 'wild oats appeared spontaneously from a piece of ground that had been ploughed but not sown'. He also remarks on the famines of 1680 when the poorer people frequented churchyards to pull 'a mess of nettles which they greedily fed upon, boiled without meal'.

The parish of Dyke and Moy was, according to the Rev. John Dunbar 'the best provided with a variety of native and planted trees, including beech'. Ships carrying coal and lime from Newcastle were reloaded with wood slabs of firs for use as pit-props. Credit for the extensive planting of trees must go to two people, Mary Sleigh wife of Alexander Brodie of Brodie and to the Earl of Moray.

For Edenkillie the Rev. John Macdonald wrote 'the natural

woods in this parish are very extensive. The banks of the River Findhorn are in general covered with trees. Along the west bank is the ancient forest of Darnaway consisting of oaks, ash, elm, birch, allars, holly, geans and aspen'. In Elgin the Rev. John Grant makes the first reference to the fact that 'the sowing of clover and rye-grass was gaining ground'. In Urquhart the Rev. William Gordon notes the extensive planting of trees and hedges by the Earl of Fife.

Later, in the *New Statistical Accounts*, published at intervals, the lists of wild plants become longer, but perhaps not always reliable.

The earliest plant list for Moray is dated 1794 and was compiled by an untraced collector and was made at Innes House, Urquhart. This list of 296 plants is in the pocket of a small leather-bound book at Elgin Museum. The book contains a small collection of pressed specimens and came into the possession of Alexander Cooper of Urquhart and is cited as Coopers collection in the text of this flora. Like many old collections there are no data slips on the specimen sheets, some of which are of garden origin and others not known to occur in Moray, so the validity of the records must be treated with some doubt.

Towards the end of the century James Brodie of Brodie (1744–1824) began to collect plants. Many were recorded from the vicinity of Brodie Castle, but the majority were from the environments of Edinburgh. His herbarium is now lodged at the Royal Botanic Gardens at Edinburgh with a few duplicate sheets at the British Museum, at Kew and elsewhere. He came in contact with many noted botanists of his day such as W. Borrer and Sir William Jackson Hooker, the latter was at that time Regius Professor of Botany at the University of Glasgow, later becoming Director of the Royal Botanic Gardens at Kew. He visited Brodie on several occasions in search of plants to illustrate *Flora Londonensis*, collecting several of the wintergreens during his visits. There are letters from Hooker to Brodie thanking him for his hospitality, in the library at Brodie Castle.

At this time James Barlow Hoy (17——1843) secretary and librarian to the Duke of Gordon at Gordon Castle, Fochabers was also interested in wild plants and had the honour in 1792 of being the first to discover *Moneses uniflora* in Moray. Brodie however, found the plant in the same year in the west of the county.

James Straith (1765–1815) son of John Straith schoolmaster at Kingussie, and later at Nairn (where James spent his boyhood) was a physician in Forres for twenty years and made a small collection of plants, but according to Gordon it was of little value to the botanist. His collection may be one of the old collections, unnamed and undated that have recently (1977) been discovered in the Falconer Museum at Forres.

Early in the nineteenth century George Gordon (1801–93) became prominent not only as a botanist, but as an eminent geologist and naturalist of great repute. His father William Gordon was minister at Glenlivet and afterwards at Urquhart and his mother was the daughter of Joseph Anderson, minister at Birnie. He entered Aberdeen University at the early age of fourteen, graduated in 1819 and was immediately appointed to the charge of Birnie where he remained for fifty-seven years. His herbarium, if he had one, cannot be traced. There are a few specimens in the herbarium at Edinburgh and in the university herbaria at Aberdeen and Glasgow.

In 1839 he published the first printed account of the flora of the Province of Moray in *Collectanea for a Flora of Moray*. The Province of Moray was a great deal larger than the political county of Moray as it stood up until 1974. It was the portion of Scotland drained on the east by the River Spey and its tributaries and on the west by the River Beauly, bounded on the north by the Moray Firth and on the south by a line running from Loch Spey to Loch Monar, thus taking in most of political Moray, a small area of Banffshire, the entire county of Nairn and a large portion of the county of Inverness as these places were known prior to the re-defining of Regions and Districts in 1975.

Gordon had many helpers, among them was William Brand (1807–69) most of whose specimens are at Aberdeen. The Rev. George Wilson, schoolmaster at Alves and the Rev. James Fraser (also from Alves) who later became minister at Grantown-on-Spey made many additions to the county flora and have specimens in the Innes herbarium at Forres.

Dr John G. Innes (1815–81) has never received the praise due to him for his study of the plants in the Forres district. His ancestors were well-known merchants in Forres and he was born in Ceylon where his father was serving with his regiment. On returning to this country he became apprenticed to Dr Adam of Forres and entered the medical profession. He later went to Edinburgh University and then to America, but soon returned to his native town where he practised until his death. His herbarium, only recently (1977) discovered in the Forres Museum is a large one, chiefly of local plants, well annotated and in good condition.

Mention must be made here of Miss Robertson who, with her brother, ran an academy for young gentlemen at Thornhill House, Forres. It was here that the brilliant young Shetland botanist Thomas Edmondston, while studying medicine at Edinburgh University, took a tutoring post during the summer vacation in 1844. It appears that Miss Robertson was also interested in botany and had shown Edmondston a specimen of *Rubus arcticus* that she had found in the Cairngorms and given to Grigor nurseryman at Forres to grow. This information came to me through the Shetland botanist Walter Scott who showed me Edmondston's letter confirming Miss Robertson's claim (for details see under *Rubus arcticus*). Edmondston himself stimulated a great deal of enthusiasm for botany in the county and was the founder of the Moray Field Club in 1844. Unfortunately the transaction of this club cannot be traced, all that is now known is that the club went into abeyance during World War

I and was revived in 1971 under the name of The Moravian Field Club.

James Grant (–1899), schoolmaster and later Provost of Lossiemouth, a lover of music, botany and geology made his mark at this time. He was not a collector of great repute but rather stimulated others to enjoy nature in all its aspects.

From the middle of the nineteenth century onwards the enthusiasm for Natural History continued and one of the outstanding figures was the Rev. James Keith (1825–1905) parish minister at Forres. Born in Keith he went to Kings College, Old Aberdeen and then became schoolmaster at Knockando until he entered the ministry of the Church of Scotland. His first appointment was to Grantown-on-Spey but he only remained there for one year and then went to Forres. He was not a robust man and his medical advisor Dr John G. Innes advised him 'that a vasculum was best suited to his needs and to use it frequently'. This advice he most certainly took to heart. Apart from his interest in botany he became a mycologist of note and two fungi were named in his honour, *Polyporus keithii* and *Peziza keithii*. His herbarium of about 1,500 specimens is in the Falconer Museum at Forres. In 1976 his collection of fungi was transferred to the Royal Botanic Gardens in Edinburgh for safe keeping.

Towards the end of the century several botanists of note visited Moray and added new records, chiefly of the critical genera. Among these were Dr G. C. Druce, often accompanied by my grandmother Mrs M. L. Wedgwood. The Rev. E. S. Marshall, W. A. Shoolbred, Charles Bailey and W. Moyle Rogers were others who made valuable contributions to the flora.

In the early twentieth century the members of the Field Club continued to record plants on a parish basis, culminating in *The Flora of Moray* edited by James J. Burgess and published just after his death in 1935. Burgess was schoolmaster at Dyke Primary School for thirty-eight years and was ably assisted by Alexander MacGregor, schoolmaster at Forres, by George Birnie, minister at Speymouth for forty years, Peter Leslie, lecturer in Forestry at Aberdeen University and many other members of the club.

In 1911 P. Ewing published a short and incomplete list of plants from the Culbin Sands but cited his specimens which are in the herbarium at Glasgow University as 'between Findhorn and Nairn' thus including two botanical vice-counties. Two years later Dr Donald Patton and E. J. A. Stewart made an extensive exploration of the area and published two important papers in the *Transactions of the Botanical Society of Edinburgh* **26** (1914) and **29** (1924).

My own work in the county commenced in 1920 until World War II (during which time no other botanical activities took place) and continued after the war until the present day. My *Check List of the Flora of the Culbin State Forest* was published in 1968 and reprinted in 1977.

Several Field meetings have been held in the county by the

Botanical Society of the British Isles and by the Wild Flower Society resulting in further additions to the known distribution of plants.

In 1954 and onwards the area has been surveyed on a ten kilometre basis of the national grid in connection with the Distribution Map Scheme that culminated in the publication of *The Atlas of the British Flora* (1962) and *The Critical Atlas* (1968) and with which work I was ably assisted by Alan and Marion Souter of Buckie and Kate Christie of Forres. The survey continues and additions are made yearly which is surprising in such a well-botanised area.

As well as the herbaria already mentioned there are several small collections of pressed specimens in the museum at Elgin. A collection of 106 plants by Dr A. Cruickshank (1836); 18 plants presented by Patrick Duff (1837) and a small collection presented by Jessie Merson (1903) probably the collection of her father Peter Merson, headmaster at Sanquhar School. Jessie was the great-aunt of W. A. Adam of Glassgreen. An unnamed and undated collection of Moray plants is in the possession of Mrs A. Dunbar of Pitgaventy.

NAIRN (Easterness)

The county of Nairn produced no botanists until the nineteenth century. In *The Statistical Accounts* the Rev. Donald Mitchell (1793) refers to firs, weeping birch, alder, hazel, ash, some oaks and much heathland in the parish of Ardclach and the Rev. John Morrison mentions the abundance of alders on the banks of the River Nairn.

William Alexander Stables (1810–90), factor to the Earl of Cawdor will however long be remembered as the most notable botanist in the county. He worked in the company of the Moray and Inverness botanists of his day and added many new records to the flora. He made a good collection of pressed specimens, well annotated, the larger amount to be found at Aberdeen University with some duplicates at Edinburgh and Glasgow.

Sheriff Alexander Falconer (mentioned by Gordon) and the Rev. J. B. Brichan, minister at Auldearn and for a short time teacher of classics at Nairn Academy (1839) were also active along with the Rev. A. Rutherford (fl 1844) who later went as minister to Kingussie.

Towards the end of the century Dr Robert Thomson, schoolmaster at Ardclach published a comprehensive list and made an excellent collection of the local plants. His herbarium is lodged in the library of the Nairn Literary Society.

In 1881 the Rev. Thomas Fraser made a long list of plants from the neighbourhood of Croy, but as Croy lies on the boundary of Nairn and Inverness it must just be guesswork as to which county his records belong.

At the turn of the century Alexander Lobban, headmaster at Nairn Academy took an interest in the local flora, and with a band of helpers Messrs Allan of Cawdor, Penny of Delnies and Swanney of Croy compiled notebooks on a parish basis. These are held with Thomson's collection at Nairn.

EAST INVERNESS (Easterness)

There is very little reference to plants, or even trees in the old *Statistical Accounts* of 1792.

In the parish of Alvie in Strathspey the Rev. John Gordon observes that 'broom will grow and whins will not' (this is still evident today the few whin bushes to be seen are confined to the railway embankment).

At Boleskine and Abertarff an 'anonymous friend' records the 'native woods chiefly of birch, allar and hazel and in the mosses trunks of oak and fir'.

In Inverness the Rev. Robert Rose notes that 'the Hill of Tomnahurich was planted with trees in 1753 and that there were no native woods except a small oak wood, but that there were many plantations. Flax was grown for private use only'.

In Kiltarlity the Rev. John Fraser wrote 'that 1,200 acres were planted in firs and that there were 4,800 acres of natural oak, birch, fir, alder and hazel etc., and flax was grown at Belladrum'.

The Rev. Alexander Fraser of Kirkhill refers to the 'scarcity of food in the parish in the seventeenth century and that four families in Kirkhill subsisted for two years on the herbs they could collect and that they ground the seed of the wild mustard to make into a meal. Flax was cultivated in small quantity'.

At Moy and Dalrossie the Rev. William McBean records 'there is a good deal of natural wood on the banks of the River Findhorn, chiefly of birch and alders. There is a tradition that before the country came into the possession of the Mackintosh of Mackintosh, it was over-run with woods and called the Forest of Strathdearn. The laird later planted firs etc.'.

In the *New Statistical Account* **14** (1845) a long list of plants was published for the parish of Kiltarlity by the Rev. C. Fraser. Many have not been refound and it is doubtful if they were ever found in the parish but they cannot be dismissed arbitrarily. They include *Dryas octapetala, Myosurus minimus, Orchis morio* and *Eriophorum alpinum*.

The Rev. John Lightfoot (1735–1788) passed through Inverness on his journey to the Highlands with Thomas Pennant in 1772, but I can find trace of only one plant in his herbarium at Kew, i.e. *Lithospermum officinale* as having been collected in vice-county 96. His *Flora Scotica* was published in 1778 but seldom refers to Easterness.

James Dickson (1738–1822), nurseryman, was probably the first collector of plants in the county. His herbarium was held at the Linnean Society London and is now at the British Museum (Natural History).

John Fraser (1750–1811) who was a nurseryman at Chelsea, and his son of the same name (1799–1852) both of Tomnacloich Inverness were both interested in plants but do not appear to have studied the local botany.

Alexander Murray (1798–1838) published the first part of *The Northern Flora* in 1836, but his death prevented the second part being published.

George Don (1764–1856) nurseryman in Forfar had two sons both collectors of plants. George (1798–1856) the better known of the two, made a large collection of alpine species now at the British Museum with a duplicate set at Kew, and David (1799–1841) whose collection is at the Linnean Society London. Both sons collected in Easterness.

There were few local botanists at this time. The Rev. James Farquharson (1781–1845) went on many excursions with Thomas Fraser of Croy and the neighbouring botanists Gordon and Stables and he was followed by Dr Hugh Clark (1836–76) a surgeon in the Bengal Army who made, when a boy, a most useful collection of plants near Inverness and these are preserved in the museum at Inverness. The collection contains many first records for the vice-county. There are two more small collections at Inverness, one is unnamed and the other has the initials P.F., both are supposed to have been collected at the end of the century.

H. C. Watson visited Easterness on at least two occasions in 1832 and 1833 and Professors Robert Graham (1786–1845) and John Hutton Balfour (1808–84) made excursions with their pupils mostly to the southern part of the county covering much new ground in the Grampian Mountains. Graham however visited Cawdor on one occasion.

Alexander Croall (1809–85) Curator of the Smith Institute at Stirling made valuable contributions near Fort George. His collection is temporarily housed at Stirling University and contains a large proportion of first records from the lowlands.

John Ball (1818–89) mountaineer, made several new records in Glen Cannich. A nurseryman from Perth, Robert Brown, was the reputed finder of *Phyllodoce caerulea* near Aviemore Strathspey. A specimen dated 1830 with Professor Graham's name as collector is at Aberdeen and the specimen at Edinburgh is ex herb. James Dickson from Professor Graham and dated 1835. This locality has not been refound but there are several hills to the south-west of Aviemore on which the plant could grow, but which are botanically dull and not visited by botanists.

The Rev. Andrew Rutherford, minister at Kingussie made a small collection, some of his specimens are at Edinburgh and towards the end of the century Alexander Somerville a Glasgow botanist, made a large collection of arctic-alpines in the Cairngorms, most of his specimens are also at Edinburgh with duplicates at the British Museum and at Glasgow University. Other well-known botanists to visit the county were William McNab, John and Paul MacGillivray, John G. Macvicar, Frederick Townsend, the brothers

Henry and James Groves, G. C. Druce and many others. At the turn of the century Professor J. W. H. Trail of Aberdeen University did much work in the Cairngorms but little in Easterness.

In the twentieth century the outstanding collection for the vice-county was made by Anthony Johnston, now Director of the Commonwealth Mycological Institute, Kew, in the region surrounding Inverness. His herbarium is held at Aberdeen University and is a very valuable contribution to the lowland flora of the north. Others who deserve mention are M. E. D. Poore, C. D. Pigott, D. N. McVean, D. A. Ratcliffe, Colin Murdoch and R. McBeath.

I have, myself, covered most of the county on a 10 km basis over the last 50 years making many new records, especially in the neglected lowland areas containing weeds of arable land and sea-coast plants etc. sadly neglected by the visitor who is, quite naturally, more interested in the plants of the mountains. In 1976 a survey of roses was undertaken with Dr Ronald Melville of Kew, resulting in the finding of six new undescribed varieties of *Rosa* to be added to the British Flora.

The Inverness Scientific Society and Field Club was inaugurated in 1885 and held periodic meetings but botany seems to have been one of the less well studied subjects. However, Dr Thomas Aitken, Medical Superintendant of the District Asylum at Craig Dunain, published in the *Transactions* a list of plants from around Inverness. Other authors from round Inverness were J. Don (1895) and G. Lang a few years later, none of these lists are of much value. The Field Club is flourishing today as is the Inverness Botany Group, formed in 1955 the first President being Dr Helen Caldwell teacher of Biology at the Academy.

TOPOGRAPHY

THE ADMINISTRATIVE COUNTIES

The botanical vice-counties of Elgin (v.c. 95) and Easterness (v.c. 96) cover an area of approximately 677,670 hectares (1,674,541 acres). This area at its widest point is 130 km (81 miles) and at its longest 90 km (56 miles). It includes eighty-nine 10 km squares, or parts of squares, of the National Grid.

It is bounded on the north by the mountain range dividing Inverness District from Ross and Cromarty District in Highland Region (v.c. 106 East Ross) and by the Moray Firth eastwards to the mouth of the River Spey in Moray District of Grampian Region. On the east it is bounded by a portion of Moray District in Grampian Region (v.c. 94 Banff) and by Kincardine and Deeside District in Tayside Region (v.c. 92 South Aberdeen). On the south it is bounded by Perth and Kinross District in Tayside Region (v.c. 89 East Perth and v.c. 88 Mid Perth) and on the west by Lochaber District (v.c. 97 Westerness) less an enclave in the Ben Alder Forest, and by Skye and Lochalsh District in Highland Region (v.c. 105 West Ross), and the western boundary of Ross and Cromarty District.

THE WATSONIAN VICE-COUNTIES

Much confusion has arisen in the past over the vice-county boundaries of Elgin and Easterness owing to the continual change and redefinition of the political boundaries, and more will now arise with the new arrangement of Districts and Regions that took place in 1975. For sequence of records the Watsonian vice-county boundaries must be adhered to. Even G. C. Druce and E. S. Marshall recorded plants as being in Easterness when they should have been placed in Elgin. These errors were probably due to the map in Druce's *Comital Flora* erring in only showing the boundaries of the then, modern County of Moray.

With the kind permission of The Ray Society the following information has been extracted from J. E. Dandy's *Watsonian Vice-Counties of Great Britain*, 32–33 (1969).

V.c. 95 **Elgin** Morayshire (Elginshire) Includes part of the county of Inverness. On the basic map published by the Society for the Diffusion of Useful Knowledge (1844) the county of Elginshire was shown with its old boundaries, by which a large detached area was separated from the main body of the county by an intervening enclave of Inverness-shire. Watson laid down 'that this enclave of Inverness-shire which cuts Elginshire into two disjointed parts, is deemed to be a portion of the latter county'; so that v.c. 95 consisted of the whole of Elginshire (as then constituted) together with the intervening portion of Inverness-shire. The administrative position of the two detached areas has since been reversed, the enclave of Inverness-shire being now united with Morayshire and the enclave of Elginshire with Inverness-shire. Thus v.c. 95 comprises the modern county of Morayshire (with some adjustments on the eastern boundary*) together with the adjacent portion of Inverness-shire which was formerly an enclave of Elginshire and which consists of the north-western part of Duthil & Rothiemurchus parish and the north-eastern part of Abernethy & Kincardine parish. The western boundary of this former enclave from Nairnshire to the River Spey, coincides with the north-western boundary of Duthil & Rothiemurchus parish, starting from the Nairnshire boundary at Càrn Glas-choire and passing over Càrn Loigste, Càrn nam Bain-tighearn, Càrn na Lair, Càrn Phris Mhoir, Càrn Coire na Caorach, Sgùman Mòr, Creag Shoilleir, Geal-chàrn Beag and Càrn Dearg Mòr to the River Spey at Kinakyle, opposite Corrour. From that point eastwards to Banffshire the boundary is not marked on modern maps, but it has been worked out and conveniently defined by W. D. Roebuck in *The Scottish Naturalist*, 80–81 (1917) as follows: River Spey downwards (from Kinakyle) to the railway bridge (by Tomachrochar) and thence the railway and then the road (B970) east-south-east to the Duack Burn (at Birchfield); Duack Burn up to the ford where a smaller burn comes in on its right bank; thence a line running up the lower part of this smaller burn and through the Abernethy Forest over the height marked 1,066 feet to the bridge over the River Nethy at Forest Lodge (above Lyngarrie), and then over Càrn a' Chnuic, Càrn Bheadhair and Càrn Tarsuinn to the Banffshire boundary at the nearest bend of the Water of Caiplich.

V.c. 96 **Easterness** Inverness-shire north-east of 'the line of watershed between the western and eastern sides of Scotland, continued along Loch Ericht to the borders of Perthshire', including Nairnshire but excluding the area (in Duthil and Rothiemurchus and Abernethy and Kincardine parishes) which was formerly an

*Adjustments on the eastern boundary
(1) The boundary crosses to the east side of the River Spey at Ballindalloch Castle and a small enclave of Banffshire is included (this was formerly part of the Province of Moray) it then returns to the course of the river downstream to Boat of Brig.
(2) The boundary again crosses to the east of the Spey and from Delfur follows the Burn of Mulben for one mile, turns south up the hill of Knock More and about one mile towards Ben Aigan to the source of the Allt Daley, then turns south-east for approximately one mile and then north-east to Mulben and on eastwards to the junction of two small burns by the road junction of the A96 with the B9103 at Fife Keith. Thence north by the A96 to Fochabers, through the policies and the loch at Gordon Castle and crossing to the west bank of the Spey two miles north of Fochabers and then following the river to its mouth at Garmouth and thus excluding the eastern portion of political Moray that is, according to Watson, part of v.c. 94 Banff.

enclave of Elginshire. The line of watershed between east and west Scotland has to cross the top level of the Caledonian Canal (i.e. the stretch between Laggan Locks and Cullochy Lock including Loch Oich) and the most convenient point for crossing is Bridge of Oich. Thus the full line of watershed separating the vice-county from the next runs as follows: from the Ross and Cromarty boundary in Loch Loyne over Mullach Coire Ardachaidh, Càrn Bàn, Clach Criche, Càrn Tarsuinn, Meall Dubh, Ceann a' Mhaim, Meall Mor, Bridge of Oich, Druim Laragan, Mullach a' Ghlinne, Eilrig, Glas Chàrn, Càrn Dearg (north), Poll-gormack Hill Càrn Leac, Creag a' Chail, Creag a' Bhanain, A' Bhuidheanach, Meall Ghoirleig, Creag Ruadh, Beinn Eilde, Meall na Sguabaich, Meall Leac na Sguabaich, Meall Beag, Meall Mòr, Beinn Bheòil, Sròn Coire na h-Iolaire and Ben Alder then following the Perthshire county boundary to Càrn an Fhidhleir in Aberdeenshire and north to the Banffshire county boundary following the Cairn Gorm summit to Bynack More and joining the Elgin boundary at the Water of Caiplich and the small burn of Caochan Bheithe near Càrn Tarsuinn.

LOWLANDS

This is name given in the Flora to the area reaching from the coast line to the '1,000 foot Peneplain' that runs in the general direction of Orton on the River Spey – by Blackhills, Pluscarden, Relugas and Cawdor – to Inverness and from thence to the Falls of Kilmorack on the River Beauly.

The coast runs from the mouth of the River Spey westwards to the Beauly Firth. It consists mainly of shingle beaches, sandy bays and esturine mud. Small areas of sandstone rock occur at Hopeman and Covesea and at the latter there is a cliff of considerable height. Raised beaches of varying heights run from Burghead to Alturlie Point and between Findhorn and Nairn lie the dune area of the Culbin Forest. Salt-marshes occur in the estuaries of the River Findhorn and the River Beauly and to a lesser degree at Garmouth, between the Culbin Forest and the Old Bar, and at Castle Stuart.

There are four shallow lochs of note, Loch Spynie, Cran Loch, Lochloy and Loch Flemington.

Inland from the coast lies the tract of land known as the Laich or Howe of Moray and The Aird on the west side of Inverness. This consists of good agricultural land, the chief crops grown are barley, potatoes, turnip and raspberries, with some oats, wheat, swedes, carrots and oil-seed rape. Very occasionally rye is grown.

HIGHLANDS

This term is applied to the area running south of the '1,000 foot Peneplain' to the Pass of Drumochter and from the Hills of Cromdale in the east to the Monar and Affric Mountains in the west. It consists of moorland, undulating hills, mountain ranges, acid lochs and lochans, river valleys, native and planted woods. The land is cultivated in the lower regions of the river straths.

Mountains

The area is bounded by mountain ranges on the south-east, the south, the west and part of the north. Between the two large massifs of the Cairngorm Range and the Glen Affric mountains in the Forest of Kilmorack, lie several lateral ranges of inferior height forming boundaries of watersheds between the river straths. The most notable of these, the Monadhliath Mountains, forms a plateau with a predominance of blanket bog at a height between 610–915 m (2,000–3,000 feet). The Cairngorms are the highest range reaching some 1,286 m (4,248 feet) on Braeriach. In the west the highest peak is Carn Eige in Glen Affric reaching 1,176 m (3,880 feet) and in the north Sgùrr na Lapaich in Glen Strathfarrar and several of the Monar mountains reach heights of over 915 m (3,000 feet). The Hills of Cromdale on the east of the area are a low range, Creggan a' Chaise Cairn reaching 717 m (2,367 feet).

Moorland

Some 80 per cent of the district consists of undulating acid moorland moulded by the retreating ice and water action during the last stages of the ice age. Peat bogs and mires are innumerable as is dry moorland on the lower hills and on moranic deposits in the river straths.

Rivers

There are five major rivers. They flow in a west-easterly direction in their upper reaches and then in a northerly direction with outlets in the Moray Firth. All have many tributaries, some of considerable size.

The River Spey (between 90 and 100 miles in length), the Findhorn (80 miles), the Lossie (30 miles), the Nairn (35 miles) and the Glass-Beauly (measuring 20 miles under that name but has two sources, the River Affric and the Abhain Deabhag, the latter of considerable length. All are fast flowing in their upper reaches becoming slower lower down especially the River Spey where it flows through the Insh marshes between Kingussie and Kincraig. The River Findhorn is noted for its ravines and gorges in its lower reaches.

Other rivers of note are the River Ness a short river with outlet to the Beauly Firth, and the rivers Tarff, Moriston and Enrick all of which run into Loch Ness. The enormous volume of water contained by these rivers and the movements of the tides forces the deposits brought down from the mountains to form shingle bars and gradually the mouths of these rivers move to the east.

Lochs

The largest loch in the district is Loch Ness, 27 miles long and

approximatly 2 miles wide for most of this distance. It reaches a
depth of 200 m (700 feet) and has the reputation of never giving up
its dead. It is also famous for being the home of the 'monster'.
Other lochs of note are Alvie Ashie, Cluanie, Ericht, Duntelchaig,
Mhor, Meikle and Morlich. Smaller lochs and lochans are too
numerous to count. Some are shallow and stone bottomed, others
deep, and the abundant moorland lochans usually steep sided,
peaty and with little surrounding vegetation. The numerous
moorland pools surrounded with sphagnum and of unknown depth
are known locally as 'murder holes'.

Forests

The counties of Moray and Nairn have long been known as the
most afforested counties in Scotland. There are native woods in
the lower reaches of the river valleys and considerable plantations
of deciduous woods at Altyre, Darnaway and Cawdor. There are
many large coniferous plantations throughout both vice-counties
and native woods of birch, oak and pine the lineal descendants of
the old Caledonian Forest at Abernethy, Rothiemurchus, Glen
Affric and Glen Strathfarrer in the highland district.

In the *Third Statistical Account for Scotland* John McGarva reckoned
that there were 82,500 acres (approx. 33,000 hectares) of plantations
in the joint counties of Moray and Nairn, or 20 per cent of the surface
area. Since then little has been lost to building, road improvements
or agriculture but something over 10,000 acres (approx. 4,000
hectares) has been planted on the moors of Knockando and Dallas.
In Inverness District the plantations are now some 67,000 acres
(approx. 26,400 hectares) and not likely to increase substantially
as the moors are retained for red deer stalking.

Population and Occupation of the People

The following figures are taken from the 1974 census. The population is concentrated in the towns, the largest are Inverness (population 36,595), Elgin 7,589, Nairn 5,890, Lossiemouth 5,903, Forres
4,169, Rothes 1,282, Grantown-on-Spey 1,577, Burghead, 1,363,
Aviemore 1,224 and Kingussie 1,067. There are several smaller
towns with a population nearing 1,000 namely Beauly, Fochabers,
Fort Augustus and Newtonmore. The population is now rapidly
expanding owing to the oil-related industries in the Moray Firth and
the growth of tourism in Strathspey. The airforce establishment at
Kinloss and the naval establishment at Lossiemouth are not
included in these figures but naturally add considerably to the
total.

The occupation of the people is mainly agriculture, forestry,
fishing, distilling and tourism. Revenue is also obtained from salmon
fishing, deer stalking and grouse shooting. Communications are by
the Highland railway from Perth to Inverness and from Aberdeen
to Inverness. There is an Airport at Dalcross. There are three main

roads, the A9 and the A95 and A96. Small harbours for fishing boats
are at Lossiemouth, Burghead, Hopeman, Nairn and Inverness and
the Caledonian Canal runs from Bridge of Oich to Inverness.

CLIMATE

The climate of the area is governed by the distribution of the
mountain masses of the Scottish Highlands which act as a barrier
to the prevailing south-west to west winds.

These moisture-laden winds, blowing from the Atlantic, give
cloudy and very wet weather over the high ground in the west of
the vice-county, while to the lee of the mountains, the low ground
bordering the Moray Firth enjoys a dry mild climate due to the
Föhn effect of the dry descending air.

From fig. 1 the mountainous west and south are seen to have a
mean annual rainfall in excess of 2,000 mm with peak values of
over 3,000 mm in the Loch Cluanie area. Rainfall decreases steadily

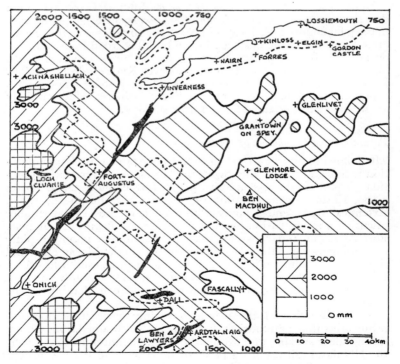

Rainfall – annual average in millimetres 1941–1970.

eastwards with much of the hilly country having annual values between 1,000 and 1,500 mm. Districts bordering the Moray Firth, being the most sheltered as well as the lowest lying, have average falls of less than 750 mm per annum. Near the coast the most favoured parts have less than 650 mm per annum. This means that in the west the annual rainfall is close to the highest in Scotland while in the Moray Firth area it is close to the minimum.

A selection of monthly rainfall averages is given in Table 1. These show that in the mountainous parts most of the rain falls in the autumn and winter months, while in the low lying coastal parts the maximum occurs in July and August. The low spring rainfall on the Moray Coast is of importance because moisture losses due to evaporation and transpiration in this period lead to a soil moisture deficit with the risk of drought (Table 2). Though the west has the highest annual rainfall, in the east heavy summer rainfall and thunderstorms may lead to severe flooding in the river valleys and flash floods in the hilly areas.

Temperature patterns are also affected by the topography, particularly by elevation, where an increase of height of 100 m leads to a falling off in the temperature of the free air by 2° to 3°C. This may be partially offset by favourable exposure, such as on south-facing slopes. In winter, on the other hand, valleys act as channels for cold air drainage, and over snow-covered surfaces very low temperatures may result.

The annual temperature range can be very wide: extreme values of 21·4° to minus 15·6°C have been recorded on Cairngorm at 1,090 m, 28·5° to minus 20·6°C at Grantown on Spey at 229 m and 30·6° to minus 16·7° at Forres at 37 m.

The waters of the Moray Firth are at their coldest in March with temperatures of around 4°C and warm but slowly towards a maximum of some 13°C in August. This results in the maximum temperatures in summer on the coast being reduced when onshore winds blow. When the warm south-east winds of spring and summer blow over this cold sea, widespread low cloud forms – a condition known on the east Scottish coast as 'haar'. Coastal areas affected by the haar remain cool with sea fog and drizzle while a few miles inland the weather may be bright and sunny.

The winter sea temperatures of around 4°C are still sufficient to have a moderating effect on minimum temperatures close to the coast. Air frost occurs on some 60 days per annum near the coast, increasing inland to over 100 days per annum in central parts of the vice-county. In the extreme west where more persistent cloudy and windy conditions prevail, this increase is not too great. Air frost may occur in any month in central parts but not in July and August on the coast. Ground frost, on the other hand, may be up to twice as frequent as air frost and occur in any month.

One method of assessing the effect of temperature on the growth of common vegetation is to use as an index of the length of the growing season, the number of days per year that the daily mean

TABLE 1

Monthly and Annual Averages of Rainfall (millimetres)

Station	Height (metres)	Jan	Feb	Mar	Apr	May	Jun	Jul	Aug	Sep	Oct	Nov	Dec	Year
Achnashellach	67	210	173	160	158	113	127	148	158	207	253	204	253	2163
Inverness	4	49	43	36	41	51	53	65	86	57	64	58	61	665
Nairn	6	45	37	33	37	48	52	60	81	67	57	54	54	625
Kinloss	5	46	39	35	35	50	55	68	89	55	60	56	54	642
Lossiemouth	6	47	40	35	36	51	52	70	89	57	61	59	55	652
Elgin	15	57	47	41	41	52	58	80	96	64	67	66	65	734
Gordon Castle	32	59	50	43	40	52	57	78	98	69	66	74	69	755
Glenlivet*	215	74	67	53	63	69	71	97	112	82	82	90	92	952
Glenmore Lodge*	341	96	76	68	68	87	73	89	106	93	110	105	106	1077
Ardtalnaig*	130	129	103	77	84	83	72	81	100	121	139	120	159	1268
Dall*	232	107	94	75	74	72	68	77	92	104	125	105	139	1132
Fascally*	82	82	65	49	51	67	62	68	90	83	87	80	100	884

The averages in the above table are for the period 1941 to 1970 but those marked * have been estimated from short-term records.

TABLE 2

Estimated Average Evapotranspiration (P.T.) in Millimetres for Coastal Districts of Moray

Jan	Feb	Mar	Apr	May	Jun	Jul	Aug	Sep	Oct	Nov	Dec	Summer Apr/Sep	Winter Oct/Mar
6	12	32	60	83	93	87	62	43	25	8	6	427	91

TABLE 3

Averages of Daily Mean Temperature in Degrees Celsius

Station	Height (metres)	Jan	Feb	Mar	Apr	May	Jun	Jul	Aug	Sep	Oct	Nov	Dec	Year
Achnashellach	67	2·6	2·7	5·0	6·9	9·7	12·4	13·3	13·3	11·7	9·0	5·3	3·7	7·9
Fort Augustus	21	2·5	2·8	5·1	7·1	9·9	12·7	13·7	13·9	11·9	9·1	5·3	3·6	8·1
Nairn	6	2·8	3·3	5·2	7·2	9·5	12·5	14·1	13·7	12·1	9·5	5·6	3·9	8·3
Forres	37	2·7	2·9	5·1	7·2	9·7	12·7	14·0	13·7	12·3	9·3	5·3	3·7	8·3
Gordon Castle	32	2·9	3·2	5·2	7·4	9·7	12·9	14·1	13·9	12·3	9·5	5·7	4·0	8·4
Grantown on Spey	229	1·9	0·9	3·1	5·1	8·3	11·7	12·9	13·0	10·5	7·5	2·8	2·6	6·7
Glenmore Lodge	341	2·0	1·0	2·7	4·6	7·8	11·3	12·3	12·4	10·2	7·5	2·8	2·6	6·4
Ardtalnaig	130	3·5	2·6	4·5	6·7	9·7	13·1	14·1	14·2	11·7	8·9	4·7	4·4	8·2
Dall	232	1·9	1·1	3·5	5·5	8·7	11·7	13·0	12·7	10·4	7·7	3·4	3·0	6·9
Fascally	82	3·0	2·3	4·5	6·5	9·4	12·8	14·1	14·1	11·7	8·8	4·2	3·8	7·9

Averages for the first five stations are for 1941–1970: the remainder for the period 1966–1975.

temperature at a site is above 5·6°C. The mean temperatures listed in Table 3 indicate that the growing season in the parts of the vice-county bordering the Moray Firth extends from the end of March to mid November and that at sites approaching 300 m above sea level it is shortened by some two months. In the much more exposed higher hills this index may approach zero, meaning that only the hardier strains of vegetation will survive in the rigorous climate.

Because of the falling off of temperature with height, snow occurs with much greater frequency in the higher parts of the vice-county than nearer the coast. The average number of days per year with snow falling varies from 40 on the coast to over 70 in the coldest parts of the central Highlands. Snow may lie on the higher hills for most of the winter and in some sheltered corries patches may linger through most summers. The persistence of snow cover depends not only on the amount that falls and on drift accumulation, but also on the chances of its removal by rain. There are many references to major flooding in the river valleys being brought about by the rapid thawing of lying snow by rain.

The comparatively mild south-west to west winds are the pre-vailing winds over the area, but spells with cold north-west winds in winter and cold easterlies in spring are also common. In summer, in the more settled periods, light and variable winds may prevail for short periods. Most gales occur with the south-west to west winds. On the Moray Coast these are not frequent, occurring on average on some 12 days per year, but gales occurring in the spring drought period may lead to severe soil blowing where soils are light and sandy. Inland in the deeper valleys there is a tendency for the winds to blow along their length and wind speeds here are often reduced by afforestation. However on the exposed higher ground the frequency of gale force winds increases rapidly and there are probably more than 100 days per year with gales in the highest parts. Very severe gustiness is frequently encountered, especially on ridges and on lee slopes. At these levels vegetation is greatly affected by buffeting from the wind and in parts such as the Cairngorm Plateau, where much of the surface is covered by coarse granitic sand the effects of sandblasting have to be considered where the loose material is not anchored by vegetation. Here too, prolonged blizzard conditions of great severity occur in winter.

From this summary, the climate of the vice-county is seen to be one of great contrasts and must exert a considerable influence on the distribution of the flora. Readers wishing more detailed information on the climate are referred to the works listed in the bibliography.

GEOLOGY AND SOILS

J. H. Gauld

Macaulay Institute for Soil Research, Aberdeen

Plant ecologists have described in detail the succession of vegetation from the first colonising plant species on an exposed surface of rock, sand or soil material to the ultimate establishment of a complex plant community. Within the soil, a similar evolutionary sequence exists by which a random mass of mineral particles can be transformed, over a varying period of time, by a succession of intricate chemical, physical and biological processes. Rather than being a static or inert substance, soil is therefore a dynamic system. Within the soil continuum, the large number of distinct soil types merely reflect the ability of the soil mass to become adjusted, through time, to the prevailing conditions of climate, topography, vegetation and parent material.

The term soil, as used by the pedologist, refers to that surface layer of the earth's crust, which has distinctive layers or horizons and which has the necessary physical, chemical and biological properties to sustain plant life. The aim of the Soil Survey of Scotland is to describe, classify and map the soils of the country. The classification is based on the nature of the soil parent material and the characteristics of the soil profile. The primary unit used in mapping is the soil series, which is defined as a group of soils having similar morphology and developed on similar parent material. Typical profiles of two different soil series may differ widely, but where the series are contiguous, it is usual for them to merge, sometimes over a considerable distance.

Differences among soils can be both local and regional. However, while it is known that soil variation exists, the distribution is not so haphazard as might be expected. Each soil reflects the environment in which it has formed, occupies a definite geographic area and occurs in certain patterns with other soils. By first recognising the main factors of soil formation and by noting the reflected characteristics in the soils themselves, we can later segregate soil geographic units.

FACTORS AFFECTING THE SOILS

I CLIMATE While it is impossible to generalise about the climate in this area due to the considerable amplitude in relief from sea level to over 1,220 m and the variations in exposure, there are certain common factors meriting discussion. These factors apply to precipitation and temperature, both of which are elements of climate known to affect soil formation.

The configuration of the landmass is such that the mountain barrier in the west and south-west of the area is capable of affording protection from the high rainfall associated with the south-westerly winds of the Atlantic depressions. Most of the rain is orographic so that the area with the highest rainfall is associated with high altitude. Thus while the Moray Firth coast may lie in a virtual rainshadow and experience only about 635 mm of rainfall per year, this figure increases to over 965 mm near Ardclach at 214 m and over 1,524 mm per annum in the highest areas of the Cairngorms. It is also generally true that the Grampian and Monadhliath Mountains, even at high altitudes, receive considerably less rain than the mountains to the west of the Great Glen.

The rainfall distribution is such that the second half of the year is generally wetter than the first. However, from the viewpoint of possible soil erosion, this distribution has two implications. In the late summer, high rainfall intensities are a notable feature of the Monadhliath and Cairngorm Mountains and may lead to catastrophic flooding. Such accelerated run-off will depend both on the absorptive capacity of the topsoil or peat and on possible changes in hill-land use, e.g. the removal of woodland. In direct contrast, drought conditions may prevail during spring in the Moray Firth coastlands where the position is compounded by the presence of light sandy soils, which become highly susceptible to windblow.

Temperatures in the west are higher than in the east at about the same altitude, with lower mean temperatures throughout the year being associated with increased altitude. The effect of altitude on the first and last incidence of frost is obviously considerable. However, relief also plays an important role in determining soil temperature, so that west of the Great Glen where there are a considerable number of east-west trending valleys, local variations in temperature between the north and south valley slopes are to be expected. Temperature controls the length of the growing season, the presence or absence of certain plant species and the degree of vegetation cover to be found in certain altitudinal zones. In upland areas, where there are lower temperatures and higher rainfall, organic matter accumulates and hill peat is formed.

2   PARENT MATERIAL   There are three main groups of soil parent material in the area.
a   Rock weathered *in situ* or transported only a short distance.
b   Pleistocene deposits: These deposits include glacial till, which consists of non-sorted material, formerly embedded in the base of glacial ice. Where the upper layers of such till have been modified by water to show some water-sorting into coarse textured material, such material has been termed water modified till. The group also includes fluvioglacial sands and gravels, where meltwater from the glacial ice during deglaciation has produced a coarse textured, completely sorted deposit, with associated rounded stones.

The nature of the rock from which the above two groups are derived affects the properties of soils considerably, in particular their colour, texture, nutrient supply, drainage, degree of stoniness and indirectly the vegetation they support.

c Recent or Post-Pleistocene deposits: These include alluvium, peat and raised beach deposits.

a ROCKS There are only isolated examples of soils directly derived *in situ* from the underlying rock. Consequently the glacial drifts are of paramount importance in soil formation studies and since these superficial deposits have originated from the local rocks, it is considered beneficial to describe the latter. These drifts have an important role in the soil survey because soils developed on parent material from the same rock type are brought together into groups, known as 'associations'.

The solid rocks, of which the high ground east and west of the Great Glen is formed, are mainly the Moine Series of the Highland Schists with practically none of the Dalradian Series of meta-morphosed sediments being represented. These rocks are mainly former sediments laid down in geosynclinal troughs, probably during Pre-Cambrian times, which then underwent extensive meta-morphism and uplift into a mountain chain during the Caledonian Orogeny. The variability of the former sediments is reflected in the range of resultant metamorphic rocks. Much of the area has been mapped as undifferentiated schist and gneiss. However, some of the metamorphosed, argillaceous sediments have been identified as pelitic gneiss and schist and from the arenaceous sediments, quartz granulites, quartzite, quartz schists and biotite granulites have been recognised. It is impossible to differentiate and map soils from such a variation in rock type, and all the soils have therefore been grouped into the Strichen Association.

The movements associated with or closely following upon the Caledonian Orogeny were accompanied by the large scale intrusion of igneous rocks. Such intrusions include the granite of the Cairn-gorms, Lethen Bar, Moy, Ardclach, Monadhliath and Aberchalder. Despite their considerable areal extent (58 sq. miles), the intrusions at Moy and Ardclach are almost entirely covered by hill peat and fluvioglacial gravels respectively and therefore are of little signifi-cance in soil formation. Where granite forms the parent material, the soils have been mapped as the Countesswells Association. Minor basic outcrops of serpentine, hornblendic rocks, crystalline limestone and other ultrabasic and basic gneisses are also to be found. However, their extent is limited, and the rock is often covered by glacial debris so that they seldom form a soil parent material in themselves. Their influence is usually reflected in the soil possessing a higher base exchange capacity and supporting a more exotic range of plant species than the surrounding soil. This explains the varied and rare plant species to be found in certain areas, e.g. Glenfeshie, Milton of Drumnadrochit, parts of the Monadhliath plateau.

The northern coastal section is underlain by rocks of the Middle and Upper divisions of the Old Red Sandstone. Between Inverness and Nairn and around Fochabers, the rocks are of the Middle division and rest unconformably on the Moine schist. The general succession begins with a basal conglomerate, followed by shales with fish-bearing beds, for example at Lethen Bar, Clune, Tynet and Dipple, and overlain by sandstones and flags. The basal conglomerates are exposed in the district south-west of Inverness, e.g. on the River Beauly at Kilmorack. The conglomerates and arenaceous sandstones form the parent material of the Kessock and Cromarty Associations respectively with the water modified till represented by the Kindeace Association. In a belt from Nairn by Forres to Elgin and beyond, beds of the Upper division rest unconformably on various members of the Middle Old Red Sandstone or upon the Moine schists. These coarse, grey and yellow sandstones give rise to soils of the Elgin Association.

North of Elgin, there are extensive sandstone outcrops, which by virtue of their containing a remarkable reptilian fauna, have been dated to be of Permian or Triassic age. However, it is impossible to deduce the stratigraphical relationship between these outcrops because of the extent of drift cover and the probable occurrence of intense faulting. Near Lossiemouth the rocks are nearly white, very soft and fine grained.

The solid geology plays a more important role in determining the relief of the area than in being a specific parent material of the soil. However, the occurrence of residual soils developed directly in deeply weathered rock have been recorded and in some cases they are large enough to map at 1:63,360 scale. Examples of deeply weathered acid schist and Old Red Sandstone sandstone have been mapped in Moray and Nairn (soils of the Leathain and Kinsteary Series respectively), while exposures of deeply weathered granite, hornblende schist and Old Red Sandstone conglomerate have also been noted.

b PLEISTOCENE DEPOSITS The details of the various glaciations, which have not only left an imprint on the pattern of relief but also covered the landscape with a variety of deposits, is too complex for brief description. However, attention is directed to certain characteristics, which influence the soil parent material.

It is an accepted fact that the area under study experienced a series of ice movements, which were associated either with major sources of ice or with smaller valley glaciers. Variations in the intensity of glacial erosion between east and west are detectable. West of the Great Glen, it is postulated that the east-west trending valleys such as Glen Urquhart, Glen Affric and Glen Moriston are the result of glacial breaching of the watershed zone due to intense glaciation in this area of high precipitation. This means that on the ridges and summits glacial erosion was severe with the glacial drift being patchily distributed and often thin or absent. Rock outcrops are numerous and the distribution of soil types extremely variable.

On the drier eastern side of the area erosion was less intense than in the mountains of the west. The imprint of glacial erosion is nevertheless strong and is reflected in the impressive scenery of the Cairngorm massif. With the retreat of the glaciers the deposition of material associated with stagnant ice masses took place, as exemplified at Glenmore on Speyside and Dava Moor on the Findhorn. The most widespread glacial deposit in the higher areas are the morainic deposits on the valley sides and floor.

The net result of the glaciation has been the scouring and denudation of the high plateau over 608 m and the deposition of a thick mantle of till over the Highland margin between this ground and the coastal Lowlands. This margin consists of the '1,000 foot peneplain', although it varies in altitude in a series of steps, which represent old erosion surfaces. The till mantle is 30 m thick in places and the complexity of its depositional history is evidenced by some exposures indicating a succession of till types, which can be associated with different fronts.

During the melt-out stage of end-glacial times, vast spreads of fluvio-glacial sands and gravels were deposited over the peneplain and across the coastal lowlands in a series of irregular mounds and ridges. It is during this period that the great mounds of sand and gravel, e.g. Tomnahurich and Torvaine, Inverness and the series of kames at Flemington and Strathnairn may have been deposited.

Following deglaciation, differential movements of sea and land resulted in the formation of four distinct raised beaches at 15, 25, 50 and 100 foot levels. The immediate post-emergence stages of these beaches were of a lagoonal nature and pockets of sandy alluvium frequently overlie the fluvio-glacial silt and clays of depressed sites, for example at Dalcross. The easily eroded deposits of these raised beaches, in combination with the prevailing direction of the tidal current has led also to the concurrent processes of erosion and deposition, such that much of the Carse of Delnies and Ardersier are of recent origin. A more spectacular change has occurred at Culbin and Maviston to the west of the Findhorn estuary. In this instance, sand, which was transported offshore in a westerly direction, has been blown back eastwards by wind to produce the present day sand dunes. These dunes have buried the old raised beach and former agricultural land. The final emergence of land also left the shallow depression of Findhorn Bay and Loch Spynie ponded back behind shingle, and while the former remains as broad mud-banks at low tide, the latter has been artificially drained, apart from a small reed-grown remnant.

c   RECENT DEPOSITS   River alluvium is present throughout the area along most river and stream courses. Lacustrine alluvium, marking former lagoonal sites (for example Dalcross, Loch Spynie central basin and between Milton Brodie and Kinloss) is of less frequent occurrence. The sediments are extremely variable both vertically and laterally, with all grades of texture from gravel to clay being encountered. In some areas, e.g. Rivers Spey and Dulnain haughlands, old meander channels are characterised by an infilling of peat alluvium or poorly drained sand overlain by a silt capping.

Mixed beach sediments are located along the coast between

Inverness and Fochabers and while varying grades of sand dominate these deposits, bands and lenses of gravel may also be present.

Hill peat covers part of the hill tops, saddles and slopes mainly on the '1,000 foot' peneplain and '2,000 foot' dissected plateau areas where the noted environmental conditions favourable to its formation prevail. Basin peat is found in low-lying hollows or depressions throughout the area.

3   RELIEF   Geomorphological processes affect both the landscape and the distribution of soil parent material. Topography and parent material together influence soil drainage and may produce repeating patterns of soil related to position on slopes.

On steep and convex slopes, freely drained soils, which are characterised by bright yellow, brown and red colours in the subsoil, tend to be found. In contrast, the soils located in depressions or flat sites are generally poorly or very poorly drained and are referred to as gley soils. In these soils where the water table remains close to the surface throughout the year, reducing conditions are favoured so that subsoil colours tend to be grey or blue-grey. Soils of poor drainage, however, may be found on good slopes where the lower soil horizons, texture of the parent material or proximity of bedrock can modify the typical drainage sequence.

The texture of the soil greatly affects the degree of vertical movement by leachates and the pattern of base saturation. More vertical leaching occurs through the more permeable soil, thereby reducing base saturation. On the coastal plain and peneplain most of the soils are freely drained, owing to the combined effect of relatively low rainfall, the generally coarse texture of the soil parent material and the possible prevalence of steep slopes in the latter area.

Relief affects soil formation indirectly through its effect on climate, particularly in the incidence of orographic rainfall and vegetation which changes according to altitudinal limits. Other features such as erosion and soil temperature, in relation to aspect, are also associated with relief and have been briefly discussed.

4   ORGANISMS   The occurrence of particular plant species reflects the interaction of parent material, relief, drainage and climate. However, besides plants, living organisms in the soil include animals, insects, fungi, bacteria and other biological forms. Together they play an important role in soil development, such as determining the kind and amount of organic matter in the soil under natural conditions. They also govern the manner in which organic matter is added, whether as leaves and twigs on the surface or as fibrous roots within the profile. The rate of organic matter decomposition is strongly influenced by the type and activity of living organisms present. In contrast to the climatically induced hill or blanket peat, basin peat deposits originate locally in poorly drained hollows where drainage is impeded and where anaerobic

conditions leading to the accumulation of organic residues prevail.

The nature of the vegetation cover is also known to have a decided influence on soil development. For example, a forest cover promotes a different soil forming process than either grass or moorland. Tree species also differ in their influence on soil development because conifers are more conducive than deciduous trees to podzol formation, especially on acid parent materials. This explains the presence of brown podzolic soils rather than iron or humus-iron podzols on the fluvioglacial gravels of the long established deciduous woods of Cawdor and Darnaway.

Man has had, and is having, a considerable influence on the soil of the area. In agricultural areas man has greatly affected profile development, indirectly by removal of the natural vegetation and directly by drainage, cultivation and the addition of nutrients. This is particularly marked in the conversion of moorland heath to permanent grassland. Through modifications in the system of agriculture, man has, in recent years, created soil erosion problems on the Moray Firth coastlands. Afforestation also influences the soil in a number of ways as a consequence of draining and ploughing. Regeneration on the moorland sites has little chance of success while heather burning is so rigorously practised. Conversely, the commercial felling of the Caledonian Pine Forest has influenced soil development and the ecological balance in those areas, where these majestic trees once grew.

5   TIME   Most of the area, except possibly parts of the Cairngorm Mountains, the Western Mountain mass and 'nunataks' within the '2,000 foot' plateau, was ice covered during the last glaciation and earlier soils have probably been destroyed or radically modified. Only one example, at Teindland, of an interglacial soil has been found. Soil forming processes were initiated at the end of the last cold period some 10,000 years ago but probably only reached their present intensity at the beginning of the Atlantic period, about 7,500 years ago.

Many exposures of peat show the stratigraphy of the deposit in which layers containing pine tree stumps, occurring, at 1 to 2 m from the surface, or examples of birch, alder or willow stems at the base, may be present. Upon drying and subsequent weathering, the zonation in the profile is often emphasised. The technique of pollen analysis, based on the fact that pollen and spores remain preserved in anaerobic conditions, enables the history of vegetational changes to be studied. From this continuous record of vegetation, major climatic changes, with possible implications in soil formation, can be assessed.

SOILS

The soils are best described in groups according to the major landform unit in which they occur. These units include:

1. Lowest Raised Beach
2. Coastal Lowlands
3. Drumossie Muir and
   Peneplain fringe
4. The '1,000 foot' Peneplain

5. The '2,000 foot' Plateau
6. Cairngorm Plateau
7. Mid Strathspey and Badenoch
8. Monadhliath Mountains

Systematic soil survey as undertaken by the Soil Survey has not yet been carried out in the rugged terrain to the west of the Great Glen and the extreme south-west of the area under discussion. It is therefore intended to exclude a generalised description of soils for these areas because there is insufficient soil information.

1   LOWEST RAISED BEACH   This unit comprises the Carses of Ardersier and Delnies, the shores of Findhorn Bay and the coastal fringe east of Findhorn to Lossiemouth and beyond to the mouth of the River Spey.

The soils show little horizon differentiation other than that due to texture. Sand and fine sand are the principal textural components but bands or lenses of fine gravel, silt, silty clay and peat have all been identified. Despite such a highly permeable parent material, the soils may not be freely drained. Within flat sites a high water table and poor surface run-off lead to conditions of poor or very poor internal drainage. Associated with this water-logging and anaerobic reducing conditions are extremely acid conditions with pH values varying between 3.5 and 4.5.

Along the coastal fringe soils are developed on dune sand, shingle bars and storm beach ridges. Between Lossiemouth and Burghead and around Nairn, wind blown sand overlies the beach deposits and has been stabilised by a relatively close vegetation, except on areas where the sand has accumulated as dunes. The soils developed on stabilised wind blown sand are classed as Links. They are mainly freely drained and consist of 10 to 15 cms of dark brown loamy sand over light brown sand. Thin buried organic stained A horizons are present in the upper part of the profile. The blown sand is usually just a veneer but at Culbin and west of Burghead it has accumulated into massive dunes. Afforestation with Corsican pine has stabilised the dune complex and beneath the litter immature iron podzols are developing on the raw sand. Occasional imperfectly drained flats, characterised by a birch-grass vegetation, indicate buried, cultivated, alluvial soils of varying texture. In the Maviston area, apart from a few isolated dunes, the terrain is flat and has a high water table. Poorly drained soils are generally underlain by peat bands and alluvium of a silty texture, reflecting the lagoonal evolution of the area.

2   COASTAL LOWLANDS   Fluvioglacial sands and gravels of the Boyndie and Corby Associations form the dominant parent materials of the coastal lowlands. Both deposits are free draining with possible exceptions being soils on the summits of gravel mounds or ridges and in depressions where excessive and poorly

drained soils respectively prevail. The parent material is water sorted with the most extensive spreads in the form of moundy terrain. A further topographic unit is represented by relatively flat upper terraces, well above the present flood plain of the Rivers Nairn, Findhorn, Lossie and Spey and their tributaries.

Within the Corby Association, the gravel deposits of the Flemington, Auldearn and Kinsteary kames tend to be coarser and less well sorted than the lower moundy deposits. Much of the Boyndie series has a characteristic deep plough horizon often exceeding 50 cm in depth, in which the physical composition is remarkably uniform with clay 10 ± 2 per cent and silt 25 ± 5 per cent. Around Castle Stuart a fluvioglacial sand has been mapped as a new series of the Boyndie Association. These soils possess a high silt (2–$50\ \mu$) content of 20 to 30 per cent throughout so that texturally the whole profile to a depth of 180 cm is a sandy loam.

Other parent materials, not associated with glacial meltwater, occur in localised areas. At Poolton and Allanfearn the parent materials are peat and lacustrine silty clay respectively, while around Duffus, on what was formerly the bed of a much larger Loch Spynie, the soils are developed on a deposit of calcareous lacustrine clay. Soils of the Carden Association are also derived from a lacustrine deposit of reddish brown silty clay and occur patchily on some of the higher mounds and ridges at Wester Alves, Hillhead and Carden. Soils with such a high silt and clay content present physical problems in relation to their cultivation. However, their chemical properties with high exchangeable bases and a base saturation of normally 100 per cent, more than compensate for these disadvantages. The soils of the Duffus Association are all calcareous with pH values up to 8·5 and, in some cases, free calcium carbonate in the surface horizons. The imperfectly drained soil has been mapped as a brown forest soil with gleyed B and C horizons while the predominating poorly drained soil is a calcareous gley. The peat deposit at Poolton seldom exceeds 1·5 m in depth and is underlain by marl deposits. A group of poorly drained soils represents the most recently formed soils in this zone. The parent material is an alluvium, which ranges in texture from sand to silty fine sand. A permeable water table is found between 72 and 100 cm from the surface and prominent iron mottles, especially in the form of iron streaks around root channels, are common in the upper part of the profile.

The majority of the soils of the coastal lowlands are cultivated humus podzols, iron podzols and humus-iron podzols. Their chemical regime has been complicated by manurial practice, but even under natural conditions their coarse texture, high porosity and low clay values mean that exchange capacity is low, except in the surface organic horizons. In these soils, clay plus silt contents in the topsoil amount to about 10 per cent with clay values seldom exceeding 5 per cent in the subsoil. Under natural conditions the iron and humus-iron podzols have an accumulation of partially and well decomposed organic matter on the surface. A sharp change occurs into a prominent, bleached A_2 horizon, which overlies a brown or yellowish red B_2 horizon. In the case of the humus-iron podzol a well defined organomineral horizon is present between these horizons. It is usually between 2 and 8 cm thick, shows a significant colour difference and a marked increase in loss on ignition values and correlated per cent organic matter as compared with the A_2 and B horizons.

The humus podzols are mainly restricted to the sandy high raised beaches and the Boyndie series to the east of Forres and in the Balloch area. Near Binsness and Cothill, 'butte' dunes with a buried well developed humus podzol have been exposed by wind erosion. Under natural conditions the A_2 horison is present and the boundary with the underlying dark brown cemented B horizon is usually abrupt. This strong cementation of the illuvial humus zone is a feature of these soils colloquially referred to as the 'Moray Pan'.

3 DRUMOSSIE MUIR AND PENEPLAIN FRINGE Drumossie Muir is a low lying ridge rising to about 150 m above sea level and effectively separating the Nairn valley from the coastal lowlands. Sandstones of the Middle Old Red Sandstone underlie this ridge and are overlain by a reddish brown, stony, sandy loam till. However, the upper part of the till is more commonly represented by a partially water sorted material so that the soils belong to the Kindeace Association, rather than the Cromarty Association. A very compact, indurated horizon, which impedes the surface drainage of the soil, is a resultant characteristic property. They have been mapped as imperfectly drained peaty podzols and poorly drained peaty gleys. A peaty topsoil, of 5 to 15 cm depth, overlies a dark brownish grey, coarse textured A_2 horizon, the latter being replaced by a gleyed A/B horizon in the case of the peaty gley. The B horizon possesses all the properties of a gley horizon. A strongly developed platy structure is typical of the indurated horizon, which is impermeable and exceedingly difficult to break or penetrate with a spade. Indurated layers seldom exceed 15 to 23 cm in depth. Pore volumes in the indurated layer are only 25–30 per cent compared with over 50 per cent in the overlying horizons. Beneath this B_3 horizon, the profile is either freely or slightly imperfectly drained. Marked changes in the surface drainage conditions are associated with slight variations in the slope of the site. The result is that the drainage in the A and B horizons above the induration may range from imperfect to very poor.

The peneplain fringe is represented by undulating hill terrain between 152 and 304 m and stretches from Drumossie Muir to Teindland in the east with an extension south along the River Spey to include Archiestown and Knockando. This fringe coincides roughly with the contact of the Old Red Sandstone and rocks of the Moine series. However, the geological formations are relatively unimportant in profile development because of the vast spreads of fluvioglacial deposits, especially gravels, which cover the underlying strata. Freely drained iron humus podzols of the Corby series are consequently dominant. Beneath the forest litter produced from long established deciduous trees, the prominent bleached A_2 horizon is replaced by a horizon where there is an intimate mixture of organo-mineral material with common bleached sand grains. The ubiquitous podzol profile is replaced by a brown podzolic soil, which has been mapped in Cawdor and Darnaway woods.

Elsewhere the soil parent materials include till derived from acid schists (Strichen Association), mixed granite and acid schists (Aberlour Association), mixed Old Red Sandstone and acid schist (Orton Association) and Middle Old Red Sandstone sediments (Elgin Association). Soils of the Elgin Association are best seen on the Monaughty Ridge and south of Loch na-Bo, while those derived from Orton till are encountered at the edge of the moor between Glen Latterach and the Glen of Rothes. At Elchies and Carron, mixed till derived from schist, granite and syenite occurs. The moorland and hill ground of the Gedloch and Rothes estates are covered by till of the Strichen Association, which is also to be found on the lower forested ground of Newtyle Forest. The soils of Teindland Forest are developed on both Elgin and Orton till.

The soils of the high moorlands are peaty podzols and peaty gleys with extensive hill peat on the flat sites and north facing slopes. On the arable land the soils include cultivated podzols, non-calcareous gleys and peaty gleys. Under the scattered afforested areas, humus-iron podzols predominate and are freely drained except, as at Teindland, where intense induration affects the internal drainage of the profile. In these cases, the soils are identical to those found on Drumossie Muir.

4   THE '1,000 FOOT' PENEPLAIN   Fluvioglacial sands, gravels and gravelly loamy sands of the Boyndie, Corby and Dulsie Associations respectively form the largest group of soils within this unit. Such deposits range from the flat wide terraces of the Rivers Nairn, Findhorn and Lossie, to the steep sloped moundy terrain around Lochindorb and south of Dallas. Rudely stratified gravelly loamy sands of the Dulsie Association are scattered across the peneplain in long low mounds, some of which may be capped by a well sorted gravel. The soils on the Dulsie and Corby Associations are identical in all respects except for their degree of sorting and topographic position.

The dominant profile is a peaty podzol with iron pan, although limited areas of a humus-iron podzol occur on the steeper slopes where peat development has been inhibited. Despite the presence of an iron pan overlain by peat up to 25 cm thick, there is little evidence of gleying in the A_2 horizon of the peaty podzol. On flatter sites where the peat may be up to 40 cm thick, the A and AB horizons are strongly gleyed above the iron pan. This group of soils has been classed as peaty gleyed podzols. The occurrence of peaty gley soils is restricted to flush sites or very gentle slopes. Within depressions or basins, there are extensive accumulations of peat, with a recorded depth of 8 m on Dava Moor. Average depths tend to be about 5 m. Outwith these basins, the peat is mainly of the hill or climatic peat variety, dark brown to black in colour and composed of *Eriophorum*, *Sphagnum* and *Calluna* residues. In general, hill peat has a low water holding capacity, high acidity, low to moderate ash content and is markedly deficient in calcium, potassium and phosphorus. Extensive deposits are located in the Larig Hill, Carn Dearg and Carn Kitty region south of Dallas, Dava Moor and the headwaters of the River Divie.

5 '2,000 FOOT' PLATEAU The Strichen Association and to a lesser extent the Countesswells Association are widespread across the flanks of this unit and on isolated hills protruding above the plateau. Derived from acid schists and granites respectively, the parent materials are a stony sandy loam and a stony gritty sandy loam. Both associations have a series formed on deeply weathered rock, which is probably the result of pre-glacial weathering. Although these profiles often display an 0·7 m layer of solifucted till-like debris, weathering and alteration often exceeds 3 m in depth and the material is easily dug by spade.

Ranker/rock complexes, sub-alpine podzols, peaty podzols with iron pan, peaty gleyed podzols (poorly drained above the iron pan) and hill peat have all been identified in this zone. Peaty gley soils are scarce and confined to concave colluvial sites. The sub-alpine podzol is developed on a very stony gritty, sandy loam derived from frost shattered and physically weathered schist. The profile is very shallow, seldom exceeding a metre in depth. Eight to ten centimetres of raw humus, produced from mosses and *Vaccinium myrtillus*, overlie a very dark brown, organic stained, sandy loam to loam, about 8 cm thick. There is little trace of an A_2 horizon except for a few bleached sand grains. The B horizon becomes progressively lighter in colour and more stony with increasing depth. Where the relief is sufficiently rugged to allow rock outcrops, ranker profiles are encountered. An A horizon of raw humus lies directly on a C horizon of either shattered or relatively unweathered rock.

On some very steep slopes, for example in the Streens area of the River Findhorn, freely drained colluvial or creep soils are to be found. The development of a mor humus or peaty surface horizon, as is found in the surrounding mineral soils, is prevented by the steepness of slope. They are replaced by moder/mull A horizon overlying a strong brown B_2 horizon.

Hill peat, 2 to 3 m in depth, covers most of the '2,000 foot' plateau. The considerable contemporary erosion is indicated by the extensive and deeply incised dendritic drainage patterns, many of which are now cutting into the mineral soil. Erosion in a few areas has exposed dense clumps of pine tree stumps rooted in the mineral soil. The courses of some underground streams may be revealed by lines of 'sink holes'.

6 CAIRNGORM PLATEAU   Residual mountain top detritus covers most of the Cairngorm plateau. The derived soils belong to the Countesswells Association and reveal a distinct altitudinal zonation.

At about 610 m the typical peaty podzol with iron pan is replaced by a profile transitional to an alpine podzol. This zone is characterised by a stunted Callunetum with *Arctostaphylos uva-ursi* colonising eroded areas on exposed or steep sites. Above 762 m this vegetation is gradually replaced by an *Empetrum-Vaccinium-Rhacomitrium* zone and the soil profile is an alpine podzol. This soil has a loose fabric due to the intense frost heaving at this altitude. The surface horizon contains numerous bleached sand grains and overlies a black layer up to 20 cm thick, with the coatings on stones and grit having a silty, greasy feel. This layer is characterised by a high organic matter content and is considerably thicker than the similar horizon found in sub-alpine podzols. It merges into a B or BC horizon where there has been an accumulation of silty material.

Around 1,067 m the increasing severity of climate leads to a patchy vegetation cover, mainly of a *Juncus trifidus-Carex bigelowii* community. Two to five centimetres of loose granitic grit overlie a thin organic A_1 horizon and rest directly on the B or BC horizon previously described. In sheltered areas of persistent snowfields there is usually an accumulation of mineralised peaty humus up to 25 cm thick under a closed vegetation mat of *Nardus stricta*. A well developed alpine podzol occurs in these sites.

7 MID STRATHSPEY AND BADENOCH   The whole area is underlain by acid schist of the Moine series and associated igneous intrusions. Whereas severe glaciation has denuded most of the hill tops, the lower hill slopes and valleys are mantled with thick deposits of fluvioglacial sands and gravels.

In the sand and gravel outwash plains, soils of the Corby and Boyndie Association have been identified. On the mounds and ridges, which support a natural or semi-natural pine forest, the dominant soils are iron podzols and humus-iron podzols. Outwith the forest canopy on open moorland sites, peaty podzols with well developed iron pan overlying a strongly cemented B_2 horizon are present. Most of the kettle holes and basins have been infilled with peat, which around Loch Garten and Loch Mor, Duthil, may be 6 m deep.

Soils developed on the alluvium of the Rivers Spey, Dulnain and Nethy vary in both texture and drainage. Old meander channels within the flood-plain often possess a silt capping up to 35 cm thick and the overlying sand is consequently poorly or very poorly drained. South of Loch Insh, the River Spey haughland consists of extensive marshes in which the alluvial material varies from sand to silt. Extensive bands of mineralised peat are also present and the whole area is liable to flooding.

Soils belonging to the Strichen, Dulsie and Aberlour Associations are found above approximately 275 m outwith the main valley systems. Intense induration characterises all these soils, the dominant type being a peaty podzol, with lesser peaty gleys and humic gleys often found below spring lines. Soils on the upper convex slopes of isolated ridges are formed mainly in shattered rock, which may be overlain by a veneer of morainic material. Humus iron and peaty podzols are dominant. On the summit of the ridges, these podzols may be associated with minor rock outcrops. Isolated occurrences of soils developed in completely weathered schist rock have been found and such soils may support a more diverse vegetation cover than is found in the immediate vicinity.

8 MONADHLIATH MOUNTAINS Although rising to approximately 915 m in the south-east, the mountains represent a broad plateau surface with a dominant level between 710 and 860 m. Such relief has been developed on rocks of the Moinean Assemblage and undergone modification through the erosive agencies of glacial and river action.

Extensive hill peat, with an average depth of 1·5 m, occupies most slopes on the plateau. Within col sites and basins, depths of 4 m have been recorded. The distinctive erosional pattern of dendritic and linear hagging is ubiquitous and often exposes iron pans within the underlying mineral profile. Of further interest is the occurrence of deeply weathered schist rock within stream sections or exposed beneath the peat. Two forms of redistributed peat have been encountered. In the eastern zone, long flush ribbons, which are characterised by a *Juncus squarrosus-Nardus stricta* vegetation, occupy concave slopes below col sites. In these cases the peat has been eroded from the col. In the slightly wetter and more exposed west, the redistributed peat is devoid of vegetation and occupies small depressions. Whereas the hill peat extends to the edge of the steep sided Findhorn valley, it is not only of limited extent but also less eroded within the Monadhliath foothills, which fringe the Spey valley.

On ridge crests and convex slopes of the plateau, alpine podzols belonging to the Strichen Association have been mapped. These soils are characterised by a stunted *Calluna, Cladonia* spp. and *Empetrum nigrum* vegetation with the boundary between the mineral soils and the surrounding hill peat being abrupt. A veneer of relatively clean fluvioglacial sand and gravel may overlie the shallow drift or shattered rock forming the parent material of these soils. Steep slopes associated with meltwater channels or deeply incised streams support peaty gleyed podzols with a closed vegetation cover dominated by *Vaccinium myrtillus* with associated *Nardus stricta* and *Juniperus communis*.

Acknowledgements

I am indebted to colleagues from the Soil Survey of Scotland for their advice in the preparation of this paper.

I am particularly grateful to Mr A. D. Walker for demonstrating many of the sites referred to in the text and for his constructive criticism of the paper.

VEGETATION

The area covered by this work contains a flora as varied as that in any county in Britain. Some 944 native species (excluding sub-species, varieties and hybrids which account for many more) and 400 introduced species have been recorded.

The area extends from the mild, dry coast of the Moray Firth, into which the gulf stream flows, to one of the highest, coldest and most wind-swept mountain plateau in the Cairngorm mountains, where snow can lie all the year round. Between these extremes lie the rich fertile agricultural lands of the Laich of Moray, and the Aird to the west of Inverness, many woodlands of both deciduous and coniferous trees, native and planted, river straths, lochs, moorland and mountain ranges of varying heights.

The distance from the shore, the elevation above sea level and the humidity seem to influence the distribution of species far more than the mineralogical character of the strata over which they grow. The nature of the vast accumulation of alluvial matter and the vegetation it supports depends mainly upon the circumstance of the deposits or subsoil being gravelly, peaty or clayey.

With three quarters of the area being moorland a calcifuge flora is dominant. Calcicoles occur on the few limestone, diorite, corn-stone and sandstone outcrops and on the calcareous lacustrine clays such as are found at Loch Spynie. Calcicoles are also found surrounding the old lime kilns of former moorland crofts where the lime was imported and the crofters burnt their own lime.

The richest flora is to be found in the river straths and ravines and can be said to contain some 80 per cent of the total flora of the two vice-counties.

Detailed accounts of the vegetation and soil for Moray, Nairn, Strathspey and the Monadhliath Mountains have been written by the Macaulay Institute for Soil Research and others are in prepara-tion, and many authors have discussed the Cairngorm Mountains but the area west of the Great Glen has been sadly neglected. Paleobotany has also been studied in the Cairngorms and it is known that before the post-glacial period juniper, dwarf birch, crowberry and other plants were present.

A large proportion of plants native to the British Isles can be regarded as introduced to North Scotland some of which are temporary, other becoming well established. Many have been introduced by the agency of man, other by migratory birds but the chief sources of introductions come with sand, gravel and cinders used to build railway embankments, canal banks etc., even whins were said by Brodie to be 'rare in the highlands' in 1800. Proprietors of some of the larger properties planted exotics in their policies such as *Luzula luzuloides, Poa chaixii* and the double-flowered *Saxifraga granulata.* This was the case with *Heracleum mantegazzianum* the Giant Hogweed planted as a gimmick at Glenferness House by Ronald, Earl of Leven and Melville about the year 1890 and from whence it has escaped to the banks of the River Findhorn and subsequently elsewhere, forming large colonies on waste land and river shingle and exterminating the native flora.

Distillery refuse and bird cage sweepings account for many annuals found on rubbish tips and railway sidings although nowadays the barley for distilleries is delivered by road and not by rail. In 1961 a sack of foreign grain was fed to hens in an allotment in Inverness producing seventy-six alien species and another sack was washed down the River Findhorn during the floods of 1957 and several species subsequently appeared in Greshop Wood, but only for a period of two years after which they were exterminated by the native vegetation. New roadside verges and carrot fields are another source of introductions, usually short-lived, for the verges soon become colonised by native species, and the spraying of herbicides destroys the weeds in the root crops although *Amsinckia calycina* found at Wester Coltfield Farm near Alves thirteen years ago seems to resist all spraying. The less-said-the-better about the planting from a Danish source of *Erica tetralix, Vaccinium myrtillus* and *V. vitis-idaea* in 1976 on the verges of the A9 near Dalwhinnie!

Some of the former cornfield weeds are not seen now but fortunately *Centaurea cyanus* (cornflower), can still be found in quantity in the fields near Kincorth and Kintessack in Moray.

Over the last fifty years very few plants can be said to have disappeared altogether, but some are diminishing owing to drainage and planting of pine forests. Industrial development has not so far accounted for the loss of species.

Some of the old records have not been refound but they cannot be arbitrarily dismissed as errors. They include the following where no voucher specimen has been seen: *Ranunculus sardous, Potentilla tabernaemontanae, P. anglica, Callitriche platycarpa, Oenanthe lachenalii, Carum verticillatum, Scrophularia aquatica, Cirsium pratense, Juncus subnodulosus* and *Sesleria caerulea.*

A mysterious plant named *Ulleriore* was supposed to have been abundant in Loch Spynie (F. M. Webb, *J. Bot.* **13**: 50 (1975)). A translation from the latin given in the Agricultural Survey of the Province of Moray from the History of Scotland by John Leslie Bishop of Ross (1578) states *Ulleriore*, a water-weed . . . moreover Moray contains a lake of fresh water denominated Spynie, greatly frequented by swans, in which there is a certain uncommon herb with which the swans are greatly allured, we call it the *Ulleriore*, it is moreover of this kind, that when fully established its roots it spreads itself so rapidly, that in my memory it hath extended its basis so far as to have rendered five miles of the lake itself of Spynie, where salmon formerly abounded, altogether shallow'.

This query elicited the following reply from the Rev. George Gordon of Birnie (*J. Bot.* **13**: 237 (1875)), 'A notice or query regarding this plant, and of similar import to that in the *Scottish Naturalist* **3**: 79 (1875–76) was sent to *Loudens Magazine of Natural History* **4**: 188 (1831). No information was given in reply. In the notice will be found an extract in Latin from Bishop Leslie where the plant is called *Olorina*. Why it was translated *Ulleriore* in the *Survey of the Province of Moray* is not known. *Typha latifolia* was suggested as likely to have been the plant meant by the Bishop. It was once abundant in the Loch of Spynie. An inroad of the sea at the time of the Moray Floods (1829) almost killed it out. It again revived and spread its roots when some years later the flow of the tide was excluded, but now it has almost disappeared, with the beautiful lake itself through the "progress of agriculture", *Scirpus lacustris*, *Arundo phragmites*, *Sparganium* and *Potamogeton* etc. were also frequent in and around the same extensive sheet of water.'

In this account the area embraced by the Flora has been divided into **Lowlands** and **Highlands** and some of the plant communities to be found discussed. Lack of space has not permitted full lists to be included.

LOWLANDS

The Lowlands consist of the Region of the Plains as described by H. C. Watson in *Remarks on the Geographical Districts of British Plants* (1835) and include Regions 1–3 as described by J. H. Gauld (1977). They reach from the shore line to the '1,000 foot' peneplain. According to Gordon (1839) Watson did not include land north of the Grampian Mountains as falling into this category but as it clearly does so, it has been included here.

From the old accounts of the land it can be assumed that the area was formerly covered with native forests of oak, pine, birch, alder and ash. Old stumps were dug from the mosses (lowland bogs) as recently as the nineteenth century. With the progress of agriculture the greater part of this area is now cultivated but there still remain areas untouched by man chiefly on the coast-line cliffs, dunes, shingle bars and salt-marshes and inland on the open moors, in oak birch and pine woods, in river straths, ravines and lochs.

1 Lowest Raised Beach

The area surveyed is NW of the Buckie Loch in the Culbin Forest, Nat. grid 28/9964 consisting of a sheltered inlet behind the Bar and formerly exposed to strong tidal scour and containing some siltings of mud, this is termed a strand. It is followed landward by low embryo mobile dunes, high mobile dunes, fixed dunes, and dune slacks.

The Culbin Forest near Binsness looking over the Findhorn estuary to Forres, Moray.

Young botanists at the dried out Buckie Loch, Culbin Forest.

Looking into the Coire an t-Sneachda from the summit plateau of Cairn Gorm,
East Inverness.

Lineal descendants of the old Caledonian Forest at Glenmore.

(a) *Strand*

An association of strand plants has developed on the seaward side of the embryo dunes and is rapidly colonising the inlet mentioned above at the rate of 15 yards a year. The earliest invaders were *Puccinellia maritima* and *Agrostis stolonifera* forming large circular patches quickly followed by *Glaux maritima, Armeria maritima, Spergularia media, S. marina, Festuca rubra* spp. and *Plantago maritima*.

(b) *Low embryo mobile dunes*

On the seaward side of these low dunes most of the vegetation consists of halophytes (plants tolerant of salt water) the most common being *Atriplex glabriuscula, A. laciniata, A. littoralis* and *Honkyena peploides* with occasional plants of *Cakile maritima, Sedum acre* and *Salsola kali*. Above this shore line grow plants that play an important part in dune consolidation such as *Carex arenaria, Agropyron junceiforme, Ammophila arenaria, Elymus arenarius* and *Poa subcaerulea*.

(c) *High mobile dunes*

The higher dunes are dominated by *Ammophila arenaria* a grass with the power of growing up to the surface when buried in sand by the wind, and possessing long underground rhizomes that stabilise the dunes. The grasses to be found on the low dunes are also to be found on the high dunes and with the shelter afforded long-rooted plants become established such as *Ligusticum scoticum, Hypochoeris radicata, Cirsium arvense* and *Senecio jacobaea* and on the more consolidated dune slopes the spring annuals of *Cerastium semidecandrum, C. diffusum, Vicia lathyroides, V. sativa subsp. segetalis, Filago minima, Myosotis ramosissima* and the small grass *Aira praecox*.

Erosion by the sea and the strong winds is taking place with alarming rapidity to the east of this area towards Findhorn, and in many places the forest trees are being washed away. It may not be long before the tides break through and the Buckie Loch becomes inundated.

(d) *Fixed dunes*

Landward of the high mobile dunes are many ridges of fixed dunes running parallel to the coast line and between which lie slacks of varying size where water formerly lay in winter but are now mostly dried out on account of the extensive planting of pines by the Forestry Commission. On the dunes (not under tree canopy) are most of the plants referred to as growing on the mobile dunes, with the inclusion of such plants as *Pilosella officinarum, Gnaphalium sylvaticum, Ononis repens, Lotus corniculatus* and *Festuca rubra* spp.

The vegetation, on the dunes under the tree canopy, consists chiefly of mosses and lichens with *Agrostis tenuis, Goodyera repens* and *Carex arenaria*, and in the damper areas *Pyrola minor, P. media* (rare), *Orthilia secunda, Moneses uniflora, Listera cordata, Corallorhiza trifida* and occasional plants of *Dryopteris austriaca*.

Other dune communities are found between Findhorn and

Burghead and are dominated by *Calluna vulgaris*, *Ulex europaeus* and *Cytisus scoparia* with occasional plants of *Erica cinerea*, *Lotus corniculatus*, *Veronica officinalis* and many lichens.

(e) *Dune slacks*

The slacks in this area are those that have suffered most with the drying out of the soil. Twenty years ago water lay in most of them during the winter months and resulted in a rich community of plant species such as *Pinguicula vulgaris*, *Drosera rotundiflora*, *Parnassia palustris*, *Eleocharis quinqueflora*, *Dactylorhiza purpurella*, *D. incarnata* and many others, now the area is overgrown with tussocks of *Salix repens* and *Empetrum nigrum* and many young trees of *Betula pubescens* subsp. *pubescens* and *Salix aurita*. In the dried out open spaces *Juncus balticus* is dominant with dwarf plants of *Juncus gerardii* (4–5 cm in height). Other species are *Sagina nodosa*, *S. maritima*, *Centaurium littorale* and *Gentianella campestris*, but it will not be long before the rank grasses, willow and crowberry have smothered these small plants.

An entirely different plant association was found (prior to the drying out process in 1959), in dune slacks known as The Winter Lochs, at the east end of the Culbin Forest. Here *Lycopodiella inundata* was abundant with *Drosera rotundifolia*, *Peplis portula*, *Apium inundatum*, *Calluna*, *Erica tetralix*, *Narthecium ossifragum* and *Deschampsia setacea*.

(f) *The Buckie Loch, Culbin Forest*

Although once holding water throughout the year but now nearly completely dried out, this area has still the richest flora in the forest. A relic of the former outlet of the River Findhorn, and subsequently hemmed in by a succession of high wind-blown dunes on each side save the east, only one small damp area remains at the west end of the loch. *Salix aurita*, *Iris pseudacorus* and *Agrostis tenuis* are encroaching rapidly into the last remaining open spaces and only just surviving are *Menyanthes trifoliata*, *Potentilla palustris*, *Lychnis flos-cuculi*, *Myosotis scorpioides*, *Parnassia palustris*, *Dactylorhiza purpurella*, *D. incarnata* and the sedge *Carex diandra*. In drier places under alders that were probably the old bank of the river are *Platanthera chlorantha*, *Listera ovata*, *Pyrola minor* and *Corallorhiza trifida*. At the east end of the loch now dominated by *Agrostis tenuis* there is a unique association of *Botrychium lunaria* and *Ophioglossum vulgatum* often reaching 18 cm in height with very tall plants of *Gentianella campestris*.

(g) *Salt-marsh*

Surveyed between the Culbin Forest and the Old Bar, Nat. grid 28/9561. The seaward side is influenced by daily tides the landward side only by the spring tides.

The muddy sand of the seaward zone is first colonised by *Salicornia europaea* followed inland by short turf intersected by small creeks and pan-holes. The turf consists of *Puccinellia maritima*,

Juncus gerardii with *Suaeda maritima* and *Glaux maritima* and on the edges of the runnels and holes *Aster tripolium*, *Plantago maritima* and *Triglochin maritima*. In the less brackish pools are found *Scirpus tabernaemontani*, *Eleocharis palustris*, *E. uniglumis*, *Blysmus rufus* and occasionally *Scirpus maritimus* and *Juncus maritimus*. As the turf becomes further removed from the tidal influence there is a sudden change in the flora. *Juncus gerardii* is still common with the addition of *Linum catharticum*, *Potentilla anserina*, *Cochlearia officinalis*, *Schoenus nigricans*, *Euphrasia foulaensis*, *Carex extensa*, *C. flacca* and *C. distans*.

In some places on the perimeter of the forest the low dunes are banked with shingle and there the common plants are *Silene maritima*, *Ligusticum scoticum*., *Matricaria maritima* subsp. *maritima*, *Agropyron repens*, *Elymus arenarius* and *Poa subcaerulea*.

(h) *Shingle beaches*

Surveyed between Ardersier and Fort George, Nat. grid 28/7755 and consisting of stable shingle with flat rounded pebbles (4–10 cm diameter) occasionally influenced by spray. The fringe flora consists of *Silene maritima*, *Matricaria maritima* subsp. *maritima*, *Agropyron repens*, *Vicia hirsuta*, *V. sativa* subsp. *nigra*, *Geranium robertianum* and *Rumex crispus*, and where more soil occurs between the pebbles are *Teucrium scorodonia*, *Pilosella officinarum*, *Agrostis tenuis*, *Poa subcaerulea*, *Hypochoeris radicata*, *Rosa pimpinellifolia*, *R. rubiginosa* and in open spaces *Plantago coronopus*, *Vicia lathroides*, *Viola canina* and a prostrate form of *Vicia sylvatica*. The landward shingle is dominated by whins.

Other shingle communities occur along the coast. Those of the Old Bar in the Culbin Forest between Nairn and Findhorn are subject to frequent disturbance by wave action and are poor in species. The seaward pebbles are covered with lichens. Strong wave action on the shingle west of Nairn has accounted for the disappearance of *Mertensia maritima* which was abundant fifty years ago.

2 Coastal Lowlands

Consisting of the fertile, rich, agricultural lands of Laich of Moray and The Aird, west of Inverness.

(a) *The calcareous lacustrine clays of Loch Spynie*, Nat. grid 38/2366

This loch, once an arm of the sea, has now only a small area of open water and is becoming smaller and smaller owing to drainage. The surrounding marshland contains a unique flora for the area. It is feared recent extensive draining (1976–77) will seriously affect the status of certain species. On the west and south sides of the loch are dense stands of *Phragmites australis* and *Salix* spp. but on the east and north sides a more open community is found extending on the north side to marshland dominated by *Iris pseudacorus*. In the loch grow *Potamogeton gramineus*, *P. perfoliatus*, *P. crispus*, *P. pectinatus* *P. x suecicus*, *Elodea canadensis* and *Ceratophyllum demersum*. The low eastern bank is formed into several zones commencing on the water side with *Scirpus lacustris*, *Iris pseudacorus*, *Filipendula ulmaria*,

Juncus effusus and *Carex rostrata*, followed inland with a lower type vegetation of *Lychnis flos-cuculi, Mimulus guttatus, Myosotis laxa* subsp. *caespitosa, Pedicularis palustris, Potentilla palustris, Dactylorhiza purpurella, Juncus balticus* and *Carex diandra* etc. Further inland the communities merge with a heath type flora of *Erica tetralix, Potentilla erecta, Achillea millefolium* and other species. In wet ditches two of the rarest plants of north Scotland occur, *Butomus umbellatus* (very rare) and *Berula erecta*. Under willows are *Pyrola minor, Eleocharis palustris, E. uniglumis* and occasional *Corallorhiza trifida*.

(b) *Cultivated land*

Surveyed in a root field at Craigfield Farm, Kintessack, Nat. grid 38/0064. Certain plants of cultivated land are found on every type of soil throughout the two vice-counties these include *Raphanus raphanistrum, Sinapis arvensis, Spergula arvensis, Anchusa arvensis, Polygonum aviculare, Chenopodium album* and *Galeopsis tetrahit*. Others are confined to the Lowlands or are extremely rare in the Highlands and these include *Fumaria* spp. *Chrysanthemem segetum, Papaver* spp.

On sandy soil at Craigfield Farm in addition to the above are *Papaver argemone, Fumaria capreolata, Euphorbia helioscopa, Thlaspi arvense, Anthemis arvensis, Centaurea cyanus* and *Galeopsis bifida*.

Veronica agrestis, V. persica and *V. sublobatum* are some of the common garden weeds. The latter appears to be the common species of Ivy-leaved Speedwell in the north although it has not yet been critically examined.

Other sandy fields near the coast line may have *Anagallis arvensis, Coronopus didymus* and *Hypochoeris glabra* but these are all very local.

3 Drumossie Muir and the Peneplain Fringe

Consisting of dry moorland, wet moorland bogs, deciduous woodlands and river banks.

(a) *Dry moorland*

Surveyed at Essich, Nat. grid 28/6538. A treeless area studded with whins and poor in species. Dominated by *Calluna vulgaris, Nardus stricta* with *Deschampsia flexuosa, Agrostis tenuis, Molinia caerulea, Galium saxatile* and *Polygala serpyllifolia*.

(b) *Wet moorland bogs*

Surveyed by the Big Burn near Essich, Nat. grid 28/6338. Underlain by calcareous sandstone and rich in species. The flushes contain *Eriophorum latifolium, Schoenus nigricans, Saxifraga aizoides, Trollius europaeus, Parnassia palustris, Sagina nodosa, Dactylorhiza fuchsii, D. purpurella, Veronica scutellata, Eleocharis quinqueflora, Carex dioica, C. lepidocarpa, C. diandra* and many other species.

(c) *Deciduous woodland*

Surveyed at Darnaway, Nat. grid 38/0052. Influenced by the river valley and containing a flora rich in species. The larger proportion of the trees are beech with some sycamore, ash and oak. Large areas of the beech woods are invaded with *Luzula sylvatica* too dense to allow natural regeneration to take place or for other species to survive, there are also large areas dominated by *Holcus mollis*. Where these two plants have not invaded the woods the common species to be found are *Anemone nemorosa, Adoxa moschatellina, Ajuga repens, Galium odoratum, Sanicula europaea* and *Mercurialis perennis*. Very rare is *Neottia nidus-avis*. On slopes leading to the river and containing many wet flushes there is an addition to the tree cover of *Prunus padus, Alnus glutinosa, Salix capraea* and *Sorbus aucuparia* and the ground flora includes *Cardamine amara, Stellaria nemorum, Geum* × *intermedium* with both parents, *Phegopteris connectilis, Gymnocarpium dryopteris, Equisetum sylvaticum* and several colonies of *Equisetum hyemale*

A small ash wood, nearby, which may well be native, growing on the cornstone outcrop opposite Cothall, is dominated by *Brachypodium sylvaticum* with *Sanicula europaea, Listera ovata, Orthilia secunda* and occasional *Primula vulgaris, Circaea* × *intermedia* and *Carex sylvatica*.

In the Kellas oak wood, Nat. grid 38/1553, the vegetation is similar to that in most Highland oakwoods in that there are several plant communities to be found. *Quercus petraea* is the dominant tree with occasional seedlings of *Sorbus aucuparia* and *Betula pubescens* subsp. *pubescens*. The ground flora consists of *Holcus mollis* (dominant), *Viola riviniana, Pteridium aquilinum, Galium saxatile, Oxalis acetosella, Conopodium majus, Melampyrum pratense* var. *hians, Trientalis europaea, Veronica chamaedrys, Lonicera periclymenum, Anthoxanthum odoratum* and *Deschampsia flexuosa*. There is also a large area of *Rhododendron ponticum*.

(d) River banks

Surveyed by the River Findhorn at Darnaway, Nat. grid 38/0052. The plant associations are numerous due to the variation in the river bank rock formations. In the steep sided ravines *Vicia sylvatica, Solidago virgaurea, Campanula rotundifolia, Festuca altissima, Melica nutans* and *Hieracium* spp. are frequent and on the grass banks at the waters edge the more interesting species are *Galium boreale, Mimulus guttatus, Geranium sylvaticum, Trollius europaeus, Geum rivale, Carex flacca* and *Equisetum pratense* only known to cone regularly in this locality.

HIGHLANDS

Formerly termed the Upland regions and including Watsons' median, sub-alpine and Alpine regions and comparative to Gaulds regions 4–8.

4 The '1,000 foot' Peneplain

(a) *Dry moorland*

Surveyed near Lochindorb on Dava Moor, Nat. grid 28/9735.

Dominated by *Calluna* and poor in species but notable for the two clubmosses, *Diphasium alpinum* and *Lycopodium annotinum* reaching a low level of 300 m.

(b) *Wet moorland*

Surveyed at Easter Limekilns on Dava Moor, Nat. grid 28/9936. A grassy hillside slope studded with juniper. Eighty-four species were noted over a very small area the more interesting being *Trollius europaeus*, *Anemone nemorosa*, *Geum rivale*, *Linum catharticum*, *Pedicularis palustris*, *Drosera rotundifolia*, *Pinguicula vulgaris*, *Galium uliginosum*, *Carex flacca*, *C. dioica* and *C. hostiana*.

5 The '2,000 foot' Plateau

Surveyed on Creagan a' chaise in the Hills of Cromdale, Nat. grid 38/1024. Steep-sloped treeless moorland with shallow soil. Vast areas in the country come into this category of vegetation. Dominated by stunted *Calluna vulgaris* with some *Erica cinerea*, *Vaccinium myrtillus*, *V. vitis-idaea*, *Arctostaphylos urva-ursi*, *Nardus stricta*, *Scirpus cespitosus*, *Genista anglica*, *Empetrum nigrum*, *E. hermaphroditum*, *Anthoxanthum odoratum*, *Antennaria dioica* and *Rubus chamaemorus*. Very few flushes or burns are encountered but when present will contain *Saxifraga stellaris*, *Epilobium anagallidifolium*, *Montia fontana* and occasional sedge species. In the shelter of burn ravines, *Sorbus aucuparia*, *Populus tremula*, *Dryopteris* spp. and oak and beech ferns are frequent.

6 The Cairn Gorm Plateau

Surveyed in the Coire an Lochain of Cairn Gorm between 900 and 1,200 m, Nat. grid 28/9802.

(a) *Moorland slopes above the tree limit*

Chiefly consisting of *Calluna vulgaris* the associated species being *Vaccinium myrtillus*, *V. vitis-idaea*, *Cornus suecicus*, *Lycopodium clavatum*, *L. annotinum* and *Empetrum hermaphroditum*. With increasing altitude more open communities occur and *Diphasium alpinum*, *Alchemilla alpine*, *Antennaria dioica*, *Loiseleuria procumbens*, *Arctostaphylos urva-ursi* and *Vaccinium uliginosum* are common and in wetter places *Erica tetralix*, *Juncus squarrosus*, and *Nardus stricta*.

Approaching the lochan in the corrie large boulder fields are encountered and many ferns find shelter, notably *Athyrium distentifolium*, *Cystopteris fragilis*, *Gymnocarpium dryopteris*, and *Phegopteris connectilis*. On the gravels between the boulders where water seeps continually are *Saxifraga stellaris*, *Tofieldia pusilla*, *Luzula spicata*, *Pinguicula vulgaris* and *Juncus triglumis*.

(b) *Grass slopes below the cliffs and the summit plateau*

Subject to snow cover until mid June or later. The grasses are *Festuca vivipara*, *Deschampsia caespitosa*, *D. flexuosa*, *Agrostis tenuis*, *A. canina* subsp. *montana*, *Phleum alpinum* and occasional *Deschampsia alpina.* Other common species are *Alchemilla alpina, Ranunculus acris, Epilobium anagallidifolium, Cerastium cerastoides, Polygonum viviparum, Veronica serpyllifolia* subsp. *humifusa* and *V. alpina.*

(c) *Cliff ledges at* 1,100 *m*

Where water seeps regularly a rich flora is encountered such as *Cerastium arctium, C. cerastoides, Cochlearia officinalis* subsp. *alpina, Saxifraga rivularis, Saussurea alpina* and *Carex lachenalfii* and in pockets with more nutrients in the soil a tall herb community is found of *Trollius europaeus, Geranium sylvaticum, Epilobium angusti-folium, Angelica sylvestris* and *Rumex acetosa.* In dry rock clefts *Poa flexuosa* is not growing in its usual habitat of wet scree. The small fern *Cryptogramma crispa,* rather rare in this corrie, grows on scree with *Saxifraga nivalis, Silene acaulis* and *Cherleria sedoides.*

(d) *The summit plateau*

The most characteristic plant of the exposed mountain top vegeta-tion along with several mosses and lichens is *Juncus trifidus* with some *Salix herbacea, Gnaphalium supinum, Carex bigelowii, Festuca vivpara* and the rare *Luzula arcuata.* In the shelter of the terraces *Huperzia selago, Cardaminopsis petraea, Silene acaulis, Armeria maritima* and *Saxifraga stellaris* are frequent.

Further to the south on the diorite cliffs and burn ravines in Gaick Forest and Glen Feshie a rich calcicole flora is found several of which are not found on Cairn Gorm, they include *Polystichum lonchitis, Asplenium viride, Dryas octopetala, Silene acaulis, Saxifraga oppositifolia, Arabis hirsuta, Draba incana, D. norvegica, Veronica fruticans, Galium sterneri, Orchis mascula* and *Carex capillaris* most of which are locally common.

The mountains in Glencannich Forest to the west have two plants so far not recorded for other localities in *Juncus castaneus* and *Juncus biglumis* and the bogs and flushes on the low summits of the hills about Dalwhinnie and in Gaick Forest quite commonly have *Carex rariflora.*

7 Mid Strathspey and Badenoch

(a) *Swamps*

Between Kingussie and Loch Insh the River Spey becomes slow-flowing and marshes usually known as 'fens' occur on each bank for a distance of some three miles. There are large areas dominated by *Phragmites australis, Potentilla palustris, Equisetum fluviatile, Meny-anthes trifoliata, Carex rostrata, C. vesicaria* and *C. aquatilis,* and some of the pools have the small yellow water-lily *Nuphar pumila.* Where the ground is less swampy the common plants are *Lychnis flos-cuculi, Pedicularis palustris, Filipendula ulmaria, Cicuta virosa, Galium palustre* and many sedges. The drainage ditches have *Sparganium erectum, Glyceria fluitans* and *Phlaris arundinacea* among other plants. The whole area is subject to extensive winter flooding.

(b) *Dwarf shrub moors*

Surveyed south of Nethybridge, Nat. Grid 28/9820. Dry, undulating ground of calcareous schist. Heavily grazed.

Dwarf *Calluna* dominant with *Helianthemum chamaecistus, Gymnadenia conopsea, Pseudorchis albida, Genista anglica, Campanula rotundifolia, Thymus praecox* subsp. *arcticus, Festuca ovina, Veronica chamaedrys,* and in one hollow *Listera ovata.*

(c) *Consolidated river alluvium*

Surveyed at Newtonmore, Nat. grid 27/7298. Among the plants to be found on this very rich consolidated shingle are *Viola lutea, Polygala vulgaris, Pimpinella saxifraga, Platanthera chlorantha, Galium boreale, Cirsium helenoides* and *Helichtotrichon pratense.* On unstable shingle, *Saxifraga stellaris* and *S. oppositifolia* are short-lived. On the banks of the river is *Mimulus guttatus × luteus* in many colour forms and in the backwaters *Myriophyllum alterniflorum* and several *Potamogeton* spp.

(d) *Pine woods*

Surveyed at Loch Garten, Nat. grid 28/9718. Mature pine wood probably lineal descendants of the old Caledonian Forest.

With the *Pinus sylvestris* are a few trees of *Betula pubescens* and *Sorbus aucuparia, Juniperous communis* is plentiful. The forest floor is dominated by *Calluna vulgaris, Vaccinium myrtillus, V. vitis-idaea* and many mosses. An interesting form of *V. myrtillus* grows in association with the type and is readily distinguished by its uniformly taller (15 cm) habit and black shining fruits without bloom. Where the *Vaccinium* is not so dense associated species are *Goodyera repens, Potentilla erecta, Equisetum sylvaticum, Blechnum spicant, Pyrola media, Trientalis europaea, Luzula pilosa, L. multiflora, Oxalis acetosella, Deschampsia flexuosa, Molinia caerulea* and *Agrostis tenuis.*

There are many birch woods in Strathspey and the vegetation is varied, west of the strath *Betula pubescens* subsp. *odorata* is the common tree and the ground flora is often herb-rich with *Trientalis europaea, Lysimachia nemorum* and *Melampyrum pratense* etc. Other birch woods have a *Vaccinium-rich* association, and in others the floor cover consists of *Pteridium aquilinum, Conopodium majus* and *Endymion nonscriptus.* The native oak woods of *Quercus petraea* can have a floor cover of *Thelepteris limbosperma, Lysimachia nemorum, Anemone nemorosa* and *Primula vulgaris* or be quite poor in species dominated by *Holcus mollis* and *Conopodium major.*

8 The Monadhliath Mountains

Surveyed at Alvie, Nat. grid 28/7911. Large areas of blanket bog with some outcrops of basic soil.

The blanket-bog community is typical with much *Sphagna, Erica tetralix, Scirpus cespitosus* subsp. *germanicus, Myrica gale, Rubus chamaemorus, Betula nana, Eriophorum vaginatum, Juncus squarrosus* and some *Calluna.*

(b) *Basic outcrops*

These outcrops are of rare occurrence in the Monadhliaths and form small dry hummocks throughout the blanket bog. The plants encountered are *Linum catharticum, Geum rivale, Selaginella selaginoides, Botrychium lunaria, Galium boreale, Thalictrum alpinum, Alchemilla vestita* ~~filicaulis~~ subsp. ~~filicaulis~~ vestita and *Carex caryophyllea* all of very small stature.

(c) *Roadside verge (A9) near Dalwhinnie*

Newly-sown roadside verge

Surveyed by the A9 north of Dalwhinnie, Nat. grid 27/6688. Open moorland at approximately 350 m. The plants regarded as native to this area are marked with an asterisk.

ABUNDANT
Agropyron repens
Cynosurus cristatus
Festuca rubra subsp. *litoralis*
Lolium perenne
Plantago major
Poa pratensis
Poa trivialis
*Sagina procumbens
Spergula arvensis

FREQUENT
Achillea millefolium
*Agrostis canina subsp. *montana*
*A. tenuis
Alopecurus pratensis
*Anthoxanthum odoratum
*Cerastium fontanum subsp. *triviale*
Cirsium vulgare
Crepis capillaris
Dactylis glomerata
*Deschampsia caespitosa
Galeopsis tetrahit
*Galium saxatile
*Holcus lanatus
*Juncus kochii
*Leontodon autumnalis
Lolium multiflorum
*Lotus corniculatus
*Luzula multiflora
Phleum pratense
*Poa subcaerulea
*Poa annua
*Potentilla erecta
*Prunella vulgaris
Ranunculus repens
*Rumex acetosa
*Trifolium repens
Veronica officinalis
V. serpyllifolia
Viola arvensis

OCCASIONAL
Alopecurus geniculatus
Anthemis arvensis
*Bellis perennis
Capsella bursa-pastoris
Chamomilla suavolens
Cirsium arvense
Leucanthemum vulgare
Matricaria maritima
Myosotis discolor
Polygonum aviculare
Rumex crispus
R. obtusifolius
R. longifolius
Stellaria media
Veronica arvensis
*V. chamaedrys

RARE
Agrostis gigantea
Anthyllis vulneraria subsp. *carpatica* var. *pseudovulneraria*
Barbarea vulgaris
Brassica napus
*Callitriche hermaphroditica
Conopodium majus
Epilobium montanum
Geranium molle
G. pusillum
Leontodon taraxacoides
Myosotis arvensis
*Myrica gale
*Ranunculus flammula
*Rubus idaeus
Senecio jacobea
S. vulgaris
Trifolium dubium
T. hybridum
T. pratense
Vicia sativa subsp. *nigra*
*Viola canina
V. tricolor

EXPLANATION OF THE TEXT

Nomenclature

The nomenclature has been checked by P. D. Sell. It follows basically J. E. Dandy, *List of British Vascular Plants* (1958) with corrections, particularly from the paper by Dandy in *Watsonia* 7: 157–178 (1969) and from *Flora Europaea* (1964–). Where the name is different from that in Dandy (1958), synonyms are given to connect the taxon with the taxon in that work. These synonyms also connect the taxa with those in A. R. Clapham, T. G. Tutin and E. F. Warburg, *Flora of the British Isles*, 2 (1962). Some critical taxa and aliens are not in any of these works, and not all nomenclature in *Flora Europaea* adhered to.

Vernacular names, as far as possible have been taken from J. G. Dony, C. M. Rob and F. H. Perring, *English Names of Wild Flowers* (1974); they are often preceded by local names taken from the old Flora's of Moray.

The area covered by the Flora falls into two botanical vice-counties according to the system devised by H. C. Watson, but to enable political Nairn to have its own identity it has been divided into three sections as follows:

M = Moray or v.c. 95 Elgin
N = Nairn or v.c. 96 Easterness in part
I = East Inverness or v.c. 96 Easterness in part

Within the above system chronological sequence of records has been followed as far as possible and the earliest record for each v.c. is always included. Locality and date are followed by the name of the collector or recorder in italics and the standard abbreviation of the herbarium where the specimen is housed in bold capitals. Where the name of the collector appears many times only initials (in italics) without punctuation are given. Records without an authority are my own and to save space my initials have been excluded except where a plant has been collected in company with another botanist or has been again collected from a locality from which the species has already been recorded. In other cases an exclamation mark following a record indicates that I have also seen it in the locality. All the records from G. Gordon's *Collectanea* are included to emphasise status at an earlier period.

When a plant is recorded as abundant or very common only a few localities are given, in most cases when a plant is *rare* all are cited, and when it is very rare only the district, with no details, are given.

List of Herbaria consulted with their standard abbreviations

Those herbaria marked as being completely checked means that every sheet in every folder in the British section has been examined. It has resulted in many early records being located. *All* the herbarium records of H. Clark, A. Croall, J. Keith, A. Johnston, R. Thomson and A. Somerville have been included.

ABD	Department of Botany, University of Aberdeen (completely checked)
BM	British Museum of Natural History (completely checked up to the end of *Rosaceae*, also *Cyperaceae* and *Gramineae* and separate collections of J. Dickson and G. Don)
CGE	Botany School, University of Cambridge (*Hieracium*, *Pilosella* and *Rubus* completely checked with numerous other records particularly those of E. S. Marshall)
UKD	Ursula K. Duncan (private collection)
DURHM	Durham University
E	Royal Botanic Garden, Edinburgh (completely checked)
ECW	Edward C. Wallace (private collection)
EWG	E. W. Groves (private collection)
ELN	Elgin Museum (completely checked)
FRS	Falconer Museum, Forres (completely checked)
GL	Glasgow University (completely checked)
INV	Inverness Museum (completely checked)
K	Royal Botanic Gardens, Kew (Rosa, Salix, and Mentha completely checked and the collections of W. Borrer, G. Don, W. J. Hooker, J. Lightfoot and H. C. Watson)
JEL	J. E. Lousley (private collection)
MANCH	The University, Manchester
MBH	Marlborough College (herb. A. & M. L. Wedgewood)
NRN	Nairn Literary Society, Nairn (completely checked)
NDS	Norman Douglas Simpson (private collection now at the BM, list supplied by NDS)
OXF	Oxford University (completely checked)
PTH	Perth Museum
STA	The University, St Andrews (completely checked)
STI	The Smith Institute, Stirling (completely checked)
S	The University of Strathclyde, Glasgow (complete list checked)

Abbreviations of collectors names

Names appearing many times in the text are reduced to initials and printed in italics without punctuation.

HC	Hugh Clark	*ESM*	Edward S. Marshall
UKD	Ursula K. Duncan	*NDS*	Norman Douglas Simpson
RAG	Rex A. Graham	*WASh*	W. A. Shoolbred
RMH	Raymond M. Harley	*WAS*	William Alexander Stables
AJ	Anthony Johnston	*AS*	Alexander Somerville
RJ	Roy Jones	*OMS*	Olga M. Stewart
JK	James Keith	*MMcCW*	Mary McCallum Webster
BAM	Beverly A. Miles		

Literature abbreviations

To save space literature abbreviations occurring many times in the text have been abridged. References to *Volume numbers* in the *Annals of Scottish Natural History* should be disregarded as they are errors. This mistake occurred owing to the volumes being numbered in the Science Library at Aberdeen University (where most of my research took place) but they are not numbered in other libraries and the references should have been referred to by a *Section number*. The date and page numbers should however be sufficient to check the records.

ABBREVIATIONS IN LITERATURE

(Aitken)   Aitken, Dr Thomas, Plant List, *Transactions of the Inverness Scientific Society and Field Club*, **1** (1875–1888).

(Atlas)   Perring, F. H. & Walters, S. M. *Atlas of the British Flora* (1962).

(Balfour, 1865)   Balfour, J. H., Report of excursion to Cairn Gorm summit, *Trans. Bot. Soc. Edin.* **8**: 220 (1865).

(Balfour, 1868)   Balfour, J. H., Notice of botanical excursions in the Highlands of Scotland during autumn 1867, *Trans. Bot. Soc. Edin.* **9**: 293–300 (1868).

(Ball, 1851)   Ball, John, List of plants in Strath Affric, *Bot. Gaz.* **3**: 42, 56 (1851).

(Boyd)   Boyd, W. B., Notes on an excursion to the district of Kingussie with the Scottish Alpine Botanical Club, *Trans. Bot. Soc. Edin.* **13**: lxix–lxxiv (1878).

(Burgess)   Burgess, James J. (ed.), *Flora of Moray* (1935).

(Critical Atlas)   Perring, F. H. & Sell, P. F., *Critical Supplement to the Atlas of the British Flora* (1968).

(J. Don, 1898)   Don, J., *Flowering Plants of Inverness-shire and of some parts of adjoining Counties* (1898).

(Druce, 1888)   Druce, G. C., Notes on the Flora of Easterness, Elgin, Banff and West Ross, *J. Bot.* **26**: 17–26 (1888).

(Druce, 1889)   Druce, G. C., First records for Moray, A. Bennett *Scott. Nat.* **4**: 104 (1889).

(Druce, 1897)   Druce, G. C., Notes of plants observed about Forres and Findhorn, *Ann. Scott. Nat. Hist.* 54–55 (1897).

(Druce, *Com. Fl.*)   Druce, G. C., *The Comital Flora of the British Isles* (1932).

(Ewing)   Ewing, P., Flora of the Culbin Sands, *Glasgow Naturalist*, **5**: 5–15 (1913).

(Gordon)   Gordon, George, *Collectanea for a Flora of Moray* (1839).

(Gordon, *annot.*)   Gordon, George, An annotated copy (1848) by J. G. Innes of *Collectanea for a Flora of Moray* and a manuscript list of additions by G. Gordon, in library N. D. Simpson, now at Botany School, Cambridge.

(Hooker, *Fl. Scot.*)   Hooker, W. J., *Flora Scotica* (1821).

(Innes)   Innes, John G., Plant list of Forres and district (*c.* 1834), in Falconer Museum, Forres.
(G. A. Lang, 1905)   Lang, G. A., A list of wild flowers and ferns of Inverness and district (1905).
(Lightfoot)   Lightfoot, J., *Flora Scotica*, edn. 2 (1792).
(Lobban)   Lobban, A., Notebooks containing lists of plants found near Nairn (1902), in Library of the Nairn Literary Society, Nairn.
(Marshall, 1887)   Hanbury, F. J. & Marshall, E. S., Notes of some plants of northern Scotland observed in 1886, *J. Bot.* **25**: 165–169 (1886).
(Marshall, 1898)   Marshall, E. S. & Shoolbred, W. A., Notes of a tour in North Scotland 1897, *J. Bot.* **36**: 166–177 (1898).
(Marshall, 1899)   Marshall, E. S. & Shoolbred, W. S., Some plants of East Scotland, *J. Bot.* **37**: 383–389 (1899).
(Moyle Rogers)   Rogers, W. Moyle, North-east Highland plants 1903, *J. Bot.* **42**: 12–21 (1904).
(Murray)   Murray, A., *The Northern Flora* **1** (1836).
(Pealling)   Pealling, R. J., The Botany of Loch Duntelchaig, *Transactions of the Inverness Scientific Society and Field Club*, **9** (1923).
(Pollock)   Pollock, J. R., *Guide to Beauly and District* (1902), botany by J. Donald.
(Stables, *Cat.*)   Stables, W. A., Marked copy of *Collectanea for a Flora of Moray*, G. Gordon (1839), no. 176 in County Catalogues of H. C. Watson in library Kew herbarium.
(Smith, *Eng. Fl.*)   Smith, J. E., *The English Flora*, **1–4** (1828–30).
(Thomson)   Thomson, Robert, *The Natural History of a Highland Parish (Ardclach)* (1900). *Nairn Telegraph.*
(Trail, 1904)   Trail, J. W. H., Alien flora of the lower part of the Spey, 1899, *Ann. Scott. Nat. Hist.* 103–106 (1904).
(Watson)   Watson, H. C., *The New Botanists Guide*, **2** (1837).
(Watson, *Loc. Cat.*)   Watson, H. C., MS lists no. 43, Inverness (1832); no. 28, Dalwhinnie Inn (1833); Kingussie to Dalwhinnie no. 44 (1833); local catalogues, H. C. Watson library Kew herbarium.
(Watson, *Top. Bot.*)   Watson, H. C. *Topographical Botany*, **2** (1847).

Other abbreviations

Punctuation and prepositions have been omitted whenever possible.
Place names such as Grantown-on-Spey are referred to as Grantown.
*   denotes the record has not been seen by the expert indicated as having named most specimens in the group
BSBI   Botanical Society of the British Isles
BSE   Botanical Society of Edinburgh
nm   nothomorph (hybrid form of sexual origin whether a member of the first generation, a later segregate or a back-cross (Melville, *Proc. Linn. Soc. London* **151**: 158 (1939)). Supplement second edition

sp.   An unidentified species of the genus indicated
subsp.   subspecies
CTW   A. R. Clapham, T. G. Tutin & E. F. Warburg, *Flora of
         the British Isles* ed. **2** (1962)
var.   variety
v.c.   vice-county
!   after a station, indicates the author has seen the plant growing
         there.

Distribution maps

Each record represents a 10 km grid square of the National Grid on
the Ordnance Survey Map. Plants confined solely to the coast and
to mountain tops and ledges over 1,000 m have not all been mapped.
Neither have the very rare or the very common species.

● = Records seen by the author
L = Records in literature and by other collectors

The metric system has been used for plant measurements and the
heights of hills, but miles have been retained for distance so as to
agree with the distances given on the old herbarium sheets and
literature records.

10 cm = 3·9 inches   1·609 km = 1 mile
5 m = 16·4 feet   0·4047 hectares = 1 acre
100 m = 330 feet

PTERIDOPHYTA

Generic sequence and nomenclature according to J. A. Crabbe, A. C. Jermy and J. T. Mickel in *Fern Gazette* 2: 141–162 (1975). Specific and infraspecific nomenclature and order according to British Museum arrangement of British Ferns.

LYCOPSIDA
LYCOPODIACEAE
Huperzia Bernh.
H. selago (L.) Bernh. ex Schrank & Mart.   *Fir Clubmoss*
Lycopodium selago L.
Native. Common in open places on moors reaching 1,220 m and descending to sea level by the shores of Lochloy in the Culbin Forest. Distinguished from the three following genera by the lack of a distinctly stalked cone. Map 1.
M Frequent (Gordon); Califer Hill near Forres (Innes); sandy ground in the Culbin Forest, 1954 **E**.
N (Stables, *Cat.*); Ardclach (Thomson); in damp ground on the north shore of Lochloy, 1954.
I Inverness, 1832 and Dalwhinnie, 1833 (Watson, *Loc. Cat.*); Badenoch, *A. Rutherford* **E**; summit of Cairn Gorm (Balfour, 1865) and 1891 *AS* **E**.

Lycopodiella Holub
L. inundata (L.) Holub   *Marsh Clubmoss*
Lycopodium inundatum L.
Native. Rare on boggy verges of lochs and in damp dune slacks. Pl 1.
M Beyond Dyke Moss, towards the sea, 1800 *Brodie* **E**; rare at Hatton, Kinloss, *G. Wilson* (Gordon)!; parish of Alves, 1834 *J. Fraser* **STI**; shire of Moray, *J. Mackay* **BM**; dune slack Winter Lochs, Culbin, 1950 **E,BM**, probably extinct there owing to forestry drainage; wide ditch on waste grassland E of Hopeman golf course, 1975 *R. Richter* **E,BM**!; a few plants in a dry slack, Culbin, 1976 *R. Richter*.
N Near Budgate, Cawdor, 1832 *WAS* **E,BM**; Lochloy, 1832 *W. Brichan* **ABD**, it was still to be seen there in 1954 but appears to have now disappeared; near Druim House, Nairn, 1937 *J. Walton* **GL**.
I Carse of Ardersier, plentiful, 1844 *A. Croall* herb. *Hooker* **K**; Loch Morlich, 1938 *E. M. Wakefield* **K**, it is still plentiful on the S shore of the loch.

Lycopodium L.
L. annotinum L.   *Interrupted Clubmoss*
Native. Common on moors over 305 m especially in Strathspey. Fig 1d, Map 2.

M Aviemore, 1833 *WAS* (Gordon); moor on the E of Lochindorb, 1950 **E**; pine wood above Fochabers, 1953 *R. Richter*; plentiful on Càrn Dearg Mòr, Aviemore, 1974 **E**.
N Cawdor and the Findhorn basin S of General Wade's road, Ardclach (Lobban); moor on the W of Lochindorb, 1950.
I Mountains N of Loch Ericht, *Brodie* **E**; Dalwhinnie, 1833 (Watson, *Loc. Cat.*); Cairn Gorm, *Hooker* **E,GL**; Fraoch-choire, Cannich (Ball, 1851); Farr, 1881 *A. G. Gregor* **K**; Cairn Gorm, 1882 *J. Groves* **BM** and 1893 *AS* **E**; frequent on the moors above Milton of Drumnadrochit, 1947.

L. clavatum L.   *Tods' Tails, Stag's-horn Clubmoss*
Native. Common on moors, in pine woods and forest breaks. Fig 1c, Map 3.
M On mountain heaths and in woods, 1800 *Brodie* **E**; frequent (Gordon); Culbin Sands, 1926 *G. C. Druce* **OXF** and 1952 *MMcCW* **ABD**; edge of a quarry at Newton Toll; forming dense patches in short grassland on the moors at Dallas.
N (Stables, *Cat.*); Cawdor, 1863 *R. Thomson* **NRN**; fixed dune on the E side of Cran Loch, 1960 **ABD**; lawn at Glenferness House where it has resisted mowing for a number of years, 1974 *Earl of Leven and Melville*.
I Dalwhinnie, 1833 (Watson, *Loc. Cat.*); Ardersier, 1843 ? *coll.* **K**; above Dochfour, 1854 *HC* **INV**; Kincraig, 1891 *AS* **BM**; lower slopes of Mealfuarvonie, 1971 **E**.

Diphasium C. Presl
D. alpinum (L.) Rothm.   *Alpine Clubmoss*
   Lycopodium alpinum L.
Native. Common on dry moors on mountains reaching 1,220 m. A straggling, densely tufted mountain plant with glaucous-green leaves. Fig 1a, Map 4.

   Var. **alpinum**
M Frequent (Gordon); Califer Hill, Forres, *J. G. Innes*; moor at the SE corner of Lochindorb, 1954; summit of Beum a' Chlaidheimh, Duthil, 1967 **E**.
N Clunas, 1833 *WAS* **ABD**; Hill of Aitnoch, Ardclach, 1884 *R. Thomson* **NRN**; common on the moor at Ballochrochin, Drynachan, 1974.
I Dalwhinnie, 1833 (Watson, *Loc. Cat.*); Badenoch, 1838 *A. Rutherford* **E**; Fraoch-choire, Cannich (Ball, 1851); heaths above Nairn valley, 1885 (Aitken); Kincraig, 1891 *AS* **BM**; An Leacainn, Blackfold, 1973 **E**; Bynack More, 1975 **BM**.

   Var. **decipiens** (Syme)
   Lycopodium issleri (Rouy) Lawalree; *L. complanatum* auct.
M Near Advie, Grantown-on-Spey, 1871 *J. S. Gamble* **K**.
I Cairngorms (Druce, 1888); SW slope of Geal-chàrn, Glen Feshie at 800 m, 1909 *A. Wilson & J.A. Wheldon* **K**.

53

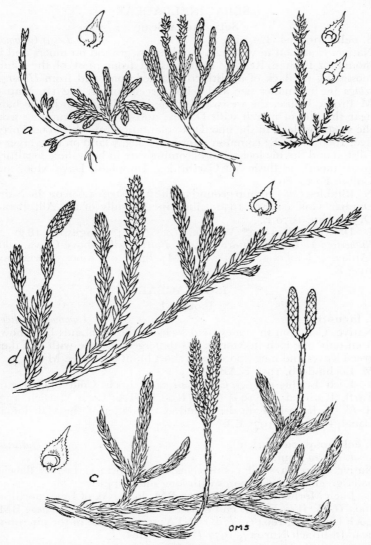

Fig 1   a, **Diphasium alpinum**; b, **Selaginella selaginoides**; c, **Lycopodium clavatum**; d, **Lycopodium annotinum**.

SELAGINELLACEAE
Selaginella Beauv.

S. selaginoides (L.) Link   *Lesser Clubmoss*
Native. Common in wet grassy and mossy ground on moors and in
mountain flushes. Readily distinguished from most of the Clubmosses by the lack of a distinctly stalked cone and from *Huperzia
selago* by its smaller size and yellow-green colour. Fig 1b, Map 5.
M Brodie House, the westward of the bog of Milldish in the bank
near the line of march with Dalvey, and in a ditch running from
the westmost loch in the new Dyke plantation into the wood Forty-
five, 1800 *Brodie* **E**; common (Gordon); boggy land on the verges of
Gilston and Spynie lochs, but becoming rare in both these localities,
1950; moorland flushes at Carrbridge, Dunphail, Dava Moor and
Culbin Forest etc.
N (Stables, *Cat.*); damp ground by the Minister's Loch on the Nairn
Dunbar Golf course, 1950; flush in the gully of the Allt Breac,
Drynachan, 1970 **E**.
I Dalwhinnie, 1833 (Watson, *Loc. Cat.*); Kingussie, 1840 *A.
Rutherford* **I**; Fraoch-choire, 1850 *J. Ball*; heaths above Clava, 1885
(Aitken); Kincraig, 1891 *AS* **BM**; flush on moor Abriachan,
1972 **E**.

ISOETACEAE
Isoetes L.

I. lacustris L.   *Common Quillwort*
Native. Common in stoney acid lochs in the highlands but known
from only one loch in Moray. A submerged aquatic with stiff dark
green leaves and megaspores with short blunt tubercles. Map 6.
M Lochindorb, 1952 **E,ABD**.
I Loch Ericht, 1867 *A. Craig Christie* **E**; Lochs Coiltry, Uanagan,
Tarff, Ruthven etc. 1904 *G. T. West* **E,STA**; Loch Morlich, 1938
A. H. G. Alston **BM**; in deep mud at the outlet of the Caledonian
Canal, Inverness, 1975 **E,BM**.

I. echinospora Durieu   *Spring Quillwort*
   I. setacea Lam.
Native. Rare in similar situations to the former. Leaves flaccid,
paler green and megaspores with long sharp spines.
I Loch Morlich, 1887 *G. C. Druce* **BM**; loch NW of Invermoriston,
1904 *G. T. West* **E**; NE end of Loch Alvie, 1922 *C. E. Salmon* **BM**;
Loch Uanagan, 1904 *G. T. West* **STA**; small lochan in the pinewood
near Inshriach Nursery, 1972 *P. Barnes* **BM**.

SPHENOPSIDA
EQUISETACEAE
Equisetum L.

E. hyemale L.   *Dutch Rush*
Native. Occasional on shady banks of rivers and ditches and in
moorland bogs. A simple stemmed dark glaucous-green plant with

winter persistent shoots. Sheaths pale with a black ring at top and bottom. Teeth are normally shed. Fig 2a.

M Abundant in a few localities, bog at Pittendreich, Hill of Monachty, *G. Wilson* and Darnaway forest (Gordon); Moray, *G. Gordon* **GL**; Dunphail, 1834 *WAS* **ABD**; burn near Forres, *J. G. Innes* **E** and 1844 *T. Edmondston* **BM**; Sanquhar, 1856 *JK* **FRS**; swamp near loch N of Boat of Garten, 1922 *C. E. Salmon* **BM**; two colonies by the river path, Darnaway woods opposite to Sluie, 1956 **BM,K**.

N Bank of ditch a few yards E of Cothill Farm near Cran Loch, 1955 **E,ABD**; small bog in pinewood between the road and Lochloy, 1956 **ABD**.

I Moss of Petty, Culblair, 1844 *A. Croall* **BM**; pinewood near the Keppoch Stone, Culloden Moor, 1956 **BM**; a few plants on the W side of a small lochan S of Essich, 1971 **E**; a large colony on a roadside bank between Easter Muckovie and Nairnside, 1975 **BM**; moor at the E end of Loch Bunachton, 1975 *M. Souter*.

E. fluviatile L.   *Paddocks' Pipes, Water Horsetail*
          E. limosum L.

Native. Common on loch margins, in bogs and in ditches. Stems simple, smooth with 10–30 shallow grooves and often with whorls of branches in the middle. Central hollow $\frac{4}{5}$ diameter of stem. Sheaths green, teeth dark and persistent. Fig 2b, Map 7.

M Innes House, Urquhart, 1794 *A. Cooper* **ELN**; frequent (Gordon); Forres, 1859 *JK* **FRS**; small bog W of Pluscarden, 1968 **CGE**.

N Cawdor, 1832 *WAS* **ABD**; the Wine Well, Auldearn, 1883 *R. Thomson* **NRN**; margin of Cran Loch, 1960 **ABD,K**.

I (Watson, *Top. Bot.*); Loch Meiklie and Inchnacardoch Bay, 1904 *G. T. West* **E,STA**; Loch nan Geadas and Loch a' Chlachain (Pealling); ditch at Drumochter, Loch Crunachdam, Glen Shirra, Loch Loyne, Glen Strathfarrar etc.

A much branched form with many cones often referred to as var. *polystachyum* Weigel may occur in any colony from time to time. It appears to be rare in the north east and has only been recorded from a ditch near the old Ford at Achavraat, 1960.

E. arvense L.   *Common Horsetail*

Native. Abundant on poor soils, on waste ground, and dune slacks reaching 900 m on mountains. A pestilential weed difficult to eradicate. Poisonous to horses and sheep when dried. Stems branched, black and wiry, erect or decumbent, grooves 6–19, deep. Central hollow about $\frac{1}{2}$ diameter of stem. The ephemeral fertile stems are simple with loose pale brown sheaths and darker teeth and appear before the vegetative shoots emerge in March–April. Fig 2d.

M Very common (Gordon); Alves, *G. Wilson* **ABD**; Forres, 1856 *JK* **FRS**; Dunphail, 1886 *ESM* **CGE**; Culbin Forest, 1953 *A. Melderis* **BM**; roadside verge Earlseat, 1974 **E**.

N Hill of Urchany, Cawdor, 1833 *WAS* **K,GL**; Ardclach (Thomson); garden path Kinsteary, 1964 **ABD**.

I Banks of the canal, Inverness, 1833 *WAS* **GL**; near Kincraig, 1891 *AS* **BM**; roadside in Glen Convinth, 1972 **E**.

E. arvense × fluviatile = E. × litorale Kühlew. ex Rupr.
I Railway embankment by Balcarse Farm, Kirkhill, 1976 **E**.

E. pratense Ehrh.   *Shady Horsetail*
Native. Frequent on river banks and by mountain burns. Branches simple, spreading, generally drooping to one side. Stem grooves 8–20, rough. Sheaths green below, teeth long, narrow, dark and free. Central hollow ½ diameter of stem. Fig 3a, Pl 1, Map 9.
M Banks of the River Findhorn, Darnaway, opposite Sluie where it cones regularly, 1952 **E,BM**; steep grassy slope by the Allt na Criche, Cromdale, 1971 **E**; grassy slope by the Allt Iomadaidh, Fae, 1973.
N Several scattered colonies upstream from Coulmony House on the River Findhorn, 1960 **ABD**; wet hillside by Geddes Reservoir, 1961 **ABD**; roadside S of Righoul, 1973 **E,BM**.
I (Druce, *Com. Fl.*); near Boat of Garten, 1956 *E. F. Warburg* (*Atlas*); on cliffs and by the burn, Coire Chùirn, Drumochter, 1966; by River Spey, Kingussie, 1972 *R. McBeath*; gully of the burn at A'Chaoirnich, Gaick, 1974 **E**; by the Allt Coire an t-Slugain, Garva Bridge; by a small burn running into Loch Ericht from The Fara, Dalwhinnie.

E. sylvaticum L.   *Wood Horsetail*
Native. Common in damp woods, open moors and banks throughout the district. Branches drooping and again branched except in very young plants. Stem with 10–18 grooves with minute, stiff spicules giving a characteristic rough feel to the finger nail. Sheaths green below, teeth united into 3–6 large, dark to pale brown lobes, Central hollow ½ diameter of stem. Fig 3b, Map 8.
M Frequent (Gordon); woods near Forres, 1856 *JK* **FRS**; damp wood by the River Findhorn, Darnaway, 1955 **BM,K**; pinewood Culbin Forest, dry bank on verge of pinewood opposite the entrance to Glenerney House, Dunphail etc.
N Cran Loch, *Brodie* **E**; Cawdor, 1833 *WAS* **E**; Ardclach (Thomson).
I Dalwhinnie and Pitmain, 1833 (Watson, *Loc. Cat.*); Kincraig and Rothiemurchus, 1891 *AS* **E,BM**; banks of the Craggie Burn, Daviot, 1971 **E**.
A form with very slender thread-like branches is referred to as var. *capillare* Hoffm. but is not considered by present day taxonomists to be a variety. Recorded from Dunphail in 1886 by *ESM* **CGE** as new to Britain (*J. Bot.* **25**: 165 (1887)); by River Dulnain, Muchrach, 1972, shady path by Randolph's Leap, Relugas, 1973 **E**, and by River Lossie at Kellas. Also from Ardclach and Dulsie Bridge, 1974 **E**; Ardbrolach, Kingussie, 1887 *G. C. Druce* **OXF**; wood by Loch Ness, Lewiston, 1975 **E**; Newtonmore and Reelig Glen.

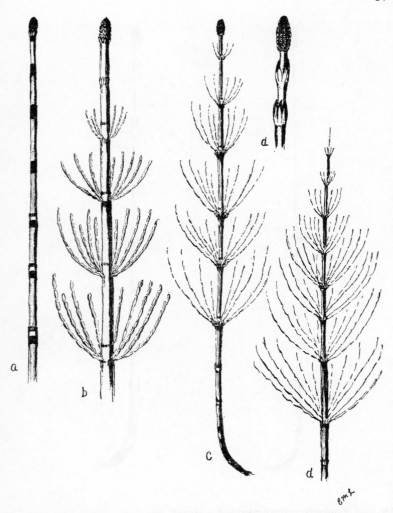

Fig 2   a, **Equisetum hyemale**; b, **E. fluviatile**; c, **E. palustre**; d, **E. arvense.**

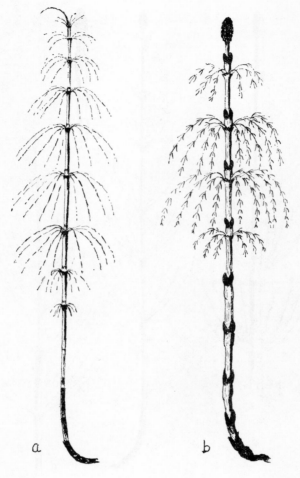

Fig 3   a, **Equisetum pratense**; b, **E. sylvaticum**.

E. palustre L. *Marsh Horsetail*
Native. Abundant on wet moorland, verges of ditches and bogs.
Usually with upward curving branches with shallow furrows and
dark-tipped clasping teeth. Grooves on stem few, 4–8, deep,
moderately rounded. Fig 2c.
M Frequent (Gordon); Balnageith, 1857 *JK* **FRS**; roadside bank,
Hatton Farm, 1918; Culbin Forest etc.
N Moss of Kinudie, 1832 *WAS* **ABD**; Ardclach, 1890 *R. Thomson*
NRN; Loch of the Clans, 1960 **ABD**; pond near Boghole, 1967 **E.**
I Dalwhinnie, 1833 (Watson, *Loc. Cat.*); Rothiemurchus, 1892
AS **E**; Strath Mashie, Laggan, 1916 *ESM* **E,BM**; throughout the
district, 1954.

OPHIOGLOSSACEAE
Ophioglossum L.
O. vulgatum L. subsp. **vulgatum** *Adder's Tongue*
Native. Very rare in short grassland, chiefly near the sea, and in
one peaty moorland depression in the Cairngorms. Pl 1.
M Near Burghead, 1834 *G. Wilson* **E** and *J. Fraser* **GL**; Culbin
Sands, *Mr Macleod* (Aitken); growing with *Botrychium* in grassland
at the E end of the Buckie Loch, Culbin Forest, 1950 **E,K.**
N Near Druim, Nairn, 1937 *J. Walton* **GL**.
I Natural pasture 2 miles from Croy, 1883 *T. Fraser* (*Trans. ISS
& FC* 3: 40 (1884)); rare on the Braes of Kilmorack, 1902 (Pollock);
over 1,000 plants in a peat depression 14 × 18 yards wide that had
evidently held water, Glen Feshie, 1974 **E,EM.**

Botrychium Sw.
B. lunaria (L.) Swartz *Unshoe the Horse, Moonwort*
Native. Frequent on dunes, short grassland, railway verges and
mountain ledges. Said to have had magical properties and been
able to unlock locks and loosen shoes from horses feet (Thomson).
Pl 1, Map 10.
M Lossiemouth, 1832 *? coll.* **ELN**: frequent at Knockando and
moor above Forres, *J. G. Innes* (Gordon); bridge of Findhorn, 1857
JK **FRS**; Nethybridge, 1876 *J. Fraser* **E**; sand dunes Findhorn,
1925; short grass Buckie Loch, Culbin Forest, 1968 **E,CGE**;
abundant on the old railway embankment above Huntly's Cave,
Dava Moor.
N Kinsteary Park, Auldearn, 1833 *WAS* **ABD**; Cawdor, *WAS*
(Gordon); on the hill beside the Bell Tower, Ardclach, 1882 *R.
Thomson* **NRN**; plentiful on the old bridge over the Tomlachin
Burn, Glenferness, 1948, grass bank Dulsie Bridge and in dune
slacks on the Nairn Dunbar golf course.
I S of Loch Ericht, 1867 *P. N. Fraser & A. Craig Christie* (Balfour,
1868); Bught and Leachkin, Inverness, 1885 (Aitken); Boat of
Garten, 1889 *G. C. Druce* **OXF**; Daviot, 1904 *J. S. Gamble* **K**;
Dalwhinnie, 1911 *ESM* **E,K**; abundant on the railway embankment
between Drumochter and Newtonmore, 1954; short turf by the

old lime kiln at Mid Morile, Strathdearn; basic outcrop on Alvie moor; mountain ledges at Gaick, Loch Killin, Sgùrr na Lapaich, Glen Feshie and The Fara etc.

OSMUNDACEAE
Osmunda L.

O. regalis L.   *Royal Fern*
Native. Very rare and probably introduced in the north-east.
M Damp dune slack, Culbin Forest, 1955 *E. C. Wallace.*
I Brought by tinkers from the west coast and planted in several places round Inverness, but not known to have survived; Glen Strathfarrar, not common 1902 (Pollock); in a burn *c.* 1½ miles up the glen from Glen Feshie Lodge, 1956 *R. A. Boniface* **BM**; among rocks by Loch Ness N of Inchnacardoch, 1974 *P. Stewart.*

HYMENOPHYLLACEAE
Hymenophyllum Sm.

H. wilsonii Hook.   *Wilson's Filmy Fern*
Native. Very rare in the north-east. A small translucent, dark-leaved fern of damp rock crevices.
I Rock face, Pass of Inverfarigaig, 1971 *R. Richter* **E**!; gully of burn on the NW slopes of Ciste Dhubh, Glen Affric and in several other places nearby, 1973 *R. McBeath et al.*; Mealfuarvonie, 1973 *R. W. M. Corner*; top of a gully on Creag nan Clag, Torness, 1975 *R. Richter* **E**!.

DENNSTAEDIACEAE
Pteridium Gled. ex Scop.

P. aquilinum (L.) Kuhn   *Brakens, Bracken*
Native. Abundant on acid moorland, woods and hillsides, but often absent from large areas.
M Very common (Gordon); about Forres, 1856 *JK* **FRS**; among pines in the Culbin Forest 1925 and throughout the county, 1954.
N (Stables, *Cat.*); Ardclach, 1886 (Thomson); throughout the county, 1954.
I Dalwhinnie and Pitmain, 1833 (Watson, *Loc. Cat.*); Kincraig, 1891 *AS* **BM**; rare in upper Glen Feshie and Gaick Forest but abundant elsewhere.

CRYPTOGRAMMACEAE
Cryptogramma R. Br.

C. crispa (L.) Hook.   *Parsley Fern*
   Allosorus crispus (L.) Röhl.
Native. A small tufted fern with differing fertile and vegetative fronds. Locally common among rocks and on scree in the higher mountains. Pl 2, Map 11.
I Dalwhinnie, 1833 (Watson, *Loc. Cat.*); Kingussie, 1841 *A. Rutherford* **E**; near Kincraig, 1891 *AS* **BM**; on the high hills of

Cannich and Glen Affric, 1902 (Pollock); Mam Sodhail, 1947 *N. D. Simpson* **BM**; rare on scree on the E side of the lochan in the Coire an Lochain of Cairn Gorm, 1953; scree by the Elrick Burn, Coignafearn; S-facing slopes of The Fara, Dalwhinnie; at 900 m on Càrn Ban, Glen Banchor, 1965 *S. Haywood.*

THELYPTERIDACEAE
Phegopteris (C. Presl) Fée
P. connectilis (Michx) Watt   *Beech Fern*
  Thelypteris phegopteris (L.) Slosson
Native. Common in shady places by rivers and burns and among rocks and boulders on mountains. Pl 2.
M Frequent (Gordon); Relugas, 1857 *JK* **FRS**; plentiful in the gorge of the River Findhorn, Darnaway, 1960 **K**; among rocks at Huntly's Cave, Dava Moor; burn gullies at Fae, Dorback and Auchness etc.
N Cawdor woods, 1833 *WAS* **E,BM**; Cawdor Burn, 1864 *R. Thomson* **NRN**; common in rock crevices by the River Findhorn at Dulsie Bridge, Drynachan, Coulmony and Glenferness, 1954; in a small burn gully by Clunas Reservoir, 1972.
I Dalwhinnie, 1833 (Watson, *Loc. Cat.*); at Dunain, by the Holm Burn and at Englishton, 1885 (Aitken); Kincraig, 1891 *AS* **BM**; common among boulders and rocks by Ben Alder Lodge, on Cairn Gorm and other mountains 1954; ditch at Tomich, 1972 **E**.

Oreopteris J. Holub

O. limbosperma (All.) J. Holub   *Mountain Fern, Lemon-scented Fern*
  Thelypteris limbosperma (All.) H. P. Fuchs;
  T. oreopteris (Ehrh.) Slosson
Native. Abundant in burn gullies, in woods and on moorland slopes.
M Calcots (Gordon); Relugas, 1856 *JK* **FRS**; among pines in the Culbin Forest, 1925.
N Cawdor, 1832 *WAS* **ABD**; Cawdor woods, 1833 *? coll.* **BM**; small birch wood by the E drive at Holme Rose, 1955 **K**; among rocks by the River Findhorn, Dulsie Bridge, 1962 **ABD**.
I Dalwhinnie, 1833 (Watson, *Loc. Cat.*); near Leys, Inverness, 1854 *HC* **INV**; Leachkin, 1885 (Aitken); Kincraig, 1891 *AS* **BM**; common on the banks of burns and moorland slopes by Ben Alder Lodge, at Creag Dhubh, Laggan and Fort Augustus etc. 1954.

ASPLENIACEAE
Asplenium L.
A. scolopendrium L.   *Hart's Tongue*
  Phyllitis scolopendrium (L.) Newm.
Doubtfully native in the north-east. Rare on old walls and on rocks.
M Innes House, Urquhart, 1794 *A. Cooper* **ELN**; Castle of Old Duffus (Gordon); garden wall Dalvey House near Forres, 1856 *JK* **FRS**!; wall by the Black Burn at Pluscarden Priory, 1938 *R. Richter*;

3 plants on the rock face of Newton Toll Quarry, 1973 *R. Richter* **E**!.
N Cawdor wood, 1834 *WAS* **ABD,GL,BM**!; formerly at Ardclach
but not seen for several years (Thomson).
I   Foyers, 1881 *F. Townsend* **INV**; on rock face below the Falls
of Foyers by the road bridge (probably the same locality) 1971
E; Glen Strathfarrar, 1902 (Pollock); a few young plants on a road
culvert in Reelig Glen, shown to him by a keeper on the Reelig
estate were recorded in 1955 by *J. W. Dyce*.

A. adiantum-nigrum L.   *Black Spleenwort*
Native. Common in rock crevices and on old walls and ruins.
Fig 4a, Map. 12.
M Pluscarden Hill (Gordon); banks of the River Findhorn, 1857
JK **FRS**; walls of Spynie Palace, 1925; rocks by the River Lossie
near Buinach, 1973 **E**.
N (Stables, *Cat.*); walls of the Castle of Rait, *J. Brichan* (Watson)!;
rare at Ardclach, 1882 *R. Thomson* **NRN**; rock crevice by the River
Findhorn below the Bell Tower, Ardclach, 1973 **E**; area walls of
Geddes House, 1975 **E**.
I Craghue and Badenoch (Gordon); Fort Augustus (Innes); rocks
by the Moniack Burn at Reelig and by the River Nairn, also at
Abriachan (Aitken); wall opposite the entrance to Newton House,
Kirkhill, 1930 *M. Cameron*!; bridges at Milton and in Strathglass;
cliffs in the Pass of Inverfarigaig and below Creag Dhubh by Loch
Ness; walls of Urquhart Castle and Dochfour House etc. 1954; rock
crevice Tom Bailgeann near Torness, 1972 **E**.

A. cuneifolium Viv.
  A. serpentini Tausch
Native. A rare diploid found on serpentine rocks. Never found on
limestone. Fig 4b.
I W of the road junction at Milton, 18.. *Miss McInnes*, a narrow
pinnuled form mentioned by Moore (1859) was named *A. adiantum*
-nigrum var. *leptorachis* (R. H. Roberts & A. McG. Stirling, *Fern
Gazette* **11**: (1974)); plentiful on serpentine outcrops which occur
over a wide area N of Polmaily, Glen Urquhart, 1973 *A. McG.
Stirling* **E**; rock crevice in small quarry N of Lochan an Torra
Bhuidhe, Upper Gortally, 1976 **E**.

A. marinum L.   *Sea Spleenwort*
Native. Sea cliffs and caves. Only known from one locality.
M Abundant Covesea (Gordon). Collected here by *G. Wilson ABD*;
1878 *J. Fraser* **E**; 1898 *ESM* **CGE** and 1956 *MMcCW*. **BM**.

A. trichomanes L.   *Maidenhair Spleenwort*
Records of the two following subspecies have been determined by
J. D. Lovis unless marked by an asterisk. They can be recognised
with some difficulty.

  Subsp. **trichomanes**
Native. Rare in dark sheltered places on serpentine and ultra basic
rocks. The essential feature is the delicacy and the separation of

Fig 4   a, **Asplenium adiantum-nigrum**; b, **A. cuneifolium**; c, **A. tricho-manes** subsp. **trichomanes**; d, **A. trichomanes** subsp. **quadrivalens**.

the pinnae, the thin wiry rachis and the few, not crowded, sori.
Fig 4c.
M Craigellachie Reserve, Aviemore, 1964 *P. H. Davis* **E**; rock face
in deep shade by the River Findhorn opposite Coulmony, 1975 **E**.
I Conglomerate cliff near Crask of Aigas, 1975 **E**; low rocks by
the road near Eskadale, 1976 **E**; conglomerate cliff below the
Plodda Falls, Guisachan, **E**.

Subsp. **quadrivalens** D. E. Meyer
Native. Common on mortared walls and base-rich rocks. Pinnae
are crowded, oblong and more robust than in subsp. *trichomanes*,
the upper pinnae inserted at right angles. Fig 4d, Map 13.
M * Abundant on the walls of Elgin Cathedral, Pluscarden Priory,
Stotfield and at Kinloss Abbey (Gordon); *banks of the River
Findhorn, 1858 *JK* **FRS**; walls of Lochindorb Castle, 1886
(Thomson) and 1974 *M.McC.W.* **E**; among stones in the rockery
at Kellas House, 1975 **E**.
N *Firwood Burn, Ardclach, 1833 *WAS* **ABD**; *Cawdor Burn,
1864 *R. Thomson* **NRN**; small bridge over the Burn of Blarandualt,
Ordbreck, 1975 **E**; very luxuriant plants on the farm wall at
Geddes House, 1976 **E,BM**.
I *Fort Augustus, *J. G. Innes* (Gordon); *Bunchrew Burn, 1854
HC **INV**; Inellan, 1860 *A. Craig Christie* **E**; rock face Plodda Falls,
Guisachan, 1972 **E**; cliffs of Creag Dhubh, Laggan, 1974 **E,BM**.

A. viride Huds.   *Green Spleenwort*
Native. Frequent on base-rich rocks in the highlands. Differs from
A. trichomanes in the green petiole and crenulate margins of the
pinnules. Map 14.
M Sluie, *J. G. Innes* (Gordon, *annot.*); diorite rocks by the Allt
Iomadaidh, Fae, 1973 **E**.
N Cawdor woods, 1833 *WAS* **E,ABD,GL**; Ardclach, *G. Wilson*
(Gordon); occurs at the bottom of the Falls at Ault n' airidh on the
River Findhorn and at Lynemore on the Torr Burn, 1886 (Thomson).
I South side of Loch Ericht, 1867 *P. N. Fraser & A. Craig Christie*
(Balfour, 1868); head of a glen 3 miles N of Truim, 1882 (Boyd);
Moniack Burn (Aitken), 5 plants seen here in 1964; plentiful on
the diorite rocks of glens Feshie and Banchor; crevices of the burn
gullies at Gaick; in mortar on Torgyle bridge, Glen Moriston, 1971
E; old lime quarry Suide, Kincraig, cliffs of Creag nan Clag, rare
in the Coire an Lochain of Cairn Gorm; cliff above Loch Killin,
1972 *M. Barron* **E**; low cliff by road near Eskadale at 60 m *A. J.
Souter* **BM**; rocks at the top of Ord Ban, Inshriach, 1972 *R. McBeath*.

A. ruta-muraria L.   *Wall Rue*
Native. Common in mortar of old walls and bridges and occasionally
in rock crevices. Map 15.
M Walls of Elgin Cathedral, 1838 *WAS* **E**; ruins of Kinloss Abbey,
G. Wilson **ABD**; abundant (Gordon); Kinloss Abbey, 1856 *JK*
FRS; garden wall Moy House, 1956 **ABD**; wall by the old Forres–
Grantown railway line at Pilmuir, 1972 **E**.

N (Stables, *Cat.*); Cawdor Castle, 1864 *R. Thomson* **NRN**; rocks at Dulsie Bridge and wall at Auldearn, 1954.
I Fort Augustus, *J. G. Innes* (Gordon); ramparts of Fort George, 1885 *R. Thomson* **NRN**!; ruins of Urquhart Castle and at Abriachan (Aitken); Kincraig, 1891 *AS* **BM**; abundant on the walls of the mausoleum at Kirkhill, 1930 *M. Cameron*!; formerly abundant on the old wall at Dores Kirk but now rare there, 1954; bridges at Struy, Milton, Truim and Dalcross, cliffs by Loch Ness, on Creag nan Clag, in Glen Feshie and a burn gully at Lurgmore; walls at Kincraig, Alvie and Aviemore, 1973 *R. McBeath*.

A. septentrionale (L.) Hoffm.   *Forked Spleenwort*
Native. Crevices of rocks. Very rare. Not seen in recent years.
I Inverfarigaig, 1885 (Aitken).

A. ceterach L.   *Rusty-back Fern*
Ceterach officinarum DC.
Native but probably introduced in the north-east.
N Cawdor Castle, 1863 *R. Thomson* **NRN**.
I Twelve plants on a wall by the A82 near Dunain House, Inverness, 1962 **E,BM**. This colony had increased to 16 plants by 1970 but owing to the imminence of road widening and the demolition of the wall five plants were removed and transplanted to three different walls nearby. Only one survived in the wall between Dunain House and the road.

ATHYRIACEAE
Athyrium Roth

A. filix-femina (L.) Roth   *Lady Fern*
Native. Abundant in damp woods, by ditches and among boulders and rocks on mountains.
M Frequent (Gordon); Relugas, 1857 *JK* **FRS**; rare in the Culbin Forest but abundant elsewhere in the vice county, 1954.
N Cawdor woods, 1833 *WAS* **BM**; Ardclach, 1886 (Thomson); birch wood by Cran Loch, 1960 **ABD**.
I Dalwhinnie, 1833 (Watson, *Loc. Cat.*); Ness Islands, Inverness, 1854 *HC* **INV**; Kincraig, 1891 *AS* **BM**; Feshie Bridge, 1904 *G. B. Neilson* **GL**.

A. distentifolium Tausch ex Opiz   *Alpine Lady Fern*
A. alpestre (Hoppe) Rylands, non Clairv.
Native. Frequent among rocks and boulders on mountains over 650 m. Differs from *A. filix-femina* in having broader, blunter leaf segments, very small orbicular sori and the indusium falling before spores are ripe. Map 16.
I Pitmain, Kingussie, 1833 (Watson, *Loc. Cat.*); Braeriach, 1833 *Backhouse* herb *Hooker* **K**; Cairn Gorm (Balfour 1868); head of a glen 3 miles N of Truim, 1882 (Boyd); Coire at t-Sneachda, 1,000 m, 1898 *ESM* **E,BM**; Sgùrr nan Conbhairean, 610 m, 1960 *H. Milne-Redhead*; boulder scree Coire an t-Sneachda, Cairn Gorm, 1955 **ABD, BM**.

A. flexile (Newm.) Druce
 A. filix-femina var. *flexile* (Newm.) Jermy
Native. Rare among boulders on mountains over 915 m. Smaller
than *A. distentifolium* with narrower leaves borne on very short
petioles of 1–2 cm, suddenly bent near the base. Sori on the lower
part of the leaf.
I Braeriach, 1880 *G. C. Druce* **BM,OXF**; Glen Einich, 1888 ex
herb. *Palmer* and *G. C. Druce*, **BM,OXF**; Cairn Gorm, 915 m, 1898
ESM & WASh **K**; Coire Dhondail, Glen Einich, 1973 *R. McBeath*
BM; among boulders below the buttress above the lochan in Coire
an Lochain, Cairn Gorm, 1973 **E**.

<h3 style="text-align:center">Gymnocarpium Newm.</h3>

G. dryopteris (L.) Newm.   *Oak Fern*
 Thelypteris dryopteris (L.) Slosson
Native. Common in damp woods, river banks and among rocks
and boulders on mountains. Pl 2.
M Frequent (Gordon); Relugas, 1860 *JK* **FRS**; Grantown, 1897
J. S. Gamble **K**; dune hollow in pine wood, Culbin Forest, 1956
ABD; among stones on the banks of the Allt Iomadaidh, Fae, 1972 **E**
and very common on the shady banks of the rivers Spey and
Findhorn.
N Cawdor woods, 1832 *WAS* **ABD,BM**; by the Cawdor Burn, 1867
R. Thomson **NRN**; in a wall (now demolished) by the A96 near
Courage Farm, Auldearn, 1967 **ABD**.
I Dalwhinnie, 1833 (Watson, *Loc. Cat.*); Fraoch-choire Cannich
(Ball, 1851); Dunain and by the Holm Burn near Inverness, 1885
(Aitken); Kincraig, 1891 *AS* **BM**; between Glenmore and Loch
an Mare, 1966 *E. Rosser* **E**.

<h3 style="text-align:center">Cystopteris Bernh.</h3>

C. fragilis (L.) Bernh.   *Brittle Bladder-fern*
Native. Common on walls and in rock crevices in shady places
reaching 915 m in the Cairngorms, also frequently found beneath
railway platforms. Map 17.
M Common (Gordon); Bridge of Brown, 1905 *ESM* **BM**; Brodie
station platform, 1950 **CGE**; walls of the old greenhouse Brodie
Castle, 1972 **E**; among rocks by the Allt Iomadaidh at Fae, 1974
BM.
N Cawdor Castle, 1832 *WAS* **E,ABD**; by the Tomnarroch Burn,
Ardclach, 1889 *R. Thomson* **NRN**; rock crevices by the River
Findhorn at Dulsie Bridge and Coulmony 1954; gully of the Allt
Breac, Drynachan, 1973.
I Dalwhinnie, 1833 (Watson, *Loc. Cat.*); Kingussie, 1838 *A.
Rutherford* **E**; Foyers, 1854 *HC* **INV**; Moniack Burn, 1885 (Aitken);
Glendoe, 1904 *G. T. West* **STA**; garden walls at Ben Alder Lodge,
1956 and Culloden House, 1975 **E**.

C. montana (Lam.) Desv.   *Mountain Bladder-fern*
Native. Very rare on shady mountain cliffs. Pl 2.

I North facing rocks of Coire Garbhlach, Glen Feshie, 1957 *D. A. Ratcliffe et al.*

ASPIDIACEAE
Polystichum Roth

P. lonchitis (L.) Roth   *Holly Fern*
Native. Locally plentiful on basic rocks in the higher mountains descending to 60 m in Strathglass. Young plants of *P. aculeatum* are often mistaken for this species. Pl 3. Map 18.
M One specimen on rocks by the River Findhorn, Edinkillie (Burgess). I suspect this record to be an error.
N Ardclach, 1833 *WAS* (Thomson). There is no specimen to confirm this record and it is probably also an error.
I Dalwhinnie, 1833 (Watson, *Loc. Cat.*); rocks on S side of Loch Ericht, 1867 *P. N. Fraser & A. Craig Christie* (Balfour, 1868)!; a few on Braeriach, 1877 (Boyd); Mealfuarvonie, 1885 (Aitken); rocks of Dirc Mhòr, Dalwhinnie and Coire Madagan, Gaick, 1949 *N. Y. Sandwith*; Marg na Craige, Glen Banchor, 1955 *C. D. Pigott*; plentiful on rocks by the waterfall in Coire Bhatain, Gaick, 1973 **E**; Glen Markie and Glen Feshie; rock crevice, 60 m, by the east road near Eskadale in Strathglass, 1973 *A. J. Souter* **E,BM**!; cliffs above Loch Gynack, Kingussie, 1973 *R. McBeath*.

P. aculeatum (L.) Roth   *Hard Shield-fern*
    P. lobatum (Huds.) Chevall.
Native. Frequent in dark ravines and wet rocks by burns and rivers. A large stiff-leaved fern, shining on top. Pinnae decreasing in size almost to the base of the stipe. Margins of upper pinnules meeting each other at an acute angle. Map. 19.
M On the Findhorn (Gordon) and 1857 *JK* **FRS**; abundant by the small burn just S of Rafford Kirk, 1925; gorge of the River Findhorn opposite Coulmony House, 1972 **E**; ravine of burn at Slochd N of Carrbridge, 1972 **BM**; plentiful in the Findhorn valley and the ravine of the Divie Burn, and in small quantity by the Allt Iomadaidh at Fae, and by the Muckle Burn at Dyke.
N Auchindown near Cawdor, 1832 *WAS* **E,GL**! and in 1833 **BM**; frequent at Ardclach, 1885 *R. Thomson* **NRN**; in shade on the banks of the River Nairn at Holme Rose, 1955 **ABD**; wall at Cawdor Castle, 1976 **E**.
I By the Bunchrew Burn, 1854 *HC* **INV**; ravine of the Moniack Burn (Aitken)!; near Drumnadrochit, 1947 *C. C. Townsend* **K**; Pass of Inverfarigaig, 1971 **E**; Falls of Divach, 1975 **BM**; frequent by the burns round Flichity and Farr, gullies in Glen Feshie, in the Ryvoan Pass and elsewhere.

P. setiferum (Forsk.) Woynar   *Soft Shield-fern*
    P. angulare (Willd.) C. Presl
Native. Very rare, Confined to the Findhorn valley. Differs from *P. aculeatum* in the flaccid yellowish-green fronds, pinnae stopping suddenly leaving a distinct length of stipe and the margins at the

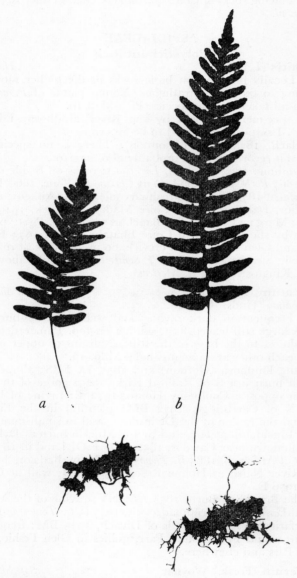

Fig 5   a, **Polypodium vulgare**; b, **P. interjectum**.

Ophioglossum vulgatum × ⅔
Adder's Tongue

Botrychium lunaria × ⅔
Unshoe the Horse, Moonwort

Lycopodiella inundata × ⅔
Marsh Clubmoss

Equisetum pratense × ⅔
Shady Horsetail

PLATE 1

Cystopteris montana × ⅔
Mountain Bladder-fern

Cryptogramma crispa × ⅔
Parsley Fern

Phegopteris connectilis × ¼
Beech Fern

Gymnocarpium dryopteris × ¼
Oak Fern

J. K. Butle

PLATE 2

Viola lutea × ⅔
Mountain Pansy

Trollius europaeus × ⅕
Lucken Gowan, Globe Flower

Silene acaulis × ½
Moss Campion

T. G. Collett

Silene vulgaris subsp. maritima ×
Sea Campion

PLATE 5

Geranium sylvaticum × ⅔
Wood Cranesbill

Oxalis acetosella × ⅔
Wood Sorrel

Rubus chamaemorus × ⅔
Aivrons, Cloudberry

R. chamaemorus (*fruit*) × ⅔

PLATE 6

base of the upper pinnules meeting each other at right angles, those of the lower at an obtuse angle.

M By the River Findhorn at Sluie, 1885 (Aitken); a single plant in a crevice in the gorge of the River Findhorn opposite Sluie, 1957 **BM** (possibly Aitken's locality), but it disappeared from there in a landslide in 1964.

N Ardclach, not common, 1886 (Thomson).

Dryopteris Adans.

D. filix-mas (L.) Schott   *Male Fern*
Native. Abundant in damp woods, banks of burns and rivers especially in the lowlands.

M Common (Gordon); Relugas, 1856 *JK* **FRS**; Culbin Forest, 1925 and throughout the county.
N (Stables, *Cat.*); Cawdor, 1865 *R. Thomson* **NRN**; birch wood at Holme Rose, 1953 **ABD**.
I Dalwhinnie and Pitmain, 1833 (Watson, *Loc. Cat.*); bank at Ben Alder Lodge, 1956 and throughout the district.

D. pseudomas (Wollaston) J. Holub & Pouzar   *Scaly Male Fern*
    D. borreri (Newm.) Tavel
Native. Abundant among rocks, on screes on mountains and in woods. Differs from *D. filix-mas* in the dense orange-brown scales on the petioles and rhachis, the truncate pinnules with a blackish mark on the upper surface of the pinnae at the point of junction with the rhachis, and, when young, for its yellow colour and stiff stems.

M Dunphail near Forres, 1859 *F. Browne* herb. *K. Moore* and at Moseley near Birnie, 1863 *N. Fraser* E (*Watsonia* 3: 1 (1953)); Culbin Forest, 1925; scree on Beum a Chlaidheimh, Duthil, 1973 **E** etc.
N Banks of the River Nairn, Holme Rose, 1953 **ABD**; wood at Cawdor Castle, 1960 **K**; by the Allt Breac, Drynachan, Dulsie Bridge, Geddes and Loch of the Clans, 1954.
I Dalwhinnie, 1839 *H. C. Watson* **K**; Dunain Hill, 1866 *N. Fraser* **E**; Drumnadrochit, 1947 *C. C. Townsend* **BM,K**; Badger Fall, Fasnakyle, 1947 *N. Y. Sandwith* **K**; cliffs above Loch Killin, 1975 **E**; scree below Creag nan Clag, **E**; quarry at Suidhe, Kincraig, **E**.

D. oreades Fomin
    D. abbreviata auct.
Native. Mountain slopes and screes. Rare or overlooked. Generally shorter than the two preceding species. Petiole with a few pale brown scales, pinnules slightly concave giving the leaf a crisped appearance. Sori very few, sometimes only one on either side of the longest pinnules. Indusium with glands on the margins.
I ? locality, 1958 *C. D. Pigott*; among rocks in the Coire nan Laogh, Newtonmore, 600 m, *R. McBeath* **E**; Creag Liath, Garva Bridge, 1973 *R. McBeath* **E**; burn gully A' Chaornich, Gaick, 1973; rocks above Allt Beithe Min, 800 m and in the Coire na Geurdain of Mulloch Fraoch-choire, 1974 *R. W. M. Corner et al.*; scree below the Fara, Dalwhinnie, 1975 **E**.

D. aemula (Ait.) Kuntze   *Hay-scented Buckler Fern*
Native. A fern of the west, very rare in the north-east.
I   Coire an Lochain of Braeriach, just above the tarn (Marshall, 1899).

D. carthusiana (Vill.) H. P. Fuchs   *Narrow Buckler Fern*
    D. lanceolatocristata (Hoffm.) Alston; *D. spinulosa* Watt
Native. Frequent in wet grassy margins of lochs and in damp birch woods. A medium sized fern, fronds generally erect, narrowish, light green, uniformly pale scales to the stipe. Pinnae flat, the fronds growing singly from creeping rhizomes. In Collectanea for a Flora of Moray, 1839 the Rev. George Gordon recorded this fern as common in error for *D. austriaca* (Burgess). Map 20.
M   Relugas, 1856 *JK* **FRS**; marshy ground Castle Grant, 1934 (Burgess); in a small quarry by the road to Outlawel, 1963 *J. W. Dyce* **JWD**; birch wood slope by the road, Hill of Mulundy, Remachie, 1968 **E,BM**; among pines Culbin Forest, 1968 **BM**; bog by the Black Burn W of Pluscarden Priory, 1968 **BM**; railway embankment beneath birches at Slochd, Carrbridge, 1974 **E**.
N   Grassy bog at the W of Loch of the Clans, 1960 **E,BM**; marsh surrounding Cran Loch, 1960 **ABD,K**; margin of Loch Litie near Auldearn, 1963 **ABD,K**; grassy bog on the Hill of Urchany, 1974 **E**.
I   Near Kincraig, 1891 *AS* **BM**; Corries Cave, Loch Ness, 1904 *G. T. West* **STA**; Reelig Glen, 1954 *J. W. Dyce* **JWD**; moorland by Lochan Uaine, Ryvoan, 1971 **E**; bog below Netherwood, Newtonmore, 1972 **E**; banks of River Beauly by Black Bridge, Kilmorack, 1972 **E**; in shade by the River Calder, Newtonmore, 1972 *J. D. Buchanan*; Inshriach 1973 *R. McBeath* **E**; bog near Creagdhubh Lodge, Laggan, bank near the Home Farm at Dochfour, below cliffs at Loch Duntelchaig **E**, bog near Cultie and in shade by the old railway line at Broomhill.

D. austriaca (Jacq.) Woynar   *Broad Buckler Fern*
    D. dilatata (Hoffm.) Gray
Native. Abundant throughout the district. A most variable fern in size and shape, typically with arching fronds, broader and darker green than *D. carthusiana*, with heavier texture and usually dark-coloured scales to the stipe, the plume generally with bent down margins. Fronds emerging from distinct crowns.
M   Burn of Denavin, Darnaway, *Brodie* **E**; by the Divie Burn, Dunphail, 1878 *R. Kidston* **GL**; railway embankment Slochd, 1972 **BM**; bank by the Allt Iomadaidh, Fae, 1972 **E**; bog at Drakemyres, 1972 **E**.
N   (Stables, *Cat.*); Ardclach (Thomson); birch wood Holme Rose, 1955 **ABD,K**; banks of the River Findhorn at Dulsie Bridge, 1960 **ABD,K**.
I   Dalwhinnie and Pitmain, 1833 (Watson, *Loc. Cat.*); Ness Islands, 1854 *HC* **INV**; Kincraig, 1891 *AS* **BM**; pine wood in the Affric Forest, Glen Affric, 1971 **E**; among boulders Dunmaglass,

1972 **E**; open moor Leachkin, Inverness, **E**; below the cliffs of Creag Dhubh, Laggan, **E**.

D. austriaca× **carthusiana** = **D.**× **deweveri** (Jansen) Jansen & Wachter
M Railway embankment by the A9 ¼ mile S of Slochd, 1972 **E,BM**; by the Allt Mòr, Bridge of Brown, 1973 **E**.
I Among rocks in a birch wood between Lochain Uvie and Creagdhubh Lodge, Laggan, 1975 **E**.

D. assimilis S. Walker
D. dilatata var. *alpina* T. Moore
Native. Common but under-recorded. Closely resembles the alpine forms of *D. austriaca* but is more yellow-green in colour especially when dry. Frond flat, the upper pinnae and pinnules more finely divided and the scales concolorous red-brown. Map 21.
M Rock scree, Craigellachie Reserve, Aviemore, 1964 *M. Cowan et al.* **E**.
I N side of Braeriach, 1,000 m 1898 *ESM* **BM,K**; Carn Eige, Cannich, 1947 *E. C. Wallace* **ECW**; N facing slope of Coire Garbhlach, Glen Feshie, 1958 *K. Goodway & M. C. F. Proctor* **CGE**; cliff ledge Sgùrr na Lapaich, Glen Strathfarrar, 1971 **E,BM**; meadow by the Moniack Burn, Kirkhill, 1972 **E**; among boulders below the cliffs by Loch Killin, **E**; felled pine wood Leachkin, **E**; cliff ledges above Loch Toll a' Mhuic, Monar, **E**; by the Allt Coire Chùirn, Drumochter, 1975 **ABD**.

D. assimilis× **austriaca**
M In woodland by the path above the gorge of the River Findhorn N of Logie House, 1967 **E,BM, CGE**.
I Cliff ledge above Loch Toll a' Mhuic, Monar, 1975 **E**; among boulders below the cliffs by Loch Killin, 1975 **E**.

BLECHNACEAE
Blechnum L.
B. spicant (L.) Roth   *Ladder Fern, Hard Fern*
Native. Abundant in woods, on moors and in mountain ravines on acid soil. A dark green stiff fern with differing fertile and vegetative fronds.
M Very common (Gordon); Relugas, 1856 *JK* **FRS**; Grantown, 1890 *E.F.* **E**; pine wood Culbin Forest, 1925 etc.
N Highland Boath, 1832 *WAS* **ABD**; Ardclach, 1888 *R. Thomson* **NRN**; Cawdor woods, 1955 **ABD**.
I Dalwhinnie, 1833 (Watson, *Loc. Cat.*); Craig Phadrig, 1854 *HC* **INV**; Kincraig, 1891 *AS* **BM**; and throughout the district, 1954.

POLYPODIACEAE
Polypodium L.
P. vulgare L.   *Common Polypody*
Native. Abundant on walls, among rocks, on tree trunks and

occasionally on sand dunes. Frond outline $\pm$ parallel sided for most of its length. Pinnae with rounded ends. The plant is more leathery than the other species. Fig 5a.

M Common (Gordon); Relugas, 1856 *JK* **FRS**; sand dunes at Hopeman and in the Culbin Forest, 1930; wall near Randolph's Leap, Relugas, 1972 **BM**; on rocks in the gorge of the River Findhorn opposite Coulmony House, 1972 **E**; Lochindorb Castle, 1974 **BM**.
N Walls at Cawdor Castle, 1832 *WAS* **ABD**; Ardclach, 1886 (Thomson); top of a dry wall near Auldearn, 1967 **ABD**.
I Dalwhinnie, 1833 (Watson, *Loc. Cat.*); Clachnaharry, 1854 *HC* **INV**; by the Allt Mullardoch, Cannich, 1947 *A. J. Wilmott* **BM**; throughout the district, 1954; road bridge at Craggie, Daviot, 1971 **E**.

P. interjectum Shivas
   P. vulgare subsp. *prionoides* Rothm.
Native. Found in the same type of habitat as *P. vulgare*, but rare in the north-east. Differs in the fronds being usually larger and the longest pinnae near the middle of the frond and the outline generally narrowing gradually both above and below this point. Fig 5b.
M Railway embankment near Slochd, 1972 **BM**; rocks Craig-ellachie Reserve, Aviemore, 1973 **E**.
I Top of a wall on the hill road Reelig Glen, Kirkhill, 1975 **E**; north facing cliffs by Loch Monar, **E**; on the bridge at Tromie Bridge, **E**; roadside cliff Crask of Aigas, **E**; walls of Urquhart Castle **E**.

P. interjectum × vulgare = P. × mantoniae Rothm.
I Among rocks and boulders by a small burn ½ mile N of Tomich, 1975 **E**.

MARSILEACEAE
Pilularia L.

P. globulifera L.   *Pillwort*
Doubtfully native, probably introduced by waterfowl. Very rare, possibly extinct.
M Alves, near the sea shore, *G. Wilson* (Gordon); in marshes, not infrequent as at a marsh near Gordon Castle, ? date ? *coll.* **GL** (there is no specimen remaining on this sheet).
N Loch of the Ord, 1926 *J. B. Simpson*! (not seen since 1956).

SPERMATOPHYTA

GYMNOSPERMAE

Most of the larger estates have a number of planted coniferous trees and the Forestry Commission have planted different species on an experimental basis, many of which have proved unsuitable to the area and have now been discontinued. To make a full list would be beyond the scope of this book.

PINACEAE
Larix Mill.

L. decidua Miller *European Larch*
    L. europaea DC.
Introduced. Regenerates freely. Twigs yellowish, drooping.
M Throughout the county, 1954; shingle of the River Findhorn by Broom of Moy, 1965 **CGE**.
N Ardclach (Thomson); moorland near Drynachan, 1963 **ABD**.
I 'Larch trees were introduced into Scotland by the Duke of Atholl shortly after the '45. At that time Fraser of Belladrum was commissioner on the forfeited Lovat estate and he bribed a man to steal some. During the process of planting the men became conscious of a thirst and went into a public house to allay the same. In their absence, the person bribed, picked up a bundle of young trees and made tracks for Beaufort. The bundle was not missed and the stolen trees were planted at Beaufort and Belladrum and can still be seen there. Some were also planted at Bruiach Farm and one of these, at a distance of 4 feet from the ground measures 12½ feet in circumference' (Pollock); Fort Augustus, 1904 *G. T. West* **STA**; throughout the district, 1954.

L. kaempferi (Lambert) Carrière *Japanese Larch*
    L. leptolepis (Sieb. et Zucc.) Endlicher
Introduced. Twigs reddish, stiff.
M Extensively planted throughout the county, 1954; plantation near Dallas, 1967 **CGE**; by the Spynie Canal, 1972 **E**.
N Throughout the county.
I Planted at Fort Augustus, in Glen Strathfarrar, at Cannich, Culloden and Daviot etc.

L. decidua × kaempferi = L. × eurolepis A. Henry
Planted at Cloddymoss in the Culbin Forest in 1948 and is now the most commonly planted larch at Inverfarigaig Forest, and doubtless elsewhere.

Pinus L.
P. sylvestris L. *Scots Pine*
Native. Abundant throughout the district. Commonly planted, regenerating freely. About 150 variants have been described.

Subsp. **scotica** (Schott) E. F. Warburg
The native Scottish plant. The crown remaining pyramidal for a
long period, rounded only in the old trees.
M Indigenous in Strathspey (Gordon).
N (Stables, *Cat.*); a few trees of the ancient Caledonian Forest, N
of Ardclach, 1886 (Thomson).
I In Strathspey going from Badenoch to Inverness, Glen Moriston,
Strathglass, Abernethy and Rothiemurchus (Lightfoot);
Rothiemurchus, 1900 *JK* **FRS**; do. 1953 *P. H. Davis* **E**;
Descendants of the old forests can be seen in glens Feshie and
Shirra, in the Cairngorms and elsewhere.

CUPRESSACEAE
Juniperus L.
J. communis L.
Subsp. **communis**   *Aitnach, Melmot Berries, Juniper*
Native. Common throughout the district on moors, hillsides and
in birch and pine woods. Berries used to flavour gin and were
formerly exported in large quantities to Holland.
M Very common (Gordon); mosses about Forres (Innes); frequent
on dunes in the Culbin Forest.
N (Stables, *Cat.*); scrub by Lochloy, 1964 **ABD**; scattered over the
moors at Cawdor, Clunas, Carse of Delnies and Glenferness.
I Dalwhinnie and Pitmain, 1833 (Watson, *Loc. Cat.*); Clava, 1855
HC **INV**; Rothiemurchus, 1891 *AS* **E** and 1900 *JK* **FRS**.

Subsp. **nana** Syme
Exposed moorland chiefly in the north-west. Differs from subsp.
communis in its prostrate habit and the upturned, or upcurved leaves.
I Cairn Gorm (Druce, 1888); near Fort Augustus and by the
shores of Loch a' Choire, 1904 *G. T. West* **STA**; Glen Afric, 1947
N. D. Simpson (*Atlas*); Càrn Bàn Mór, Cairngorms, 915 m *C.
Murdoch*; Sgùrr na Lapaich, Glen Strathfarrar, 1971; exposed
summit of Meall Dubh, Glen Moriston.

ANGIOSPERMAE
DICOTYLEDONES
RANUNCULACEAE
Caltha L.
C. palustris L.   *Marsh Marigold, Kingcup*
  C. radicans auct.
  Subsp. **palustris**
Native. Common in bogs, ditches and by burns and lochans.
Flowers golden-yellow. Poisonous to cattle. 'Known by some as
Bull's Eye, a corruption of Pool's Eye or Drunkard from the excess
of water it is able in a short time to pass through its system'
(Thomson).
M Innes House, Urquhart, 1794 *A. Cooper* **ELN**; very common
(Gordon); Forres, 1898 *JK* **FRS**.

N (Stables, *Cat.*); Ardclach, 1886 *R. Thompson* **NRN**; bog by the River Findhorn, Dulsie Bridge, 1961 **ABD**.
I Inverness, 1832 (Watson, *Loc. Cat.*); near Inverness, 1854 *HC* **INV**; Lower Tullochgrue, Rothiemurchus, 1892 *AS* **E**; near Colyumbridge, 1907 *R. S. Adamson* **BM**.

Subsp. **minor** (Mill.) Clapham
Mountain flushes and burns. Smaller than subsp. *palustris*, often rooting at the nodes. Flowers yellowish-green.
I Rothiemurchus, *JK* **FRS**; stream Cairngorm mountains, 1884 *J. Groves* **BM**!; Braeriach, 1898 *WASh* **BM,K**; by the Allt Coire Chùirn, Drumochter, 1948 *N. Y. Sandwith* **K**!.

Trollius L.

T. europaeus L.   *Lucken Gowan, Butter Balls, Globe Flower*
Native. In hill pastures, river banks and mountain ledges. Rare in Moray and Nairn owing to cultivation and drainage but common in the highlands. Pl 5, Map 22.
M Common in the upper districts (Gordon); Greshop, 1875 *JK* **FRS**; frequent by the rivers Spey and Findhorn, 1954; in a bog on the hill at Douneduff, meadow by Grantown station and by a small burn at Lettoch.
N (Stables, *Cat.*); by the River Findhorn at Ardclach, 1881 *R. Thomson* **NRN**; River Nairn, 1885 (Aitken); a small colony on the raised beach near Maviston, 1964 **E,ABD**; meadows at Cantraydoune, Dulsie Bridge, Coulmony and Glenferness.
I Inverness, 1833 (Watson, *Loc. Cat.*); Culloden Moor, 1833 *WAS* **GL**; Dulnain wood and Abriachan, 1885 (Aitken); Kincraig, 1891 *AS* **BM**; Balnafoich, Strathnairn, 1947 *AJ* **ABD**; frequent on the banks of the rivers Spey and Beauly and at 1,000 m on rock ledges on Cairn Gorm.

Helleborus L.

H. viridis L. subsp. **occidentalis** (Reut.) Schiffn.   *Green Hellebore*
Garden escape or planted. Reported from Cawdor, 1900 (Lobban).

Eranthis Salisb.

E. hyemalis (L.) Salisb.   *Winter Aconite*
Garden escape. Greshop wood, Forres, 1953 *K. Christie*; Cawdor, 1900 (Lobban); under a hedge by the Home Farm at Clunes, Kirkhill, 1954 *M. Cameron* **E**!.

Aconitum L.

A. napellus L.   *Common Aconite, Wolf's Bane, Monkshood*
Garden escape. Very poisonous. Cromdale and Lhanbryde (Burgess); railway embankment N of Rothes; near Kinchurdy, 1954; Ardclach, 1900 (Thomson), and on a bank below the Bell Tower at Ardclach (probably Thomson's locality), 1973; banks of the River Beauly, 1902 (Pollock); Kincraig, 1930 *R. Rymer Roberts* **ABD**; shingle of River Enrick, Drumnadrochit, 1947 *N. D. Simpson* **BM**; shingle verge of Loch Alvie, 1952, by the road at Easter Brae

of Cantray, near the farm of Tullochgrue and in a wood by Loch Ness at Lewiston.

A. bicolor Schultes
Garden escape. Flowers blue and white. Roadside N of Braemoray opposite a cottage, 1954; out-cast near Grantown tip, 1971 **E**; waste land at Dulnain Bridge in Easterness.

Delphinium L.
D. ambiguum L.   *Larkspur*
   D. ajacis auct.
Casual.
M Occurred on a mass of cinders and rubbish from a sulphuric acid manufactory, Waterford near Forres, 1871 *JK* **FRS**; Forres rubbish tip, 1971 **E**.

Actaea L.
A. spicata L.   *Baneberry, Herb Christopher*
Casual. Berry ovoid, black and shining. Very poisonous.
M Appeared as a garden weed, Crofts of Buinach, Kellas, 1956 **ABD**; garden weed Rose Cottage, Dyke, 1970.

Anemone L.
A. nemorosa L.   *Wind Flower, Wood Anemone*
Native. Abundant in deciduous woods, on river banks and on open moorland.
M Very common (Gordon); Grantown, 1843 *J. B. Brichan*; Greshop Wood, 1856 *JK* **FRS**; a large colony with lilac flowers, near Sluie, 1900 (Lobban); with deep pink flowers in the wood at Green Gates, Brodie, 1964 **E**.
N (Stables, *Cat.*); Ardclach, 1880 (Thomson); common in all deciduous woods and on the open moors at Clunas etc., 1954.
I Inverness, 1832 (Watson, *Loc. Cat.*); moor at Abernethy, *R. Meinertzhagen* **BM**; throughout the district, 1954; wood by the Falls of Foyers, 1964 **CGE**.

A. apennina L.   *Blue Anemone*
A garden escape at Forres (Burgess).

Clematis L.
C. vitalba L.   *Old Man's Beard, Traveller's Joy*
Planted in hedges at Elgin (Burgess).

Ranunculus L.
R. acris L.   *Meadow Buttercup, Crowfoot*
Native. The common plant of meadows, waste land etc. is referable to subsp. **acris** var. **acris**. Dwarf plants of the Cairngorms and other mountains with shallowly lobed leaves which are glabrous in early summer are referable to subsp. **acris** var. **pumilus** Wahlenb. They have not been separated here.
M Innes House, Urquhart, 1794 *A. Cooper* **ELN**; abundant (Gordon); Forres, 1900 *JK* **FRS**; the Leen, Garmouth, 1953 *A. Melderis* **BM**; grass bank by the Muckle Burn, Dyke, 1974 **E**.

N (Stables, *Cat.*); Ardclach, 1882 *R. Thomson* **NRN**; grass field near Newton of Park, 1962 **ABD**.
I Inverness, 1832 and Pitmain, 1833 (Watson, *Loc. Cat.*); near Inverness, 1854 *HC* **INV**; Kingussie, 1879 *F. J. Lyne* **E**; Kincraig, 1891 *AS* **E,BM**; shingle of the Allt Fearn, Glen Mazeran, 1968 **CGE**.

R. repens L.   *Sit siccar, Creeping Buttercup*
Native. Abundant throughout the district on cultivated ground, in meadows and on sand dunes.
M Innes House, 1794 *A. Cooper* **ELN**; very common (Gordon); frequent, 1856 *JK* **FRS**; in a wood at Cothall, 1953 *A. Melderis* **BM**; garden weed Dyke, 1974 **E**.
N (Stables, *Cat.*); Ardclach, 1880 (Thomson); pasture near the River Nairn, Kilravock Castle, 1962 **ABD**.
I Inverness, 1833 (Watson, *Loc. Cat.*); near Inverness, 1854 *HC* **INV**; Kincraig, 1891 *AS* **BM**; garden weed, Kirkhill, 1954 **E**.

R. bulbosus L.   *Bulbous Buttercup*
Native. Frequent in short turf and on sand dunes. Easily recognised in the field by its tuberous roots and reflexed sepals. The very hairy form of dunes is referable to var. **dunensis** Druce. It has not been separated here. Map. 23.
M Frequent (Gordon); near Waterford, 1856 *JK* **FRS**; Grantown, 1903 *T. A. Sprague* **K**; banks of the River Spey, Aviemore, short turf by the lime kiln, Fae, plentiful on the grass banks in Dyke Kirkyard, dunes in the Culbin Forest etc., 1954; sandy bank, Burghead, 1972 **E**.
N (Stables, *Cat.*); a curious variety by the schoolhouse road, 1882 *R. Thomson* **NRN**; dunes on the Nairn Dunbar golf course, 1962 **ABD**; bank at Clephanton, 1976 **BM**.
I Pitmain, 1833 (Watson, *Loc. Cat.*); by the River Ness, Inverness, 1854 *HC* **INV**; grass bank below Creag Dhubh, Laggan, roadside verges Tomdow and Garva Bridge, lower slopes of Mealfuarvonie, 1954; roadside bank near Mains of Balnagowan, Ardersier, 1970 **E**; abundant on the grass banks by the Caledonian Canal from Cullochy Lock to Fort Augustus; meadow by Beauly football field, 1976 **BM**.

R. arvensis L.   *Corn Buttercup*
Casual, introduced with foreign grain. A rare annual with lemon-yellow coloured flowers and spiny achenes.
M Shingle of River Spey, Garmouth, 1953 *G. Haines*!; also in 1964 *M.McC.W.*; near the Plasmon Flour Mill Forres, 1973 *K. Christie* **E**!.
I Allotment near a distillery, Inverness, 1961 **E**.

R. sardous Crantz   *Hairy Buttercup*
Recorded by Druce for v/c 95 but I have been unable to locate a specimen or any other reference to this species.

R. auricomus L.   *Goldilocks*
Native. Rare in shady places by burns and in deciduous woodlands;

occasionally on open moorland in shelter of rocks and boulders.
An aggregate of apomictic taxa. Distinguished by the long-stalked
roundish, hardly lobed lower leaves and the few petals, sometimes 0,
rarely 5.
M Between Knockando and Ballintomb (*New Stat. Acc. Scot.* 14: 66
(1845)).
N Blackhills, Nairn, *J. B. Brichan* (Gordon).
I Side of Loch Ness, *G. Anderson* (Gordon); a small colony near
the bottom of the path at the Falls of Foyers, 1971 **E**; in a hazel
wood near the waterfalls at Farley, 1972 **E**; a small colony by the
Allt Garbh, Eskadale, 1973 *M. Barron*!.

R. lingua L.   *Greater Spearwort*
Native. Very rare on loch margins and bogs in the lowlands.
M Abundant in one or two localities. In the loch below Lesmurdie
Cottage, Elgin! (now filled in with rubbish), Loch of Spynie! and
at Leuchars (Gordon); Loch Spynie, 1860 *JK* **FRS**, 1886 *G. C.
Druce* **OXF** and 1898 *ESM* **CGE**!; Dead Water, Elgin, 1881 *? coll.*
ELN; Buckie Loch, Culbin, 1899 *JK* **FRS**! (it was still to be
found there in 1960 but extensive draining has dried up the habitat
and it is feared the plant has been lost); bog on the S side of
Lochinvar, Elgin, 1960 **ABD,K**.
I ? locality, *AS* (Trail, *ASNH* **14**: 224 (1905)).

R. flammula L.   *Wil Fire, Lesser Spearwort*
  Subsp. **flammula**
    Var. **flammula**
Native. Abundant in bogs, ditches, among stones on loch margins
etc. throughout the district. Very variable in size and habit. The
small plants of mountain lochs, with very narrow leaves and rooting
at the nodes are often mistaken for *R. reptans* L.
M Innes House, 1794 *A. Cooper* **ELN**; very common (Gordon);
Dava Moor, 1900 *JK* **FRS**; by the Buckie Loch, Culbin, 1953
A. Melderis **BM** and 1968 *MMcCW* **E**.
N (Stables, *Cat.*); Ardclach (Thomson); by the Minister's Loch,
Nairn, 1900 *R. Thomson* **NRN**!.
I Inverness, 1832 (Watson, *Loc. Cat.*); banks of the River Ness,
1854 *HC* **INV**; shores of Loch Uanagan, 1904 *C. T. West* **STA**;
among stones by Loch Affric, 1947 *N. Y. Sandwith* **K**; peaty pool
by the dam at Mullardoch, 1971 **E**.

  Var. **ovatus** Pers.
A very robust plant with ovate basal leaves occurs in a pool on the
east side of the road between Auldearn and Blackhills Farm near
Nairn, 1967 **E,BM**.

  Var. **pseudoreptans** Syme
Stony margins of mountain lochs and pools. A slender very small
form, creeping and rooting at the nodes and forming arches with
tufts of leaves at each node.
M Winter Lochs, Culbin, 1963 **CGE**; stony shores of Avielochan

near Aviemore, 1964 **CGE**; pool on the moor by Beum a' Chlaid-
heimh, 1973 **E,BM**.
N Marshy ground, Lochloy, 1974.
I Margins of Loch Morlich (Druce, 1888); shores of Loch Beinn
a' Mheadhoin, Glen Affric, 1947 *N. D. Simpson* **BM**; shores of
Loch Toll a' Mhuic, Glen Strathfarrar, 1974 *D. Kingston* **E**!.

Subsp. **scoticus** (E. S. Marshall) Clapham
Similar to subsp. *flammula* but always erect, more robust and
flowers usually solitary. Probably under recorded.
M Buckie Loch, Culbin, 1953 *N. D. Simpson* **BM**!; bog by Loch
Mòr, Duthil, 1973 **E**; backwater of the River Spey, Craigellachie.
N Bog at the north-west end of Lochloy, 1974.
I Stony shore of Loch Alvie, 1898 *ESM & WASh* **BM**!; bog by
Loch Morlich, 1898 *ESM* **OXF**; Shores of Lochs Uanagan and
Tarff, 1904 *G. T. West* **STA**; margin of the pond at Tomatin
House; shores of Loch Insh, Kincraig; bog at Muirtown Basin,
Inverness, 1976 **E**.

R. sceleratus L.   *Celery-leaved Buttercup*
Native. Scattered in muddy pools in the lowlands but nowhere
abundant.
M Frequent (Gordon); bridge at Kinloss (Innes); Loch Spynie,
1860 *JK* **FRS**!; the Leen, Garmouth, 1933 *G. Birnie* (MacGregor,
Fl. of NE); by the edge of a pool in a field at Snab Farm, 1954 **E**;
pools at Netherton near Forres, Gilston Lochs and Newton of
Dalvey, ditches at Kincorth, Sanquhar and Duffus Castle, pond
at Knock of Alves, 1954; farmweed (not typical) at Kirkhill near
Elgin, 1967 *D. Gill* **E**!
N Not seen in Nairn by Thomson who states 'the bruised foliage
applied to the skin produces ugly blisters and beggars are said to
take advantage of this and use poultices to raise sores in order to
stimulate compassionate benevolence'.
I Near Beauly, 1892 *ESM* **BM**; bog in a field between the railway
and the A9 near Lentran, 1925; pool near Culloden, 1956; salt
marsh at Castle Stuart; abundant in the mud of the lagoons at
Muirtown Basin, Inverness, 1975 **E**.

The following species are white flowered.
R. hederaceus L.   *Ivy-leaved Crowfoot*
Native. Frequent in muddy pools and in shallow waters of ditches
and burns. Fig 6b, Map 24.
M Common (Gordon); Berrigeith, 1857 *JK* **FRS**; uncommon at
Advie and Grantown, 1903 (Moyle Rogers); backwater of the
River Spey at Pitcroy, 1967 **E**; ditches at Carrbridge, Lochindorb,
Dyke, Snab of Moy, Cromdale and Dava Moor etc., 1954.
N (Stables, *Cat.*); common in marshy places at Ardclach, 1886
R. Thomson **NRN**; in mud at the edge of Loch of the Clans, 1954;
runnel by the dam at Blackhills Farm, Auldearn, 1971.
I (Watson, *Top. Bot.*); Loch Tarff, 1904 *G. T. West* **STA**; ditches
at Boblainy and Coignafearn 1954; pool by a track on the moor

at Coillenaclay, Farley, 1972 **E**; in a small burn by the Faille
Bridge, Daviot where it was still in flower on 12 December 1975;
runnel by Loch an Ordain near Torness; very robust plants in a
ditch by the golf course at Newtonmore.

R. trichophyllus Chaix   *Thread-leaved Water-crowfoot*
  R. drouetti F. W. Schultz ex Godron
  Subsp. **trichophyllus**
Native. Frequent in ditches, by small burns and muddy margins
of pools. Floating leaves absent.
M (Druce, *Com. Fl.*); margin of Gilston Loch, near Elgin, 1953;
muddy margin of Dyke pond, 1963 **BM**; small burn at Darnaway
Castle; ditch by the road ¼ mile W of Duffus Castle, 1976 **E**.
N In a little pool not far from the sea, *c.* 1 mile E of Nairn (as
R. drouetii) habit quite typical but carpels almost glabrous, 1898
ESM & WASh **BM**; pool by the golf course, Nairn, 1947 *UKD*
UKD; margin of Loch of the Clans, 1960 **ABD**; Loch Flemington,
1962 **ABD**; pool in a field between Auldearn and Blackhills Farm,
1967 **E,CGE**.
I Loch Laide, Abriachan, 1961 **K**; Loch Flemington, 1956;
muddy pool at Crelevan, Strathglass 1925 **BM** and at Eskadale;
by the small burn running through the Inverness golf course,
1976 **E**.

R. aquatilis L.   *Common Water-crowfoot*
Native. Rare. Floating leaves present.
M Recorded by Gordon as common but this status must refer
to all the white-flowered species as no others were listed save *R.
hederaceous*; pool on the Lossiemouth golf course (as *R. heterophyllus*
var. *submersus*) 1954.
N Pool near Nairn, 1954; pool at Dallaschyle, 1955.
I Culloden, 1854 *HC* **INV**; Beauly river, 1892 *ESM* **BM**; Loch
Ashie, 1934 *E. S. Todd*; near Dunain Park, 1940 *AJ* **ABD**; muddy
margin of a pool on the moor by Mains of Bunachton, Dunlichity,
1976 **E**.

R. peltatus Schrank
  R. aquatilis subsp. *peltatus* (Schrank) Syme
Native. Rare.
M Muddy backwater of River Lossie, Elgin, 1941 *? coll.* **STA**.
Pool by the River Spey, Garmouth, 1950; do. 1953 *A. Melderis* **BM**;
ditch by Loch Spynie, 1956.
N Pool near Budgate, Cawdor and margin of Loch Flemington,
1954; small pool by the road at Dallaschyle, 1956; Loch of the
Clans, 1957 **K**.
I Loch Flemington, 1936 *A. Stewart Sandeman* **BM**; do. 1976
MMcCW **E**.

R. penicillatus (Dumort.) Bab.
  R. aquatilis subsp. *pseudofluitans* (Syme) Clapham
M Small burn by Darnaway Castle, 1956; burn running from

Auchnagach House, near Grantown, 1970 *A. J. Souter* **E,BM**!; in very fast flowing stream of the River Spey, Rothes, 1973 *A. J. Souter* **E**!. In the latter locality it has been known by local fishermen only for the last six years. It forms large clumps and extends for about 2 miles upstream from Rothes on both sides of the river. It probably originated from Auchnagach House.

R. baudotii Godr. *Brackish Water-crowfoot*
Native. Rare in brackish pools by the sea.
M Loch Spynie, 1898 *ESM* **E,CGE**; pool on the Lossiemouth golf course, 1954; near the estuary of the River Findhorn N of Invererne, 1968 **E,BM**; deep ditch between the airfield and Duffus Castle, 1972 **E**.
N Sands by the sea shore, Nairn, 1874 *A. Ley* **CGE**.
I Ditch near the River Beauly at Windhill (as var. *marinus*), 1947 *E. C. Wallace* **ECW**!.

The following species has yellow flowers.
R. ficaria L. *Common Pilewort, Lesser Celandine*
    Subsp. **ficaria**
Native. Abundant in woods, by rivers and in meadows but absent from the higher hills. Frequent (Gordon); Sanquhar House, Forres, 1856 *JK* **FRS**; woods at Brodie Castle, 1963 **CGE**; banks of the River Spey at Grantown, 1972 **E**.
N (Stables, *Cat.*); Ardclach (Thomson); Newmills near Auldearn *G. A. Hosking* **K**. The form with double flowers grows on the lawn at Budgate House, Cawdor, 1960 **E,BM**.
I Drumchree, Kirkhill, *c.* 1880 *? coll.* **INV**; banks of the Big Burn, Ness Castle, 1971 **E**; cliffs by Loch Killin, 1972 **E**. The form with double flowers grows at Auchnagairn, Kirkhill.

    Subsp. **bulbifer** Lawalree
Probably introduced. Shady places generally near habitations. Closely resembles subsp. *ficaria* but it is more slender with smaller flowers, has petals that do not overlap, and small bulbils in the axils of the leaves.
M Abundant in the Kirkyard and in the garden at Rose Cottage, Dyke, 1962 **CGE**; small wood by Rothes rubbish tip; Greshop wood near Forres, 1967; edge of the salt marsh at Kingston, 1972 **E**; Fochabers, 1973 *A. J. Souter*; ditch near Dunphail Station, 1974.
N Garden weed Lochloy House near Nairn, 1967 **ABD,CGE**.
I Abundant by the burn and in the Hotel garden, Kingussie, 1954; banks of the Big Burn Glen Cottage, Inverness, 1970 **E,BM**; Culcabock and Dunain, 1970 *M. Barron*; bank at Inshriach Nursery; ditch by the farm of Viewhill, Gollanfield; by the potting sheds at Flichity House; beneath the garden wall at Culloden House; banks of the burn at Farr House, and by the River Beauly below Lovat Bridge.

Adonis L.
A. annua L.   *Pheasant's-eye*
    A. autumnalis L.
Casual. A single plant in a beet field at the Royal Infirmary,
Inverness, 1971 **E**.

Myosurus L.
M. minimus L.   *Mousetail*
I   Recorded by *J. Fraser* for the parish of Kiltarlity (*New Stat. Acc.
Scot.* **14**: 493 (1845)). Probably introduced with grain.

Aquilegia L.
A. vulgaris L.   *Columbine*
Originally a garden escape which has become established on river
shingle and elsewhere.
M Innes House, 1794 *A. Cooper* **ELN**; doubtfully native, Dunphail
woods (Gordon); water side Forres, 1857 *JK* **FRS**; banks of River
Findhorn below Sluie, 1948; established in Greshop wood, 1954;
garden outcast on moor S of Carrbridge.
N Muckle Burn above Craighead, 1832 *WAS & W. Brichan* **ABD**;
shingle of River Findhorn at Glenferness and Coulmony, 1954;
banks of River Nairn near Kilravock Castle, 1967.
I   Kinrara, 1831 (Gordon); Falls of Foyers, 1832 *G. Anderson &
W. Brand* **ABD**; by the Bunchrew and Moniack burns and by the
Kaim below the asylum at Inverness (Aitken); woods at Bunchrew,
1940 *AJ* **ABD**; well established on the shingle of the River Enrick
at Milton, 1947 **E**; railway embankment N of Kingussie, 1973 **E**;
Balnahaun wood, Cannich, 1973 *A. J. Souter*; banks of the River
Enrick at Corrimony School and abundant on shingle on both
shores of Loch Ness from Lochend to Fort Augustus.

Thalictrum L.
T. alpinum L.   *Alpine Meadow Rue*
Native. Common in grassy flushes, on mountain ledges and by sides
of burns in the highland areas. Fig 6a, Map 25.
M Gully of a small burn in the Cromdale Hills, 1925; flush in the
Craigellachie Reserve, Aviemore, 1940 *E. Crapper* **STA**; grass flush
on Càrn Dearg Mòr, Aviemore.
I   Cairn Gorm, 1861 *J. Roy* (Watson); Kingussie, 1890 *A.
Rutherford* **E**; Kincraig, 1891 *AS* **BM**; Larig Pass, 1892 *AS* **E,GL**.
Abundant in suitable habitats on all the hills over 610 m. Minute
plants up to 2 cm high occur on a dry, base-rich rock outcrop in the
Monadhliath Mountains at Alvie.

T. minus L. subsp. **arenarium** (Butcher) Clapham
                                         Lesser Meadow Rue
Native. Occasional on dunes by the sea.
M Dunes in the parishes of Drainie, Kinloss and Urquhart
(Burgess); dunes near Covesea Lighthouse, 1925; edge of the
Roseisle Forest at Burghead, 1954; now rare on the dunes at
Findhorn, the Bar on Culbin and at Lossiemouth.
I   *Macvicar* sp. (*J. Bot.* **43** (1905)).

T. aquilegiifolium L. occurred as a garden relic on a roadside at Easter Galcantry, 1961 **E**, and a member of the **T. minus** agg. occurs as a garden outcast on the steep banks of the River Calder opposite Banchor Mains, Newtonmore.

BERBERIDACEAE
Epimedium L.

E. alpinum L. *Barren-wort*
Garden escape.
M Gone wild about Gordon Castle, Fochabers (Gordon); Relugas, 1867 *JK* **FRS**.
I Near Loch an Eilein, cultivated, 1891 *AS* **E**.

Berberis L.

B. vulgaris L. *Barberry*
Introduced, probably by birds.
M Certainly introduced, Old Duffus Castle (Gordon); Greshop near Forres (Innes); S of the River Findhorn, Forres (Druce, 1888); parishes of Dyke, Alves and Speymouth (Burgess); single bush by the old road bridge over the River Spey at Aviemore, 1973 **E**; roadside near the entrance to Glenerney House, Dunphail.
N Glenferness, 1887 *R. Thomson* **NRN**; Cawdor (Lobban).
I Midmills, Inverness, 1854 *HC* **INV**; Kinrara, Aviemore, 1894 *AS* **E**; by the River Spey 2 to 3 miles below Kingussie, probably bird sown, 1898 *ESM* **E**; Ness Islands (Aitken); planted in Reelig Glen and by the roadside at Tomich; single bushes by the A9 at Alvie, in a field at Achnagairn, Kirkhill and by a small burn at Inchnacardoch 1972 **E**.

B. darwinii Hook. occurs as a garden relic in the old shrubbery at Binsness near Forres, and in the policies of Ness Castle, Inverness.

Mahonia Nutt.

M. aquifolium (Pursh) Nutt. *Oregon Grape*
Garden escape or planted for pheasant food.
N Woods at Cawdor Castle, Holme Rose, Coulmony and Lethen House, 1954; a few plants by the ford at Fornighty, 1973 **E**.
I A single bush in a rock crevice near the road at Eskadale, 1972 **E**.

NYMPHAEACEAE
Nymphaea L.

N. alba L. *White Water-lily*
Native. Frequent in lochs. The plant of some highland lochs has been called subsp. **occidentalis** Ostenf. It is smaller in all its parts, but grades into subsp. **alba**. Map 26.
M Loch near Rinniner, Dallas, *Mr Dick* (Gordon) (this loch may refer to the artifical pond at Dallas Lodge, the present day name for Rinniner); Avielochan near Aviemore, 1925; planted in the parish of Dyke (Burgess), and in the Knock Pond at Newton Toll; also in

Loch an t-Sithein, Lochindorb, Loch Mòr near Duthil, a small
lochan at Brokentore near Kellas etc. 1954.
N Loch of Litie, near Auldearn (Gordon)!; Loch Belivat
(Thomson)!; pool by the River Nairn at Kilravock Castle and Loch
Flemington, 1954.
I Lochs of Badenoch, 1831 (Gordon); loch near Fort George,
1832 *W. Brand* **E,ABD**; between Fort George and Delnies, *WAS*
(Gordon); pond near Castle Hill, Inverness, 1885 (Aitken); Loch
Meiklie, 1904 *G. T. West* **STA**; Loch Duntelchaig, 1940 *AJ* **ABD**;
abundant in most of the small lochans in Strathspey and Strath-
glass.

Nuphar Sm.

N. lutea (L.) Sm.   *Yellow Water-lily*
Introduced in north Scotland.
M Planted in the Knock Pond, Newton Toll, 1930; in the parishes
of Dyke, Alves and Elgin (Burgess).
N Cawdor, 1900 (Lobban); pool by the River Nairn, Kilravock
Castle, 1954.
I Raigmore Pond, Inverness, 1954.

N. pumila (Timm) DC.   *Least Yellow Water-lily*
Native. Frequent in lochans in the Spey valley and in Strathglass.
M In two ponds, one large, one small in the Rape Park at Brodie
House, to which it was brought from Loch Puladdern at Aviemore,
Brodie **E**; collected again (the name of the loch being variously
spelt, Loch Belladeran etc.) in 1811 by *Brodie* **E,K**; 1824 *Dr
Bainbridge* **ABD**; 1837 *W. Brand* **E**; 1865 *JK* **FRS,BM**; and *H. C.
Watson* **GL**; Loch Vaa, 1887 (Druce, 1888)!; Avielochan, 1889
G. C. Druce **OXF**!.
I Loch below Milton of Badenoch, 1831 (Gordon); pool by the
road in Glen Cannich, 1867 *J. Farquharson & T. Fraser* (*Trans.
BSE* 9; 474 (1868)); peaty pool among the Spey marshes between
Kincraig and Kingussie, 1898 *ESM & WASh* **BM**!; Loch Meiklie,
1904 *G. T. West* **STA**!; Strathglass, 1936 *A. J. Wilmott* **BM**!;
Lochan Uvie near Laggan, *Y. Heslop Harrison* (*Watsonia* 1953)!;
Loch nan Bonnagh and Loch an Ordain 1976.

CERATOPHYLLACEAE
Ceratophyllum L.

C. demersum L.   *Hornwort*
Native. A rare submerged aquatic plant with stiff dark green filiform
leaves in whorls. Fruit a nut with two spines at the base.
M Spynie Loch, 1950 and Gilston Lochs 1972 **E,BM**.

PAPAVERACEAE
Papaver L.

P. rhoeas L.   *Field Poppy*
Introduced in the north-east. A rare weed of cultivated and waste
places. Flowers deep red, capsule obovoid.

M Forres, 1863 *JK* **FRS**; casual in the parishes of Dyke and Speymouth (Burgess); distillery yards at Carron, 1957 **ABD** and Knockando; Elgin tip, 1960 **ABD,K**; garden weed Dyke, 1967 **BM**; river shingle Fochabers, 1971 **E**; cornfield at Wester Manbeen, 1973 *R. Richter*; waste ground in the Cooper Park, Elgin, 1973 *K. Christie.*
N Recorded by Lobban for Auldearn, Ardclach, Cawdor and Nairn in 1900 but these may be an error for *P. dubium* which he did not list.
I Introduced with foreign barley from a distillery, in an allotment at Inverness, 1961 **E,K**.

P. dubium L. *Blavers, Long-headed Poppy*
Native or introduced. An abundant weed in cornfields, roadside verges and waste places in the lowland areas, rare in the highlands. Flowers brick red, capsule obovoid-oblong. Map 27.
M Very common '*Blavers*' (Gordon); cornfields, frequent, 1856 *JK* **FRS**; Nethybridge and Grantown (Moyle Rogers); arable field near Greshop, 1962 **CGE**.
N (Stables, *Cat.*); Ardclach, not common (Thomson); waste land near Nairn and throughout the county, 1954.
I Inverness, 1832 (Watson, *Loc. Cat.*); mouth of the Caledonian Canal, 1889 *J. Stirling Smith & R. Kidston* **STI**; near Inverness, 1940 *AJ* **ABD**; garden weed Ben Alder Lodge, 1956; rare at Tomich, Tomatin, Fort Augustus and Drumnadrochit.

P. hybridum L. *Round Prickly-headed Poppy*
Casual. Very rare. Flowers pale crimson, capsule globose covered with yellow bristles.
M Sandy fields in the parishes of Dyke, Speymouth and Urquhart (Burgess); introduced with wool shoddy, Greshop House, Forres, 1958.
I Introduced with foreign barley from a distillery, in an allotment at Inverness, 1961 **E**.

P. argemone L. *Long Prickly-headed Poppy*
Native or introduced. An occasional weed in cornfields and on waste ground. Flowers deep red, the petals with a dark base, capsule obovoid-oblong, ribbed with a few dark bristles.
M Very common (Gordon); frequent in cornfields, Forres 1856 *JK* **FRS**; sandy ground near Elgin and Forres (Druce, 1897); parishes of Dyke and New Spynie (Burgess); distillery yard, Carron, 1956 **ABD**; appears yearly in the cornfields at Craigfield Farm, Kintessack where it is well established along with Cornflowers,1963 **CGE**; farm yard near Crook of Alves, 1968; abundant on a new roadside verge by Tearie crossroads, Dyke, where it only appeared for one year, 1972 **E**.
N (Stables, *Cat.*); on the steep bank by the railway bridge on the E of Nairn, 1961 **E,BM**.
I Inverness, 1832 (Watson, *Loc. Cat*); cornfields at Clunes, Kirkhill, 1930 *M. Cameron*; causeway between Clachnaharry and Kessock, 1955 *J. M. Whyte*.

4

P. somniferum L.   *Opium Poppy*
Casual on waste ground and new roadside verges. Flowers large,
lilac or pink. The glaucous-green capsule contains opium.
M Rare on walls about Pluscarden Priory (Gordon); garden weed
at Fochabers, 1921 *? coll.* **GL**; casual in the parishes of Lhanbryd
and Urquhart (Burgess); waste ground at Dyke, 1963 **E**; roadside
verge at Nethybridge; waste ground Burghead, 1973 *R. M.
Suddaby.*
N Nairn tip, 1970; new roadside verges at Blackcastle and Brackla
and waste ground by Nairn golf course.
I On the raised shingle beach at Ardersier, 1952; waste ground
by Loch Morlich, on the Longman tip Inverness, at Kingussie and
Allanfearn, and new roadside verge at Gollanfield.

P. orientale L. occurs as a garden out-cast opposite Dalvey Cottage
near Forres and on the Elgin tip in Moray, and on the raised beach
at Ardersier in Easterness.

P. lateritum C. Koch with pale brick red flowers occurs as a
garden out-cast in the car park at Inshriach Nursery and on the
raised beach at Ardersier.

Meconopsis Vig.

M. cambrica (L.) Vig.   *Welsh Poppy*
Casual. Naturalised in a few places. Flowers yellow.
M Certainly introduced near the gate of Gordon Castle (Gordon);
in rocky places in the parishes of Elgin, Bellie, Speymouth and at
Pluscarden Priory (Burgess); river shingle Garmouth and Blacksboat, 1954; ditch by a small burn W of Pluscarden, 1968 **E**; waste
ground between Whiterow and Dallas Dhu Distillery near Forres,
1971 *K. Christie.*
N By the River Nairn, Holme Rose, 1954; woods at Lethen and by
the Red Burn Cawdor Castle.
I Near the Schoolhouse, Rothiemurchus, 1879 *JK* **FRS**; abundant
in the woods at Balblair near Beauly, 1930 *M. Cameron*!; river bank
near Scaniport and at Bunchrew, 1940 *AJ* **ABD**; introduced with a
load of coal, Ben Alder Lodge, 1956; garden escape on the moor by
Glen Feshie Lodge and waste ground by the hill road between
Inverfarigaig and Foyers, where it has orange flowers.

Glaucium Mill.

G. flavum Crantz   *Yellow Horned-poppy*
Casual. Reported by Dr Gordon from Elgin but not seen since
(Burgess).

G. corniculatum (L.) Rudolph   *Red Horned-poppy*
Casual. Introduced with foreign grain.
M Waste ground in Greshop wood, Forres where it was washed
down the River Findhorn during a spate in a sack of grain from a
distillery above, 1957 **E**.

Chelidonium L.

C. majus L.   *Greater Celandine*
Casual. Walls and banks near habitations. Poisonous.

M Dunphail and Kinloss Abbey (Gordon)!; Cherry Grove, 1856
JK **FRS**; on old walls at Pluscarden Priory, 1925; waste ground at
Findhorn, Forres and Blacksboat, 1954; a few plants in the Elgin
Car Park, 1971.

N Croft of Manoch, 1826 *WAS* **E**; Cawdor (Gordon) and 1885
R. Thomson **NRN**; garden out-cast Brackla; edge of a field opposite
houses, Nairn 1972 **E**.

I Falls of Kilmorack, 1870 ? *coll.* **ELN**; Moniack Burn, 1878
? *coll* **INV**; roadside verge a few yards W of Lovat Bridge, Beauly,
1947 **E**; waste ground Ardersier, 1972 **E**; wood by Pityoulish House,
Rothiemurchus.

Eschscholzia Cham.

E. californica Cham.   *Californian Poppy*
Casual. Garden escape.

M Roadside opposite Wellhill cottages, Kintessack, 1964 **E**; Elgin
tip.

N Tip at Newton of Park, 1975.

I Waste ground by the Longman tip, Inverness, 1975 **E**.

FUMARIACEAE

Corydalis Medic.

C. claviculata (L.) DC.   *Climbing Corydalis*
Native. A climbing, delicate much branched plant with the rhachis
ending in a branched tendril. Inflorescence *c.* 6-flowered, petals
cream coloured. Map 28.

M Among nettles in the Oaken Nursery at Brodie House, 1797
Brodie **E**; frequent at Conerack, Rothes, *J. S. Bushman*, at Knockando, Relugas and Lynleish in Strathspey, *J. G. Innes* (Gordon);
locally abundant at Advie, 1903 (Moyle Rogers); banks of the
River Spey near Carron House, 1957 **K**; old wall at Kellas House,
1967 **E**; woods at Darnaway and in the gorge of the burn by
Huntly's Cave, Dava Moor.

N Cawdor wood, *WAS* **E** and herb. *Hooker* **K**!; by the River Nairn
and the Delny Burn (Aitken); Ardclach, in one spot by the River
Findhorn between a clump of Blackthorn and the precipice opposite
the Schoolhouse, 1887 *R. Thomson* **NRN**!; climbing over brambles
and whins above the quarry 3 miles S of Nairn on the A939, 1954.

I Inverness, 1821 *G. Anderson* (Hooker, *Fl. Scot.*); banks of the
River Nairn near the Manse of Daviot, 1860 *JK* **FRS**; island in
Loch Knockie (Aitken); near Beauly, 1892 *ESM* **CGE**; island in
Loch Tarff, 1904 *G. T. West* **E,STA**; roadside near Struy, 1946
M. S. Campbell **BM**!; hedgerow near Groam of Annat and among
brambles at the junction of the A9 and B9164 at Dunballoch near
Beauly, 1950 *M. Cameron*!.

C. lutea (L.) DC. *Yellow Corydalis*
Introduced. Rare on walls near habitations.
M Certainly introduced on garden walls at Elgin (Gordon)!; Manse of Rafford, 1860 *JK* **FRS**; escape at Dyke (Burgess); walls at Forres, Milton Brodie, Kinloss Abbey and Greyfriars House, Elgin, 1954.

Fumaria L.

F. capreolata L. subsp. **babingtonii** *White Ramping Fumitory*
(Pugsley) Sell
Native. The most robust of our Fumitories. Flowers 10–14 mm, white, the wings of the upper petal and tips of the inner petals blackish-red; bracts equalling the fruiting pedicels which are strongly arcuate-recurved in fruit. Sepals 4–6 × 2–3 mm. Fruit more or less rectangular in outline, smooth when dry. Frequent on cultivated land and shingle by the sea in the lowlands, rare in the highlands. Fig 6f, Map 29.
M Doubtfully native, abundant in one or two localities, Pluscarden and Kinloss (Gordon); Chapelton near Forres, 1860 *JK* **FRS**; Elchies Distillery, 1899 ? *coll.* **ABD**; colonist on banks in the parishes of Dyke, Speymouth, Elgin, New Spynie and Bellie (Burgess); Lossie Bridge, Elgin, 1960 **ABD**; arable field to the S of Fochabers; rubbish heap on the dunes at Findhorn; garden weed Moy House, Forres, 1960 **ABD,K**; weed at Dyke, 1976 **E,BM**.
N (Stables, *Cat.*).
I Inverness, 1832 (Watson, *Loc. Cat.*); Campbelltown, *A. Croall* (Watson)!; waste ground near Kirkhill, 1930 *M. Cameron*; root field at Dores; on shingle by the sea at Alturlie Point and Longman Point, 1955; raised beach Ardersier, 1971 **E,CGE**.

F. purpurea Pugsley *Purple Ramping Fumitory*
Native. Rare on waste and cultivated land. Rather similar to *F. capreolata* but with pinkish-purple flowers, wings and tips dark purple, pedicels less recurved. Fruit faintly rugose when dry.
M Root field by Kinloss Abbey, 1950 *MMcCW* herb. **JEL**.
I Several plants on waste ground at Longman Point, Inverness, 1976 **E,BM**.

F. muralis Sond. ex Koch *Common Ramping Fumitory*
Subsp. **boraei** (Jord.) Pugsley
Var. **boraei**
Native. Frequent on cultivated and waste land. Flowers 10–12 mm, pink, wings and tips dark red. Bracts about half as long as the more or less erect pedicels. Sepals 3–5 × 1·5–3 mm. Fruit obovoid, more or less smooth when dry. Fig 6d, Map 30.
M Sandy ground near Elgin, 1896 (Druce, 1897); waste ground at Findhorn, among rubble at Kinloss Abbey, garden weed at Logie House, Dallas, Kellas and Newton Houses, 1954; root field at Rafford, 1967 **CGE**.
N Garden weed Cawdor Castle, Coulmony House and Nairn, 1954; arable field by Cran Loch, 1955; farm yard at Maviston, 1973 **E**.

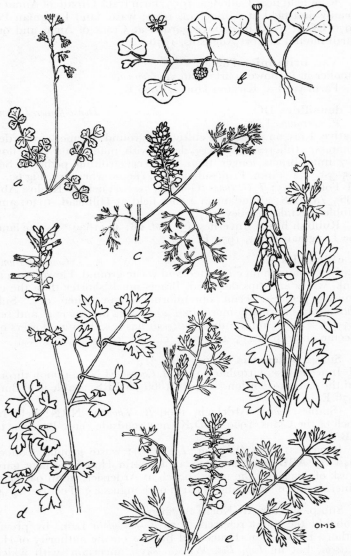

Fig 6 a, **Thalictrum alpinum**; b, **Ranunculus hederaceus**; c, **Fumaria densiflora**; d, **F. muralis** subsp. **boraei**; e, **F. officinalis** subsp. **officinalis**; f, **F. capreolata**.

I Root field near Culloden, 1955; farm track Groam of Annat near Beauly and field at Abriachan, 1962; waste land Longman Point, 1970 **CGE**; weed in cottage garden at Crask of Aigas and on an earth mound at Fort Augustus, 1970.

Var. **britannica** Pugsley
Smaller in its flower parts than var. *boraei*.
I Farm yard at Kirkton House, 1976 **E**.

F. **densiflora** DC.   *Dense-flowered Fumitory*
F. *micrantha* Lag.
Native. Rare on waste and cultivated ground. Formerly recorded as common. Inflorescence very dense with many pale pink flowers 6–7 mm. Bracts longer than the erect fruiting pedicels. Sepals 2·5–3·5 × 2–3 mm. Fruit subglobose, rugose when dry. Fig 6c.
M Forres, 1844 *J. G. Innes* (Gordon, *annot.*); field at Kinloss Abbey, 1865 *JK* **FRS**; abundant in a root field at Hillhead, 1930; among rubble at Kinloss Abbey, 1970.
I Rubbish heaps, Inverness (*J. Bot.* **28**: 40 (1890)); waste land on the Longman Point, 1975 **E**.

F. **officinalis** L.   *Common Fumitory*
Native. Common on cultivated and waste ground. Flowers 7–9 mm, pink, wings and tips dark red. Bracts much shorter than the erect fruiting pedicels. Fruit obreniform, rugose when dry. Subsp. *officinalis* has normally more than 20 flowers in a raceme and sepals 2·5–3·5 × 1–1.5 mm. Subsp. *wirtgenii* has less than 20 flowers on a raceme and sepals 1·5–2 × 1 mm (cf. Sell, 1963a).

Subsp. **officinalis** Fig 6e, Map 31.
M Innes House, Urquhart, 1794 *A. Cooper* **ELN**; common throughout the district (Gordon); Forres, 1866 *JK* **FRS**; arable field Burgie, 1976 **E**.
N (Stables, *Cat.*); Ardclach, 1899 *R. Thomson* **NRN**; field N of Loch of the Clans, 1960 **ABD,K**; new roadside verge Brackla, 1976 **E,BM**.
I Inverness, 1832 (Watson, *Loc. Cat.*); waste ground Culloden, 1940 *AJ* **ABD**; garden weed at Tomatin House, cottage garden Crask of Aigas, on the raised beach at Ardersier, in an oat field at Kincraig and on a mound of earth on Kingussie golf course, 1954.

Subsp. **wirtgenii** (Koch) Arcangeli
Probably the plant referred to as *F. parviflora* Lam. by previous authors, i.e. Gordon, Druce and Burgess on the authority of H. C. Watson, but in *Top. Bot.* Watson says 'uncertain' with which I agree. The distribution of this subsp. has not been thoroughly investigated.
M Cloddach quarry near Elgin, 1976 **E**.

CRUCIFERAE
Brassica L.

B. oleracea L.   *Wild Cabbage*
Casual, rarely naturalised. Waste land and cliffs.
M Rare casual Dyke (Burgess); long established on the cliffs by
Lossiemouth harbour, 1966 **E,ABD**.
I Castlehill, Inverness, 1854 *HC* **INV**; rubbish tip Longman
Point, waste land by Kilmartin House, Glen Urquhart, 1954; tip at
Muirtown Basin, Inverness, 1958 *E. W. Groves* **EWG**.

B. napus L.   *Lochindorb Kail, Swede, Swedish Turnip*
Casual or relic of cultivation. Leaves glaucous-green, flowers pale
yellow the buds overtopping the open flowers.
M 'On a small island, partly artificial, in the northern half of the
lake are the ruins of Lochindorb Castle, once the stronghold of the
Wolf of Badenoch. In 1303 the castle was reduced by Edward I and
in 1336 by Edward III. Among the ruins is found a peculiar plant
resembling red cabbage, known locally as Lochindorb Kail. The
country people transplant it to their gardens and use it as a vegetable' (*Cambridge County Geographies*, **20** (1915)); collected here by
Brodie **E**, and mentioned by Gordon; abundant, 1974 **E,BM**;
scattered throughout the lowlands in suitable habitats but only as a
casual.
N Bank by the railway bridge just E of Nairn, and on roadside
verges at Achavraat, Clunas and Newton of Park, 1954.
I Distillery yard, Tomatin, farm yard Knockie, railway embankment Culloden, waste ground at Foyers, Kincraig and Laggan, 1954.

B. rapa L.   *Yellow Turnip*
  B. campestris L.
Casual or relic of cultivation. Leaves grass-green, flowers bright
yellow overtopping the buds. Cultivated varieties known as *Oil-seed
Rape* are now being grown as a crop.
M Certainly introduced (Gordon); waste land Grantown, 1959 **E**;
cornfield at Muchrach.
N Nairn, 1900 (Lobban).

B. nigra L.   *Black Mustard*
  Sinapis nigra L.
Casual. Rare on waste land and tips.
M Certainly introduced, Deanshaugh, Elgin (Gordon); casual at
Dyke and Speymouth (Burgess); from foreign grain, Forres tip,
1954 and in 1956 **BM**.
I Introduced with foreign grain, allotment near a distillery,
Inverness, 1961 **E,CGE**; waste land on the Longman Point, 1971 **E**.

B. juncea (L.) Czern & Coss.
Casual. Introduced with foreign grain and bird seed.
M Waste land near Longmorn Distillery near Elgin, 1954; Rothes
tip, from distillery refuse, 1956 **E**; on the Elgin tip, 1975.
I Longman Point tip, 1970 and 1971 **E**.

B. tournefortii Gouan
Casual. Introduced with foreign grain.
M Rothes tip, from distillery refuse, 1954 **E,ABD,BM**; railway
siding at Knockando Distillery, 1967 **ABD,CGE**.
I Allotment Inverness, 1961 **E,CGE**; distillery yard, Tomatin,
1961 **E,BM,K**.

Rhynchosinapis Hayek
R. cheiranthos (Vill.) Dandy   *Tall Wallflower Cabbage*
Introduced. Rare on railway embankments and river shingle.
M Railway embankment between Carrbridge and Slochd, 1930.
N Many plants on shingle by the River Nairn at Little Kildrummie,
1974 **E**.
I Railway embankment from Slochd to Culloden Moor station,
1930 and 1971 **E,BM**.

Sinapis L.
S. arvensis L.   *Skelloch, Charlock, Wild Mustard*
Native. Abundant on cultivated and waste land throughout the
district.
M Doubtfully native, frequent (Gordon); Forres, 1860 *JK* **FRS**;
root field, Dyke, 1974 **E**.
N (Stables, *Cat.*); very common in cultivated fields, 1890 *R.
Thompson***NRN**; roots fields at Cawdor, Glenferness, Coulmony and
Nairn, 1954.
I Dalwhinnie (Watson, *Cybele Brit.* 1: 161 (1847))!; cornfield,
Dochgarroch, 1940 *AJ* **ABD**; abundant in fields etc. in the lowland
areas and extending to Glen Moriston, Strathdearn and Cluny
Castle, Laggan.

S. alba L.   *White Mustard*
Casual. Rare on waste and cultivated land.
M Sandy ground near Elgin, 1897 (Druce, 1888); parishes of
Dyke, Kinloss, Speymouth and New Spynie (Burgess); track in the
Culbin Forest, 1953; Forres and Elgin tips, waste ground at Burg-
head and Longmorn, 1954; arable field at Darnaway, 1964 **CGE**;
waste land near Mosset Park, Forres, 1974 *K. Christie* **E**.
I Campbelltown, 1844 *A. Croall* (Watson); Longman tip, 1970
E; roadside verge S of Dores by Loch Ness, 1971 **E**.

Hirschfeldia Moench
H. incana (L.) Lagr.-Foss.   *Hoary Mustard*
Casual. Introduced with bird seed and wool shoddy.
M Greshop House garden, from wool shoddy, 1958 **E**.
N Wild Goose Cottage, Nairn, from 'Swoop', 1969 **E,ABD**.

Carrichtera DC.
C. annua (L.) DC.
  Vella annua L.
Casual. Introduced with foreign grain. A small, hairy annual with
pale yellow, violet-veined flowers and deflexed silicula in two
distinct segments.
I Waste ground in a distillery yard, Inverness, 1961 **E**.

Diplotaxis DC.
D. muralis (L.) DC.   *Wall Rocket, Stinkweed*
Introduced. Sandy ground near the sea.
M Plentiful round buildings by the quay at Findhorn, 1925 **E**; river shingle, Garmouth, waste ground near Forres, 1953; gravel drive, Moy House, near Forres, 1956 **BM**; garden weed at Rose Cottage, Dyke, 1968; sandy ground in Cloddach quarry, 1976.

D. tenuifolia (L.) DC.   *Perennial Wall Rocket*
Introduced.
M Waterford, Forres, 1880 *J. W. H. Trail* **ABD**.

Eruca Mill.
E. sativa Mill.
Casual. Rare on waste ground and tips.
M Waste ground, Moy House, Forres, 1956 **ABD**.
I Several plants on the Longman Point tip, Inverness, 1971 **E**.

Raphanus L.
R. raphanistrum L.   *Runches, Wild Radish*
Doubtfully native. Very common on cultivated land throughout the district. Flowers yellow. Map 32.
M Doubtfully native, very common (Gordon); near Forres, 1860 *JK* **FRS**; Rothes tip, 1966 **E**; arable land Sanquhar House, Forres, 1961 **BM,K** etc.
N (Stables, *Cat.*); Cawdor (Aitken); root field Ordbreck, 1974 **E**.
I Inverness, 1832 (Watson, *Loc Cat.*); abundant by the Spey at Kingussie, 1903 (Moyle Rogers); fields at Craggie (Aitken); allotment at Inverness, 1961 **E**; roadside verge S of Dores, 1971 **E**.
    The white-flowered form appears to be introduced with foreign grain and is rare.
M Siding at Knockando Distillery, 1967 **E**; river shingle, Fochabers, 1971 **E**.
I Distillery yard, Inverness, 1961 **E**; distillery yard Tomatin, 1961, **BM**; roadside verge S of Dores and on waste land at Longman Point.

R. sativus L.   *Radish*
Garden out-cast on tips and waste land.
M Elgin tip, 1954 **E**.
I Waste land at Longman Point, 1971.

Rapistrum Crantz
R. rugosum (L.) All. subsp. **rugosum**   *Bastard Cabbage*
Casual. Introduced with foreign grain and bird seed.
M Tip at Grantown, 1971 **E**; Elgin tip 1974 **E**.
N From bird cage sweepings, Newton of Park tip, 1975 **E**.
I Distillery yard, Inverness, 1961 **E,K**; on the Longman tip, 1971 **E,BM**.

R. hispanicum (L.) Crantz
Casual. Introduced with foreign grain.
M Rothes tip, 1960 **E**.

Var. **hirsutum** (Cariot) O. E. Schultz
M Distillery yard, Carron, 1954; Rothes tip, 1960 **E** and in 1966
E,BM.

Cakile Mill.
C. maritima Scop. subsp. **maritima**   *Sea Rocket*
Native. Common on dunes and on shingle along the whole length
of the coast.
M Abundant (Gordon); dunes at Findhorn, 1899 *JK* **FRS**; dunes
in the Culbin Forest, 1953 *A. Melderis* **BM** and 1968 *MMcCW* **E,
ABD**
N (Stables, *Cat.*); sea shore, 1900 (Lobban)!.
I Carse of Ardersier, 1950 and abundant on the shingle bar,
Whiteness Head, 1976 **E**.

Conringia Adans.
C. orientalis (L.) Dumort.   *Hare's Ear Cabbage*
Casual. Rare.
M Rubbish tip, Waterford, 1892 *JK* **FRS**.

Lepidium L.
L. sativum L. var. **sativum**   *Garden Cress*
Casual. Waste land and tips.
M Rothes tip, 1954 **E,ABD**; Elgin tip, 1961 **ABD**; waste land near
Hillhead, S of Elgin.
I Tip at Longman Point, 1970 **ABD**.

Var. **crispum** L.
A variant with pale mauve flowers and parsley-like leaves which
remain on the plant longer than in var. *sativum* is often introduced
with foreign grain.
M Rothes tip, 1966 **E**.
I Frequent on the Longman tip, Inverness, 1970 **E,CGE**.

L. campestre (L.) R. Br.   *Pepperwort*
Native but probably introduced in north Scotland. Rare.
M Old quarry, Lossiemouth, *G. Wilson* **ABD**; Forres (Innes);
Inverugie, 1830, Ardgay, 1830, and near Brodie House, 1837
P. Cruickshank (Gordon); garden weed Dyke, 1974 **E,BM**.
N Nairn, 1900 (Lobban).

L. heterophyllum Benth.   *Smooth Pepperwort, Smith's Cress*
    L. smithii Hook.
Native. Common on dry banks, river shingle and railway embank-
ments. Fig 7f, Map 33.
M Dalvey (Innes)!; Dyke, 1867 *JK* **FRS**; siding at Boat of Garten
railway station, 1930; shingle of River Spey, Garmouth, 1953
A. Melderis **BM**!; dry bank at Fochabers, 1974 **E**.
N Nairn, 1900 (Lobban); roadside verge by the Nairn Dunbar golf
course, 1962 **ABD**.
I East approach to Holme, Inverness, 1833 *WAS* **E,K**; Beauly,
1870 ? *coll.* **ELN**; Culcabock dam (Aitken); near Kincraig, 1891

AS **BM**; abundant on stable shingle by the River Enrick, Milton, 1947; very common on the cinder verges of the railway line from Drumochter to Inverness, 1954; waste land at Daviot, 1971 **E**.

L. ruderale L. *Narrow-leaved Pepperwort*
Casual. Rare.
M Fochabers, 1923 *S. Bisset* **GL**.

Coronopus Zinn

C. squamatus (Forsk.) Aschers. *Swine Cress, Wart Cress*
Senebiera coronopus (L.) Poir.
Native. Rare on sandy ground in farmyards and in gateways.
M Findhorn village, *G. Wilson* (Gordon) and in herb. *Hooker* **K**; Findhorn, 1866 *JK* **FRS**; in a gateway at Cullerne House, Findhorn, 1953; farm yard at Netherton, Forres, 1973 *R. M. Suddaby*.

C. didymus (L.) Sm. *Lesser Swine Cress*
Senebiera didyma (L.) Pers.
Introduced. A rare weed of cultivation. Fig 7b.
M Parishes of Urquhart and Speymouth, *G. Birnie* (Burgess); Newmill, Elgin, *J. Harrison* **E**; garden weed, Innes House, Urquhart, 1967 **CGE**; two plants in the garden at Rose Cottage, Dyke, 1966 **E,CGE**.
I Waste land, Longman Point, 1975 **E,BM**; arable field by Abbar Water, Lochend, 1975 **E**.

Cardaria Desv.

C. draba (L.) Desv. *Hoary Cress, Hoary Pepperwort*
Lepidium draba L.
Introduced. Waste ground.
M Long established on a roadside bank near the bridge to the golf course at Garmouth, *G. Birnie* (Burgess) and 1953 *MMcCW* **E**; waste land at Findhorn, 1954; railway embankment by the viaduct at Slochd, N of Carrbridge, 1973 **E**; edge of a sand quarry just W of Hopeman, 1975 *K. Christie*!.

Iberis L.

I. umbellata L. *Candytuft*
Casual. Garden escape.
M Shingle of River Spey, Garmouth, 1953; by the Grantown tip, 1973 **E**; waste ground near Forres, 1974.
I Roadside verge S of Dores, 1956; waste ground at Longman Point; roadside in new cutting, Milton, 1971 **E**; dunes by the sea at Fort George.

Thlaspi L.

T. arvense L. *Field Pennycress*
Doubtfully native. A common weed of cultivated and waste land in the lowlands. Fig 7e.
M Elginshire, 1837 *G. Gordon* **E**; Greshop, 1878 *JK* **FRS**; fields at Aviemore, Grantown, Culbin Forest, Rothes etc., 1954; river shingle, Fochabers, 1971 **E**.

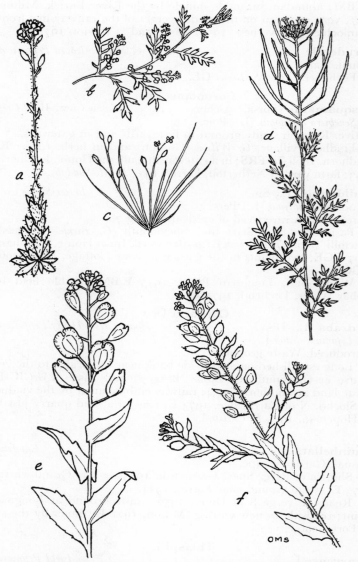

Fig 7 a, **Draba incana**; b, **Coronopus didymus**; c, **Subularia aquatica**;
d, **Descurania sophia**; e, **Thlaspi arvense**; f, **Lepidium heterophyllum**.

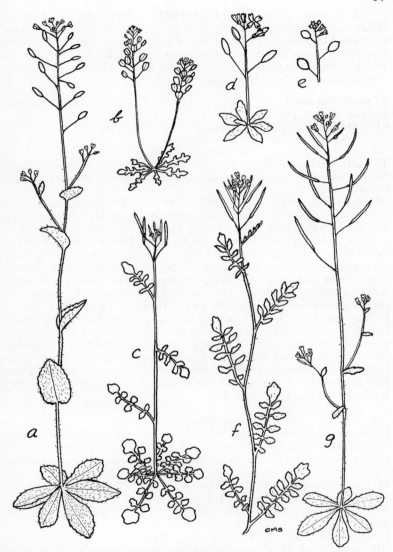

Fig 8   a, **Draba muralis**; b, **Teesdalia nudicaulis**; c, **Cardamine hirsuta**; d, **Erophila verna**; e, **E. spathulata**; f, **Cardamine flexuosa**; g, **Arabidopsis thaliana**.

N Fields about Nairn, 1839 *WAS* **ABD**; Cawdor, 1887 *R. Thomson*
NRN; waste land at Achavraat and in a root field at Ferness, 1954.
I Fort Augustus, 1820 herb. *Borrer* **K**; Inverness, 1832 (Watson,
Loc. Cat.); near Inverness, 1887 herb. *Tarras* **E**; fields at Dores,
Kirkhill and Allanfearn, 1954; garden weed, Mains of Banchor,
Newtonmore, 1975 *S. Haywood*.

T. perfoliatum L.   *Perfoliate Pennycress*
Casual. Rare.
M Rubbish tip, Waterford, Forres, 1872 *JK* **FRS**.

Teesdalia R.Br.
T. nudicaulis (L.) R.Br.   *Shepherd's Cress*
Native. Common on coastal dunes and often on river shingles inland.
Fig 8b, Map 34.
M Linkwood, 1824, Bishopmill, Alves and Findhorn, *G. Wilson*
(Gordon); sands at Findhorn, 1832 *WAS* **E,GL**; Alves, 1833
W. Brand **ABD**; near Elgin, 1833 *WAS* **E,GL**; Glassgreen, 1836
J. Shier **E**; Clunyhill, Forres, 1856 *JK* **FRS**; Binsness, Culbin, 1953
A. Melderis **BM**; dunes near the Buckie Loch, Culbin, 1961 **E,BM**.
N Nairn grove, 1836 *WAS & J. B. Brichan* (Gordon); near Nairn,
1898 *WASh* **BM** and *ESM* **CGE**; sandy bank at Cran Loch, 1960
ABD; dunes by the Nairn Dunbar golf course, 1972 **E**; inland on
the shingle of the Allt Lorgaidh, Dulsie Bridge.
I Dry bank by Loch Flemington, 1955; abundant on a small
island in the River Glass near Eskadale, 1956; gravel pit by River
Spey, Kingussie, 1959 *F. W. Knaggs* (*Atlas*); plentiful on shingle by
the rivers Spey and Calder at Newtonmore; roadside verge E of
Fort George, shingle spit Whiteness Head and sandy ground near
Kincraig station, 1960; on a heap of gravel by the bridge over the
River Doe, Glen Doe, 1972 **E**.

Capsella Medic.
C. bursa-pastoris (L.) Medic.   *Witches' Pouches, Shepherd's Purse*
Native. Abundant in cultivated and waste land. 'Named *Witches'*
Pouches in Moray in the belief that the siliculae were the secret
repository for their enormous wealth' (Thomson).
M Innes House, Urquhart, 1794 *A. Cooper* **ELN**; very common
(Gordon); Forres, 1871 *JK* **FRS**; garden weed, Dyke, 1974 **E**.
N (Stables, *Cat.*); Ardclach, 1886 *R. Thomson* **NRN**; garden at
Nairn, 1974 **E**.
I Inverness, 1832 and Pitmain, 1833 (Watson, *Loc. Cat.*); Newton
Hill, 1880 ? *coll.* **INV**; near Windmill, Beauly, 1947 *A. J. Wilmott*
BM.

C. rubella Reut.
Casual. Introduced with foreign grain. Differs from the former in
the smaller flowers, concave-sided siliculae with a shallow apical
notch, and bright red flower buds.
I Distillery yard Tomatin, 1962 **E,CGE** det. J. E. Lousley.

Hornungia Reichb.

H. petraea (L.) Reichb. *Rock Hutchinsia*
Casual.
M On shingle of River Spey, Speymouth, *G. Birnie* (Burgess).

Cochlearia L.

C. officinalis L. *Scurvygrass*
Native. A common, but variable plant of salt-marshes and on both coastal and mountain cliffs. Formerly eaten by sailors for its sharp taste, being a source of ascorbic acid.

Subsp. **officinalis**
Leaves fleshy, the basal cordate.
M Very common (Gordon); rocks at Covesea, *G. Wilson* **ABD**; Findhorn Bay, 1856 *JK* **FRS**; do. 1953 *A. Melderis* **BM**; salt-marsh 1 mile W of the Buckie Loch, Culbin, 1968 **E**.
N (Stables, *Cat.*); salt-marsh E of Nairn, 1960 **ABD**.
I Lentran, ? *coll.* **INV,CGE**; abundant on mud by the River Beauly, 1892 *ESM* **CGE**; salt-marsh by Castle Stuart, Petty, 1970 **E**.

Subsp. **alpina** (Bab.) Hook. *Alpine Scurvygrass*
Differs in the not fleshy leaves being generally reniform in outline. Mountain ledges. According to *Flora Europaea* **1**: 314 (1964) *C. alpina* (Bab.) H. C. Watson and *C. micacea* E. S. Marshall are nearer to *C. pyrenaica* DC. Both have narrower fruits than *C. officinalis* subsp. *officinalis*.
I *Gordon ms* (Watson, *Top. Bot.*); Braeriach, 1877 (Boyd); Glen Einich, 1884 *J. Groves* **BM**; Coire an t-Sneachda of Cairn Gorm, 1898 *ESM* **BM**; head of Glen Einich and in the Larig Pass, 1893 *AS* **E**; Coire Bogha-cloiche, 1949 *P. S. Green* **E**; cliffs at 850 m Sgùrr na Lapaich, Glen Strathfarrar, 1971 **E**; abundant in gullies of burns on Bynack More, Cairngorms.

C. officinalis × **scotica** Druce
N Salt-marsh 1 mile E of Nairn Dunbar golf course, 1960 **K** det. E. Pobedimova.

C. scotica Druce *Scottish Scurvygrass*
Native. Frequent on salt-marshes and the margins of shingle bars, basal leaves usually truncate. It shows much variation in the size of the seeds and sometimes forms sterile hybrids with *C. officinalis*. It is possible that many of the following records are referable to this hybrid.
M Shore line the length of the Culbin Forest, 1952 **BM,K**.
N Salt-marsh at Carse of Delnies, 1960 **ABD,K**; shingle margin of the Old Bar, Culbin, 1968 **E**.
I Hills in Glen Feshie (as *C. groenlandica*) 1831 (Gordon); sea shore Carse of Ardersier, 1844 *A. Croall* **GL**; Loch Ness, 1921 *Webb* (*BEC Report* **9**: 333 (1930)).

C. micacea E. S. Marshall (see note under *C. officinalis*)
I Coire an t-Sneachda, Cairn Gorm, 1942 *Chaworth Masters* **BM**.

C. danica L.   *Danish Scurvygrass*
Native. A rare annual with pale mauve flowers and stalked leaves, the lower palmately 3- to 7-lobed. Below walls and on rocks by the sea.
M Culbin Sands (Gordon); sandy ground below buildings at Findhorn, 1954; plentiful near Hopeman Harbour, 1967 **E** (now gone owing to the tarring of the paths); Lossiemouth Harbour, 1972 *K. Christie*; a few patches among rocks near the Lossiemouth golf course, 1973 **E,BM**.

C. anglica L.   *Long-leaved Scurvygrass*
Native. Rare on banks of rivers and burns with tidal waters. Basal leaves cuneate at the base, silicula much compressed laterally, its septum three times as long as wide.
    'In the east coast of Scotland *C. anglica* is not known to occur. It is entered in *Top. Bot.* for the counties of Edinburgh and Elgin but with a query after each county name, cf. Report of the Curator (J. T. Boswell) of the *BEC Report* 1875' (*J. Bot.* **14**: 275 (1876)). 'The Rev. J. B. Brichan observes—may not Cochlearia in all its varied forms, from the most luxuriant state at Burghead to its dwarfish state on the sea coast at Nairn include *C. officinalis*, *C. anglica*, *C. groenlandica* and *C. danica*? The soil situation and quantity of fresh water that is present likely have sufficient influence to effect all these changes. It assumes the form *anglica* at Kincorth on account of fresh water present' (Watson).
M Kincorth, 1830 (Gordon); outlet of the Mosset Burn near Forres, 1976 **E**; salt-marsh by Garmouth golf course, 1976 **E**.
I (Druce, *Com. Fl.*); Windhill near Beauly, 1947 *A. J. Wilmott* **BM**!, confirmed by E. Pobedimova; banks of the River Beauly by the football pitch, Beauly, 1976 **E,BM**.

Subularia L.
S. aquatica L.   *Awlwort*
Native. A dwarf aquatic annual, often flowering under water. Frequent on muddy margins of acid lochs and pools in the highlands. Fig 7c, Map 35.
M Introduced by waterfowl, Bellie, *G. Birnie* (Burgess); margin of Loch Romach, 1937 *R. Richter*.
I Loch Ericht (Balfour 1868); pool by the River Beauly (Buchanan White, *J. Bot.* **8**: 129 (1870)); margins of Loch Morlich (Druce, 1888) and 1907 *W. Edgar Evans* **E**!; Kincraig, 1891 *AS* **BM**; N shore of Loch Mullardoch, 1947; abundant with large plants of up to 15 cm high in a roadside pool 1½ miles E of Dundreggan, Glen Moriston, 1971 **E,BM**; in an unusual habitat in a flush below cliffs on Sgùrr na Lapaich, Glen Strathfarrar, 1971 **E**.

Alyssum L.
A. alyssoides (L.) L.   *Small Alison*
Casual. A rare annual of cultivated ground. Not seen in recent years.
M Near Linkwood, 1840 *Mr Martin* (Gordon, *annot.*); weed in a

field at Drumduan, Forres, 1861 *JK* **FRS**; railway embankment
Boat of Garten, 1888 *R. S. Wishart* **GL**.
N Piperhill near Cawdor, 1902 *Mr Allan* (*Trans. ISS & FC* 6: 214
(1899))
I Inverness, 1871 ? *coll.* **ELN**.

Lobularia Desv.

L. maritima (L.) Desv.   *Sweet Alison*
   Alyssum maritimum (L.) Lam.
Casual. A well known white-flowered border plant escaping from
gardens and becoming established for short periods on river shingle
and railway embankments.
M On tips at Elgin and Forres, 1954; shingle of River Spey at
Fochabers; rubbish tip at Rothes 1966 **E**.
N Waste ground Newton of Park, 1970.
I River shingle, Inverdruie, 1956; waste ground at Longman
Point and on the raised beach at Ardersier, 1970.

Draba L.

D. norvegica Gunn.   *Rock Whitlowgrass*
   D. rupestris R. Br.
Native. A rare plant of mountain scree and rocks.
I Rocks on the summit of Cairn Gorm (Hooker, *Fl. Scot.*); rare
in the Cairngorms (Gordon); Coire Garbhlach in Glen Feshie,
1957 *E. C. Wallace*, found flowering at the latter locality at the end
of April 1974 by *D. Hayes*.

D. incana L.   *Hoary Whitlowgrass*
Native. Rare on mountain ledges and scree and sometimes washed
down and becoming established on river shingle and overspills of
burns. Fig 7a.
I Cairngorms, herb. *Borrer* **K**; county of Inverness, *H. C. Watson*
(Gordon); a few plants on the NE face of Mealfuarvonie, 1971; in
several places on gravel by burns and on scree in the gullies near
Gaick Lodge, 1973 *MMcCW & R. McBeath* **E**.

D. muralis L.   *Wall Whitlowgrass*
Casual. Garden weed often attaining 40 cm in height. Fig 8a.
I Garden weed Dalwhinnie Station, 1964; hotel garden Newton-
more, 1973 *R. McBeath*; abundant in the vegetable garden at
Dochfour House, 1975 **E,BM**.

Erophila DC.

E. verna (L.) Chevall.   *Whitlowgrass*
   Draba verna L.
Native. Abundant on gravel paths, on walls and railway embank-
ments and on sand dunes throughout the district.
   Subsp. **verna** Fig 8d, Map 36.
M Very common (Gordon); Forres, 1856 *JK* **FRS**; abundant at
Carrbridge, 1903 (Moyle Rogers); gravel drive at Brodie Castle,
1963 **BM**; among cinders on the railway line at Pitcroy, 1967 **E**.

N Walls at Cawdor Castle, 1833 *WAS* **ABD**; dunes at Nairn, 1970 **E**.

I (Watson, *Top. Bot.*); moorland Abernethy, 1928 *R. Meinertz-hagen* **BM**; gravel at Ben Alder Lodge, 1956; on cinders by the railway at Dalwhinnie, 1970 **CGE**; raised beach at Ardersier, 1970 **E**.

Subsp. **spathulata** (Láng) Walters. Fig 8e.
Differs from subsp. *verna* in the obovoid silicula and in the short, fine, forked stellate hairs on the leaves, simple longer hairs on the margins and fewer (18–40) seeds. Rare on sand dunes.

M Dunes to the NW of the Buckie Loch, Culbin Forest, 1930; on dunes a short distance to the W of Lossiemouth golf clubhouse, 1970 **E**; sandy turf by the sea E of Hopeman, 1973 *R. Richter* **E**!.

Armoracia Gilib.

A. rusticana Gaertn., Mey. & Scherb.   *Horse-radish*
Garden out-cast.
M Roadside verge near Forres, in an old quarry at Lossiemouth, Elgin rubbish tip and grassy roadside verge to the S of Rothes, 1954.
I Waste ground at Longman Point, Inverness, 1948.

Cardamine L.

C. pratensis L.   *Cuckoo Flower, Lady's Smock*
Native. Abundant in wet pastures and by rivers and burns through-out the district. A variable taxon probably containing more than one entity.

M Very common (Gordon); Dunphail, 1898 *JK* **FRS**; the Leen, Garmouth, 1953 *A. Melderis* **BM**; policies of Brodie Castle (with white flowers), 1962 **CGE**; meadow by the River Spey at Advie, 1972 **E**.

N (Stables, *Cat.*); common at Ardclach, 1887 *R. Thomson* **NRN**; by the river at Nairn, 1963 **ABD**. A form with double flowers is found in the meadows at Kilravock Castle.

I Loch na Shanish, 1854 *HC* **INV**; by Lochan Càrn a' Chuilinn, 760 m, 1904 *G. T. West* **E**; near Kincraig, 1891 *AS* **BM**; bog by Loch Conagleann, Dunmaglass, 1972 **ABD**.

C. amara L.   *Wood Bitter-cress, Large Bitter-cress*
Native. Frequent in ditches and by burns generally in shade. Appears to be absent from Easterness. Similar in size to *C. pratensis* but always has white flowers and the basal leaves do not form a rosette.

M Elginshire, 1837 *G. Gordon* **E**; Dunphail and Relugas, 1901 *JK* **FRS**; banks of the Mosset Burn, Forres, in the alder wood by Kellas House, banks of the Muckle Burn at Dyke, woods by the River Spey at Cromdale, bog at Lochinvar near Elgin, 1954; ditch in the wood by the River Findhorn at Darnaway, 1963 **BM**; in a backwater by the River Spey at Fochabers, 1972 **E**; by a small burn in an open habitat by Mains of Allanbuie near Keith, 1974.

N Nairnshire, *Alex. Falconer* (Gordon); Ardclach (Thomson)!; in a flush in the beech wood at Cawdor Castle, banks of the River Nairn,

Holme Rose, 1954. The record for v/c 96 cited by Watson in *Top. Bot.* refers to Falconer's record for Nairnshire.

C. flexuosa With.   *Wavy Bitter-cress*
Native. Common by burns and in ditches, generally in wet places but often frequent on dry scree on mountains reaching 915 m, and as a weed of cultivation. Distinguished from *C. hirsuta* by its larger size, flexuous leafy stem and loose rosette of basal leaves. Fig 8f.
M On the River Divie at Relugas, 1837 *J. G. Innes* (Gordon, annot.); near Dyke, 1885 *R. Thomson* **NRN**; shingle of the River Spey at Fochabers, 1972 **E**; among stones in the ruins of Lochindorb Castle, 1974.
N (Stables, *Cat.*); common at Nairn, Holme Rose and Coulmony, 1954; bog near Dulsie Bridge, 1960 **ABD**; by the Muckle Burn at Lethen, 1961 **K**.
I (Watson, *Top. Bot.*); flush on the moor near Boat of Garten, 1922: among stones below Plodda Falls, Guisachan, 1954; by the burn behind Ben Alder Lodge, 1956.

C. hirsuta L.   *Hairy Bitter-cress*
Native. An abundant weed of cultivated land throughout the district. Fig 8c.
M Very common (Gordon); Forres, 1870 *JK* **FRS**; Grantown, 1903 (Moyle Rogers); garden weed, Dyke, 1970 **E**.
N (Stables, *Cat.*); Ardclach, 1887 *R. Thomson* **NRN**; garden weed Drynachan Lodge, 1954.
I Invermoriston, 1800, herb. *Borrer* **K**; Reelig garden, 1880 ? *coll.* **INV**; Bunchrew, 1940 *AJ* **ABD**; bankKi W oflmorack Kirk, 1972 **E**.

C. bulbifera (L.) Crantz   *Coralroot*
Introduced. Flowers pinkish-purple. Propagated by small brownish-purple bulbils in the axils of the upper leaves.
M Planted or introduced with other plants in the policies of Blackhills House, Lhanbryde, 1965 **E,CGE**. A single plant appeared at the back door of Rose Cottage, Dyke in 1967, probably from this source.

C. kitaibelii Becherer
Dentaria polyphylla (Waldst. & Kit.) O. E. Schulz, non D. Don
M Planted or introduced in the policies of Burgie House, Forres, 1955 **K**.

Barbarea R. Br.
B. vulgaris R. Br.   *Yellow Rocket, Winter-cress*
Native. Common, and variable, on waste ground and river shingle in the lowlands. The widest part of the basal leaves falls below the rounded terminal lobe. Map 37.
M Rare by the Gas House, Elgin and at Ardgay, 1830 *WAS* (Gordon); Balnageith, 1898 *JK* **FRS**; shingle of River Spey, Garmouth, 1953 *A. Melderis* **BM**; waste ground near Elgin station, 1959 **K**; banks of the Muckle Burn between Dalvey and Dyke, 1972 **E**.

N Auldearn, *J. B. Brichan* (Watson); shingle of River Nairn at
Kilravock Castle and at Nairn 1954; waste ground near Cawdor
Castle, 1960 **ABD,K.**
I Rare near Inverness, 1829 (Gordon); Castlehill, 1854 *HC* **INV**;
Culcabock dam (Aitken); shingle of River Enrick, Drumnadrochit,
1945 *M. S. Campbell* **BM**; roadside verge 1 mile S of Nethybridge,
1972 **E**; waste ground behind the Badenoch Hotel, Newtonmore,
1975 *J. Clark.*

Var. **arcuata** (Opiz ex J. & C. Presl) Reichenb.
A yellowish-green variant having a cuneate base to the terminal
lobe of the lower leaves, and upward-curving siliquae.
N Shingle of River Nairn at Holme Rose, 1955 **ABD.**
I Bank at Learag, Tomatin, 1974 **E**; waste ground at Skye of
Curr, 1975 **E.**

B. intermedia Bor.   *Intermediate Winter-cress*
Introduced. Occasional on new roadside verges and waste ground.
Basal leaves with a large terminal lobe and 3 to 5 pairs of lateral
lobes. Flowers darker yellow than in *B. vulgaris.*
M Roadside verge ½ mile W of Cardow Distillery, Knockando,
1955; roadside verge near Lochinvar, Elgin, 1960 **ABD**; near
Knockando, 1966 **CGE**; near Whitemire, 1967 **BM**; waste ground
by the Cairn Gorm petrol station at Aviemore, 1973 *R. McBeath* **E,**
BM; shingle of River Spey below the road bridge at Fochabers,
1976 **E.**
N Waste ground at Holme Rose, 1955 and shingle of River Nairn
near Nairn, 1960.
I Shingle of River Enrick, 1945 *M. S. Campbell*; shingle of River
Ness, Inverness, 1947 *UKD* **E**; shingle of River Enrick, Milton, 1972
E; waste ground at Struy Bridge in Strathglass, yard by a cottage at
Aldourie Pier, raised beach at Ardersier, abundant on a bank on the
hill road between Beauly and Resaurie, garden weed at Drumkin-
herras and waste ground at Smithtown.

B. verna (Mill.) Aschers.   *Early Winter-cress*
Introduced. Rare on cultivated and waste land. Distinguished by
the long siliquae 3–6 cm (twice as long as the other species), curving
upwards on thick stalks, and by the basal leaves having a small
terminal lobe and 6–10 pairs of lateral lobes.
N New roadside verge opposite Brackla Distillery, Cawdor, 1976 **E.**

Cardaminopsis (C. A. Mey.) Hayek
C. petraea (L.) Hiit.   *Northern Rock-cress*
  Arabis petraea (L.) Lam.
Native. Occasional on rocks and scree on the Cairngorm Mountains.
Often washed down rivers becoming established on shingle. Not
recorded from the western hills. Pl 3.
M (Trail, *ASNH* **14**: 226 (1905)); washed down the Spey at
Craigellachie and Knockando, (Burgess)!; shingle of River Spey
at Garmouth, Orton and Blacksboat, 1950.

I Braeriach, 1822 herb. *Greville* **E**; Larig Pass, 1884 *J. Groves* **BM**; Coire an t-Sneachda and Braeriach, 1888 *G. C. Druce* **OXF**!; summit of Cairn Gorm, 1953; Glen Feshie, 1973 *R. McBeath*.

The following records have been named var. **hispida** DC.
I On granite, 1880 *W. West* **BM**; do. 1886 *H. Ward* **BM**; Braeriach, 1888 *G. C. Druce* **OXF**; near the top of the Larig Pass, 820 m, 1892 *AS* **E,BM**.

Arabis L.

A. caucasica Willd. *Garden Arabis*
Casual. Garden escape.
M Rubbish tip at Forres, 1960; site of old cottage between Altyre and Clashdhu bridge, 1967 **ABD**; dunes by Lossiemouth golf course, 1973.

A. hirsuta (L.) Scop. *Hairy Rock-cress*
Native. Rare on base-rich rocks on mountains, on walls and often washed down rivers becoming naturalised on shingle. Map 38.
M Abundant in one or two places, Inverugie (Gordon); shingle of River Findhorn N of Greshop wood near Forres, 1954; shingle of River Spey near Baxter's Factory, Fochabers, 1973 **E**.
N Shingle of River Nairn at Nairn, 1956.
I Reelig Glen, 1902 (Pollock); shingle of River Ness opposite the Islands, Inverness, 1956; rare on the cliffs of Mealfuarvonie, 1971; conglomerate cliffs of Tom Bailgeann, 1972 **E**; buttress on Creag nan Clag, 1972 *M. Barron*!; plentiful on overspills and in the burn gullies at Gaick Lodge, 1973 *MMcCW & R. McBeath* **E**; scree and by the track in the lower regions of Glen Feshie; abundant on a low wall by the Kilmorack power station.

A. muralis Bertol
    A. rosea DC.
Casual. Flowers pinkish-lilac.
M Introduced at Blackhills House, Lhanbryde, 1955 **ABD,BM**.

Lunaria L.

L. annua L. *Honesty*
Garden escape.
M Shingle of River Spey at Garmouth, 1953, waste ground at Smallburn, on Forres tip, 1968 **E**.
N Bank near Drumbeg, Nairn, 1966, wood by Cawdor Castle and bank at Woodend.
I Waste ground Tomatin, 1952; shingle of River Ness, Inverness, by the old Kirk at Ardersier and a few plants with dark purple flowers on the new roadside verge by Loch Ceo Glais, Torness, 1972 **E**.

Rorippa Scop.

R. nasturtium-aquaticum (L.) Hayek *Watercress*
Nasturtium officinale R. Br.
Native. Frequent in ditches, ponds and small burns. Fruit 13–18 mm,

seeds distinctly 2-rowed with *c.* 25 depressions in each face of the seed-coat. Map 39.

M Very common (Gordon); Balnageith, 1878 *JK* **FRS**; Gilston lochs, 1972 **E**; bog at Drakemyres, 1972 **E**; pond at Sanquhar House; ditch behind Knockomie, Forres and by burns at Spynie and Pittendreich etc.

N (Stables, *Cat.*); Ardclach (Thomson); pool by the River Nairn, Kilravock Castle, 1954.

I Burn near Inshes E of Castlehill and about Bogroy (Aitken); ditches between Longman Point and the railway, near Dalcross, outlet of a small burn at Ardersier, Raigmore Pond and Ness Castle loch, 1954; ditch near Lewiston; burn by Balloan road Inverness, 1976 **E**.

R. microphylla (Boenn.) Hyland.   *One-rowed Watercress*
    Nasturtium microphyllum (Boenn) Reichenb.
Native. In similar habitats to *R. nasturtium-aquaticum* which it closely resembles, but generally smaller, save the flowers which are larger, and the seeds being in one row. The plant turns purple-brown in the autumn.

M Burn at Garmouth, 1953 *A. Melderis* **BM**!; river shingle at Fochabers, and ditches at Gilston, Forres, Darnaway, Rothes and Culbin Forest.

N Nairn, 1911 *P. Ewing* **GL**; edge of the Minister's Loch, Nairn, 1960 **ABD**; Geddes Reservoir, 1961 **ABD**; Loch of the Clans and in a small burn behind Boath House, Auldearn.

I Drumnadrochit, 1947 *A. J. Wilmott* **BM**; ditches at Beauly, Inverness, Bunchrew and Ardersier, 1954.

R. microphylla × nasturtium-aquaticum = R. × sterilis
Airy-Shaw
The following records have been determined by H. K. Airy-Shaw.
M Bog at Lochinvar near Elgin, 1961 **CGE**; Sanquhar pond, 1961 **CGE**; by the River Lossie near Kellas House, 1968 **E**.
N Shingle of River Nairn by the cemetery, 1960 **ABD,K**; small pond in a field between Auldearn and Blackhills Farm, 1967 **ABD**.
I Allanfearn, 1943 *UKD* **UKD**; pool in Carse Wood near Fort George, 1964 **E,K**.

R. sylvestris (L.) Besser   *Creeping Yellow-cress*
    Nasturtium sylvestre (L.) R.Br.
Doubtfully native in north Scotland. Probably introduced with foreign grain. A persistent weed of cultivated ground and river shingle. Difficult to eradicate on account of its wiry stoloniferous roots.
M Garden weed at Milton Brodie, 1965 **E,BM**, also at Logie House and Newton House, Elgin; shingle of River Spey from Fochabers to Garmouth, 1965 *A. J. Souter*!; garden at Cotterton Cottage, Brodie, 1967 **E**; forestry track, Knockando House, 1970 **CGE**; shingle of the River Spey at Kingston 1972 **E**.

R. islandica (Oeder) Borbás   *Marsh Yellow-cress*
Nasturtium palustre (L.) DC.
Native. Rare on loch margins and dried up pools.
M Speymouth, *G. Birnie* (Burgess); dried up loch behind the school at Boat of Garten, 1973 **E**; shores of Loch Vaa, Aviemore, 1973 **E,BM**; muddy shores of Avielochan, 1973; a single plant on river shingle below the rifle range at Rothes, 1973 *A. J. Souter.*
N South shore of Loch Flemington, 1954 **E,BM**.
I North shore of Loch Flemington, 1954 **E,BM**.

Matthiola R.Br.

M. longipetala (Vent.) DC.   *Night-scented Stock*
  subsp. **bicornis** (Sibth.) P. W. Ball
M. bicornis (Sibth. et Sm.) DC.
A rare casual on rubbish tips.
M Forres tip, 1968 **E,CGE**.

Malcomia R.Br.

M. maritima (L.) R.Br.   *Virginian Stock*
Garden escape.
M Elgin and Forres tips, 1954.
I Longman Point tip, Inverness, 1971 **E**.

Hesperis L.

H. matronalis L.   *Dame's Violet*
Introduced. Often escaped from gardens. River shingles and waste ground.
M Certainly introduced (Gordon); Speymouth, *G. Birnie* (Burgess); sand dunes, Findhorn, 1930; bank by the Burnie Path, Dyke, waste ground at Kinloss, wood at Sanquhar House, Forres, river shingle, Blacksboat and Fochabers, 1973 **E**.
N Cawdor glebe, 1837 *WAS* **E**; roadside verge, Kingsteps, 1962 **ABD**; river bank at Coulmony and roadside verge near Newton of Park.
I Waste ground at Kirkhill, 1930 *M. Cameron*!; shingle of River Enrick, Drumnadrochit, 1944 *M. S. Campbell*!; frequent in the lowland areas extending to Kingussie, Tomatin, Invermoriston and Newtonmore.

Erysimum L.

E. cheiranthoides L.   *Treacle Mustard*
Introduced. Locally plentiful on cultivated and waste ground. Map 40.
M Doubtfully native, rare in a ploughed field at Grantown, *Mr Fraser* (Gordon); Boat of Garten (Druce, 1888)!; two or three plants on shingle at Rothes (Trail, 1905); cornfield by the River Spey at Advie and waste ground at Aviemore, 1903 (Moyle Rogers); river shingle at Relugas, cultivated fields at Dunphail and Garmouth, 1954; waste ground, Grantown, 1957 **E**; roadside verge E of Carrbridge; introduced at Dyke.
N Shingle of River Findhorn by Coulmony House, 1961 **ABD**.

I Tomatin Distillery yard, 1961 **E**; shingle of the Moniack Burn, Kirkhill, 1967 **E**; waste ground S of Nethybridge, Newtonmore station yard, cornfield at Inverdruie and new roadside verge E of Boat of Garten.

E. linifolium Pers.
Casual. Flowers lilac.
I Several plants on the new roadside verge of the Beauly to Milton road junction, at Milton, 1971 **E,BM**.

Cheiranthus L.
C. cheiri L. *Wallflower*
Introduced. Old walls and quarries and as a garden out-cast on rubbish tips.
M Elgin Cathedral, 1837 *G. Gordon* **E**; Greyfriars and Kinloss (Gordon); long established on cliffs in the old quarry near Lossiemouth harbour, 1970 **E,BM**.
N Walls of the old Kirk at Auldearn, 1975.
I (Druce, *Com. Fl.*); garden out-cast on Longman Point and on the raised beach at Ardersier.

Alliaria Scop
A. petiolata (Bieb.) *Garlic Mustard, Jack-by-the-Hedge*
Cavara et Grande
Native. Common in hedges and on margins of woods in the lowland areas, rare in the highlands. Smells strongly of garlic when crushed. Map 41.
M Rare in the old Tanyard at Elgin, common at Sheriffmill bridge, Altyre and Darnaway (Gordon); Forres, 1901 *JK* **FRS**; ½ mile S of Forres station, 1953 *A. Melderis* **BM**; waste ground, Brodie Castle, 1963 **BM**; hedgerow at Dyke, 1973 **E**.
N Rare at Cawdor Castle, *WAS* (Gordon); common on waste ground at Holme Rose, Nairn, Geddes and Auldearn, 1954.
I Common on wood margins and in shady places by roads, Kirkhill, 1930 *M. Cameron*; wood margins at Bunchrew, Ardersier and Inverness, 1954; ditch at Culloden House, 1968; roadside verge opposite entrance to Cluny Castle, Laggan and about farm buildings at Pityoulish.

Sisymbrium L.
S. officinale (L.) Scop. *Hedge Mustard*
Native. Common on cultivated and waste land.
Var. **officinale**. Map 42.
M Very common (Gordon); Forres, 1898 *JK* **FRS**; ½ mile S of Forres station, 1953 *A. Melderis* **BM**; garden weed, Dyke, 1962 **E**.
N (Stables, *Cat.*); common by roadsides in the lower district, Ardclach, 1886 *R. Thomson* **NRN**; Dulsie Bridge, Cawdor, Nairn etc., 1954.
I Inverness and Dalwhinnie, 1832 (Watson, *Loc. Cat.*); dyke at Englishton (Aitken); roadside verges at Milton, S of Aviemore and Tomatin, 1954; garden weed at Kirkhill, 1968.

Var. **leiocarpum** DC.
A variant with smooth fruits. Not common.
M Roadside near Brodie, 1898 (Marshall, 1899); Findhorn tip,
1953 *A. Melderis*, **BM**; shingle of River Spey at Garmouth, 1973 **E**.
I New roadside verge by Loch Ness, S of Dores, 1971 **E**.

S. irio L.   *London Rocket*
Casual. Rare.
M Rubbish tip at Waterford, 1891 *JK* **FRS**; Elgin tip, 1961 **E,BM**.

S. orientale L.   *Eastern Rocket*
Casual. Introduced with foreign grain and bird seed.
M Rothes tip, 1954 **E**; station yards at Blacksboat, and Knockando,
1956; station yard, Forres, 1961 **BM**; river shingle at Fochabers,
1971 **E**; common on the Elgin tip, 1976 **BM**.
N Waste ground by Nairn habour, 1954 **ABD**.
I Inverness docks, 1955; distillery yard, Inverness, 1961 **E,BM**;
distillery yard at Tomatin, 1963 **E,BM**.

S. altissimum L.   *Tall Rocket*
Casual. Waste ground.
M Urquhart, *G. Birnie* (Burgess); tip at Elgin 1961; two plants on
waste ground in the village of Carrbridge, 1973 **E**.
I (Druce, *Com. Fl.*); waste land on Longman Point, Inverness,
1955 and in 1961 **E**.

Arabidopsis (DC.) Heynh.
A. thaliana (L.) Heynh.   *Thale Cress*
  Sisymbrium thalianum (L.) Gay
Native. A common weed of cultivated and waste land, often on river
shingle, sea cliffs and mountain ledges. Fig 8g, Map 43.
M Very common (Gordon); Forres, 1856 *JK* **FRS**; Carrbridge,
1903 (Moyle Rogers); Longmorn, 1934 *L. Clarke* **BM**; garden weed,
Dyke, 1970 **E**.
N (Stables, *Cat.*); Ardclach, 1887 *R. Thomson* **NRN**; garden weed,
Nairn, 1960 **ABD**.
I (Watson, *Top. Bot.*); wall near Cantray (Aitken); to the W of
Kincraig, 1891 *AS* **BM**; frequent at Kingussie, 1903 (Moyle Rogers);
bank near Kilmorack Kirk, 1972 **E**; waste ground at Ardersier,
1972 **E,BM**; on the railway embankments from Drumochter to
Inverness.

Camelina Crantz
C. sativa (L.) Crantz   *Gold of Pleasure*
Casual. Introduced with foreign grain chiefly on rubbish tips and
in railway yards, occasionally in arable fields.
M Gordon Castle, Fochabers, 1812 *Brodie & J. Hoy* **E**; among flax
at Alves, 1833 *WAS & W. Wilson* **ABD**; Rashcrook, 1846 *G. Birnie*
(Watson); Forres station, 1861 *JK* **FRS**; Rothes tip (and also in
a root field) 1954 **ABD**; rubbish tip, Forres, 1957 **K**; waste ground
near Longmorn distillery, 1966. A variant with simple as well as

stellate hairs (var. *pilosa* DC.) was collected from gravel at Moy
House, Forres, 1953 *A. Melderis* **BM**!.
N Roadside bank by a carrot field at Broombank, Auldearn, 1957
E,ABD; from 'Swoop', Nairn, 1969.
I Longman Point rubbish tip, 1970 **E,BM**; new roadside verge
between Charleston and Inverness, 1973 *M. Barron*.

C. microcarpa Andrz. ex DC.
Differs from *C. sativa* in being densely hairy, the silicula slightly
longer than wide and the seeds not more than 1 mm.
I Longman Point rubbish tip, 1971 **E,ABD**.

Neslia Desv.
N. paniculata (L.) Desv.
Casual. Annual plant resembling a small *Camelina*. Introduced with
foreign grain.
M Forres station yard, 1954 **E**.
I Allotment at Inverness, 1961 **E**.

Descurania Webb & Berth.
D. sophia (L.) Webb ex Prantl   *Flixweed*
 Sisymbrium sophia L.
Colonist. Frequent on waste and cultivated ground in the lowland
areas, chiefly near the sea. Fig 7d, Map 44.
M Innes House, Urquhart, 1794 *A. Cooper* **ELN**; at Lossiemouth,
Elgin, Kinloss and Forres, *J. G. Innes* (Gordon); Greshop, 1898
JK **FRS**; sandy ground at Findhorn, 1953 *A. Melderis* **BM** !; garden
weed, Dyke, 1976 **E**.
N Waulkmill (Gordon); Nairn, 1844 *A. Croall* (Watson); outskirts
of Nairn, 1898 (Marshall, 1899); near the harbour, Nairn, 1936
G. Taylor **BM**; shingle of the River Nairn near Howford bridge,
1960 **ABD,K**.
I Inverness, 1832 (Watson, *Loc. Cat.*)!; near Inverness, 1833
W. Brand **ABD**; Ruthven Barracks, Kingussie, 1877 (Boyd); waste
ground at Ardersier and Dalcross, 1954; garden weed, Kirkhill,
1976 *M. Cameron*.

RESEDACEAE
Reseda L.
R. luteola L.   *Dyer's Rocket, Weld*
Native. Local on waste ground, chiefly in the lowlands.
M Frequent but doubtfully native, Lossiemouth, Cothall and
Dunphail (Gordon)!; bridge at Findhorn (Innes); Cothall, 1870
JK **FRS**; railway embankment, Rothes, quarry at Lossiemouth,
grass verge at Newton Toll, sandy ground by the fishermen's bothy,
in the Culbin Forest, river shingle at Fochabers and Garmouth,
1954; station yard, Forres, 1960 **E**; railway embankment, Knock-
ando, 1967 **CGE**.
N Auldearn, *J. B. Brichan* (Watson); river shingle Nairn, 1962
ABD; waste ground by the old ford at Achavraat.
I Ardersier, 1844 *A. Croall* (Watson); by the smithy at Moniack

and at Englishton, 1902 (Pollock); dry ground at Torvaine, 1940
AJ **ABD**; near Holm Mill, Inverness and waste land at Allanfearn,
1972.

R. lutea L.   *Wild Mignonette*
Introduced. Rare on railway embankments, river shingle and
rubbish tips.
M Certainly introduced, rare at Lossiemouth and Greshop, *J. G.
Innes* (Gordon); Forres station yard, 1875 *JK* **FRS** and 1886 *ESM*
CGE; Waterford, 1880 *J. W. H. Trail* **ABD**; station yards at Blacks-
boat and Carron and rubbish tip at Findhorn, 1954; Forres station,
1961 **E,BM**; siding at Knockando, 1967 **CGE**; river shingle,
Fochabers, 1972 **E**.
N Rare at Nairn, *A. Falconer* (Gordon); Kinsteary near Auldearn,
J. B. Brichan (Watson); river shingle, Nairn 1954; railway embank-
ment, Auldearn, 1956 **CGE**.
I Canal banks at Inverness, Bught Dyke and Englishton, 1885
(Aitken); near Campbelltown, 1898 *ESM* **E**; track by Allanfearn
station, 1955; introduced with foreign grain from a distillery, in an
allotment at Inverness, 1961 **E**; railway siding, Inverness, 1962
ABD; rubbish tip at Longman Point, 1964 **E,CGE**; and in 1976
BM.

R. alba L.   *Upright Mignonette*
Casual. Rare.
M A single plant in Greshop Wood near Forres, 1957; many plants
appeared as a garden weed at Rose Cottage, Dyke for several years
but have now disappeared, 1964 **E,CGE**.

VIOLACEAE
Viola L.

V. odorata L.   *Sweet Violet*
Doubtfully native in the north-east. Naturalised by houses, in woods
and on banks. Both the white and violet colour forms occur.
M Innes House, Urquhart, 1794 *A. Cooper* **ELN**; Delfur near Rothes
(with a small flower) (Burgess); garden escape, Newton House,
Alves, 1954; beechwood by the Muckle Burn between Barley Mill
farm and Brodie station, 1966 **E**; in a small wood by the River
Lossie near Hillhead, 1967 **E**.
N Below the garden wall at Kilravock Castle, *Cosmos Innes* (Gordon,
annot.); in a small wood by the river at Nairn, 1954.
I Bank at Clunes Farm, Kirkhill and roadside bank by the A9
opposite the road to Kiltarlity, 1930 *M. Cameron*!; railway embank-
ment by the bridge near Stratton, 1954 **E**; in shade by the River
Beauly near Beauly, 1957 and outside the garden wall at Beaufort
Castle, 1972.

V. riviniana Rchb.   *Common Dog Violet*
Native. Abundant on banks, in woods, on moorland and in rocky
places ascending to 1,000 m.

Subsp. **riviniana**
M (Watson, *Top. Bot.*); Carrbridge, Aviemore and Nethybridge,
1903 (Moyle Rogers); policies of Brodie Castle, 1963 **CGE**.
N (Stables, *Cat.*); Cawdor, 1867 *R. Thomson* **NRN**; Ardclach,
1891 *R. Thomson* **NRN**; bank by the road at Blackhills farm, 1962
ABD.
I (Watson, *Top. Bot.*); Dalwhinnie, and by Loch Insh, 1903
(Moyle Rogers); Glen Feshie, 1950 *C. D. Pigott*; dry bank at
Foyers, 1964 **CGE**.

Subsp. **minor** (Gregory) Valentine
The flowers are smaller and have narrower petals than subsp.
riviniana. The spur is always violet-coloured and the plant comes
into flower 2 to 3 weeks earlier. Under-recorded. The old records
for *V. reichenbachiana* probably belong here.
M Bank by the old railway bridge in the policies of Altyre House,
Forres, 1967 **CGE**; dry bank near Lower Derraid, Grantown, 1975
E; bank by a track in Monaughty Forest, 2 miles W of Pluscarden,
1976.
I Bank by Kyllachy House, Strathdearn, 1974; on the raised
beach at Alturlie Point, 1975; roadside bank a few hundred yards
north of the road to Breakachy in Strathglass, 1975 **E**.

V. canina L. subsp. **canina**   *Heath Dog Violet*
Native. Common on dry moorland and sand dunes. Rather similar
to *V. riviniana* but the leaves are darker green, thick, and shallowly,
not deeply cordate at the base. The coastal plants have almost
blue flowers, those on the moorland deep purple-blue, both have
a yellow or greenish-white spur. Map 45.
M Very common (Gordon); bed of the Spey, Aviemore, 1892 *AS*
E,GL; common in sandy tracks by the sea in the Culbin Forest,
1920; by the Buckie Loch, Culbin, 1953 *A. Melderis* **BM**; station
yard Boat of Garten, 1973 **E**.
N (Stables, *Cat.*); Ardclach (Thomson); dunes near Lochloy, 1964
ABD; track on the moor near Glenferness House.
I Inverness, 1832 and Pitmain, 1833 (Watson, *Loc. Cat.*); moor
by Loch Ruthven, 1904 *G. T. West* **STA**; shingle of River Enrick,
Milton, 1947 **E**; fixed dunes at Fort George, 1972 **E**.

V. palustris L. subsp. **palustris**   *Marsh Violet*
Native. Common in turfy bogs, by burns and in mountain flushes.
Flowers pale lilac with dark veins.
M Moray, 1836 (Murray); frequent (Gordon); Aviemore, 1892
AS **E**; Cothall, 1898 *JK* **FRS**; under alders by the Buckie Loch,
Culbin, 1964 **CGE**; bog in meadow by the River Spey at Grantown,
1972 **E**.
N (Stables, *Cat.*); Ardclach and Glenferness, 1891 *R. Thomson*
NRN; bogs at Cawdor, Drynachan and Cran Loch, etc. 1954.
I Marsh at Englishton (Aitken); Cherry Island, Loch Ness, 1904
G. T. West **STA**; bog below the Plodda Falls, Guisachan, 1972 **E**
and throughout the district.

V. cornuta L.   *Horned Pansy*
Introduced. Garden escape on banks and waste places. Flowers
large, pale violet on very long peduncles.
M By the roadside near Dunphail, *A. MacGregor* (Burgess) (it is
still plentiful there); roadside near Orton, hedgerow near Dallas,
railway embankment at Blacksboat and banks of the River Spey at
Aviemore and Grantown, 1954; verge of a field near Grantown
station, 1963 **E**.
N Ardclach, 1900 (Lobban); roadside bank near cottages Galcantry, 1962 **ABD**.
I Garden relic on the railway embankment at Dalwhinnie, 1955;
bank by the A9 S of Tomatin bridge and waste ground near
Kingussie golf course, 1964; by a ditch away from habitation on
the hill road at Ruilick above Beauly, 1975 **E**.

V. lutea Hudson   *Mountain Pansy*
Frequent on base-rich grassland in the highlands. Flowers large,
purplish-blue in var. **amoena** Syme, or yellow in var. **lutea**. Pl 5,
Map 46.
M Moray, *WAS* (Murray); abundant in one or two places
(Gordon); Dava, scarce, 1898 *ESM & WASh*; banks of the River
Spey at Blacksboat and Advie, grassy moorland verges at Carrbridge
and Nethybridge, verge of moorland track at Torbreck near Slochd,
1954; grass field at Cromdale, Lochindorb and Duthil, 1973; by
the old lime kiln at Rychorrach on Dava Moor, 1974 **E**. All var.
amoena Syme.
N (Stables, *Cat.*); Ardclach, the purple variety (Thomson); grass
field to the W of Lochindorb (var. *lutea*) 1960; by the river Findhorn
at Glenferness, 1960 **ABD**.
I Badenoch, 1830 (Gordon); Inverness, 1832 (Watson, *Loc. Cat.*);
low ground at Kingussie, 1877 (Boyd); Kincraig, 1891 *AS* **E**; in
short turf by Ben Alder Lodge, at Gaick, in Glen Feshie, Strathdearn
and Shenachie etc.; abundant on the golf course at Newtonmore.

V. tricolor L.   *Heartsease, Wild Pansy*
Native. Common on cultivated and waste ground and river shingle.

Subsp. **tricolor**
M Very common (Gordon); Aviemore, 1898 and Forres, 1900 *JK*
FRS; turnip field, Kinloss, 1925; the Leen, Garmouth (as var.
lloydii (Bor.)), 1953 *A. Melderis* **BM**.
N (Stables, *Cat.*); not uncommon on the Ord Hill (Aitken); dunes
by the Nairn Dunbar golf course and shingle of River Findhorn,
Glenferness, 1961 **ABD**; shingle of the River Nairn at Little
Kildrummie.
I Frequent at Culloden, Kingussie and Dalwhinnie, 1903 (Moyle
Rogers)!; shores of Loch Ness at Fort Augustus, 1904 *G. T. West*
STA; pasture at Bunchrew, 1940 *AJ* **ABD**; arable fields in glens
Moriston, Urquhart and Strathglass and all the coastal areas, 1954;
dry bank by Mains of Balnagowan, Ardersier, 1970 **E**.

Subsp. **curtisii** (E. Forster) Syme
A yellow-flowered pansy of sand dunes; chiefly in the north-west.
M Dunphail, 1850 herb. *J. A. Power*. C. E. Salmon states 'I cannot
find that *V. curtisii* is on record for v/c 95, Elgin, a specimen of this
exists in the Herbarium of the Holmesdale Natural History Club,
Reigate' (*J. Bot.* **49**: 276 (1911)).

V. arvensis Murr.   *Field Pansy*
Native. An abundant weed of cultivated and waste land. Flowers
small cream-coloured.
M Sandy ground near Elgin, 1896 (Druce, 1897); Advie, Avie-
more, Carrbridge and Nethybridge, 1903 (Moyle Rogers); Lossie-
mouth, 1907 *G. B. Neilson* **GL**; garden weed at Dyke, 1974 **E**.
N (Stables, *Cat.*); Ardclach (Thomson); on an old wall at Dulsie
Bridge, 1960 **ABD**; carrot field at Broombank, Auldearn, 1969
ABD.
I Pitmain, 1833 (Watson, *Loc. Cat.*); by the road to Leys, ? date
? *coll* **INV**; Windhill near Beauly (as var. *obtusifolia* Jord.) 1947
A. J. Wilmott; arable fields at Tomatin, Fort Augustus, Kincraig
and Glen Urquhart, waste land by Loch Morlich, 1954.

POLYGALACEAE
Polygala L.
P. vulgaris L.   *Common Milkwort*
Native. Frequent on base-rich grassland, on moors and on dunes.
Leaves all alternate; flowers usually purplish-blue, sometimes white
or pink but never the clear dark blue of *P. serpyllifolia*. Map 47.
M Innes House, Urquhart, 1794 *A. Cooper* **ELN**; very common
about Forres (Gordon) (this presumably included *P. serpyllifolia*
which was not mentioned by Gordon); near Forres, 1881 *JK* **FRS**;
Garmouth golf course, 1953 *A. Melderis* **BM**; Findhorn sands
(flowers slaty-blue) 1938 *A. H. G. Alston* **BM**; short turf by the old
lime kiln at Fae near Dorback, bank at Muchrach Castle and
railway embankment near Forres, etc. 1954.
N (Stables, *Cat.*); river bank Glenferness, 1961 **ABD**; dunes at
Nairn, grass outcrop on the moor at Ballochrochin near Drynachan,
grassland at Lochindorb and by Cran Loch.
I Inverness, 1832 Pitmain, 1833 (Watson, *Loc. Cat.*); Culloden
Moor, 1834 ? *coll.* **ABD** Kincraig, 1891 *AS* **E,BM**; below Creag
Bheag, Feshie, 1951 *E. C. Wallace* **BM**; grass verge of the road by
Spey Dam, Laggan, 1972 **E**; dry bank below the conglomerate
cliffs of Tom Bailgeann near Torness, 1972 **E**.

P. serpyllifolia Hose   *Heath Milkwort*
Native. The common plant of acid moorland. Abundant throughout
the district. Lower leaves opposite and crowded. Flowers dark blue,
pale blue, pink or white.
M Boat of Garten (Druce, 1888); short grass in the Culbin Forest,
1962 **CGE** and throughout the county.

N Ardclach, 1883 *R. Thomson* **NRN**; moor near Holme Rose, 1962
ABD; dunes on the Nairn Dunbar golf course, 1962 **CGE**.
I Glen Einich and Rothiemurchus, 1892 *AS* **E,BM**; Fort Augustus,
1904 *G. T. West* **STA**; bank on the moorland road at Blackfold,
1962 **CGE**; moor at Abriachan, 1972 **E**.

GUTTIFERAE
Hypericum L.

H. androsaemum L.   *Tutsan*
Garden escape, probably bird sown. Rare in damp woods and rock
crevices.
M Policies of Brodie Castle, 1961 **CGE**.
I Banks of River Ness opposite the Islands, Inverness, 1954.

H. calycinum L.   *Rose of Sharon*
Garden escape, generally near cottages. Flowers very large.
M Planted by the North Lodge, Altyre House, 1867 *JK* **FRS**!;
wood above the River Findhorn near Cothall, 1973 *R. M. Suddaby*.

H. perforatum L.   *Common St. John's Wort*
Native. Frequent on grass verges, waste places and open woods
in the lowlands. Rare and generally confined to railway embankments in the highlands. Leaves covered with translucent glandular
dots. Map 48.
M Clunyhill, Forres, 1856 and railway line near Greshop, 1898 *JK*
FRS!; Bareflat Hills, Elgin, 1882 ? *coll.* **ELN**; station yards at Boat
of Garten and Aviemore, 1954; railway siding at Forres, 1961 **E,K**;
roadside verge on the Invererne road near Kinloss and banks of
the River Findhorn at Waterford.
N (Stables, *Cat.*); Ardclach (Thomson); waste ground by the river
at Nairn, 1954.
I (Watson, *Top. Bot.*); Bunchrew, 1940 *AJ* **ABD**; roadside verge
near Beauly and near the Holm Mill, Inverness, 1954; waste ground
S of Dores and at Farr; railway embankments at Culloden, Tomatin,
Moy and Dalwhinnie.

H. maculatum Crantz   *Imperforate St. John's Wort*
  H. dubium Leers
Recent records of the two following subspecies have been deter-
mined by N. K. Robson.
  Subsp. **maculatum**
Native. Banks and roadside verges. Very rare. Stems quadrangular,
not winged; leaves without pellucid glands. Branches of the
inflorescence making an angle of 30 degrees with the stem. Sepals
ovate.
M Policies of Moy House near Forres, 1955 **ABD**.
I Top of a bank by the road near Polmaily, Glen Urquhart, 1973
E,ABD.

  Subsp. **obtusiusculum** (Tourlet) Hayek
Leaves often with pellucid glands. Branches of the inflorescence

making an angle of 50° with the stem. Sepals lanceolate. This is the common plant of the area. Map. 49.

M Cothall, 1898 *JK* **FRS**; banks of the River Spey, Grantown, 1954 **ABD,BM**; waste ground by the golf course at Nethybridge, 1972 **E**; banks of River Spey by Knockando station, 1956 **K**; policies of Brodie Castle, 1961 **K,CGE**.

N Hedgerow near Kinsteary, Auldearn, 1954; by the river at Nairn, 1960 **ABD**; roadside verge just north of Glenferness House, 1972 **E**.

I Near Beauly, 1942 *UKD* **UKD**; by the River Beauly near Groam of Annat, 1947 **E,CGE**; roadside bank near Ferrybrae, 1973 **E**; wall at Culloden House, 1973 **E**; abundant in a small meadow and by the road at Suie Hotel, Kincraig, 1974 **ABD**; embankment by the Caledonian Canal, Muirtown Basin, Inverness, 1975 **BM**.

H. tetrapterum Fr.   *Square-stemmed St. John's Wort*
Native. Frequent in wet meadows in the lowlands. Stems quadrangular with angled wings. Petals paler yellow than the other species. Map 50.

M Frequent at Loch Spynie (Gordon)!; by the River Findhorn below Sluie and at Cothall (Aitken); in bogs at Brodie, Conicavel, Findhorn, Fochabers, Carrbridge and on the Leen, Garmouth, 1954; bog at the edge of Lochinvar, Elgin, 1960 **K**; edge of a burn by the old Spey Bridge at Grantown, 1975 **E**.

N (Stables, *Cat.*); Meikle Geddes, 1864 *R. Thomson* **NRN**; bogs by the Loch of the Clans, Holme Rose, near Geddes Reservoir and by the River Findhorn at Coulmony, 1954; ditch at Newton of Park, 1973 **E**.

I Freetown, 1844 *A. Croall* (Watson); Culcabock and Dunain (Aitken); Dochgarroch, 1940 *AJ* **ABD**; shores of Loch Flemington, 1956; rough pasture by the Moniack Burn, Kirkhill, 1961 **K**; banks of the River Beauly by Groam of Annat; ditch by the bridge over the Breakachy Burn, Strathglass; edge of the salt-marsh, Castle Stuart, Petty, 1975 **E**.

H. humifusum L.   *Trailing St. John's Wort*
Native. Occasional on open gravelly places on moors and waste ground.

M Frequent at Garmouth, Urquhart and Alves, 1830 *G. Wilson* (Gordon); Aviemore, 1881 *JK* **FRS**; railway yard at Boat of Garten, 1925; path at Kellas House and waste ground by the station at Grantown.

N Track on the moor by the Loch of the Clans, 1954.

I Culblair, Petty, *WAS* and Insh in Badenoch (Gordon); Dunain (Aitken); Inverness, 1903 *F. & H. C. Palmer* **OXF**; Dochgarroch, 1940 *AJ* **ABD**; shingle of River Glass at Eskadale, station yard at Kincraig and gravel path at Inchnacardoch, 1955; in gravel by a small burn at Corrimony, 1964 **CGE**; open moorland by a roadside pool, Glen Moriston, 1971 **E**.

H. pulchrum L.   *Slender St. John's Wort*
Native. A striking plant, often with red stems. On moors, roadside
banks etc. on acid soil. Abundant throughout the district.
M Very common (Gordon); Grantown, 1903 (Moyle Rogers);
Garmouth golf course, 1953 *A. Melderis* **BM**!; throughout the county,
1954.
N (Stables, *Cat.*); *R. Thomson* **NRN**; moorland bank, Coulmony,
1961 **ABD**.
I Inverness, 1832, Pitmain and Dalwhinnie, 1833 (Watson, *Loc.
Cat.*); Westhill moor, 4 miles SE of Inverness, 1937 *J. Walton* **GL**;
one plant with lemon-yellow flowers at Drumnadrochit, 1947 *A. J.
Wilmott*; throughout the district, 1954.

H. hirsutum L.   *Hairy St. John's Wort*
Native. Rare on base-rich soil in woods and on banks.
M Lower Craigellachie, 1842 *Mr Martin* (Gordon, *annot.*)!; woods
in the parishes of Dyke, Edinkillie, Urquhart, Speymouth, Rothes,
Bellie and Knockando (Burgess); policies of Brodie Castle, 1961
E,BM; waste ground at Bradbush, Darnaway, 1965 **E**.
N Wood by Kinsteary House, Auldearn, 1973.
I Policies of Creagdhubh Lodge, Laggan (probably introduced
there) 1975.

CISTACEAE
Helianthemum Mill.
H. nummularium (L.) Mill.   *Sol Flower, Common Rockrose*
   H. chamaecistus Mill.
Native. Frequent on base-rich grassland. Map 51.
M Near Pitgaveny, 1830 *G. Gordon* **ABD**; raised beach near
Brigges, *G. Wilson* (Gordon); Aviemore, 1900 *JK* **FRS**; locally
abundant at Carrbridge, 1903 (Moyle Rogers); bank E of the River
Spey at Grantown, 1971 **E**; banks on verges of moorland at Lettoch
and Cromdale, by the River Spey at Rothes and Pitcroy, abundant
on the golf course at Boat of Garten, verge of the moor at Torbreck
and grass slope in the gully of the Allt Iomadaidh at Dorback.
N (Stables, *Cat.*); the Druim, *J. B. Brichan* (Watson); bank by
Keppernach Farm, Ardclach, 1900 *J. B. Simpson* (this locality is
now overgrown with whins and broom and the plant has not been
seen since 1962).
I Pitmain, 1833 (Watson, *Loc. Cat.*); common between Campbelltown and Fort George, 1846 *J. Tolmie* (Watson)!; Drumnadrochit,
1885 (Aitken)!; Kincraig, 1891 *AS* **E,BM**; locally abundant at
Kingussie and Culloden Moor, 1903 (Moyle Rogers); bank at
Dalwhinnie, below the cliffs of Tom Bailgeann near Torness, edge
of the moor at Tomatin, abundant in short turf below Creag
Dhubh, near Loch Coulan and on a steep bank at Ruthven Farm,
Shenachie, 1954.

CARYOPHYLLACEAE
Silene L.

S. vulgaris (Moench) Garcke   *Bladder Campion*
  S. cucubalis Wibel
  Subsp **vulgaris**
Native. Scattered throughout the lowlands on roadside verges and
railway embankments, but never abundant. Appears to be absent
from Badenoch. Map 52.
M Frequent (Gordon); roadside Greshop, 1856 *JK* **FRS**; near
Forres (Druce, 1888); near Carron Distillery, 1907 *W. G. Craib*
ABD; banks at Knockando, Elgin, Lossiemouth, Forres and Boat
of Garten, 1954; roadside verge N of Alves, 1974 **E**.
N (Stables, *Cat.*); roadside near Foynesfield, Nairn, 1883 *R.
Thomson* **NRN**; bank by the road near Lethen, by the path near
Delnies School, near Brackla Distillery and at Littlemill, 1954; a
few plants on a bank at Kingsteps, 1974 **E**.
I (Watson, *Top. Bot.*); railway line near Culloden, 1885 (Aitken);
decidedly uncommon, Tomatin, 1903 (Moyle Rogers)!; railway
embankment at Allanfearn, 1930; waste land on Longman Point,
plentiful by the road at Gollanfield station and several clumps in a
wall at Cannonbank near Kirkhill.

Subsp. **maritima** (With.) Á. & D. Löve   *Sea Campion*
Native. Common on shingle on the coast and frequently on mountain
rocks and in ravines, often washed down and established on river
shingle. Pl 5, Map 53.
M Frequent (Gordon); by the River Spey, in great quantity at
Aberlour and Advie, 1903 (Moyle Rogers); dunes at Culbin and
Findhorn, 1925; shingle of River Spey, Garmouth, 1953 *A. Melderis*
BM; abundant on river shingle from Aviemore to Garmouth,
1954.
N (Stables, *Cat.*); shore at Nairn, 1900 (Lobban); Carse of Delnies,
1960 **ABD**; the old Bar, Culbin, 1968 **E**.
I Inverness, 1832 (Watson, *Loc. Cat.*); Culloden, ? *coll.* **INV**;
Longman Point, 1885 (Aitken)!; Rothiemurchus, 1888 *JK* **FRS**;
dry bed of the Spey at Aviemore, 1892 *AS* **E**; shores of Loch Ness
at Aldourie and Fort Augustus, 1904 *G. T. West* **E,STA**!; cliffs
in the burn gullies at Gaick and Coire Chùirn at Drumochter, 1954
and abundant on river shingle in Strathspey, Strathglass, Glen
Markie and Glen Cannich.

S. conica L.   *Striated Catchfly*
Native. Very rare on sand dunes and roadside verges. Not seen in
recent years.
M Barefleet Hills near the railway, 1857 *Mr Martin* (Watson);
native near Elgin, Moray,(Druce, 1897); not common on a roadside
at Lossiemouth, *G. B. Neilson* **GL**; between the railway and
Ramsay MacDonald's house, Lossiemouth, 1935 *A. Stewart Sandeman*
BM.
I Fort George, 1840 Painting by *Sarah Bland*.

S. dichotoma Ehrh. *Forked Catchfly*
Introduced. Recorded for Nairn and Ardclach by Lobban and for
v/c 96, without locality, by Trail.

S. gallica L. *Small-flowered Catchfly*
  S. anglica L.
Casual. Introduced with foreign grain and bird seed.
M Alves, *J. B. Brichan* and Duffus, *G. Wilson* (Gordon); Dyke,
J. G. Innes (Gordon, *annot.*); Waterford, 1876 *JK* **FRS**.
N (Stables, *Cat.*); Inshoch, *J. B. Brichan* (Gordon); Cawdor and
Nairn, 1900 (Lobban); from 'Swoop', Nairn, 1969 **E,ABD**.
I Culloden, 1856 *W. R. McNab* **E**.

S. acaulis (L.) Jacq. *Moss Campion*
Native. A dwarf moss-green plant of mountain scree and rock faces.
Flowers deep pink to white. Pl 5, Map 54.
I Glen Feshie and Mam Soul (Gordon)!; Cairn Gorm, 1863. *J. P.
Bisset* **ABD**!; NE of Dalwhinnie (Balfour, 1868); Glen Urquhart
(Aitken); Coire an Lochain of Cairn Gorm, 1953 *A. Melderis* **BM**!;
cliffs in the Coire Chùirn, summit of Ben Alder, Sgùrr na Lapaich,
Glen Strathfarrar, and on all the high hills at the head of Glen
Afric, 1954.

S. armeria L. *Sweet William Catchfly*
Garden escape. Tips and waste ground.
M The Manse garden, Forres, 1858 *JK* **FRS**; rubbish tips at Elgin
and Forres, 1959 **E**.

S. noctiflora L. *Night-flowering Catchfly*
Casual. Introduced with foreign grain.
M Waste ground at Carron Distillery and several plants on the
railway siding, 1956; siding by Knockando Distillery, 1966 **CGE**.
I Rubbish tip, Longman Point, Inverness, 1956.

S. pendula L. *Drooping Catchfly*
Introduced.
M Garden weed Greshop House, near Forres, 1954 **E**.

S. dioica (L.) Clairv. *Red Campion*
  Melandrium rubrum (Weigel) Garcke
Native. Common in woods, hedgerows, on river banks and shingle,
and on sea cliffs in the lowlands; occasionally on mountain ledges.
M Innes House, Urquhart, 1794 *A. Cooper* **ELN**; very common
(Gordon); Forres, 1861 *JK* **FRS**; S of Forres station, 1953 *A.
Melderis* **BM**; Craigellachie Nature Reserve, 1964 *P. How* **E**.
N (Stables, *Cat.*); Ardclach, 1887 (Thomson); river banks at
Coulmony, Dulsie Bridge and Glenferness, 1954.
I Inverness, 1832 (Watson, *Loc. Cat.*); by the river Spey at Boat
of Garten, 1925; in Strath Glass, by the Bunchrew Burn, small wood
at Fort Augustus and ledge near the summit of Ben Alder etc.,
1954.

S. alba (Mill.) Krause   *White Campion*
    Melandrium album (Mill.) Garcke
Native. Abundant on cultivated and waste land and new roadside
verges; chiefly in the lowlands.
M Innes House, 1974 *A. Cooper* **ELN**; Forres, 1861 *JK* **FRS**;
Aviemore, Boat of Garten and waste ground near Forres (Druce,
1888); ½ mile S of Forres station, 1953 *A. Melderis* **BM**.
N (Stables, *Cat.*); Ardclach (Thomson); waste land at Drynachan,
Cawdor and Glenferness etc., 1954.
I Near Kincraig, 1891 *AS* **BM**; roadside at Bunchrew, 1920 *AJ*
ABD; waste land at Dalwhinnie, near Laggan, at Fort Augustus
and railway embankment at Daviot, etc., 1954.

S. alba×dioica
Under-recorded. Often fertile, back crossing with both parents.
M Roadside verge at Earnhill, 1930; ½ mile S of Forres station, 1953
A. Melderis **BM**; grass banks near Grange Hall, Dalvey Cottage
and by Essle Cemetery.
N Cothill, 1970 **E**; roadside verges at Ardclach, Howford Bridge
Foynesfield and Nairn.
I Grass bank to the W of Lewiston village, 1972 **E**; farm track,
Groam of Annat, 1972 **E**; raised beach at Ardersier, bank opposite
the Kirk at Kilmorack and roadside verge at Loch Flemington.

Lychnis L.
L. flos-cuculi L.   *Ragged Robin*
Native. Common in wet grassy meadows and bogs. Map 55.
M Very common (Gordon); Greshop, 1900 *JK* **FRS**; Advie, 1903
(Moyle Rogers); in the bog at the W end of the Buckie Loch,
Culbin, 1925; the Leen, Garmouth, 1953 *A. Melderis* **BM**.
N (Stables, *Cat.*); Ardclach (Thomson); bog on the raised beach
at Maviston, 1968 **E**.
I (Watson, *Top. Bot.*); Parkes of Inshes, ? *coll.* **INV**; bogs at
Dalwhinnie, Loch Meiklie, Fort Augustus, Strathdearn and Loch
Flemington etc., 1954.

L. coronaria (L.) Desr.
Garden escape. Forres rubbish tip, 1970 and roadside verge near
Kiltarlity in Easterness, 1964.

Agrostemma L.
A. githago L.   *Papple, Corn Cockle*
Introduced. Formerly common in cornfields, now very rare. The
seeds are poisonous to animals.
M Doubtfully native (Gordon); frequent, 1850 *JK* **FRS**; sandy
ground near Elgin (Druce, 1888); common in the cornfields at
Kinloss, 1925; casual among rye, wheat and tares in the parishes
of Dyke, Forres, Elgin, Duffus and New Spynie (Burgess); railway
yard at Boat of Garten, 1944 *L. Riddell-Webster*; rubbish tip at
Rothes, 1954; roadside verge near Milton Brodie, 1968 *R. M.
Suddaby*.

N (Stables, *Cat.*); cornfield at Logie bridge, Ardclach, 1883 *R. Thomson* **NRN**; Cawdor, 1900 (Lobban).
I   Inverness, 1832 (Watson, *Loc. Cat.*); cornfield at Culloden, ? *coll.*
INV.

Dianthus L.

D. armeria L.   *Deptford Pink*
Garden escape. River bank near Coulmony House, 1960 and by the River Glass at Glassburn, 1954.

D. barbatus L.   *Sweet William*
Garden escape. Near Binsness, Culbin, 1973 *R. M. Suddaby* and roadside verge opposite Hardmuir, Auldearn, 1962.

D. deltoides L.   *Maiden Pink*
Native. Rare on dunes and short grassy banks.
M Oakwood, Elgin, 1839 *G. Taylor* (Gordon); dunes at Findhorn, 1950 *J. Harrison* (seen again in 1967 by *K. Christie*, but now appears to be extinct); shingle of River Spey at Pitcroy, 1955 *H. Du Pre.*
I   Cromwell's Mount, Campbelltown *WAS* (Gordon) (seen again in 1974 in very small quantity and grazed to the roots by rabbits!); Crown, Inverness, 1871 ? *coll.* **ELN**; grass verge near Kincraig, 1971 *J. D. Buchanan*; do., 1972 *MMcCW* **E,BM.**

Vaccaria Medic.

V. pyramidata Medic.   *Cowherb*
Casual. Introduced with foreign grain and bird seed.
M Flaxfields at Alves, 1842 *G. Wilson* (Gordon, *annot.*); rubbish tip, Elgin, 1956; from 'Swoop', Dyke, 1975 *A. Munro* **E.**
N From 'Swoop', Nairn, 1968 **E,ABD**; garden at Househill, Nairn, 1974.

Saponaria L.

S. officinalis L.   *Soapwort*
Introduced. Banks and roadsides near habitations. Usually the double-flowered form.
M Mill of Brodie, Dyke, *Brodie* **E**; Deanshaugh, Elgin (Gordon); in a hedge at Forres Nurseries and at Cothall (Innes); Forres, 1861 *JK* **FRS**; in several places about the village of Alves, 1920; railway embankment at Elgin, bank near Rothes and near Boat of Garten, 1954; waste land at Findhorn, 1970 *K. Christie.*
N Ruins of Penick Castle, 1832 *WAS & J. B. Brichan* **ABD**; bed of the river at Firhill, 1885 *R. Thomson* **NRN**; between Nairn and Forres, 1955 *J. Milne* **E**; banks of the River Nairn at Holme Rose, 1955; near the town of Nairn, 1960 **ABD**; waste land at Wester Delnies.
I   Roadside W of Bunchrew, 1920 (it remained there until 1966 when the road was widened and then disappeared); collected there in 1940 *AJ* **ABD**; by the railway station and on the embankment at Gollanfield, 1954; roadside E of Croy and garden relic at Broomhill station.

Petrorhagia (Ser. ex DC.) Link
P. nanteuilii (Burnat) P. W. Ball & Heywood   *Proliferous Pink*
Kohlrauschia prolifera auct.
Casual. Introduced with wool shoddy at Rose Cottage Dyke, 1967.
It reproduced itself for 8 years.

Cerastium L.

C cerastoides (L.) Britton   *Starwort Mouse-ear Chickweed*
C. trigynum Vill.
Native. A rare arctic-alpine of grassy slopes and scree on the higher
mountains. Styles usually 3. Pl 4.
I Glen Feshie, 1831 and Mam Soul, 1836 *Messrs Anderson,
Stables and Gordon* (Gordon); Braeriach, 1867 herb. *Balfour* **E**;
springs on Cairn Gorm, 1886 *H. & J. Groves* **E,OXF**!; rocky scree,
900 m on Tom a' Choinich, 1947; scree near the summit of Ben
Alder, 1956; A' Bhuideanach, Laggan, 830 m, 1957 *D. A. Ratcliffe*;
wet grassy slope of Meall a' Chaoruinn Mòr near Garva Bridge,
700 m; grassy slope below the cliffs on Sgùrr na Lapaich, 1971 **E**;
gully on Bynack More and in most of the corries of the Cairngorm
range.

C. arvense L.   *Field Mouse-ear Chickweed*
Native. Common on sandy banks and grass verges in the lowlands.
Rare in the highlands. Map 56.
M Moray, *G. Gordon* **GL**; abundant in one or two places at Crofts,
Gallowhill, Grangehall, Coltfield and Kinloss, *G. Wilson* and near
Clunyhill, Forres *J. G. Innes* (Gordon); Boat of Garten, 1888 *R. S.
Wishart*; roadside bank between Dalvey Cottage and the Smithy,
1920 **E**; Knockando, Dunphail and Garmouth etc., 1954.
N Viewfield and Kinsteary, 1833 *WAS* **GL**; Hill of Penick, 1833
W. Brand **ABD**; local round Nairn, 1885 *R. Thomson* **NRN**;
abundant by the Nairn Dunbar golf course, 1963 **ABD,BM**; dry top
of a wall near Altonburn, 1972 **E**.
I Kingussie, 1842 ? *coll.* **ABD**; abundant on wall tops and banks
round Beauly, 1947; on sandy soil at Ardersier, Fort Augustus, Skye
of Curr and Tomatin village, 1954; wall on the hill road by Falls
of Divach, at Allanfearn and Ferrybrae; grass bank near Mains of
Balnagowan, 1970 **E,CGE**.

C. alpinum L. subsp. **lanatum** (Lam.) Ascherson et Graebner
Alpine Mouse-ear Chickweed
Native. A rare arctic-alpine of the higher mountains. Occasionally
washed down rivers.
M Brought down the River Spey from the hills, 1935 *G. Birnie*
(Burgess).
I County of Inverness (Gordon); north face of Cairn Gorm, 1874
A. Ley **BM**!; Glen Urquhart (Aitken); S side of Glen Einich (Druce,
1888); Larig Ghru, 1907 *W. Edgar Evans* **E**; Coire Garbhlach, 1955
E. C. Wallace **ECW**; A' Bhuidheanach, Laggan, 1957 *D. A. Ratcliffe*;
scree on the summit of Ben Alder, 1956; rocks on Mealfuarvonie,
1961 *D. A. Ratcliffe*; washed down from the hills above the burn in

Glen Mazeran, 1968 **E**; Creag na Caillich Glen Feshie and head
of Loch Einich, 1972 *R. McBeath.*

C. alpinum×fontanum
I (as *c. alpinum* var. *pubescens* Syme) (Trail, *ASNH* **14**: 230 (1905));
on rocks of mica schist in Coire Garbhlach in Glen Feshie, 1951
E. C. Wallace **ECW**; below the buttress in Coire an Lochain of
Cairn Gorm, 1953.

C. arcticum Lange   *Arctic Mouse-ear Chickweed*
Native. Very rare in rock crevices and scree on the higher mountains
of the Cairngorms.
I Cairn Gorm, 1881 *H. Groves* **BM**!; collected there in 1886 by
G. C. Druce **BM**, in 1898 by *WASh* **BM**, in the Coire an t-Sneachda
of Cairn Gorm in 1889 by *H. Groves* **BM**, in 1920 by *G. C. Druce,*
OXF and 1936 by *JEL* **JEL** and *ECW*, and in the Coire an Lochain
1953 by *J. Souster* **K**; Coire Ruadh of Braeriach, 1944 *R. Mackechnie*
GL,ECW.

C. fontanum Baumg. subsp. triviale (Murb.) Jalas
   C. holosteoides Fr.   *Common Mouse-ear Chickweed*
Native. Abundant in short grass and cultivated places throughout
the district, reaching 1,100 m in the Cairngorms. Fig 9f.
M Innes House, Urquhart, 1794 *A. Cooper* **ELN**; frequent (Gordon);
Findhorn, 1857 *JK* **FRS**; the Leen Garmouth, 1953 *A. Melderis* **BM**;
garden weed Dyke, 1973 **E**.
N (Stables, *Cat.*); Ardclach, 1882 *R. Thomson* **NRN**; Nairn Dunbar
golf course, 1962 **K**; sandy field at Maviston, 1968 **E**.
I Canal bank, Inverness, ? *coll.* **INV**; Tomatin and Dalwhinnie,
1903 (Moyle Rogers); Glen Affric, 1947 *E. Milne-Redhead* **K** and
E. Vachell **BM**; short grassland in Glen Mazeran, 1968 **E**; in grass
on mountain slope of Sgùrr na Lapaich, Glen Strathfarrar, 1971 **E**.

C. glomeratum Thuill.   *Sticky Mouse-ear Chickweed*
   C. viscosum auct.
Native. Abundant on cultivated and waste ground, especially in
farmyards and gateways. Fig 9e.
M Very common (Gordon); field at Brodie, 1962 **K**; garden weed
Dyke, 1973 **E**.
N (Stables, *Cat.*); Ardclach, 1891 *R. Thomson* **NRN**; by the River
Nairn (Aitken), and in 1963 *MMcCW* **E**.
I Inverness, 1832 and Pitmain, 1833 (Watson, *Loc. Cat.*); English-
ton Muir (Aitken); Tomatin, 1903 (Moyle Rogers); yard at Ben
Alder Lodge, 1956 etc.

C. diffusum Pers.   *Dark-green Mouse-ear Chickweed*
   C. atrovirens Bab.; *C. tetandrum* Curt.
Native. Very common on sandy places by the sea and occasionally
on cinders at railway stations inland. Bracts herbaceous, fruit stalks
erect and parts of flowers usually in fours but sometimes in fives.
Fig 9h. Map. 57.

M Abundant, Burghead and Ladyhill (Gordon); Findhorn, 1857 *JK* **FRS**; Culbin, 1923 *D. Patton* **GL**; below the lighthouse at Burghead, 1972 **E**.
N Nairn, 1874 *A. Ley* (*J. Bot.* **14**: 276 (1876)); Nairn, 1911 *P. Ewing* **GL**; on dunes on the Nairn Dunbar golf course, 1962 **K**; dunes on the old Bar, Culbin, 1968 **E**.
I Kincraig, 1951 *E. C. Wallace* **BM**; sandy ground by the sea at Bunchrew and at Castle Stuart, 1954; gravel of the River Glass, Erchless, 1956; shingle bank S of Fort George, 1972 **E**; in cinders at Culloden station.

C. semidecandrum L.   *Little Mouse-ear Chickweed*
Native. In similar situations to *C. diffusum*. Upper bracts with broad scarious margins and tips, the fruit stalk at first recurved or sharply deflexed, the parts of the flowers in fives. Fig 9g, Map 58.
M Innes House, Urquhart, 1794 *A. Cooper* **ELN**; frequent (Gordon); Carrbridge and Nethybridge, 1903 (Moyle Rogers); Chapelton, Forres, 1901 *JK* **FRS**; Culbin, 1953 *A. Melderis* **BM**; dunes at Lossiemouth, 1970 **CGE**; bare ground at Nethybridge, 1973 **E**.
N (Stables, *Cat.*); Ardclach, 1888 *R. Thomson* **NRN**; dunes at Nairn, 1962 **ABD,K**; sandy outcrop in a field at Lynemore, Dulsie Bridge, 1973.
I Top of an old wall by Castle Stuart, Petty, 1970 **E**; dunes N of Fort George, 1970 **E,CGE**; dry bank at Alturlie Point, among cinders at Broomhill station, railway line at Culloden, below the parapet at Struy bridge in Strathglass and edge of a rabbit scrape near Groam of Annat near Beauly.

C. biebersteinii DC.
Garden escape. A handsome well-known grey-leaved plant of rockeries with flowers two to three times as long as the calyx.
M Roadside verge near Rothes, 1954; dunes at Findhorn; roadside verge at Aviemore; railway yard at Blacksboat, 1963 **E**; dunes near the bathing huts to the W of Hopeman, 1967 **E,CGE**; shingle of River Spey at Fochabers, dunes at Lossiemouth and roadside near Essle Kirk.
I Roadside verge near Bunchrew station and railway embankment at Broomhill, 1954.

C. tomentosum L.
Garden escape. An untidy plant with smaller flowers than the former.
I Waste ground Ardersier, 1975 **E**.

Stellaria L.
S. nemorum L. subsp. **nemorum**   *Wood Stitchwort*
Native. Damp woods, river banks and in ditches. Frequent in the lowlands but only recorded from one locality in Easterness. Map 59.
M Rare at Slaginan near Dunphail, 1827 *William Gordon* (Gordon); Darnaway Forest, 1847 *J. G. Innes* (Gordon, *annot.*); Innes's garden, Forres, 1856 *JK* **FRS**; Dunphail, 1886 *ESM* **CGE**; do. (Aitken):

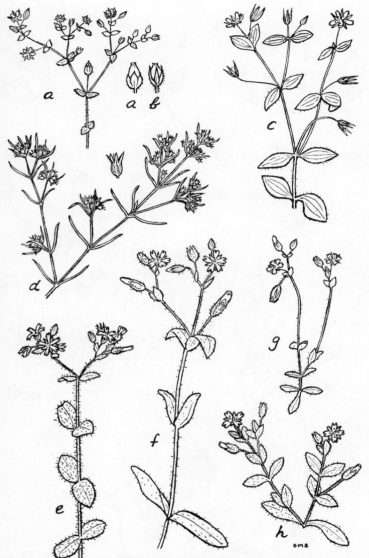

Fig 9   a, **Arenaria serpyllifolia**; b, **A. leptoclados**; c, **Moehringia trinervia**; d, **Scleranthus annuus**; e, **Cerastium glomeratum**; f, **C. fontanum** subsp. **triviale**; g, **C. semidecandrum**; h, **C. diffusum**.

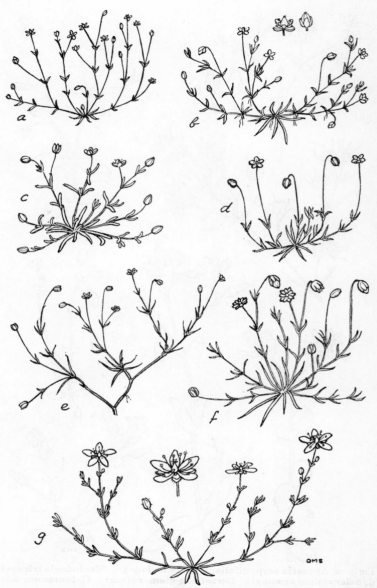

Fig 10   a, **Sagina apetala** subsp. **erecta**; b, **S. procumbens**; c, **S. maritima**; d, **S. subulata**; e, **S. × normaniana**; f, **S. saginoides**; g, **S. nodosa**.

Greshop wood, 1950; banks of the River Findhorn a few yards upstream from Randolph's Leap, Relugas, 1973 **E**; plentiful on the banks of the River Spey from Garmouth to Craigellachie; ditch at Orton, 1973 **BM**.
N Cawdor woods, 1834 *WAS* **E**!; do. 1839 *G. Gordon* **E**; Ardclach, *G. Wilson* (Gordon)!.

S. media (L.) Vill. *The Hen's Inheritance, Chickweed*
Native. Abundant in cultivated land throughout the district. Seeds reddish-brown, 0·9–1·3 mm diam.
M Innes House, Urquhart, 1794 *A. Cooper* **ELN**; very common (Gordon); Forres, 1901 *JK* **FRS**; ½ mile S of Forres station, 1953 *A. Melderis* **BM**; verge of a sandy track by Lossiemouth golf course, 1970 **E,CGE**.
N (Stables, *Cat.*); Ardclach (Thomson); sandy field, Kingsteps, 1960 **ABD**.
I Inverness, 1832 and Pitmain, 1833 (Watson, *Loc. Cat.*); the Manse garden, Rothiemurchus, 1903 *H. Groves* **BM**; waste ground, Inverness, 1940 *AJ* **ABD**.

S. pallida (Dumort.) Piré *Lesser Chickweed*
    S. apetala auct.
Native. Rare. Sandy soil generally near the sea. Differs from *S. media* in being smaller in all its parts, in having no petals and in the seeds being pale yellowish-brown, 0·6–0·8 mm diam.
M Sandy soil by houses in the village of Findhorn, 1972 **E**.
N Track by the sea E of Nairn Dunbar golf course, 1970 **E,ABD**.
I Shingle beach, Campbelltown, 1898 **ESM,E**; dunes opposite the old kirkyard E of Fort George, 1972 **E**.

S. neglecta Weihe *Greater Chickweed*
Very rare. Not seen in recent years. Seeds reddish-brown, larger than *S. media*, never less than 1·3 mm diam.
M By the river Findhorn near Logie (Druce, 1888).

S. holostea L. *Snap Stalks, Break Bones, Greater Stitchwort*
Native. Common on grassy banks, edges of woods and by rivers and burns.
M Innes House, 1794 *A. Cooper* **ELN**; very common (Gordon); Greshop wood, 1856 *JK* **FRS**.
N (Stables, *Cat.*); Ardclach, 1889 *R. Thomson* **NRN**; banks of the River Findhorn at Coulmony, 1961 **ABD**; bank of the Geddes Reservoir, 1961 **K**.
I (Watson, *Top. Bot.*); Ness Islands, Inverness ? *coll.* **INV**; Glendoe, 1904 *G. T. West* **STA**; by the River Spey at Newtonmore, at Coignafearn and among rocks and boulders under juniper in the Ryvoan Pass etc., 1954.

S. graminea L. *Lesser Stitchwort*
Native. Abundant on roadside banks, grassy moorland and by rivers and burns throughout the district.

M Frequent (Gordon); Greshop wood, 1956 *JK* **FRS**; shingle of the River Spey at Garmouth, 1953 *A. Melderis* **BM**.
N (Stables, *Cat.*); Ardclach, 1891 *R. Thomson* **NRN**; banks of the river Findhorn upstream from Dulsie Bridge, 1961 **E,ABD**; grassy bank by the Geddes Reservoir, 1961 **K**.
I Inverness, 1832 (Watson, *Loc. Cat.*); Dalwhinnie, 1903 (Moyle Rogers). A dwarf plant collected by George Don from the sides of rivulets on the mountains of Badenoch (and by Loch Nevis in Westerness) is named *S. scapigera*, **BM,K**. It is probably a stunted form of *S. graminea*.

S. alsine Grimm   *Bog Stitchwort*
    S. uliginosa Murr.
Native. Abundant in ditches, bogs and flushes on moorland throughout the district.
M Innes House, Urquhart, 1794 *A. Cooper* **ELN**; frequent (Gordon); Waterford, 1898 *JK* **FRS**; bog by the Muckle Burn, Dyke, 1973 **E**.
N (Stables, *Cat*); Ardclach, 1881 *R. Thomson* **NRN**; by Geddes Reservoir, 1961 **E,K**.
I Pitmain, 1833 (Watson, *Loc. Cat.*); Leys quarry and Englishton (Aitken); SE of Loch Morlich, 1938 *J. Walton* **GL**.

Sagina L.
Notes on *S.* × *normaniana* Lagerh. and *S. subulata* (Swartz) C. Presl var. *glabrata* Lange are by P. Harrold.

S. apetala Ard.   *Ciliate Pearlwort*
    Subsp. **apetala**: *S. ciliata* Fries
Native. Rare on sandy soil near the sea. Outer pair of sepals pointed.
M Sandy ground near the sea at Hopeman, 1892 *ESM* **BM**; fixed dune in the Roseisle Forest at Burghead, 1954 **E**.
I Culloden Moor station, 1907 *J. W. H. Trail* **ABD**.

Subsp. **erecta** (Hornem.) F. Hermann   *Common Pearlwort*
Native. On bare ground below walls and on gravel paths and cinder tracks. Common in the lowlands, less so in the highlands. Leaves awned; flowers 4-merous; sepals ovate, hooded and blunt; stamens 4. Fig 10a, Map 60.
M Moray, 1836 *WAS* (Murray); near Elgin (Druce, 1897); lay-by on Dava Moor and gravel path at Dyke, 1954; shingle of River Spey at Garmouth, 1964 **CGE**; in cinders at Knockando station, 1966 **E**; waste ground near Carrbridge and garden weed at Boat of Garten etc.
N Nairn, quite rare (Marshall, 1899); banks of the River Nairn (Aitken); common in the garden paths at Cawdor Castle, Glenferness House and Holme Rose, 1954; track near Nairn, 1960 **ABD**; path in the kirkyard at Auldearn, 1975 **E**.
I Sea shore near Inverness, 1793 *James Dickson* **E,BM!**; gravel paths at Inshriach Nursery, Kirkhill and Culloden, 1954; soil heap by the road near Nuide E of Newtonmore, 1975 **E**; waste ground by the old kirk at Ardersier, 1975 **E**; waste ground at Inverness,

Alturlie Point, lay-by by Loch Ceo Glais and cinder track at Broomhill station etc.

S. maritima Don *Sea Pearlwort*
Native. Frequent on fixed dunes, mud flats and occasionally in rock pockets by the sea. Leaves fleshy, blunt or apiculate, not awned; flowers 4-merous; sepals hooded, blunt; stamens 4. Fig 10c.
M Coast of Moray, 1836 *WAS* (Murray); abundant in one or two places (Gordon); near the Old Bar at Findhorn, 1843 *J. B. Brichan & J. G. Innes* (Gordon, *annot.*); salt-marsh Findhorn Bay, 1953 *A. Melderis* **BM!**; fixed dunes NW of the Buckie Loch, Culbin, 1965 **E**; rock pocket by the Sculptor's Cave at Covesea, 1973 **E**.
N (Stables, *Cat.*); dune slack between Nairn Dunbar golf course and the sea, 1960; salt-marsh on the Carse of Delnies, 1960 **ABD**.
I Sea shore near Beauly, 1808 *Miss Hutchins* herb. *Dawson Turner* **K!**; Fort George, 1826 (Gordon)!; Inverness, *c.* 1833 *W. Borrer* herb. *Forster* **BM**; muddy hollow on the raised beach at Ardersier, 1956.

S. procumbens L. *Procumbent Pearlwort*
Native. Abundant in paths, tracks and short grassland throughout the district. Leaves shortly awned; flowers usually 4-merous; sepals ovate, hooded and blunt; stamens 4. Fig 10b.
M Very common (Gordon); Forres, 1865 *JK* **FRS**; by the Muckle Burn near Moy House, 1953 *A. Melderis* **BM**. A sterile double-flowered form known as var. **daviesii** (Druce) Druce grows in abundance on shingle in the upper reaches of the River Lossie in the Moss of Bednawinny near Dallas, 1976 **E**.
N (Stables, *Cat.*); Nairnshire, *J. B. Brichan* (Watson); the school-house garden at Ardclach, 1882 *R. Thomson* **NRN**; throughout the county 1954.
I Inverness, 1832 and Pitmain, 1833 (Watson, *Loc. Cat.*); near Inverness, 1940 *AJ* **ABD**; Dalwhinnie and Tomatin, 1903 (Moyle Rogers); on scree in the Coire an Lochain of Cairn Gorm at 1,150 m 1953; throughout the district 1954.

S. saginoides (L.) Karst *Alpine Pearlwort*
Native. A rare, and often overlooked, arctic-alpine of grassy slopes and gravel by burns. Leaves mucronate; flowers 5-merous; stamens 10 (cf. note under *S.* × *normaniana*). Fig 10f.
M Shingle of the River Spey, 1972 **E**.
I (Watson, *Top. Bot.*); below the cliffs of Coire an t-Sneachda and the Coire an Lochain of Cairn Gorm, 1947; Coire Garbhlach in Glen Feshie, 1957 *E. C. Wallace* (*Atlas*); grass slope below the cliffs of Sgùrr na Lapaich in Glen Strathfarrar, 1971; by a small burn at Gaick Lodge, 1973 **E**. The specimen collected by E. S. Marshall in 1911 at **E** from the Coire Chùirn, Drumochter has been re-determined as *S.* × *normaniana* Lagerh.

S. × **normaniana** Lagerh. *Scottish Pearlwort*
  S. saginoides subsp. *scotica* (Druce) Clapham
Native. Frequent on gravel by burns, in flushes and short grass on

mountain slopes. It is probably a hybrid between *S. procumbens*
and *S. saginoides*. It can be distinguished from *S. saginoides* in (a) the
presence of 4-merous and 5-merous flowers and the number of
stamens being between 5 and 8 (usually 10 in *S. saginoides*); (b) the
forming of extensive creeping mats (whereas *S. saginoides* forms
single rosettes from which arise leafy flowering shoots each year);
(c) the lack of capsules in many flowers in the wild, but if so, much
shorter and containing fewer seeds than *S. saginoides*; (d) the normally low pollen fertility (occasionally 10 per cent) in the wild
(usually above 65 per cent fertile in *S. saginoides*). Fig 10e.

I Coire an t-Sneachda of Cairn Gorm, 1889 *H. Groves* **BM**; Coire
Chùirn, Drumochter (as *S. saginoides*) 1911 *ESM* **E,BM**!; Glen
Einich, *G. C. Druce* (*BEC Report*, 1912); Glen Feshie, 1944 *Rex A.
Graham*; grassy slope on Sgùrr na Lapaich in Glen Strathfarrar, 1972
E; gully of small burn at Gaick, 1974 **E**; scree by a burn on Bynack
More, 1975 **E**; grass slope on Sgùrr na Muice, Monar, 1975.

S. subulata (Swartz) C. Presl   *Awl-leaved Pearlwort*
Native. Frequent on sandy tracks near the coast and on cinder
verges of railway lines and bare open places on moors inland.
Leaves awned; flowers 5-merous; sepals hooded, blunt at apex;
stamens 10. Fig 10d. Map 61.

M Birdsyards near Forres, 1834 *J. G. Innes* **E**; abundant in one or
two places (Gordon); Culbin, 1867 *JK* **FRS**; between Brodie and
Forres, 1896 *G. C. Druce* **OXF**; by the coast at Garmouth, 1898
ESM & WASh **BM,CGE**; Glen Brown, 1905 *ESM* **E,BM**; verge
of a track in the Culbin Forest, 1964 **E,BM**; bare ground on Dava
Moor near Lochindorb and track on top of the cliffs at Hopeman.
N (Stables, *Cat.*); near Nairn, 1898 *ESM* **CGE**; forest track N of
Lochloy, 1975.
I Dalwhinnie, 1903 *F. A. Rogers* **K**!; track near Kincraig station,
1956; small passing place on the hill road at Kinachyle, 1972;
track in Glen Feshie, 1974 **E**; moorland road above the Stag Hut
on Alvie moor, 1974 **E**; hillside near Kingussie and shingle of the
River Spey at Newtonmore, 1974 *J. D. Buchanan*!

Var. **glabrata** Lange
This plant has often been confused with *S. saginoides* but may
readily be distinguished by the distinctly aristate leaves.
I Bare ground on moorland below the Cluanie Dam in Glen
Moriston, 1971 **E**.

S. nodosa (L.) Fenzl   *Knotted Pearlwort*
Native. Common on damp dune slacks by the coast, but rare inland
where it grows on tussocks by lochans and in bogs. The flowers are
5-merous, the petals twice as long as the sepals; stamens 10. Fig 10g,
Map 62.
M Frequent (Gordon); Findhorn Bay, 1856 *JK* **FRS**; ½ mile from
Kincorth, 1938 *J. Walton* **GL**; abundant on fixed dunes in the
Culbin and Roseisle forests and in dune slacks at Lossiemouth,
1954; on tussocks by the lochs at Gilston and in bogs on the Crom-

dale Hills, 1954; flush by the Allt Iomadaidh at Dorback; sandy track in the Culbin Forest, 1961 **E,K**.
N (Stables, *Cat.*); fixed dunes at Kingsteps, 1973 **E**; shores of Lochloy, 1973 **BM**
I Inverness, 1832 (Watson, *Loc. Cat.*); Inverness, 1871 ? *coll.*
ELN; Essich, 1940 *AJ* **ABD**!; flush on the moor 2 miles N of Milton, 1947 and quite common on tussocks surrounding the small lochs above Upper Gartally; bog in a birch wood at Tullochgrue; small bog to the E of Essich Farm, 1975 **E** and by the pond at Achvraid near Essich; bog in Strathglass, on tussocks by the lochan at Mains of Bunachton and in flushes running into Loch Bunachton at Dunlichity.

Var. **moniliformis** (C. F. Mey.) Lange
Stems procumbent, the upper axillary buds and leaf-fascicles become detached as bulbils and propagate the plant vegetatively. All coastal records are probably referable to this variant.
M Elgin coast, 1912 (Druce, *Trans. BSE* **26**: 147 (1913)).

Minuartia L.

M. verna (L.) Hiern   *Vernal Sandwort*
Arenaria verna L.
Native. Rare on serpentine and base-rich rocks. Washed down from the hills.
M Speymouth on dry rocks and banks, *G. Birnie* (Burgess).

M. sedoides (L.) Hiern   *Cyphel*
Cherleria sedoides L.
Native. A rare alpine forming a yellow-green mossy cushion on mountain scree and rock ledges.
I On cliffs of mica schist, Coire Garbhlach, Glen Feshie, 1951 *C. D. Pigott* (*Watsonia*, **2**; (1952)); on scree below the buttress on Coire an Lochain, Cairn Gorm, 1953.

Honkenya Ehrh.

H. peploides (L.) Ehrh.   *Sea Sandwort*
Arenaria peploides L.
Native. Common on dunes and shingle along the whole length of the coast.
M Frequent (Gordon); Findhorn, 1897 *JK* **FRS**!; do. 1953 *A. Melderis* **BM**; shingle on the margin of the Culbin Forest, 1925.
N (Stables, *Cat.*); seashore at Nairn, 1890 *R. Thomson* **NRN**; shingle bank of the old Bar, Culbin, 1968 **E**.
I Raised shingle beach Ardersier, 1925; salt-marsh verge, Bunchrew and shingle spit at Whiteness Head, 1954; shingle N of Fort George, 1976 **E**.

Moehringia L.

M. trinervia (L.) Clairv.   *Three-nerved Sandwort*
Arenaria trinervia L.
Native. Common in deciduous woods and shaded river banks.

Often mistaken for chickweed, from which it can be distinguished by the three prominent veins on the under surface of the leaves. Fig 9c, Map 63.

M Common (Gordon); Greshop wood, 1871 *JK* **FRS**; common in the woods at Brodie, Kellas, Fochabers and in the gorge at Huntly's Cave on Dava Moor etc. 1954; wood at Relugas, 1973 **BM**.
N (Stables, *Cat.*); woods at Geddes, Cawdor Castle, Dulsie Bridge etc. 1954; by the Muckle Burn at Lethen House, 1961 **K**; in shade by the River Findhorn, Ardclach, 1973 **E**.
I The Islands, Inverness, 1833 ? *Coll.* **INV**; at Quarry Hill, Mound and Petty (Aitken); beneath alders by the River Enrick, Milton, 1947; woods by the River Nairn at Daviot, by Loch Ness at Fort Augustus, Dochfour House, by the Falls of Divach and near the school at Corrimony, 1954; birch wood at Farley, 1972 **E**; in shade of juniper in the Ryvoan Pass.

Arenaria L.
A. serpyllifolia L.
    Subsp. **serpyllifolia**   *Thyme-leaved Sandwort*
Native. Frequent on bare ground on dunes and gravel paths. Capsule flask-shaped with firm curving sides. Fig 9a, Map 64.
M Very common (Gordon); Castlehill, Forres, 1856 *JK* **FRS**; Aviemore, 1891 *AS* **E**; on shingle on the west bank of the River Findhorn near Forres, 1953 *A. Melderis* **BM**; by the lime kiln at Fae, Dorback, 1972 **E**.
N (Stables, *Cat.*); Ardclach, by Redburn bridge and by the Free Kirk, 1886 *R. Thomson* **NRN**; Milltown of Kilravock (Aitken); dunes on the Nairn Dunbar golf course, 1961 **ABD**; waste ground by the ford at Achavraat, 1973 **E**.
I (Watson, *Top. Bot.*) Millarton (Aitken); roadside verge at Milton, 1947; station yards at Dalwhinnie, Kingussie and Kincraig, car park at Fort Augustus etc. 1954; among the ruins of Urquhart Castle and by a new road cutting at Spey Dam, Laggan, 1971; gravel path at Gaick Lodge, 1973 **E**.

    Subsp. **leptoclados** (Reichb.) Nyman
                     Lesser Thyme-leaved Sandwort
More slender than subsp. *serpyllifolia* and smaller in all its parts. Capsule conical, straight-sided and not firm. Fig 9b.
M Grantown and Cromdale, 1903 (Moyle Rogers); waste ground by the River Findhorn near the Waterford tip, 1974 **E**; disused railway line at Rafford, 1976 **E**.
I Banks of River Oich near Fort Augustus, 1883 *C. Bailey* **BM**; near Kincraig, 1891 *AS* **BM**.

A. balearica L.   *Mossy Sandwort*
Garden escape. Rare on walls near habitations.
M Wall by the entrance to Newton House near Elgin, 1954; garden walls at Milton Brodie and Kellas House.
N Garden wall Kinsteary House, Auldearn, 1964 **ABD**.

laminopsis petraea × 1
hern Rock-cress

M. C. F. Proctor

ystichium lonchitis × ½
ly Fern

S. J. Heyward

PLATE 3

Cerastium cerastoides × 1
Starwort Mouse-ear Chickweed

Alchemilla glomerulans × 1

PLATE 4

Spergula L.

S. arvensis L.   *Yarr, Corn Spurrey*
Native. An abundant and troublesome weed of cultivated and
waste land throughout the district.
M Very common (Gordon); Forres, 1898 *JK* **FRS**; shingle of the
River Spey, Garmouth, 1953 *A. Melderis* **BM**.
N (Stables, *Cat.*); Ardclach in all the fields, 1883 *R. Thomson* **NRN**;
sandy field at Kingsteps, 1966 **ABD**.
I Inverness, 1832 and Pitmain, 1833 (Watson, *Loc Cat.*); Kincraig,
1891 *AS* **BM**: Manse garden at Rothiemurchus, 1903 *H. Groves*
BM; Bunchrew, 1940 *AJ* **ABD**; on waste ground at Glen Feshie
Lodge, among stones at Dalwhinnie and on a moorland track at the
head of Glen Strathfarrar etc., 1954.

Spergularia (Pers.) J. & C. Presl

S. rubra (L.) J. & C. Presl   *Sand Spurrey*
Native. Frequent on bare sandy ground, on tracks and among
cinders in railway yards. Map 65.
M Innes House, Urquhart, 1794 *A. Cooper* **ELN**; frequent (Gordon);
railway siding, Forres, 1875 *JK* **FRS**; among cinders in the station
yard at Knockando, 1963 **CGE**; yard of an old sawmill at Carr-
bridge, 1972 **E,BM**.
N (Stables, *Cat.*); Cawdor, 1885 *R. Thomson* **NRN**; waste ground
near Nairn harbour, 1960 **ABD,K**.
I Inverness, 1832 (Watson, *Loc. Cat.*); by the River Oich, Fort
Augustus, 1883 *C. Bailey* **E**; near Kincraig, 1891 *AS* **E,BM**; by the
River Spey at Newtonmore, 1897 *R. Corstorphine* **STA**!; railway at
Culloden (Aitken)!; station yard at Dalwhinnie, track at Moy Hall,
shingle of burn at Corriegarth in Stratherrick, etc. 1954.

S. media (L.) C. Presl   *Greater Sea Spurrey*
   S. marginata Kittel
Native. Common in salt-marshes and dune slacks. Seeds all broadly
winged. Fig 11b.
M Findhorn, 1900 *JK* **FRS**; salt-marsh near Kinloss school, 1925;
salt-marsh at Findhorn, 1951 **E**.
N (Stables, *Cat.*); a compact form with glandular pedicels, 3 miles
E of Nairn, 1898 (Marshall, 1899); salt-marsh on the old Bar,
Culbin, 1968 **E**.
I Salt-marsh at Lentran, 1925; Bunchrew, 1940 *AJ* **ABD**; salt
marshes at Castle Stuart and Carse of Delnies, 1954.

S. marina (L.) Griseb.   *Lesser Sea Spurrey*
   S. salina J. & C. Presl
Native. Salt-marshes, but not as frequent as *S. media*. Seeds generally
not winged. Fig 11a.
M Frequent (Gordon); Kinloss (Innes); salt-marsh near Findhorn,
(Marshall, 1899); in all the coastal parishes (Burgess); dune slack
NE of the Buckie Loch, Culbin, 1968 **E**.
N (Stables, *Cat.*); dune slack on the Carse of Delnies, 1954; salt-
marsh ½ mile E of the Nairn Dunbar golf course, 1960 **ABD**.

Fig 11   a, **Spergularia marina**; b, **S. media**; c, **Radiola linoides**; d, **Montia fontana** subsp. **fontana**; e, **Linum catharticum**; f, **Montia perfoliata**; g, **Montia sibirica**

I Inverness, 1832 (Watson, *Loc. Cat.*); Inverness, 1853 *A. Atkins*
BM; Easterness (as var. *neglecta* Kindb.) (Trail, *ASNH* 14: 231
(1905)); salt-marsh at Castle Stuart, 1976 **E**; by the River Beauly
at Tomich near Beauly.

Polycarpon L.

P. tetraphyllum (L.) L.   *Four-leaved Allseed*
Introduced with wool shoddy and established in the garden at Rose
Cottage, Dyke where it appears yearly in the gravel paths, 1967
BM,CGE.

ILLECEBRACEAE
Scleranthus L.
S. annuus L. subsp. **annuus**   *Annual Knawel*
Native. In dry sandy fields, waste ground and river shingle. Fig 9d,
Map 66.
M Frequent (Gordon); fields about Forres (Innes); Forres, 1899
JK **FRS**; shingle of River Spey at Garmouth, 1964 **CGE**; arable
field at Calcots, 1967 and roadside cutting near Earlsmill, 1968 **E**;
shingle of River Spey at Orton, 1973 **E**.
N (Stables, *Cat.*); common at Ardclach, 1887 *R. Thomson* **NRN**;
cornfield at Little Kildrummie, 1960 **ABD**; root field by Loch of the
Clans, 1960 **ABD,K**; cornfield at Bankhead E of Nairn, 1966,
ABD,BM.
I Inverness, 1832 (Watson, *Loc. Cat*); near Kincraig, 1891 *AS* **BM**;
shingle of River Enrick, Drumnadrochit, 1947; arable fields at
Inverfarigaig, Fort Augustus and Farr etc., 1954; fields at Newtonmore and Knappach, 1973 *R. McBeath*; root field near Farley,
1975 **E**.

PORTULACACEAE
Montia L.
Recent records have been determined by Dr S. M. Walters. **M.
fontana** L. sensu lato has been recorded for every 10 km grid square
in the district,
M. fontana L.
Subsp. **fontana**   *Blinks*
Native. Abundant in wet places, on moorland, by burns and in
mountain flushes. Ripe seeds smooth and shining. Fig 11d.
M Innes House, Urquhart, 1794 *A. Cooper* **ELN**; common
(Gordon); Forres, 1857 *JK* **FRS**; sandy backwater of the River
Lossie near Kellas House, 1968 **E**; edge of a small burn by Dal-
rachney Beg, Carrbridge, 1972 **BM**; by the Allt Iomadaidh,
Dorback, 1972 **ABD,K**.
N (Stables, *Cat.*); Ardclach, 1891 (Thomson); bog by Loch of the
Clans, 1960 **ABD,K**; by the Minister's Loch Nairn Dunbar golf
course, 1961 **ABD,BM**; runnel by the Ord Loch, 1967 **E,CGE**; bogs
at Glenferness and Geddes Reservoir.
I Dalwhinnie and Pitmain, 1833 (Watson, *Loc. Cat.*); near

Kincraig, 1891 *AS* **BM**; by a burn in Glen Mazeran, 1968 **CGE**; pool by Loch Killin, 1972 **E**; runnel on moor, Coillenaclay, 1972 **BM**.

Subsp. **chondrosperma** (Fenzl) Walters

Plant yellowish-green with short erect branches. Seed dull and entirely covered with coarse tubercules.

M Schoolhouse, Lossiemouth, 1875 *James Grant* herb. *Babington* **CGE**.

Subsp. **variabilis** Walters

Plant usually loose in habit. Seed more or less smooth but not as shining as subsp. **fontana**.

N Ardclach, 1898 *R. Thomson* **NRN** det. S. M. Walters; flush below the 8m (25 ft) raised beach **E**.

I Maviston farm, 1968 **E**.

M. perfoliata (Donn ex Willd.) Howell   *Perfoliate Claytonia*
   Claytonia perfoliata Donn ex Willd.   Spring Beauly
Introduced. A pestilential weed of cultivated ground. Native of S. America. Fig 11f, Map 67.

M Pinefield Nursery, Elgin (Gordon); Forres Nurseries, 1898 *JK* **FRS**; near Sueno's Stone Forres, 1898 *R. Thomson* **NRN**; garden weed at Sheriffston, Brodie Castle, Pitcroy Lodge, Kellas House, Newton House near Elgin and at Rothes; rubbish tip at Forres, 1962 **E,ABD,CGE**; Culbin Forest, 1976.

N Cawdor (Thomson); arable field, Nairn, 1954; garden weed Brackla, 1975 *I. A. Robertson*.

I Allotment, Inverness, 1954; dry bank by the road near Kilmorack Kirkyard, 1972 **E**; garden weed Nairnside House, 1975.

M. siberica (L.) Howell   *Spring Beauty*
   Claytonia siberica L.   Pink Purslane
Introduced. Damp woods and by burns and rivers. Occasional, but locally abundant. Fig 11g.

M Fochabers, 1923 *S. Bisset* **GL**; garden escape in the parishes of Edinkillie, Lhanbryd and Urquhart (Burgess); Forres, 1938 *R. Richter*; bank by Sluie near Forres, banks of the River Spey at Blacksboat, Rothes and Knockando; garden weed Dyke, 1967 **E,BM**; by a small burn in the ravine at Slochd near Carrbridge, 1973.

N Waste ground at Littlemill, 1954; near Glenferness school, 1963 **ABD**; road junction W of Nairn, 1963 *Mrs Allardyce* **ABD**.

I Woods by Tomatin House and near Inverness, 1954; ditch at Knockie Lodge, by the burn at Kingussie, small wood at Aberhalder; garden weed, Banchor Mains, Newtonmore, 1975 *S. Haywood*.

Calandrinia Humb., Bonp. et Kunth
C. menziesii Hook.
Introduced.

I Cornfield at Blackpark near Aviemore, 1907 *W. Edgar Evans* **E**.

TETRAGONIACEAE
Tetragonia L.
T. tetragonoides (Pallas) O. Kuntze   *New Zealand Spinach*
  T. expansa Murr.
Garden escape.
M Rubbish tip, Elgin, 1975 **E**.

AMARANTHACEAE
Amaranthus L.
A. retroflexus L.   *Common Amaranth*
Introduced with foreign grain or bird seed.
M Rubbish tip at Forres, 1970 **E**.

A. albus L.
Introduced with 'Swoop' at Wild Goose Cottage, Nairn, 1968
ABD.

A. blitoides S. Watson
Introduced with foreign grain or bird seed.
M Rubbish tip at Elgin, 1958 **E,K**, and also in 1975 **E**.

A. caudatus L.
Garden escape. Much cultivated for ornament and has long pendant
red or green inflorescences.
M Rubbish tip, Elgin, 1961 **E,K**.

CHENOPODIACEAE
Chenopodium L.
C. bonus-henricus L.   *Smear Dock, Good King Henry*
Introduced. By farm buildings and old cottages. Formerly eaten as a
vegetable.
M Frequent (Gordon); churchyard at Forres, 1856 *JK* **FRS**; about
Forres, *J. G. Innes* **ABD**; parishes of Dyke, Forres, Rafford, Crom-
dale, Alves, Duffus, Urquhart, New Spynie, Rothes and Knockando
(Burgess); lane by the old ruined 'Stake and Rice' cottage at
Muirtown, Dyke, 1968 **E,BM,CGE**; by Kinloss Abbey farm and at
Rothes.
N (Stables, *Cat.*); near Auldearn and Cawdor, 1900 (Lobban);
roadside verge near the saw-mill at Geddes, 1961 **E,ABD,K**; by
Budgate Farm, Cawdor, 1970 **E**.
I Kirkhill kirkyard ? *coll.* INV; Kincraig, 1891 *AS* **E**; Ruthven
Barracks, Kingussie, 1904 *G. B. Neilson* **GL**!; ditch near the bridge
over the River Calder, Newtonmore 1911 *ESM* **E,BM** (this is by the
farm of Banchor Mains where the plant is still to be found); by the
River Enrick, Drumnadrochit, 1947; waste land at Fort Augustus,
1950 *M. Cameron*!; by the farm buildings at Croft near Inverdruie,
1975.

C. album L.   *Melgs, Fat Hen*
Native. Abundant on cultivated and waste land throughout the
district. Map 68.

M Frequent (Gordon); Pilmuir, Forres, 1856 *JK* **FRS**; rubbish tip, Forres, 1970 **E**.
N (Stables, *Cat.*); Ardclach (Thomson); field at Kingsteps, 1966 **ABD**; from 'Swoop' Nairn **E,CGE**.
I (Watson, *Top. Bot.*); near Kincraig, 1891 *AS* **BM**; Kingussie (as var. *incanum* Moq.-Tand.) (Druce, 1888); rubbish tip at Longman Point, 1970 **E,CGE**; track at Garva Bridge, cultivated land at Tomich, Kincraig, Dalwhinnie and Newtonmore etc.

C. ficifolium Sm.   *Fig-leaved Goosefoot*
Casual. Cultivated and waste land.
M Garden weed, Moy House, 1953 *A. Melderis* **BM**!; arable land St. John's Mead, Darnaway, 1953 *UKD* **UKD**!.

C. murale L.   *Nettle-leaved Goosefoot*
Casual. Waste land.
M Several plants on a railway siding W of Forres station, 1968 **E**.

C. rubrum L.   *Red Goosefoot*
Casual. Waste land.
M Parishes of Urquhart and Speymouth (Burgess).

C. capitatum (L.) Aschers.   *Strawberry-blite*
Casual. Waste land. The dense, sessile heads of the inflorescence become scarlet at maturity.
M Garden weed at Moy House, 1954 *M. Murray*!.
I In a root field and in the garden of Clunes, Kirkhill, 1955 and 1956 *M. Cameron*.

C. probstii Aellen
Casual. Waste ground.
I Rubbish tip at Longman Point, Inverness, 1970 **E**.

Atriplex L.
A. littoralis L.   *Grass-leaved Orache*
Native. Common on shingle, verges of salt-marshes and dunes by the sea.
M (A. Bennett, *Scot. Nat.* 1888); coast at Urquhart (Burgess); verge of the salt-marsh on the Findhorn estuary, 1925; verge of the sand dunes NE of the Buckie Loch, Culbin, 1968 **E,CGE**; on shingle at the mouth of the River Lossie at Lossiemouth and on mud banks at Garmouth.
N Coast E of Nairn, 1898 *ESM & WASh* **CGE**; verge of the salt-marsh between the old Bar and the Culbin Forest, 1954; shingle bank by the sea, Carse of Delnies, 1976 **E**.
I Near Lentran, 1892 *ESM* **BM**; raised shingle beach between Ardersier and Fort George, 1925; salt-marshes at Clachnaharry and Whiteness Head; verge of the salt-marsh at Castle Stuart, 1976 **E,BM**.

A. patula L.   *Spreading Orache, Common Orache*
Native. Common in cultivated and waste land and on shingle near the sea. Map 69.

M Frequent (Gordon); Pilmuir, 1856 *JK* **FRS**; Findhorn Bay, Lossiemouth and Kingston (maritime forms) (Marshall, 1899); cultivated ground Forres (var. *erecta* Huds.) (Marshall, 1899); waste land at Aviemore, Grantown, Kellas etc. 1954; station yard at Knockando, 1966 **ABD**.
N (Stables, *Cat.*); on the beach at Nairn and as a weed at Ardclach, 1890 *R. Thomson* **NRN**; garden weed at Cawdor, Holme Rose, Auldearn, Nairn and Dulsie Bridge, 1954.
I (Watson, *Top. Bot.*); near Loch an Eilein, 1891 *AS* **E**; Beauly and Lentran, 1892 *ESM* **BM**; cultivated land at Newtonmore, Kincraig, Tomatin and Tomich etc. 1954.

A. hastata L. *Hastate Orache*
Native. Sea coast. Very variable and difficult to distinguish from *A. glabriuscula* Edmond. Plant erect to 100 cm. Bracteoles sessile, sometimes cuneate at the base, thin not inflated. Nothing on the north coast compares with *A. hastata* from S. England.
M East side of Findhorn Bay, 1898 *ESM* **CGE**, determined by A. J. Wilmott as var. *oppositifolia* 'seeds rather small, black and shining'; Culbin Sands, 1923 (Stewart & Patton, *Trans. BSE*, **29**: 40 (1924)).
N Nairn coast, 1898 *ESM & WASh* (Lobban).
I Inverness, 1881 *G. C. Druce* **BM**.

A. glabriuscula Edmondst. *Babington's Orache*
    A. babingtonii Woods
Native. The common plant of the north coast, on shingle banks and sandy verges of salt-marshes. Plant usually procumbent or up to 20 cm. Bracteoles thick and inflated.
M Findhorn, 1860 *JK* **FRS**; in all the coastal parishes (Burgess); verge of the salt-marsh NE of the Buckie Loch, Culbin, 1954.
N Beach at Nairn, 1898 *ESM* **BM,CGE**; abundant from the Carse of Delnies to the Culbin Forest, 1954.
I Shore of the Beauly Firth at Englishton (Aitken); Lentran (as var. *virescens*, a prostrate dark blue-green form), 1892 *ESM* **BM**; the length of the coast, 1954; salt-marsh at Clachnaharry, Inverness, 1975 **E**; verge of shingle bar, Whiteness Head, 1976 **E**.

A. laciniata L. *Frosted Orache*
    A. sabulosa Rouy
Native. Common on the shore line; generally in sand. A mealy, silvery-white, decumbent plant, often with red stems.
M Very common (Gordon); Findhorn, 1899 *JK* **FRS**; do. 1953 *A. Melderis* **BM**; sand on the tide line, Culbin Forest, 1968 **E**.
N Nairn, 1884 *J. Groves* **BM**; abundant on the beach E of Nairn, 1898 *ESM* **BM,CGE**; the length of the coast, 1954.
I Ness mouth, 1930 ? *coll.* **STA**; Bunchrew, 1940 *AJ* **ABD**; verge of the salt-marsh at Lentran Point 1947; dunes at Whiteness Head, 1976 **E**; shore line Camp Sands, Fort George.

A. hortensis L.　　　　　　　　　　　　　　　*Garden Orache*
Garden escape. Waste land and rubbish tips. Plant green or maroon
coloured.
M Waste ground near the Mosset Park, Forres, 1974 *R. Richter* **E**!;
rubbish tip by the River Findhorn, 1974 *K. Christie.*

Suaeda Forsk. ex Scop.

S. maritima (L.) Dumort.　　　　　　　　　*Annual Sea-blite*
Native. Common on salt-marshes along the whole length of the
coast. The plants of the north-east are probably referable to var.
macrocarpa (Desv.) Moq. being prostrate or decumbent, rarely
erect (and then very small) and with seeds *c.* 2 mm in diameter.
M Lossiemouth, 1867 *J. Grant* FRS; Findhorn (as var. *procumbens*
Syme) (Marshall, 1899); salt-marsh NW of the Buckie Loch in the
Culbin Forest, 1964 **E,CGE.**
N East of Nairn (as var. *procumbens* Syme) (Marshall, 1899); salt-
marsh N of Lochloy, 1975 **E.**
I Fort George, 1832 *G. Gordon* herb. Shuttleworth **BM**; Longman,
Inverness, 1871 ? *coll.* **ELN**; abundant on the Carse of Ardersier,
1954; salt-marsh at Whiteness Head, 1976 **E.**

Salsola L.

S. kali L.　　　　　　　　　　　　　　　　*Prickly Saltwort*
Native. Frequent on sand dunes along the whole length of the coast.
M Abundant in one or two places, Findhorn, Stotfield and Burg-
head (Gordon); Findhorn, 1899 *JK* **FRS**!; dunes NW of the
Buckie Loch in the Culbin Forest, 1965 **E,CGE.**
N Sands E of Nairn, 1832 *WAS* **ABD,GL**!.
I Abundant on the dunes at Camp Sands, Fort George and on
Whiteness Head, 1975 **E.**

Salicornia L.

S. europea L.　　　　　　　　　　　　　　　　*Glasswort*
　　S. herbacea (L.) L.
Native. Abundant on mud and sand on the salt-marsh verges.
M Below Brodie House, *Brodie* **E**; the old Bar of Findhorn, *J. Strath*
and in Kinloss Bay *J. G. Innes* (Gordon); Findhorn Bay, 1857 *JK*
FRS; Spynie, 1865 *JK* **FRS**; in mud NW of the Buckie Loch, 1968
E,CGE.
N On the Moray coast E of Nairn harbour, 1832 *WAS* (Murray);
Carse of Delnies, 1954.
I Shore near Fort George, 1844 *A. Croall* (Watson); Clachnaharry,
Inverness, 1854 *HC* **INV**; salt-marsh at Bunchrew, 1940 *AJ* **ABD**;
abundant in the Beauly Firth, 1954.

TILIACEAE
Tilia L.

T. platyphyllos Scop.　　　　　　　　　*Large-leaved Lime*
Planted. Under-recorded. Leaves pubescent beneath.
M (Burgess); Mill Cross near the railway bridge S of Millbuies,
1976 **E.**

N Park at Cawdor Castle, 1832 herb. *Shuttleworth* **BM**!; small wood between Cawdor and Newton in Inchnacaorach, 1976 **E**.
I By the loch at Dochfour House, 1940 *AJ* **ABD**; planted as an avenue by the road near Kinchyle, Scaniport, 1976 **E,BM**.

T. cordata Mill. *Small-leaved Lime*
Planted. Cymes erect. Leaves glabrous beneath except for tufts of rusty hairs in the axils of the veins.
M Policies of Gordon Castle, Fochabers, 1967 *A. J. Souter*.
I Planted in the avenue by the road at Kinchyle, Scaniport, 1976 **E,BM**.

T. cordata× **platyphyllos** = **T.**× **vulgaris** Hayne *Lime*
T.× *europaea* auct., non L.
Planted. Cymes pendulous. Leaves glabrous beneath except for tufts of whitish hairs in the axils of the veins.
M Certainly introduced (Gordon); planted throughout the county, 1954.
N Lawn at Cawdor Castle, 1832 *WAS* **ABD**; Glenferness and Ardclach, 1886 *R. Thomson* **NRN**; Kilravock Castle, Holme Rose and Geddes House etc. 1954.
I Kingussie, 1903 (Moyle Rogers)!; planted in the policies of the larger houses and in towns and villages, 1954.

T. petiolaris auct. is planted at Newton House near Elgin and **T. tomentosa** Moench by the road approaching Reelig Glen near Kirkhill, and probably elsewhere.

MALVACEAE
Malva L.
M. moschata L. *Musk Mallow*
Introduced in the north-east. Waste ground generally near houses.
M Doubtfully native, rare at Pluscarden, 1830 and at Linksfield, 1839 *John Lawson* (Gordon); railway bank near Forres towards Dunphail (Marshall, 1899); railway siding Aviemore, 1907 *W. Edgar Evans* **E**; Forres rubbish tip, 1970 **E**; roadside at Mundole, at Carrbridge, Garmouth and Rothes etc., 1954; near Binsness, Culbin, 1968 *P. Gordon*.
N The church brae Ardclach, 1883 *R. Thomson* **NRN**!; roadside verge near Brackla distillery, and river bank Coulmony, 1954; by the River Findhorn below the Bell Tower, Ardclach, 1973 **E**.
I Inverness, 1832 (Watson, *Loc. Cat.*); Rothiemurchus, 1876 *JK* **FRS**; roadside verge at Ardersier, Invermoriston, and Tomich, 1955; waste land at Kincraig; Culcabock, 1970 *M. Barron*; single plant by Loch Morlich, 1973; on shingle by the River Glass at Cannich, Broomhill station yard and between the viaduct and the village of Tomatin.

M. sylvestris L. *Common Mallow*
Native. Frequent on waste ground near buildings, and by roadsides.
M Doubtfully native (Gordon); near Elgin and Forres (Druce,

1897); Waterford, 1900 *JK* **FRS**!; waste ground by the Palace of
Spynie, Burghead, Gateside of Alves, Garmouth, Lossiemouth,
Grantown, Findhorn, Kinloss Abbey and Longmorn, 1954; under
the hedge at Dyke Kirk, 1969 **E,BM**.
N (Stables, *Cat.*); Cawdor, 1900 (Lobban); waste ground in a
field near Achnacloich, 1960 **ABD,K**; near a cottage at Clunas
and by the tip at Newton of Park.
I (Watson, *Top. Bot.*); banks of the River Ness, Englishton Burn
and at Alturlie Point (Aitken); waste ground at Ardersier, 1954;
Longman Point and at Tornagrain near Petty, 1973.

M. nicaeensis All.
Casual.
M Findhorn, 1953 *A. Melderis* **BM**!; introduced with *Molucella* seed
in the garden at Rose Cottage, Dyke, 1969.

M. neglecta Wallr.   *Dwarf Mallow*
　　M. rotundifolia auct.
Native. Rare on sandy ground, chiefly near the sea.
M Doubtfully native, rare about Westpark and the Cathedral at
Elgin and at Forres, *J. G. Innes* (Gordon); Findhorn, 1878 *R.
Kidston* **GL**!; Lossiemouth, 1913 *D. Patton* **GL**!; below the walls of
the boat house at Findhorn, 1966 **E,CGE**; sandy soil in a quarry
near Upper Hempriggs, 1974 **E**.
N Nairn, *J. B. Brichan* (Watson).
I Campbelltown, 1844 *A. Croall* (Watson); Inverness, 1859 *JK*
FRS; the Bught, Inverness (Aitken); waste ground at Ardersier,
by Castle Stuart and by the sea on Longman Point, 1954.

M. parviflora L.
Casual. Introduced with foreign grain.
M Carron Distillery yard, 1956 **ABD**; rubbish tip at Rothes 1954
and 1966 **E,K**.
I Allotment near a distillery at Inverness, 1961 **E,ABD,BM**.

Lavatera L.
L. arborea L.   *Tree Mallow*
Native. Sandy ground near the sea.
M In a garden at Findhorn, 1900 *JK* **FRS** (it was abundant on
the sea wall at Findhorn in 1920, but was cleaned out from there
and is now rare and chiefly found in cottage gardens); Duffus
(Burgess); waste land to the E of the harbour at Burghead, 1973
R. M. Suddaby.

LINACEAE
Linum L.
L. usitatissimum L.   *Cultivated Flax*
Casual. Introduced with foreign grain and bird seed on waste
ground and rubbish tips.
M Rare, certainly introduced (Gordon); Forres, 1859 *JK* **FRS**;
casual at Dyke (Burgess); rubbish tips at Elgin in 1953, Rothes in
1966 **E**, and Forres in 1969 **E**.

N Cawdor, 1900 (Lobban); casual from bird seed at Newton of Park, 1975.
I Distillery yard, Tomatin, 1958; rubbish tip at Longman Point, 1970 **E,ABD** and on the raised shingle beach at Ardersier, 1975.

L. catharticum L.   *Purging Flax*
Native. Abundant on base-rich grassland, on verges of moorland roads, and in sand dunes. Fig 11e.
M Very common (Gordon); Findhorn, 1900 *JK* **FRS**; sand dunes by the Buckie Loch, Culbin, 1925; banks of the River Findhorn, 1953 *A. Melderis* **BM**; throughout the county, 1954.
N (Stables, *Cat.*); common round the Free Kirk at Ardclach and on the banks of the River Findhorn, 1882 *R. Thomson* **NRN**; dunes by the Nairn Dunbar golf course, 1961 **ABD**; in short turf at Maviston, 1968 **E**.
I Inverness, 1832, Dalwhinnie and Pitmain, 1833 (Watson, *Loc. Cat.*); Bunchrew, 1940 *AJ* **ABD**; abundant throughout the district, 1954.

Radiola Hill

R. linoides Roth   *All-seed*
Native. Frequent in damp dune slacks and on moorland cart tracks. A minute, much branched annual 2–8 cm high with numerous flowers. Fig 11c.
M Leverach Loch, 1824 (Gordon); banks of the River Spey between Fochabers and Orton, 1832 *WAS* (Murray); Loch of Spynie, Pluscarden and Stotfield (Gordon); Seapark at Kinloss, 1857 *JK* **FRS**; in profusion by the Buckie Loch, Culbin and at Garmouth *ESM & WASh* !; abundant in damp slacks in the old shingle workings at Kingston, 1966 **E**; track at Covesea, 1968.
N Nairnshire, *WAS* (Gordon); damp sandy ground near Nairn, 1898 *ESM* **CGE** !; muddy verge of Loch of the Clans, 1960 **ABD**; cart track on the moor at Ord Loch near Littlemill, 1973 **E**; bare ground at Lochloy and by the Nairn Dunbar golf course.
I ? locality, 1844 *A. Croall* (Watson, *Top. Bot.*); on a boulder at Torr Riach near Inverness (Aitken); track to the west of Loch of the Clans, 1960; track near Culburnie, 1976 *M. Cameron* **E** !; track on the hill between Kirkton House and Inchberry Hill, 1976 **E**.

GERANIACEAE
Geranium L.

G. pratense L.   *Meadow Cranesbill*
Doubtfully native to the west of the Spey valley although it appears native in runnels leading to the sea in v/c 94 Banff.
M Innes House, Urquhart, 1794 *A. Cooper* **ELN**; seen only by the River Dulnain at Carrbridge where it looks like a garden escape, 1903 (Moyle Rogers); sand dunes at Lossiemouth, banks near Rothes and S of Grantown and several plants by the lake at Gordonstoun College, 1954; near cottages at Cromdale, 1968.

N Cawdor, Nairn and Ardclach, 1900 (Lobban); garden escape by the river at Nairn and Holme Rose, 1954.
I Railway embankment just W of Clachnaharry, 1925 and 1972 **E**; meadow at Clachnaharry, 1940 *AJ* **ABD**; garden out-cast by the River Beauly near Lovat Bridge, 1947; Dirr Wood, Dores, 1973 *M. Barron*; garden escape on waste ground at Struy, on the raised shingle beach at Ardersier, by the roadside ½ mile W of Bunchrew and by the farm of Croft at Tullochgrue.

G. sylvaticum L.   *Wood Cranesbill*
Native. Abundant on river banks, in wet meadows and by ditches, and often on mountain ledges, up to 915 m in the Cairngorms. Pl. 6, Map 70.
M Greshop woods, *J. G. Innes* (Gordon) and 1900 *JK* **FRS**; throughout the county, 1954.
N (Stables, *Cat.*); banks of the River Findhorn, Ardclach (Thomson); by the river at Nairn, 1973 **ABD,CGE**.
I Dalwhinnie, 1833 and the lower Falls of Foyers (Watson, *Loc. Cat.*); Strath Affric, 1850 (Ball, 1851); by the Moniack Burn, 1878 (*Trans. ISS & FC* 1: 170 (1878)); wood at Kinrara, 1882 *J. Groves* **BM**; near Kincraig, 1891 *AS* **BM**; rock ledge near the summit of Ben Alder, 1956; by the Plodda Falls at Guisachan, on rock ledges on Sgùrr na Lapaich in Glen Strathfarrar etc.

G. endressii Gay
Garden escape.
M Bank of the old road over the River Lossie on the W of the town of Elgin, 1953 **K**; rubbish tip at Rothes, 1954; shingle of the River Spey near the top of Long Pool, Knockando, 1955 *G. Shepherd*.
I Path by the river at Beauly, 1947 ('not the hybrid usually so-called', det. E. F. Warburg) **BM**; river shingle at Drumnadrochit, 1956; well established by the River Enrick near Corrimony school, Glen Urquhart, 1975 **E,BM**.

C. versicolor L.
M Garden escape near Rothes, 1954.

G. phaeum L.   *Dusky Cranesbill*
Introduced.
M Near the lake Gordon Castle, 1844 (Gordon, *annot.*); parishes of Dyke, Forres and Speymouth (Burgess); well established by the road a few yards N of Earnhill, Kintessack, and waste ground at Forres, 1954.
N Cawdor and by the Muckle Burn at Ardclach, 1900 (Lobban); garden escape outside the walled garden at Geddes House, 1960 **E,ABD K,**
I Garden escape in the policies of Achnagairn House, Kirkhill, 1954 **E,K**.

G. sanguineum L.   *Bloody Cranesbill*
Native. Very rare on base-rich rocks. Occasionally occurs as a garden escape.

M Near the Heronry on the River Findhorn, Darnaway, *J. G. Innes* (Gordon) and 1902 *JK* **FRS**, (it is still there on an inaccessible cliff); Aviemore, 1894 *R. S. Adamson* **BM**; cliff in a quarry near Lossiemouth Harbour, 1954.

N Ardclach in the Findhorn basin S of General Wade's road, 1900 (Lobban); garden escape near Nairn, 1973 *Ruth Mackay*.

I About Inverness, *G. Anderson* (Gordon); near Fort George, scarce, 1844 *A. Croall* (Watson); Falls of Kilmorack near Beauly, 1899 *G. B. Neilson* **GL**.

G. pyrenaicum Burm. fil.   *Mountain Cranesbill*
Doubtfully native. Frequent in the lowlands on waste ground by farms and railway stations.

M Near the Manse at Dyke, 1840 *J. G. Innes* (Gordon, *annot.*) (it remained there until 1962 but it was destroyed when the road was widened); Sanquhar House, Forres, 1857 *JK* **FRS**; denizen in the parishes of Dyke, Forres, Lhanbryd and Elgin (Burgess); roadside opposite the old Manse, Dyke, 1962 **E**; roadside verge near Tearie cross-roads, waste ground at Findhorn, junction of the A96 and the road to Coxton Tower and in the goods yard and on the rubbish tip at Elgin.

N Rare, Millfield near Nairn, 1833 *WAS* (Gordon); by the distillery near Nairn, 1898 *ESM* **BM,CGE**; roadside verge by Achnacloich, E of Nairn, 1960 **ABD,K**; railway embankment Nairn station, 1961 **BM,CGE**; waste ground by Wester Delnies, 1972 **E**; roadside verge E of Geddes, 1976 **BM** and at Newton of Budgate.

I (Druce, *Com. Fl.*); by the mausoleum at Clunes, Kirkhill, 1930 *M. Cameron*!; waste ground at Longman Point and on the raised beach at Ardersier, 1954.

G. columbinum L.   *Long-stalked Cranesbill*
Introduced with foreign grain. Very rare on river shingle.

M Shingle of River Spey at Garmouth, 1953; shingle of River Findhorn at Relugas, Darnaway, 1954, and waste ground at Findhorn, 1954.

G. dissectum L.   *Cut-leaved Cranesbill*
Native. Common on waste ground, in pastures and on the margins of cultivated fields in the lowlands; rare in the highlands.

M Frequent but doubtfully native, grass fields by the Spey and at Forres, *J. G. Innes* (Gordon); Forres, 1900 *JK* **FRS**!; pasture at St. John's Mead, Darnaway, 1953 *A. Melderis* **BM**; waste ground at Knockando station, 1966 **ABD**.

N (Stables, *Cat.*); Cawdor and Ardclach, 1900 (Lobban); pastures at Holm Rose, Nairn, Geddes and Kinsteary, 1954; waste ground by the ford at Achavraat, 1973.

I Inverness, 1833 (Watson, *Loc. Cat.*); Erchless Castle, 1836 (Gordon); Campbelltown, 1844 *A. Croall* (Watson); near Kincraig, 1891 *AS* **BM**; Dochgarroch, 1940 *AJ* **ABD**; meadow to the S of Loch Meiklie, waste land at the road junction to Invergarry in Glen Moriston and pasture at Tomatin etc.

G. molle L.   *Dove's-foot Cranesbill*
Native. Abundant on waste ground, dunes and cultivated land.
M Very common (Gordon); Forres, 1900 *JK* **FRS**; throughout the
county, 1954.
N (Stables, *Cat.*); Ardclach (Thomson); sand dunes on the Nairn
Dunbar golf course, 1962 **ABD,CGE**.
I Inverness, 1833 (Watson, *Loc. Cat.*); waste ground Ceannocroc
Bridge, Glen Moriston, 1950; garden weed at Ben Alder Lodge,
1956; garden weed at Glen Feshie, Newtonmore and Kincraig etc.

G. pusillum L.   *Small-flowered Cranesbill*
Native. Waste and cultivated land. Frequent in the lowlands, rare
in the highlands. Rather similar to **G. molle** in general appearance
but the plant is usually more robust, the flowers smaller and of a
dull lilac colour, and the carpels are pubescent. Map 71.
M Doubtfully native (Gordon); roadside between Spynie and
Elgin, 1898 *ESM* **E,BM**; waste ground at Carrbridge, Culbin,
Grantown, Rothes and Burghead etc., 1954; weed of cultivation,
Dyke, 1971 **E,CGE**.
N (Stables, *Cat.*); rare at Ardclach, 1887 *R. Thomson* **NRN**; near
Nairn, 1898 *ESM* **BM,CGE**; sandy field E of Nairn, 1960 **E**; bank
near the farm of East Milton, 1967 **E,ABD,BM**.
I Inverness, 1833 (Watson, *Loc. Cat.*); garden weed Kirkhill,
waste ground at Inverness railway embankment at Culloden and on
the raised beach at Ardersier, 1954; car park at Drumnadrochit,
1973 **E**; by the garden gate at Flichity House, waste ground by the
golf course at Kingussie, garden weed at Newtonmore and railway
station at Kincraig etc.; farm lane Groam of Annat, 1973 **E**; waste
land at Longman Point, 1976 **BM**.

G. lucidum L.   *Shining Cranesbill*
Native. Locally common on base-rich rocks and old walls.
M Huntly's Cave, Dava Moor, 1876 *Edward Stewart* **FRS**; Sim's
Nursery, Forres, 1896 *JK* **FRS**; rocks in the gorge of the burn at
Huntly's Cave, 1973 **E**; by the Allt Iomadaidh, Dorback, in small
quantity, 1974.
N (Stables, *Cat.*).
I Urquhart Castle, *G. Anderson* (Gordon)!; Falls of Kilmorack,
1870 ? *coll.* **ELN**; Newton Hill, ? date and *coll.* **INV**; Craigie wood
and Dunain (Aitken); cliffs above Loch Killin, 1954; cliffs on
Creag nan Clag and Inverfarigaig, 1962 *D. A. Ratcliffe*!; abundant
in the lane opposite the kirkyard at Kilmorack, 1963 **E**; Creag a'
Chlachain Dunlichity, 1971 *D. A. Ratcliffe*; a few plants on the cliffs
of Creag Dhubh, Laggan, 1971; conglomerate cliffs of Tom
Bailgeann and under juniper in the Pass of Ryvoan, Glenmore.

G. robertianum L.   *Herb Robert*
Native. Abundant in woods and shady places, by burns and rivers
and among rocks and boulders throughout the district.
M Very common (Gordon); Greshop wood, 1856 *JK* **FRS**; St.
John's Mead, Darnaway, 1953 *A. Melderis* **BM**!.

N (Stables, *Cat.*); Ardclach (Thomson); wood at Lethen House, 1961 **ABD**.
I Newton Hill, 1880 ? *coll.* **INV**; Kincraig, 1891 *AS* **BM**; among rocks in upper Glen Feshie, in a birch wood at Gaick Lodge, on cliffs at Loch Killin, and in a burn gully at Dalwhinnie etc.

G. × magnificum Hylander
G. platypetalum auct.
Garden escape. Flowers large, dark purplish-blue.
M Waste ground at Lossiemouth and Elgin, 1963.
N By the river at Nairn, 1962 **E**; policies of Holme Rose, 1962 **ABD**.
I Waste ground Clachnaharry, 1954; established by the River Enrick near Corrimony school, Glen Urquhart, 1975 **E**.

Erodium L'Hérit.
E. cicutarium (L.) L'Hérit.   *Common Storksbill*
Native. Common on waste and cultivated ground especially on sandy soil near the sea.
M Innes House, Urquhart, 1794 *A. Cooper* **ELN**; very common (Gordon); bed of the Spey, Aviemore, 1892 *AS* **E**; Drumduan, Forres, 1900 *JK* **FRS**; Findhorn Bay, 1953 *A. Melderis* **BM**; arable field Kinloss, 1957 **K**; garden weed Greshop House, Forres, 1964 **E**.
N (Stables, *Cat.*); Auldearn, Nairn and Cawdor, 1900 (Lobban); shore at Nairn, 1916 *D. Patton* **GL**; station yard at Auldearn, arable fields at Geddes and new roadside verge at Clunas, 1954.
I Inverness, 1833 (Watson, *Loc. Cat.*); Millarton and Longman Point (Aitken); Abriachan, 1940 *AJ* **ABD**; on a new verge S of Dores, waste land at Kirkhill and Culloden, 1955; abundant on the old railway line at Tomachrochan near Nethybridge and in an arable field at Balnafoich near Dunlichity, 1975; dunes at Ardersier and Fort George.

TROPAEOLACEAE
Tropaeolum L.
T. majus L.   *Garden Nasturtium*
Garden escape on most of the rubbish tips and on shingle W of Kingston in Moray, 1954.

OXALIDACEAE
Oxalis L.
O. acetosella L.   *Wood Sorrel*
Native. Abundant in deciduous woods, on banks of burns and rivers, on tree stumps and among rocks and boulders on mountains up to 915 m in the Cairngorms. Pl 6.
M Frequent (Gordon); Greshop, 1856 *JK* **FRS**; throughout the county, 1954.
N (Stables, *Cat.*); common at Ardclach, 1887 *R. Thomson* **NRN**; wood at Kinsteary, Auldearn, 1954 **ABD**.
I Dalwhinnie, 1833 (Watson, *Loc. Cat.*); Reelig Burn and Craig

Phadrig, ? *coll.* **INV**; Kincraig, 1891 *AS* **BM**; Dochgarroch, 1940 *AJ* **ABD**; by the Falls of Foyers, 1964 **CGE**.

O. corniculata L. var. **microphylla** Hook. fil.
Garden escape. A small procumbent plant forming dense mats and with small yellow flowers. An escape from greenhouses occasionally found growing between pavements and on waste land.
M By the greenhouse at Brodie Castle, 1971 **BM**; gravel drive of Sheriffston House, Lhanbryde.
N Paths in the walled garden at Househill, Nairn, 1967 **CGE**; abundant at Kilravock Castle.
I Garden weed at Moy Hall, among paving stones by Aldourie Pier and in a cottage garden at Dores, 1975.

O. europaea Jordan
Casual. Garden weed at Rose Cottage, Dyke, 1964 **E**.

O. articulata Savigny
Garden escape. Station yard, Knockando, 1954.

O. corymbosa DC.
Garden escape.
M By the greenhouse at Brodie Castle, 1965 **E,ABD,BM**; garden weed at Burghead, 1973.

LIMNANTHACEAE
Limnanthes R. Br.
L. douglasii R. Br.   *Butter and Eggs*
Casual. A well known garden annual with flowers of pale and dark yellow, much visited by bees.
M Garden escape near Cromalt, Dyke, 1975 *M. Cameron*!.
I Rubbish tip at Longman Point, 1971 **E,CGE**.

BALSAMINACEAE
Impatiens L.
I. noli-tangere L.   *Touch-me-not Balsam*
M (Druce, *Com. Fl.*).

I. parviflora DC.   *Small Balsam*
Introduced. River shingle and waste land, local but abundant where it occurs.
M Shingle of the Muckle Burn near Moy House, Forres, 1953; waste ground at Sanquhar House, Forres, 1954; abundant on the roadside verges near the Spey bridge at Fochabers, 1954 **E,K,CGE**; garden weed, Dyke, 1975.

I. glandulifera Royle   *Policeman's Helmet*
Introduced. River banks and shingle.
M Shingle of the River Spey at Kingston and by the Muckle Burn near Moy House, 1953; banks of the River Findhorn at Forres, 1954 **CGE**; banks of the River Spey at Carron, 1968 **E**; by a small burn at Auchinroath near Rothes and in a wood near St Mary's Well, Orton.

N Shingle of River Nairn near Brackla, 1969; by a small burn at Boath, Auldearn, 1975.
I By the Big Burn at Glen Cottage, Ness Castle, 1970.

ACERACEAE
Acer L.

A. pseudoplatanus L.   *Sycamore*
Introduced in the sixteenth century. Regenerates freely on rich soils but is absent from acid moorland unless planted.
M Aviemore, 1891 *AS* **BM**; planted everywhere, the timber was used for calico-rollers (Burgess); throughout the county, 1954.
N Shrubbery at Ardclach, 1885 *R. Thomson* **NRN**.
I Dochgarroch, 1940 *AJ* **ABD**; naturally regenerating in all the lowland areas but planted at Ben Alder Lodge, Dalwhinnie and Glenmore etc.

A. platanoides L.   *Norway Maple*
Introduced.
M Planted at Forres, 1930; by the railway bridge at Alves, by the Muckle Burn at Brodie, wood by the Moss of Meft, Calcots, policies of Altyre House and at Aviemore etc.
N Wood near Howford Bridge, Nairn, 1965 **ABD**; policies of Cawdor Castle, Holme Rose etc.
I Planted in the Reelig Glen, 1925; in the policies of Tomatin House, at Fort Augustus, Invermoriston, Newtonmore and Tomich etc.

A. campestre L.   *Common Maple*
Introduced in north Scotland.
M Planted at Dyke, Rafford and Forres (Burgess).
N Policies of Holme Rose, 1954.
I Reelig Glen, 1925; small tree on the embankment by the lagoons at Clachnaharry, Inverness, 1975 **E,BM**.

HIPPOCASTANACEAE
Aesculus L.

A. hippocastanum L.   *Horse Chestnut*
Planted throughout Moray and Nairn and the lowland areas of Easterness, reaching Kyllachy in Strathdearn, Kincraig and Kingussie.

AQUIFOLIACEAE
Ilex L.

I. aquifolium L.   *Holly*
Native. Often planted or bird sown. In woods, by hedges on cliffs and among rocks on moors.
M Abundant in one or two places, Oakwood, Elgin (Gordon); Culbin Forest, 1950 and throughout the county.
N Cawdor (Gordon); Ardclach (Thomson); woods at Kilravock Castle, Dulsie Bridge, and Glenferness etc., 1954.

I Dunain (Aitken); Nairnside, 1940 *AJ* **ABD**; wood at Drum-nadrochit, 1947; among rocks on Mealfuarvonie, Inverfarigaig, Fort Augustus and by Loch Ness at Dores, 1954; cliffs by Loch Killin, near Eskadale and in the Ryvoan Pass; several plants on the cliffs of Creag Dhubh, Laggan, 1973 *D. Weir.*

CELASTRACEAE
Euonymus L.
E. europaeus L.   *Spindle Tree*
Planted.
M Certainly introduced, Gordon Castle (Gordon); in the parishes of Urquhart and Bellie (Burgess); planted at Sanquhar House, Forres, 1960 *K. Christie.*
I Planted in the lane by Achnagairn House, Kirkhill, 1954 **E,CGE.**

E. latifolius (L.) Mill.   *Broad-leaved Spindle Tree*
I Several trees at the entrance to Reelig Glen, 1925 **CGE.**

RHAMNACEAE
Frangula Mill.
F. alnus Mill.   *Alder Buckthorn*
M Below the Heronry, Darnaway, *J. G. Innes* (Gordon, *annot.*).

LEGUMINOSAE
Lupinus L.
L. nootkatensis Donn ex Sims   *Lupin*
Introduced. Locally abundant on river shingle. Flowers blue-pink-white.
M Casual by the river at Forres (Burgess)!; frequent on the shingle of the River Spey from Aviemore to Garmouth, 1954; shingle of the Spey at Knockando, 1967 **K,CGE.**
N Sandy bank by the river at Nairn, 1962; new roadside verge at Wester Delnies, 1975.
I Strathglass (Buchanan White *J. Bot.* **8**:129 (1870))!; an escape from cultivation in great profusion on the banks of the River Beauly, especially at the railway bridge, the Cruives and several places in Strathglass, 1902 (Pollock); riverside at Beauly, 1919 *G. C. Druce* **BM,OXF!.**

L. arboreus Sims   *Tree Lupin*
Garden escape. Flowers yellow.
M Frequent on banks by the sea at Kinloss and about the houses and dunes at Findhorn village, 1925.
I Raised beach at Ardersier, 1956; waste ground at Milton, 1972; by the sea at Longman Point, Inverness.

L. polyphyllus Lindley   *Garden Lupin*
Garden escape. Flowers pink, purple, white or yellow.
M Pine plantation at Conicavel, 1967 **E**; roadside verge S of Carrbridge and by the Mains of Allanbuie, Keith; plentiful on the

railway embankment at Slochd and Kinveachy; abundant on shingle of River Spey on the island below Aviemore.
N Track by Loch Belivat, 1974.
I Abundant on the railway embankment north of Kingussie, 1925; track to Clunes station, 1930 *M. Cameron*; new roadside verge at Milton, station yard at Tomatin, on the raised beach at Ardersier, waste ground in Daviot village, island in the River Spey at Broomhill and in a small wood by Loch Ness at Lewiston.

Laburnum Medic.

L. anagyroides Medic.   *Laburnum*
Planted on forest margins and by roadsides, regenerates naturally.
Pods appressed-pubescent. Seeds poisonous.
M Planted extensively (Burgess); planted as a hedge opposite Sheriffston school; by the River Findhorn at Randolph's Leap, Relugas, 1973 **E**.
N Introduced at Ardclach (Thomson); by the river at Nairn, 1962 **ABD**.
I Common on the margins of pine woods at Port Clair, Invermoriston, 1950; woods at Dochfour, near Bunchrew, at Kinnianan kirkyard by Loch Ness and at Kirkhill etc.

L. alpinum (Mill.) Berchtold et J. Presl   *Alpine Laburnum*
Introduced. Differs from *L. anagyroides* in the more slender panicles of deeper yellow flowers and in the rhachis and pods being glabrous.
M Less common (Trail, *ASNH* **14**: 233 (1905)); planted by the road between Craigroy and Dallas Kirk, 1975 **E**.
N Roadside near Airdrie, Ferness, 1976 **E,BM**.
I In the small wood at the outlet of the River Tarff at Fort Augustus, 1975 **E**; planted by the road at Kinchyle near Scaniport.

Genista L.

G. anglica L.   *Carline's Spurs, Needle Furze, Petty Whin*
Native. Common on dry open moorland. Fig 12e, Map 72.
M Innes House, Urquhart, 1794 *A. Cooper* **ELN**; frequent (Gordon); Califer Hill near Forres, *J. G. Innes*; Culbin Forest, 1925; moors at Aviemore, 1904 *R. Meinertzhagen* **BM**; between Grantown and Tomintoul, 1954 *A. W. Exell* **BM**; 4 miles N of Aviemore, 1955 *J. Milne* **E**; moorland bank at Cromdale, 1970 **E**.
N (Stables, *Cat.*); heath near Nairn, 1898 *ESM* **E**; moor at Dulsie Bridge, 1961 **ABD**; by the Minister's Loch on the Nairn Dunbar golf course, 1962 **CGE**.
I Dalwhinnie and Pitmain, 1833 (Watson, *Loc. Cat.*); Culloden, 1880 ? *coll.* **INV**; Rothiemurchus, 1892 *AS* **E** and 1901 *JK* **FRS**; Dunlichity, 1940 *AJ* **ABD**; moors at Newtonmore, 1972 **E**.

Var. **subinermis** Rouy et Fouc.
A form without spines.
M Trailing among heather, Boat of Garten, 1916 (H. W. Pugsley, *J. Bot.* **60**: 201 (1922))!.
I Kincraig, 1891 *AS* **BM**.

Ulex L.

U. europaeus L.   *Whins, Gorse*
Native. Abundant on waste ground, on moors and by roadsides and
railway lines throughout the district. According to Brodie it was
'rare' in the highlands in 1800.
M Very common (Gordon); Forres, 1898 *JK* **FRS**; abundant
throughout the county, 1954; near Elgin, 1975 **E**.
N (Stables, *Cat.*); by the golf course Nairn Dunbar, 1976 **E**.
I Rare near Inverness, *Brodie* **E**; Inverness, 1832 and Pitmain,
1833 (Watson, *Loc. Cat.*); Craig Phadrig, 1940 *AJ* **ABD**; railway
embankment at Drumochter, 1954; by Ben Alder Lodge, 1956 etc.

U. gallii Planch.   *Western Gorse*
Introduced in the north-east. The former records for *U. minor* Roth
may all belong here. Differs from *U. europaeus* in having deeper
yellow flowers, the primary spines weakly grooved, and minute
bracteoles 0·6–1·3 mm long. It flowers in late autumn. Recent
records have been determined by P. M. Benoit.
M Hopeman Harbour, 1836 (Gordon); Oakwood, Elgin and three
bushes in widely separate localities in Speymouth, 1932 *G. Birnie*
BM (Burgess); one bush by Cullerne House, Findhorn, 1935;
abundant on both sides of the A96 to the S of Fochabers, by Leitch
Wood, 1961 **E,K,CGE**.
I A small patch by Auchgourish Farm, Boat of Garten, 1959
G. B. Morison, **E,ABD**; do. 1971 *MMcCW* **E,BM**; top of a roadside
bank by the A96 100 yards W of Morayston near Dalcross, 1975
E.

Cytisus L.

C. scoparius (L.) Link
  Sarothamnus scoparius (L.) Wimm. ex Koch   *Broom*
Native. Abundant on waste and scrub land, hedgerows and railway
embankments.
M Very common (Gordon); Forres, 1898 *JK* **FRS**; abundant on
rides and under pines in the Culbin Forest, 1925; ½ mile S of Forres
station, 1953 *A. Melderis* **BM**.
N (Stables, *Cat.*); by the river at Nairn, 1967 **ABD**.
I Inverness, 1832, Pitmain and Falls of Foyers, 1833 (Watson,
Loc. Cat.); by Loch Ness, 1904 *G. T. West* **STA**; Caledonian Canal,
1940 *AJ* **ABD**; throughout the county, 1954. White flowered culti-
vars often escape, e.g. on the dunes at Fort George.

C. striatus (Hill) Rothm.
Planted on new roadside verges. Pods with long white hairs.
I Roadside banks at Milton and Polmaily in Glen Urquhart, 1970

Ononis L.

O. repens L.   *Rest Harrow*
Native. Common on dunes, roadside verges and banks, chiefly in the
lowlands. Map 73.

M Common (Gordon); riverside at Greshop, 1856 *JK* **FRS**; railway at Aviemore, 1904 *R. S. Adamson* **BM**; pink and white flowered forms occur near the Buckie Loch, Culbin, 1925.
N Broadley, Nairn, 1833 *WAS* **E**; Ardclach (Thomson); Cawdor (Aitken); very tall plants on the roadside verge opposite Lochloy House, 1961 **ABD,K**; abundant by the track to Rait Castle.
I Inverness, 1832 (Watson, *Loc. Cat.*); near Inverness, 1850 *P. H. MacGillivray* **ABD**; Culcabock and Balloan (Aitken); Allanfearn, 1940 *AJ* **ABD**; grass bank at Milton, 1947; roadside verge at Farr and in all the coastal areas.

O. spinosa L.
Casual. Reported by *Charles Watt* from Knockando, Moray (Burgess) and from the Asylum grounds, Dunain, Easterness (Aitken).

O. baetica Clemente
  O. salzmanniana Boiss. et Reuter
N Introduced with 'Swoop'. Wild Goose Cottage, Nairn, 1969 **E,ABD,CGE**.

Medicago L.

M. sativa L.   *Lucerne*
  Subsp. **sativa**
Introduced. A relic of cultivation which has become established on grass banks and waste ground. Flowers blue-violet.
M Findhorn, *J. E. Allan* and Ardgay, *WAS* (Gordon, *annot.*); Spynie, 1860 *JK* **FRS**; Bareflathills, Elgin, 1882 ? *coll.* **ELN**; escape from a crop, Covesea, 1954; waste land Lossiemouth; roadside bank E of Lesmurdie, Elgin, 1967 **E,ABD**; old railway line at Hopeman, 1975 *A. H. A. Scott*.
I On the raised beach at Ardersier, 1955.

  Subsp. **falcata** (L.) Arcangeli   *Sickle Medick*
Flowers yellow. Pod falcate, sometimes nearly straight.
M Waterford, 1876 *JK* **FRS**; banks of the River Findhorn near Forres, 1880 *J. W. H. Trail* **ABD**; Forres and Speymouth (Burgess); washed down the River Findhorn in a sack of foreign grain during a flood to Greshop wood near Forres, 1957 **E,K**; grass verge at the top of the quarry near the harbour at Lossiemouth, 1957 *J. Harrison*!; waste ground by Forres station, 1975 *K. Christie* **E**.

M. lupulina L.   *Black Medick*
Native. Frequent on waste ground and short grassland, chiefly near the sea. Fig 12g, Map 74.
M Frequent (Gordon); Castlehill, Forres, 1856 *JK* **FRS**; waste land at Dyke, Dunphail, Forres, Findhorn, Rothes and Elgin, etc., 1954; dunes at Lossiemouth, 1966 **CGE**; distillery yard, Knockando and roadside verge at Mains of Craigmill.
N (Stables, *Cat.*); Ardclach, Cawdor and Nairn, 1900 (Lobban); dunes by the golf course at Nairn, waste ground at Holme Rose, Coulmony, Loch of the Clans and Cran Loch, 1954.

I Inverness, 1832 (Watson, *Loc. Cat.*); grain fields about Inverness (Aitken); Kincraig, 1891 *AS* **BM**; Bunchrew, 1940 *AJ* **ABD**; railway yard at Kingussie, waste ground at Kirkhill, Castle Stuart and Ardersier, 1955; waste ground near the Kirk at Dores on the towpath at Clachnaharry and on a heap of earth by Kilmartin House, Glen Urquhart, 1975; sandy soil by the sea, Longman Point, 1976 **E**.

M. polymorpha L.   *Hairy Medick*
   M. denticulata Willd.
Casual. Introduced with foreign grain.
M Speymouth, *G. Birnie* (Burgess); introduced with wool shoddy, Greshop House, Forres 1958.
I In an allotment near a distillery, Inverness, 1961 **E,ABD,K**; distillery yard Tomatin, 1962 **CGE**.

   Var. **apiculata** (Willd.)
I Allotment near a distillery, Inverness, 1961 **E,K**.

M. praecox DC.
Casual. Introduced with wool shoddy, Greshop House, Forres, 1968 **E**.

M. minima (L.) Bartal   *Bur Medick*
Casual. Introduced with wool shoddy, Greshop House, Forres, 1958.

M. truncatula Gaertn.
   M. tribuloides Desr.
Casual. Introduced with foreign grain to an allotment near a distillery, Inverness, 1961 **E,K**.

Melilotus Mill.

M. altissima Thuill.   *Tall Melilot*
Casual.
M (Druce, *Com. Fl.*)

M. officinalis (L.) Pall.   *Common Melilot*
Casual. Waste ground, rubbish tips and newly sown roadside verges.
M Near Findhorn, 1838 *J. E. Allan*, Linksfield, 1839 *J. Lawson*, Urquhart, 1843 (Gordon, *annot.*); the Manse garden, Forres, 1870 *JK* **FRS**; Waterford, 1880 *J. W. H. Trail* **ABD**; Brodie, 1940 *AJ* **ABD**; farmyard, Rosehaugh, 1957 **E,K**; rubbish tip, Elgin, 1961 **E,BM**; Carron station, 1961 **ABD**; new roadside verges at Boghole and Barley Mill, Brodie; cornfield at Mains of Allanbuie, 1975.
N Millhill, *J. B. Brichan* (Gordon, *annot.*); newly-sown verge at Bankhead, 1970 **E,CGE**.
I Fields below Castle Heather near Inverness (Aitken); distillery yard, Tomatin, 1961 **K**; allotment at Inverness, 1961 **E**.

M. alba Medic.   *White Melilot*
Casual. Waste land, rubbish tips and newly-sown roadside verges.
M Parishes of Dyke, Forres, Alves and Speymouth (Burgess); bank by the Muckle Burn at Whitebridge, Kintessack, 1960 **ABD**; roadside verge at Boghole, 1970 **E**.

N Nairn, rare, 1891 *Dr Sclanders* **NRN**; Cawdor, 1900 (Lobban); roadside verge near Bankhead, 1970 **E**.

M. indica (L.) All.   *Small-flowered Melilot*
Casual. Introduced with foreign grain.
M Garden weed at Grange Hall, Forres, 1952; rubbish tip at Findhorn, 1953; station yards at Knockando and Blacksboat, 1954; waste ground near the distilleries at Rothes and Carron, 1955; distillery yard, Smallburn near Rothes, 1956 **ABD**; rubbish tip, Elgin 1961 **E,K**; distillery yard at Carron, 1968 **E,BM**.
I Allotment near a distillery, Inverness, 1961 **ABD,K,CGE**; several plants on the tip at Longman Point, Inverness, 1975 **E**; root field at Dalveallan W of Daviot, 1975.

Trigonella L.
T. caerulea (L.) Ser.
Casual. Waste land.
M Frequent on shingle of the River Spey (Trail, *ASNH* **13**: 103 (1904)); Carron district, 1907 *W. G. Craib* **ABD**; ? locality and date, *J. P. Bisset* **E**; distillery yard, Carron, 1968 **E**.

Trifolium L.
T. pratense L.   *Red Clover*
Native. Abundant in grassy places throughout the district. Often sown as a crop.
M Doubtfully native (Gordon); common, Forres, 1900 *JK* **FRS**; throughout the county, 1954; Aviemore, 1963 *P. F. Wilson* **E**.
N (Stables, *Cat.*); Ardclach, naturalised from cultivated fields (Thomson); meadow at Cawdor, 1961 **ABD**.
I Inverness, 1832 and Pitmain, 1833 (Watson, *Loc. Cat.*); Kincraig, 1891 *AS* **BM**; Leys, 1940 *AJ* **ABD**.

T. medium L.   *Zigzag Clover*
Native. Common on grassy banks and by rivers and burns. Map 75.
M Frequent (Gordon); common in the uplands, 1900 *JK* **FRS**; throughout the county, 1954.
N (Stables, *Cat.*); Nairn and Cawdor, 1900 (Lobban); banks of the River Findhorn at Coulmony, 1961 **ABD**; bank N of Maviston, 1968 **E**.
I Inverness, 1832 (Watson, *Loc. Cat.*); Badenoch, 1838 *Hughes* **E**; road to Leys, ? date and *coll.* **INV**; Bunchrew, 1940 *AJ* **ABD**; banks at Dalwhinnie, Newtonmore and Fort Augustus etc. 1954.

T. arvense L.   *Dogs and Cats, Hare's-foot Clover*
Native. Frequent on sandy soil and among cinders on railway sidings. Map 76.
M Very common (Gordon); Findhorn, 1856 *JK* **FRS**; Findhorn Bay, 1898 *ESM* **CGE**; Hopeman, 1898 *ESM*; Findhorn, 1953 *A. Melderis* **BM**; Covesea, 1963 *V. Mackinnon* **E**; sandy soil near Burghead, 1963 **CGE**; on a new roadside verge at Craigmill near Dallas, 1973 and abundant in the station yards at Aviemore and Boat of Garten.

N Nairn Links, 1832 *WAS* **ABD**; seaside at Nairn, *E. M. Thomson*
NRN; Nairn, 1898 *ESM* **E**; do. 1961 *MMcCW* **E,K**; station yard
at Auldearn; shingle of the Old Bar, Culbin, 1968 **E**.
I Abundant on the shingle bar at Whiteness Head and on dunes
at Fort George, 1954; in an allotment with grain aliens, Inverness,
1961 **E,K**; among cinders on the railway line at Gollanfield station,
1976 **E,BM**.

T. striatum L.   *Soft Trefoil, Knotted Clover*
Introduced with wool shoddy, Rose Cottage, Dyke, 1972 **E**.

T. scabrum L.   *Rough Clover*
Doubtfully native in the north-east. Dry sandy banks near the sea.
M Bank near the lighthouse at Burghead, 1966 *A. J. Souter*,
E,ABD,CGE!.

T. hybridum L.   *Alsike Clover*
Introduced. Planted as a crop and becoming naturalised on grass
verges and waste ground. Map 77.
M Near Forres (Druce, 1897); introduced with farm seed (Burgess);
throughout the county, 1954; roadside near Rothes, 1966 **CGE**;
railway yard at Boat of Garten and in the car park at the Aviemore
Centre, 1975.
N Station yard at Auldearn and in hay fields at Drynachan and
Holme Rose, 1954; river shingle at Nairn, 1961 **ABD,K**; sandy
field at Kingsteps, waste ground by the ford at Achavraat and on
roadside verges at Ordbreck and Wester Delnies.
I Near Kincraig, 1891 *AS* **BM**; Tomatin, 1901 *JK* **FRS**; roadside
verge near Beauly, 1947 and S of Dores; railway yards at Dalwhinnie
and Newtonmore; waste ground at Kingussie, Milton and Culloden
etc.; root field above the Falls of Divach, 1975.

T. repens L.   *White or Dutch Clover*
Native. Abundant in short grassland, roadside verges and on sand
dunes. Many varieties are cultivated for hay.
M Very common (Gordon); Forres, 1900 *JK* **FRS**; ½ mile S of
Forres station, 1953 *A. Melderis* **BM**; throughout the county, 1954;
Garmouth, 1968 *E. S. Harrison* **E**.
N Nairn Links and Cawdor, 1832 *WAS* **ABD**; Ardclach (Thom-
son); Cawdor road, 1961 **ABD,BM**; on a bank by the Milk Market-
ing Board offices, Nairn (form with deep pink flowers) 1961
ABD,K.
I Inverness, 1832, Dalwhinnie and Pitmain, 1833 (Watson, *Loc.
Cat.*); road to Leys, ? date and *coll.* **INV**; Hilton, 1940 *AJ* **ABD**;
throughout the county reaching the car parks in the Cairngorms,
1954.

T. campestre Schreb.   *Hop Trefoil*
    T. procumbens auct.
Native. Common on waste land, on cinders in railway sidings and
by roadside verges. Fig 12f, Map 78.

M Abundant in one or two places (Gordon); Forres, 1860 *JK* **FRS**; Findhorn, 1953 *A. Melderis* **BM**; waste land by the Elgin rubbish tip, 1961 **ABD,BM**; dunes at Lossiemouth, 1966 **CGE**; station yard at Forres, 1967 **E,K**.
N Near Nairn, 1832 *WAS & J. B. Brichan* **ABD**; Ardclach (Thomson); waste ground at Holme Rose, Geddes and in the station yard at Auldearn, 1954; sandy ground by the Nairn Dunbar golf course, 1961 **ABD,K**.
I Inverness, 1832 (Watson, *Loc. Cat.*); waste land Drumnadrochit, 1947; waste ground at Kingussie, Kincraig, Invermoriston, Fort Augustus and all the coastal areas, 1954.

T. aureum Poll.
Casual. Introduced with foreign grain. Similar in general appearance to *T. campestre* but usually larger and the flowers a deeper yellow in colour.
M Near Dunphail, 1887 *G. C. Druce* **BM**; Boat of Garten, 1889 *G. C. Druce* **BM**; near Aviemore, 1891 *AS* **E**; introduced with clover seed at Dyke and Kinloss (Burgess).
N Nairn, 1900 herb. *Murchison* **E**; Nairn and Ardclach, 1900 (Lobban).

T. dubium Sibth.   *Lesser Trefoil*
Native. Abundant in short grassland and on waste land and dunes. Fig 12d.
M Very common (Gordon); near Aviemore, 1892 *AS* **E**; Garmouth, 1898 *ESM & WASh*; Findhorn, 1899 *JK* **FRS**; by the Muckle Burn, Moy House, 1953 *A. Melderis* **BM**; throughout the county, 1954; dunes at Lossiemouth, 1966 **E,CGE**.
N Near Boath, Auldearn, 1832 *WAS & J. B. Brichan* **ABD**; Ardclach and Nairn, 1900 (Lobban); waste land at Drynachan, Cawdor, Dulsie Bridge and Glenferness etc., 1954.
I Beauly (*J. Bot.* **28**: 39 (1890)); Scaniport, 1940 *AJ* **ABD**; Dalwhinnie and Tomatin, 1903 (Moyle Rogers); very common at Newtonmore, Fort Augustus, Inverness etc. 1954; forest track Glen Affric, in an old quarry in Glen Moriston, on the track in upper Glen Feshie and in all the lowland areas.

T. micranthum Viv.   *Slender Trefoil*
     T. filiforme L. nom. ambig.
Doubtfully native in north Scotland. Very rare in lawns and garden paths. Fig 12b.
M Castle Hill, Forres, 1856 *JK* **FRS**; Speymouth, *G. Birne* (Burgess).
N Lawn at Cawdor Castle, 1960 **E,ABD,BM**.
I *A. Somerville* sp. (A. Bennett, *ASNH* **7**: 228 (1898)) (there is no specimen to confirm this record in either **E,GL** or **BM**); gravel path Dochfour House, 1975 **E,BM**.

T. lappaceum L.
Casual. Introduced with 'Swoop' at Wild Goose Cottage, Nairn, 1968 **E**.

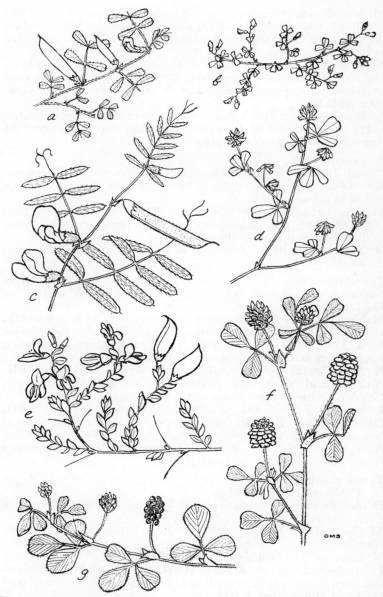

Fig 12   a, **Vicia lathyroides**; b, **Trifolium micranthum**; c, **Vicia angusti-folia**; d, **Trifolium dubium**; e, **Genista anglica**; f, **Trifolium campestre**; g, **Medicago lupulina**.

Astragalus L.

A. danicus Retz   *Purple Milk-vetch*
Native. Frequent on fixed dunes and turf on the edge of moorland along the whole length of the coast.
M Frequent at Hopeman and Burghead (Gordon)!; Hopeman, 1860 *JK* **FRS**; Lossiemouth, 1879 ? *coll.* **ELN**; short turf by the sea at Garmouth, Covesea and Findhorn, 1950; dunes at Findhorn, 1964 **E,BM**.
N (Stables, *Cat.*); dunes on the Nairn Dunbar golf course, 1954.
I Campbelltown (Gordon); raised beach between Ardersier and Fort George, 1954 **E**.

A. glycyphyllos L.   *Wild Liquorice, Milk-vetch*
Native. Very rare.
M Elginshire, 1837 *G. Gordon* **E**; abundant in one or two places, Lossiemouth and Cothall, *G. Wilson* (Gordon); Cothall, 1860 *JK* **FRS**; sandy waste ground near the sea, Lossiemouth harbour, 1962 **E,CGE** (this habitat disappeared when the waste land was made into a promenade, but there still remain one or two plants on the cliffs of the old quarry nearby).

Ornithopus L.

O. perpusillus L.   *Bird's-foot*
Native. Rare in dry sandy and gravelly places.
M Garmouth road opposite Urquhart, *G. Wilson* (Gordon); parishes of Dyke, Elgin, Lhanbryd, Speymouth and Birnie (Burgess); sandy bank by the road near Coxton Tower 1954; Innes Links, 1960 *A. J. Souter* **ABD**; gravel tracks by the sea at Kingston, 1966 **E,ABD**; sandy roadside verge on Dava Moor, 1974 **E,BM**.
I Many plants on the gravel drive at Inchnacardoch Hotel, Fort Augustus, 1957 but it now seems to have disappeared.

Coronilla L.

C. varia L.   *Crown Vetch*
Introduced. Rare, not seen in recent years.
M Waterford, 1877 *JK* **FRS**; escape at Forres and Cromdale (Burgess).

Onobrychis Mill.

O. viciifolia Scop.   *Sainfoin*
Introduced. Formerly sown as a crop.
M (Watson *Top Bot.*); reported from Lesmurdie, Elgin but not observed sine (Burgess).

Vicia L.

V. hirsuta (L.) Gray   *Hairy Tare*
Native. Common on waste and cultivated land. Map 81.
M Frequent (Gordon); Waterford, 1856 *JK* **FRS**; waste ground at Findhorn, Knockando, Grantown and Dunphail, 1954; grassy bank by Dalvey Cottage, 1963 **BM,CGE**; waste ground at Lossiemouth, 1966 **E**.

N (Stables, *Cat.*); Ardclach (Thomson); grass verges at Dulsie Bridge and Glenferness, 1954; garden weed at Holme Rose and waste ground by the ford at Achavraat, 1973.
I (Watson, *Top. Bot.*); at Englishton and by the Moniack Burn (Aitken); between Campbelltown and Fort George, 1898 *ESM* **BM**; mouth of the Caledonian Canal, Inverness, 1889 *R. Kidston & J. S. Stirling* **STI**; a weed in the Manse garden, Rothiemurchus, 1903 *H. Groves* **BM**; Bunchrew, 1940 *AJ* **ABD**; railway embankments at Allanfearn and near Nethybridge; garden weed in Inshriach Nursery, 1974 *R. McBeath*; waste ground by Kingussie golf course and in Lewiston kirkyard, 1973, etc.

V. tetrasperma (L.) Schreb.   *Smooth Tare*
Native. Grass verges and waste ground.
M Waste ground at Greshop House, Forres, 1948; near Brodie station and at Elgin, 1954; by Rothes distillery, 1956 **ABD** and grass verge of road at Dallas.

V. cracca L.   *Mice Peas, Tufted Vetch*
Native. Common in rough and grassy places by roads and rivers.
M Very common (Gordon); roadside near Hillhead (a white-flowered variant), 1830 *G. Gordon* (Gordon, *annot.*); ½ mile S of Forres, 1953 *A. Melderis* **BM**; throughout the county, 1954.
N (Stables, *Cat.*); riverside at Ardclach, 1888 *R. Thomson* **NRN**; near Nairn (a form with pretty pinkish-white flowers), 1898 *ESM* **E,BM,CGE**; throughout the county, 1954; bank near Nairn, 1962 **ABD**.
I Inverness, 1832 Pitmain and Dalwhinnie, 1833 (Watson, *Loc. Cat.*); Kincraig, 1898 *AS* **BM**; Rothiemurchus, 1903 *H. Groves* **BM**; Fort Augustus, Corrimony, Kingussie etc., 1954.

V. tenuifolia Roth   *Fine-leaved Vetch*
Introduced. Rather similar to *V. cracca* but with more lax racemes, the flowers larger, pale blue or violet with whitish wings.
M Waste ground near Brodie station, 1963 **E**, but now feared lost owing to road widening.

V. villosa Roth   *Fodder Vetch*
Introduced. Waste land.
I Rubbish tip at Longman Point, Inverness, 1971 **E**.

V. orobus DC.   *Bitter Vetch*
Native. On rocks and in woods. Recorded by Lobban for Cawdor, Croy and Dalcross but must be considered doubtful without a voucher specimen.

V. sylvatica L.   *Wood Vetch*
Native. Frequent on mountain scree and on rocks in shaded places by rivers. Map. 82.
M Abundant in one or two places on the Findhorn above Sluie, 1830 (Gordon)!; sandstone rocks near the Heronry, Darnaway, 1857 *JK* **FRS**; parish of Bellie (Burgess).

N (Stables, *Cat.*); Cawdor, 1885 *R. Thomson* **NRN**.
I Near Inverness at the foot of Beinn a' Bha'ach Ard and at
Kilmorack, 1829 (Gordon); Moniack Burn and Glen Urquhart
(Aitken); by the side of the road by Loch Ness, 1938 *M. L. & T. A.
Sprague* **K**; Glen Einich, 1892 *AS* **E**; by Loch Ness, 1947 *A. J.
Wilmott* **BM**; near Castle Urquhart, 1940 *AJ* **ABD**; cliffs by Loch
Killin, scree in Glen Feshie, burn gullies at Gaick, by the Plodda
Falls at Guisachan etc., 1954.

   Var. **condensata** Druce
Variant making prostrate, compact mats on shingle.
I Stoney ground between Campbelltown and Fort George, 1826
R. Graham (Trans. BSE **1**: 26 (1840))!; do. 1898 *ESM* **BM**; do. 1898
WASh **OXF,K**, still to be found there in small quantity.

V. sepium L.   *Bush Vetch*
Native. Abundant in grassy places by roadsides and rivers.
M Castles of Old Duffus and Spynie (Gordon); Rafford, *G. Wilson*
ABD; Nethybridge, 1903 (Moyle Rogers); throughout the county
1954.
N Cawdor woods, 1833 *WAS* (Gordon); by Lord Leven's girder
bridge at Ardclach, 1886 *R. Thomson* **NRN**; banks of River Findhorn
at Coulmony, 1961 **ABD**.
I (Watson, *Top. Bot.*); Glen Einich, 1886 *H. Groves* **BM**; Inverallan
(as var. *ochroleuca* Bast.), 1918 *Mr Nicholson* herb. *NDS* **BM**;
Drumnadrochit, 1947 *R. D. Graham & C. Wickham* **BM**; by Loch
Ericht, banks of the Spey at Newtonmore, cliffs by Loch Killin
etc., 1954; birch wood on the hillside by Gaick Lodge, gully of the
burn at Coignashie, among boulders in the Ryvoan Pass etc., 1972.

V. sativa L.
   Subsp. **sativa**   *Fitchacks, Cultivated Tare, Common Vetch*
Escape from cultivation, naturalised on the grassy margins of fields
and on waste land. Stipvles often with a dark blotch.
M Innes House, Urquhart, 1794 *A. Cooper* **ELN**; very common
(Gordon) (this probably refers to subsp. *nigra* (L.) Ehrh.); Forres,
1856 *JK* **FRS**; in a field of oats at Kinloss, 1926; barley field at
Dyke, 1963 **CGE**; rubbish tip, Rothes, 1966 **E**; root field at Rafford,
1967 **E,ABD**.
N Grass verge of a field at Nairn, waste ground Auldearn station
and near Whitemire, 1954.
I (Watson, *Top. Bot*); cultivated at Kincraig, 1891 *AS* **E**; Kiltar-
lity, 1940 *AJ* **ABD**; waste land at Dores, near Inverness and in the
station yard at Kincraig, 1956; among grass by the sea at Alturlie
Point, farm lane at Kirkhill and in a root field at Mid Craggie near
Daviot, 1975.

   Subsp. **nigra** (L.) Ehrh.   *Narrow-leaved Vetch*
   Subsp. *angustifolia* (L.) Gaud.; *V. angustifolia* L.
Native. Common in short turf on moors and in grassy banks and
sand dunes. Very variable. Fig 12c, Map 83.

M Rare (Gordon) (this probably refers to subsp. *sativa*); Forres, 1900 *JK* **FRS**; grass banks at Dyke and Brodie, 1964 **CGE**; dunes at Lossiemouth, 1966 **CGE**; roadside verge on Dava Moor, railway embankment near Spynie and river bank opposite Coulmony, etc., 1974.

N (Stables, *Cat.*); Nairnshire, *J. B. Brichan* (Watson); Ardclach, 1887 *R. Thomson* **NRN**; grass bank at Drynachan, near Nairn, Holme Rose and Cawdor, 1954; bank by the road ½ mile E of Geddes, 1976 **E**.

I Near Kingsmills (Aitken); meadows at Kincraig, Glen Affric, Fort Augustus and Tomatin, etc., 1954; near the garden at Flichity House, by a small burn by Loch nan Lann, Knockie, 1975.

Subsp. **segetalis** (Thuill.) Gaud.
V. angustifolia var. *bobartii* Koch

Native. A very slender plant of sand dunes and dry waste ground. Probably more widespread than the records suggest.

M Common on the sand dunes at Findhorn and Culbin, 1954; dry moor on top of the sea cliffs near Hopeman; railway embankment at Spynie, 1973.

N Ardclach, 1884 *R. Thomson* **NRN**; among cinders in the railway yard at Auldearn, 1962 **ABD**; roadside bank on sandy soil W of the Druim near Nairn, 1964 **ABD**.

I Near Kincraig, 1892 *AS* **E**; dunes at Fort George, 1970 **CGE**.

V. lathyroides L. *Spring Vetch*

Native. Frequent on fixed dunes on the coastal belt and on sandy banks, road verges and wall tops inland. Fig 12a.

M Frequent (Gordon); sea coast, Moray, *G. Wilson* **ABD**; roof of castle (?) 1856 *JK* **FRS**; Culbin sands, 1920; Findhorn, 1953 *A. Melderis* **BM**; Lossiemouth golf course, 1966 **E**; on cinders on the old railway line at Spynie; on top of the kirkyard wall at Dyke, 1968 **E**; roadside verge on Dava Moor, 1968 **E**.

N (Stables, *Cat.*); Ardclach, not common (Thomson); abundant on the fixed dunes by the Nairn Dunbar golf course, 1970 **E,ABD**.

I Top of a wall by the old Kirk, Castle Stuart, 1970 **E**; common on the dunes at Fort George and on a sandy bank by the road junction A96-B3039 near Newton, 1976 **BM**.

Lathyrus L.

L. aphaca L. *Yellow Vetchling*

Casual. Among grass at the Leen, Kingston, 1934 *E. T. Dawson*.

L. pratensis L. *Meadow Vetchling*

Native. Common in hedgerows and on grassy banks. Map 84.

M Birnie, 1837 *G. Gordon* **ABD**; Greshop, 1856 *JK* **FRS**; throughout the county, 1954; ½ mile S of Forres, 1953 *A. Melderis* **BM**.

N (Stables, *Cat.*); Blackpark, 1884 *R. Thomson* **NRN**; bank at Drynachan and throughout the county, 1954.

I Inverness, 1832 Pass of Drumochter and Pitmain, 1833 (Watson, *Loc. Cat.*); Kincraig, 1891 *AS* **BM**; railway embankment at Beauly,

m rivale × $\frac{2}{3}$
er Avens

Dryas octopetala × $\frac{2}{3}$
Mountain Avens

lum villosum × 1
ry Stonecrop

Saxifraga oppositifolia × $\frac{2}{3}$
Purple Saxifrage

PLATE 7

Rosa pimpinellifolia × ½
Scots Rose, Burnet Rose

R. pimpinellifolia (*fruit*) × ½

Rosa canina var. sylvularum × ⅔
Dog Rose

Rosa dumetorum × ⅔

Rosa afzeliana var. reuteri × ⅔

Rosa coriifolia var. lintoni × ⅔

PLATE 8

a villosa var. mollis × 1
ıny Rose

Rosa mollis var. relicta × ⅔

a rubiginosa × ⅔
eet Briar

Rosa micrantha var. nov. × 1   *D. A. Lawrie*

sa pimpinellifolia × sherardii × ⅔

Rosa canina × afzeliana × ⅔

PLATE 9

Hippuris vulgaris × ⅕
Marestail

M. Coll

Cornus suecica × ½
Dwarf Cornel

Cornus suecica (*fruit*) × ⅔

PLATE 10

1892 *ESM* **E,BM**; throughout the district except on the higher mountains.

L. tuberosus L. *Tuberous Pea*
Introduced with garden refuse and well established by the rubbish tip at Newton of Park near Nairn, 1975 **E,BM**.

L. sylvestris L. *Narrow-leaved Everlasting Pea*
Introduced. A single plant, nowhere near habitations, on dunes very near the sea in the Culbin Forest, 1970 *R. K. Smith* **E**!

L. latifolius L. *Everlasting Pea*
Introduced.
M Near Forres, *JK* **FRS**; for many years on a bank by the road in the Oakwood near Elgin, now gone with road widening, 1950 **K**; rubbish tip at Elgin, 1957.

L. montanus Bernh. *Gnapperts, Carmile, Bitter Vetch*
Native. Abundant on wet banks and by rivers and burns chiefly in the highlands. The tubers were formerly eaten. Map 85.

Var. **montanus**
M Very common (Gordon); Clunyhill, Forres, 1900 *JK* **FRS**; throughout the county, 1954.
N Cawdor woods, 1837 *WAS* **E**; Ardclach (Thomson); banks of the River Findhorn at Dulsie Bridge and Coulmony, 1960 **ABD**.
I Inverness, 1832 Pitmain and Dalwhinnie, 1833 (Watson, *Loc. Cat.*); S side of Loch Affric, *A. J. Wilmott* **BM**; Rothiemurchus, 1940 *E. Crapper* **STA**; throughout the district, 1954.

Var. **tenuifolius** (Roth) Garcke
A common variant with very narrow leaflets.
M In the Oakwood at Elgin and the 'Blasted Heath' at Brodie (Gordon); parishes of Dyke, Forres and Edinkillie (Burgess).
I Tomatin (W. Moyle Rogers *J. Bot.* **42**: 13–21 (1904)); by Loch Affric, 1947 *A. J. Wilmott* **BM**; cliffs by Loch Killin, 1958; in shingle on the raised beach at Ardersier, 1972 **E**; wet bank by Loch Tarff, among rocks in Glen Feshie, bank by the river Enrick W of Loch Meiklie, verge of a pine wood at Inverdruie and on a steep bank by the hill road to Abriachan, 1973.

L. niger (L.) Bernh. *Black Pea*
I Rare at Craiganain near Moy, Inverness (Gordon).

L. inconspicuus L.
Introduced. Shingle of the River Spey at Rothes and Aberlour (Trail, *ASNH* **13**: 103 (1904)).

Pisum L.

P. sativum L. *Garden Pea*
Garden escape.
M Rothes, (Trail, *ASNH* **14**: 234 (1905)); field near Aviemore, 1907 *R. S. Adamson* **BM**.

Var. **arvense** (L.) Poir.   *Field Pea*
Flowers purple. Casual on shingle at Rothes, Moray (Trail, 1905).

ROSACEAE
Spiraea L.
S. salicifolia L.   *Willow Spiraea, Bridewort*
Introduced. Garden relic.
M Certainly introduced, Gordon Castle etc. (Gordon); Invererne,
Forres, 1898 *JK* **FRS**; waste ground near old cottage, Kintessack,
1963 **CGE**; field in the village of Carrbridge, lane at Auchinroath,
roadsides at Cromdale, Dunphail, Boat of Garten, Knockando,
etc.; Spynie, 1941 ? *coll.* **STA**; site of old cottage at Tilliglens near
Relugas, 1973 **E**.
I Banks of the Ness and at Laggan (Aitken); Kingussie (Druce,
1888); Kincraig, 1891 *AS* **BM**; Carphick, 1940 *AJ* **ABD**; near old
buildings at Dores, Alvie, Bunloit, Culloden, Ardersier and
Tomatin etc., 1954.

S. douglasii Hook.
Introduced. Similar to *S. salicifolia* but the leaves are tomentose
beneath.
I Frequent by the road between Kingussie and Dunachton, 1954.

Filipendula Mill.
F. vulgaris Moench   *Dropwort*
   Spiraea filipendula L.
Garden escape.
M Parishes of Dyke and Forres (Burgess).
N Found on rubbish near the schoolhouse at Ardclach, 1883
R. Thomson **NRN**.
I Loch an Eilein, Rothiemurchus, cult, 1891 *AS* **E**.

F. ulmaria (L.) Maxim.   *Meadowsweet*
   Spiraea ulmaria L.
Native. Abundant on river banks, in ditches and on margins of
lochs and bogs.
M Very common (Gordon); by the railway at Forres, 1898 *JK*
FRS; ditch by the Burnie Path, Dyke and throughout the county,
1954.
N (Stables, *Cat.*); Ardclach, 1886 *R. Thomson* **NRN**; by the River
Findhorn at Drynachan, etc., 1954.
I Inverness, 1832, Pitmain and Dalwhinnie, 1833 (Watson, *Loc.
Cat.*); Parks of Inshes, ? date and *coll.* **INV**; Kincraig, 1891 *AS* **BM**;
banks of the River Tarff, ? date *Coles* **BM**; Inchnacardoch Bay,
1904 *G. T. West* **E**; Loch Dochfour, 1940 *AJ* **ABD**; ledge at 823 m
Sgùrr na Lapaich, 1971, and throughout the district.

Rubus L.
R. chamaemorus L.   *Aivrons, Cloudberry*
Native. Common on wet acid moorland often covering large areas,
usually over 600 m. Pl 6.

M (Watson); parishes of Edinkillie and Cromdale (Burgess); peat bog in the Moss of Bednawinny at 305 m, 1954; bogs to the SW of Lochindorb; Càrn Dearg Mòr above Aviemore, 1974 **E**; dominant over a large area where it fruits regularily on Càrn Tuairneir by Bridge of Brown.

N Rare on the hill behind Drynachan Cottage, 1833 *WAS* (Gordon); abundant on the hills S of Ballochrochin near Drynachan, 1974 **E**.

I Dalwhinnie, 1833 (Watson, *Loc. Cat.*); hills near Loch Ness, *G. Anderson* (Gordon); Fraoch-choire, Cannich (Ball, 1851); hills above Dunmaglass and on Mealfuarvonie (Aitken); head of Glen Shirra, 1916 *ESM* **CGE**; Creag Coire Doe, 1904 *G. T. West* **STA**; throughout the district, 1954; abundant in the Cairngorms where, in the higher areas, the plant only fruits when the snows have lain until May, thus protecting the flowers from the late frosts.

R. saxatilis L.   *Dog's Berries, Stone Bramble*

Native. Common on base-rich soil on banks, among rocks and in ravines of burns and rivers. Map 86.

M Elginshire, 1837 *G. Gordon* **E**; frequent at Glenlatterach and Dunphail (Gordon); banks of the River Findhorn at Cothall, 1857 *JK* **FRS**; Dunphail, 1886 *ESM* **CGE**; grassy bank by the road between Grantown and Cromdale, 1971 **E**; among rocks in the gorge at Huntly's Cave on Dava Moor, in a birch wood at Kellas and in the ravines of the rivers Spey, Findhorn and Divie etc.

N Cawdor woods, 1839 *WAS* **E,ABD**; banks of the River Findhorn at Ardclach, 1887 *R. Thomson* **NRN**; in the gorge of the Reireach Burn at Cawdor and on the banks of the River Nairn at Holme Rose, 1954; among rocks at Glenferness House, 1961 **K**.

I Dalwhinnie, 1833 (Watson, *Loc. Cat.*); by the Falls of Foyers, 1838 *C. Babington* **CGE**; Moniack and Holm burns (Aitken); Glen Einich and near Kincraig, 1887 *G. C. Druce* **OXF**; Glen Doe (with a variegated leaf) 1904 *G. T. West* **STA**; near Dores, 1937 *NDS* **BM**; Milton 1940 *AJ* **ABD**; cliff ledge at 820 m on Sgùrr na Lapaich in Glen Strathfarrar, on rocks below the Plodda Falls at Guisachan; burn ravine at Gaick etc.

R. arcticus L.

The following extracts are from letters written by Thomas Edmondston to Sir W. J. Hooker in 1844, now in the possession of Lt. Col. L. D. Edmondston of Buness, Shetland, and copies of which are in the possession of Walter Scott of Scalloway from whom the information was obtained. They read as follows—23 August 1844—'while in Morayshire lately I observed this plant in the garden of a nurseryman (John Grigor) at Forres and he assured me that the original plant from which his stock had been reared, was brought from Cairn Gorm by a lady who has now unfortunately left the country, but who was well acquainted with Botany, and gave the plant to Mr Grigor under its proper name'. In the second letter dated 1 October 1844 'you may be sure if the reputed discoverer of *Rubus arcticus* on Cairn Gorm had been accessible, I would have endeavoured to have ascertained more fully regarding the plant – the lady was Miss Robertson sister of a Mr Robertson who formerly was master of an Academy near Forres – I believe the lady knew Botany well and she gave the plant to Mr Grigor as a *great prize* under its *true name*'.

R. idaeus L.   *Raspberry*
Native. Abundant in waste places, by rivers and on railway embankments throughout the district. A yellow-fruited form is found occasionally.
M Frequent (Gordon); abundant in the Spey basin, 1903 (Moyle Rogers); in several places in the Culbin Forest and throughout the county, 1954. The yellow-fruited form is recorded from a bank near the railway bridge at Garmouth and from Duffus and by the River Lossie at Craigroy near Dallas.
N (Stables, *Cat.*); common on heaths and in thickets at Ardclach, 1891 *R. Thomson* **NRN**; banks of the River Nairn near the town, 1962 **E,ABD**.
I Inverness, 1832 Pitmain and Dalwhinnie, 1833 (Watson, *Loc. Cat.*); Englishton, 1880 ? *coll.* **INV**; near Torvaine, 1940 *AJ* **ABD**; on scree at Gaick Lodge and throughout the district, 1954.

Var. **rotundifolius** Bab.
M In a small wood by the old Kirk at Balnabreich near Mulben, 1974 **E**.

R. spectabilis Pursh   *Canadian Raspberry*
Introduced as pheasant food. More robust than *R. idaeus* with large purple-red flowers that flower in early spring before the leaves are mature, and fruits that are at first yellow eventually turning dull purple when ripe.
M Brodie pond, Binsness and Innes House near Urquhart (Burgess); banks of the River Lossie by the Cooper Park at Elgin, 1965 **CGE**; abundant in woods to the N of Innes House, 1971 **E**.
N Policies of Lethen House, 1961.
I Banks of the River Beauly below Lovat Bridge, 1930 *V. Cameron*.

R. caesius L.   *Dewberry*
Reported by Dr Gordon from Forres but not observed since (Burgess) and from Kincraig in 1891 by *A. Somerville* **E**, but both are probably errors.

R. fruticosus L. *sensu lato*   *Bramble*
In the following account herbarium specimens have been determined by B. A. Miles or E. S. Edees unless marked with an asterisk. The pocket descriptions and up-to-date nomenclature have been supplied by E. S. Edees. The first four taxa have erect stems like raspberry canes, which do not bend over and root at the tip.

R. nessensis W. Hall
    R. suberectus Anders. ex Sm.
Damp woodlands. Stems angled with few, very short, purple,

conical prickles; frequently with seven leaflets; petals white; stamens having the curious habit of hanging down when they begin to wither; fruits dark red when ripe. Fig 13a.

M Kinchurdy, near Boat of Garten (Druce, 1888).

I Banks of Loch Ness, 1787 *William Hall* (*Trans BSE* **3**: 21 (1794)), the original reference and first record; on a small island in the River Enrick below the kirk at Milton, 1973 **E**.

R. scissus W. C. R. Wats.

Moors among heather and bracken. Prickles slender, numerous, slightly curved but not falcate; frequently with seven leaflets which are sometimes pleated; petals white; fruits black when ripe. Fig 13b.

M Bank by the car park at Randolph's Leap at Relugas, 1961 **E**; railway embankment at Avielochan N of Aviemore, 1973 **E**.

N Moorland at Cantraydoune, 1954 **CGE**.

I Kingussie, 1883 *A. E. Lomax* **CGE**; thicket by the River Beauly near Groam of Annat and in a pine wood by Loch Garten, 1972 **E**.

R. plicatus Weihe & Nees

On moors among heather and bracken. Prickles falcate; terminal leaflets plicate; petals white sometimes tinged pink; stamens about equal to the styles in length; young carpels with a fringe of hairs at the base; fruits black when ripe. Fig 14a.

M Roadside bank at Brodie (Marshall, 1899); rare at Advie, 1903 (Moyle Rogers); edge of a ditch W of Dallas Lodge, 1976 **E**.

N Near Nairn, 1898 *ESM & WASh* **E,CGE**; *Holme Rose, 1955; 100 yards S of Righoul School, 1974 **E**.

I By the road near the outlet burn at E end of Loch Meiklie, 1976 **E**; roadside bank by Dalcross Castle, 1976 **E**.

R. fissus Lindl.

R. rogersii E. F. Linton

On moors among heather and bracken. Prickles strong; terminal leaflets ovate, finely toothed; petals white; fruits dark red when ripe. Fig 14b.

M By the River Spey at Advie, 1903 *W. M. Rogers* **CGE**; Blacksboat, 1903 *ESM* **CGE**; plentiful at Alves junction and at Dunphail (Marshall, 1899); small wood near Rosehaugh, 1961 **E**; thicket in the north drive at Brodie Castle, 1961 **E**; roadside near Cloddymoss, 1965 **CGE**.

N Near Nairn, 1898 *ESM* **CGE**; S of Nairn, 1966 *E. S. Edees* **ESE**.

I *Loch Pityoulish, 1887 *G. C. Druce* **OXF**; by Loch Insh at Kincraig and at Culloden, 1903 (Moyle Rogers).

R. conjungens (Bab.) Warren

This and the following species are closely related, but, whereas *R. conjungens* has leaflets which are more or less evenly toothed and pink petals, *R. latifolius* has deeply incised leaflets and white petals. Fig 15a.

N Bank by railway bridge at Auldearn station, 1976 *OMS & MMcCW* **E**.

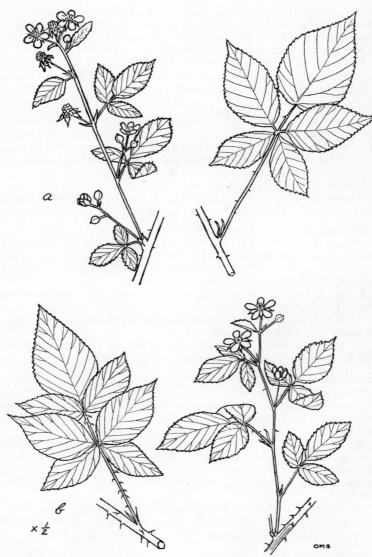

Fig 13   a, **Rubus nessensis**; b, **R. scissus**.

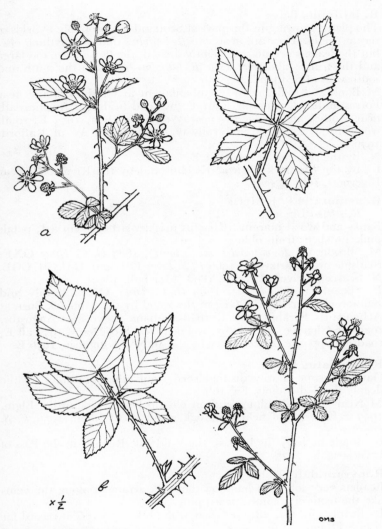

Fig 14   a, **Rubus plicatus**; b, **R. fissus**.

R. latifolius Bab.

The plants growing in this part of Scotland are probably hybrids of recent origin. They are nearer to *R. latifolius* than to anything else, but the leaflets are not sufficiently incised, the flowers are too large and the fruit is abortive. True *R. latifolius* occurs in the south and south west of Scotland. Fig 15b

M Bank at Brodie, railway embankment at Greshop House near Forres and at Garmouth, 1961 **E**; by road bridge at Auchinroath near Rothes and hedgerow near Westfield, Elgin, 1973 **E**; small wood by Dyke Kirk and railway embankment W of Rafford, 1976 **E**.

N By the river at Nairn, 1961 **E**.

I By the Moniack Burn at Kirkhill and by the River Ness near Inverness, 1961 **E**.

R. nemoralis P. J. Muell.
    R. selmeri Lindeb.

Banks and wood margins. Rhachis prickles strongly curved; petals pink, notched; fruits oblong. Fig 16a.

M *Brodie (as *R. villicaulis* var. *selmeri*), 1887 *G. C. Druce* **OXF**; roadside near Darnaway, 1966 *E. S. Edees* **ESE** and *MMcCW* **CGE**.

N Scarce by river at Nairn, 1898 (Marshall, 1899).

I *Beauly (as *R. laetus*), 1889 *G. C. Druce* **OXF**; Beauly and Kilmorack, 1892 *ESM* **CGE**; in the wood by the Moniack Burn at Achnagairn, Kirkhill, 1961 **E**; roadside bank above Fort Augustus, roadside Glendoe, steep bank by Loch Ness S of Dores, 1972 all **E**; roadside opposite the cemetery by the A9 N of Kingussie, 1976 **E**.

R. laciniatus Willd.

Garden escape. Easily distinguished from other brambles by the deeply divided leaflets. Fig 16b.

M Near Newmill, Elgin, and on old tip below Lesmurdie, Elgin, 1957 **E,K**; edge of field near oak wood W of Elgin, 1959 *E. S. Harrison*!

I Single bush by the burn in the Wild Life Reserve in the Pass of Inverfarigaig, 1975.

R. pyramidalis Kalt.

Leaflets very soft with hairs on the back arranged along the veins like the teeth of a comb; petals pink.

I Beauly (*J. Bot.* **28**: 39 (1890)), doubtful, no recent record for N of Scotland.

R. leptothyrsos G. Braun
    R. danicus Focke ex Frider. & Gelert

Common. The snowy white (sometimes faintly pinkish) petals combined with hairy anthers should be enough to distinguish it. Fig 17a.

M Dunphail and near Elgin (Marshall, 1899); railway embankment near Greshop Farm and on a bank leading to the Muckle Burn at Tearie cross-roads, Dyke, 1961 **E**; roadside verge between

Greengates House and the entrance to Brodie Castle, 1965 *B. A. Miles* **CGE**; lay-by at Randolph's Leap, Relugas, 1966 **E** and *E. S. Edees* **ESE**; near Darnaway Forest, 1966 *E. S. Edees* **ESE**; Findhorn, 1966 *E. S. Edees* **ESE**.

N Common round Nairn, 1898 (Marshall, 1899); by the River Nairn near the cemetery, 1961 **ABD**; Nairn, 1966 *E. S. Edees* **ESE**; by railway bridge at Auldearn station, 1976 *OMS & MMcCW* **E**.

I Banks of the Caledonian Canal, Inverness, 1881 ? *coll.* **ESE**; Kilmorack and above Beauly, 1892 *ESM* **CGE**; Fasnakyle at Cannich, 1947 *E. Milne-Redhead* **CGE**; by the Moniack Burn at Kirkhill and by the River Ness, Inverness, 1961 **E**; roadside bank on moor N of Loch Killin and by Loch Ness S of Dores, 1972 **E**; roadside near Polmaily in Glen Urquhart, 1976 *OMS & MMcCW* **E**.

R. septentrionalis W. C. R. Wats.
This is the correct name for the Scottish bramble which was formerly called *R. villicaulis* or *R. insularis*. Common. Stems with spreading white hairs; prickles straight and long; petals white or pinkish; stamens glabrous. Fig 17b.

M Garmouth, Dunphail, Alves station and near Forres etc. (Marshall, 1899); roadside between Dyke and Brodie Castle, 1965 *B. A. Miles* **CGE,ESE**; S of Forres and on a track near Binsness, 1965 *B. A. Miles* **CGE**; pinewood near Lossiemouth, Findhorn and near Darnaway Forest, 1966 *E. S. Edees* **ESE**.

N Nairn (Marshall, 1899); ditch near Ardclach, 1966 **E**; Coulmony, 1966 *E. S. Edees* **ESE**; bank by Ferness cross-roads, 1972 **E**; river bank at Little Kildrummie, 1974 **E**; roadside bank a few yards W of entrance to Lochloy House, 1976 *OMS & MMcCW* **E**.

I Bank NW of Loch Meiklie in Glen Urquhart, 1973 **E**; roadside at Leishmore, 1974 *D. Kingston* **ESE**.

R. cardiophyllus Muell. & Lefèv.
  R. rhamnifolius auct.
Leaflets becoming concave; terminal leaflets with unusually long stalks, half the length of the blade; petals white, raised to make the flower cup-shaped instead of flat like a saucer.

M Near Forres (Druce, 1888), unconfirmed and doubtful.

R. mucronulatus Bor.
A very beautiful bramble with several distinctive features. Terminal leaflets broad, heart-shaped at the base and remarkably cuspidate; petals pink; anthers densely hairy. Fig 18a.

M Near Forres (Druce, *ASNH* **6**: 55 (1897)).

N Roadside E of Nairn, 1966 *E. S. Edees* **ESE**.

I Achnagairn, Kirkhill, 1961 **E**.

R. radula Weihe ex Boenn.
Barren stems rough like a file with short bristles; terminal leaflets ovate; petals pink. Fig 18b.

M Wood near Forres station, 1898 *ESM* **E**; near Spynie, more

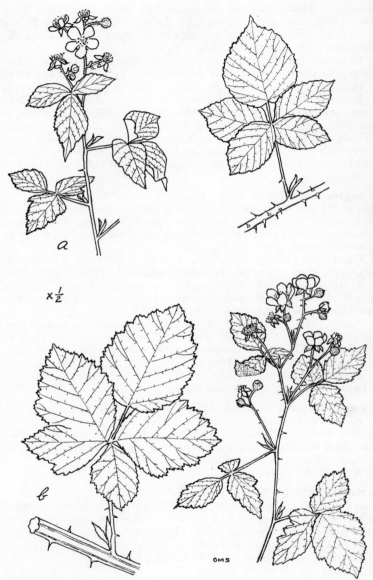

$\times \frac{1}{2}$

Fig 15 a, **Rubus conjungens**; b, **R. latifolius**.

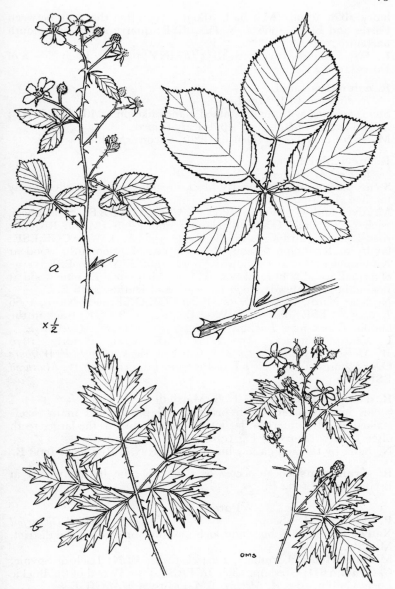

Fig 16   a, **Rubus nemoralis**; b, **R. laciniatus**.

hairy than usual (Marshall, 1899); wood by the A96 between Forres and Elgin, 1966 *E. S. Edees* **ESE**; quarry near Lossiemouth harbour, 1972 **E.**

I   *Ness Islands, Inverness, 1854 *HC* **INV**; bank by Loch Ness S of Dores, 1972 **E.**

R. echinatus Lindl.
> *R. discerptus* P. J. Muell.

Stems grooved with many stalked glands like pins with dark heads; incised leaflets; petals red; sepals leafy-tipped.

M Altyre near Forres (Druce, 1888), unconfirmed.

R. furvicolor Focke
> *R. melanoxylon* auct.

Stems dark, sometimes nearly black in sunny places; leaflets finely toothed. Fig 19a.

M Alves and Garmouth, 1898 *ESM* **E,CGE,OXF**; *Fochabers, 1926 *G. C. Druce* **OXF**; Blacksboat and Elgin (Moyle Rogers); roadside near entrance to Brodie Castle, 1965 *B. A. Miles* **CGE,ESE**; lay-by at Randolph's Leap, Relugas, near Lossiemouth, wood at Kintessack and near Darnaway Forest, 1966 *E. S. Edees* **ESE**; verge of a small wood N of railway at Moss of Meft, 1968 **E**; pine wood at Lochnabo, Lhanbryde, 1972 **E**; wood S of Duffus, 1976 **E.**

N Near Nairn, 1898 *ESM & WASh* **CGE,OXF**; near Nairn, 1966 *E. S. Edees* **ESE**; pine wood W of Delnies, 1968 **CGE**; track in the Culbin Forest near Lochloy House, 1976 *OMS & MMcCW* **E.**

I Near Kilmorack, 1892 *ESM* **CGE**; Craig Phadrig, 1870 *W. Mathews* herb. *Babington* **BM**; Crask of Aigas, 1947 *E. F. Warburg* **OXF**; shores of Loch Ness E of Invermoriston, 1969 *C. W. Muirhead* **ESE.**

R. dasyphyllus (Rogers) E. S. Marshall

Stems heavily armed with prickles and pricklets of many sizes; stalked glands numerous; leaflets unevenly toothed, the larger teeth often slightly recurved. Fig 19b.

N Bank by the A939 a few hundred yards N of Redburn, 1968 **E.**

R. giraldianus Focke occurs as a garden relic in the woods at Brodie Castle, 1972 **E.**

Potentilla L.

P. palustris (L.) Scop.   *Marsh Cinquefoil*

Native. Common in bogs and loch margins throughout the district. Map 87.

M Innes House, Urquhart, 1794 *A. Cooper* **ELN**; Loch of Spynie, *G. Gordon* **ABD**; Greshop, 1895 *JK* **FRS**; at the W end of the Buckie Loch, Culbin, 1953 *A. Melderis* **BM** and 1973 *MMcCW* **E.**

N (Stables, *Cat.*); in a moss near the Free Kirk, Ardclach, 1885 *R. Thomson* **NRN**; the Minister's Loch, Nairn, 1960 **ABD**; bog below the raised beach at Maviston, 1968 **E.**

I Dalwhinnie, 1833 (Watson, *Loc. Cat.*); Kincraig, 1891 *AS* **BM**;

Inchnacardoch and Fort Augustus, 1904 *G. T. West* **E,STA**; throughout the district, 1954.

P. sterilis (L.) Garcke *Barren Strawberry*
Native. Very common on dry banks and in deciduous woods in the lowlands, less so in the highlands. Map 88.
M Abundant on the banks of the Spey at Fochabers, at Newmill of Alves, *G. Wilson* and at Birdsyards, Forres, *J. G. Innes* (Gordon); Sanquhar, 1871 *JK* **FRS**; Aviemore, 1892 *AS* **E**; bank in the Cromdale Hills etc., 1954.
N (Stables, *Cat.*); Auldearn 1837 *J. B. Brichan* (Watson); banks of the Findhorn at Ardclach and at Cawdor, 1888 *R. Thomson* **NRN**; banks of the River Nairn at Nairn, 1963 **ABD**; wood by Lochloy House, 1968 **E,CGE**.
I (Watson, *Top. Bot.*); bank by the road to Beauly, Milton, 1947; common in Strathglass, woods at Foyers and round Inverness, 1954; lower slopes of Mealfuarvonie, 1964 **CGE**; railway embankment at Allanfearn, on dry grassy slope below Creag nan Clag, rare on a bank at Moy Hall, by the River Spey at Inverdruie etc.

P. anserina L. *Mascorns, Silverweed*
Native. Common on fixed dunes, farmyards, waste ground, among stones on loch margins and by roadside verges. Map 89.
M Innes House, 1794 *A. Cooper* **ELN**; very common (Gordon); Forres and Culbin, 1898 *JK* **FRS**; dunes by the Buckie Loch, Culbin, etc., 1954.
N (Stables, *Cat.*); Ardclach, near the schoolhouse, 1885 *R. Thomson* **NRN**; dunes on the Nairn Dunbar golf course, 1962 **ABD**.
I Inverness, 1832 (Watson, *Loc. Cat.*); by Loch Ness, Fort Augustus, 1904 *G. T. West* **STA**; near Kincraig, 1891 *AS* **BM**; waste ground in several places in Glen Moriston, farm yards at Newtonmore and Pityoulish and among stones by Loch Flemington etc., 1954.

P. argentea L. *Hoary Cinquefoil*
Doubtfully native in north Scotland. Not seen in recent years.
M Foot of Moss Wynd, Elgin, 1826 *G. Gordon* **GL**; field west of Glassgreen, Elgin, 1847 *J. Robertson* **E**.
N Rare at Nairn, *A. Falconer* (Gordon).
I Not uncommon in Glen Urquhart, 1885 (Aitken).

P. recta L. *Sulphur Cinquefoil*
Introduced. River shingle and waste land.
M Farm track, Sandyhillock, Calcots, 1971 *E. R. Grigor* **E**!; shingle of River Spey between Dandaleith and Craigellachie, 1974 *G. Shepherd* **E**!

P. norvegica L. *Ternate-leaved Cinquefoil*
Introduced.
M Forres, 1871 *JK* **FRS**; station yard Knockando, 1956 **ABD**.
I (Druce, *Com. Fl.*)

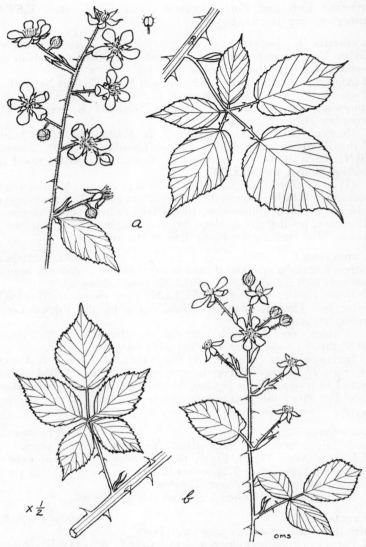

Fig 17 a, **Rubus leptothyrsos**; b, **R. septentionalis**.

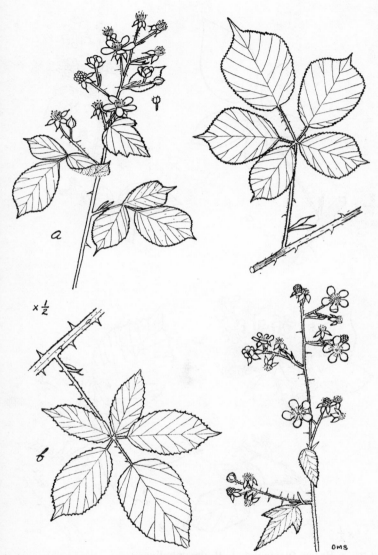

Fig 18   a, **Rubus mucronulatus**; b, **R. radula**.

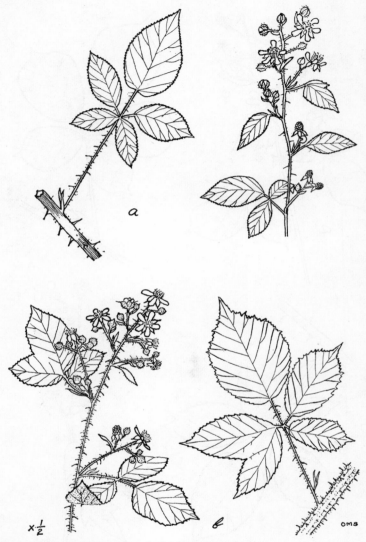

$x \cdot \frac{1}{2}$

Fig 19 a, **Rubus furvicolor**; b, **R. dasyphyllus**.

P. inclinata Vill.
P. canescens Besser
Garden out-cast on the sea bank by the golf clubhouse at Nairn,
1930, collected 1961 **E,ABD**.

P. tabernaemontani Aschers.   *Spring Cinquefoil*
Recorded for Ardclach, Cawdor and Nairn by Lobban but must be
considered a doubtful record in the absence of a voucher specimen.

P. crantzii (Crantz) G. Beck ex Fritsch   *Alpine Cinquefoil*
Native. Rare on mountain rock ledges on basic soil.
I   Very few plants on a ledge of mica schist above Glen Feshie,
1950 *C. D. Pigott* **CGE**; SE side of Mealfuarvonie, 1961 *D. A.
Ratcliffe*!; low base-rich rocks on the Fara, Dalwhinnie, 1973
R. McBeath et al.

P. erecta (L.) Rauschel   *Tormentil*
    Subsp. **erecta**
Native. Abundant on moors, in short grassland and birch woods,
ascending to 1,000 m in the Cairngorms.
M Innes House, Urquhart, 1794 *A. Cooper* **ELN**; Darnaway Forest,
herb. *Borrer* **K**; very common (Gordon); Forres, 1871 *JK* **FRS**;
throughout the county 1954; Càrn Dearg Mòr near Aviemore,
1974 **E**; shingle of River Spey at Aviemore, 1974 **CGE**; steep bank
under birch in the gorge at Huntly's Cave, Dava Moor, 1974 **BM**.
N (Stables, *Cat.*); common, Ardclach, 1883 *R. Thomson* **NRN**;
pine wood at Milton of Kilravock, 1974 **E**; banks of the River
Findhorn at Dulsie Bridge, 1974 **BM**.
I   Inverness, 1832 Pitmain and Dalwhinnie, 1833 (Watson, *Loc.
Cat.*); Culloden, 1832 *I. K.* **E**; Coire an Lochain of Cairn Gorm,
1953 *A. Melderis* **BM**; moorland in Coire Laogh of Cairn Gorm
and by Gaick Lodge, 1974 **E** etc.

    Subsp. **strictissima** (Zimm.) Richards
Differs from subsp. *erecta* in being usually more robust, stiffly erect,
exceeding 150 mm in length; radical leaves coarsely and obtusely
serrate, the stem-leaves coarsely and acutely serrate to base and
exceeding 20 mm in length; leaflets oblong-lanceolate to obovate;
stipules long, usually exceeding 10 mm, divided almost to base;
peduncles exceeding 20 mm in fruit, erect, stiff; flowers sparse,
exceeding 11 mm in diameter; petals exceeding 4·5 mm in length
(Richards, *Watsonia*, **9**: 301–317 (1973)). Map 90.
M Boat of Garten, 1888 *G. C. Druce* **OXF**; woods at Castle Grant,
1895 *C. Bailey* **OXF**; birchwood at Aviemore, 1904 *R. S. Adamson*
BM; moor near Bridge of Brown, 1974 **E**; on the island in Lochin-
dorb, 1974 **E**; in grass by the Dorback Burn at Glenerney House,
Dunphail, 1974 **E**; on tussocks in Dyke Moss near Darklass Farm,
Dyke, 1974 **BM**; grass field by the Westerton Burn, Orbliston, 1974
CGE.
N Moor at Glengeoullie, 1974 **E**; bank by Fornighty Ford, 1974
E,K; near a small moorland pool at Urchany, 1974 **CGE**.

7

I Pine wood at Loch an Eilein, 1955 *E. R. Laundon* **BM**; Coire an Lochan of Cairn Gorm, 1953 *A. Melderis* **BM**; near the White Lady shieling on Cairn Gorm, *A. J. Richards* **OXF**; moor near Ceannocroc Bridge, Glen Moriston, 1974 **E**; moor below Creag Lundie, Glen Moriston, 1975 **E**; Sgùrr na Muice, Monar, 1975 **BM**; dry bank near Kiltarlity, 1975 **CGE**; on moorland by Creag nan Clag, by Loch nan Lann at Knockie and in Glen Markie.

P. anglica Laichard has been recorded for Moray by Burgess and for Easterness by Trail. In the absence of voucher specimens these records must be treated with some doubt. It is possible there was confusion with *P. erecta* subsp. *strictissima*.

P. reptans L.   *Creeping Cinquefoil*
Doubtfully native in north Scotland.
M Dune bank E of Hopeman Harbour, 1966 *A. J. Souter* **E**.
I Waste land near the rubbish tip at Longman Point, Inverness, 1954.

Sibbaldia L.
S. procumbens L.   *Sibbaldia*
    Potentilla sibbalii Haller fil.
Native. A dwarf compact arctic-alpine with bluish-green leaves and dull yellow flowers. Frequent on scree, grassy slopes and rock crevices on the higher mountains.
I Dalwhinnie, 1833 (Watson, *Loc. Cat.*); abundant in one or two places, Badenoch and Mamsoul (Gordon); Braeriach, Glen Einich, 1892 *AS* **E,BM**; near Kincraig, 1891 *AS* **BM**; Coire Chùirn, Drumochter, 1911 *ESM* **E**; Tom a' Chòinich, Glen Affric, 1947 *C. C. Townsend* **K**; common in all the corries of Cairn Gorm, also in Coire nan Laogh in Glen Banchor, Sgòr Bhothain at Gaick, Sgùrr na Lapaich in glens Affric and Strathfarrar, Meall a' Chaoruinn Mòr near Garva Bridge, Coire a' Bhèin and The Fara etc.

Fragaria L.
F. vesca L.   *Wild Strawberry*
Native. Common on dry banks and among stones and rocks, usually on base-rich soil. Map 91.
M Frequent but more abundant in the upper district of the Province (Gordon); Greshop, 1850 *JK* **FRS**; Dunphail (Aitken); wood at Brodie Castle, 1963 **CGE** and throughout the county.
N (Stables, *Cat.*); Ardclach, Cawdor and Nairn, 1900 (Lobban); by the River Findhorn at Dulsie Bridge and Glenferness, woods at Lethen and Geddes, gorge of the Allt Dearg at Cawdor, 1954; gully of the Allt Breac on the moor at Drynachan, 1973.
I Dalwhinnie, 1833 (Watson, *Loc. Cat.*); Meadowbank woods, 1855 *HC* **INV**; common (Aitken); Kincraig, 1891 *AS* **BM**; Dores, *AJ* **ABD**; railway embankment Dalwhinnie, cliffs at the Plodda Falls, Guisachan, steep bank by the Falls of Doe, Glen Doe etc., 1954; on scree in a birch wood near Gaick Lodge, 1973.

F. moschata Duchesne *Hautbois Strawberry*
Garden escape or planted.
N Ardclach and Cawdor, 1900 (Lobban).
I Planted in the woods at Achnagairn House, Kirkhill, 1955.

F. × ananassa Duchesne *Garden Strawberry*
Garden escape, often naturalised on railway embankments.
M By the railway line near Elgin and by the bridge at Garmouth, 1953; 1 mile E of Findhorn on the dunes, 1955; railway embankment near Rothes, 1967 **ABD**; in shade by the river at Nethybridge.
N Banks of the River Findhorn at Coulmony, 1954.
I Railway embankment at Dalwhinnie, 1955 and in a wood at Moy Hall, 1975.

Geum L.

G. urbanum L. *Herb Bennet, Wood Avens*
Native. Common in woods and by rivers in the lowlands, less so in the highlands. Map 92.
M Frequent at Pluscarden, on the N side of Ladyhill and at Cothall, *J. G. Innes* (Gordon); Waterford, 1900 *JK* **FRS**; S of Forres station, 1953 *A. Melderis* **BM**; wood by the Muckle Burn, Brodie, 1963 **BM,CGE**, and throughout the county.
N (Stables, *Cat.*); common at Ardclach, 1887 *R. Thomson* **NRN**; woods at Cawdor Castle, Glenferness, Lethen etc., 1954.
I Ness Islands, Inverness, 1854 *HC* **INV**; in shade by the Plodda Falls, Guisachan, 1947; wooded slopes below the cliffs of Creag Dhubh, Laggan, cliff at Loch Killin, woods by Loch Ness etc., 1954.

G. macrophyllum Willd.
Introduced. Plant more robust than *G. urbanum* and flowers larger of a deeper yellow in colour.
M Waste land at Delfur, Boat o' Brig, 1976 **E**.
I Roadside verge near an old saw-mill near Balblair, Kirkhill, 1961 *M. Cameron*; do. 1962 *MMcCW* **E,BM**; track by the old kirkyard at Kilmorack, opposite the keeper's lodge on the roadside and behind the walled garden of Beaufort Castle, 1972; track by the Black Bridge at Kilmorack, woodland path near the road junction at Dunballoch near Beauly and by the track to the Ferry at Lochend, 1975.

G. rivale L. *Water Avens*
Native. Common by burns and rivers and on mountain ledges up to 1,100 m near the summit of Ben Alder. Pl. 7, Map 93.
M Innes House, Urquhart, 1794 *A. Cooper* **ELN**; frequent (Gordon); Greshop, Forres, *J. G. Innes* **ABD**; woods at Sanquhar House, Forres, 1856 *JK* **FRS**; Craigellachie, 1940 *E. Crapper* **STA**; by the Muckle Burn, Brodie etc., 1954.
N (Stables, *Cat.*); banks of the River Findhorn at Daltra, 1882 *R. Thomson* **NRN**; wood by the Muckle Burn at Lethen, 1961 **ABD,BM**; flush on the raised beach at Maviston, 1963 **E**.
I Inverness, 1832 and Dalwhinnie, 1833 (Watson, *Loc. Cat.*);

Kingussie, 1841 *A. Rutherford* **E**; Ness Islands, 1854 *HC* **INV**;
Bunchrew and Holm (Aitken); summit of Ben Alder, 1956; rock
ledges on Sgùrr na Lapaich, among boulders in the Ryvoan Pass
etc., 1954.

C. rivale× urbanum = G.× intermedium Ehrh.
Frequent where both parents occur.
M Moray, *G. C. Druce* (*ASNH* **26**: 111 (1888)); Dyke and Forres
(Burgess); by the River Findhorn at Darnaway, 1952 **BM,CGE**;
wood at Burgie, 1963 **CGE**; banks of the Findhorn opposite
Coulmony House, 1972 **E**; small belt of woodland on the path to
the River Spey by Baxter's Factory, Fochabers, 1973 **E**; wood near
Rothes, and in the gorge by Huntly's Cave on Dava Moor.
N Cawdor, 1887 *R. Thomson* **NRN**; by the Muckle Burn, Lethen,
1961 **ABD**.
I Wood by the River Enrick, Drumnadrochit, 1947 *E. J. Gibbons*
BM!; by the track near the Moniack Burn, Achnagairn, Kirkhill
1962 **E,CGE**; in shade by the Breakachy Burn in Strathglass and
in Reelig Glen, 1975.

Dryas L.
D. octopetala L.   *Mountain Avens*
Native. On ledges, scree and grassy slopes on base-rich soil. Pl 7.
I Rocks NE of Dalwhinnie (Balfour, 1868); head of a glen 3 miles
N of Truim (Boyd); Coire Chùirn, Drumochter, 1911 *ESM* **E,BM!**;
Glen Feshie, 1950 *C. D. Pigott* **CGE!**; rock ledges at the head of
Loch Einich, 1972 *R. McBeath!*; burn gully of A' Chaornich, Gaick,
1973.

Agrimonia L.
A. eupatoria L.   *Common Agrimony*
Native. Rare on grassy banks and roadside verges. Leaves slightly
glandular beneath, fruit receptacle obconic, deeply grooved, the
basal spines spreading laterally.
M Abundant in one or two places, at Linksfield and Cothall
(Gordon); Cothall, 1860 *JK* **FRS**; parishes of Forres, Rafford and
Rothes (Burgess); bank by the Spey bridge Craigellachie, 1930;
near Dallas Dhu distillery, Forres, 1968 *R. M. Suddaby*; single plant
in rough roadside vegetation near Seapark, Kinloss, 1972 *R. M.
Suddaby*.
N (Stables, *Cat.*); waste ground Blackhills, Auldearn, *J. B. Brichan*
(Watson); bank under pines by the track to Lochloy, 1975 **E**.
I Near the Manse, Ardersier, 1844 *A. Croall* (Watson); Ness
Islands and by the Falls of Divach, Lewiston (Aitken); roadside
verge Crask of Aigas, 1930 *M. Cameron*; do. 1975 *MMcCW* **E**.

A. procera Wallr.   *Fragrant Agrimony*
  A. odorata auct.
Native. Leaves with many sessile glands beneath; fruit receptacle
companulate, shallowly grooved and basal spines deflexed. Plant
fragrant.

I Near Milton of Drumnadrochit, 1947 *A. Campbell & Mrs Campbell of Jura* **BM**; grass verge of the hill road to Achmony at Drumnadrochit, 1976 **E**.

Alchemilla L.

A. alpina L.   *Alpine Lady's Mantle*
Native. Abundant on rocks and scree, by burns and in mountain grassland. Often washed down rivers and becoming established on shingle. Map 94.
M Abundant in one or two places, by the Spey opposite the Kirk at Speymouth, in 1825, and on Dava Moor and at Daltulich (Gordon); shingle at St. John's Mead, Darnaway, 1953 *A. Melderis* **BM**; on shingle and rocks the length of the rivers Spey and Findhorn, 1954; gravel bed in the Culbin Forest, 1964 **CGE**.
N (Stables, *Cat.*); Ardclach by the River Findhorn, 1881 *R. Thomson* **NRN**; on rocks by the River Findhorn at Glenferness, 1960 **ABD**, and on the moors at Drynachan.
I Falls of Foyers, 1829, Badenoch, 1831 and the sources of the Spey, *Dr Bostock* (Gordon); Dalwhinnie and Pitmain, 1833 (Watson, *Loc. Cat.*); by the falls of Foyers, 1837 *WAS* **E**; Kincraig, 1882 *J. Groves* **BM** and in 1891 *AS* **BM**; Bunchrew and by the Dochfour Burn (Aitken); throughout the highlands, 1954.

A. conjuncta Bab.
Garden escape
M Roadside verge near Grantown station, 1968 *A. Langton*.
N Roadside verge by the old ford at Achavraat, 1960 **ABD**; planted on the walls of Cawdor Castle.

A. filicaulis Buser
Native. Common on base-rich grassland and mountain slopes. At least the lower part of stem and petioles with spreading hairs, leaves hairy on both surfaces. All *MMcCW* records have been determined by S. M. Walters.

Subsp. **vestita** (Buser) M. E. Bradshaw
Stems and pedicels hairy throughout. Fig 20e, Map 95.
M Banks of the Spey at Aviemore, 1922 *C. E. Salmon* **BM**; grass field at Grantown station, 1963 **CGE**; banks near Smallburn, Longmorn and Rothes; in short turf at Bridge of Brown, 1967 **CGE**; by the moorland track at Slochd, 1973 **E**; pasture at Avielochan, 1973 **BM**.
N Bank at Cawdor Castle, 1956 **ABD**; in short grass at Dallaschyle, 1956; Holme Rose, 1962 **ABD,CGE**; grass verge of the moor on Lethen Bar, 1964 **ABD**.
I East bank of the Spey near Kincraig and by Loch an Eilein, 1922 *C. E. Salmon* **BM**; birch wood, Inverdruie, 1922 *C. E. Salmon* **OXF**; between Mam Suim and Creag nan Gall in the Cairngorms, 1939 *E. F. Warburg* **BM**; railway embankment at Culloden, 1962 **K,CGE**; grass slope below Creag Dhubh near Laggan, 1973 **E**;

base-rich grassland at Essich, 1975 **E**; on scree in the Ryvoan Pass and in short grass at Upper Gartally, 1975 **BM**.

Subs. filicaulis

Differs from subsp. *vestita* in the leaves being less hairy and the upper part of the stem and pedicels being glabrous. Fig. 20c, Map 96.

M East side of Loch Ban at Boat of Garten and at Avielochan, 1922 *C. E. Salmon* **BM**; grass bank at Kellas and near Dunphail, 1956 **ABD**; Lynemore (inter *vestita* et *filicaulis*) 1957 **E**; banks of the River Spey at Grantown, 1972 **ABD**; short turf by the lime kiln at Fae near Dorback, 1972 **BM**; grass field above Huntly's Cave on Dava Moor, 1973 **K**.

I East bank of the Spey below Kincraig, 1922 *C. E. Salmon* **BM**; Larig Ghru, 1938 *E. F. Warburg* **BM**; Glen Strathfarrar, 1954 **CGE**; Coire an Lochain of Cairn Gorm, in Coire Chùirn at Drumochter, at Kincraig and Invermoriston etc., 1954; grass bank of a small burn on Sgùrr na Lapaich at 1,000 m, 1971 **E**; field at Newtonmore, 1972 **ABD**; gully of the burn at Gaick Lodge, 1973 **E**; grass slope in Coire nan Laogh in Glen Banchor, 1973 **BM**; recorded for all the 10 km grid squares at the head of Glen Affric, 1974 *R. W. M. Corner et al.*

A. xanthochlora Rothm.
A. pratensis auct.
Lady's Mantle

Native. Frequent in grassland, on roadside verges and river banks. Stems and petioles clothed with spreading hairs; leaves glabrous (or nearly so) above, hairy on the veins (and sometimes surface) beneath. Fig. 20d, Map 97.

M Morayshire, *G. Wilson* **ABD**; frequent by the side of the Blackburn, 1824 and by the Spey below the bridge at Fochabers (as *A. vulgaris* agg.) (Gordon); banks of the River Spey at Aviemore, 1922 *C. E. Salmon* **BM**; Garmouth, 1953 *NDS* **BM**; grass verge Kinloss, by the Spey at Knockando 1954; and in Greshop Wood near Forres, 1956 **ABD**.

N Lawn at Cawdor Castle, *WAS* (Gordon); banks of the River Findhorn at Coulmony, 1960 **K**; by the river at Nairn, roadside at Righoul, river bank at Ardclach and roadside verge at Galcantray.

I Easterness (as *A. vulgaris* var. *pratensis* Schmidt (Trail, *ASNH* **15**: 33 (1906)); Larig Ghru, 1936 *E. F. Warburg* **BM**; Milton of Drumnadrochit, 1947 *C. C. Townsend* **K**; roadside verge by the east lodge of Ness Castle, among rocks by the road above Balcraggan, by the A95 at Skye of Curr, river bank at Kincraig, roadside verges at Culloden, Erchless in Strathglass, Tomatin and Ardersier etc.

A. glomerulans Buser

Native. Mountain ledges. Stems and petioles densely clothed with appressed silky hairs; leaves more or less hairy on both surfaces although hairs sometimes confined to veins on lower surface. Fig 20b, Pl. 4.

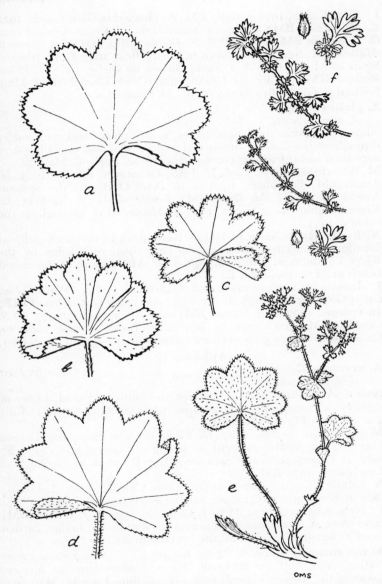

Fig 20 a, **Alchemilla glabra**; b, **A. glomerulans**; c, **A. filicaulis** subsp.
filicaulis; d, **A. xanthochlora**; e, **A. filicaulis** subsp. vestita; f, **Aphanes
arvensis**; g, **A. microcarpa**.

I Glen Einich, 1916 *J. Roffey* **BM**; A' Phocaid in Glen Einich, 1922 *C. E. Salmon* **BM**; Coire an t-Sneachda of Cairn Gorm, 1922 *C. E. Salmon* **BM** and 1938 *E. F. Warburg* **BM** and 1942 *Chaworth Masters* **BM**; Sgùrr na Lapaich in Glen Affric, 1947 *S. M. Walters*; Sgùrr na Muice, 1956 *A. A. Slack*; Gleann na Ciche, 1961 *H. Milne-Redhead* **HMR**; Coire na Cralaig and Coire Ghaidheil of Mam Sodhail in Glen Affric, 1974 *R. W. M. Corner et al.*

A. glabra Neygenf.

A. alpestris auct.

Native. Abundant by rivers, burns, grassy places and rock ledges throughout the district. Stems and petioles glabrous or with sparse appressed hairs; leaves glabrous or nearly so. Fig 20a, Map 98.

M Waterford, Forres, 1898 *JK* **FRS**; Grantown, 1925 herb. *A. H. Maude* **BM**; Fochabers, 1926 *G. C. Druce* **OXF**; by the Spey at Aviemore, 1927 *I. M. Roper* **BM**; shingle of the W bank of the River Findhorn near Forres, 1953 *A. Melderis* **BM**; throughout the county, 1954.

N By the River Nairn at Holme Rose, flush in the burn gully at Cawdor, Glenferness and Ardclach etc., 1954; rock ledge by the River Findhorn at Dulsie Bridge, 1961 **ABD**; bank of the raised beach at Maviston, 1968 **E**.

I Inverness, 1832 Pitmain and Dalwhinnie, 1833 (Watson *Loc. Cat.*); Glen Einich, 1893 *AS* **E**; Kingussie, 1906 *G. C. Druce* **OXF**; Inverdruie, 1922 *C. E. Salmon* **BM**; near Affric Lodge, 1947 *M. S. Campbell* **BM**; grass slope by Ben Alder Lodge, among rocks on Creag Dhubh, burn gully at Gaick Lodge and throughout the district.

Aphanes L.

A. arvensis L.   *Parsley Piert*

Alchemilla arvensis (L.) Scop.

Native. A common annual of waste and cultivated land. Lobes of stipules surrounding the inflorescence triangular-ovate; fruit 2·2–2·6 mm. Fig 20f, Map 99.

M Chapelton, Forres, 1887 *JK* **FRS**; track at Sanquhar House, Forres, 1956 **ABD**; cornfield at Lochinvar near Elgin, 1957 **E**; field at Lynemore near Grantown, 1957 **E**; track at Slochd N of Carrbridge, 1973 **BM**.

N (Stables, *Cat.*); common at Nairn and Cawdor, 1900 (Lobban); field at Kinsteary, 1954; waste land at Kingsteps, 1966 **ABD**.

I Near Kincraig, 1891 *AS* **BM**; Cannich, 1936 *A. J. Wilmott* **BM**; farm track at Newtonmore, in gravel at Glen Feshie Lodge, garden weed at Kingussie and in all the lowland areas.

A. microcarpa (Boiss. & Reut.) Rothm.   *Slender Parsley Piert*

Native. Abundant on acid soil. More slender than the former. Lobes of the stipules oblong; fruit 1·4–1·8 mm. Fig 20g, Map 100.

M Garden wall Fochabers, 1922 ? *coll.* **GL**; near Moy House, Forres, 1953 *NDS* **BM**; station yard at Carron, 1956 **ABD**; station yard at Knockando, 1963 **BM,CGE**; shingle of the River Spey at Garmouth, 1964 **CGE**; by the River Dulnain W of Carrbridge, 1972 **E**.

N Moorland tracks at Drynachan, Cawdor and Cran Loch etc., 1954; roadside verge at Dulsie Bridge, 1960 **ABD**; sandy ground on the Nairn Dunbar golf course, 1962 **CGE**; waste ground by the Ford at Achavraat, 1973 **E**.
I Kincraig, 1891 *AS* **BM**; Loch Ness, 1904 *G. T. West* **STA**; Milton of Drumnadrochit, 1956 *UKD* **E**; track by the River Spey at Newtonmore, 1973 **BM**; shingle verge of Loch Ness by Urquhart Castle, 1971 **E**.

Sanguisorba L.

S. minor Scop. subsp. **muricata** Briq.   *Fodder Burnet*
   Poterium polygamum Waldst. et Kit.
Introduced with foreign seed; formerly sown as a crop in pastures. Erroneously recorded by Gordon and Burgess as *Poterium sanguisorba* L.
M Introduced on Bareflat hills near Elgin, 1857 *Mr Martin* and N side of Loch Spynie near the railway, 1865, *Mr Grant* (Gordon, *annot.*); railway line Spynie, 1867 *Mr Grant* **FRS**; Lossiemouth, 1869 ? *Grant* **E**; waste ground between the harbour and the cemetery at Lossiemouth, 1957 **K**.
N Dry bank by the A96 at Broombank, Auldearn, 1960 **ABD**.

Acaena Mutis ex L.

A. ovalifolia (J. R. & G. Forst.) Druce
Introduced by Canadian lumbermen. The record for *A. anserinifolia* in the *Atlas* should be referred to this species.
I Abundant on the hill road to Blackfold, Dochgarroch, 1956 *N. M. Pritchard & J. C. Newbould* **OXF**; do 1961 *MMcCW* **E,ABD,K,CGE**; by a small burn on the hill above Lurgmore, Lochend, 1975.

Rosa L.

In the following account the *Key* to Rosa has been supplied by Dr R. Melville in accordance with A. H. Wolley-Dod, *A revision of the British roses*, *J. Bot. Lond.*, **68** and **69** *suppl.* (1930–31).
   In general the order of species follows that of Dandy (1958) except that *R. dumetorum* is recognised in addition to *R. canina* and *R. coriifolia* in addition to *R. afzeliana*. The last named is used instead of *R. dumalis* Bechst. Hybrids are in alphabetical order.
   Owing to the great taxonomic confusion which exists within the genus only records determined by R. Melville have been included with the exception of a few reliable records which are preceded by an asterisk. Records cited by G. C. Druce (*J. Bot.* 1888) and W. Moyle Rogers (1904) have been omitted.
   In a recent (1976) survey of the roses of the district by R. Melville and the author, several unnamed new variants were collected. Specimens have been lodged in the herbaria of the Royal Botanic Gardens at Edinburgh and Kew. The species shown in the distribution maps include all the variants within the species.
   Leaflet shape and serrature are shown in Fig 21.
   Prickle types and hip shapes are shown in Fig 22.

Key to British Rosa species

1. Styles exserted and united into a column.
    2. Weak and scrambling shrub, petals chalky white, column
       slender as long as the stamens, on a flat disc   . *R. arvensis*
    2. Strong arching shrub, petals pink or white, colum plump,
       shorter than the stamens on a conical disc   .   *R. stylosa*
1. Styles not united into a column although sometimes long
   exserted.
    3. Flowers solitary, without bracts, sepals entire, leaflets small
       (5–12 mm) serration simple (3)–4–5 pairs of leaflets, stems
       densely prickly and bristly with acicles   *R. pimpinellifolia*
       Flowers with 1 or more bracts, serration simple or double,
       leaflets 2–3(–4) pairs, stems with stout prickles, rarely
       unarmed or if with dense acicles then leaves with fruity
       fragrance   4
    4. Leaflets glabrous below, sometimes with a few non-odorous
       glands along the midrib.   5
        5. Styles glabrous, hispid or rarely woolly forming a ± loose
           group, orifice of disc about $\frac{1}{5}$ its width   .   .   *R. canina*
        5. Styles forming a dense rounded woolly mass, orifice of
           the disc about $\frac{1}{3}$ its width, sepals spreading to erect after
           flowering, finally ± erect .   .   .   .   *R. afzeliana*
    4. Leaflets tomentose or hairy below, at least on the midrib   6
        6. Leaflets with prominent glands on the lower surface,
           giving a sharp, fruity (sweetbriar) odour when
           rubbed   7
            7. Leaflets cuneate at the base, pedicels glabrous
                                                        R. agrestis
            7. Leaflets rounded at the base, pedicles glandular
               hispid   8
                8. Stems erect, prickles unequal, primary stems
                   often ± densely acicular, styles hispid *R. rubiginosa*
                8. Stems arching, prickles stout, uniform, styles
                   glabrous, fruit usually contracted below the
                   disc .   .   .   .   .   .   *R. micrantha*
        6. Leaflets glandless or with ± inconspicuous glands which may
           produce an aromatic, turpentine like odour on rubbing   9
            9. Leaflets with simple hairs only, (rarely a few non-odorous
               glands on the midrib) simply or bi-serrate   .   .   10
                10. Styles glabrous, hispid or ± woolly, orifice of disc
                    about $\frac{1}{5}$ its width. Sepals reflexed (erect in one
                    var.) .   .   .   .   .   .   *R. dumetorum*
                10. Styles woolly in a dense flat head, orifice of disc
                    about $\frac{1}{3}$ its width. Sepals erect or spreading after
                    flowering .   .   .   .   .   .   *R. coriifolia*
            9. Leaflets usually glandular biserrate and usually with
               some glands below at least on midrib .   .   .   11
11. Orifice of the disc about $\frac{1}{5}$ of its width   12

12. Stems with stout, broad, strongly hooked prickles, sepals short broad and pinnate, generally not glandular on the back . . . . *R. obtusifolia*

12. Stems with slender prickles, usually straight or occasionally arching. Sepals elongated not strongly pinnate, sometimes simple, glandular setose on the back. Pedicels long, slender glandular hispid . . . . . *R. tomentosa*

11. Orifice of the disc about ⅓ of its width. Leaflets densely soft tomentose. Stigmas densely villous 13

13. Sepals erect, simple, persistent, the tapering tips expanded into a small blade. Prickles slender, straight. *R. villosa*

13. Sepals erect or spreading, not persistent generally with small lateral pinnae.
Stems often zigzag. Prickles slender, straight, inclined or slightly curved . . . . . . *R. sherardii*

Warning: This key applies to the species only. In Wolley-Dod's treatment *R. canina* sections *andegavenses* and *scabratae*, *R. dumetorum* sections *deseglisei* and *mercicae* and certain 'varieties' included under other species are hybrids. The inclusion of these hybrid forms within the species has obscured the limits of the species. R. Melville (1974).

R. arvensis Huds.   *Field Rose*
Doubtfully native in north Scotland.
M Parishes of Urquhart, Speymouth and Bellie (Burgess). Probably an error. No voucher specimen.

R. pimpinellifolia L.   *Scots Rose, Burnet Rose*
R. spinosissima auct.
Native. Frequent in bushy places and on dunes near the sea. Rare inland. Map. 101. Pl 8a, 8b.

Var. pimpinellifolia
M Frequent at Bankhead near Garmouth (Gordon); Forres, 1856 *JK* **FRS**; roadside verge by Dalvey Cottage near Forres, 1920; *roadside near Cothall quarry, 1963 **CGE**; waste land by the River Spey at Garmouth, 1964; by a small burn at Crossley near Pluscarden; roadside E of Elgin near Waulkmill, 1976 *RM & MMcCW* **E,K**; bank by the River Nethy at Nethybridge, 1976 *RM & MMcCW* **E,K**; shingle bed at Pitgavenny and by the Orton road, *R. Richter*.
N (Stables, *Cat.*); rare at Achagour (Thomson); *Nairn, 1911 *P. Ewing* **GL**; *dunes on the Nairn Dunbar golf course, 1960 **ABD**; roadside verge and in scrub near Lochloy House and by the fishing hut at Loch Belivat; roadside verge at Cothill Farm, 1976 *RM & MMcCW* **E,K**.
I By Loch Ness (Gordon); sandy bank by the A96 at Tomhomie W of Nairn, 1925; Creag nan Eun, Invermoriston, 1961 *D. A. Ratcliffe*; by the sea in scrub, Alturlie Point, 1975; garden relic by an old cottage in the Pass of Inverfarigaig and by an old cottage near Skye of Curr, 1976 **K**.

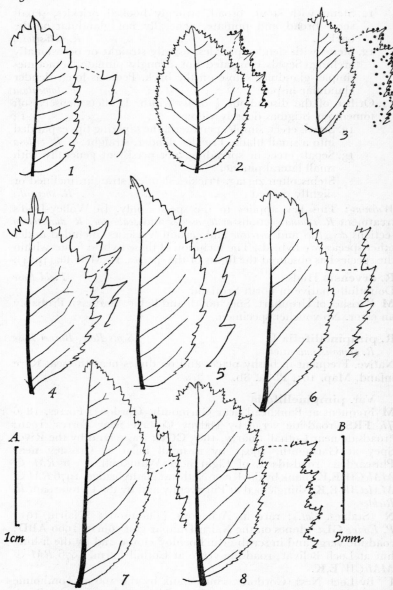

Fig 21 **Leaflet shape and serrature.** On account of the variation in the shape of the leaflets only leaflets of the distal pair are illustrated.

1 **Rosa pimpinellifolia**	5 **R. afzeliana** var. **afzeliana**
2 **R. rubiginosa**	6 **R. coriifolia** var. **watsonii**
3 **R. micrantha**	7 **R. villosa** var. **mollis**
4 **R. canina** var. **sylvularum**	8 **R. sherardii** var. **suberecta**

A   Scale for leaflets   B   Scale for serrature

Fig 22   **Prickle types and hip shapes.**

1–5 *Main stem armature*

1 **Rosa pimpinellifolia**
2 **R. rubiginosa**
3 **R. rubiginosa**
4 **R. sherardii** var. **omissa**
5 **R. villosa** var. **mollis**

6–11 *Branch armature*

6 **R. afzeliana** var. **afzeliana**
7 **R. afzeliana** var. **jurassica**
8–9 **R. coriifolia**
10–11 **R. canina** var. **sylvularum**

12–14 *Rhachis armature*

12 **R. canina** var. **sylvularum**
13 **R. sherardii** var. **suberecta**
14 **R. rubiginosa**

15–23 *Hip shapes* (not to scale)

15 globose. 16, subglobose. 17, ellipsoid. 18, narrow ellipsoid. 19, obovoid. 20, ovoid. 21, urceolate. 22, pyriform. 23, turbinate.

24–29 *Hip shapes and sepal types*

24 **R. canina** var. **flexibilis.** Sepals reflexed, pinnate, disc conical
25 **R. pimpinellifolia** var. **hispidula.** Sepals simple, erect, pedicels glandular hispid
26 **R. coriifolia** var. **watsonii.** Sepals spreading-erect, outer pinnate, glands ciliate
27 **R. coriifolia** var. nov. Sepals erect, pinnate, hips long ellipsoid
28 **R. villosa** var. **mollis.** Sepals simple erect, backs densely glandular, a few glandular setae on pedicel
29 **R. villosa** var. nov. Sepals erect, glandular on backs, terminal lobes enlarged, hips turbinate with a few glandular setae

1–14 Scale A   24–29 Scale B

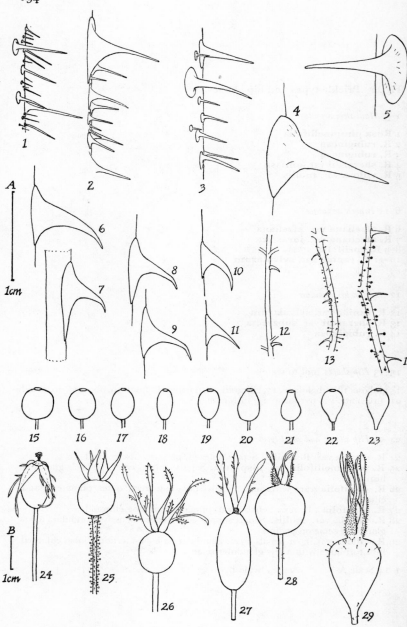

Fig 22

Var. **hispidula** Rouy
A variant with acicles on the pedicels.
M Hedgerow W of Cotterton Cottage near Brodie, 1976 **K**.
I Steep bank of the raised beach between Kirkton and Baddock
E of Fort George, 1976 **K**.

R. pimpinellifolia L.× **canina** L. = **R.**× **hibernica** Templeton
var. **grovesii** Baker
M *Woods at Dunphail, 1886 *ESM* **CGE**.

R. pimpinellifolia L.× **rubiginosa** L. = **R.**× **cantiana**
(W.–Dod) W.–Dod
M Roadside verge on the Nethybridge road a short distance from
the river bridge at Broomhill, 1975 **E**.

R. pimpinellifolia L.× **rubiginosa** L.× **sheradii** Davies = **R.**×
perthensis Rouy
M Near Forres, pre-1924 *J. W. Heslop-Harrison* (J. R. Mathews,
BSE News **18**: 15 (1976)).

R. pimpinellifolia L.× **sherardii** Davies = **R.**× **involuta** Sm.
Pl 9e.
M Woods near Forres, 1834 *J. G. Innes* and in 1856 *JK* **FRS**.
N *Near Nairn (as *R. pimpinellifolia*× *tomentosa*) 1898 *ESM* **E**;
hillside N of Ferness Bridge near the road to Cairnglass (as *R.
pimpinellifolia* var. *hispidula* Rouy), 1972 **E** and collected there in 1976
RM & MMcCW **E,K**; in scrub by the road opposite Lochloy House,
1976 *RM & MMcCW* **E,K**.

R. pimpinellifolia L.× **villosa** var. **mollis** Sm = **R.**× **sabinii**
Woods
M *By the outlet of the Fochabers Burn to the River Spey at
Fochabers (as *R. arvensis*) 1909 ? *coll.* **E**; banks of the Fochabers
Burn (probably the same locality and the same plants), 1976
RM & MMcCW **E,K**.
I Lane at Achnagairn Farm, Kirkhill, 1963 **K**.

R. rugosa Thunb.   *Japanese Rose*
Introduced. A garden relic.
N *Grass verge near the sea by the Nairn golf clubhouse, 1962
E,ABD.
I Bank by the A9 at Craggie, 1930; roadside verge at Foyers, 1956;
planted by the A9 near Newtonmore, 1958; at Bunchrew and
Dunain, 1970 *M. Barron*.

R. canina L.   *Dog Rose*
Map 102.

Var. **andegavensis** (Bast.) Desf.
A mixture of *R. canina* hybrids has been placed under this name.
I Roadside by Loch Ness S of Invermoriston, 1976 *RM &
MMcCW* **E,K**.

Var. **biserrata** (Mér.) Baker forma **sphaeroidea** (Rip.)
W.–Dod.
I   Banks of Loch Ness S of Bunloit near Drumnadrochit, 1976
RM & MMcCW **E.**

Var. **dumalis** W.–Dod, non (Bechst.) Dum.
M Aviemore (with sub-globose fruits) (Druce, 1888); roadside a
short distance N of Coxton Tower, Llanbryde, 1976 *RM & MMcCW*
E.
I   S side of a small burn at Mid Crocail in Strathglass, 1976
RM & MMcCW **E,K**; banks of Loch Ness N of Inverfarigaig, 1976
RM & MMcCW **E.**

Forma **viridicata** (Pug.) Rouy
I   Drumnadrochit, in scrub on the hill road to Achmony, 1976 **E.**

Var. **flexibilis** (Déségl.) Rouy
I   Grass meadow S of Struy, Strathglass, 1976 *RM & MMcCW*
E,K.

Var. **fraxinoides** (H. Bor.) W.–Dod forma **recognita** Rouy
M Roadside near Brodie, 1961 **E.**

Var. **globularis** (Franch) Dum.
I   Raised beach N of the rubbish tip between Ardersier and Fort
George, 1976 *RM & MMcCW* **E,K**; banks of Loch Ness N of
Inverfarigaig, 1976 *RM & MMcCW* **E.**

Var. **lutetiana** (Lem.) Baker
M Banks of the River Findhorn at Broom of Moy, 1954 **K.**
N   Road junction near the ford at Fornighty, 1976 *RM &
MMcCW* **E,K.**
I   Waste ground on the raised beach near the village of Ardersier,
1976 *RM & MMcCW* **E.**

Forma **lasiostylis** Borb.
I   Between the road and Loch Ness, in thick scrub, S of Lurgmore,
1976 *RM & MMcCW* **E,K.**

Var. **medioxima** (Déségl.) Rouy
M By the entrance to the Culbin Forest at Wellhill, 1965 **K.**

Var. **rhynchocarpa** (Rip.) Rouy
M By the old Spey Bridge road, opposite Tignacoile, at Grantown,
1976 *RM & MMcCW* **E,K.**

Var. **sphaerica** (Gren.) Dum.
M Boat of Garten (Druce, *J. Bot.* **26**: 17 (1888)).

Var. **sylvularum** (Rip.) Rouy
Map 102, Pl 8c.
M Bank opposite the entrance to Leuchars House, Lhanbryde, 1976
RM & MMcCW **E,K**; bank of road at Foresterseat near Pluscarden
Priory, 1976 *RM & MMcCW* **K**; scrub at Broom of Moy, 1976

RM & MMcCW **E,K**; moorland verge on the Cromdale road E of
Grantown, 1976 *RM & MMcCW* **E,K**; roadside opposite Brodie
pond, 1976 **K**.
N By a wall on the S side of the road just east of the entrance to
Lethen House, 1976 *RM & MMcCW* **E,K**.
I Top of a wall near Reelig House, 1976 *RM & MMcCW* **E,K**;
hillside on the road to Achmony at Drumnadrochit, 1976 *RM &
MMcCW* **E**; banks of Loch Ness N of Inverfarigaig and opposite
Urquhart Castle, 1976 *RM & MMcCW* **E,K**; grassy hillside in
Glen Liath above Foyers, 1976 *RM & MMcCW* **E**.

Var. **verticillacantha** (Mér.) Baker
M Hedgerow between the village and Cromalt at Dyke, 1976
RM & MMcCW **E**.

R. canina L. × **afzeliana** Fr.
Pl 9f.
I By Loch Ness N of Inverfarigaig, 1976 *RM & MMcCW* **E,K**;
in the scrub by the hill road to Achmony at Drumnadrochit, 1976
RM & MMcCW **E**; shingle beach at Ardersier, by the old Manse
at Bailebeag in Stratherrick and below the wall near the S lodge of
Ness Castle, Inverness, all *RM & MMcCW* **E,K**.

R. canina L. × **coriifolia** Fr.
M In scrub at Broom of Moy and in a ditch opposite Rafford Kirk,
1976 *RM & MMcCW* **E,K**.
N Steep bank above Furness Bridge by the road to Cairnglass, 1976
RM & MMcCW **E,K**.
I In scrub by the hill road to Achmony at Drumnadrochit, at
Crinaglack in Strathglass and by a small burn at Mid Crocail, all
RM & MMcCW **E,K**.

R. canina L. × **sherardii** Davies
I Roadside W of Kiltarlity, 1976 *RM & MMcCW* **E,K**.

R. dumetorum Thuill.
Map 103, Pl 8d.

Var. **dumetorum** forma **semiglabra** (Rip.) W.–Dod
M By the old railway at Millcross near Millbuies (with glandular
petioles), 1976 *RM & MMcCW* **E**; hedgerow between the village
and Cromalt at Dyke, 1976 *RM & MMcCW* **E**; near Sheriffston,
1976 *RM & MMcCW* **E,K**.
I Balmacaan road near Lewiston, 1976 *RM & MMcCW* **E**.

Forma **urbica** (Lem.) W.–Dod
M Aviemore (as *R. canina* var. *urbica*) (Druce, 1888).
I Edge of a birch wood at the junction of two tracks on the
Divach road W of Balmacaan, 1976 *RM & MMcCW* **K**; wood by
Loch Ness S of Lurgmore, 1976 *RM & MMcCW* **E**.

Var. **gabrielis** (F. Ger.) R. Kell.
N Roadside bank near Cothill Farm on the Lochloy road, 1976
RM & MMcCW **E**.

Var. **sphaerocarpa** (Pug.) W.–Dod forma **spinetorum** (Déségl. et Ozan) W.–Dod

I Roadside bank near Ness Castle, 1976 *RM & MMcCW* **E**.

R. dumetorum Thuill. × **afzeliana** Fr

M Top of the bank above the 'island' in the River Spey at Craigellachie (with very small hips), 1976 *RM & MMcCW* **E,K**.

R. coriifolia Fr.
Map 104.

Var. **coriifolia**

M Kinchurdy (as *R. canina* var. *coriifolia*) (Druce, 1888); river bank at Broom of Moy (as *R. caesia* Sm.) 1954 **K**.

N Auldearn, 1835 *J. B. Brichan* (Watson); Holme Rose, 1955; Lethen 1960; Nairn Dunbar golf course, 1960 *I. Blewitt* **K**; by the river at Nairn, 1962 **ABD,K**.

I Kingussie (as *R. canina* var. *coriifolia*) (Druce, 1888).

Forma **frutetorum** (Bess.) W.–Dod

M Near Greshop House, Forres, 1954 **K**; Pluscarden road, 1976 *RM & MMcCW* **E**.

N Banks of the river at Nairn, 1962 **K**.

Forma **oblonga** Chr.

I Field by the keeper's cottage at Stray in Strathglass, 1976 *RM & MMcCW* **E,K**.

Var. **bakeri** (Déségl.) W.–Dod

M Roadside bank on the N side of the road to Nethybridge near Broomhill, 1976 *RM & MMcCW* **E**.

N Rough ground by Loch of the Clans (forma *floribus albus*), 1960 *I. Blewitt* **K**.

I Roadside bank N of the road junction to Foxhole in Glen Convinth, 1976 *RM & MMcCW* **E**; by Loch Ness S of Dores, 1976 *RM & MMcCW* **E**; by Loch Ness N of Inverfarigaig, 1976 *RM & MMcCW* **K**.

Forma **cryptopoda** (Baker) W.–Dod

M Roadside by the A95 near the road to Boat of Garten and by the old Spey Bridge at Grantown, 1976 *RM & MMcCW* **E**.

Var. **caesia** (Sm.) W.–Dod

M Near Greshop House, Forres, 1954 **K**.

Var. **celerata** (Baker) W.–Dod

M Roadside by the A95 near the junction to Boat of Garten, 1976 *RM & MMcCW* **E**.

Var. **pastoralis** R. Kell.

M Near Bishop Croft above Blacksboat in Strathspey, 1976 *RM & MMcCW* **E,K**.

Var. **subcollina** Chr.
M Longmorn, 1929 *K. D. Little* **K**; banks of the River Findhorn at Broom of Moy, 1954 **K**; bank near Greshop House, Forres, 1954 **K**.

Var. **subcoriifolia** (Barclay) W.–Dod.
M Hedgerow opposite Kinloss Nursery (probably this), 1976 *RM & MMcCW* **E,K**; railway embankment by Loch Vaa, Aviemore, 1976 *RM & MMcCW* **E**.
N Kingsteps near Nairn, 1976 *RM & MMcCW* **E,K**.
I Roadside near Reelig House, Kirkhill, 1976 *RM & MMcCW* **E**.

Forma **perpubescens** W.–Dod
M North side of Sheriffston cross-roads, 1976 *RM & MMcCW* **E,K**.

Var. **watsonii** (Baker) W.–Dod
I Kingussie (Druce, 1888); roadside near Pitmain, 1975 **E**.

The following undescribed variants were collected by *RM & MMcCW* in September 1976. Specimens are at **E** and **K**.

Var. nov.
I Passing place by Loch Mhòr near the road to Fenecreich *RM* (76.123) *& MMcCW* (19,174).

Var. nov.
M Near road junction on waste land, between Todholes and Rafford, *RM* (76.140) *& MMcCW* (19,199).

Var. nov. (aff. *subcollina*)
M Roadside E of Rafford, *RM* (76.141).

Var. nov.
I Station road Tomatin, *RM* (76.201) *& MMcCW* (19,272).

R. coriifolia Fr. × **rubiginosa** L.
M Among scrub by the River Findhorn near Seafield near Forres, 1976 *RM & MMcCW* **E,K**.
N Among whins at Kingsteps near Nairn, 1976 *RM& MMcCW* **E,K**; roadside near Milltown N of Ferness, 1976 *RM & MMcCW* **E,K**.

R. coriifolia Fr. × **sherardii** Davies
I Edge of a field near Reelig House, Kirkhill, 1976 *RM & MMcCW* **E,K**; under birch trees at Old Town of Aigas in Strathglass, 1976 *RM & MMcCW* **E,K**.

R. coriifolia Fr. × **villosa** L. var. **mollis** Sm.
I Several bushes by the road at Borlum House E of Tarff bridge at Fort Augustus, 1976 *RM & MMcCW* **E,K**.

R. afzeliana Fr.
Map 105

Var. **afzeliana** Pl 8e.

Var. *reuterii* (God.) W.-Dod

M Roadside by the A95 near the road junction to Boat of Garten, 1976 *RM & MMcCW* **E**; Dell road, Nethybridge (forma), 1976 *RM & MMcCW* **E,K**; in juniper scrub W of Carrbridge, 1976 *RM & MMcCW* **E,K**.

N By the river at Nairn, 1962 **K**.

I Roadside bank Scaniport, 1976 *RM & MMcCW* **E,K**; bank between Lewiston Hotel and Balmacaan, 1976 *RM & MMcCW* **E**.

Forma **crepiniana** (Déségl.) W.-Dod

M Steep bank above the 'island' in River Spey at Craigellachie, 1976 *RM & MMcCW* **E,K**.

Forma **transiens** (Gren.) W.-Dod

I Verge of the moorland at Etteridge, Glen Truim, 1976 **K**.

Var. **glandulifera** (R. Kell.) W.-Dod

N Burn gully by the River Findhorn upstream from Drynachan Lodge, 1970 **K**.

I Meadow on the right bank of the Moniack Burn at Kirkhill, 1972 **E**.

Var. **glaucophylla** (Winch) W.-Dod

M Forres (as *R. canina* var. *subcristata*) (Druce, 1888); roadside opposite Firknowe at Nethybridge, 1976 *RM & MMcCW* **E,K**; by the A95 near the road junction to Boat of Garten at Nethybridge, 1976 *RM & MMcCW* **E,K**; by the A95 near the road junction to Boat of Garten 1976, *RM & MMcCW* **E**.

I Achnagairn, Kirkhill, 1963 **K**; among boulders below the cliffs of Loch Killin, 1972 **E**.

Forma **adenophora** (Gren.) W.–Dod

M Dunes in the Culbin Forest, 1964 **K**; roadside bank by Advie bridge, 1976 *RM & MMcCW* **E,K**.

Forma **jurassica** (Rouy) W.–Dod

M Roadside by the A95 near Boat of Garten and by a lay-by on the Elgin–Rothes road near Millcotts, 1976 *RM & MMcCW* **E,K**; railway embankment by Loch Vaa N of Aviemore, 1976 *RM & MMcCW* **E**.

I Roadside near Reelig House, Kirkhill and by the cemetery N of Kingussie on the A9, 1976 *RM & MMcCW* **E,K**.

Forma **myriodonta** (Chr.) W.–Dod

M Field by the hill road above Cromdale village, 1976 *RM & MMcCW* **E,K**.

I Raised shingle beach near the rubbish tip between Ardersier and Fort George, 1976 *RM & MMcCW* **E,K**.

Var. **oenensis** (R. Kell.) W.–Dod

M Railway embankment by the Rothes rubbish tip (probably this, but midribs sometimes pubescent), 1976 *RM & MMcCW* **E**.

N River bank upstream from Nairn, 1962 **K**.

I West of Kirkhill, 1960 **K**.

Var. **pseudohaberiana** (R. Kell.) W.–Dod
M Dunes on the Bar, Culbin Forest (aff. this var.), 1967 **K**.
I Lane by Achnagairn Farm, Kirkhill (aff. this var.), 1963 **K**.

Var. **stephanocarpa** (Déségl. et Rip.) W.–Dod
M Scrub by the River Findhorn at Broom of Moy, 1976 *RM &
MMcCW* **E,K**.
I Roadside verge on moorland W of Achlain in Glen Moriston,
1976 **K**.

Var. **subcanina** (Chr.) W.–Dod forma **latifolia** W.–Dod
M On the W side of the old railway line opposite the N entrance
to Castle Grant, Grantown, 1976 *RM & MMcCW* **E,K**.
I Top of a wall N of the Torbreck road junction on the Inverness–
Dores road, 1976 *RN & MMcCW* **E,K**.

R. afzeliana Fr. × **rubiginosa** L.
M Scrub by the River Findhorn at Broom of Moy, 1976 *RM &
MMcCW* **E,K**.
I Achnagairn, Kirkhill, 1963 **K**; roadside W of Kiltarlity, forest
track at Milton of Balnagowan E of Ardersier, 1976 both *RM &
MMcCW* **E,K**; moorland by Dundreggan Kirk, Glen Moriston,
1976 **K**.

R. afzeliana Fr. × **sherardii** Davies
M Scrub by the River Findhorn at Broom of Moy, gully of a
small burn below Lagg Farm near Grantown, both 1976 *RM &
MMcCW* **E,K**.
N River bank at Nairn, 1962 **K**.
I Towpath of the Caledonian Canal at Muirtown Basin, Inverness,
1958 *E. W. Groves* **BM**; hedge by the hill road to Leachkin near
Inverness, grassy verge to the road between Balvraid and Tomatin,
1976 *RM & MMcCW* **E,K**; in rough ground at the top of the hill
N of road junction to Foxhole in Glen Convinth, 1976 *RM &
MMcCW* **K**.

R. afzeliana Fr. × **villosa** L. var. **mollis** Sm. = **R.** × **glaucoides**
W.–Dod
M Wall top opposite a pine wood at the road junction S of Elgin
golf course at Duffushillock, 1976 *RM & MMcCW* **E,K**.
I Field just E of Tarff bridge Fort Augustus, roadside near alder
wood at Glendoe bridge and roadside verge by the A9 near the
entrance to Dunachton, Kincraig, all 1976 *RM & MMcCW* **E,K**.

R. tomentosa Sm. var. **pseudocuspidata** (Crép.) Rouy
M *Forres, Dunphail, Aviemore and Kinchurdy (Druce, 1888).
I *Loch an Eilein and Kingussie (Druce, 1888); Caledonian
Canal, Inverness, 1947 *UKD* **E**.

R. sherardii Davies
Map 106.

Var **sherardii**
I Roadside between Kingussie golf course and Pitmain, 1976
RM & MMcCW **E,K**.

Var. **cinerescens** (Dum.) W.–Dod
I Moorland by a ruined cottage at Grubenmore bridge N of
Dalwhinnie, 1976 **K**.

Var. **omissa** (Déségl.) W.–Dod
N Kingsteps near Nairn, 1976 *RM & MMcCW* **E**.
I Roadside between Lewiston and Balmacaan, 1976 *RM &
MMcCW* **E**.

Forma **resinosoides** (Crép.) W.–Dod
M Banks of the River Findhorn at Broom of Moy, 1954 **K**; hedgerow
near Brodie station 1961 **ABD**; near Greshop House, Forres,
1954 **K**; roadside near the bridge over the Muckle Burn N of Moy
House, 1961 **K**; roadside in the village of Nethybridge, 1976 *RM &
MMcCW* **E,K**.

Var. **schulzei** R. Kell.
I Roadside a short distance up Glen Affric near Fasnakyle
(cf. this var.), 1976 **K**.

Var. **suberecta** (Ley) W.–Dod
M Dunphail and Balnaferry, Forres (as *R. tomentosa*), 1856 *JK* **FRS**;
bank at 'elbow' of road below Lagg Farm near Grantown, 1976
RM & MMcCW **E,K**.
I Rothiemurchus (as *R. tomentosa*), 1856 *JK* **FRS**; roadside at
Leachkin, Inverness, by the River Glass at Breakachy in Strathglass
and by a small burn at Mid Crocail, all 1976 *RM & MMcCW* **E,K**;
roadside between Kingussie and Pitmain, 1976 *RM & MMcCW* **K**;
bank N of Tomich, near Achlain, at Ceannocroc Bridge and
opposite Dundreggan Kirk in Glen Moriston, 1976 **K**.

R. sherardii Davies × **rubiginosa** L. = **R.** × **burdonii** W.–Dod
M Scrub by Cothall quarry near Forres, 1968 *A. Berens & MMcCW*
K.
N In several places by the road between Lochloy House and
Cothill Farm, 1976 *RM & MMcCW* **E,K**.
I Top of a grassy bank on the W side of the bridge over the River
Spey at Kingussie, 1975 **E**; roadside near Reelig House, Kirkhill,
near Balmacaan Farm, Lewiston, margin of a field by Loch Ness
at Lurgmore, near the entrance to Culloden station, bank by the
road W of Nairnside House in Strathnairn, near the Boars Stone,
Essich, all 1976 *RM & MMcCW* **E,K**; edge of a birch wood near
Fasnakyle in Glen Affric and on moorland opposite Mackenzie's
Cairn in Glen Moriston, 1976 **K**.

R. sherardii Davies × **villosa** L. var. **mollis** Sm. = **R.** ×
shoolbredii W.–Dod

M Roadside bank on the N side of the Nethybridge road near Broomhill, 1976 *RM & MMcCW* **E,K**.

R. villosa L. var. **mollis** Sm.   *Downy Rose*
Map 107, Pl 9a.

Var. **mollis**
M Dunphail, 1896 *JK* **FRS**; banks of the River Findhorn at Broom of Moy, 1954 **K**; near Greshop House, Forres, 1954 **K**; sand dune by the Moray Firth, 1963 *P. A. Briggs* **K**.
N Nairn Dunbar golf course, 1960 **ABD**.
I *By the River Calder at Newtonmore (with white flowers), 1911 *ESM* **CGE**; Achnagairn, Kirkhill, 1963 **K**; by a wall near the farm of Balmacaan near Lewiston, 1976 *RM & MMcCW* **E,K**; by the old Manse of Bailebeag in Stratherrick, roadside by the A9 at Dunachton, Kincraig and on the railway embankment by Newtonmore golf course, all 1976 *RM & MMcCW* **E,K**.

Forma **annesiensis** (Déségl.) R. Kell.
I Railway embankment near the bridge on the Newtonmore golf course, 1976 *RM & MMcCW* **K**; by the River Moriston at Ceannocroc Bridge in Glen Moriston, 1976 **K**.

Forma **caerulea** (Woods) W.–Dod
I Roadside between Kingussie golf course and Pitmain, 1975 **E**; on the E side of the railway bridge on Newtonmore golf course, 1976 *RM & MMcCW* **K**.

Forma **glandulosa** W.–Dod
M Culbin Sands, 1934 *W. B. Turril* **K**; by the A95 near Boat of Garten, 1976 **K**.
I Roadside bank of ditch W of Nairnside House, 1976 *RM & MMcCW* **K**.

Var. nov. (with pendulous globular fruits)
I Roadside near Reelig House, 1976 (*RM* 76.86, *MMcCW* 19,126) **E,K**.

Var. nov. (with large turbinate fruits)
M By the N Lodge of Castle Grant near Grantown, 1976 (*RM* 76.134, *MMcCW* 19,190) **E,K**.
I Roadside bank by Borlum House, Fort Augustus, 1976 (*RM* 76.117, *MMcCW* 19,169) **E,K**; roadside near Gorthleck House, Errogie, 1976 (*RM* 76.153) **K**.

R. villosa L. var. **mollis** Sm. × **rubiginosa** L. = **R.** × **molliformis** W.–Dod
M Rough ground between the road and the river on the old Spey Bridge road at Grantown, 1976 *RM & MMcCW* **E,K**; bank by the railway bridge at Mains of Allanbuie near Keith, 1976 **K**.
I Rough ground between Lewiston and Balmacaan, 1976 *RM & MMcCW* **E,K**.

R. rubiginosa L.   *Sweet Briar*
Map 108, Pl 9c.

Var. **rubiginosa**
M Forres, 1895 *JK* **FRS**; *near Forres, 1898 *ESM* **CGE**; dunes
NW of the Buckie Loch in the Culbin Forest, 1954; scrub at
Findhorn, 1954 **K**; by the River Spey at Fochabers, 1971 **E**; bank
by Millbuies loch, 1972 **E**; roadside near Kinakyle, Aviemore, 1973.
N Rare, doubtfully native near Penick, *J. B. Brichan* (Gordon);
Ardclach, introduced (Thomson); *near Nairn, 1898 *ESM* **CGE**;
near Holme Rose 1954; by the river at Nairn, 1962 **K**; roadside
verge W of Cothill Farm on the Lochloy road, 1976 *RM &
MMcCW* **E,K**.
I River bank opposite Bught, Inverness (Aitken); lane at
Achnagairn, Kirkhill, 1963 **K**; near the railway station at Kincraig
and by the sea at Alturlie Point, 1973; roadside W of Kiltarlity, 1976
RM & MMcCW **E,K**; shingle beach near the rubbish tip between
Ardersier and Fort George, 1976 *RM & MMcCW* **K**.

Var. **rotundifolia** Rau
M Edge of Millbuies loch, 1972 **E**.
I Lane by Achnagairn at Kirkhill, 1963 **K**.

R. micrantha Borrer ex Sm.
Var. nov. Pl 9d.
I Hedge by Loch Ness Camp Site S of Invermoriston, 1976
(*RM* 76.114, *MMcCW* 19,165) **E.K**; roadside hedge by the Foyers–
Inverfarigaig road opposite the pier, 1976 (*MMcCW* 19,211) **E**.

R. pendulina L.
R. alpina L.
I Planted in Reelig Glen (*Trans. ISS & FC* **1**: 170 (1878)).

R. rubrifolia Vill
M Several plants by the River Spey at Pitcroy, 1973.

Prunus L.
P. spinosa L.   *Blackthorn, Sloe*
Native. Common in scrub and by ditches and burns. Map 109.
Var. **spinosa**
M Frequent (Gordon); Sanquhar, 1898 *JK* **FRS**; Advie, 1903
(Moyle Rogers); by the Muckle Burn at Dyke, 1967 **CGE**; through-
out the county 1954.
N (Stables, *Cat.*); Ardclach, 1887 (Thomson); common by the
River Findhorn at Coulmony, Glenferness and Dulsie Bridge;
scrub-land by Rait Castle, Cawdor and Nairn; rare at Drynachan.
I (Watson, *Top. Bot.*); Tomich, Milton and Fort Augustus, 1947;
plentiful in Strathglass and the coastal areas, 1954; on the moor
near Tomatin distillery; roadside banks near Ruthven Barracks,
Kingussie, Invermoriston and Kincraig etc.

Var. **angustifolia** Wimmer et Grab.
I Hill road to Achmony, Drumnadrochit, 1976 *R. Melville &
MMcCW* **E**.

P. domestica L.   *Bullace*
    Subsp. **insititia** (L.) C. K. Schneid.
Introduced. Planted in hedges and by houses. Fruit globose.
M Near Ashgrove on the Linkwood road, Elgin (Gordon, *annot.*);
near Binsness, Culbin 1954 **ABD,CGE**; common in hedges round
Urquhart and Garmouth; hedge at Cluny Hill, Forres and by the
road near Blervie farm, Rafford, 1971.
I Planted about Inverness, 1972; hedge at Kingussie, 1975.

    Subsp. **domestica**   *Plum*
Leaves large. Fruit oblong-ovoid. Introduced, usually near old
cottages.
M Castle of Old Duffus (Gordon).
I On the hillside on the W of Daviot, 1974; steep bank on hill
at Cabrich near Beauly, and roadside bank S of Struy in Strathglass, 1976.

P. avium (L.) L.   *Gean, Wild Cherry*
Native, but often planted.
M Rocks and cliffs by the Divie Burn, Dunphail, 1897 *ESM*
BM,CGE; banks of the Findhorn at Cothall, 1898 *JK* **FRS**;
throughout the county, 1954.
N (Stables, *Cat.*); Ardclach (Thomson); throughout the county,
1954; hedgerow by East Milton, 1967 **CGE**; planted on the moor
by the croft of Ballochrochin, Drynachan.
I Inverness, 1832 (Watson, *Loc Cat.*); Kilmorack and by the
Culcabock Burn etc. (Aitken); Beauly river by the Falls of Kil-
morack, 1892 *ESM* **BM**; by Loch Ness at Dores, 1904 *G. T. West*
STA; Kilmorack, 1940 *AJ* **ABD**; wood E of Castle Stuart, 1972 **E**;
common in the lowlands and reaches Dalwhinnie where it was
probably planted.

P. cerasus L.   *Sour Cherry, Dwarf Cherry*
Introduced. Planted as hedges and near houses.
M Certainly introduced (Gordon); by the River Lossie at Arthur's
bridge, Lossiemouth, 1970 **E,CGE**; by the farm of Balnallan,
Cromdale, 1971 **E,CGE**; by cottages at Advie and Knockando,
Nethybridge and Boat of Garten.
N Near a house to the W of Tradespark, Nairn, 1972 **E**.
I Bunchrew, 1885 (Aitken); planted in Reelig Glen, 1954.

P. padus L.   *Hagberry, Bird Cherry*
Native. Very common by rivers, burns and in woods. Map 110.
M Abundant, sides of the rivers Spey and Findhorn and at Ald-
roughty (Gordon); Sanquhar, 1857 *JK* **FRS**; abundant in the
Spey basin near Grantown, 1903 (Moyle Rogers); by the bridge
at Aviemore, *J. Milne* **E**; wood at Brodie Castle, 1963 **CGE**;
throughout the county 1954.
N (Stables, *Cat.*), Ardclach by the Findhorn, 1887 *R. Thomson*
NRN; throughout the county reaching Drynachan, 1954.
I Ness Islands, 1885 (Aitken); Loch an Eilein, 1891 *AS* **E,BM**;

Lovat Bridge, Beauly, 1922 *C. E. Salmon* **BM**; Milton, 1940 *AJ* **ABD**; abundant in the Spey valley reaching Dalwhinnie; a few small trees at Gaick.

P. laurocerasus L.   *Cherry Laurel*
Planted.
M Policies of Brodie Castle, 1925; woods at Altyre House and by the Moss of Meft, 1970, and doubtless elsewhere; wood above the River Findhorn at Cothall, 1973 *R. M. Suddaby*.
I Planted in Reelig Glen, Kirkhill, 1954, and in the policies of Ness Castle, 1956; Culcabock, 1970 *M. Barron*.

Cotoneaster Medic.
C. simonsii Baker   *Himalayan Cotoneaster*
Planted in hedges and gardens; readily bird-sown and found in rock crevices and woods.
M Shingle of River Spey at Kingston, 1953 *J. Souster*!; felled wood at Binsness and in the pine woods of the Culbin Forest, 1967 **E**; roadside verge at Conicavel, Darnaway.
N Pine wood below Drumbeg near Nairn, 1962; regenerating in the policies of Achareidh, Nairn, 1965 *D. McClintock;* on the old walls of Auldearn Kirk and by the roadside near Wester Delnies.
I River bank above Cannich, 1947 *J. Raven* **BM**; by the River Ness, Inverness, 1947 ? *coll.* **E**; edge of a pine wood at Culloden, several plants in the Wild Life Reserve at Inverfarigaig, by the hill road at Polmaily in Glen Urquhart, rock crevice by Loch Ness S of Lochend; roadside banks at Daviot, Moy and Kirkhill etc.; Culcabock, 1970 *M. Barron*.

C. microphyllus Wall. ex Lindl.   *Small-leaved Cotoneaster*
Garden relic near old buildings and occasionally bird-sown.
M By the ruins of an old cottage near Earlsmill, Darnaway, 1967 **ABD,CGE**.
I Roadside verge by Loch Ness near Urquhart Castle, 1956; foot of the bridge of the Moy Burn at Moy Hall, 1975.

C. frigidus Wall. ex Lindl.
Garden escape.
N Planted by the road on the edge of the Laiken Forest S of Nairn, 1954; wood in the policies of Boath House, Auldearn, 1968 **ABD**.

Crataegus L.
C. laevigata (Poir.) DC.   *Midland Hawthorn*
    C. oxyacanthoides Thuill.
Doubtfully native in the north east.
I Planted at Nairnside, 1940 *AJ* **ABD**.

C. laevigata×**monogyna** = **C**×**media** Bechst.
I Single tree among sloes, elder, brambles and bracken etc., on the hillside above Drumnadrochit, 1976 *R. Melville & MMcCW* **E**.

C. monogyna Jacq.   *Chaws, Hawthorn*
Native and extensively planted for hedges.

M Frequent (as *C. oxyacantha*) (Gordon) and abundant (Burgess); throughout the county, 1954; a yellowish-red-fruited form in a hedge S of Moy House, 1972 *M. Hunter* **E**!

N (Stables, *Cat.*); Ardclach, 1887 (Thomson); by the river at Nairn, 1962 **ABD**; river bank at Dulsie Bridge, 1962 **ABD**, and throughout the county.

I Indigenous about Loch Ness (Gordon); Inverness, 1832 (Watson, *Loc. Cat.*); opposite Urquhart Castle, 1935 *A. J. Wilmott* **BM**; hedge by the River Beauly, 1947 **BM**; common in the lowlands and reaching Newtonmore, Glen Strathfarrar, Tomich and Kincraig etc. 1954.

Amelanchier Medic.

A. confusa Hyland.   *Juneberry*
Planted by the pond at Househill, Nairn, 1967 **E,CGE**.

Sorbus L.

S. aucuparia L.   *Roddan Tree, Rowan, Mountain Ash*
Native. Abundant, in woods, by rivers, on rocks and mountain ledges up to 1,000 m. Often planted near houses and sheilings as it was supposed to have had magical properties and contain a remedy against spells and witchcraft (Thomson). In Strathspey the people were said to make hoops of the wood of this tree and on the first day of May cause all their sheep and lambs to pass through them. Said to be a druid superstition.

M Abundant (Gordon); throughout the county, 1954.

N (Stables, *Cat.*); Ardclach (Thomson); river bank Nairn, 1962 **ABD**, and throughout the county.

I Pitmain and Dalwhinnie, 1833 (Watson, *Loc. Cat.*); Culloden, 1837 *J. Forbes*, ex herb. *J. Macnab* **E**; Loch Ness near Dores (var. *flava* Druce) 1937 *NDS* **BM**; by the River Ness, Inverness (var. *glabra* Gilib.), 1930 *M. L. Wedgewood* **MBH**; cliff crevice Ben Alder, 1,000 m, 1956, and throughout the district.

S. domestica L.
Resembles *S. aucuparia* but is a larger tree. Recorded for Culloden, Easterness, 1837 *J. Macnab* **E**.

S. hybrida L.
Recorded from Elgin by *P. Leslie* (Burgess).

S. intermedia (Ehrh.) Pers.   *Swedish Whitebeam*
Planted.
N Plantation by the road at the entrance to Garblies farm, Auldearn, 1968 **BM,CGE**.

I Banks of the River Ness, Inverness, 1930 *M. L. Wedgewood* **MBH**; do. *G. C. Druce & R. H. Corstorphine* **OXF**; South Kessock, 1946 *D. F. Young*; policies of Achnagairn, Kirkhill; roadside verge at Englishton, 1971 **E**; roadside at Balnagowan, policies of Kyllachy House, Strathdearn, by Loch Flemington and large single trees at Glen Affric Lodge and by farm buildings at Farley.

S. aria (L.) Crantz sensu lato   *Henapple, Whitebeam*
Planted and occasionally bird-sown.
M Planted extensively by James Earl of Fife in the eastern part
of the province (Gordon); near Forres (Druce, 1897); planted
throughout the county, 1954; bird-sown in a birch wood between
Hazelbank and Pluscarden, 1968 **E**; in a wall by Spynie Kirk, 1976.
N Planted at Nairn and Lethen, 1954.
I Planted at the Bught, Inverness, 1889 (Aitken); near Ferrybrae,
Beauly, 1947 *S. M. Walters & E. F. Warburg*; Glen Strathfarrar,
Corrimony, Strathglass, by Loch Mhòr at Errogie, Reelig Glen,
Inverness and Culloden, 1954; Inshriach, 1963 *F. Brooke* **E**; bird-
sown in a dense birch wood on Inchberry Hill above Kirkton
House, Bunchrew, 1976.

S. aria × aucuparia = S. × thuringia (Isle) Fritsch
I Old pine wood, 770 m at Dunain, Inverness, 1972 *M. Barron* **E**.

S. latifolia (Lam.) Pers. sensu lato   *Broad-leaved Whitebeam*
N Single tree by the lane from Croy to Holme Rose, 1955 **ABD**.
I Ness Islands, Inverness, 1930 *M. L. Wedgewood* **BM,MBH**; Ness-
side, planted, *G. C. Druce & R. H. Corstorphine* **OXF** 'yes, the true
latifolia – I have it from the same station', Fraser August 1931;
Islands Inverness, 1947 *UKD* **E**; single tree in a grass field at
Essich, 1975 **E**; single tree on the old railway bridge over the canal
at Fort Augustus **E**.

S. rupicola (Syme) Hedl.
Native. Base-rich rocks. Rare.
I (Druce, *Com. Fl.*); 20–30 trees on the cliffs of Creag Dhubh,
Laggan, 1955; crags of Beinn a' Bhacaidh, Loch Ness, 1961 *D. A.
Ratcliffe*; cliff at Inverfarigaig, 1962 *D. A. Ratcliffe*!.

Pyrus L.
Pyrus communis L. agg.   *Pear*
Cultivated varieties are found as garden relics etc., in the lowland
areas.

Malus Mill.
M. sylvestris Mill sensu lato   *Apple*
Planted or garden relics. Not the true *Crab-apple*.
M Certainly introduced (Gordon); Aviemore (Druce, 1888);
hedge near Forres, 1954; policies of Darnaway Castle, 1973.
N Planted in a hedge near Geddes House on the Nairn road, 1954
E.
I Inverness, 1832 (Watson, *Loc. Cat.*); Ballindarroch, Scaniport,
1940 *AJ* **ABD**; Crinaglack Strathglass, 1947 *M. S. Campbell*; waste
land by Broomhill station, edge of a field near Dalcross Airport
and by the railway N of Kincraig.

Stranvaesia davidiana Decne
This shrub with conspicuous red berries is planted on forest verges.
I Strathglass, 1973 **E**; by the road to Tomich near Cannich.

Neillia sinensis Oliver
Planted.
I   Policies of Glen Cottage, Ness Castle, 1971 **E**, and planted by
the path to the lochan.

Holodiscus discolor (Pursh) Maxim.
Shrub with inflorescences resembling a *Spiraea*.
M   Bank between Dallas Kirk and Craigroy and by the bridge over
the Divie Burn near the S entrance to Logie House.

CRASSULACEAE
Sedum L.

S. rosea (L.) Scop.   *Rose-root*
Native. A glaucous-green fleshy arctic-alpine of mountain ledges
and rock crevices, descending to sea level.
M   On the sea cliffs at Covesea (Burgess); seen there in 1954, but
appears to have now disappeared. It is still common on the Banffshire coast to the east of Macduff.
I   Dalwhinnie, 1833 (Watson, *Loc. Cat.*); SE of Dalwhinnie, herb.
Borrer **K**; Badenoch and the Falls of Foyers (Gordon); Cairn Gorm,
1867 *J. Roy* (Watson); rock crevice near the summit of Ben Alder,
cliffs of Sgùrr na Lapaich, Bynack More, Glen Feshie and Gaick
etc.; low cliff by Loch Killin, 1972 **E**.

S. telephium L.   *Orpine, Livelong*
   Subsp. **fabaria** Syme
Native. Banks and waste grassy places.
M   Near Dalvey, *J. B. Brichan* (Watson); near Forres, *J. G. Innes*
(Gordon, *annot.*); parishes of Edinkillie and Urquhart (Burgess);
roadside verges at Boat of Garten, Dunphail, Kintessack, near
Mundole, at Findhorn, Kellas and Whitemire, 1954; bank near
Lochinvar, 1975 **E**; railway embankment Clashdhu near Altyre.
N   Woods near Cawdor, 1855 *HC* **INV**!.
I   Stone wall by Kingussie bridge, 1833 (Watson, *Loc. Cat.*);
roadside verge N of Croy; raised beach at Ardersier, 1954; roadside
bank to the W of Kilmorack Kirk and a few plants on the edge of
the pine wood at Loch Garten.

S. spurium Bieb.
Garden escape.
M   In a rabbit scrape in a rough field at Hamlets, Spynie, 1966 **E**;
established on the river bank near Inverallan, Grantown, 1970 **E**.

S. anglicum Huds.   *English Stonecrop*
Native. Rare on river shingle, rocks and gravel flats near the sea.
M   Raised beaches near Garmouth on the Moray Firth, Elginshire,
1837 *G. Gordon* **E**!. Garmouth, 1900 *JK* **FRS**; gravel flats near
Hatton loch near Kinloss, 1923; shingle of River Findhorn at
Waterford, 1974 **E**.
N   (Stables, *Cat.*); shingle of River Nairn at Holme Rose, among
stones by Loch of the Clans and paths in the garden at Glenferness

House where it was probably introduced, 1954; shingle of River Findhorn at Drynachan, 1970.

I Planted SW of Dores, 1955 *J. M. Whyte*; shingle of River Findhorn at Tomatin and on the gravel drive of Moy Hall, 1975.

S. album L. subsp. **album** *White Stonecrop*
Introduced. Walls, and dunes and waste land.

M Escape in the parish of Dyke (Burgess); by the sea at Covesea, 1936 *R. Richter*; shingle of the River Spey at Orton and Garmouth, 1953; waste land at Lossiemouth, 1955; sand dunes at Findhorn, 1964 **CGE**; walls at Fochabers and Longmorn.

N On a wall away from habitation by the road junction to Righoul near Geddes, planted on walls at Piperhill and Nairn, 1954; waste ground by the ford at Achavraat, 1973 **E**.

I Shingle of the River Enrick at Milton, 1947; dry bank near Alvie, quarry at Mid Coul, raised beach at Ardersier, walls of Dunlichity Kirk and among rocks and stones by Loch Ness 1 mile S of Lurgmore.

S. acre L.   *Biting Stonecrop*
Native on dunes and shingle on the coast; probably introduced on railway embankments and river shingle.

M Very common (Gordon); Castlehill, Forres, 1856 *JK* **FRS**; Findhorn, 1953 *A. Melderis* **BM**; shingle of the River Spey from Aviemore to the sea, 1954; railway line at Brodie, Dunphail, Elgin and Rothes and the whole length of the coast, 1954; Culbin Forest, 1964 *P. How* **E**.

N (Stables, *Cat.*); Ardclach on walls (Thomson); Culbin sands, Nairn, 1911 *P. Ewing* **GL**; river shingle Holme Rose, walls at Cawdor, Nairn and Auldearn kirkyard, 1954; shingle beach, the Bar, Culbin, 1968 **E**.

I Longman, Inverness, 1869 ? *coll.* **ELN**; banks of the canal, Inverness (Aitken); abundant among cinders on the railway embankments at Allanfearn, Dalcross, Culloden and Nethybridge, 1954; canal bank at Fort Augustus, planted on walls at Kingussie, Lewiston, Insh and Daviot etc.; shingle of River Findhorn at Tomatin and the raised beaches at Ardersier and Whiteness Head.

S. forsteranum Sm.   *Rock Stonecrop*
Garden escape or planted.

    Subsp. **forsteranum**
M By a small burn at Conicavel, Darnaway, 1963.

    Subsp. **elegans** (Lej.) E. F. Warb.
I Planted on the walls of Dunlichity kirkyard, 1954 **E**.

S. reflexum L.   *Reflexed Stonecrop*
M Planted on a wall at Greyfriars, Elgin (Gordon).

S. villosum L.   *Hairy Stonecrop*
Native. Rare in grassy flushes and bogs, usually associated with *Galium uliginosum*. Pl 7.

M Near the station at Garmouth, 1837 *G. Gordon* **E**; near Bridge of Brown (Marshall, *J. Bot.* **44**: 156 (1906))!; 3 miles from Grantown, on moors among moss, 1915 *Miss Wilkinson* **OXF**; base-rich flush by the small burn ½ mile to the W of Wester Ryndballoch on the Haughs of Cromdale, 1971 **E**; flush at Fae, Dorback, 1972 and in several flushes running into the Allt Iomadaidh between Fae and Bridge of Brown.

Sempervivum L.

S. tectorum L.   *Fouse, House-leek*
Introduced. Formerly planted on roofs. Now found as garden escapes on old walls and river banks.
M Frequent, certainly introduced (Gordon); parishes of Dyke and Kinloss (Burgess).
N (Stables, *Cat.*); Ardclach (Thomson); wall of the embankment of the River Nairn by Little Kildrummie, 1974 **E**.
I Gable of old ruined cottage at Blackfold 1971, and on a rock outcrop far from habitation on the moor by the River Dulnain downstream from Caggan bridge, 1975 *M. Barron*.

Crassula L.

C. tillaea L.-Garland   *Mossy Stonecrop*
Tillaea muscosa L.
Introduced. Cinder tracks and gravel paths.
M Introduced with wool shoddy, Dyke, 1968.
I Extending for ½ mile on the cinder track by the Caledonian Canal, Inverness, 1970 *M. Barron*!; abundant on the garden paths at Ballindarroch House, Scaniport, 1975 **E**; in gravel by the lock at Dochgarroch, 1975.

Umbilicus DC.

U. rupestris (Salisb.) Dandy   *Pennywort, Navelwort*
Planted on walls at Kellas House in Moray, 1967, and near Campbelltown in Easterness (Hooker, *Fl. Scot.*).

SAXIFRAGACEAE
Saxifraga L.

S. nivalis L.   *Alpine Saxifrage*
Native. Rare on mountain ledges and scree.
I Beinn a' Bha'ach Ard, Erchless Forest, 1902 (Pollock); Drumochter ? date and *coll.* (*Atlas*); on scree above the 'Apron' in Coire an Lochain, Cairn Gorm, 1953; Coire Garbhlach, Glen Feshie, 1972 *R. McBeath*.

S. nivalis × **stellaris** = **S.×crawfordii** E. S. Marshall
I Coire an t-Sneachda of Cairn Gorm, 1902 *F. C. Crawford* (Marshall, *J. Bot.* **47**: 98 (1909)).

S. stellaris L.   *Starry Saxifrage*
Native. Common in mountain flushes, by burns and on wet rocks and gravel. Often washed downstream and becoming established on shingle and river banks.

M Moy Carse and Relugas, *J. G. Innes* (Gordon, *annot.*); river shingle at Darnaway and in flushes further up the River Findhorn, 1954; common in the Cromdale Hills and in runnels and flushes by the Allt Iomadaidh between Fae and Bridge of Brown.
N (Stables, *Cat.*); Ardclach and Dulsie, *G. Wilson* (Gordon); rare at Ardclach, 1885 *W. Gordon* **NRN**; shingle of the River Findhorn at Glenferness and by the Allt Breac at Drynachan.
I Dalwhinnie, 1833 (Watson, *Loc. Cat.*); Badenoch (Gordon); Rothiemurchus and Strathnethy, 1892 *AS* **E**; Cairn Gorm (var. *fontana* Dr.), 1919 *M. L. Wedgewood* **MBH**; Coire an Lochain, 1953 *A. Melderis* **BM**; Mam Sodhail, Glen Affric 1940 ? *coll.* **STA**; in all the hill districts, 1954.

S. spathularis Brot. × **umbrosa** L. = **S.** × **urbium** D. A. Webb
London Pride
Garden out-cast. Banks and woods.
M Pine wood Grantown, 1954 **E**; Aviemore, 1964 *M. Cowan et al* **E**; by the burn in the wood at Burgie House, 1967 **CGE**.
N Bank by the reservoir at Geddes, 1961 **ABD**; roadside ditch near Brevall, Lethen, 1967 **ABD**.
I Roadside ditch just N of Fort Augustus, 1950; established by the Moniack Burn, Kirkhill, 1963 **K**.

S. cymbalaria L.
A form of this small annual yellow-flowered saxifrage is established in the Nursery at Inshriach and has been introduced with plants to the garden of Rose Cottage, Dyke.

S. granulata L.   *Meadow Saxifrage*
Native. Common on dunes by the sea, less so on grassy banks inland
M Elginshire, 1837 *G. Gordon* **E**; Mundole near Forres, 1857 *JK* **FRS**; in all the coastal parishes (Burgess); dunes and grassy cliffs at Burghead, 1967 **CGE**; near Hopeman, 1967 **E**; dunes at Lossiemouth, 1972 **ABD**.
N Moyness, 1900 (Lobban).
I Kilmorack, 1885 (Aitken)!; wood at Achnagairn, Kirkhill, 1930 *M. Cameron*!; abundant for a short distance on the grassy bank of the roadside to the W of Kilmorack kirk, 1972 **E**.
The form with double flowers was planted in the policies of many of the larger houses.
M Innes House, Urquhart, 1794 *A. Cooper* **ELN**; wood Brodie Castle, 1963 **BM,CGE**, also at Dalvey and Altyre.
N Policies of Geddes House, 1954.

S. rivularis L.   *Brook Saxifrag*
Native. Rare in rock crevices and wet places in the Cairngorms usually over 1,000 m.
I Cairn Gorm, 1874 *A. Ley* **ABD**; Cairngorm mountains, 188*?* *H. & J. Groves* **E,GL** Larig Pass, 1893 *AS* **E**; Coire an Lochain, 195*?* *A. Melderis* **BM**!; also in the Coire an t-Sneachda and in the corrie of Braeriach in the Cairngorms.

S. cespitosa L.   *Tufted Saxifrage*
Native. Very rare in one of the corries of Cairn Gorm, ? date and
coll.!.

S. hypnoides L.   *Mossy Saxifrage*
Native. Frequent on rock ledges and scree in the highlands and on
grassy banks in the lowlands, preferring base-rich soil.
I   Glen Feshie, 1831 (Gordon); Dalwhinnie and Pitmain, 1833
(Watson, *Loc. Cat.*); SE of Dalwhinnie, herb. *Borrer* **K**; Bunchrew,?
date and *coll.* **INV**; Glen Einich, 1887 *G. C. Druce* **BM**; Pass of
Inverfarigaig, 1904 *G. T. West* **STA**; Coire Chùirn, Drumochter,
1911 *ESM* **E,BM**; cliffs above Loch Killin, cliffs of Creag Dhubh,
Laggan, Creag nan Clag and Tom Bailgeann near Torness etc.

S. aizoides L.   *Yellow Saxifrage*
Native. Common in bogs, flushes and wet mountain ledges. Often
washed downstream and becoming established on river shingle and
by the sides of burns. Map 112.
M   Darnaway, *Brodie* **E**; Elginshire, 1837 *G. Gordon* **E**; by the River
Findhorn, 1844 *A. Croall*; Cothall, 1900 *JK* **FRS**; flushes in the Hills
of Cromdale and shingle of the burn at Bridge of Brown.
N   (Stables, *Cat.*) on the banks of the Nairn at Broadley, 1835
J. B. Brichan **ABD**; by the Findhorn at Levrattich, 1885 *R. Thomson*
NRN; rocks by the River Findhorn at Glenferness and Dulsie
Bridge, 1954; flush in a gully on Creag a' Chròcain W of
Drynachan.
I   Dalwhinnie and Pitmain, 1833 (Watson, *Loc. Cat.*); Dalwhinnie,
1841 herb. *Borrer* **K**; Badenoch (Gordon); Bunchrew Burn, 1854
HC **INV**; Moniack Burn (Aitken); Inverfarigaig, 1940 *AJ* **ABD**;
flush near Dalwhinnie, 1963 **CGE**; on wet gravel by a small moor-
land pool in Glenmoriston, 1971 **E**.
A curious form with orange-coloured flowers var. **aurantia**
Hartm. was found near Dalwhinnie in 1877 by *Mr Potts* (Boyd,
Trans. BSE **13**: appendix lxix (1878)).

S. oppositifolia L.   *Purple Saxifrage*
Native. Frequent on base-rich rocks and scree, often washed down-
stream and becoming established for short periods on river banks
and shingle. Pl 7, Map 113.
I   Glenfeshie, 1831 (Gordon); on the N shore of Loch Ericht, 1833
(Watson, *Loc. Cat.*); Loch Einich, 1882 *JK* **FRS**; Mealfuarvonie
(Aitken); by the River Spey near Ralia, Newtonmore, 1951
C. Murdoch; on most of the higher mountains in glens Affric, Cannich,
and Strathfarrar, 1954; locally common in Glen Feshie and at
Gaick; low conglomerate cliffs of Tom Bailgeann near Torness,
1972 **E**; Creag nan Clag near Torness, 1972 *M. Barron* !.

Tellima R. Br.
T. grandiflora (Pursh) Dougl. ex Lindl.
Introduced.
M   Policies of Brodie Castle, 1961; near Duffus House, 1962 *S. M.*

8

Laurensen; by the Muckle Burn at Dyke, 1972 where it was found by two pupils of Dyke Primary School, *Catherine Finlayson* and *Robert Smith Jnr.* **E.**
N Abundant in the policies of Lethen House, 1961 **ABD,BM.**

Tolmiea Torr. et Gray
T. menziesii (Pursh) Torr. et Gray   *Pick-a-back-plant*
Introduced.
M Policies of Duffus House, 1962.
N Wood opposite the Home Farm Cawdor, 1970 **E.**
I Policies of Achnagairn House, Kirkhill, 1963 **CGE.**

Chrysosplenium L.
C. oppositifolium L.   *Opposite-leaved Golden Saxifrage*
Native. Abundant by burns and in flushes, wet woods and mountain springs.
M Frequent (Gordon); Greshop, 1876 *JK* **FRS**; bog in the wood at Greengates, Brodie Castle 1963 **CGE**, and throughout the county.
N (Stables, *Cat.*); Edgehill, Auldearn, 1843 *J. B. Brichan* (Gordon, *annot.*); common at Ardclach, 1890 *R. Thomson* **NRN**; flush by Geddes reservoir, 1964 **ABD**, and throughout the county.
I Dalwhinnie, 1833 (Watson, *Loc. Cat.*); S of Dalwhinnie, herb. *Borrer* **K**; Dulnain, Holm and Moniack etc. (Aitken); near Drumnadrochit, 1940 *AJ* **ABD**; flush by the Falls of Foyers, 1964 **CGE**; spring in Reelig Glen, 1967 **E,CGE**; bog by the Big Burn, Ness Castle, 1971 **E,CGE**; flushes on Sgùrr na Lapaich, Glen Strathfarrar at 1,000 m, on Mealfuarvonie and Bynack More etc.

C. alternifolium L.   *Alternate-leaved Golden Saxifrage*
Native. Frequent in similar situations to *C. oppositifolia* often growing together. Distinguished by being slightly more robust, the cauline leaves being alternate and the bracts more yellow than green. Map 114.
M Dunphail, Alves and Newmill, *J. B. Brichan* (Gordon); Greshop, 1870 *JK* **FRS**; Dunphail, 1886 *ESM* **CGE**; banks of the River Findhorn, Darnaway, by a ditch at Longmorn, wet wood at Huntly's Cave on Dava Moor, by the River Spey at Grantown and Rothes, 1954; Greshop Wood, 1964 **CGE.**
N Cawdor *WAS* (Gordon); Auldearn, 1837 *J. B. Brichan* (Watson); bank by Ferness bridge, 1964; by the River Nairn at Kilravock Castle, 1966 **ABD.**
I Badenoch (Gordon); Strathglass, Mealfuarvonie, Inverfarigaig and Daviot, 1956; Reelig Glen, 1967 **E**; by the Big Burn at Glen Cottage, Ness Castle, 1971 **E**; wet wood by the River Enrick at Milton.

Bergenia Moench
B. crassifolia L.
Planted in Reelig Glen, Easterness, 1964.

PARNASSIACEAE
Parnassia L.

P. palustris L. *Grass of Parnassus*
Native. Frequent in grassy bogs and flushes on moors; also in dune
slacks. Frontispiece, Map 115.
M Elginshire, 1827 *G. Gordon* **E**; Morayshire, *G. Wilson* herb.
Borrer **K**; Spynie, 1860 *JK* **FRS**; rare on the fixed dunes NW of the
Buckie Loch, Culbin, 1930; grassy bogs at Grantown, Relugas,
Kellas, edge of Lochinvar and Longmorn, 1954; among willows on
the S edge of the Buckie Loch, Culbin, 1968 **E**; grass field by the
small burn at Auchinroath near Rothes, 1972; very rare among
rushes and heather on the steep banks of the River Lossie near
Torwinny, Dallas, 1965 *J. Mitchell*!
N Lochloy, 1833 *WAS & J. B. Brichan* **ABD**; in a boggy hollow
on the lower slope of the Aitnoch Hill, Ardclach, 1885 *R. Thomson*
NRN; abundant on the sandy margin of Lochloy, 1956; flushes
by the Allt Breac at Drynachan, 1974 **E**.
I By the road to Foyers, 1885 (Aitken); ? locality, date and *coll.*
INV; Struy in Strathglass, 1902 (Pollock) SE corner of Loch
Morlich, 1938 *J. Walton* **GL**; near Dores, 1940 *AJ* **ABD**; abundant
in grassy flushes running into Loch Ness on the W side, 1925; bog
at Dalballoch in Glen Banchor, 1951 *C. Murdoch*; grass bank by the
Big Burn, Essich, 1965 *M. Cuthbert*; common in bogs by lochs
Coulan, Bunachton and Tarff; meadow at Dunmaglass, grassy
moorland Glen Mazeran, slopes of Sgùrr na Lapaich etc.

HYDRANGEACEAE
Philadelphus L.

P. coronarius L. *Syringa, Mock Orange*
Planted at Glenferness, Nairn and in Reelig Glen, Easterness, 1960.

GROSSULARIACEAE
Ribes L.

R. rubrum L. *Red Currant*
   R. sylvestre (Lam.) Mert. et Koch
Doubtfully native. Common in woods and by rivers.
M Knockando (Burgess); by the Muckle Burn at Dyke, by the
River Spey at Boat of Garten, Rothes and Fochabers, wood at
Wester Elchies, etc., 1954; small wood at Delfur, Boat o Brig,
1967 **E**.
N (Stables, *Cat.*); Ardclach (Thomson); woods at Holme Rose and
Kilravock, 1954; by the River Findhorn at Coulmony and Glen-
ferness; wood to the W of Kinsteary House, Auldearn, 1964 **ABD**;
wood near Cawdor, 1972 **E**.
I Bridge near Essich, 1885 (Aitken); roadside bank near Tomich,
1947; woods at Tomatin and Invermoriston 1952; by Loch Morlich
and by the burn near Flichity House, among rocks below the cliffs
of Loch Killin, in the small wood at the outlet of the River Enrick at
Lewiston and by the River Enrick at Corrimony school etc.

R. spicatum Robson   *Downy Currant*
  R. petraeum auct.
Native. Frequent in Strathspey, and nearby on the banks of the
river and its tributaries. Map 116.
M In many places on the Spey from Craghue to Grantown, 1827
(Gordon); banks of the Spey above Grantown, *WAS* (Murray);
Blacksboat and Knockando, 1903 (Moyle Rogers); native by the
Spey from Doune (v/c 96) to Kinchurdy (Druce, 1888); on the E
side of the River Spey by the old bridge at Grantown, 1972 **E**; by
the River Nethy a short way upstream from the village.
I *Gordon sp.* (Watson, *Top. Bot.*); lochs Garten and Mallachie
(Druce, 1888); by Loch Morlich, 1953; wooded slope of the River
Dulnain at Skye of Curr; shingle island by the golf course at
Newtonmore; in the small wood at the exit of the River Enrick to
Loch Ness, Lewiston.

R. nigrum L.   *Black Currant*
Doubtfully native. Woods and river banks.
M Parishes of Dyke, Forres and Edinkillie (Burgess); Culbin
Forest, in a small wood at the E end of the Burnie Path, Dyke, 1972
E; banks of the Spey at Blacksboat and Knockando, gully in the
Hills of Cromdale and in the gorge at Huntly's Cave on Dava Moor
etc.
N (Stables, *Cat.*); Ardclach (Thomson); in the woods at Holme
Rose, Kilravock Castle and Cawdor Castle, 1954; on the raised
beach near the Druim E of Nairn and by the River Findhorn at
Coulmony and Glenferness etc.; wood at Kinsteary House,
Auldearn, 1964 **ABD**.
I Near Tomatin station, 1952; Ness Islands, Inverness, in the Pass
of Inverfarigaig and in several places by the River Glass in Strath-
glass, 1954; among boulders below the cliffs of Creag Dhubh,
Laggan, by the river at Newtonmore, and generally scattered over
the district.

R. sanguineum Pursh   *Flowering Currant*
Garden escape. Frequent in woods and waste land.
M Railway embankment at Dunphail and hedge at Longmorn,
1954; beech wood by Barley Mill, Brodie, 1966 **E,CGE**, and in
many other places.
N Banks of the River Findhorn at Coulmony and in a small wood
at Cantraydoune, 1954; by the river at Nairn and by the River
Findhorn at Glenferness etc.; policies of Kinsteary House, Auldearn,
1964 **ABD**.
I In a wood at Invermoriston and on the railway line at Tomatin,
1950; cliff face at Aigas, in the wood at Reelig Glen and at the
outlet of the River Tarff at Fort Augustus etc.; roadside near
Kilmorack, 1972 **E**.

R. alpinum L.   *Mountain Currant*
Planted in the policies of Darnaway Castle, Moray, 1950 **E**, and in

the policies of Achnagairn House and Reelig Glen, Kirkhill, Easterness, 1952.

R. uva-crispa L. *Gooseberry*
Doubtfully native. Common at the foot of walls and in woods and waste places.
M Certainly introduced (Gordon); near Forres (Druce, 1897); bird sown (Burgess); Fochabers, 1925 ? *coll*. **GL**; quarry at Lossiemouth, 1970 **ABD**, and throughout the county.
N (Stables, *Cat.*); Ardclach (Thomson); woods at Dallaschyle and Kinsteary, Cawdor and Glenferness etc., 1954; river bank at Kilravock Castle, 1961 **ABD**.
I Inverness, 1832 (Watson, *Loc. Cat.*); Culcabock Dam and Craig Phadrig (Aitken); Dalwhinnie, Fort Augustus, Tomich and Beauly etc., 1954; hedgerow near Kilmorack, 1972 **E**.

DROSERACEAE
Drosera L.
D. rotundifolia L. *Round-leaved Sundew*
Native. Abundant in moorland bogs throughout the district.
M Very common (Gordon); Pilmuir and Mundole (Innes); rare in dune slacks NW of the Buckie Loch, Culbin, 1925; Kincorth, 1938 *J. Walton* **GL**; winter lochs, Binsness, 1953 *A. Melderis* **BM**; throughout the county, 1954.
N (Stables, *Cat.*); frequent in marshes at Ardclach, 1888 *R. Thomson* **NRN**; bogs on the moors at Drynachan, Glenferness etc., 1954.
I Dunain, 1871 ? *coll*. **INV**; Glen Urquhart (Aitken); Drumashie Moor, 1940 *AJ* **ABD**; abundant throughout the district, 1954.

D. anglica Huds. *Long-leaved Sundew*
Native. Frequent in moorland bogs and locally plentiful. The scape is twice as long as the leaves the flowers more robust than *D. rotundifolia*. Map 117.
M Abundant in one or two places from Culbin to Fort George (Gordon); Dyke Moss, 1832 *WAS* **E**; bogs near Dava station, 1959 *D. A. Ratcliffe*; common in the sphagnum bogs round Lochan Dubh, Aviemore, 1974 **E**.
N (Stables, *Cat.*); not common at Levrattich, Ardclach, 1888 *R. Thomson* **NRN**.
I East of Fort George, 1836 *WAS* **E**; Invermoriston, herb. *Borrer*
K; Badenoch (Gordon); Loch Ericht (Balfour, 1868); marshy places in Glen Urquhart and Strathglass (Aitken); Rothiemurchus, 1886 *JK* **FRS**; wet gravel by a pool near Loch Loyne, Glen Moriston, 1971 **E**; very local on the lower slopes of Mealfuarvonie, 1973 **E**; plentiful in bogs at the head of Glen Strathfarrar, in Strathglass, at Milton, Fort Augustus and in Glen Affric etc.

D. rotundifolia × anglica = D. × obovata Mert. et Koch
Frequent where both parents occur.
M Loch Vaa (Druce, 1888); marsh near Boat of Garten, 1887

G. C. Druce **OXF**; with both parents by Lochan Dubh, Aviemore, 1974 **E**.
I   Strathaffric (Ball, 1851); bog, with both parents, on the shores of Loch Mullardoch, 1947 *S. M. Walters* **CGE**; Loch Beinn a' Mheadhoin, 1947 *E. F. Warburg* **OXF**; bogs by the Monar Dam and on Sgùrr na Muice in Glen Strathfarrar, 1954; boggy slopes by Lundie in Glen Moriston, small bog on the lower slopes of Meal-fuarvonie and in a small bog on the E side of the Spey at Boat of Garten.

D. intermedia Hayne   *Oblong-leaved Sundew*
Native. Rare. Often confused with *D.*×*obovata* and recorded as such. In the absence of herbarium material the following records must be treated with some doubt. Scapes a little longer (sometimes shorter) than the leaves and curved or decumbent at the base.
M   Delete from v/c 95 (Trail, *ASNH* **15**: 40 (1906)).
I   Moors above Glen Urquhart (Watson); bog at the foot of Craig Dhu (*Trans. BSE* **13**: appendix lxx (1878)); Badenoch, *AS* (*ASNH* **2**: 123 (1892)). G. T. West's record from the hills N of Fort Augustus has been re-determined as *D.*×*obovata*.

LYTHRACEAE
Lythrum L.
L. salicaria L.   *Purple Loosestrife*
Doubtfully native in north Scotland. Rare in marshes.
M   Abundant in a meadow by Elgin station, 1920; garden out-cast in a small pond opposite Mudhall Farm at Dyke, 1953 **E,ABD**; in a small quarry opposite cottages near Milton Brodie and in a ditch near Kintrae, 1954.
N   Garden out-cast by the distillery pond at Brackla, 1968.
I   Very scarce between Kingussie and Kincraig (Marshall, 1899); banks of the River Beauly between the town and Lovat Bridge, 1947; abundant, and perhaps native lower down the river opposite Ferrybrae, 1975 **E**.

L. meonanthum Link
Casual on the Elgin rubbish tip, 1957 **E**.

Peplis L.
P. portula L. var. **portula**   *Water Purslane*
Native. Frequent on the muddy margins of acid lochs and pools. Map 118.
M   Elginshire, 1837 *G. Gordon* **E**; near the Manse of Urquhart at Sunbank and at Alves, G. Wilson (Gordon); Kintessack, 1843 *J. B. Brichan & J. G. Innes* (Gordon, *annot.*); abundant at the Buckie Loch, Culbin (Marshall, 1899) (now dried up owing to drainage); pool in the wood at Conicavel, verge to the lochan at Avielochan and on the muddy margins of Gilston lochs, 1954; abundant by Millbuies loch, the loch at Crofts of Buinach and in depressions in the shingle beds at Kingston; pool at Speyslaw, 1972 **E**.

N (Stables, *Cat.*); pond at Newton of Park, *J. B. Brichan* (Gordon); on the 'Blasted Heath' near Auldearn (Watson); Culbin Sands, Nairn, 1910 *P. Ewing* **GL**; pool on the moor between Easter Lochend and the Loch of the Clans, 1960 **ABD**; pool on the Carse of Delnies, 1960 **ABD**.

I Pitmain, 1833 (Watson, *Loc. Cat.*); by the River Nairn (Aitken); Kincraig, 1891 *AS* **E**; backwater of the river at Coiltry Loch and small loch near Aldourie Pier by Loch Ness, 1904 *G. T. West* **E**; pool between Cannich and Balmore, 1954, and in several bogs in Strathglass; small bog near the river bridge at Boat of Garten, pool by Flichity Inn, in a ditch at Tomachrochar near Nethybridge, plentiful in boggy fields near Coulan, on the E shore of Loch Flemington and on the margins of Loch Doirb above Inverfarigaig; bog at Essich, 1972 *M. Barron*.

THYMELAEACEAE
Daphne L.
D. laureola L.   *Spurge Laurel*
Doubtfully native in north Scotland. Rare in woods; usually near houses.
M Policies of Newton House, near Elgin, 1948; wood between Tearie Cottages and the Lodge of Darnaway Castle, 1954; very well established in a wooded ravine by a small burn S of the Manse at Rafford, 1967 **E,CGE**.
N Policies of the Druim E of Nairn and abundant about Cawdor Kirk, 1950.
I Woods at Holme House, Inverness, 1956.

ELAEAGNACEAE
Hippophaë L.
H. rhamnoides L.   *Sea Buckthorn*
Introduced. Rare in hedges and on dunes.
M Introduced, 1900 *JK* **FRS**; planted by the old curling pond at Seapark, Kinloss, 1918; on dunes by the drain running through Roseisle forest near Burghead, 1964; several bushes on the railway embankment S of Lossiemouth and covering a large area on the dunes near the golf clubhouse, 1972 **E**.
N Planted in the policies of the Druim near Nairn, 1930.

ONAGRACEAE
Epilobium L.
E. hirsutum L.   *Great Hairy Willow-herb, Codlins and Cream*
Native. Rare in ditches and by small burns.
M Ditch behind Brodie, 1840 *J. G. Innes* (Gordon, *annot.*); parishes of Urquhart and Speymouth (Burgess); ditch to the N of Lochinvar, 1960 **K**; abundant in the mill stream at Pittendreich, 1975 **E,BM**. The record in *Proceedings* **2**: 4 (1957) from the railway bridge between Garmouth and Spey Bay is in political Moray but in v/c 94 Banff.

N (Stables, *Cat.*); in the Findhorn basin N of Wades road, Ardclach, 1900 (Lobban).
I Marsh in the glebe of Rothiemurchus, 1872 *JK* **FRS**; probably introduced, on a dry bank at the entrance to Culloden station, 1971 *J. B. Dougan* **E**; by a quarry pool near Blacktown, Strathnairn, plentiful in runnels to the sea at Bunchrew and in a ditch near Balcarse farm, Kirkhill, 1975. The records shown in the (*Atlas*) for grids 28/43 and 28/44 are in error.

E. parviflorum Schreb.   *Small-flowered Hairy Willow-herb*
Native. In wet places usually in ditches. Formerly frequent but appears to be diminishing.
M Ditches round Loch Spynie (Gordon); Cothall, 1898 *JK* **FRS**; Culbin Sands, 1942 *UKD*, **UKD**; railway siding at Mosstowie, 1950; ditches at Boat of Garten, Dyke, Knockando and in dune slacks at Lossiemouth etc., 1954.
N (Stables, *Cat.*); Cawdor and Auldearn, 1900 (Lobban); wet track by the Loch of the Clans, 1954.
I (Watson, *Top. Bot.*); dune slack at Ardersier, 1954; ditch at Kirkhill, 1956; wood near Glen Feshie, 1966 *E. Rosser* **E**; bog in Glen Moriston, 1971 **E**.

E. montanum L.   *Broad-leaved Willow-herb*
Native. Abundant in woods, as a garden weed, on waste land and sometimes on mountain slopes. Fig 23g
M Darnaway, *Brodie* **E**; very common (Gordon); Greshop, 1856 *JK* **FRS**; ½ mile S of Forres station (albino form) and St. John's Mead, Darnaway, 1953 *A. Melderis* **BM**; Greshop wood, Forres, 1957 **K**; and throughout the county.
N (Stables, *Cat.*); by Lord Leven's bridge, Ardclach, 1882 *R. Thomson* **NRN**; garden weed at Drynachan, Geddes and Lethen etc., 1954; waste ground by the Allt Dearg, Cawdor Castle, 1961 **ABD,K**.
I Ness, 1854 *HC* **INV**; Kingussie, 1893 *J. Buchanan* **GL**; Inverness, 1940 *AJ* **ABD**; garden at Ben Alder Lodge, waste ground Dalwhinnie and throughout the district, 1954; scree slope at Gaick Lodge, 1973 **E**.

E. montanum × palustre = E. × montaniforme Knaf ex Čelak.
I Kilmorack near Beauly, 1892 *ESM* **CGE**; (Trail, *ASNH* **15**: 37 (1906)).

E. roseum Schreb.   *Small-flowered Willow-herb*
Native. Rare in waste places and as a weed of cultivation.
M Garden weed Westfield House near Elgin, 1953 **BM**; garden weed at Newton House, 1954; railway yard at Boat of Garten, 1956; rubbish tip at Elgin, 1957 **K**; distillery yard at Carron, 1968 **E,CGE**.
N Garden weed Cawdor Castle, 1960.
I Below walls about the town of Fort Augustus, 1950.

E. ciliatum Raf.   *American Willow-herb*
   E. adenocaulon Hausskn.
Introduced. Frequent on waste land and as a weed of cultivation.
Under-recorded. Fig 23a.
M Garden weed Dyke, 1972 **E**; in a flower bed Tolbooth street
Forres, abundant in Legge's Nursery Kinloss and waste ground at
Pluscarden, 1975.
N By the burn behind Boath House, Auldearn, 1975.
I Rubbish tip Longman Point, Inverness, 1971 **E**; by the pond
at Culloden House, garden weed Nairnside, Dores and Fort
Augustus, waste ground at Kirkhill, Daviot village and by Loch
Flemington, verge of a moorland track at Knockie, 1975.

E. tetragonum L.   *Square-stemmed Willow-herb*
Not seen in recent years. In the absence of herbarium specimens
former records must be treated with some doubt. They probably
referred to *E. obscurum*.
M Frequent (Gordon) (Moyle Rogers, *ASNH* **14**: 237 (1905)).
I Inverness, 1832 (Watson, *Loc. Cat.*); Englishton burn (Aitken).

E. obscurum Schreb.   *Short-fruited Willow-herb*
Native. Abundant in ditches, woods and as a weed of cultivation.
Fig 23e.
M Cliffs between Lossiemouth and Hopeman, at Garmouth and
Dunphail (Marshall, 1899); throughout the county, 1954; edge of a
bog at Sanquhar loch, Forres, 1961 **K**.
N (Stables, *Cat.*); between Findhorn and Nairn, 1912 (Ewing);
ditches at Drynachan, Cawdor, Dulsie Bridge etc., 1954.
I Beauly (*J. Bot.* **28**: 39 (1890)); Feshie Bridge, 1904 *G. B.
Neilson* **GL**; garden at Ben Alder Lodge, 1956; waste ground
Dalwhinnie, Glen Moriston etc., 1954.

E. obscurum × palustre = E. × schmidtianum Rostk.
M Moray (Trail, *ASNH* **15**: 33 (1906)).

E. obscurum × parviflorum = E × dacicum Borbás
M By the River Spey, Grantown, 1954 and near Forres, 1955,
det. G. M. Ash.

E. palustre L.   *Marsh Willow-herb*
Native. Very common in bogs and by rivers and burns throughout
the district. Fig 23f.
M Frequent (Gordon); Spynie, 1941 ? *coll.* **STA**; the Leen,
Garmouth, 1953 *B. Welch* **BM**; bog at the W end of the Buckie
Loch, Culbin and throughout the county, 1954.
N (Stables, *Cat.*); Ardclach, 1892 *R. Thomson* **NRN**; flush at
Drynachan, bogs by Loch of the Clans, Dulsie Bridge and Cawdor
etc., 1954.
I Dalwhinnie and Pitmain, 1833 (Watson, *Loc. Cat.*); Rothie-
murchus, 1872 *JK* **FRS**; Feshie Bridge, 1904 *G. B. Nielson* **GL**;
Drumashie Muir, 1940 *AJ* **ABD**; abundant throughout the district
1954.

Fig 23   a, **Epilobium ciliatum**; b, **E. anagallidifolium**; c, **E. alsinifolium**;
d, **E. brunnescens**; e, **E. obscurum**; f, **E. palustre**; g, **E. montanum**.

E. anagallidifolium Lam.   *Alpine Willow-herb*
   E. alpinum auct.
Native. Common in flushes by burns and on wet scree in the
mountain districts, often washed down rivers and becoming estab-
lished on loch margins and river shingle. Fig 23b, Map 119.
M Brought down the Spey from the uplands in the parishes of
Urquhart and Speymouth (Burgess); flushes in the Hills of Crom-
dale, 1954; flush by the Allt Iomadaidh at Dorback and on the
slopes of Càrn Dearg Mòr above Aviemore, 1972.
N Streens near Nairn, 1886 *R. Thomson* **NRN**.
I Dalwhinnie, 1833 (Watson, *Loc. Cat.*); Badenoch (Gordon);
Cairn Gorm, 1839 *J. Johnstone* **K**; Kingussie, 1840 *A. Rutherford* **E**;
Fraoch-choire, Cannich (Ball, 1851); Cairngorms, 1874 *A. Ley* **E**;
Coire an t-Sneachda, 1893 *AS* **K**; Creag Coire Doe and Càrn a'
Chuilinn, 1904 *G. T. West* **STA**; on a shingle spit at the outflow
of the River Tarff to Loch Ness, 1904 *G. T. West*; common on all
the higher mountains, 1954.

E. anagallidifolium × palustre
I Shingle in a stream bed Glen Affric, 1947 *E. Vachell* **E** (*Watsonia*
1 : 1 (1949)); on shingle of a small burn not far from the track on
the N side of Loch Affric and *c.* ½ mile from Affric Lodge (perhaps
the same locality), 1971 **E**.

E. alsinifolium Vill.   *Chickweed Willow-herb*
Native. Frequent in flushes and by mountain burns. Fig 23c,
Map 120.
M Flush in the Hills of Cromdale, 1954; by a small waterfall by
the Allt Iomadaidh near Dorback, 1972 **E**.
N Abundant in one or two flushes by the Allt Breac Drynachan,
1974 **E**.
I Dalwhinnie, 1833 (Watson, *Loc. Cat.*); Cairn Gorm, 1839
J. Johnstone **K**; Glen Einich, 1893 *AS* **E,GL,K**; by the burn in
Glen Mazeran, flushes in glens Feshie and Banchor etc., 1954;
by a burn by Loch Affric, 1971 **E**; in the burn gully at Gaick
Lodge, 1973 **E**.

E. brunnescens (Cockayne) Raven & Englehorn   *New Zealand*
   subsp. **brunnescens**   *Willow-herb*
   E. nerterioides auct.; *E. pedunculare* auct.
Introduced. Abundant on river shingle, rocks and waste places.
Fig 23d, Map 121.
M Speymouth (as *E. alsinifolium*) an escape multiplying rapidly in
some places (Burgess); quarry on the Knock Hill near Elgin, 1950;
shingle of the River Findhorn near Forres, 1953 *A. Melderis* **BM**;
distillery yard at Knockando, 1966 **E,ABD**, etc.
N On shingle of the River Findhorn at Banchor, 1951; shingle of
the River Nairn at Nairn, 1962 **ABD**; and wet gravel bank at
Drynachan, 1963 **E,BM**; and elsewhere.
I A few miles from Inverness, 1944 *R. J. Pealling* **E**; gravel drive
Ben Alder Lodge, 1956; shingle by a small burn Corrimony, 1964

BM; by the Elrick Burn, Coignafearn, 1970 **E,BM**; Dunmaglass, 1972 **E** etc.

E. angustifolium L. *Rosebay Willow-herb, Fireweed*
Chamerion angustifolium (L.) Scop.
Native. Abundant in wood clearings, waste land and on mountain ledges and among rocks.
M Frequent, on rocks by the Divie above Relugas and by the Falls of Lossie (Gordon); Relugas, 1857 *JK* **FRS**; abundant throughout the county, 1954.
N (Stables, *Cat.*); rare by the Findhorn at Ardclach, 1888 *R. Thomson* **NRN**; rare in a burn gully on Drynachan moor, but abundant throughout the county elsewhere.
I Falls of Foyers and woods by Loch Ness, 1821 (Hooker, *Fl. Scot.*); Dalwhinnie, 1833 (Watson, *Loc. Cat.*); Slochmuichd, *WAS* (Gordon); fields at Diveach (Aitken); head of Glen Einich, 1893 *AS* **E**; Petty, 1940 *AJ* **ABD**; cliff ledge at 850 m Sgùrr na Lapaich, Glen Strathfarrar, by Ben Alder Lodge and in a burn gully at Gaick etc., 1973.

Oenothera L.
O. biennis L. *Evening Primrose*
Garden escape.
M Railway cutting, Speymouth, *G. Birnie* (Burgess).

O. erythrosepala Borbás
Garden escape.
M Station yard, Forres, 1967 **CGE**.

Circaea L.
C. lutetiana L. *Enchanter's Nightshade*
Native. Woods. Very rare, formerly confused with *C. intermedia*. Fig 24a.
M Policies of Brodie Castle, 1961 **ABD,BM,K**.
N In shade of the River Findhorn at Daltra, *Dr Sclanders* **NRN**.
I Inverness and parishes of Urquhart and Glen Moriston (Murray); Dochfour wood and Muckovie Quarry (Aitken); among rocks in shade by the shores of Loch Ness near Brackla, 1975 **E**.

C. alpina L. *Alpine Enchanter's Nightshade*
Native. Very rare among rocks and woods in the mountain districts. Fig 24c.
I Under juniper above the old lime quarry on the hillside of Suidhe, Kincraig, 1973 *R. McBeath* **E!**, confirmed by P. H. Raven.

C. alpina × lutetiana = C. × intermedia Ehrh.
Intermediate Enchanter's Nightshade
Native. Frequent in shade by rivers and burns and among stones and rocks on the shores of lochs. Many of the old records for *C. alpina* belong here (cf. P. H. Raven, *Watsonia* 5: 262–272 (1963)). Fig 24b, Map 122.
M Gordon Castle woods, 1829, by the burn at Burgie, *G. Wilson* and by the Findhorn at Sluie (as *C. alpina*) (Gordon); Speyside in

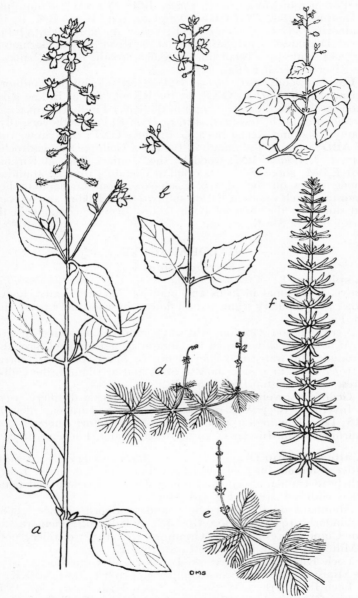

Fig 24 a, **Circaea lutetiana**; b, **C. intermedia**; c, **C. alpina**; d, **Myrio phyllum alterniflorum**; e, **M. spicatum**; f, **Hippuris vulgaris.**

the Province of Moray (as *C. alpina*) *WAS* (Murray); damp gully at Dunphail, 1886 *ESM* **CGE**; Dunphail, 1897 *WASh* **BM**; by the Findhorn, 1897 *G. C. Druce* **OXF**; below the cliffs at St. John's Mead, Darnaway, 1950; backwater of the River Spey at Garmouth, 1953; in shade by a small burn at Auchinroath, Rothes; banks of the River Lossie at Kellas, 1973 **E**.

N Cawdor woods (as *C. alpina*) *WAS* (Gordon); by the Findhorn, Nairn, 1889 *G. C. Druce* **OXF**; by the River Nairn at Holme Rose, woods at Cawdor and the length of the River Findhorn, 1954.

I   Shores of Loch Ness (as *C. alpina*), *WAS* **E**; Falls of Foyers, 1870, Beauly, 1889 and Reelig burn, *G. C. Druce* **OXF**; Bunchrew, 1940 *AJ* **ABD**; wood at the mouth of the River Coiltie, Drumnadrochit, 1947 *A. J. Wilmott* **BM**; wood by the Moniack Burn at Kirkhill, 1961 **E,BM**; gorge by the waterfall at Glendoe, 1972 **E**; abundant among stones on the edge of Loch Ness from Lochend to Fort Augustus; birch woods at Farley and by Loch Meiklie; among rocks and stones on the shore of Loch Killin, 1972 **E**; common in the burn gullies of Strathglass etc.

HALORAGACEAE
Myriophyllum L.

M. spicatum L.   *Spiked Water-milfoil*
Native. Occasional in lochs and pools of base-rich water. Leaves with 13–35 capillary segments; all flowers in whorls and petals dull red. Fig 24e.

M Very common (Gordon) (the status must be considered an error for *M. alternifolium* not listed by Gordon); stagnant pool, probably introduced by waterfowl, Speymouth, *G. Birnie* (Burgess); Forres, 1871 *JK* **FRS**; in a water hole at Snab and fire dam in the Culbin Forest, 1954; Gilston Loch, 1972 **E**.

N Geddes reservoir, 1961 **ABD**; pond at Brackla distillery, 1975.

I   Inverness, 1833 (Watson, *Loc. Cat.*); Loch Ruthven, 1860 *JK* **FRS**; Loch Flemington, pool by Loch Ness at Fort Augustus, and Muirtown Basin, Inverness, 1954; Abbar Water, Lochend, 1975.

M. alterniflorum DC.   *Alternate-flowered Water-milfoil*
Native. Common in pools, lochs and burns in acid water. Leaves with 6–18 capillary segments, upper flowers alternate and petals yellow with red streaks. Fig 24d, Map 123.

M Abundant on the rivers Spey and Findhorn (Druce, 1888); Lochindorb, 1955 *J. Milne* **E**; Loch Puladdern, Aviemore, fire dam Culbin Forest, pool at Grantown etc., 1954; distillery pond at Miltonduff, 1972 **E**; Spynie Loch, 1972 **E**.

N Loch Kirkaldy, Glenferness, Lochloy and Cran Lochs, 1954; the Minister's Loch on Nairn Dunbar golf course, 1961 **ABD,K**.

I   Loch Ericht, 1833 (Watson, *Loc. Cat.*); Loch Alvie, 1891 *AS* **E**; Lochan Càrn a' Chuilinn, Loch nan Eun, Loch Tarff, Loch Oich, Loch Ashie etc., 1904 *G. T. West* **E,STA**; Torvaine, 1940 *AJ* **ABD**; abundant in all the lochans from Drumochter to Inverness, 1954.

HIPPURIDACEAE
Hippuris L.
H. vulgaris L.   *Mare's-tail*
Native. Common in base-rich water in pools, lochans and slow-running burns in the lowlands, less so in the highlands. Fig 24f, Pl 10, Map 124.

M Spynie Loch, 1823 Loch of Cotts, the Leen, Garmouth and Abbeywell at Urquhart (Gordon); the Moss of Chapelton, Forres, Moy Carse, Loch of Blairs, 1842 and Dyke Moss, 1843 *J. G. Innes* (Gordon, *annot.*); Netherton, 1871 *JK* **FRS**; pool in the old lime quarry at Cothall, 1930; ditch at Kinloss, 1925; fire dam Culbin Forest, moat at Duffus Castle, burn on Dava Moor etc., 1954; Gilston Loch, 1972 **E**.

N (Stables, *Cat.*); Nairn, *A. Falconer* (Gordon); Moss of Litic and Inshoch, *J. B. Brichan* (Gordon, *annot.*); Ardclach, 1886 *R. Thomson* **NRN**; lochs Flemington, Lochloy, Kirkaldy, Cran Loch and Loch of the Clans, 1954; the Minister's Loch, Nairn, 1960 **ABD**; Geddes reservoir, 1961 **ABD,K**.

I Marsh in Dunain wood and Culcabock burn, 1885 (Aitken); Loch Garten and the curling pond at Achnagairn, Kirkhill, 1954; pool by the River Glass at Eskadale; small lochan at Essich, 1972 *M. Barron*; bog at Kingussie, 1972 *R. McBeath*; loch N of An Lea-cainn, Blackfold and in the farm pond at Achvraid near Essich, 1973; bog 6 miles W of Tomatin, 1974 *J. A. Burgess* **STA**; tiny plants submerged in a peat bog pool at 700 m, near Fort Augustus, 1974 *H. A. McAllister et al.*

CALLITRICHACEAE
Callitriche L.
C. stagnalis Scop.   *Water-Starwort*
Native. Abundant in small burns, flushes and pools etc., throughout the district. A very variable species. Fig 25b. The terrestial forms with thyme-like leaves have been named var. **serpyllifolia** Lönnr. and are found in dried out places on mud and occasionally on arable land.

M Very common (Gordon); Greshop wood, 1857 *JK* **FRS**; by the Muckle Burn, Moy, 1953 *A. Melderis* **BM**; margin of the pond at Sanquhar House Forres, 1961 **K**; ditch at Boat of Garten, 1962 **ABD**; pond at Dallas Lodge, 1967 **CGE**; ditch at Auchinroath near Rothes, 1973 **E**.

N Ardclach, 1900 (Lobban); throughout the county, 1954; pool by the Allt Breac at Drynachan, 1973 **E**; muddy verge of the Ord Loch, 1973 **E**.

I Pitmain and Dalwhinnie, 1833 (Watson, *Loc. Cat.*); abundant throughout, 1954; runnel in Glen Mazeran, 1968 **E,BM**; pool by the River Enrick, Milton, 1973 **ABD**; pool W of Newtonmore, 1973 **CGE**; runnel on Sgùrr na Lapaich at 850 m, and throughout the district.

C. platycarpa Kütz.   *Various-leaved Water-starwort*
   C. verna auct.
Both Gordon and Burgess recorded *C. verna* as common, which, in
the absence of herbarium specimens must be treated as errors for
C. stagnalis.
M Ornamental pond at Altyre House, 1954 det. A. Melderis.
N Ardclach (Thomson), probably an error.
I Inchnacardoch Bay, 1904 *G. T. West* **E,STA** (det R. D. Meikle
as *C. stagnalis*); Daviot, 1940 *AJ* **ABD**.

C. hamulata Kütz. ex Koch   *Intermediate Water-starwort*
   C. intermedia Hoffm.
Native. Common. Leaves linear, widened and deeply emarginate at
the apex resembling a bicycle spanner. Fig 25c, Map 125.
M Logie (Druce, 1888); burn at Dava, 1890 *JK* **FRS**; by the
Allt Iomadaidh at Dorback, 1972 **E**; small burn by Lochindorb,
1972 **E**; scattered throughout the county.
N Ardclach, 1900 (Lobban); small burn at Drynachan, 1954;
ditch at Dunearn, 1961 **ABD**; Geddes reservoir, 1961 **E,K**; Loch
Kirkaldy by Glenferness, 1972 **E**.
I (Watson, *Top. Bot.*); Loch Uanagan, 1904 *G. T. West* **E**, also
collected from lochs Tarff, Ness, and the lochs north of Inver-
moriston and in Balmacaan forest etc.; salt marsh at Beauly, 1943
UKD **E**; overflow of Loch Conagleann at Dunmaglass, 1972 **E**;
burn above Gaick Lodge, 1973 **E**; pool by the River Spey W of
Newtonmore, 1973 **E**.

C. hermaphroditica L.   *Autumnal Water-starwort*
   C. autumnalis L.
Native. Frequent. Leaves linear-lanceolate, widest at the base
tapering to an emarginate apex. Fig 25a, Map 126.
M Frequent, Alves, *G. Wilson* (Gordon); Loch Spynie, *P. Leslie*
(Burgess); ditch on moor at Boat of Garten, pools at Grantown,
Dallas, Loch Romach etc., 1954; in the Allt Iomadaidh at Dorback,
Millbuies lochs, Lochnabo at Lhanbryde and Lochindorb all in
1972 **E**.
N In the Coulmony burn at Ardclach, 1894 *R. Thomson* **NRN**;
Loch Kirkaldy near Glenferness, 1972 **E**; Clunas reservoir, 1975.
I Loch Gynack, Kingussie (Druce, 1888); pool in Glen Mazeran,
1968 **CGE**; flush at 915 m. Sgùrr na Lapaich **E**; backwater of the
River Calder at Newtonmore, Loch Carrie, Glen Cannich, Doch-
four and Moy Lochs, pool by Loch Toll a' Mhuic, muddy shores
of Loch Ness, Loch Flemington and Loch Bunachton etc.

LORANTHACEAE
Viscum L.
V. album L.   *Mistletoe*
Introduced.
M Two plants, one male the other female grew at Duffus House
until 1936, *R. Richter.*

Ligusticum scoticum × ⅕
Scots Lovage

Phyllodoce caerulea × 2   *J. Thorne*
Blue Heath

Polygonum viviparum × 1
Alpine Bistort

Loiseleuria procumbens × ⅔
Trailing Azalea

PLATE 11

Arctostaphylos urva-ursi × ½
Gnashacks, Bearberry

Arctostaphylos urva-ursi (*fr*

Vaccinium vitis-idaea × ¼
Scots Cranberry, Cowberry

Vaccinium vitis-idaea (*fruit*) × ½

Arctostaphylos alpinus × ⅔
Alpine Bearberry

PLATE 13

Moneses uniflora × ⅔
One-flowered Wintergreen

ola minor × 1
mmon Wintergreen

Pyrola rotundifolia × 1   *S. J. Heyward*
Larger Wintergreen

Pyrola media × 2   *D. A. Lawrie*
Intermediate Wintergreen

PLATE 14

Orthilia secunda × ⅔
Serrated Wintergreen

Trientalis europaea × ⅔
Trientale, Chickweed Wintergreen

Linnaea borealis × 1
Linnaea, Twin-flower

Monotropa hypopitys × 1
Yellow Bird's-nest

PLATE 15

I Beaufort Castle Gardens (*Stat. Acc. Scot.* 1794); Springfield, Inverness (believed by Dr Simpson to be self-sown) (Aitken); planted on an apple tree 15 years ago at Inverness, *James A. Gossip* (*Trans. ISS & FC* **6**: 75 (1902)).

CORNACEAE
Cornus L.

C. sanguinea L.   *Dogwood*
   Swida sanguinea (L.) Opiz: *Thelycrania sanguinea* (L.) Fourr.
Introduced.
M Certainly introduced, rare in old shrubberies such as at Innes House and Gordon Castle (Gordon).
I About Inverness (Gordon); Kilmorack, 1940 *AJ* **ABD**; Reelig Glen, 1925; Alvie and Kincraig, 1952; railway embankment near Moy station.

C. sericea L.
   Swida sericea (L.) Holub; *Thelycrania sericea* (L.) Dandy
Introduced.
M Near Carron station, 1965, planted by the lake at Gordonstoun, 1972 **E**, relic of an old cottage garden at Tilliglens near Relugas.
I By the railway S of Daviot and by the path to the lochan at Ness Castle, 1954; edge of a wood at Cruives.

C. suecica L.   *Dwarf Cornel*
   Chamaepericlymenum suecicum (L.) Aschers. & Graebn.
Native. Common on acid moorland over 610 m. Pl 10, Map 127.
M Among blaeberries and heather on the N slope of Geal-chàrn Beag, Aviemore, 1974 **E**.
I About Inverness, *G. Anderson* (Hooker, *Fl. Scot.*); Dalwhinnie, 1833 (Watson, *Loc. Cat.*); Mamsoul, 1836 *Messrs Anderson, Stables & Gordon* (Gordon); Kingussie, 1841 *A. Rutherford* **E**; Glen Strathfarrar and Mealfuarvonie (Aitken); Strathnethy and the Larig Pass, 1892 *AS* **E** etc.

ARALIACEAE
Hedera L.

H. helix L.   *Ivy*
   Var. **helix**
Native. Climbing on trees, walls and rocks. Very variable in leaf shape. Bears fruit only in the lowland areas.
M Frequent (Gordon); woods at Forres (Innes); throughout the county, 1954.
N (Stables, *Cat.*); Ardclach (Thomson); woods at Holme Rose, Cawdor Castle etc., 1954; on the bridge at Dulsie; small wood by Cran Loch etc., 1954.
I Inverness, 1855 *HC* **INV**; Strathglass, 1940 *AJ* **ABD**; on cliffs by Loch Duntelchaig and Loch Killin, rocks in the Pass of Inverfarigaig and at Eskadale etc.; on scree and rocks (with very small leaves) above Lurgmore, 1975 etc.

Var. **hibernica** Kirchin   *Irish Ivy*
Planted on walls, perhaps native in woods. Differs from var. *helix*
by its larger 5-lobed leaves of a soft pliant texture, paler green
colour and long pinkish-green petioles. The leaves are not quite
flat and do not turn purple-black in winter.
M Garden out-cast opposite Dalvey Cottage near Forres, extending
over a large area on the railway embankment, 1975 **E**; policies of
Newton House near Elgin, 1975 **E**; abundant at Pitgaveny.
N Wood by the lodge in the policies of Geddes House, near
the shop and by the Kirk in Cawdor village, on the wall and in the
wood of Firhall House at Nairn, 1976.
I On the lodge wall at Newton House, Kirkhill, on the Kirk wall
at Eskadale, by an old cottage at Torgormack by Kilmorack and
on a wall by the pond at the keeper's cottage Beaufort Castle, 1976.

UMBELLIFERAE
Hydrocotyle L.
H. vulgaris L.   *Marsh Pennywort*
Native. Common in bogs and marshes on acid soil. Map 128.
M About Elgin, *G. Gordon* **ABD**; Chapelton, Forres, 1856 *JK* **FRS**;
by the Buckie Loch, Culbin 1913 *D. Patton* **GL**!; the Leen at
Garmouth, 1953 *A. Melderis* **BM**; by Loch Puladdern at Aviemore,
Loch Spynie, Loch Vaa, Blairs Loch near Forres, at Grantown,
Lochinvar, Kellas etc. 1954.
N (Stables, *Cat.*); by the Free Church Moss at Ardclach, 1882
R. Thomson **NRN**; bogs by Lochloy, Cran Loch and near Littlemill,
1954 etc.; Geddes Reservoir, 1961 **ABD,K**.
I (Watson, *Top. Bot.*); behind Dunain, ? date and *coll.* **INV**;
Scaniport, Lentran and Nairnside (J. Don, 1898); bog by Loch
Ashie, 1904 *G. T. West* **E,STA**; Loch Flemington, 1940 *AJ* **ABD**;
common round lochans in Strathspey and Strathglass, 1954; in
mud by the River Beauly, on a steep grassy hillside at Breakachy
and by the farm pond at Essich etc.

Sanicula L.
S. europaea L.   *Sanicle*
Native. Frequent in deciduous woods, on river banks and occasion-
ally among rocks in the mountain areas. Map 129.
M Wood at Grantown, 1834 *J. G. Innes* **FRS**; frequent in Glen-
latterach and Dunphail (Gordon); Darnaway woods, 1857 *JK*
FRS; in the gorge at Huntly's Cave on Dava Moor, banks of the
River Findhorn at Dunphail and Relugas, under hazel at the
Cothall quarry and by the River Spey at Rothes and Wester
Elchies, 1954; small wood by the old kirk at Balnabreich near
Mulben, 1974 **E**.
N Cawdor Woods, *WAS* (Gordon)!; woods by the River Findhorn
at Glenferness, Dulsie Bridge and Coulmony, 1954; banks of the
River Nairn at Holme Rose and in a small wood by the ford at
Fornighty.

I (Watson, *Top. Bot.*); Dunain and by the Holm Burn near Inverness (Aitken); Reelig (J. Don, 1898); near Dores, 1940 *AJ* **ABD**; common in the woods by the shores of Loch Ness and by the River Glass in Strathglass, 1930; gully on the cliffs of Tom Bailgeann near Torness, among rocks in the Ryvoan Pass, in Glen Banchor, Glen Doe and on Creag na h-Iolaire above Aviemore, 1954; among rocks below the cliffs at Creag Dhubh near Laggan, 1972 **E**.

Astrantia L.
A. maxima L.
Introduced. Well established on the Islands at Inverness, 1961 **E**.

Chaerophyllum L.
C. temulentum L. *Rough Chervil*
Native. Frequent on roadside verges and in hedgerows in the lowlands. Map 130.
M Doubtfully native, Forres 1830 and Alves 1833 (Gordon); Forres, 1878 *JK* **FRS**; hedges near Forres, 1898 *ESM* **CGE**; plentiful by the roadside near Milton Brodie, 1925; ½ mile S of Forres station, 1953 *A. Melderis* **BM**; in all the coastal areas reaching Dallas and Kellas.
N (Stables, *Cat.*); hedgerows at Holme Rose, Cawdor, Nairn and Auldearn, 1954; roadside bank near Kinsteary, Auldearn, 1962 **ABD**.
I Lochdhu (Aitken); near Bellfield (J. Don, 1898); roadside bank near Ardersier, 1955; Culloden, Kirkhill, Croy, near Beaufort Castle, by the Longman tip Inverness, and by the railway bridge ¼ mile W of Bunchrew, 1956.

Anthriscus Pers.
A. caucalis Bieb. *Bur Chervil*
   A. neglecta Boiss; *A. vulgaris* Pers.
Native. Frequent in the coastal areas on sandy soil in waste places. In Collectanea for a Flora of Moray the Rev. G. Gordon cites *A. caucalis* as 'frequent' and *A. sylvestris* as 'doubtfully native'. This is an error and his records, should be transposed. Map 131.
M Forres, 1834 *J. G. Innes* **FRS**; doubtfully native, Elgin Cathedral (Gordon); Castlehill, Forres, 1856 *JK* **FRS**; Knockando, *C. Watt* **ABD**; roadside near Forres (Marshall, 1899); Lossiemouth, 1907 ? *coll.* **GL**; waste sandy ground at Findhorn, 1925; Rothes and Kingston, 1954; waste land at Lossiemouth, 1966 **CGE**; garden weed at Dyke, 1966 **E,BM**.
N (Stables, *Cat.*); close to Nairn, 1898 *ESM & WASh*; Ardclach and Cawdor, 1900 (Lobban); waste ground at Auldearn, 1958 **ABD,BM**; roadside bank by the road turning to Geddes on the Cawdor road, 1976 **E**.
I Ruthven Barracks, Kingussie (Gordon) sandy soil at Ardersier, 1925; waste land at Kirkhill, 1930 *M. Cameron*; on Longman Point, Inverness and Whiteness Head, 1954; waste land by Kilmorack Kirk, 1975; bank by Ruthven Barracks, 1973 **E,BM**.

A. sylvestris (L.) Hoffm.   *Queen Anne's Lace, Cow Parsley*
Native. Abundant on roadside verges, in waste places, on banks and
by ditches. Map 132.
M Innes House, Urquhart, 1794 *A. Cooper* **ELN**; frequent (Gordon);
Greshop, 1856 *JK* **FRS**; abundant on the island in Lochindorb,
1882 (Thomson)!; throughout the county, 1954.
N (Stables, *Cat.*); roadside verges at Holme Rose, Nairn, Cawdor,
Lethen and Auldearn etc., 1954.
I Dalwhinnie, 1833 (Watson, *Loc. Cat.*); Dalcross (Aitken); railway
embankment S of Newtonmore and in all the lowland areas, 1954;
between Inverness and Beauly, 1962 *J. Lamond* **E**.

A. cerefolium (L.) Hoffm.   *Garden Chervil*
Garden escape.
M Speymouth, *G. Birnie* (Burgess); near Forres, 1975 *K. Christie*.
I (Druce, *Com. Fl.*); planted at Dochfour House, 1975 **E,BM**.

Scandix L.
S. pecten-veneris L.   *Shepherd's Needle, Venus's Comb*
Introduced. Formerly abundant in cornfields, now only as a casual
in railway station yards.
M Cornfield at Alves, 1834 *J. Fraser* **FRS**; doubtfully native,
cornfields at Drainie, Duffus and Alves (Gordon); garden weed
Forres, 1856 *JK* **FRS**; introduced with farm seed in the parishes of
Forres, Elgin, Urquhart, Knockando and Bellie (Burgess); shingle
of the River Spey at Kingston, 1953; railway siding at Forres, 1961
E,ABD.
N (Stables, *Cat.*); Nairnshire, *J. B. Brichan* (Gordon, *annot.*).
I Found once or twice at Inverness (J. Don, 1898).

Myrrhis Mill.
M. odorata (L.) Scop.   *Myrrh, Sweet Cicely*
Introduced, but now well established on roadside verges, river
banks and waste places. Smells strongly of aniseed. Map 133.
M Certainly introduced, rare at Bilboahall, Elgin and above
Cothall near Forres, *J. G. Innes* (Gordon); Dalvey, 1857 *JK* **FRS**;
abundant in the lowland areas, 1954; by the Burnie Path, Dyke
and by ditches at Kinloss etc., reaching Carrbridge station, Boat
of Garten and the hills of Cromdale in the highlands.
N (Stables, *Cat.*); Lethen, 1833 *J. B. Brichan* **ABD**; Holme Rose,
Cawdor, Geddes, Glenferness and Dulsie Bridge, 1954; roadside
bank near Auldearn, 1962 **ABD**.
I Kingussie, 1840 *A. Rutherford* **E**; near Inverness, 1854 *HC* **INV**;
frequent in the lowland areas but often near cottages and farms,
1954; planted at Tomatin; rare at Newtonmore near houses and
reaching Skye of Curr, Kincraig, Invermoriston and by the Lynwilg
Hotel at Alvie.

Torilis Adans.
T. japonica (Houtt.) DC.   *Upright Hedge Parsley*
   T. anthriscus (L.) C. C. Gmel; *Caucalis anthriscus* (L.) Huds.
Native. Common in the lowlands on roadside banks and verges, and

in open deciduous woods, rare below cliffs and on steep banks in the highlands. Map 134.

M Forres, 1834 *J. G. Innes* **FRS**; between Rothes and Sourden, at Pluscarden and Old Duffus (Gordon); Waterford, near Forres, 1856 *JK* **FRS**; common in all the lowland areas, 1954; verge to a track in the Culbin Forest, 1968 **E**.

N By the River Findhorn at Daltra, 1888 *R. Thomson* **NRN**; roadside verges at Holme Rose, Cawdor, Dulsie Bridge and Glenferness etc., 1954.

I Inverness, 1833 (Watson, *Loc. Cat.*); Lentran, 1892 *ESM* **CGE**; Lower Leachkin, 1940 *AJ* **ABD**; grass verge of small wood at Milton, 1947; in all the lowland areas, 1954; waste ground by an old cottage in the Pass of Inverfarigaig, near Kingussie station and below the cliffs of Creag Dhubh near Laggan etc.

Coriandrum L.
C. sativum L. *Coriander*
Casual on waste land and rubbish tips. Smells strongly of bed-bugs.
M Speymouth (Burgess).
I Rubbish tip on the Longman Point at Inverness, 1971 **E**.

Conium L.
C. maculatum L. *Hemlock*
Native. Frequent on waste ground and open places in the lowlands, rare in the highlands where it is probably introduced. Deadly poisonous. Readily recognised in the field by the purple-blotched stems. Map. 135.
M Doubtfully native, frequent at Waulkmill, in the churchyards at Spynie, Duffus and Birnie, waste ground at Lossiemouth, Ardgay and several other places in Alves, at Castlehill, Forres, *J. G. Innes* (Gordon); Castlehill, 1857 *JK* **FRS**; waste ground at Kinloss 1930; in all the coastal areas, 1954; waste ground at Grantown, 1956; bank near Rafford Kirk, 1976.
N Cawdor woods, *WAS* (Gordon); Auldearn, 1837 *J. B. Brichan* (Watson)!; roadside verge near Penick, waste ground at Cawdor, by Howford Bridge, and by the Newton Hotel, Nairn, 1954.
I Inverness, 1833 (Watson, *Loc. Cat.*)!; near Inverness, 1854 *HC* **INV**; waste ground at Ardersier and Kirkhill; tow path by the river at Beauly, roadside verges at Gollanfield, at Petty and Skye of Curr; ditch at Dalwhinnie, 1974.

Bupleurum L.
B. lancifolium Hornem. *False Thorow Wax*
Casual. Introduced with bird seed.
M Rubbish tip at Elgin, 1960 **E,K** and 1973 **E**.
N From 'Swoop', Brackla, 1968 *H. M. Finlay* **ABD** and Nairn, 1968, *MMcCW* **E,CGE**.

Apium L.
A. graveolens L. *Wild Celery*
M Certainly introduced near Kinloss Abbey, 1831 *G. Wilson*

(Gordon); bridge at Kinloss, 1834 *J. G. Innes* **FRS**; Kinloss, 1837 *G. Gordon* **ABD** and 1856 *JK* **FRS**.

A. nodiflorum (L.) Lag. *Fool's Watercress*
Native. Very rare in ditches and runnels.
M (Druce, *Com. Fl.*)
I (Druce, *Com. Fl.*); roadside ditch near Tornagrain Farm, Croy, 1954; runnel on moor by Loch an Ordain near Torness, 1976 **E,BM**.

A. inundatum (L.) Reichb. fil. *Lesser Marshwort*
Native. Frequent in shallow pools and margins of lochans on acid moorland. Map 136.
M Alves, *J. B. Brichan* (Watson); Spynie Loch (Gordon); moss near Kintessack, *J. G. Innes* (Gordon, *annot.*); Garmouth, 1900 *JK* **FRS**; formerly plentiful in the Buckie Loch, Culbin and Hatton Loch near Kinloss but probably extinct there owing to drainage; pool near Relugas and in the Dyke pond, 1954; abundant on the muddy margin of Millbuies Loch S of Elgin, 1972 **E**.
N East of Nairn, *WAS* 1833 **ABD**; Merrytown near Nairn, *WAS* (Gordon); pool on moor S of Loch of the Clans, 1960 **ABD**; bog near Easter Lochend, 1960 **K**.
I Pitmain, 1833 (Watson, *Loc. Cat.*); Inches near Inverness, 1871 ? *coll.* **ELN**; parish of Croy, 1881 *T. Fraser*; ditch at Bunchrew (Aitken); Loch Dochfour, 1904 *G. T. West* **E,STA**; frequent in all the pools in Strathglass, at Abbar Water Lochend, margin of lochans near Newtonmore and Kincraig and at Lochan Dhubh, Inverfarigaig etc., 1954; pool in Glen Banchor, 1973 **BM**.

Petroselinum Hill

P. crispum (Mill.) Nyman *Parsley*
Garden escape on the sand dunes at Lossiemouth in Moray, and on waste land near Foyers in Easterness.

Cicuta L.

C. virosa L. *Water Hemlock, Cowbane*
Native. Rare in marshes.
I Ruthven Barracks near Kingussie, 1898 *JK* **FRS**; do., 1973 *MMcCW* **E**; marsh by small pool SE of the railway at Balavil near Kingussie, 1973 **E,BM**; marsh by small lochan at Inverdruie, 1975

Carum L.

C. verticillatum (L.) Koch *Whorled Caraway*
Native. Rare in acid grassy bogs. Not seen in recent years.
I (Druce, *Com. Fl.*).

C. carvi L. *Caraway*
Introduced. Formerly planted by farms and crofts. The seeds were used to flavour cheese.
M Certainly introduced, near Waterton on the Duffus road and at Drumbain (Gordon); around farmsteads in the parishes of Dyke, Elgin, Duffus, Rothes, Knockando and Cromdale (Burgess); farm

lane E of Archiestown, 1955; grass bank at the edge of a field at
Tomlea near Knockando, 1956 **ABD**.
N (Stables, *Cat.*); plentiful by the Free Church at Ardclach, 1892
R. Thomson **NRN**.
I Deserted gardens, Inverness (J. Don, 1898); Longman (G. A.
Lang, 1905).

Conopodium Koch
C. majus (Gouan) Loret   *Lousey Arnuts, Knotty Meal, Cronies, Pignut*
Native. Abundant in birch woods, on grassy banks by burns and
rivers and in meadows. The tubers are edible. Map 137.
M Very common (Gordon); Greshop, 1898 *JK* **FRS**; by the Muckle
Burn at Moy, 1953 *A. Melderis* **BM**; grass verge at Dyke, 1963
CGE, and throughout the county.
N (Stables, *Cat.*); Ardclach (Thomson); by the river at Nairn,
1962 **ABD**; and throughout the county.
I Dalwhinnie and Pitmain, 1833 (Watson, *Loc. Cat.*); Ness
Islands, 1854 *HC* **INV**; grassland at Ben Alder Lodge, meadow by
Loch an t-Seilich, at Gaick Lodge and throughout the district.

Pimpinella L.
P. saxifraga L.   *Burnet Saxifrage*
Native. Frequent in short grassland on base-rich soil. It flowers
later than the other umbellifers. Map 138.
M Innes House, Urquhart, 1794 *A. Cooper* **ELN**; abundant in one
or two places, Alves, *G. Wilson* and Grantown *WAS* (Gordon);
grassy roadside verge at Scotsburn, 1950; river pasture at Grantown,
1970 **E,ABD**; railway embankment near Rafford, pasture by the
river Dulnain at Muchrach at Boat of Garten and by the River
Spey at Fochabers.
N (Stables, *Cat.*); Pigeon Hill, Auldearn, *J. B. Brichan* (Watson).
I Pitmain, 1833 (Watson, *Loc. Cat.*); abundant in one or two
places in Badenoch (Gordon); pasture at Drumnadrochit, 1947;
at Achnagairn House, Kirkhill, 1961 **BM,K**; bank by the road
S of Nethybridge, 1971 **E**; locally common in Strathglass and the
Spey valley, grass field at Essich and by the River Dulnain at Skye
of Curr etc.

Aegopodium L.
A. podagraria L.   *Goutweed, Bishop's Weed, Ground Elder*
Introduced, now well naturalised in waste places and a persistent
weed in gardens.
M Doubtfully native, frequent (Gordon); Greshop, 1856 *JK* **FRS**;
banks of the River Lossie at Elgin, 1941 ? *coll.* **STA**; throughout
the county 1954.
N (Stables, *Cat.*); Ardclach (Thomson); throughout the county,
1954 but rare at Drynachan and on the Carse of Delnies.
I Inverness, 1832 (Watson, *Loc. Cat.*); Midmills, Kingsmills and
Ness etc. (Aitken); Lower Leachkin, 1940 *AJ* **ABD**; garden at
Ben Alder Lodge and only absent from the moorland areas without
cultivated land.

236

Berula Koch
B. erecta (Huds.) Coville   *Narrow-leaved Water-parsnip*
Sium erectum Huds.; *S. angustifolium* L.
Native. Rare in marshes.
M Spynie Loch, 1835 *G. Gordon* **E**!; do. *WAS* **E,ABD**; W of Spynie
Loch, 1898 *ESM* **E,CGE**; pool opposite the Boar's Head in the
Lossie Forest, 1975 *R. Richter*.

Crithmum L.
C. maritimum L.   *Rock Samphire*
Casual. Shingle by the sea. Recorded by Lobban in 1900 from the
coast at Auldearn. Not seen in recent years.

Oenanthe L.
O. fistulosa L.   *Water Dropwort*
Native. Marshes. Reported from the Leen at Garmouth by Gordon
in 1828 and later from the same locality by G. Birnie. Not seen in
recent years.

O. lachenalii C. C. Gmel.   *Parsley Water Dropwort*
Native. Reported by Druce in *Com. Fl.* for v/c 96. No other record.

O. crocata L.   *Hemlock Water Dropwort*
Native. Ditches chiefly near the sea. Root tubers poisonous.
M Near Gordon Castle, *J. Hoy* herb. *G. Don* **BM,K**; abundant in
one or two places, Leuchars and Speyside from the bridge to
Garmouth (Gordon); the Leen at Garmouth, 1900 *JK* **FRS**;
do. 1953 *UKD* **E**; runnel in Culbin Forest, ditch near the sea at
Kinloss, by the Mosset Burn near Forres and at Spynie and Gilston
lochs, 1954.
N (Stables, *Cat.*); inland at Nairn, 1900 (Lobban).
I Campbelltown, dug up and eaten by sailors two of whom died
the same day (*Phyt.* **2**: 414–415 (1858)); banks of River Beauly, not
common (J. Don, 1898); Temple Bay and Urquhart Bay, Loch
Ness, 1904 *G. T. West* **STA**; Drumnadrochit, 1940 *AJ* **ABD**; river
marshes at Beauly and salt-marshes at Castle Stuart and Muirtown
Basin at Inverness, 1954; on a small island in the River Ness at
Dochgarroch, 1971; plentiful in runnels by the sea to the E of
Bunchrew and in a ditch at Lewiston behind the old graveyard,
1975; bog at Clachnaharry, Inverness, 1975 **E**; salt-marsh by the
River Beauly at Tomich Farm near Beauly.

Aethusa L.
A. cynapium L. subsp. **cynapium**   *Fool's Parsley*
Introduced. Rare on rubbish tips and as a garden weed.
M Doubtfully native, gardens about Elgin and Alves, 1833
(Gordon); near Elgin, 1843 *R. Beveridge* **ABD**; Dr Innes's garden,
Forres, 1874 *JK* **FRS**; garden weed at Greshop House and Milton
Brodie, track in the Culbin Forest and waste ground at Burghead,
1953; rubbish tip at Forres, 1973.
N Garden weed at Cameron Crescent, Nairn, 1968.

I   Garden weed at Kirkhill, 1950; railway embankment at
Tomatin, 1952; rubbish tip on Longman Point, Inverness, 1971
E,CGE.

Foeniculum Mill.
F. vulgare Mill.   *Fennel*
Casual. Garden escape or outcast. An aromatic herb with yellow
flowers.
M Escape at Speymouth, *G. Birnie* (Burgess); roadside verge near
the railway W of Elgin, 1954; single plant at Findhorn, 1966
A. J. Souter.

Anethum L.
A. graveolens L.   *Dill*
Introduced with foreign seed. Annual resembling *Fennel*.
M Carrot field at Milton Duff, 1975 *R. Richter* **E**.
I From distillery refuse on the Longman tip at Inverness, 1970
E,BM, CGE.

Meum Mill.
M. athamanticum Jacq.   *Spignel, Meu, Baldmoney*
Native. Rare in mountain pastures. An aromatic herb with dark
green feathery leaves.
M Innes House, Urquhart, 1794 *A. Cooper* **ELN** (perhaps cultivated).
I Rare on the N side of Loch Ness, *A. Murray* (Hooker, *Fl. Scot.*);
Longman (G. A. Lang, 1905); pastures near Alvie (*Atlas*); grass
verge of a field at Easter Lynwilg near Alvie, 1974 *D. Hayes* **E,BM**!

Ligusticum L.
L. scoticum L.   *Scots Lovage*
Native. Common on sand dunes, shingles and rocks by the sea. Pl 11.
M Rare at Covesea and Cummingstown, seashore at Alves, *Fraser*
(Gordon); Covesea, 1902 *JK* **FRS**; Burghead, 1915 *T. McGrouther*
GL; Findhorn Bay, 1953 *A. Melderis* **BM**; Culbin, 1964 **CGE**;
Findhorn, 1964 *P. H. Davis* **E**.
N Coast at Nairn, 1900 (Lobban); shingle bar and dunes at
Nairn, 1954.
I Among shingle and sand dunes at Whiteness Head, 1954; sand
dunes on Camp Sands, Fort George, 1975 **E**.

Angelica L.
A. sylvestris L.   *Ait-Skeiters, Wild Angelica*
Native. Abundant in damp places by rivers, lochans and roadside
ditches and often on cliff ledges in the higher mountains.
M Frequent (Gordon); Waterford, 1956 *JK* **FRS**; by a small ditch
by Kinloss Kirk, 1925 and throughout the county, 1954.
N (Stables, *Cat.*); Ardclach (Thomson); throughout the county,
1954.
I Inverness, 1832 Pitmain and Dalwhinnie, 1833 (Watson, *Loc.*
Cat.); banks of the River Ness, Inverness, 1855 *HC* **INV**; Culcabock
Burn (Aitken); Inverfarigaig, 1940 *AJ* **ABD**; throughout the district
1954.

Peucedanum L.
P. ostruthium (L.) Koch   *Master-wort*
Introduced. Naturalised on roadsides, and by rivers and old
buildings.
M Certainly introduced, escape at the Manse of Rothes (Gordon);
on the left bank of the River Spey not far from the old Spey Bridge
at Grantown, 1972; banks of the River Dulnain at Balnaan, 1973.
I Roadside near Lovat Bridge near Beauly, 1902 (Pollock), and
in 1963 *MMcCW* **BM,CGE**; roadside verge at Leachkin near
Inverness, 1907 *M. Barron*!; on a bank and by the River Foyers in
the village of Foyers, 1971 **CGE**; banks of the River Dulnain at
Skye of Curr, 1975.

Pastinaca L.
P. sativa L.   *Parsnip*
Carden outcast.
M Escape in a lane near Covesea Farm, 1954; distillery yard at
Smallburn, Rothes, 1956 **ABD**; car park in Forres, 1976.
I Rubbish tip on the Longman Point, Inverness, 1971 **E**.

Heracleum L.
H. sphondylium L.   *Bear-Skeiters, Cow Parsnip, Hogweed*
Native. Abundant on grassy banks, roadside verges and often on
mountain cliff ledges.
M Very common (Gordon); Greshop, 1856 *JK* **FRS**; throughout
the county, 1954.
N (Stables, *Cat.*); Ardclach (Thomson); banks of the River
Findhorn at Drynachan and throughout the county, 1954.
I Inverness, 1832 Dalwhinnie, 1933 (Watson, *Loc. Cat.*); through-
out the district, 1954.
   Var. **angustifolium** Huds.
A variant with linear-lanceolate leaf segments occurs in the woods
at Burgie House, Forres, 1967 **ABD**.

H. mantegazzianum Somm. et Lev.   *Giant Hogweed*
Introduced. River banks, ditches and waste ground. The hairs on
the stems are very poisonous and cause blisters if touched. It was
introduced by Ronald Earl of Leven and Melville *c* 1890 to an
ornamental garden pool at Glenferness House, in the county of
Nairn, from whence it has escaped to the banks of the River
Findhorn and other open places. Map 139.
M Abundant by the banks of the River Findhorn near Forres,
1925; by the River Lossie at Kellas, 1935; shingle of the River Spey
at Orton, Knockando and Garmouth, by the Broad Burn near
Rothes, roadside verge at Dunphail, by the Roseisle Forest, and
well established on the dunes of the Bar in the Culbin Forest.
N Glenferness House *c* 1890 *Lord Leven*; escape from a garden at
Cawdor and Nairn, 1900 (Lobban); by the Allt Dearg at Cawdor
Castle, by the River Nairn at Nairn, by the River Findhorn at
Coulmony, 1954; bank of the burn at Millhill near Auldearn and
roadside verge at Ferness.

I By a ditch near Seafield E of Inverness, 1925; planted at Milton, 1947; bank at Ardersier, 1954; near Loch Flemington and planted at Kilmorack, 1954; planted at Tomatin, 1974 and at Culloden House; Abriachan and one of two plants near Aldourie Pier, 1970 *M. Barron.*

H. mantegazzianum × sphondylium
M Roadside between Invererne and Kinloss, 1970; by the Muckle Burn between Moy House and Earnhill, 1976 *F. Stewart.*

Daucus L.
D. carota L. *Wild Carrot*
Introduced in north Scotland. Railway embankments and waste ground.
M Doubtfully native (Gordon); railway bank near Elgin (Druce, 1897); tips at Elgin, Forres and Rothes where it is a garden outcast; shingle of the River Spey at Garmouth, 1972 **E**; railway embankments near Burghead and near Rafford, 1976 **E**.
N Banks of the River Nairn (Aitken); Nairn and Cawdor, 1900 (Lobban); railway embankment at Auldearn, 1967 **ABD**.
I Bunchrew Burn, ? date and *coll.* **INV**; Englishton Burn (Aitken); wayside at Culloden and banks of a stream at Englishton (J. Don, 1898) near Tomich *A. Radford (Atlas)*; abundant on the river embankment between Tomich Farm, Beauly and Tarradale, 1976 **E**.

D. glochidiatus (Fisch. & C. A. Mey.) Avé Lall
Introduced with wool shoddy at Greshop House, Forres, 1958 **K**.

ARISTOLOCHIACEAE
Asarum L.
A. europaeum L. *Asarabacca*
Introduced. Firmly established in a wood at Wester Elchies House, Knockando, *Mr Ross* (Burgess). Not seen in recent years.

Aristolochia L.
A. clematitis L. *Birthwort*
Introduced, supposedly by the Monks from Fountains Abbey for its medicinal properties.
M Innes House, Urquhart, 1794 *A. Cooper* **ELN**; under the garden walls at Milton Brodie, 1965 **E**.

EUPHORBIACEAE
Mercurialis L.
M. perennis L. *Dog's Mercury*
Native. Common in woods and shady places on rich soil in the lowlands and among rocks and in gorges of burns in the highlands. Map. 140.
M Frequent (Gordon); Greshop Wood near Forres, 1856 *JK* **FRS**; throughout the county 1954; small wood by the River Spey at Delfur, 1967 **E**.

N (Stables, *Cat.*); Howford near Nairn, 1882 *R. Thomson* **NRN**;
banks of the river at Nairn, 1962 **ABD**.
I Ness Islands, Inverness, 1854 *HC* **INV**; by the Holm Burn and
Dunain etc. (Aitken); Bunchrew, 1940 *AJ* **ABD**; among rocks on
the cliffs of Mealfuarvonie, 1964 **CGE**; wood at Clunes, Kirkhill,
1972 **E**; on cliffs at Loch Killin, on Tom Bailgeann and in the
Ryvoan Pass in the Cairngorms etc.

Euphorbia L.

E. lathyrus L.   *Caper Spurge*
Introduced. Garden weed. A drastic purge and also supposed to
be a preventative from moles, which appears to be a fallacy!
M Gordon Castle grounds (Burgess); waste ground near Garmouth
station, 1954; garden weed at Greshop House and at Dyke; Elgin
tip, 1975.
I Garden weed at Clunes, Kirkhill, 1930 *M. Cameron*.

E. dulcis L.   *Sweet Spurge*
Introduced. Garden escape.
M Certainly introduced (var. *purpurata* Thuill.). Escaped about
Gordon Castle and Grant Lodge gardens (Gordon); Sanquhar,
Forres, 1880 *J. W. H. Trail* **ABD**; grounds of Gordon Castle, 1921
G. Birnie **ABD**; bank of a small burn at Altyre House, Forres, 1949;
bank at Dunphail, 1954.
N By the River Nairn at Holme Rose, 1954; about 1 mile S of
Nairn on the Carrbridge road, 1955 *J. B. Simpson*; in the grounds of
the Nairn cemetery, 1961 **ABD**; roadside bank at Cothill, 1963
CGE; banks of the river at Nairn, 1966 **E**; grass verge of road not
near habitations ½ mile E of Cantraydoune, 1974.
I (Druce, *Com. Fl.*); banks of the River Ness at Inverness, 1954
E,K; garden at Achnagairn, Kirkhill, policies of Tomatin House
and by Loch Ness near Aldourie Castle; dry bank by a lay-by near
Aigas in Strathglass, 1972 **E**.

E. helioscopia L.   *Sun Spurge*
Native. Common in cultivated land. Map 141.
M Very common (Gordon); frequent at Forres, 1856 *JK* **FRS**;
throughout the county especially in root fields, 1954.
N (Stables, *Cat.*); common at Ardclach, 1890 *R. Thomson* **NRN**;
throughout the county 1954.
I Common in fields at Charleston and Culduthel, Inverness
(J. Don, 1898); Torvaine, 1940 *AJ* **ABD**; waste ground at Inverness,
Ardersier and Culloden, 1954; common in the lowlands reaching
Fort Augustus, Boat of Garten, Kincraig and Farr; garden weed at
the Badenoch Hotel, Newtonmore, 1975 *J. Clark*; root field at
Laggan, 1975.

E. peplus L.   *Petty Spurge*
Native. Frequent in cultivated ground rare in the highlands.
Map. 142.
M Doubtfully native, garden weed at Elgin (Gordon); in the

Manse garden Forres, 1856 *JK* **FRS**; Fochabers, 1815 *D. Patton* **GL**; garden weed at Kinloss, Newton House near Elgin, and at Garmouth, Fochabers, Conicavel and Forres, 1954.
N Nairn, *J. B. Brichan* (Watson); Cawdor, 1900 (Lobban); garden weed at Holme Rose, Cawdor Castle, Nairn and Kinsteary House at Auldearn, 1954.
I Fields and gardens at Inverness (J. Don, 1898); garden weed Holm House, Inverness, 1954; Fort Augustus, 1956 *M. T. T. Phillips* (*Atlas*); garden weed at Myrtle Cottage, Reelig and at Ballindarroch at Scaniport, cottage garden at Allanfearn, waste ground at Ardersier and at Banchor Mains at Newtonmore.

E. exigua L.   *Dwarf Spurge*
Introduced with foreign grain.
M Single plant in the garden at Moy House near Forres, 1952; railway siding by Knockando distillery, 1968 *G. Shepherd* **E**.

E. esula L. × **virgata** Waldst. & Kit. = **E.** × **pseudovirgata** (Schur) Soó
Introduced. Rare on railway embankments and waste ground. This is the *Spurge* erroneously called *E. virgata*, *E. virgata* var. *esulifolia* or *E. uralensis* Fisch. & Link.
M Railway embankment near Loch Spynie (as *E. esula* L.), 1898 *ESM & WASh* **E**; railway embankment behind Greshop Farm near Forres, 1961 **E,K**.

E. cyparissias L.   *Cypress Spurge*
Introduced.
M Wood near Darnaway Castle, 1877 *W. Brown* (*Trans. BSE* **13**: appendix lxix (1878); Darnaway, 1888 *JK* **FRS**; escape in the parishes of Dyke and Elgin (Burgess).
Records for **E. hyberna** L. from Fochabers, **E. pilosa** L. from Spynie and **E. amygdaloides** L. from the banks of the River Findhorn by Burgess must be treated with some doubt.

POLYGONACEAE
Polygonum L.
P. patulum M. Bieb.
Introduced with foreign grain.
M Rubbish tip at Elgin, 1958 **E**.

P. aviculare L.   *Rabbit Sugar, Knotgrass*
   P. heterophyllum Lindm.
Native. Abundant on cultivated ground and bare open waste places. Branch leaves smaller than the stem leaves, perianth divided almost to the base, fruit trigonous with three concave sides. Fig. 25e.
M Stotfield, 1831 *G. Gordon* **E**; Forres, 1856 *JK* **FRS**; throughout the county, 1954; stubble field at Lochinvar near Elgin, 1961 **K**; garden weed at Brodie Castle, 1961 **E**.
N (Stables, *Cat.*); everywhere at Ardclach, 1889 *R. Thomson* **NRN**;

throughout the county, 1954; arable field at Cran Loch, 1960
ABD,K; dunes on the Nairn Dunbar golf course, 1966 **ABD**.
I   Inverness, 1832 and Dalwhinnie, 1833 (Watson, *Loc. Cat.*);
near Inverness, 1854 *HC* **INV**; Bunchrew, 1940 *AJ* **ABD**; Glen
Affric Lodge, 1947 and throughout the district, 1954.

P. rurivagum Jord. ex. Bor.
Native. Plant very slender with narrow linear-lanceolate leaves,
perianth segments narrow and reddish, fruit slightly exserted.
M   Boat of Garten (Druce, *J. Bot.* **28**: 43 (1890)).

P. arenastrum Bor.
    P. aequale Lindm.
Native. In similar situations to *P. aviculare*. Differs in the branch
and stem leaves being of equal size, the perianth divided to half
its length and the fruit having two sides convex and one concave.
Fig 25d. Map 143.
M   Farmyard at Cullerne House, Findhorn, 1953; sandy ground
in the Culbin Forest, 1961 **E**; roadside verge at Auchterblair near
Carrbridge, 1972 **E**; in the car park at the Aviemore Centre and
roadside verges at Duthil and Roseisle Forest etc.
N   Dunes by Cran Loch, 1960 **ABD**; gateway at Rait Castle and
root field at Ordbreck near Clunas.
I   Raised shingle beach at Ardersier, 1956; track on the Newtonmore
golf course, 1975 **E**; track above the falls of Divach, 1975 **E**; station
yard at Dalwhinnie, car park at Inshriach Nursery, waste ground
at Kingussie, Kilmorack, Tomatin, Knockie and Daviot station etc.

P. viviparum L.   *Alpine Bistort*
Native. Common in mountain pastures, rock ledges and by burns,
often washed down rivers and becoming established on river banks
in the lowland areas. The upper part of the inflorescence has
small white flowers, the lower part has reddish-purple bulbils.
Pl 11. Map 144.
M   Moray, *G. Gordon* **GL**; near Aviemore, Strathspey, 1839 *J. A.
Hankey* **OXF**; Carrbridge, 1894 *JK* **FRS**; Inverallan near Grantown,
1895 *C. Bailey* **OXF**; banks of the River Findhorn at Broom of Moy,
1930; and in all the highland areas, 1954.
N   (Stables, *Cat.*); Lochloy, *J. B. Brichan* and Lethen, 1840 *J. G.
Innes* (Watson); flushes at Drynachan, at Coulmony and Dulsie
Bridge, 1954; banks of the River Findhorn, 1961 **ABD**; flush
above the small gorge by Clunas Dam, 1973 **E**.
I   Dalwhinnie, 1833 (Watson, *Loc. Cat.*); near Balloan Farm,
Inverness (Aitken); Kincraig, 1891 *AS* **E**; Kiltarlity and Kilmorack,
1902 (Pollock); Moy, 1919 *G. C. Druce* **OXF**; golf course at
Newtonmore, 1914 *T. McGrouther* **GL**; throughout the highland
district, 1954.

P. bistorta L.   *Snakeweed, Bistort*
Native. Meadows, ditches, and grassy banks of rivers and burns.
Map 145.

Fig 25   a, **Callitriche stagnalis**; b, **C. hermaphroditia**; c, **C. hamulata**;
d, **Polygonum arenastrum**; e, **P. aviculare**; f, **P. hydropiper**.

M Pluscarden and Muchrach, *J. G. Innes* (Gordon); Moray,
G. Gordon **GL**; Dalvey near Forres, 1860 *JK* **FRS**; meadows at
Boat of Garten, Forres, Kellas and Rothes, 1954; by the Muckle
Burn near Barley Mill, Dyke, 1963 **ABD,BM**; ditch by the path to
the old curling pond at Grantown, 1971 **E,CGE**; ditch near
Mulben, 1974.
N (Stables, *Cat.*); Nairn, *A. Falconer* (Gordon); bank by the river
at Nairn, 1954.
I Inverness, 1867 ? *coll.* **ELN**; banks of the River Nairn (Aitken);
Holm burn (G. A. Lang, 1905); bank on the north side of the
railway bridge at Stratton, 1925; Bunchrew, 1950 *M. Cameron*;
small ditch between Beauly and Muir of Ord, 1950; pasture by
railway at Kincraig, 1951 *E. C. Wallace*!; wood at Moy Hall, bog
near the river bridge at Tomich, banks of the river at Foyers and in
Boleskine kirkyard; bank at Loch Flemington and ditch at Culburnie near Kiltarlity.

P. amphibium L.   *Amphibious Bistort*
  Amphibious form
Native. Common in lochans and moorland pools and at the shallow
edges of the larger lochs. Floating leaves rooting at the nodes.
Map 146.
M Frequent (Gordon); Moy Carse, 1856 *JK* **FRS**; Loch Spynie,
1941 ? *coll.* **STA**; pond by the Muckle Burn at Dyke, pools at Boat
of Garten, Dava Moor, at Forres and Garmouth etc., 1954; Loch
Pulladern at Aviemore, 1964.
N (Stables, *Cat.*); Ardclach, not common, 1889 *R. Thomson* **NRN**;
Cran Loch, 1954; Loch Flemington, 1972 **E**; small burn N of
Boath House, Auldearn, 1975.
I Lentran, ? date and *coll.* **INV**; Culcabock Dam (Aitken);
Cherry Island in Loch Ness, 1904 *G. T. West*; Loch Flemington,
1940 *AJ* **ABD**!; pool by River Beauly, Groam of Annat, 1947;
near Ardesier, Alturlie point and Inchnacardoch Bay, 1954;
Raigmore pond, Inverness, bog by Lewiston kirkyard and in a small
lochan in the wood by Aldourie Pier, etc.

  Terrestrial form
In the terrestrial form the stems are erect, rooting only at the lower
nodes, the leaves are oblong-lanceolate and hispid. Frequent on
banks and in arable fields.
M River bank at Waterford near Forres, 1850 *JK* **FRS**; rubbish
tip at Forres, 1954; arable field at Crosslets near Duffus, bank at
Broom of Moy, hedge by ditch at Forsterseat, cross-roads near
Urquhart and bank by a burn in the village of Kingston.
N Bank by the burn at Boath House, Auldearn, 1975.
I Roadside bank ¼ mile E of Fort Augustus, bank near the football
field at Beauly and in a root field at Farley near Beauly, 1975.

P. persicaria L.   *Redleg, Persicaria*
Native. Abundant in cultivated ground and waste places. Map 147.
M Very common (Gordon); Forres, 1899 *JK* **FRS**; throughout the

county, 1954; arable field near Brodie station, 1963 **E**; garden weed, Knockando, 1967 **ABD**; Dyke, 1969 **E,BM**.
N   (Stables, *Cat.*); schoolhouse garden at Ardclach, 1883 *R. Thomson* **NRN**; root field at Drynachan and throughout the county, 1954; waste land at Lethen, 1961 **ABD,K**.
I   Pitmain, 1833 (Watson, *Loc. Cat.*); by the Moniack Burn, ? date and *coll.* **INV**; Bunchrew, 1940 *AJ* **ABD**; in all the lowland areas, 1954; waste ground on Longman Point, Inverness, 1971 **E,CGE**; root field at Laggan, 1974.

P. lapathifolium L.   *Pale Persicaria*
Introduced. Waste ground on rubbish tips and near distilleries.
M Abundant in one or two places (Gordon); Netherton of Grange, 1867 *JK* **FRS**; near Forres (Druce, 1897); roadside verge near Knockando distillery, 1967 **E,ABD**; Elgin rubbish tip with bird seed aliens, 1975 **E**.
N Rubbish tip at Newton of Park, 1975 **E**.
I Fort Augustus, 1957 *J. Milne* (*Atlas*); with sweepings from a distillery on the Longman tip, Inverness, 1971 **E**.

P. nodosum Pers.
Introduced with foreign seed. Peduncles and under sides of the leaves densely dotted with yellow glands.
N Carrot field at Broombank Farm, Auldearn, 1967 **E,ABD,BM**.

P. hydropiper L.   *Water-pepper*
Native. Frequent in wet places, among stones by burns and on the shallow margins of lochs. Leaves lanceolate, inflorescence very slender, the plant burning to the taste. Fig 25f, Map 148.
M Pond at Pittendreich (Gordon); abundant in the parishes of Dyke, Edinkillie and Speymouth (Burgess); ditch at Kinloss, 1925; by the pond at Sanquhar House, Forres, bog at Lochinvar and backwater of the River Spey at Garmouth, 1954; wet field at Brodie station, 1963 **E**; bog near the Black Burn, Pluscarden, 1975; very large plants with glands absent on the Elgin rubbish tip, 1975 **E**.
N Newton of Park, 1832 *WAS & J. B. Brichan* **ABD,FRS**; rare at Millhill, *J. B. Brichan* (Gordon); pool near Holme Rose, gravel of burn at Cawdor and margin of Loch Flemington, 1954.
I Banks of the River Nairn (Aitken); shore of Urquhart Bay, Loch Ness, 1904 *G. T. West* **E,STA**; Scaniport, 1940 *AJ* **ABD**; common by pools in Strathglass and among stones on the margin of Loch Ness, among stones in a dried up burn at Corrigarth and at Loch Meiklie; by the River Spey at Laggan, 1974; verge of pool by the River Glass at Cannich, 1975 **E**; dried up pool by Durris Farm near Dores, 1976 **E**.

P. convolvulus L.   *Black Bindweed*
Native. Common in waste places and cultivated ground. Map 149.
M Very common (Gordon); Forres, 1900 *JK* **FRS**; in arable fields throughout the county, 1954; cornfield at Dyke, 1963 **CGE**.

9

N Moyness, Nairn, 1832 *WAS* **ABD**; common in fields and gardens at Ardclach, 1889 *R. Thomson* **NRN**; throughout the county, 1954; sandy field at Kingsteps, 1966 **ABD**.

I Inverness, 1832 (Watson, *Loc. Cat.*); Tower of Culloden, 1855 *HC* **INV**; Bunchrew, 1940 *AJ* **ABD**; common in the lowland areas, less so in the highlands but reaching Newtonmore, Strathglass and Tomatin, 1954.

P. cuspidatum Sieb. et Zucc.   *Japanese Knotweed*
Introduced. A tall plant reaching 2 m with broad ovate glabrous leaves that are cuspidate at the tips and truncate at the base. Waste land, by ditches and near old cottages.

M By the River Spey at Fochabers, 1950; west bank of the River Findhorn, on shingle, near Forres, 1953 *A. Melderis* **BM**; garden outcast at Conicavel, waste land at Rothes, by the River Spey at Essle, by the River Lossie at Elgin, and in a lay-by at Longley Farm near Dyke, 1954; by the burn at Smallburn N of Rothes, 1972 **E**.

N Roadside verge near cottages at Galcantray, 1954; by the old saw-mill at Dalaschyle, 1962 **E,BM**; shingle of the River Nairn at Little Kildrummie.

I Waste land at Clachnaharry, 1954; Strathglass, 1958; abundant by the kennels at Culloden House, in a wood by Loch Ness near Lewiston, by Dell Farm at Inverdruie, by a ruined cottage at Corrich in Glen Convinth, river bank at Beauly and waste ground by the sea at Allanfearn.

P. sachalinense F. Schmidt   *Giant Knotweed*
Introduced. Usually taller than *P. cuspidatum*. Leaves ovate, acute at tip, weakly cordate at base.

M By the Muckle Burn at Moy House, 1950; garden outcast near Kellas House and in the policies of Brodie Castle.

I Garden escape near Boleskine House, Foyers, 1954.

P. polystachyum Wall. ex Meisn.   *Himalayan Knotweed*
Introduced. Leaves lanceolate, acuminate, cordate to cuneate at base.

M Planted by the pond at Sanquhar House, Forres, 1925; bank by the bridge over the River Lossie near Dallas, 1954; grass verge of road by a cottage near Pitcroy, Strathspey, 1968; planted by the lake at Gordonstoun College, Duffus, 1972 **E**; grass bank in the village of Relugas, 1973.

I Inverness, 1952; grass verge of the A9 by Mealmore Lodge S of Daviot, 1964.

P. campanulatum Hook fil.   *Lesser Knotweed*
Introduced. Leaves whitish tomentose beneath.

M By the River Lossie near the Cooper Park, Elgin, 1955; garden outcast on the Elgin rubbish tip and in Greshop wood, Forres.

N Roadside verge between Easter Galcantray and Dalaschyle, 1962 **E,ABD**.

I Roadside verge near Hardhill, Croy, 1955; garden outcast by a small burn at Torbreck near Inverness, 1975.

P. compactum Hook. fil.
Introduced.
I A large patch just N of the railway crossing at the junction of
the A9 and A96 Inverness, 1967 **E,ABD,BM,CGE**.

Fagopyrum Mill.

F. esculentum Moench *Buckwheat*
Introduced with foreign grain.
M Neighbourhood of Forres, 1828 *J. B. Brichan* (Gordon, *annot.*);
introduced with tares in the parishes of Dyke and Edinkillie
(Burgess); waste ground in Greshop Wood, Forres, 1957 **E**; rubbish
tip at Elgin, 1973 **E**.

Oxyria Hill

O. digyna (L.) Hill *Mountain Sorrel*
Native. Locally common by burns and on wet rock ledges in the
highlands; often washed down rivers and becoming established in
rock crevices and shingle in the lowlands. Map 150.
M Side of the Spey at Speymouth, at Aviemore, *J. G. Innes* and
banks of the River Findhorn at Relugas, *WAS* (Gordon); in the
parishes of Dyke, Edinkillie, Forres, Urquhart, Speymouth, Bellie
and Rothes (Burgess); in rock crevices by the River Findhorn
opposite Coulmony, 1973.
N Shingle of the River Findhorn at Glenferness House, 1973.
I By the River Truim and at Pitmain, 1833 (Watson, *Loc. Cat.*);
Fraoch-choire in Glen Cannich (Ball, 1851); Abriachan (J. Don,
1898); bed of the River Spey, 1893 *AS* **E**; by the Falls of Divach,
by Loch Ruthven and on the banks of the River Findhorn (Aitken);
in all the mountain areas, 1954; rocks by the Plodda Falls at
Guisachan, 1972 **E**.

Rheum L.

R. rhabararum L. (*Rhubarb*) is found as a garden relic by old ruined
cottages and on railway embankments.

Rumex L.

R. acetosella L. *Sheep's Sorrel*
Native. Abundant on moorland and on open waste and cultivated
land on acid soil.
M Innes House, Urquhart, 1794 *A. Cooper* **ELN**; very common
(Gordon); Forres, 1898 *JK* **FRS**; on river shingle at Garmouth and
at Forres, 1953 *A. Melderis* **BM**; throughout the county, 1954;
garden path at Brodie Castle, 1961 **BM,K**; shingle of the River Spey
at Fochabers, 1972 **E**.
N (Stables, *Cat.*); Ardclach (Thomson); throughout the county,
1954; moor near Lethen, 1961 **ABD,K**; sand dunes at Nairn, 1962
CGE.
I Inverness, 1832 Dalwhinnie, 1833 (Watson, *Loc. Cat.*); Bunchrew, 1940 *AJ* **ABD**; throughout the district, 1954; shingle of the
River Spey at Spey Dam near Laggan, 1972 **E**.

R. tenuifolius (Wallr.) Löve   *Narrow-leaved Sheep's Sorrel*
Native. Frequent on very poor soil chiefly among stones and shingle
of rivers and burns. Plant usually decumbent with very narrow
leaves 7–10 times as long as broad and of a glaucous-green colour.
Many intermediates occur with *R. acetosella*. Recent records have
all been confirmed by J. E. Lousley. Map 151.
M Culbin, 1911 *P. Ewing* **GL**; shingle of the River Findhorn at
St John's Mead, Darnaway, 1953 *UKD* **UKD**; dunes by the Culbin
Forest, 1957 **K**; shingle of the River Spey at Fochabers, sand dunes
at Findhorn and shingle of the River Dulnain near Carrbridge,
1972 **E**; track near Grantown station and shingle at Bridge of
Brown, 1973 **E,BM**.
N Abundant on the dunes of the Nairn Dunbar golf course,
1962 **ABD,CGE**; shingle of the River Findhorn at Glenferness
House, 1973 **E,BM**.
I River shingle at Spey Dam, Laggan, 1972 **E**; shingle of the
River Killin between the loch and Sronlairig Lodge, 1972 **E**;
sandy path at Glen Shirra Lodge and by the Markie burn in Glen
Markie, 1972 **E**; abundant on the fixed shingles by the River Spey
at Newtonmore, 1973 **E,BM**; gravel overspill of the burn at Gaick
Lodge, 1973 **E,ABD**.

R. acetosa L.   *Souracks, Sorrel*
Native. Abundant in grassland, open woods, waste places and
mountain ledges throughout the district. Very variable in leaf
shape. Narrow-leaved forms occur on mountain ledges and river
shingle.
M ? locality and *coll.*, 1830 **FRS**; very common (Gordon); bank of
the River Findhorn, on shingle, 1953 *A. Melderis* **BM**; grass bank
at Dyke, 1963 **CGE**.
N (Stables, *Cat.*); Ardclach (Thomson); throughout the county,
1954.
I Inverness, 1832 and Dalwhinnie, 1833 (Watson, *Loc. Cat.*);
near Inverness, 1854 *HC* **INV**; summit rocks on Ben Alder, 1956;
cliff at 820 m Sgùrr na Lapaich in Glen Strathfarrar, 1972 **E**;
cliffs by Loch Killin, 1972 **E**.

R. hydrolapathum Huds.   *Great Water Dock*
Native. Ditches. Not seen in recent years.
M Only a few plants in marshy ground near Kinloss Kirk (Marshall,
1899).
I Near Beauly, *ESM* (*J. Bot.* **31**: 233 (1893)).

R. alpinus L.   *Butter Blades, Monk's Rhubarb*
Introduced. Ditches and roadside verges usually near cottages.
Formerly used to wrap butter.
M Certainly introduced and escaped from gardens where it was
formerly cultivated (Gordon); parishes of Dyke, Forres, Rafford,
Urquhart, Rothes and Knockando (Burgess); abundant in a ditch
by Woodside, Burgie, 1950; roadsides at Fochabers, Dallas, Dava

and Rothes, 1954; near Knockando, 1964 **E**; roadside by the Manse at Rafford.
N By old buildings at Ferness bridge, 1964 **ABD**; ditch at Tomloan near Achavraat, 1970.
I Carse of Ardersier, 1846 *A. Croall* **STI**; railway embankment W of Dalcross, 1968.

R. longifolius DC. *Northern Dock*
   R. domesticus Hartm.; *R. aquaticus* auct., non L.
Native. Common on grassy roadside verges and waste ground. Recognised in the field by its dense compact inflorescence and lack of tubercules on the fruiting perianth. Map 152.
M Frequent, about Elgin, 1837 *W. Brand* (Gordon); Aviemore meadows, 1893 *AS* **E**; not uncommon at Brodie, Dunphail, Garmouth and Aviemore (Marshall, 1899); grass verges at Kinloss and Findhorn, 1925; Findhorn, 1953 *A. Melderis* **EM**; throughout the county, 1954.
N Frequent near Cawdor Castle, 1839 *Messrs Braid & Campbell* (Graham, *Trans. BSE* **1**: 25 (1840)); bank at Drynachan, roadside verges at Holme Rose, Cawdor and Geddes, 1954; sandy field at Kingsteps near Nairn, 1966 **ABD**.
I Dalwhinnie, 1833 (Watson, *Loc. Cat.*); at Holm, Beauly and Culloden etc. (J. Don, 1898); grass verges at Tomatin, Glenmore Lodge, by Loch Flemington, near Inverness, Ardersier etc., 1954; damp grassy place near a burn, Sronlairig by Loch Killin, 1972 *M. Barron.*

R. longifolius × obtusifolius L. = **R. × arnottii** Druce
M (Trail, *ASNH* **15**: 180 (1906)).
I Dalwhinnie, 1911 *ESM* **E**.

R. crispus L. *Dockans, Curled Dock*
Native. Abundant on waste and cultivated land and on dunes and shingle by the sea. Map 153.
M Frequent (Gordon); near Forres, 1856 *JK* **FRS**; throughout the county, 1954.
N (Stables, *Cat.*); Ardclach (Thomson); garden weed at Drynachan and throughout the county, 1954; arable field at Maviston, 1968 **E**.
I Common at Inverness (J. Don, 1898); Tomatin, 1903 (Moyle Rogers); Lower Leachkin, 1940 *AJ* **ABD**; garden at Ben Alder Lodge, at Gaick Lodge and newly sown roadside verge by Loch Loyne in Glen Moriston, and in all the lowland areas, 1954; shingle bar by Fort George, 1975 **E**.

   Var. **subcordatus** Warren
M Aviemore, 1893 *AS* **E**.

R. crispus × longifolius = R. × propinquus Aresch.
N By the river *c* 2½ miles from Nairn (with parents) (Marshall, 1899).
I Dalwhinnie, 1912 *ESM & WASh* (Trail, *ASNH* **15**: 180 (1906)).

R. crispus× obtusifolius = R× acutus L.
M Blacksboat near Knockando, 1903 *Moyle Rogers* and at Spey-
mouth, *G. Birnie* (Burgess).
N By the river *c* 2½ miles from Nairn (Marshall, 1899).
I Dalwhinnie, 1912 (Marshall, *J. Bot.* **51**: 166 (1913)).

R. obtusifolius L.   *Broad-leaved Dock*
Native. The most common species in north Scotland. On roadside
verges and waste and cultivated land.
M Frequent (Gordon); Relugas, 1857 *JK* **FRS**; throughout the
county 1954.
N (Stables, *Cat.*); Ardclach (Thomson); by the river at Drynachan
and throughout the county, 1954.
I Inverness, 1832 and Dalwhinnie, 1833 (Watson, *Loc. Cat.*);
Lentran, 1940 *AJ* **ABD**; roadside verge at Drumochter and
throughout the district save the higher mountains, 1954.

R. obtusifolius× sanguineus L. var **viridis** Sibth.
M Elgin rubbish tip, 1957 **K**.

R. sanguineus L. var. **viridis** Sibth.   *Wood Dock*
Native. Rare on the margins of deciduous woods.
M Abundant in one or two places, Findrassie and Gordon Castle
woods (Gordon); woods at Gordonstoun, 1937 *R. Richter*; Greshop
Wood near Forres, 1957 **K**; by the pond at Gordon Castle, 1972 **E**.
N Woods at Holme Rose and Cawdor Castle, 1954.
I Tomatin, 1903 (Moyle Rogers); near Beauly, 1954; wood at
Farr, 1968; small wood by the road at Daviot and by Loch Meiklie;
by the river at Tomich, 4 or 5 plants, 1971 *A. J. Souter*; woods at
Culloden House, 1975 **E,BM**.

R. conglomeratus Murr.
Doubtfully native in north Scotland.
M Near Forres (Marshall, 1899); in the parishes of Forres and
Bellie, *P. Leslie* and Knockando, *C. Watt* (Burgess); on rubbish tips
at Elgin and Forres, 1954. All other records for north and north-east
Scotland in the *Atlas* must be considered errors.

R. maritimus L.   *Golden Dock*
M The Leen Garmouth, *G. Birnie* (Burgess).

R. pulcher L and **R. dentatus** L. subsp. **halacsyi**(Rech.)Rech.
fil. were introduced into the garden of Rose Cottage, Dyke and
regenerate freely but have so far not escaped further afield.

URTICACEAE
Parietaria L.
P. judaica L.   *Pellitory*
    P. diffusa Mert. et Koch
Native. Rare in crevices of old walls.
M Moray but not indigenous, *WAS* (Murray); certainly introduced,
rare at Pans Port, Elgin, at Pluscarden Priory, at Kinneddar, and

on the ruins of the Castle of Spynie, Old Duffus and Kinloss Abbey (Gordon); Drainie, 1860 *JK* **FRS**; roadside verge near the kirk at Dyke, 1960 **E,ABD**.
N Penick Castle E of Auldearn, *J. B. Brichan* (Watson).
I In walls at Bunchrew station, 1930. The record for v/c 96. Watson and Druce probably refers to the Nairn locality.

Soleirolia Gaud.

S. soleirolii (Req.) Dandy   *Mind-your-own-business*
   Helxine soleirolia Req.
Introduced.
N Escape from the greenhouse at Kinsteary House, Auldearn, 1964 **ABD**.

Urtica L.

U. urens L.   *Small Nettle*
Native. Abundant in cultivated ground. Map 154.
M Very common (Gordon); Forres, 1900 *JK* **FRS**; abundant in the garden at Brodie Castle and throughout the county, 1954.
N (Stables, *Cat.*); Ardclach (Thomson); Nairn, 1911 *P. Ewing* **GL**; throughout the county, 1954; sandy field E of Nairn, 1966 **ABD**.
I Inverness, 1832 and between Dalwhinnie and Truim, 1833 (Watson, *Loc. Cat.*); Dochgarroch, 1940 *AJ* **ABD**; in all the lowland areas reaching Newtonmore and Achlean in Glen Feshie.

U. dioica L.   *Stinging Nettle*
Native. Abundant on waste land, in woods and about buildings often reaching 800 m among rocks on mountains.
M Very common (Gordon); Forres, 1856 *JK* **FRS**; throughout the county, 1954.
N (Stables, *Cat.*); Ardclach, the young shoots highly valued for making 'nettle kale' to keep the system healthy (Thomson); by farm buildings at Drynachan and throughout the county, 1954.
I Inverness, 1832 Pitmain and Dalwhinnie, 1833 (Watson, *Loc. Cat.*); Dores, 1840 *AJ* **ABD**; yard at Ben Alder Lodge, among rocks in the Pass of Ryvoan and on cliffs at 800 m on Sgùrr na Lapaich etc.

CANNABIACEAE
Humulus L.

H. lupulus L.   *Hop*
Garden escape
M Hedgerow at Elgin station, 1954; bank between Duthil and Carrbridge, 1975.
N Near Cawdor Castle, 1855 *HC* **INV**.
I Near Kingussie, 1904 *G. B. Neilson* **GL**; on a fence in the village of Kirkhill, 1954; bank by a cottage in the lane above Kilmorack Kirk, 1964; garden wall at Alvie Kirk, at Moy and relic of a cottage garden by the viaduct at Clava; opposite houses at Tomatin, 1974 *E. Bulloch*; bank N of Aigas in Strathglass.

Cannabis L.

C. sativa L.   *Hemp*
Introduced with foreign grain. Occasional on rubbish tips. The
plant contains the drug *Marijuana*.
M Elgin rubbish tip, 1958 **E,K**; Cloddach quarry where rubbish
had been tipped, 1976 **E,BM**.

ULMACEAE
Ulmus L.

U. glabra Huds.   *Scotch Elm, Wych Elm*
Native. Common in the lowland areas but less so in the highlands.
M Abundant in ravines in one or two places (Gordon); Forres, 1898
JK **FRS**; throughout the county, 1954; ditch near Roseisle Forest,
1972 **E**.
N (Stables, *Cat*); Ardclach (Thomson); everywhere, but absent
from the moors, 1954; rough ground W of Geddes Reservoir, 1960
ABD.
I Fort Augustus, 1833 (Watson, *Loc. Cat.*); Inverness, common
(J. Don, 1898); Drummond, 1940 *AJ* **ABD**; scattered throughout
the district reaching Dalwhinnie and Newtonmore but probably
planted there.

U. procera Salisb.   *English Elm*
    U. campestris auct.
Planted as hedges and in the policies of some of the larger houses.
M Parishes of Dyke, Forres, Edinkillie and Bellie, the timber used
for making cart and carriage wheels and coffins (Burgess); planted
at Sanquhar House, Forres and at Elgin, 1954.
N Planted at Ardclach (Thomson).
I Hedge at Kirkhill and by Reelig House, 1954 **E**.

U.×vegeta (Loud.) A. Ley, *Huntingdon Elm* is planted on the south
side of the A96 by the Elgin Central Engineers garage on the
outskirts of Elgin, and **U.×sarniensis** (Loud.) Ley, *Jersey or
Wheatley Elm* is planted on the northern approach to Rothes, at
Innes House and by the A96, on the north side by the Coach garage
on the outskirts of Elgin. For parentage of these trees see (C. A. Stace,
Hydridization and the Flora of the British Isles, 294–6 (1975)).

JUGLANDACEAE
Juglans L.

J. regia L.   *Walnut*
Planted.
M Parishes of Forres and Bellie, at Gordon Castle and Dellachaple,
Garmouth (Burgess); planted at Moy House, Brodie Castle,
Gordonstoun and by the Mosset Burn at Forres.
I Dochfour House and probably elsewhere.

MYRICACEAE
Myrica L.
M. gale L.   *Bog Myrtle*
Native. Common in wet acid moorland. Map 155.
M Frequent (Gordon); Aviemore, 1898 *JK* **FRS**; Grantown, 1901
AS **E**; occasional in the Culbin Forest and throughout the county,
1954.
N (Stables, *Cat.*); Ardclach (Thomson); throughout the county,
1954; bog by Cran Loch, 1960 **ABD**.
I Inverness, 1832 Pitmain and Dalwhinnie, 1833 (Watson, *Loc.
Cat.*); Bunchrew Burn, 1855 *HC* **INV**; Errogie, 1940 *AJ* **ABD**;
throughout the district, 1954; bog near Essich, 1971 **E**.

BETULACEAE
Betula L.
B. pendula Roth.   *Silver Birch*
  B. verrucosa Ehrh.; *B. alba* auct.
Native. Woods and river banks. Common. Bark smooth and silvery-
white above, black and fissured at the base of the trunk. Branches
drooping.
M Very common (Gordon); Forres, 1852 *JK* **FRS**; Culbin Forest,
1925 and throughout the county, 1954.
N (Stables, *Cat.*); Ardclach (Thomson); throughout the county,
1954 but not reaching the moorland areas at Drynachan.
I Inverness, 1832 Pitmain and Dalwhinnie, 1833 (Watson, *Loc.
Cat.*); Tromie Bridge, 1904 *G. B. Neilson* **GL**; Abriachan, 1940
AJ **ABD**; throughout the district, 1954 but rare at Ben Alder
Lodge, at Laggan and in the extreme west.

B. pendula × **pubescens** = **B.** × **aurata** Borkh.
I Cannich, 1947 *A. J. Wilmott* (*Proc.* **5**: 407 (1964)). Probably
quite common but not recorded with certainty.

B. pubescens Ehrh.   *Downy Birch*
Native. Very common. Tolerant of wet places.
  Subsp. **pubescens**
Tree with a single stem. Bark smooth, brown or grey. Branches
not drooping. Very variable in leaf shape and size.
M By the River Findhorn at Logie (Druce, 1888); about Aviemore
(as var. *microphylla* (Hartm.) E. S. Marshall) (Marshall, 1899);
Culbin Sands, *D. Patton* (Burgess); throughout the county, 1954.
N Between Findhorn and Nairn, 1912 (Ewing); by the River
Findhorn at Drynachan and throughout the county, 1954.
I Craig Phadrig, 1855 *HC* **INV** and (Aitken); by the Allt an
t-Sluie by Dalwhinnie and at Laggan Bridge, 1911 *ESM & WASh* **E**;
natural birch wood at 250 m, 1 mile S of Laggan Bridge (as var.
microphylla) (Marshall, *J. Bot.* **51**: 166 (1913)); by the Allt an
t-Sluie (as var. *sudetica* (Reichenb.) E. S. Marshall) *ESM & WASh*;
wood at Gaick Lodge, Glen Feshie etc., 1954; by Lochan Uvie E
of Laggan (as var. *microphylla*), 1973 **E,BM**.

Subsp. **carpatica** (Willd.) Aschers & Graeb.
Subsp. *odorata* (Bechst.) E. F. Warb.
A shrubby tree often with several stems. The leaves have a sweet
resinous smell when young.
M Fixed dunes NW of the Buckie Loch, Culbin Forest, 1964 **CGE**;
in the gorge of Huntly's Cave on Dava Moor, on the banks of the
River Findhorn at Daltulich bridge, plentiful by the road to
Glenbeg at Grantown and by the track on the hill at Kinakyle.
N Wood by Kinsteary House at Auldearn, 1955; woods at Glenfer-
ness, at Loch Belivat and at Lochloy.
I Cannich, 1947 *A. J. Wilmott*!; near Kincraig station, 1954;
Inshriach, 1961 *P. H. Davis* **E**; dominant in some of the glens to the
west of the district, e.g. Garva Bridge etc.

B. nana L.   *Dwarf Birch*
Native. Dwarf shrub up to 1 m with small orbicular, crenate,
glabrous leaves. Locally common in wet peaty bogs and moors,
usually over 500 m, to the west of Strathspey, only found in one
or two places in the east.
M *W. J. Hooker* (Watson, *Top. Bot.*); moor on the NE slope of
Geal-chàrn Beag above Aviemore, 1974 **E**; several colonies to the
N of Carn Sleamhuinn, 1974 *D. Hayes*.
I North of Alvie, *G. A. W. Arnott* **E**; glen N of Alvie, 1830 *W. J.
Hooker* **GL**; rare in Glen Feshie (Gordon); Dalwhinnie, 1833
(Watson, *Loc. Cat.*); bogs to the N of Fort Augustus and Inver-
moriston, shores of Loch Kemp and by Loch na Bà Ruaidhe at
Milton, 1904 *G. T. West* **STA**; by the Uisge Geal near Dalwhinnie,
500–800 m, *ESM & WASh* **E,BM,OXF**; bogs NW of Beinn a'
Chrasgaid, 1957 *D. A. Ratcliffe*; moors at Creag Dhubh near Laggan
and Creag Dhubh above Pitmain and on The Fara, 1970 *C. Murdoch*;
plentiful at Druim na h-Aimhne on the moors at Corrimony, 1964
CGE; slopes of Creag na h-Iolaire at Aviemore, 1974 **E**.

Alnus Mill.
A. glutinosa (L.) Gaertn.   *Allar, Arn, Alder*
Native. Abundant by river banks, sometimes forming small woods.
Bark dark brown, fissured.
M Frequent (Gordon); used for making clogs, abundant (Burgess);
by the Muckle Burn at Dyke and throughout the county, 1954.
N Cawdor, 1837 *WAS* **E**; Ardclach (Thomson); banks of the
River Findhorn, becoming scarce at Drynachan, and by all the
small burns, 1954; boggy ground by Cran Loch, 1960 **ABD**.
I Inverness, 1832 Pitmain and Dalwhinnie, 1833 (Watson, *Loc.
Cat.*); Kincraig, 1904 *G. E. Neilson* **GL**; Cherry Island in Loch Ness,
1904 *G. T. West*; throughout the district, 1954; margins of Loch
Craskie in Glen Cannich, 1972 **E**.

A. incana (L.) Moench   *Grey Alder*
Planted on pine wood verges and in small plantations. Bark pale
grey and smooth.
M Introduced (Burgess); Low Wood in the Culbin Forest, 1937

FAGUS 255

J. Walton **GL**; planted in several rides in the Culbin Forest, 1966
E,BM; margin of the Roseisle Forest near Burghead, 1967 **E**; small
plantation at Castle Grant, Grantown, and edge of quarry at
Netherglen N of Rothes.
I Planted in Reelig Glen, 1967; plantation near Aigas in
Strathglass, also at Glencoe, near the dam at Kilmorack, by the
small bridge above the Falls of Divach, roadside verge by the A9 in
Daviot Wood and at Flichity House.

CORYLACEAE
Carpinus L.
C. betulus L. *Hornbeam*
Introduced.
M Certainly introduced (Gordon); parishes of Forres, Urquhart,
Bellie and Rafford (Burgess); single tree by the road to Cluny Hill
near Forres Police Station, 1972 **E**.
N Kinsteary House, Auldearn, 1964 **ABD**.
I Single tree at the Bogroy Inn at Kirkhill, 1925 and planted in
many of the hedges nearby between the village and the Beauly
Firth, 1925; garden in Holm Road, Inverness and by the old
stables at Culloden House.

Corylus L.
C. avellana L. *Hazel, Cob-nut*
Native. Common in riverside woods and burn gullies on base-rich
soil. Map 156.
M Very common (Gordon); Sanquhar House Forres, 1898 *JK*
FRS; by the Muckle Burn near Brodie Station, 1930; in the gorge
of Huntly's Cave on Dava Moor etc., 1954.
N Cawdor, 1837 *WAS* **E**; Ardclach (Thomson); plentiful on the
steep banks of the River Findhorn and throughout the county, 1954;
in a small burn gully on the moors at Drynachan.
I Inverness, 1832 Pitmain, 1833 (Watson, *Loc. cat.*); Abriachan,
1940 *AJ* **ABD**; rare at Ben Alder Lodge, 1956 but frequent in most
burn gorges where there is some shelter; by the road in Glen
Cannich, 1972 **E**.

FAGACEAE
Fagus L.
F. sylvatica L. *Beech*
Planted throughout the district. Regenerates naturally in some
woods.
M Certainly introduced (Gordon); planted in the Culbin Forest,
at Boat of Garten, and in the Spey and Findhorn valleys etc., 1954;
Aviemore, 1964 *P. F. Wilson* **E**.
N (Stables, *Cat.*); Ardclach (Thomson); Cawdor Wood, Coulmony
and at Drynachan etc., 1954.
I (Watson, *Top. Bot.*); Torbreck, 1940 *AJ* **ABD**; in all the policies
of the larger houses reaching Dalwhinnie and Ben Alder Lodge.

The form with deeply incised leaves is planted at Darnaway Castle, Invererne House near Forres, Dalvey Cottage near Dyke and St Mary's Well at Orton.

Castanea Mill.

C. sativa Mill.   *Sweet Chestnut, Spanish Chestnut*
Planted.
M In the parishes of Dyke, Forres, Edinkillie, Bellie, Elgin and Birnie with some very large trees at Gordon Castle (Burgess); policies of Brodie Castle, small wood by the Moss of Meft, as a hedge at Orton etc.
N Ardclach (Thomson); planted at Drynachan, Cawdor, Holme Rose and Kilravock Castle.
I Bunchrew, 1855 *HC* **INV**; Dores, 1940 *AJ* **ABD**; planted at Inverness, Ballindarroch, Croy and Pityoulish, 1954; plentiful in the woods at Balblair near Kirkhill where it occasionally forms nuts.

Quercus L.

Q. cerris L.   *Turkey Oak*
Introduced.
M Parishes of Dyke, Forres, Urquhart and Bellie (Burgess); Drumduan Wood at Forres and by a lay-by just N of Edinkillie Kirk, Dunphail.
N Single tree by the Glengeouillie bridge at Cawdor, 1970 **E**.
I Planted at Bruiach near Kiltarlity, in the policies of Guisachan House, by the Dores road at Scaniport and by the roadside at Hughtown.

Q. ilex L.   *Evergreen Oak, Holm Oak*
Introduced.
M Planted at Gordon Castle and Pluscarden Priory (Burgess); in the policies of Brodie Castle and at Findrassie near Elgin.
I Policies of Ness Castle, Inverness, 1956.

Q. robur L.   *Pedunculate Oak*
    Q. pedunculata Ehrh. ex. Hoffm.
Native. Leaves glabrous beneath with small reflexed auricles at the base; peduncle 2–8 cm.
M Frequent (Gordon); by the River Findhorn at Logie and Altyre (Druce, 1888); Forres, 1898 *JK* **FRS**; indigenous remains of oak forest at Darnaway, *P. Leslie* (Burgess); throughout the county but often planted, 1954.
N (Stables, *Cat.*); Ardclach (Thomson); policies of Geddes House and at Drynachan Lodge, 1954.
I Craighills Castle, 1855 *HC* **INV**; Rothiemurchus (Druce, 1888); common at Inverness (J. Don, 1898); Dores, 1940 *AJ* **ABD**; planted extensively but perhaps native in Strathglass and near Inverness.

Q. petraea (Mattuschka) Lieb.   *Durmast Oak, Sessile Oak*
    Q. sessiliflora Salisb.
Native. Common on acid soils. Leaves pubescent along the midrib

beneath, cuneate to cordate at the base; peduncle very short (1 cm) or absent.

M Logie (Druce, 1888); parishes of Dyke, Edinkillie and Forres, used for furniture etc., and the bark for tanning, possibly introduced now common as at Darnaway (Burgess); throughout the county, possibly native in the Kellas wood on the Hill of Wangie.

N Wood by Geddes reservoir; banks of the River Findhorn at Glenferness, Dulsie Bridge and Drynachan etc., 1954.

I Scattered throughout the district, Fort Augustus, Drumnadrochit, shores of Loch Ness, Kingussie, Pityoulish, Tomatin and Strathglass etc., 1954; among birch by the River Calder in Glen Banchor at Newtonmore.

SALICACEAE
Populus L.

P. alba L. *White Poplar*

Introduced. Catkin scales dentate.

M Certainly introduced (Gordon); planted at Dyke and Forres (Burgess).

I Ness Islands (Aitken); Drumnadrochit, 1940 *AJ* **ABD**.

P. alba × tremula = P. × canescens (Ait.) Sm. *Grey Poplar*

Introduced. Catkin scales laciniate. There is much confusion over *P. × canescens* and *P. alba* and it is probable that many of the old records for *P. alba* should be placed here.

M Planted by the Findhorn bridge near Dalvey Smithy, 1972 **E**; bank near Sheriffston school, by the old Spey Bridge at Grantown, at Duffus House and at Forres, Rothes and Blacksboat etc.

N By the river at Nairn, 1954 **E**.

I Beauly road (Aitken); planted at Newtonmore, Inverness, Strathglass, Daviot etc., 1954; by the Moniack Burn near Bogroy Inn at Kirkhill, 1972 **E**.

P. tremula L. *Quaking Ash, Aspen*

Native. In rocks of ravines and by rivers. Often planted as in the Culbin Forest. Map 157.

M Frequent (Gordon); common in Strathspey and in the Findhorn valley, 1954; small wood near Blacksboat, 1968 **E,BM**; by the River Spey at Grantown, 1972 **E**.

N Cawdor (Stables, *Cat.*); Ardclach (Thomson); common in the Findhorn valley, by the River Nairn at Holme Rose and on Lethen Mill hill, 1954.

I Pitmain and Dalwhinnie, 1833 (Watson, *Loc. Cat.*); Glen Convinth (Aitken); Inverfarigaig, 1940 *AJ* **ABD**; abundant in the Strathspey, 1954; cliff face at Errogie, by the River Moriston at Ceannocroc, among rocks in the gorge of the Allt Lorgaidh in Glen Feshie and in rocks at Gaick; by the River Spey at Laggan, 1972 **E**.

P.× canadensis Moench   *Black Italian Poplar*
   P. serotina Hartig
Introduced. Trunk of tree without bosses.
M Planted, *P. Leslie* (Burgess); single trees at Forres, Dalvey,
Kellas, Blacksboat and Grantown etc., 1954.
N Planted at Drynachan Lodge, Holme Rose and Tradespark at
Nairn and doubtless elsewhere, 1954.
I Planted at Beauly, 1947; at Farley, Daviot, Moy Hall, Inver-
druie and in Eskadale kirkyard etc., 1954.

P. gileadensis Rouleau   *Balsam Poplar*
Introduced. Frequently planted by roadside verges and on the
edges of woods. The leaves are strongly scented when unfolding.
M Planted at Seapark, Kinloss, in the Culbin Forest, at Cromdale,
Orton, Knockando etc.; by a ditch at Roseisle Forest, 1972 **E**.
N Ferness village (Thomson); Drynachan Lodge and Holme Rose,
1954.
I Plantations in Glen Feshie, Glen Mazeran, at Farley, Errogie,
S of Aviemore, at Tomatin House, at Eskdale, Drumnadrochit and
Tomich and by an old cottage at Truim Bridge etc.

Salix L.
This difficult group have not been thoroughly investigated and
require much further study. Many hybrids occur. Recent records
have been determined by R. D. Meikle.
S. pentandra L.   *Sauchs, Palms, Bay Willow*
Doubtfully native.

M Elchies and Waulkmill (Gordon); river banks in the parishes
of Dyke, Edinkillie, Forres and Knockando (Burgess); two or three
trees on the E bank of Loch Spynie, 1972 **E**; single trees at Blacksboat and near a farm at Cromdale.
N Bognafouran Wood, Nairn, *J. B. Brichan* (Gordon, annot.).
I Single tree in a field by a small burn N of Essich, 1975 **E**; one
or two trees in a bog by Antfield S of Scaniport, 1976.

S. alba L.   *White Willow*
Introduced. Planted by farm ponds and by burns.
M Doubtfully native, frequent (Gordon); planted in the parishes
of Dyke, Forres, and Edinkillie and Speymouth (Burgess); by the
farm pond at Rosehaugh, banks of the River Lossie at Elgin, at
Cromdale and Grantown, in a ditch S of Duffus, etc.; common by
Loch Spynie, 1973 **E**.
N Ardclach (Thomson); quarry-hole at Kingsteps, by the River
Nairn at Holme Rose, 1954.
I Urquhart Bay, 1904 *G. T. West*; Lentran, 1940 *AJ* **ABD**;
planted at Dalwhinnie, Kingussie, by Loch Meiklie in Glen
Urquhart, at Ness Castle, Inverness and at Culloden and Kirkhill
etc., 1954.

   Var. **coerulea** (Sm.) Sm.   *Cricket-bat Willow*
I Planted by the River Beauly near the football field at Beauly,
1976 **E**.

Var. **vitellina** (L.) Stokes
M Waterford near Forres, 1897 *JK* **FRS**.

S. alba×**fragilis**=**S.**×**rubens** Schrank
M Pond at Rosehaugh Farm near Elgin, 1954; shingle of the River
Spey at Fochabers, 1972 **E**; by the Spey at Boat o'Brig, 1972 **E**;
Spey shingle at Garmouth, 1973 **ABD**.
N By the Mill of Lethen, 1973 **E,ABD**.

S. babylonica L.×**fragilis**=**S.**×**blanda** Anderss.
M By the Mosset Burn on the Invererne road at Forres, 1956.

S. fragilis L.   *Crack Willow*
Introduced in north Scotland.

Var. **fragilis**
M Ditch at Coltfield near Alves, 1973 **E**.
I Near the football field by the river at Beauly, 1976 **E**.

Var. **russelliana** (Sm.) Koch   *The Duke's Willow*
M Gordon Castle (Gordon); by the River Lossie at Bishopmill,
1956; banks of the River Spey N of Rothes and at Loch Spynie, 1973
E,BM.
I Near Inverness, 1832 (Watson, *Loc. Cat.*); by a small burn at
Allanfearn, 1954 **E**.

Var. **decipiens** (Hoffm.) Koch
    S. fragilis auct. mult.
Frequently planted. Flowers usually male in Britain.
M Banks of the River Spey at Boat of Garten, 1973 **E,ABD**; margin
of a pool in Urquhart village and on shingle of the Spey below
Baxter's Factory at Fochabers, 1973; river shingle at Garmouth,
1973 **E**.
I Banks of the river at Beauly, 1976 **E**.

S. triandra L.   *Almond Willow*
Introduced.
M Planted at New Spynie, *G. C. Druce* (Burgess); banks of the
River Lossie at Bishopmill, 1956; shingle of the River Spey at
Garmouth.
N Roadside ditch just S of Righoul school, 1973 **E**.
I Planted by a small lochan near Aldourie and by the river
at Beauly, 1975.

S. purpurea L.   *Purple Willow*
Native. Frequent by burns and rivers, and by ditches.
M Side of the railway at Thornhill crossing at Forres, 1897 *JK* **FRS**;
Waterford (as var. *ramulosa* Borrer ex Bab.), 1898 *JK* **FRS**; a large
patch N of Culbin and not far from the Buckie Loch, *ESM & WASh*
(*J. Bot.* **37**: 387 (1899)); by the Black Burn near Pluscarden, 1925;
by a fire dam in the Culbin Forest, 1967 **ABD,BM**; by the Muckle
Burn at Earlsmill, 1969 **BM**; dune slack on Lossiemouth golf course,
1972 **E**; bog W of Glenerney, perhaps planted there, 1972 **E**; also

at Lochinvar, near Boat of Garten, at Relugas and Dava etc.,
planted at Carrbridge.

N By the River Findhorn at Ardclach (Thomson); by the River
Nairn near Bowford Bridge, 1956; by the ford at Achavraat, 1965
ABD; near Glenferness, 1966.

I By the River Nairn at Culloden and by the River Enrick at
Drumnadrochit, 1954; by Loch Ness at Invermoriston, 1956;
planted in a pond at Clunes, Kirkhill, 1963 **CGE**; plantation at
Ceannocroc Bridge and on roadside verges in other places in Glen
Moriston; planted at Raigmore Pond, Inverness, by a small
lochan at Aldourie; native by Loch Ness at Lewiston; plantation
near the A9 at Daviot, 1975 **E**.

S. purpurea × viminalis L. = **S. × rubra** Huds.
M Loch Spynie (Druce, 1888).
I Planted by the road W of Torgyle Bridge, Glen Moriston, 1964.

S. daphnoides Vill.
Introduced. Planted at the forest verge at Woodside, Blervie, in
Moray, 1970 **E**; by the railway at Nairn and in plantations at
Lethen and at Druim House in Nairn.

S. viminalis L   *Common Osier*
Doubtfully native on river banks, and frequently planted.
M Frequent (Gordon); Waterford, 1898 *JK* **FRS**; banks of the
Muckle Burn at Dyke, 1964 **E**; roadside ditch E of Loch Spynie,
1972 **E,BM**; shingle of the River Spey at Garmouth, 1973 **E,ABD**;
also at Grantown, Dunphail, Kellas and Rothes etc.
N Ardclach (Thomson); by the river at Nairn, 1900 (Lobban).
I (Trail, *ASNH* **15**: 181 (1906)); by the River Enrick at Drumna-
drochit, 1947; by Loch Ness at Invermoriston, at Kirkhill and
Culloden, 1954; verge of Loch Mhòr at Errogie; shores of Loch
Uanagan at Fort Augustus; planted by the track to Loch nan Lann
at Knockie, etc.

S. viminalis × ? = S. × stipularis Sm.
Of doubtful parentage.
M Spynie Loch (Druce, *ASNH* **6**: 53–54; (1897)). About *S. ×
stipularis* the Rev. E. F. Linton remarks 'it has a great look of
caprea × viminalis on the east side of Loch Spynie where plants of
undoubted *caprea × viminalis* were growing' (Burgess).

S. caprea L.   *Sallow, Goat Willow*
   Subsp. **caprea**
      Var. **caprea**
Native. Very common throughout the district reaching 500 m in
some mountain ravines.
M Frequent (Gordon); Thornhill at Forres, 1898 *JK* **FRS**;
policies of Innes House, Urquhart, 1967 **CGE**.
N Cawdor woods, 1833 *WAS* **GL**; Ardclach (Thomson); by the
River Findhorn at Drynachan and throughout the county, 1954.

. myrsinites × 1
...rtle-leaved Willow

M. C. F. Proctor

...x lapponum × ½
...olly Willow

M. C. F. Proctor

PLATE 12

I (Watson, *Top. Bot.*); Ness Islands (Aitken); Westhill 4 miles S. of Inverness, 1937 *J. Walton* **GL**; common by rivers and burns in the lowland areas becoming scarce in the highlands mostly in sheltered places below cliffs or in ravines; birch wood at Daviot, 1971 **E**; Glen Urquhart, 1974 *J. A. Macdonald* **STA**; cliffs of Creag Dhubh S of Loch Ericht 500 m, 1976 **E**.

Subsp **sericea** (Anderss.) Flod.
    S. coaetanea (Hartm.) Flod.
A grey-leaved variant often found by mountain burns and ravines. Under-recorded.
M Bog by a railway bridge over the A9 about 2 miles N of Carrbridge, 1976 **E**.
I By the Allt Targe flowing from Loch Tuill Bhearach to Loch Mullardoch in Glen Cannich, 1947 *E. Milne-Redhead & C. C. Townsend* **K**; ravine of the Allt Fionndairnich in Strathdearn (as var. *sphacelata* (Sm.) Wahl.) 1974 **E**; ravine of the Allt Doe in Glen Doe, 1976 **E**.

S. caprea× cinerea = S.× reichardtii A. Kerner
N By the River Findhorn at Dulsie Bridge, 1960 **K**.
I Near a waterfall ½ mile W of Loch Beinn a' Mheadhoin in Glen Affric, 1947 *E. Milne-Redhead* **K**.

S. caprea× myrsinifolia Salisb. = **S.× latifolia** Forbes
I River shingle near Drumnadrochit, 1947 *C. C. Townsend* **K**.

S. caprea× viminalis L. = **S.× laurina** Sm.
M East side of Loch Spynie (Druce, *ASNH* **6**: 54 (1897)).

S. cinerea L.   *Grey Willow*
    Subsp. **cinerea**
Native. Rare. Leaves pubescent beneath.
M Banks of the River Spey below the golf course at Boat of Garten, 1973 **E**.

    Subsp. **oleifolia** Macreight
    Subsp. *atrocinerea* (Brot.) Silva et Sobr.
Native. Common throughout the lowland areas, less so in the highlands. Leaves with short, curved rust-coloured hairs beneath.
M Frequent (Gordon); Waterford near Forres, 1898 *JK* **FRS**; throughout the county, 1954; quarry at Lossiemouth, 1967 **CGE**; banks of the River Findhorn N of Greshop, 1968 **E**; by the loch of Speys Law in the Lossie Forest, 1972 **E**; banks of the River Spey at Boat of Garten, 1973 **E,BM**.
N Ardclach (Thomson); at Cantraydoune, Holme Rose and by the River Nairn at Howford Bridge, 1954; banks of the Muckle Burn at Achavraat, 1960 **K**; dried up bog on the Hill of Urchany, 1974 **E**.
I Inverness, 1833 (Watson, *Loc. Cat*); Drumnadrochit, 1947; Inshriach, 1964 *P. H. Davis* **E**; by the old ice pond near Lovat Bridge, Beauly, 1976 **E**.

S. cinerea × **viminalis** = **S.** × **smithiana** Willd.
M By Sanquhar House pond at Forres, 1956; by the River Findhorn
opposite Greshop Wood; banks of the River Lossie at Bishopmill;
dune slack at Lossiemouth, 1972 **E**; banks of the River Spey below
Knockando station, 1973 **E**; banks of the River Findhorn N of the
road bridge W of Forres.

S. cinerea subsp. **oleifolia** × **myrsinifolia** = **S.** × **strepida** Forbes
M Banks of the River Findhorn between the road bridge and the
railway bridge at Broom of Moy, 1956.

S. aurita L.   *Eared Willow*
Native. Very common in wet places on moors, by rivers and burns
reaching 820 m on mountains.
M Frequent (Gordon); Loch of Blairs, 1859 *JK* **FRS**; Grantown,
1897 *J. S. Gamble* **K**; throughout the county, 1954; fixed dunes in
the Culbin Forest, 1966 **E**.
N (Stables, *Cat.*); banks of the River Findhorn at Dulsie Bridge
1960 **K**; bog by Cran Loch, 1960 **ABD,K** and throughout the
county.
I Pitmain and Dalwhinnie, 1833 (Watson, *Loc. Cat.*); Loch
Ruthven, Loch Ness and Loch Bran, 1904 *G. T. West* **STA**; by
Loch an Laghair and Loch Beinn a' Mheadhoin, 1947 *E. Milne-
Redhead* **K**; throughout the district, 1954.

S. aurita × **cinerea** subsp. **oleifolia** = **S.** × **multinervis** Doell
M Culbin Sands, 1942 *UKD* **UKD**; by Sanquhar pond at Forres,
1956.

S. aurita × **phylicifolia** L. = **S.** × **ludificans** F. B. White
I With both parents by the Markie Burn near Crathie at Laggan,
1916 *ESM* **E,K,CGE** det A. Neumann.

S. aurita × **repens** L. = **S.** × **ambigua** Ehrh.
I (Trail, *ASNH* **15**: 181 (1906)).

S. aurita × **viminalis** L. = **S.** × **fruticosa** Doell
I Shingle of the Markie Burn just above where it joins the Spey
at Crathie, near Laggan, 1916 *ESM* **E**.

S. myrsinifolia Salisb.   *Dark-leaved Willow*
  S. nigricans Sm.
Native. Frequent on mountain cliffs, ravines and river banks.
Twigs pubescent, leaf shape very variable; leaves deep green and
dull above, pale but not glaucous beneath, pubescent on the veins;
usually turning black when dried. Map 158.
M Sanquhar, Forres (as var. *fosteriana*), *JK* and riverside at Spey-
mouth, *G. Birnie* (Burgess); by the Spey at Kinchurdy (Druce,
1888); banks of the River Findhorn 1 mile N of Greshop Wood,
Forres, 1956; by the River Spey at Wester Elchies, 1974 **E**.
I Cairn Gorm, *W. J. Hooker* **OXF**; Coire Madagan Mor in
Gaick Forest, 1948 *N. Y. Sandwith* **K**; gully of the burn at Gaick
Lodge, 1975 **E**; burn gully of A' Chaorinich in Gaick Forest, 1974

ABD,BM; cliffs by the Allt Lorgaidh in Glen Feshie, 1974; cliffs of Creag Dhubh S of Loch Ericht, 1976 **E**.

S. myrsinifolia × myrsinites
I   Gully of the burn at Gaick Lodge, 1974 **E**.

S. myrsinifolia × phylicifolia = S. × tetrapla Walker
M Banks of the River Findhorn N of the road bridge on the A96 W of Forres, 1956; banks of the River Spey below Knockando station, 1973 **E**.
I   Banks of the River Spey at Crathie near Laggan (with both parents) 1916 *ESM* **K**; ledge of the Allt na Faing at 610 m, rock ledge in upper Glen na Ciche and in a gully on the W side of Tigh Mór na Seilge all in Glen Affric, 1974 *R. W. M. Corner et al.*

S. phylicifolia L.   *Tea-leaved Willow*
Native. Common on river banks and in burn ravines. Twigs glabrous, leaves glabrous shining above, glaucous and glabrous beneath. Map. 159.
M Aviemore on Speyside (Druce, 1888); Waterford near Forres, 1897 *JK* **FRS**; river bank St John's Mead, Darnaway, 1953 *B. Welch* **BM**; common in the Spey and Findhorn valleys, 1954; banks of the River Dulnain near Carrbridge, 1972 **E**; railway embankment by the A9 N of Carrbridge, 1976 **E,BM**.
N By the Muckle Burn at Ardclach, 1900 (Lobban); common the whole length of the River Findhorn, 1954; on shingle by the river at Glenferness House, 1960 **K**; river bank at Ardclach, 1973 **E,BM**.
I   Glen Tarff near Fort Augustus (Gordon); Rothiemurchus, 1890 *JK* **FRS**; banks of the Spey at Crathie, 1916 *ESM*; by a burn near Loch Beinn a' Mheadhoin in Glen Affric, 1947 *E. Milne-Redhead* **K**; abundant the whole length of the River Spey, 1954; by the Elrick Burn in Strathdearn, 1970 **E**; cliffs of Sgùrr na Lapaich in Glen Strathfarrar, 1972 **E**; in burn gullies at Gaick and Glen Feshie; burn ravine of the Allt Doe in Glen Doe, 1976 **BM**.
A hybrid possibly referable to **S. phylicifolia × lapponum** occurs on the cliffs of Sgùrr na Lapaich in Glen Strathfarrar, 1972 **E**.

S. repens L.   *Creeping Willow*
Subsp. **repens**
Native. Common on moors and dune slacks. Map 160.
M Common (Gordon); Culbin Sands, cult. Surrey, 1932 *O. Warburg* **K**; throughout the county, 1954; very common on dunes in the Culbin Forest, 1968 **E**; moor W of Loch Mòr, Duthil, 1973 **E**.
N (Stables, *Cat.*); Hill of Aitnoch (Thomson); Culbin, 1904 *G. T. West* **STA**; dunes by Cran Loch, 1960 **ABD,K**; dunes on the Nairn Dunbar golf course, 1972 **E**.
I   Braeriach (Gordon); Cairn Gorm, 830 m *J. Johnstone*, 1839 **K**; on the moor above Holm Burn near Inverness (Aitken); Rothiemurchus (forma), 1892 *AS* **E**; railway embankment at Daviot, 1971 **E**; bog at Newtonmore near the station, a tall plant of nearly 2 m with very narrow leaves, 1975 **E**.

Subsp. **argentea** (Sm.) G. & A. Camus
Leaves and catkins larger than subsp. *repens* the leaves being usually silky on both sides. Dune slacks.
M Fixed dunes in the Culbin Forest, 1942 *UKD*, **UKD**.
N Fixed dunes at Kingsteps near Nairn, 1964 **ABD**.
I Recorded from Loch Garten, railway line at Daviot, roadside near Flichity Inn and Kincardine, but not confirmed.

S. lapponum L.   *Downy Willow*
Native. Frequent on wet rock ledges in the higher mountains. Pl 12.
I Mountains near Dalwhinnie, herb. *Borrer* **K**; Braeriach, 1877 *W. B. Boyd* **E**; S side of Glen Einich (Druce, 1888); Coire Chàis and by the Allt Coire Bhathaich near Dalwhinnie, 610 m 1911 *ESM* **E**; by the Allt Coire Chùirn, 1911 *ESM & WASh* **K**; Geal Chàrn in Glen Markie, 1916 *ESM* **E**; cliffs of Sgùrr na Lapaich in Glen Strathfarrar, 1971 **E,CGE**; cliffs on Sgùrr na Muice at Monar, burn gully at Gaick Lodge, summit of Ben Alder etc.

S. lapponum × repens = S. × pithoensis Rouy
I By the Allt Coire Bhataich, 610 m 'with revolute edges to the leaves, a good sign of repens', 1911 *ESM & WASh* **OXF**.

S. lanata L.   *Woolly Willow*
Native. Rare on cliff ledges in the higher mountains.
I Sgùrr nan Conbhairean, Glen Moriston, 610 m, 1960 *H. Milne-Redhead*; seen in the Coire Garbhlach of Glen Feshie, 1967 *J. G. Roger et al.*; on an inaccessible cliff ledge in Coire a' Bhèin near Garva Bridge, 1975.

S. myrsinites L.   *Whortle-leaved Willow*
Native. Rare on base-rich rocks in the higher mountains. Twigs pubescent when young becoming glabrous, leaves bright green and shining on both sides at maturity. Pl 12.
I Braeriach (Gordon) and in 1877 *W. B. Boyd* **E,BM**; rocks NE of Dalwhinnie, 1867 *A. T. Coore* (Balfour); head of Loch Einich, 1899 *JK* **FRS** and in 1890 *H. & J. Groves* **BM**; Coire Chùirn near Drumochter, 1911 *ESM* **E,CGE** and in 1948 *N. Y. Sandwith* **K**; rock ledge on Sgùrr na Lapaich in Glen Strathfarrar, 1971; cliffs in a burn ravine in Glen Markie, by the Allt Lorgaidh in Glen Feshie and at Gaick Lodge, 1972.

S. myrsinites × phylicifolia
I Cliffs at 610 m on the north face of Creag Dhubh by Loch Ericht, 1976 *O. M. Stewart & MMcCW* **E**.

S. herbacea L.   *Dwarf Willow*
Native. Common on bare mountain tops and occasionally on rock ledges and grass slopes.
M Summit of Geal-chàrn Beag at Aviemore, 1974.
I Dalwhinnie, 1833 (Watson, *Loc. Cat.*) and herb. *Borrer* **K**; Badenoch (Gordon); Cairn Gorm, 1867 *JK* **FRS**; Sgòran Dubh, 1891 *AS* **E**; Toll Easa in Glen Affric, 1947 *E. Milne-Redhead* **K**;

Coire Garbhlach in Glen Feshie, 1957 *E. C. Wallace*; summit
plateau of Cairn Gorm, 1953 *A. Melderis* **BM**; at Gaick, The Fara
and on all the higher mountains.

S. reticulata L.   *Reticulate Willow*
Native. Very rare on base-rich mountain ledges.
I   Cairngorms, Scotland, 1839 *J. Johnstone* ex herb. *B. T. Lowne* **K**;
Coire Garbhlach in Glen Feshie, 1972 *R. McBeath*.

ERICACEAE
Rhododendron L.
R. ponticum L.   *Rhododendron*
Introduced. Naturalised in woods, on railway embankments and
rough ground.
M Common in the Culbin Forest near Binsness, 1954; railway
embankment by Loch na Bo near Lhanbryde and in the policies
of most of the larger houses.
N Rough ground at Dulsie Bridge, 1954 and at Lethen Mill,
Cran Loch etc.
I   On an island in the River Glass near Glassburn in Strathglass,
1954; pine wood at Dunmaglass, woods at Guisachan and Glen-
mazeran Lodge etc.

Loiseleuria Desv.
L. procumbens (L.) Desv.   *Trailing Azalea*
  Azalea procumbens L.
Native. Common on bare mountain tops over 500 m. Pl 11, Map 161.
M Summit of Geal-chàrn Beag above Aviemore, 1974 **E**.
I   Dalwhinnie, 1833 (Watson, *Loc. Cat.*); Cairngorms, herb.
Borrer **K**; Fraoch-choire, Cannich (Ball, 1851); Mam Sodhail and
Beinn a' Bha'ach Ard, *Messrs Anderson, Stables & Gordon* (Gordon);
Sgòran Dubh in Glen Einich, 1892 *AS* **E**; Sgùrr na Lapaich in
Glen Strathfarrar (with white flowers) 1904 (Boyd); abundant in the
Cairngorm range but rare on the hills in Gaick Forest.

Phyllodoce Salisb.
P. caerulea (L.) Bab.   *Blue Heath*
  Menziesia caerulea (L.) Sw.
Native. Rare on moorland slopes of 700 m. Very difficult to dis-
tinguish from *Empetrum* unless in flower. Leaves linear, serrulate.
Flowering time early June. Pl 11.
I   Near Aviemore, 1812 *Mr Brown* of Perth (Smith, *Eng. Fl.* **2**: 222
(1824)); near Aviemore in Strathspey, 1821 (Hooker, *Fl. Scot.*);
near Aviemore, 1830 *R. Graham* **ABD** and in 1835 *R. Graham* herb.
James Dickson **E**; slopes of Sròn Dréineach near Beinn Bheòil at
750 m *R. E. Groom* **E**!.

Daboecia D. Don
D. cantabrica (Huds.) C. Koch   *St Dabeoc's Heath*
Garden escape.
M Among heather on White Ash Hill near Fochabers, probably

from seed carried by birds from a neighbouring Nursery (Burgess);
garden escape in a pine wood near Cathy House, Forres, 1950
Mr Clark!.

Gaultheria L.
G. shallon Pursh   *Shallon*
Introduced. Planted for pheasant food.
M Abundant in the pine woods at Blervie House, 1925; policies
of Binsness House, Culbin Forest, 1962.
N Policies of Glenferness House, 1954.
I Roadside verge by Loch Dochfour, 1954; policies of Kyllachy
House in Strathdearn, and at Flichity House; abundant by a track
between Balnagaig and Dunain House near Inverness, 1971
M. Barron!.

Pernettya Gaudich
P. mucronata (L. fil.) Gaud. ex Spreng.   *Pernettya*
Introduced. Rare.
M Among heather near the Buckie Loch in the Culbin Forest, 1952;
pine wood at Fochabers, 1970 *A. J. Souter*.
N Pine wood near Drumbeg in the Culbin Forest, 1967; sand dune
on the Bar, Culbin (lost by erosion in 1969), 1968 **E**.
I On a small island to the SW of Loch Knockie, 1974 *J. A. Murray*
BM; top of a bank by the Neaty Burn, Glen Strathfarrar, 1974
D. Kingston **E**.

Arctostaphylos Adans.
A. uva-ursi (L.) Spreng.   *Gnashacks, Bearberry*
Native. Common on open moorland forming large prostrate mats of
small evergreen leaves. Pl 13, Map 162.
M By the burn at Fochabers, 1831 *G. Gordon* **GL**; Lochindorb, 1867
JK **FRS**; Carrbridge, *E. M. MacAlister Hall* **E**; ½ mile W of Boat
of Garten, 1939 *J. Walton* **GL**; Dava Moor, on the Cromdale Hills
and on the Moss of Bednawinny near Dallas etc., 1954.
N Moors above Cawdor, 1832 *WAS* **ABD**; common on the moors
at Ardclach, 1888 *R. Thomson* **NRN**; moors at Dulsie Bridge,
Drynachan and Glenferness etc., 1954.
I Dalwhinnie, 1833 (Watson, *Loc. Cat.*); Kingussie, 1838 *A.
Rutherford* **E,OXF**; Englishton Muir, Dunain and Dunlichity
(Aitken); Dores, 1940 *AJ* **ABD**; very common on the moors at
Newtonmore etc., 1954.

A. alpinus (L.) Spreng.   *Alpine Bearberry*
    Arctous alpinus (L.) Neidenzu
Native. Rare in the eastern districts on bare summits of mountains
over 600 m. Frequent in the west at 300 m. Pl 13, Map 163.
M Summit of Geal-chàrn Beag above Aviemore, 1974 **E**.
I Beinn a' Bha' ach Ard in Glen Strathfarrar, 1867 *J. Farquharson
& T. Fraser* (*Trans. BSE* 9: 474 (1868)); Saddle Hill, Strathglass
(Aitken); Mam Sodhail (Buchanan White, *J. Bot.* **8**: 129 (1870))
and 1936 *A. J. Wilmott* **BM**; rare at Tomatin and Strathglass

(J. Don, 1898); Carn an Tuairneir, Invermoriston, 1904 *G. T. West* **STA**; on An Meallan in Glen Affric and Mullach Tarsuinn in Glen Cannich, 1958 *D. A. Ratcliffe*; common over a large area on rocky hummocks on the moors of Corrimony in Glen Urquhart, 1950 **E**; Geal-chàrn Mòr, 1963 *C. Murdoch*; lower slopes of Braeriach, 1966 *G. E. Woodroffe & E. Pelham-Clinton* **E!**; common on the higher hills of glens Moriston, Cannich, Strathfarrar and Affric.

Calluna Salisb.

C. vulgaris (L.) Hull   *Heather, Ling*
Native. Abundant on acid moorland throughout the district.
M Gordon Castle, Fochabers, *G. Don* **BM**; very common (Gordon); Boat of Garten (as var. *glabrata*), 1898 *JK* **FRS**; Aviemore, 1904 *G. B. Neilson* **GL**; Culbin Forest and throughout the county, 1954.
N (Stables, *Cat.*); Ardclach (Thomson); near the sea E of Nairn, 1930 and throughout the county, 1954.
I Rothiemurchus, 1832 *W. Brand* **ABD**; Dalwhinnie, 1833, (Watson, *Loc. Cat.*); Bunchrew Burn, 1855 *HC* **INV**; Braeriach Cairngorms, 1893 *AS* **E**; by Loch Ness, 1940 *AJ* **ABD**; throughout the district, 1954.

Var. **hirsuta** S. F. Gray
Var. *incana* Reichb.; var. *pubescens* auct.
A densely grey-tomentose variant more frequent near the coast than in the highlands.
M Brodie House ? date *coll.* **GL**; sent by *J. Hoy* (probably from Fochabers) herb. *G. Don* **K**; Grantown, 1872 ? *coll.* **E**; Forres, 1899 *JK* **FRS**; Garmouth, 1914 *B. E. White* herb. *ESM* **CGE**; Culbin Forest, 1913 *D. Patton* **GL**, in 1948 *Fraser* **E** and in 1968 *MMcCW* **E**; Blackhills, 1932 *R. M. Adam* **E**.
N Cawdor (R. Graham, *Trans. BSE* 1: 26 (1840)) and in 1865 *R. Thomson* **NRN**; Culbin, 1910 ? *coll.* **GL** and in 1939 *J. Walton* **GL**; pine wood by Lochloy, 1974.

Erica L.

E. tetralix L.   *Cross-leaved Heath*
Native. Abundant in wet moorland bogs. Flowers in terminal clusters, usually pinkish-purple, sometimes white.
M Very common (Gordon); Califer, 1871 *JK* **FRS**; Culbin Sands, 1910 *P. Ewing* **GL**; Heldon Hill, Pluscarden, 1913 *D. Patton* **GL**; Culbin, 1953 *A. Melderis* **BM**; Lossiemouth, 1963 *V. Mackinnon* **E**, and in all moorland bogs.
N (Stables, *Cat.*); Ardclach (Thomson); in all the moorland bogs, 1954; by the Leonach Burn near Dulsie Bridge, 1974 *J. A. Burgess* **STA**.
I Rothiemurchus woods, 1832 *W. Brand* **ABD**; Dalwhinnie, 1833 (Watson, *Loc. Cat.*); on the graves at Culloden, 1841 *WAS* **GL**; abundant in moorland bogs throughout the district, 1954; planted on the new verges by the A9 near Dalwhinnie from a Danish source, 1975.

E. cinerea L.   *Bell Heather*
Native. Abundant on dry moors. Flowers usually dark crimson-purple, occasionally white.
M Very common (Gordon); Califer, 1871 *JK* **FRS**; on heathland by the banks of the River Findhorn near Forres, 1953 *A. Melderis* **BM**; on all the moors but not as common as *Calluna*, 1954.
N (Stables, *Cat.*); Macbeth's moorland W of Brodie (with white flowers), 1835 *J. G. Innes* **E**; Ardclach (Thomson); throughout the county, 1954.
I Dalwhinnie, 1833 (Watson, *Loc. Cat.*); by Reelig Burn, ? date ? *coll.* **INV**; Falls of Foyers, 1883 *C. Bailey* **E**; Abriachan, 1940 *AJ* **ABD**; Sgùrr na Lapaich (with white flowers) 1940 ? *coll.* **STA**; throughout the district, 1954.

E. vagans L.   *Cornish Heath*
Introduced.
I On a moor near Inverness, 1836 *G. Dodds* **E,GL**; planted in the Fairy Glen, Reelig, 1964 **E**.

Vaccinium L.
V. vitis-idaea L.   *Scots Cranberry, Cowberry*
Native. Common in woods on acid soil, and on moorland up to 1,000 m in the Cairngorms. Pl 13, Map 164.
M Frequent in the Oakwood, Elgin and abundant in the upper part of the province (Gordon); Grantown, 1856 *JK* **FRS**; Dunphail, 1886 *ESM* **CGE**; Culbin Forest and throughout the county, 1954.
N (Stables, *Cat.*); Dulsie, 1888 *R. Thomson* **NRN**; near Dulsie Bridge, 1960 **ABD**, and on all the moors.
I Dalwhinnie and Foyers, 1833 (Watson, *Loc. Cat.*); Strath Affric (Ball, 1851); Rothiemurchus, 1892 *AS* **E,GL**; Loch an Eilein, 1904 *G. B. Neilson* **GL**; Glenmore Forest, 1948 *E. M. Burnett* **ABD**; throughout the district, 1954.

V. myrtillus L.   *Blaeberry, Whortleberry*
Native. Abundant on acid moors and in woods, up to 1,200 m in the Cairngorms often forming dominant zones above the heather line.
M Very common (Gordon); Dunphail, 1898 *JK* **FRS**; throughout the county, 1954 reaching sea-level in the Culbin Forest.
N (Stables, *Cat.*); Ardclach (Thomson); pine wood at Glenferness House, 1961 **ABD** and throughout the county.
I Dalwhinnie, 1833 (Watson, *Loc. Cat.*); Craig Phadrig near Inverness, 1854 *HC* **INV**; Creag Coire Doe, hills to the N of Fort Augustus and moor by Loch Tarff, 1904 *G. T. West* **STA**; Daviot, 1940 *AJ* **ABD**; throughout the district, 1954; planted on the new verges of the A9 S of Dalwhinnie from a Danish source 1975.

V. uliginosum L.   *Bog Blaeberry*
Native. Frequent on the higher mountain slopes, usually with blaeberry and heather. Leaves glaucous-green. Reluctant to flower in the Cairngorms. Map 165.

M Dry hillock near Carrbridge, 1896 *JK* **FRS**; slope near the summit of Geal-chàrn Beag, 1974 **E**.
I Badenoch, 1810 herb. *Borrer* **K**; Dalwhinnie, 1833 (Watson, *Loc. Cat.*); Fraoch-choire Cannich (Ball, 1851); Cairn Gorm, 1863 *J. P. Bisset* **ABD**; Pityoulish, 1883 *JK* **FRS**; at an unusually low altitude of 200 m in a bog below Netherwood at Newtonmore, 1972 **E**.

V. oxycoccos L. Cranberry
Oxycoccus palustris Pers.
Native. Rare in moorland bogs. A delicate trailing plant with leaves of 4–8 mm, oblong-ovate, equally broad for some distance from the base or broader towards the middle, dark green above, glaucous green beneath and strongly revolute. Flowers somewhat resembling a tiny Cyclamen, pink, on long puberulous pedicels. Fruit globose or pyriform.
M Cromdale and Knockando, *C. Watt* (Burgess).
I Moy, 1839 *Miss McLauchlan*, Strathglass hills, *Miss Fraser*, and at Belladrum (Gordon); Mealfuarvonie (Aitken); Beinn a' Bha'ach Ard 1867 *J. Farquharson & T. Fraser* (*Trans. BSE* 9: 474 (1868)); S of Loch Affric, 1947 *A. J. Wilmott* **BM**.

V. microcarpum (Rupr.) Hook. fil. Small Cranberry
Native. Smaller and more delicate than the former. Leaves 3–5 mm, triangular-ovate, widest near the base; pedicels glabrous; fruit pyriform. This is the common species of north Scotland and the old records for *V. oxycoccos* may belong here. Usually in sphagnum bogs.
M In sphagnum at the edge of the lochan at Lochnellan on Dava Moor, 1950; by a small lochan below Beum a' Chlaidheimh near Duthil, 1963 **E,CGE**; on the moor at Rochuln, Edinkillie, summit plateau of the Hills of Cromdale, on the slopes of Càrn Dearg Mòr, at Aviemore and on the E side of Loch Mòr, Duthil; abundant by Loch Dallas, at Kinchurdy, 1973 *D. Hayes*!.
N Bog at the SW end of Loch Belivat, 1973 **E,BM**.
I Bog by the River Truim at Drumochter, 1956; pool edge at Pollan Buidhe near Loch Affric, 1971; in a birch wood at Ballindarroch near Scaniport, 1963 *Lady Maud Baillie*!; S side of Loch Dubh in Glen Banchor, 1973 *R. E. Groom*; near Auchgourish, 1973 *D. Hayes*; by Loch Gharbh-choire, Ryvoan, 1975.

PYROLACEAE
Pyrola L.
P. minor L. Common Wintergreen
Native. Common in woods, on rock ledges and damp banks. Flowers pink; style and stigma shorter than the petals. Pl 14, Map 166.
M Darnaway, 1810 herb. *Borrer* **K**; woods at Brodie House, 1821 (Hooker *Fl. Scot.*); in the oakwood at Elgin, 1857 *T. Watt* (Watson); Dunphail, Brodie and Newmill of Alves (Gordon); Cluny Hill,

Forres, 1869 *N. E. Brown* **ABD**; by Blairs Loch near Forres, 1882
JK **FRS**; Grantown, 1925 *A. H. Maude* **BM**; Findhorn Gorge
(as *Orthilia secunda*× *P. minor*) 1922 *E. G. Beach* **OXF**; pine woods
Culbin Forest, 1968 **E,BM** etc.
N Cawdor woods, 1834 *WAS* **ABD,FRS**!; near Dulsie, *G. Wilson*
(Gordon); Kilravock, 1846 *A. Croall* **GL**!; Culbin Forest E of Nairn,
1960 **K**; pine wood Glenferness House, 1961 **ABD**.
I Culloden wood, 1871 ? *coll.* **ELN**; by the Reelig Burn, ? date
? *coll.* **INV**; woods at Belladrum (Aitken); Rothiemurchus, 1892
AS **E,BM,GL**; Englishton Muir, 1930 *M. Cameron*; birch woods at
Farley, Milton, Inverfarigaig and Kincraig etc.; pine woods by
Loch Mallachie, Culloden Wood, near Dalcross and Tomatin etc.;
open moor by Loch Knockie, S of Nethybridge in the Cairngorms
etc.; rock ledges by a burn at Corrimony, on Creag nan Clag and
by the Allt Coire Fhàr at Drumochter.

P. media Sw.   *Intermediate Wintergreen*
Native. Common in pine woods in the highlands. Flowers white,
tinged pink; stigma exerted, straight. Pl 14, Map 167.
M Woods near Brodie House, herb. *Borrer* **K**; Darnaway woods,
1822 ? *coll.* **E**; Moray, 1827 *P. Cruickshank* **STA**; fir wood at Grantown, 1837 *WAS* **ABD**; frequent, but more so in the upland districts,
Oakwood at Elgin (Gordon); Grantown, 1877 *JK* **FRS**; Craigellachie, 1925 *A. H. Maude* **BM**; bank under birch on the Glenbeg
road, Grantown, 1973 **E**; well distributed throughout the county
but now rare in the Culbin Forest.
N Fir woods at Dulsie Bridge, 1810 *Brodie* herb. *Borrer* **K**; Clunas,
1833 *WAS* **ABD**; Ord Hill (Aitken); pine woods at Glenferness
House, 1950; pine wood at Sunnyhillock W of Nairn, 1968 **E**.
I Dalwhinnie, 1833 (Watson, *Loc. Cat.*); Bona Ferry (Aitken); by
the Green Loch, Glenmore, in Glen Einich and by Loch Morlich,
all 1892 *AS* **E**; Craig Dunain, 1940 *AJ* **ABD**; Fasnakyle in Glen
Affric, 1947 *M. S. Campbell & A. J. Wilmott* **BM**; Feshie Bridge, 1950
C. D. Pigott; Reelig Glen, Inverfarigaig, pine wood by Loch
Mallachie, moor by Loch Neaty etc.; moor at Shenachie, 1974
M. Barron.

P. rotundifolia L. subsp. **rotundifolia**   *Larger Wintergreen*
Native. Frequent on open moorland. Flowers white, stigma exerted,
decurved then curving back. Probably under-recorded as it is
reluctant to flower and difficult to distinguish from the other species
unless it is in flower. Pl 14.
M Open pine wood near the Dell at Nethybridge, 1954; planted
at Kellas House, 1973; burn gully on open moorland N of Beum a'
Chlaidheimh, Duthil, 1973.
I Culloden woods, *A. Murray* (Hooker, *Fl. Scot.*); Kingsmill,
Inverness, 1836 *G. Gordon* **E**; by the Reelig Burn, ? date ? *coll.* **INV**;
open moor between Coylumbridge and Tullochgrue, 1954; near
Drumguish, 1972; by the burn above the Alltbeithe Hostel in
Glen Affric, 1974 *R. W. M. Corner et al.*

Orthilia Raf.

O. secunda (L.) House   *Serrated Wintergreen*
   Pyrola secunda L; *Ramischia secunda* (L.) Garcke
Native. Frequent in pine woods, on rock ledges and ravines of
burns. Flowers greenish-white, secund. Pl 15, Map 168.
M Woods near Brodie, 1807 *Brodie* herb. *D. Turner* **K**; Darnaway,
1808 *Brodie* **E** and herb. *Borrer* **K**; Gordon Castle woods, in the
Oakwood at Elgin, on the raised beach near Briggses, and on the
banks of the rivers Findhorn and Divie (Gordon); Falls of Lossie,
1847 herb. *Murchison* **E**; banks of the River Findhorn, 1859 *JK* **FRS**;
Dunphail, 1886 *ESM* **E,ABD,CGE**; wood near Dunphail, 1886
F. J. Hanbury **BM,OXF**; Culbin Forest 1953 *A. Melderis* **BM**; on
rock ledges by the old railway line near Huntly's Cave on Dava
Moor, bank at Nethybridge, by burns at Fae and Kinveachy Lodge,
pine wood at Boat of Garten etc.; Craigellachie Reserve at Aviemore, *D. A. Ratcliffe.*
N Cawdor woods, 1837 *WAS* **E,ABD,STA,BM,K,CGE**!; by the
bridge over the burn at Auchneim, Cawdor, 1865 *R. Thomson* **NRN**;
woods at Glenferness and banks of the River Findhorn at Dulsie
Bridge, 1954.
I Culloden woods, *A. Murray* (Hooker, *Fl. Scot.*); Dochfour Burn,
1855 *HC* **INV**; Dalwhinnie, 1867 *A. Craig Christie* **E**; by the Falls
of Foyers, 1883 *G. V. C. Last* **K**; by the Allt Coire and t-Sneachda,
Cairn Gorm, 1887 *G. C. Druce* **OXF**; by the Green Loch, Glenmore,
1892 *AS* **E**; Glenmore forest Rothiemurchus, 1898 *ESM* **E,BM**;
roadside verge at Kirkhill, 1930 *M. Cameron*; Fasnakyle, 1936
A. J. Wilmott **BM**; rock ledges in the Coire Garbhlach Glen Feshie,
birch wood at Gaick, 1973 **E**; burn gully at Corrimony etc.

Moneses Salisb.

M. uniflora (L.) A. Gray   *One-flowered Wintergreen*
   Pyrola uniflora L.
Native. Occasional in old pine woods. Flowers large, white and
delicately fragrant. Pl 14.
M Pine wood near Brodie House, 1792 *Brodie* **E,** and herb. *Borrer* **K**;
near Gordon Castle, 1792 *J. Hoy* (Hooker, *Fl. Scot.*); discovered by
J. Lawson about 20 years ago in the oak wood at Aldroughty. It was
afterwards lost sight of until 1836 when a few specimens were
gathered by *J. Shier* and pupils (Gordon); Burgie near Forres, 1870
JK **E,ABD,FRS,BM** and in 1871 *G. Gordon* **ABD**; Snab Wood near
Dyke, 1923 *D. Patton* **GL**!; locally abundant in several places in
the Culbin Forest, 1948 **E,BM**; pine woods at Castle Grant,
Grantown; Roseisle Forest near Burghead, 1974 *J. Cuthbertson*!.
N Pine wood at Lochloy House, 1890 *Mary Baillie* **BM**.
I Glen Strathfarrar, 1867 *Miss Fraser of Lovat* (*Trans. BSE* **9**: 474
(1868)); at the upper end of Loch an Eilein, Rothiemurchus, 1882
JK **FRS**; Rothiemurchus Forest, 1886 *J. Groves* **BM**; do. 1892 *AS*
GL,BM,OXF; field near the dam at Kingsmill, Inverness, *Mr
Galloway* our oldest botanist (Aitken); Strathdearn (*ASNH* **2**: 257

(1902)); in several places in the pine woods at Loch Mallachie
and Loch an Eilein (Trail, *ASNH* **19**: 257 (1910)); woods near
Kincraig and on the E side of Loch Morlich, 1954.

MONOTROPACEAE
Monotropa L.
M. hypopitys L.   *Yellow Bird's-nest*
Native. Rare in pine woods. A saprophytic herb without chlorophyll.
Recent study includes the glabrous plants, var. **glabrata** Roth
(*M. hypophegea* Wallr.) (*Fl. Europaea* **3**: 5 (1972)). Pl 15.
M In two places in the parish of Dyke, 1925 *Mrs J. B. (Daisy)
Simpson*, and at Cothall near Forres (Burgess)!; 20 plants under
pines by a track in the Culbin Forest, 1975 *A. H. A. Scott* **E,BM**.
N Cawdor woods, 1841 *J. H. Balfour & T. Fraser* (Watson) and in
1842 *WAS* **E,ABD,GL,BM**; woody den or ravine at Galcantray,
T. Fraser (Aitken).
I In a birch wood 4 miles W of the east end of Loch Ness, 1813
P. H. MacGillivray & WAS (Watson).

EMPETRACEAE
Empetrum L.
E. nigrum L.   *Croupans, Crowberry*
Native. Common on sand dunes, cliff tops by the sea and on dry
moorland up to 800 m. Young stems reddish, leaves 3–4 times as long
as broad with parallel margins. Map 169.
M Very common (Gordon); Dallas, 1860 *JK* **FRS**; sea cliffs at
Hopeman, 1913 *D. Patton* **GL**; fixed dunes in the Culbin Forest,
1967 **E,CGE**; moor at Easter Limekilns on Dava Moor, 1973 **ABD**,
etc.
N (Stables, *Cat.*); common at Ardclach, 1888 *R. Thomson* **NRN**;
Ord Hill (Aitken); throughout the county, 1954; moor at Lochloy,
1964 **ABD**.
I Dalwhinnie, 1833 (Watson, *Loc. Cat.*); summit of Cairn Gorm
(Balfour, 1865); Dunain Hill (Aitken); N of Fort Augustus, 1904
G. T. West **STA**; woods SE of Loch Morlich, 1938 *J. Walton* **GL**;
Drumrossie, 1940 *AJ* **ABD**; moors by Blackfold lochs, near Tomatin
station, in the Ryvoan Pass etc.

E. hermaphroditum Hagerup   *Mountain Crowberry*
Native. Abundant on dry moorland in the mountain districts.
Young stems greenish, leaves 2–3 times as long as broad with
rounded margins. The flowers are hermaphrodite, the stamens
remaining round the black, bitter fruit for some time. The older
mountain records for *E. nigrum* may belong here. Map 170.
M Moors at Boat of Garten, Aviemore and Carrbridge, etc., 1954.
N Common at Drynachan, Glenferness and W of Lochindorb,
1954; moor near Dulsie Bridge, 1961 **ABD**.
I Mam Sodhail in Glen Affric, 1936 *A. J. Wilmott* (*BSBI Year
Book*, **38** (1949)); Coire an t-Sneachda of Cairn Gorm, 1953

A. Melderis **BM**; abundant on all the higher mountains Ben Alder, The Fara at Dalwhinnie, Sgùrr na Lapaich in Glen Strathfarrar and at Gaick, Glen Feshie and Glen Moriston etc., 1954.

PLUMBAGINACEAE
Armeria Willd.

A. maritima (Mill.) Willd. *Cassan Carricht, Thrift, Sea Pink*
Native. Common on sand dunes, rocks and cliffs by the sea and occasional on bare mountain tops. Map 171.
M Frequent (Gordon); Findhorn, 1899 *JK* **FRS**; salt-marsh in Kinloss Bay and dunes in the Culbin Forest, 1925; dunes by the Buckie Loch, Culbin, 1953 *A. Melderis* **BM** and in 1968 *MMcCW* **E**; Kinloss Bay, 1964 *M. Cowan et al.* **E**.
N (Stables, *Cat.*); dunes by the Nairn Dunbar golf course, 1962 **ABD**.
I Inverness, 1832 (Watson, *Loc. Cat.*); Longman Point, Inverness, 1854 *HC* **INV**; between Glenmore burn and the Snow Coire, 1877 (Boyd); Glen Einich, 1891 *AS* **E**; cliff on Sgùrr na Lapaich in Glen Strathfarrar at 915 m, 1972 **E**; summit plateau of Coire Laogh of Cairn Gorm and on Sgùrr na Lapaich in Glen Affric.

PRIMULACEAE
Primula L.

P. scotica Hook. *Scottish Primrose*
Reported by the farmer at Gladhill near Garmouth to Norman Webster of Knockomie, Forres as having been seen in short turf by the sea between Kingston and Lossiemouth about the year 1920 where it had remained for several years.

P. veris L. *Cowslip*
Native. Rare on sea cliffs and grassy banks on base-rich soil and introduced on railway embankments.
M Abundant in one or two places, at Cothall *J. G. Innes* and at Relugas (Gordon)!; Pluscarden 1857 *JK* **FRS**; Covesea, 1844 ? *coll.* **ELN**!; Boat of Garten, *Miss Sangster* (*J. Bot.* **28**: 39 (1890)); probably planted in the policies of Sanquhar House, Forres, 1953; grass bank in a gully at Easter Kellas Farm, 1963 *Mrs P. Campbell*!.
N Nairnshire, *A. Falconer* (Gordon); one specimen apparently wild near Moss Side at Ardclach (Thomson).
I On the railway embankment 1 mile S of Daviot station, probably introduced with soil, 1925 **E,BM**; garden escape in the woods at Flichity House, 1975.

P. elatior (L.) Hill *Oxlip*
Introduced. In a wood at Speymouth very likely from seed carried by birds, *G. Birnie* (Burgess). This record is more likely to refer to *P. veris × vulgaris*.

P. vulgaris Huds. *Primrose*
Native. Abundant on grassy sea cliffs, moorland banks by burns and rivers, grass slopes on mountains and in deciduous woods. Map 172.

M Frequent in the Oakwood and on the banks of the rivers Findhorn, Lossie and Divie (Gordon); Forres, 1898 *JK* **FRS**; abundant on the sea cliffs at Hopeman and Covesea, small wood by Brodie Station, gullies in the Hills of Cromdale and throughout the county, 1954.
N (Stables, *Cat.*); banks of the River Findhorn at Ardclach, 1887 (Thomson); by the Allt Breac at Drynachan, banks of the River Findhorn at Glenferness, on the raised beach of the Culbin Forest at Druim near Nairn, etc., 1954.
I Dalwhinnie, 1833 (Watson, *Loc. Cat.*); valley of the Nairn, ? date ? *coll* **INV**; grass banks at Ben Alder Lodge, on Mealfuarvonie and in Strathdearn, well distributed throughout the district, 1954; abundant on the slopes by Loch Ness and in Strathglass; bank in Glen Moriston, 1972 **E**.

P. florindae Ward
Introduced to the raised beach by Druim House near Nairn in 1910 by David Brodie of Brodie where it has greatly increased, and planted by a small burn in the policies of Farr House in Easterness.

Hottonia L.
H. palustris L. *Water Violet*
Doubtfully native, probably introduced on the feet of water fowl.
M In an artificial pool N of Binsness House in the Culbin Forest, 1923 *D. Patton & E. J. A. Stewart* **GL**!. It remained there until 1958 when the pool dried up.

Lysimachia L.
L. nemorum L. *Wood Loosestrife, Yellow Pimpernel*
Native. Abundant in deciduous woods and wet places by burns and rivers and in moorland flushes. Map 173.
M Frequent, at Edinkillie, 1825, at Dunphail, Knockando and a bog at Pittendreich (Gordon); by the River Findhorn at Cothall, 1871 *JK* **FRS**; St John's Mead Darnaway, 1953 *A. Melderis* **BM**; under alder by the Buckie Loch, Culbin Forest, under hazel at the Slochd N of Carrbridge and throughout the county, 1954.
N Cawdor woods *WAS* (Gordon); occasional by the River Findhorn at Ardclach, 1886 *R. Thomson* **NRN**; flush at Drynachan and throughout the county, 1954; wood at Lethen House, 1961 **ABD,K**.
I Inverness, 1832 and Dalwhinnie, 1833 (Watson, *Loc. Cat.*); Ardersier, *J. Tolmie* (Watson); Reelig Burn, 1880 ? *coll*. **INV**; Dores, 1904 *G. T. West* **STA**; Inshriach, 1963 *E. A. Younger* **E**; cliffs by Loch Killin, flush at Ben Alder Lodge and on Sgùrr na Lapaich at 800 m etc.

L. nummularia L. *Moneywort, Creeping Jenny*
Doubtfully native in north Scotland.
M Recorded from Pittendreich bogs near Elgin by *Dr Keith* (Burgess).

L. vulgaris L. *Yellow Loosestrife*
Doubtfully native. Rare by rivers and lochs. Leaves dotted with orange or black glands; corolla lobes not glandular.

M Sanquhar House pond at Forres but probably planted, 1954.
N Planted by the distillery pond at Brackla, Cawdor, 1975.
I In two separate localities by the River Spey between Aviemore
and Boat of Garten, near Pityoulish, 1882 *Dr McTier* **E,FRS**;
by a small loch in the birch wood by Aldourie Pier, 1904 *G. T.
West* **E,STA**!; a few plants by a small lochan in the village of
Newtonmore, 1975 *J. D. Buchanan* **E**!.

L. punctata L. *Dotted Loosestrife*
Introduced. Leaves with ciliate margins; corolla-lobes glandular-
ciliate.
I Garden escape by the roadside at Tomich, 1964; escape at
Tomatin and on a bank near Lochend; a few plants by a small
lochan in the village of Newtonmore, 1973 *J. D. Buchanan* **E**!.

Trientalis L.

T. europaea L. *Trientale, Chickweed Wintergreen*
Native. Abundant in birch and pine woods and on open moorland.
Pl 15, Map 174.
M In Oakwood, Elginshire, 1837 *G. Gordon* **E**; Dunphail, 1886
ESM **CGE**; by the river Findhorn, 1888 *G. C. Druce* **OXF**; Castle
Grant at Grantown, 1895 *C. Bailey* **OXF**; Dunphail, 1901 *JK* **FRS**;
throughout the county, 1954; open moorland in the Culbin Forest,
1968 **E**.
N Cawdor woods, 1832 *WAS* **ABD** and in herb. *Borrer* **K**; abundant
on the moors at Ardclach, 1882 *R. Thomson* **NRN**; pine wood at
Lochloy, 1960 **ABD**.
I Fields of Culloden, *A. Murray* (Hooker, *Fl. Scot.*); Inverness,
1832 Dalwhinnie, 1833 (Watson, *Loc. Cat.*); Strathaffric (Ball.
1851); Creag Phadrig, 1854 *HC* **INV**; Rothiemurchus, 1892 *AS* **E**;
throughout the district, 1954.

Anagallis L.

A. tenella (L.) L. *Bog Pimpernel*
Native. Rare in bogs and peaty margins of pools.
M Near Brodie House, *G. Don* **BM**; abundant in one or two places
Gordon); below the Keeper's House, Castle of Spynie, 1864
J. Gordon (Watson); ditch on the Leen, Kingston, 1954 *A. E.
Dawson*!; bog by a burn near the old Forres-Dava railway, Altyre,
1970 *R. M. Suddaby*; bog at Darnaway, 1970 *R. M. Suddaby*.
N Lochloy, 1832 *J. B. Brichan & WAS* **E** and in 1857 *JK* **FRS**
it is still there); bog below the raised beach at Maviston, 1968 **E**.
Near Loch Ness, 1954 *B. Ing (Atlas)*.

A. arvensis L. *Scarlet Pimpernel*
Native. Rare in cultivated ground.
M Cornfields at Cothall, Westfield, Drainie and Meft (Gordon);
Brodie, 1867 *JK* **FRS**; in the Manse garden at Fochabers, 1923
coll. **GL**; in the Forestry Commission Nurseries at Newton, 1950;
garden weed at Elgin, 1964; Dyke Nursery, 1969 *A. G. Rait* **E**;
weed at Sanquhar House and on the tip at Forres, 1970 *K. Christie*.

N (Stables, *Cat.*); in the Academy garden at Nairn, 1888 *R. Thomson* **NRN**.
I (Watson, *Top. Bot.*); Alturlie (G. A. Lang, 1905); garden weed at Achnagairn, Kirkhill, 1930 *M. Cameron*; waste ground near the Holm Mill, 1950; on the Longman tip, 1970 **E**; plentiful in a cornfield at Alturlie Point, 1975.

Var. **caerulea** Ludi
A dark-blue flowered variant introduced with foreign grain.
M Low ground in Greshop Wood, Forres, 1957, washed down the River Findhorn during a spate in a sack of foreign grain from a distillery above, **E,K**; Carron Station (as *A. foemina*), a rare casual probably introduced with barley from the distillery *W. M. Watt & J. Burgess* (Burgess).

A. minima (L.) E. H. L. Krause   *Bastard Pimpernel, Chaffweed*
 Centunculus minimus L.
Native. Rare in damp sandy places where water has lain during the winter. Fig 26a.
M Sides of Loch Spynie, *WAS* (Murray); abundant in one or two places, on the flats on the W side of Stotfield, 1828; by a pool at Sunbank, sides of Loch Spynie at Ardivot and near the schoolhouse at Kinloss (Gordon); Spynie, ex herb. *Hanbury* **BM**; Kinloss, *G. Wilson* ex herb. *Linn. Soc.* **BM**; abundant near the Leen at Garmouth, *G. Birnie* and W of Culbin Sands, 1901 (Burgess); formerly common in cart tracks by the Buckie Loch, Culbin but not observed since 1960 when extensive road-making took place.
N By the sea E of Nairn, 1833 *WAS* **E,GL,BM** and herb. *Borrer* **K**; Maviston, 1897 *JK* **FRS**; cart track on the edge of the salt-marsh by Kingsteps, Nairn, 1967 **ABD**.
I Dune slack on the raised beach between Ardersier and Fort George, 1975, **E,BM**.

Glaux L.

G. maritima L.   *Sea Milkwort*
Native. Abundant in salt-marshes and on muddy and dry sand along the whole length of the coast.
M Frequent (Gordon); Findhorn, 1900 *JK* **FRS**; dunes Culbin 1953 *A. Melderis* **BM**; Kinloss salt-marsh, 1964 *D. W. Lindsay* **E**; salt-marsh NW of the Buckie Loch, Culbin Forest, 1968 **E**.
N (Stables, *Cat.*); salt-marshes at Kingsteps, on the Bar, Culbin and on the Carse of Delnies, 1954.
I Sea coast by the Beauly Firth (Aitken); Phopachy, ? date ? *coll* **INV**; coast at Bunchrew, 1940 *AJ* **ABD**; the whole length of the coast, 1954.

Samolus L.

S. valerandi L.   *Brookweed*
Native. Rare, not seen in recent years.
M Morayshire, 1831 *WAS* **ABD**; by the side of a canal dividing the Gordonstoun and Kinnedar estates on the E side of the Lossiemouth–Elgin road (Gordon).

BUDDLEJACEAE
Buddleja L.
B. davidii Franch.   *Butterfly-bush*
Introduced.
M Garden escape by the road near Orton, 1973; single bush, far from habitation, on dunes in the Culbin Forest, 1973 **E,BM**.
I Waste land by the road E of Drumnadrochit, 1975 *K. Christie*!.

OLEACEAE
Fraxinus L.
F. excelsior L.   *Ash*
Native. Frequent on base-rich soil and forming woods in the river valleys. Often planted elsewhere.
M Native by the River Findhorn at Darnaway, in the gorge of Huntly's Cave on Dava Moor and in the oak wood at Kellas etc.
N Ardclach, 1890 *R. Thomson* **NRN**; probably native by the River Findhorn at Glenferness and Dulsie Bridge, 1954; planted at Drynachan.
I To all appearances indigenous on the N bank of Loch Ness, *WAS* (Gordon); Inverness, 1832 Pitmain, 1833 (Watson, *Loc. Cat.*); by Loch Ness at Dores, 1940 *AJ* **ABD**; native by the waterfall at Glendoe and in many places in burn ravines in Strathglass and Glen Mòr etc.; extensively planted in the lowland areas.

Var. **diversifolia** Aiton
   F. monophylla Desv.
Introduced. A tree with simple leaves.
N Planted by the Glengeoullie bridge and in the policies of Cawdor Castle, 1841 *WAS* **E**, and in 1970 *MMcCW* **E**.
I A single tree in the old kirkyard at Kiltarlity, 1975 **E,BM**.

Syringa L.
S. vulgaris L.   *Lilac*
Garden relic usually by old ruined buildings.
M By ruined cottages at Shougle, and Phorp 1954; as a hedge at Dalvey House, 1954; by the bridge over the Muckle Burn near Earlsmill, 1967 **CGE**.
N Hedgerow at Cantraydoune, 1954.
I By the A9 N of Kingussie, bank near Skye of Curr, by the Lynwilg Hotel at Alvie and by a ruin at Bunloit, 1954; in a wood ¼ mile E of Cantray House, 1975.

Ligustrum L.
L. vulgare L.   *Common Privet*
Introduced. Planted as hedges and as relics by ruins.
M Occasionally wild in the older shrubberies, as at Gordon Castle (Gordon); hedges in the parishes of Dyke, Forres, Edinkillie, Urquhart and Bellie (Burgess); at Garmouth, Rothes, Kellas and by the railway at Dunphail etc., 1954.
N Introduced at Ardclach (Thomson)

I   Hedge near Culloden station and by the road to Moniack House near Kirkhill (Aitken); Bunchrew, 1940 *AJ* **ABD**; hedges at Erchless, near Tomatin distillery, by Lovat Bridge at Beauly and roadside verge N of Kingussie, 1954; on the raised beach at Ardersier, 1956.

APOCYNACEAE
Vinca L.

V. minor L.   *Lesser Periwinkle*
Introduced. In woods and waste places usually near habitations.
M   Certainly introduced at Dunphail, 1824 and a variegated form at Innes House, 1829 (Gordon); parishes of Dyke, Elgin, Edinkillie and Bellie (Burgess); woods in the policies of Sanquhar House and Whiterow near Forres and at Wester Elchies, etc., 1954; railway line at Bush of Mulderie, bank by the road at Garbity and on waste ground on Kinchurdy road at Boat of Garten; wood at Dalvey House, 1964 **CGE**; small wood by Rafford Manse, 1967 **E**.
N   Lethen woods, 1882 *R. Thomson* **NRN**; abundant on the banks of the hill road above the Mill at Cantray near Croy, 1954.
I   By the Moniack Burn (Aitken); Dunain Park, Inverness, 1940 *AJ* **ABD**; in the beech wood near the junction of the A96 and the A9 at Raigmore; policies of Dunain House, 1972 **E**; wall at Skye of Curr, wood above Kingussie and on a bank (not near habitation) on the Kirkhill–Beauly road at Dunballoch.

V. major L.   *Greater Periwinkle*
Introduced; always near gardens.
M   Innes House, Urquhart, 1794 *A. Cooper* **ELN**; parishes of Dyke, Forres, Knockando (Burgess); outside cottages at Garmouth, 1953; disused quarry at Lossiemouth, grass verge by Essil Kirk and on the banks of a small burn near Longhill.
I   Millarton (Aitken); on Ness Islands Inverness, 1952.

GENTIANACEAE
Centaurium Hill

C. erythraea Rafn   *Common Centaury*
Native. Rare on dry grassland. The early records were probably confused with *C. littorale* and must be treated with some doubt.
M   By the seaside, Brodie House, *Brodie* herb. *G. Don* **K**; abundant in one or two places at Springfield near Elgin and in a limestone quarry in Main Wood at Garmouth, *J. Gillan* (Gordon); casual on a path at Greshop House near Forres, 1953; short turf by Spynie Loch, 1954.
N   Nairn, 1898 *WASh* **STA**.
I   Field near Aldourie (Aitken).

C. littorale (D. Turner) Gilmour   *Sand Centaury*
Native. Frequent on short turf by the sea.
M   Shores of the Moray Firth as below Brodie House, and between Burghead and Findhorn, 1826 and on the Culbin Sands, 1830

(Gordon); Findhorn Bay, 1898 *ESM* **E** and 1900 *JK* **FRS**; in short turf Culbin, 1953 *A. Melderis* **BM**!.
N Sea coast E of Nairn, *J. B. Brichan & WAS* **ABD**!; Nairn, 1898 *ESM* **E**.
I Sea shore 2–3 miles E of Fort George on the Carse of Ardersier (as *E. pulchella*) 'gathered by the common people and put in whisky to drink as bitters', 1844 *A. Croall* **E,GL**!.

Gentiana L.

G. nivalis L. *Alpine Gentian*
Native. Very rare on mountain rock ledges.
I Grassy slope below rock outcrop in Gaick Forest, 1955 *E. A. Cameron* (the specimen was held by Mrs M. D. Cameron for several years but is now unfortunately lost!).

Gentianella Moench

G. campestris (L.) Börner *Field Gentian*
Native. Common in short acid grassland, on moorland road verges and on sand dunes. Map 175.
Calyx-lobes 4, the two outer larger than the inner and overlapping and enclosing them.
M Not uncommon (Murray); very common (Gordon); Aviemore, 1871 *JK* **FRS**; coast near Garmouth, 1898 *ESM* **CGE**; Boat of Garten, 1839 *J. Walton* **GL**; in long grass by the Buckie Loch in the Culbin Forest, 1953 *B. Welch* **BM**; throughout the county, 1954; in short turf by the old lime kiln at Fae near Dorback, 1972 **E**.
N Cawdor, 1833 *WAS* **ABD**; an extraordinary branched, compact form was found in plenty at intervals between Nairn and Fort George and with unusually dingy, pale lilac-hued flowers (Marshall, 1899); damp ground E of Nairn, 1898 *ESM* **CGE**; a very slender annual plant was found in abundance a short mile E of Nairn, not far from the sea, being taken at the time for a peculiar *G. baltica* and subsequently named *G. uliginosa* (Willd.) Borner, 1898 *ESM* **BM,CGE**; Ardclach moors, 1891 *R. Thomson* **NRN**; short grassland by Loch of the Clans, 1960 **ABD**; moorland tracks at Glenferness and Dulsie Bridge etc.
I Dalwhinnie, 1833 (Watson, *Loc. Cat.*); Kincraig and Bunchrew, 1876 *Agnes Thomson* **OXF**; near Aviemore, 1898 *ESM* **E**; Rothiemurchus, 1893 *AS* **E**; Daviot, 1940 *AJ* **ABD**; Muir of Ord golf course, 1941; common throughout the district, 1954; in short grassland in Glen Affric (with white flowers), 1971 **E**; hillside at Ruisaurie, 1975 **BM**.

G. amarella (L.) Börner *Felwort*
Native. Very rare, not seen in recent years. Calyx-lobes usually 5, equal, not overlapping.
M Pastures at St Andrew's, at Calcots, at Inverugie lime quarry, near the millpond E of the new Manse of Duffus and at Westfield, *WAS* (Murray); E of Duffus Manse, and near Westfield, *G. Wilson* (Gordon) (apparently Stables' localities).

MENYANTHACEAE
Menyanthes L.
M. trifoliata L.   *Water Triffle, Bogbean*
Native. Common in bogs, loch margins and slow-running burns.
Pl 16, Map 176.
M Innes House, Urquhart, 1794 *A. Cooper* **ELN**; Loch Spynie and
Mosstowie etc. (Gordon); Manachy near Forres, 1871 *JK* **FRS**;
throughout the county, 1954; bog at the W end of the Buckie Loch
in the Culbin Forest, 1963 **E**; pool at Cothall, 1963 **BM**.
N (Stables, *Cat.*); Ardclach (Thomson); common on the loch
verges of Loch Flemington, Lochloy, Loch Belivat and Cran Loch,
1954; bogs at Geddes, Littlemill and Glengeoullie near Cawdor etc.
I Dalwhinnie, 1833 (Watson, *Loc. Cat.*); Loch-na-Shanisch near
Inverness etc. (Aitken); Lochs Tarff and Coiltry and lochan Creag
Dhubh etc.; Inchnacardoch Bay of Loch Ness, 1904 *G. T. West*
E,STA; near Inverness, 1940 *AJ* **ABD**; common throughout the
district, 1954.

Nymphoides Hill
N. peltata (S. G. Gmel.) O. Kuntze   *Fringed Water-lily*
Doubtfully native, probably introduced by water-fowl.
M Abundant in one quarry-hole pool at Hamlets by Loch Spynie,
1964 **E,CGE**.

POLEMONIACEAE
Polemonium L.
P. caeruleum L.   *Jacob's Ladder*
Introduced; a garden escape.
M Escape in the parishes of Dyke, Forres and Rothes (Burgess);
river shingle at Fochabers and roadside N of Rothes, 1954; shingle
of the River Findhorn at Waterford near Forres, 1975 *K. Christie*.
N Roadside scrub opposite Lochloy House, 1930; near Glen-
ferness, 1961.
I Garden outcast near Affric Lodge, 1947; waste land by the
Longman tip, Inverness, 1971 **E**; railway embankment at Tomatin,
near a cottage at Daviot station and on a bank between Flichity
House and the burn.

Gilea Ruiz et Pav.
G. capitata Sims.
Introduced with seed in a garden at Brackla, Nairn, 1975 *H. M.
Finlay* **E**, and Elgin tip, 1977 *OMS & MMcCW* **E**.

BORAGINACEAE
Cynoglossum L.
C. officinale L.   *Hound's Tongue*
Doubtfully native in north Scotland. Formerly cultivated by
herbalists.

M Speymouth and Bellie (Burgess).
N Certainly introduced at Auldearn Church yard, *A. Falconer* (Gordon); at the Castle of Inshoch, *J. B. Brichan* (Gordon, *annot.*).
I By the Bunchrew Burn (Aitken).

Omphalodes Mill.

O. verna Moench   *Blue-eyed Mary*
Introduced in the policies of some of the larger houses.
M Innes House, Urquhart, 1794 *A. Cooper* **ELN**; policies of Brodie Castle, 1925.
N Woods at Geddes House, 1964 **E,ABD**.
I Achnagairn, 1880 ? *coll.* **INV** (still there 1976).

Asperugo L.

A. procumbens L.   *Madwort*
Introduced. A rare casual on old wall tops and waste ground. Not seen in recent years.
M Near Burghead, a supposed Danish Fort on the Moray Firth, also from *J. Hoy, G. Don* **BM,K**; old garden dykes at Burghead, *Brodie* (Gordon); Waterford near Forres, 1871 *JK* **FRS**; Grantown, 1909 *T. Wise* **GL**.
N Castle of Inshoch, *J. B. Brichan* (Watson).

Symphytum L.

S. officinale L.   *Comfrey*
Introduced in north Scotland. Cultivated for its medicinal properties and as a food for goats and cattle. Flowers usually purple. Waste ground, banks and railway embankments. Cauline leaves strongly decurrent.
M Cultivated at Sanquhar House, Forres, 1871 *JK* **FRS**; parishes of Dyke, Forres, Rafford, Drainie, Bellie, Speymouth and Rothes (Burgess) (some of these records probably refer to *S.×uplandicum*); Orton (as var. *patens*), 1941 ? *coll.* **STA**; railway yard (with creamy-white flowers) Grantown, 1930 and in 1973, **BM**; waste land at Brodie station, 1963 **CGE**; records for Dunphail, Forres, Kinloss, Knockando, Dallas, Kellas, Elgin and Fochabers must be treated with some doubt; station yard, Aviemore, 1964.
N Cawdor and Nairn, 1900 (Lobban); by the River Nairn and at Househill at Nairn, 1962 **ABD,CGE**.
I Bunchrew (with white flowers), 1930 *M. Cameron*; waste ground at Fort Augustus, Erchless in Strathglass, Urquhart and Inverness, 1954.

S. asperum Lepech.   *Rough Comfrey*
Introduced. Formerly cultivated. Leaves stalked, the upper only shortly so; stems covered with short, stout, hooked bristles; flowers at first pink becoming a clear blue.
I Outside the walled garden at Beaufort Castle, Kiltarlity, 1972 **E.**

S. asperum × officinale = S. × uplandicum Nyman   *Russian*
                                                               Comfrey
Introduced. This is the common plant of waste places, roadsides and farmyards. Very variable. Flowers purple to blue. Map 177.

M Shingle of the River Spey at Kingston, 1953 *NDS* **BM**; by the
Smithy at Dunphail, waste ground at Fochabers, roadside verge
near Brodie Station, 1954; shingle of the River Spey at Rothes,
1973 **E**; waste ground at Brodie Castle, 1963 **CGE**; river bank at
Orton, 1973 **BM**; by the River Spey at Knockando, farmyard at
Kinloss and in the pine wood near Grantown tip etc.

N Cawdor (as *S. officinale* var. *patens*), 1885 *R. Thomson* **NRN**;
waste ground by Geddes House garden, 1961 **ABD,CGE**; roadside
verge near Lochloy House, 1972 **E**; waste ground at Newton of
Park.

I Waste land by the Glen Urquhart road at Drumnadrochit,
1947; Tomatin station yard, near Bunchrew, near the kennels at
Moy Hall and near the garden at Creagdhubh Lodge at Laggan.

S. orientale L.   *White Comfrey*
Introduced.

M Bank by the Forres tip at Waterford, 1968 *K. Christie* **E!**.

S. tuberosum L.   *Tuberous Comfrey*
Native. Common in woods and on river banks in the lowland areas.
Map 178.

M Rare, doubtfully native in Gordon Castle woods and at Speymouth (Gordon); Altyre, 1857 *JK* **FRS**; by the Earlston Burn,
Birnie (Watson); Brodie station, 1898 (Marshall, 1899); roadside
bank in Dyke, 1950; throughout the county 1954; roadside verge
at Brodie station, 1963 **CGE**.

N Banks of the River Nairn at Holme Rose, wood at Glenferness
and in the woods at Cawdor Castle, 1954; banks of the River Nairn
near the town, 1962 **ABD**.

I Ness Islands, Inverness and by the Bunchrew Burn (Aitken);
woods at Kirkhill, 1930; Strathglass, 1940 *K. N. G. Macleay*; waste
ground at Ardersier, woods at Culloden House, by Loch Ness at
Lurgmore, ditch near Alvie Kirk, by the ruins of a cottage at
Truim Bridge and on a roadside verge near Wester Tulloch in
Rothiemurchus.

S. tuberosum × officinale (or **S. × uplandicum**)
Growing with both parents on the grass verge of the Dyke road at
Brodie station. Plant with the habit of the second parent but with
tuberous roots and flower colour of *S. tuberosum*, 1963 **E,CGE**.

Borago L.

B. officinalis L.   *Borage*
Introduced. Garden escape on waste ground.

M Burgie, *A. Duff* and at Drainie (Gordon); parishes of Dyke,
Drainie, Knockando and Bellie (Burgess).

N Nairn inland, 1900 (Lobban); rubbish tip near Nairn, 1970 **E**.

I Inverness, 1871 ? *coll*. **ELN**; Bught mill (Aitken); waste ground
in the car-park at Fort Augustus, 1957.

Pentaglottis Tausch
P. sempervirens (L.) Tausch   *Evergreen Alkanet*
   Anchusa sempervirens L.
Introduced. Waste land usually near buildings.
M Abundant in one or two places, at Pluscarden, in Love Lane at
Elgin and by the Old Castle of Duffus (Gordon); Forres, 1858 *JK*
FRS; Cothall (Aitken) and in 1953 *A. Melderis* **BM**; beneath the
garden wall at Brodie Castle, roadside at Moy House, at Dunphail,
Burgie Castle, Blacksboat, Dallas, Garmouth and in several places
by the River Spey, 1954.
N (Stables, *Cat.*); Lethen, 1888 *R. Thomson* **NRN**; by the river at
Nairn, near Coulmony House, at Glenferness and on a bank below
the Bell Tower at Ardclach, 1954.
I Ness Islands, Inverness, 1854 *HC* **INV**; bank near Kilmorack,
1940 *AJ* **ABD**; roadside E of Drumnadrochit, 1947; about the
town of Beauly and at Beaufort Castle, 1947; Culloden House,
Inverbrough Lodge near Tomatin, bank near Loch Meiklie, waste
ground just N of Cannich, at Invermoriston, railway embankment
at Broomhill etc., 1954; near Kilmorack Kirk, 1972 **E**.

Anchusa L.
A. undulata L. subsp. **hybrida** (Ten.) Coutinho was recorded by
Trail in 1899 (as *A. hybrida* Ten.) from shingle of the River Spey
at Rothes, and **A. officinalis** L. from the policies of Gordon Castle
by *G. Birnie* (Burgess).

A. arvensis (L.) Bieb.   *Bugloss*
   Lycopsis arvensis L.
Native. Abundant in cultivated ground throughout the lowlands.
M Innes House, Urquhart, 1794 *A. Cooper* **ELN**; very common,
doubtfully native (Gordon); Chapelton, 1871 *JK* **FRS**; Lossiemouth, 1907 *G. B. Neilson* **GL**; Cothall, 1953 *A. Melderis* **BM**;
throughout the county, 1954.
N (Stables, *Cat.*); Ardclach (Thomson); Nairn shore, 1916 *D.
Patton* **GL**; arable field Glenferness, 1961 **ABD,K** and throughout
the county.
I Inverness, 1832 (Watson, *Loc. Cat.*); in arable fields in all the
lowland areas, 1954; oat field at Kincraig, root field at Divach
and Foyers, waste ground at Fort Augustus, etc.

Pulmonaria L.
P. officinalis L.   *Lungwort*
Introduced. Garden escape.
M Innes House, Urquhart, 1794 *A. Cooper* **ELN**; escaped from
Gordon Castle gardens, near the lake (Gordon); in a small wood
outside the walled garden at Newton House near Alves, 1948.
N Cawdor, 1900 (Lobban).
I In a wood on the hillside behind the hotel at Invermoriston,
1950.

Myosotis L.

M. scorpioides L.   *Water Forget-me-not*
  M. palustris (L.) Hill

Native. Common in ditches, by burns and loch margins. Fig 26e,
Map 179.

M Common (Gordon); Knockando (var. *strigulosa* Reichb.), 1902
(Moyle Rogers); on shingle on the W bank of the River Findhorn
near Forres, 1953 *A. Melderis* **BM**; throughout the county 1954;
margin of Sanquhar House pond at Forres, 1961 **K**.

N (Stables, *Cat.*) common in marshes at Ardclach, 1891 *R. Thomson*
NRN; throughout the county reaching Drynachan, 1954; by the
River Findhorn at Glenferness House, 1961 **ABD**.

I Inverness, 1832 (Watson, *Loc. Cat.*); by the road to Leys Castle,
c 1880 ? *coll.* **INV**; Aldourie, 1904 *G. T. West* **STA**; by the Bunchrew
Burn, 1940 *AJ* **ABD**; in a small burn at Dalwhinnie, ditch on the
gold course at Newtonmore, by the River Spey at Spey Dam and
throughout the district.

M. secunda A. Murr.   *Creeping Forget-me-not*
  M. repens auct.

Native. Common in wet peaty places. Lower part of the stem with
many spreading hairs. Fig 26g, Map 180.

M Moray, 1889 *G. C. Druce* (*Scot. Nat.* 1889); by the River Spey
near Garmouth (Marshall, 1899); by the burn at Huntly's Cave
on Dava Moor etc., 1954; ditch in a birch wood between Hazelbank
and Pluscarden, 1968 **E**; flush in the Hills of Cromdale, 1971 **E**;
peaty bog at Drakemyres, 1972 **E**.

N Flushes on the moors at Drynachan and Lochindorb, by Geddes
Reservoir and in a bog by the Loch of the Clans, 1954; flush on
the moor at Dunearn, 1961 **ABD**; on a wet bank at Carnoch, 1963
ABD.

I About ½ mile from Aldourie Pier, 1904 *G. T. West* **E**; moorland
flushes at Garva Bridge, Farley, Tomatin, Strathglass, Newtonmore
and Alvie etc.; 1954; ditch by Loch Bran, runnels on the moors at
Dores and Torness etc.

M. laxa Lehm.   *Tufted Forget-me-not*
  Subsp. **caespitosa** (C. F. Schulz) Hyl. ex Nordh.
    M. caespitosa C. F. Schulz

Native. Common on loch margins, by burns and in ditches. Flowers
smaller than in the two preceding species. Fig 26d, Map 181.

M Moray (*Scot. Nat.* 1888); Buckie Loch, Culbin and at Garmouth
(Marshall, 1899); Culbin, 1911 *P. Ewing* **GL**; common throughout
the county, 1954; bog on the margin of Lochinvar near Elgin,
1961 **CGE**; Buckie Loch and bog at Drakemyres, 1972 **E**.

N Ardclach, 1900 (Lobban); throughout the county, 1954; by
the river at Nairn, 1962 **ABD,CGE**; bog below Maviston in the
Culbin Forest, 1968 **E**.

I Bunchrew, 1940 *AJ* **ABD**; in runnels by the River Truim at
Drumochter and Dalwhinnie, and common throughout Badenoch,

1954; bog at Achnagairn, Kirkhill, 1961 **CGE**; margin of Raigmore pond, Inverness etc.

M. sylvatica Hoffm.   *Wood Forget-me-not*
Doubtfully native in north Scotland. Frequent in woods (usually near houses) and on river banks.
M Orton, *G. Birnie* (Burgess); woods near Cromdale, Relugas and Forres, 1954; in the policies of Brodie Castle, 1963 **E,CGE**; wood by the river at Fochabers, 1972 **E**.
N Banks of the River Nairn at Holme Rose, 1954; pine wood verge by the lodge at Glenferness, river bank at Nairn and in a small wood by the farm at Cawdor Castle.
I Banks of the River Beauly near Lovat Bridge, 1947; railway embankment at Drumochter, woods at Bunloit and Culloden, 1956; calcareous moine cliffs on Creag nan Eun near Invermoriston, 1961 *D. A. Ratcliffe*; woods at Farley, Farr House, Fort Augustus, by Loch Flemington and banks of the burn at Kingussie.

M. arvensis (L.) Hill   *Field Forget-me-not*
Native. Abundant in cultivated ground and waste places. Fig 26c, Map 182.
M Frequent (Gordon); Greshop and Forres, 1869 *JK* **FRS**; on shingle by the River Findhorn near Forres, 1953 *A. Melderis* **BM**; throughout the county 1954; garden weed at Brodie Castle, 1963 **CGE**; waste land by Boat o' Brig, 1967 **E**.
N (Stables, *Cat.*); throughout the county, 1954; arable field at Glenferness, 1961 **ABD**.
I Inverness, 1832 and Pitmain, 1833 (Watson, *Loc. Cat.*); Kingsmill road Inverness, 1880 ? *coll.* **INV**; Ben Alder Lodge garden, 1956; waste land at Guisachan, Fort Augustus, Farr, Dunmaglass, Kincraig etc. and in all the lowland areas.

M. discolor Pers.   *Changing Forget-me-not*
    M. versicolor Sm.
Native. Abundant in open dry places, on banks, dunes and on tracks. Fig 26f, Map 183.
M Very common (Gordon); Forres, 1869 *JK* **FRS**; the Leen, Garmouth (forma *dubia* Arrondeau) 1953 *NDS* **BM**; throughout the county, 1954; gravel path at Brodie Castle, 1963 **BM,CGE**; in short turf by the lime kiln at Fae, dunes in the Culbin Forest etc.
N (Stables, *Cat.*); Ardclach (Thomson); track at Drynachan, roadside verge at Cantraydoune, garden weed at Cawdor, Geddes, Lethen and Glenferness, 1954; shingle of the river at Nairn, 1962 **ABD**.
I Dalwhinnie, 1833 (Watson, *Loc. Cat.*); Wester Lovat, 1880 ? *coll.* **INV**; garden weed at Laggan, 1922 ? *coll.* **GL**; throughout the district, 1954 reaching Drumochter, Ben Alder Lodge and Monar Lodge in Glen Strathfarrar etc.; wall top at Castle Stuart, 1970 **E**.

286

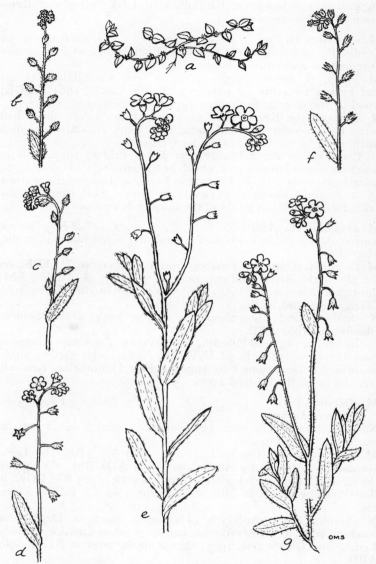

Fig 26 a, **Anagallis minima**; b, **Myosotis ramosissima**; c, **M. arvensis**; d, **M. caespitosa**; e, **M. scorpioides**; f, **M. discolor**; g, **M. secunda**.

M. ramosissima Rochel   *Early Forget-me-not*
   M. collina auct.
Native. Occasional on sand dunes and dry banks. Fig 26b.
M Abundant in one or two places (Gordon); Ladyhill, 1832 *WAS*
GL; Castlehill, Forres, 1856 *JK* **FRS**; dunes N of Buckie Loch,
Culbin, 1964 **E,CGE**; plentiful on sandy banks at Burghead, 1972 **E**;
inland on a dry bank near Longmorn and sparingly on dunes at
Lossiemouth.
N Rare on a sandy track to the sea at Kingsteps, Nairn, 1964
E,ABD.
I Inverness, 1832 (Watson, *Loc. Cat.*); Cromwell's Mount,
Campbelltown, *WAS* (Murray) and in 1841 **GL**; dry wall top at
Castle Stuart, 1954; inland on a dry bank 1 mile W of Kilmorack
in Strathglass, 1975 **E**.

Lithospermum L.
L. officinale L.   *Gromwell*
Native. Rare on base-rich soil in dry open places.
M Frequent at the Bridge of Nethy, 1834 *J. Fraser* **FRS**; by the
River Spey on the rocks above the bridge at Fochabers, 1838 and
at Cothall, *J. G. Innes* (Gordon); Elginshire, 1837 *G. Gordon* **E**;
Gothall, 1900 *JK* **FRS** where it is still to be found, only in small
quantity and not appearing every year.
I Waste uncultivated places near the Monastery at Beauly,
Dr Parsons (Lightfoot); Culloden, *Gilbert McNab* (Gordon); in
gravel at Viewhill, Kirkhill, 1958 *M. Cameron* **E**!.

Buglossoides Moench
B. arvensis (L.) I. M. Johnston   *Corn Gromwell*
   Lithospermum arvense L.
Introduced in north Scotland with foreign grain or bird seed.
M Doubtfully native but plentiful in cornfields, 1824 (Gordon);
cornfields, 1856 *JK* **FRS**; shingle of the River Spey at Kingston,
1953; station yard at Blacksboat, 1957; rubbish tip at Rothes from
distillery waste, 1966 **E,K**; station yard at Knockando, 1967
E,ABD.
I Rubbish heap on Longman road Inverness (Aitken); distillery
yard at Tomatin, 1954 and 1961 **BM,CGE**; allotment near a
distillery at Inverness, 1961 **E,BM** and in 1962 **ABD**.

Mertensia Roth
M. maritima (L.) S. F. Gray   *Oysterplant*
Native. Formerly abundant on shingle by the sea, now rare and
only found in one or two places. Pl 16.
M Covesea (Gordon) (there were about 50 plants there in 1975);
below the cliffs E of Hopeman, 1925 (this colony disappeared during
the great storms of 1953); 5 plants between Hopeman and Burghead,
1953 (disappeared in 1971); a single plant in pure fine sand on
the Bar, Culbin, 1973.
N Nairn, 1895 *JK* **FRS** (abundant there in 1930 but now gone).

I   Seashore near Inverness, 1793 *James Dickson* **BM**; Campbelltown, 1847 *WAS* **BM**; abundant on the shingle spit at Whiteness Head, 1930, now reduced to three or four plants.

Echium L.

E. vulgare L.   *Viper's Bugloss*
Introduced in north Scotland. Occasional on railway embankments and river shingle.

M   Doubtfully native, 1823 (Gordon); ? locality, 1837 *G. Gordon* **E**; Waterford, 1900 *JK* **FRS**; St John's Mead, Darnaway (Aitken); abundant on the railway embankment N of Grantown, 1948; shingle of the River Findhorn N of Waterford, 1930, seen there in 1975 by *K. Christie*.

N   (Stables, *Cat.*); Ardclach (Thomson).

I   Field near Kessock, 1872 ? *coll.* **INV** (this may refer to North Kessock, v/c 106 (G. A. Lang, 1905)); Parks of Inshes near Inverness (Aitken); occasional at Culcabock and Beaufort (J. Don, 1898); canal bank at Clachnaharry, 1930; garden outcast at Inshriach, 1964.

Amsinckia Lehm.

A. calycina (Moris) Chater
Introduced. Locally abundant in cornfields.

M   Introduced with foreign barley at Wester Coltfield about 1965, *J. C. MacIver* and 1970 *MMcCW* **E,BM**.

A. intermedia Fischer & C. A. Meyer
Introduced with foreign grain.

I   Distillery yard at Tomatin, 1961 **E**.

Cerinthe minor L. was recorded from Speymouth, introduced with barley, by *G. Birnie* (Burgess).

CONVOLVULACEAE
Convolvulus L.

C. arvensis L.   *Lesser Bindweed, Field Bindweed*
Introduced in north Scotland. Railway embankments, below walls and on sandy banks.

M   Doubtfully native, rare, field west from Gray's Hospital, Elgin, 1826, Alves, *G. Wilson*, near Sueno's Stone at Forres, *J. G. Innes* (Gordon); railway at Forres, 1856 *JK* **FRS**; railway near Spynie, 1864 (Watson); railway embankments at Boat of Garten, Grantown, near Barley Mill, Brodie, 1954; sandy bank at Roseisle and beneath a wall at Garmouth, 1956; railway embankment at Forres, 1956 **E,CGE**.

N   Nairn, *A. Falconer* (Gordon); Ardclach, 1900 (Lobban); railway embankment at Nairn, 1961 **E,BM**.

I   On the Longman at Inverness, *G. Anderson* (Gordon); near Fort George station (Aitken); Bunchrew, 1940 *AJ* **ABD**; embankment at Muirtown Basin, Inverness, 1954; below a wall at the Ferry House at Lochend, 1975; Gollanfield station, 1976 **E,BM**.

Calystegia R. Br.
C. sepium (L.) R.Br.   *Hedge Bindweed*
Native. In hedges and bushy places, also as a garden weed. Frequent
in the lowlands. Map 184.
M Doubtfully native, rare between Relugas and Dunphail, *WAS*
(Gordon); garden weed at Dyke, Elgin, Alves and Fochabers etc.,
1954; waste ground near Moy House, Kintessack, 1960 **E**; at
Aviemore sewage works, near Carrbridge and at Boat of Garten.
N Near Nairn station, at Holme Rose, Cawdor, Auldearn and by
the tip at Newton of Park, 1954.
I Dochfour, 1940 *AJ* **ABD**; bushy place by the River Beauly
between the town and Lovat Bridge, 1947; by the shop at
Dalwhinnie, hedge at Milton, at Fort Augustus and by the Post
Office at Invermoriston etc., 1954; gateway at Kirkhill, 1963 **CGE**;
in the old garden at Knockie Lodge, by the Dell Hotel at
Inverdruie and at Banchor Mains at Newtonmore, 1975.

C. pulchra Brummitt & Heywood   *Hairy Bindweed*
Introduced. Occasional in similar habitats to the last species.
Flowers bright pink.
M Side of the railway at Forres, 1901 *JK* **FRS**; near Moy House,
1954 **LIV**; railway by Greshop House at Forres, 1960 **LIV**; banks
of the River Lossie at Elgin, 1960; waste ground by Kinloss Kirk,
1960 **LIV**; Grantown tip, 1973 **E**; also at Aviemore, Rothes and
Carrbridge.
N Rubbish tip at Newton of Park, 1975 **E**.
I Inverness, 1959 *V. Gordon* **VG**; roadside at Milton 1964; policies
of Achnagairn House at Kirkhill, 1963 **CGE**; railway E of Gollan-
field, cottage fence at Allanfearn, hedge at Dores and in Strathglass,
Kingussie tip, 1975; farm yard at Kirkton House, Bunchrew, 1976
E.

C. silvatica (Kit.) Griseb.   *Large Bindweed*
Introduced. Rare.
M Waste ground opposite Kinloss Kirk, 1925; railway embankment
at Alves, 1941 ? *coll.* **STA**; banks of the River Lossie by the Cooper
Park at Elgin, 1960 **E**; garden weed in Tolbooth Street, Forres and
waste ground at Lossiemouth, 1973.
I Near a cottage at Struy, Strathglass, 1976.

C. soldanella (L.) R.Br.   *Sea Bindweed*
Reported for Moray by *A. MacGregor* (*Brit. Assoc. Report* **26** (1934)).
Not seen in recent years.

Cuscuta L.
C. epilinum Weihe   *Flax Dodder*
Introduced. A parasite growing on flax. Stems red.
M Two plants in a flax field at Alves, along with *Camelina sativa*
from American seed imported by a vessel at Burghead, 1841
G. Wilson, at Forres, 1843 *J. B. Brichan* and at Inererne, 1844
J. G. Innes (Gordon, *annot.*).

C. epithymum (L.) L.   *Common Dodder*
Native. Stems reddish, very slender.
M Among Lucerne at Dyke, but never maintaining its ground
(Burgess).

C. campestris Yunker
Introduced. Stems orange.
I Forestry garden at Lewiston, growing on *Godetia*, 1975 *Janet
Smith* **E,BM**!.

SOLANACEAE
Nicandra Adans.
N. physalodes (L.) Gaertn.   *Shoo-fly*
Introduced. A tall plant with light blue flowers. Supposed to repel
flies.
M Garden weed at Brodie Castle, 1971 **E**; rubbish tip at Elgin,
1972 **E**.
I Rubbish tip on Longman Point, Inverness, 1971 **E,CGE**.

Lycium L.
L. barbarum L.   *Duke of Argyll's Tea-plant*
   L. halimifolium Mill.
Introduced. Waste ground chiefly near the sea. A spiny shrub; stems
greyish-green; leaves lanceolate; flowers rose-purple, eventually
turning brown, lobes shorter than tube.
M On waste ground at Lossiemouth, Burghead and Hopeman,
1954.
N Near the sea at Nairn, 1954.
I Raised beach and among houses at Ardersier, 1952.

L. chinense Mill.
Introduced. Leaves ovate or rhomboid, bright green; corolla-lobes
equalling or longer than the tube.
M Burghead and Duffus (Burgess); waste land at Lossiemouth,
1975 *A. H. A. Scott* **E**.

Atropa L.
A. bella-donna L.   *Sleepy Nightshade, Deadly Nightshade*
Introduced in north Scotland. A very poisonous plant containing the
alkaloids atropine and hyoscyamine. Flowers a dingy purplish-
green; fruit black. Pl 16.
M Kinloss Abbey, *WAS* (Murray) who states 'on the authority
of Buchanan several writers upon this plant notice that a victory
of Macbeth over the Danes was obtained chiefly by mixing it with
a donation of wine and ale sent by the Scots to Sueno' (or Swain)
King of Norway during a truce. They sent both wine pressed out
of the grape and also strong drink made of barley malt mixed with
the juice of a poisonous herb abundance of which grows in Scotland
called Sleepy Nightshade'; Kinloss Abbey, 1844 ? *coll.* **BM** and
1970 *MMcCW* **E,ABD** (it is still to be found in small quantity there

despite an annual dose of weed-killer); orchard in a Nursery at
Kinloss, 1972 *E. M. Legge.*
I   Near Raigmore Hospital, Inverness, 1956 *V. Norris.*

Hyoscyamus L.
H. niger L.   *Henbane*
Introduced in north Scotland. Rare, usually near old Kirks and
Abbeys, and occasionally as a weed in gardens and tips. All parts
of the plant are poisonous and narcotic. Contains the alkaloids
hyoscyamine and scopolamine. Pl 16.
M Kinloss Abbey 1824, Gallowgreen 1825, churchyards at Alves
and Duffus, lime quarry at Sheriffmill 1826, Spynie and Castlehill
at Forres, *J. G. Innes* (Gordon); Forres Churchyard, 1871 *JK* **FRS**;
Ladyhill at Elgin, 1947 *R. Richter* and in 1972 *A. H. A. Scott*;
garden weed at Garmouth, 1950; waste ground where garden refuse
tipped at Mosset Park, Forres, 1974 **E.**
N Auldearn, *A. Falconer* (Gordon); Cawdor and Nairn, 1900
(Lobban).
I   Cornfield verge by the A9 near Lentran by the Beauly Firth,
1960 *Mrs B. H. S. Russell,*

Lycopersicon Mill.
L. esculentum Mill.   *Tomato*
Introduced. Rubbish tips and river shingle. Recorded by *Trail* from
river shingle at Rothes in 1899 and by *MMcCW* from shingle near
Fochabers and from the Elgin and Forres tips in 1954. At Newton
of Park tip, Nairn, 1975 and the Longman tip at Inverness in 1954.

Solanum L.
S. dulcamara L.   *Mad Dog's Berries, Bittersweet,*
                                                    Woody Nightshade
Native. Frequent in hedges and waste ground in the lowlands.
Map 185.
M Abundant between the Bridge of Spey at Fochabers and the
fish beds at Dipple, 1838, at Allarburn 1823, between Craigellachie
and Rothes and by the Burn of Burgie (Gordon); in many places
about Forres, *J. G. Innes* (Gordon, *annot.*); in an alder wood by the
salt-marsh, Culbin, hedgerow at Dyke and near Craigfield, by the
River Spey at Garmouth and by the farm steading at St Mary's
Well at Orton, 1954.
N (Stables, *Cat.*); Nairn, *A. Falconer* (Gordon); at Holme Rose,
by the burn at Cawdor, verge of the pond at Househill near Nairn
and among willows on the edge of Lochloy, 1954.
I   Kinrara 1830, common about Inverness, *G. Anderson* (Gordon);
Glen Urquhart (Aitken); Rothiemurchus, 1884 *JK* **FRS**; Bogroy
and Lentran (J. Don, 1898); near Dores, 1940 *AJ* **ABD**; in bushy
places at Kirkhill and Beauly, 1947; on the shingle beach at Ardersier and at Kincraig, 1956.

S. nigrum L.   *Black Nightshade*
Introduced in north Scotland. Rare on waste land.

M Casual at Knockando, *C. Watt* (Burgess); garden weed and on waste ground near the station at Forres, 1955; rubbish tip at Rothes; garden weed at Dyke, 1970 **BM**; waste ground Elgin, 1972 *K. Christie* and 1976 *A. H. A. Scott.*
N Cawdor, 1886 *R. Thomson* **NRN.**
I (Druce, *Com. Fl.*); Lentran (G. A, Lang, 1905); garden weed at Inverness, 1954.

S. sarrachoides Sendtner
Introduced.
N Carrot field at Broombank Farm, Auldearn, 1967 **E,BM,CGE.**
I Roadside verge near the sea and near the rubbish tip on Longman Point at Inverness, 1975 **E.**

Datura L.
D. stramonium L.   *Thorn-apple*
Introduced. A narcotic and very poisonous plant containing the alkaloids hyoscyamine, hyoscine and scopolamine.
M Certainly introduced. One or two plants near gardens on the Bilboahall road and near Hay Street at Elgin, on a dyke near Linksfield, 1831 and at Drainie Manse, 1831 (Gordon); garden weed Dyke, 1974.
I Found growing wild at Ness Castle, Inverness, '*Medicus*' (*Northern Chronicle*, **9**: 10 (1889)).

SCROPHULARIACEAE
Verbascum L.
V. thapsus L.   *Great Mullein, Aaron's Rod*
Native. Waste places and banks, frequent in the lowlands, rare in the highlands. Map 186.
M Elgin Cathedral, Kinloss Abbey, and at Castlehill, Forres *J. G. Innes* and on banks N of the old kirk at Dundurcus (Gordon); near Pluscarden, *J. B. Brichan* (Gordon, *annot.*); waste ground at Newton Toll, quarry at Lossiemouth, station yard at Boat of Garten, 1954; railway embankment at Knockando, 1967 **ABD.**
N Cawdor Castle, *WAS* (Gordon); waste land at Holme Rose, 1954.
I Lentran station (J. Don, 1898); near Drumnadrochit, 1940 *AJ* **ABD**; waste ground at Longman Point, Tomatin station yard, roadside verge by Loch Loyne in Glen Moriston, 1954; waste ground by Ruthven Barracks at Kingussie, 1966 *C. Murdoch*; roadside banks near Alvie and Kincraig; on the shingle beach at Ardersier, by Loch Morlich and at Beauly etc., 1973.

V. phlomoides L.   *Orange Mullein*
Introduced.
M Roadside verge 1 mile W of Dallas Lodge, 1953 **BM.**

V. pulverulentum Vill.   *Hoary Mullein*
Introduced.
M Bank by the railway at Orbliston station, *G. Birnie* (Burgess).

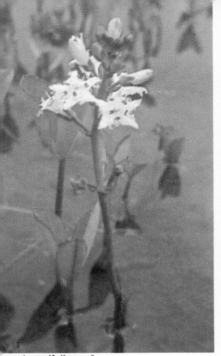

yanthes trifoliata × ⅔
er Triffle, Bogbean

Mertensia maritima × ⅔
Oysterplant

scyamus niger × ⅔
bane

Atropa bella-donna × ⅔
Deadly Nightshade

PLATE 16

Mimulus guttatus × luteus ×⅕

M. guttatus × luteus × 1   J. Whitc

M. luteus × 1 A. Silverside
Blood-drop-emlets

M. guttatus × 1 ×
Monkey-flower

Prunella vulgaris × 1
Selfheal

Petasites hybridus × ½
Butterbur

PLATE 17

Cirsium helenioides × ¼
Melancholy Thistle

phalium norvegicum × ⅔
land Cudweed

lina vulgaris × ⅔
line Thistle

Cirsium vulgare × ⅙
Scots Thistle, Spear Thistle

T. G. Collett

PLATE 18

Cephalanthera longifolia × ⅔
Long-leaved Helleborine

Corallorhiza trifida × 1　　　　　*T. G. C*
Coral-root Orchid

Luzula spicata × ½　　　　*T. G. Collett*
Spiked Woodrush

Carex buxbaumii × 4　　　　*S. J. Heyw*
Club Sedge

PLATE 19

V. nigrum L.   *Dark Mullein*
Introduced.
N Shingle of River Nairn at Holme Rose, 1954.

V. blattaria L.   *Moth Mullein*
Introduced.
M Knockando, a casual, *C. Watt* (Burgess).

V. virgatum Stokes   *Twiggy Mullein*
Introduced.
M Waterford, Forres, 1881 *JK* **FRS**; rubbish tip at Waterford,
1975 **E**.

Misopates Rafin
M. orontium (L.) Raf.   *Weasel's Snout*
   Antirrhinum orontium L.
Introduced in north Scotland.
M Garden weed at Dyke, 1970 **E**.

Antirrhinum L.
A. majus L.   *Snapdragon*
Introduced. A garden escape often on rubbish tips and occasionally
on walls and rock crevices.
M Rubbish tips at Elgin and Forres and in crevices on the quarry
walls near Lossiemouth harbour, 1954.
I Waste ground at Longman Point, Inverness, 1954.

Linaria Mill.
L. purpurea (L.) Mill.   *Purple Toadflax*
Introduced. Waste ground, wall tops and river shingle.
M Garden wall at Milton Brodie, 1930; shingle of the River Spey
at Fochabers and Garmouth, 1953; rubbish tip at Elgin, waste
ground at Findhorn and Carrbridge; old wall at Rothes; rubbish
tip at Forres, 1970 **E**.
N Top of a wall at Auldearn, 1974.
I Shingle beach at Ardersier and waste ground on Longman Point
at Inverness, 1952; wall at Drumnadrochit, at Pityoulish and Moy
Hall; waste ground at Beauly and by Aldourie Pier; in a ditch
at Inchnacardoch and on waste ground near the shop at Glenmore;
Dunain and Culcabock near Inverness, 1970 *M. Barron*.

L. repens (L.) Mill   *Pale Toadflax*
Introduced in north Scotland. Locally abundant on railway
embankments and station yards, rare on banks and waste ground
elsewhere.
M On dry banks on the cross-road between Rafford road and
Elgin road at Forres, 1901 *C. Bailey* herb. *ESM* **CGE** (is still to be
found there); station yard at Aviemore and Boat of Garten, 1930;
by the railway line near Loch Vaa N of Aviemore, 1973.
I Doubtfully native, rare at Aultnaskiach near Inverness, *P.
Nicholson* (Gordon); roadside bank S of Kincraig, 1952; railway
embankment at Tomatin, 1962 **E,CGE**; bank by the cemetery

N of Kingussie and by the railway from Phopachy to Newtonmore;
waste ground at Tomatin Distillery.

L. vulgaris Mill.   *Common Toadflax*
Native. Frequent on dry banks, roadside verges and waste ground.
Map 187.
M Elginshire, 1837 *G. Gordon* **E**; Aviemore and Mundole near
Forres, 1856 *JK* **FRS**; station yards at Aviemore, Boat of Garten
and Grantown, 1954; roadside verges at Carrbridge, Findhorn,
Rothes and Dava Moor; river shingle at Fochabers and Garmouth;
track at Dyke; access road to married quarters at Kinloss, 1973
R. M. Suddaby.
N (Stables, *Cat.*); Piperhill near Cawdor, 1886 *R. Thomson* **NRN**;
roadside banks at Holme Rose and Lethen, 1954; railway yards
at Nairn and Auldearn, 1954.
I East of Loch Meiklie in Glen Urquhart, 1947 *M. S. Campbell &
A. J. Wilmott*; railway yards at Kingussie, Inverness, Tomatin,
Newtonmore, Gollanfield and Culloden, 1954; shingle beach at
Ardersier, newly sown verge in Glen Moriston and bank by a
cottage at Loch Flemington, 1954; by a ruined cottage at Kil-
morack, 1973 *M. Barron*; Beauly, *J. M. Whyte*.

L. maroccana Hook.
Garden outcast on rubbish tips.
M Rubbish tip at Elgin, 1958; tip at Grantown, 1962 **E**; tip at
Rothes, 1967 **E,CGE**.

Chaenorhinum (DC.) Reichb.
C. minus (L.) Lange   *Small Toadflax*
Introduced in north Scotland. Rare on railway tracks and station
yards.
M Railway siding at Mosstowie, 1954 **BM**; station yard at Forres,
1961 **K,CGE** and in 1966 **E**; station yard at Knockando, 1967 **ABD**.
N Auldearn station, 1954.
I Railway track in cinders near Clunes, Kirkhill and in the
station yard at Inverness, 1961; station yards at Tomatin and
Kingussie, 1971.

Kickxia Dum.
K. elatine (L.) Dum.   *Fluellen*
Introduced with foreign grain.
M Shingle of the River Spey at Aberlour and Rothes (Trail, 1904).

Cymbalaria Hill
C. muralis Gaertn., Mey. & Scherb.   *Ivy-leaved Toadflax*
    Linaria cymbalaria (L.) Mill.
Introduced. Frequent on old walls and occasionally on river shingle
and waste ground.
M Doubtfully native, rare on a wall at Gordon Castle (Gordon);
in rubble at Kinloss Abbey, 1925; walls at Brodie Castle, Earlsmill,
Forres, Wester Elchies, Lossiemouth and Pluscarden etc., 1954.

N (Stables, *Cat.*); walls at Croy, Cawdor Castle and Nairn station, 1954; abundant on the walls by the harbour at Nairn, 1961 **ABD,K**.
I Wall at Newton House, Kirkhill (J. Don, 1898); shingle beach at Ardersier, 1955; walls at Hughton, Beaufort Castle garden, Inverness, Bunchrew station, Tomich and Kirkton Farm near Bunchrew; waste land at Kingussie, 1975.

Scrophularia L.

S. nodosa L.   *Common Figwort*
Native. Common in damp woods and by river banks. Map 188.
M Frequent (Gordon); Relugas, 1856 *JK* **FRS**; St John's Mead at Darnaway, 1953 *A. Melderis* **BM**; in sandy ground by a track in the Culbin Forest, 1953; damp woods Brodie, Dunphail, Dallas, Kellas and Fochabers etc., 1954; common by the rivers Spey and Findhorn.
N (Stables, *Cat.*); Ardclach (Thomson); wood at Geddes and banks of the River Findhorn at Glenferness and Coulmony, 1954; by the Allt Dearg at Cawdor, 1961 **E,CGE**; wood by the Muckle Burn at Lethen, 1961 **BM,K**; roadside near the East Gate at Kilravock, 1971.
I Inverness, 1832 (Watson, *Loc. Cat.*); Bught at Inverness, 1854 *HC* **INV**; Horse Shoe by Loch Ness, 1904 *G. T. West* **STA**; Bunchrew, 1940 *AJ* **ABD**; river shingle at Milton, woods in Glen Affric, at Inverfarigaig, Guisachan in Strathglass and Glen Moriston etc., 1954.

S. auriculata L.   *Water Figwort*
   S. aquatica auct.
Recorded by *G. T. West* about the shores of Loch Ness in Urquhart Bay, but there is no voucher specimen to confirm this record.

S. vernalis L.   *Yellow Figwort*
Introduced. Locally abundant in woods and on waste land.
M Speymouth, *G. Birnie* (Burgess); abundant in the policies of Brodie Castle, 1961 **BM,CGE,K**; plantation at Binsness House in the Culbin Forest, 1970.
N Cawdor, 1885 *R. Thomson* **NRN**; Nairn, 1900 (Lobban); roadside bank near Cawdor and on a wall at Little Budgate, 1954 **ABD**.
I Near Guillin, 1855 *HC* **INV**; waste ground on the raised beach at Ardersier, 1971 **E,BM**; policies of Creagdhubh Lodge, near Laggan, 1975.

Mimulus L.

M. guttatus DC.   *Monkey-flower*
Introduced to the British Isles in 1830. Now common by burns and rivers in the lowlands, less so in the highlands. Calyx and pedicels pubescent. Pl 17, Map 189.
M Dunphail, 1860 ? *coll.* **E**; by the Spey at Garmouth (a small flowered cleistogamous state) (Marshall, 1899); river shingle at Darnaway and by Sanquhar House pond at Forres, 1925; very common in shingle backwaters of the River Spey from Aviemore to

the sea, bog at the W end of the Buckie Loch in the Culbin Forest, by the river Lossie at Kellas and a ditch at Kinloss etc., 1954; roadside ditch at Pluscarden, 1968 **E,CGE**.

N By the river at Nairn (Marshall, 1899) and in 1962 *MMcCW* **ABD**; ditch on the Nairn golf course, by the Allt Dearg at Cawdor, shingle of the River Nairn at Holme Rose, pool at Kinsteary near Auldearn and common on the banks of the River Findhorn at Coulmony, Glenferness and Dulsie Bridge, 1954; in a small burn N of Boath House at Auldearn and in the distillery pond at Brackla, 1975.

I Parks of Inshes, ? date *P.F.* **INV**; river shingle Milton, 1947 *C. C. Townsend* **K**; backwater of the River Findhorn at Tomatin, 1952; ditch at Kincraig, by a small burn on the Inverness golf course, frequent in Strathglass and by the River Dulnain at Skye of Curr, 1954; rare by the river Truim between Dalwhinnie and Drumochter where the hybrid *guttatus* × *luteus* is the common plant, 1956; ditches at Farley, Culloden, Ardersier and Abriachan, shores of Loch Ness at Scaniport, in a bog at Torness, a quarry pool near Blackstone in Strathnairn and in the bogs at Muirtown Basin at Inverness.

M. guttatus × luteus

A very variable plant in the flower colour; the lobes with few or many red spots. The following have been determined by R. H. Roberts. Pl 17.

I Shingle of the River Truim at Dalwhinnie, 1954; shingle of the River Spey at Newtonmore, 1956; by the Bogroy Burn, 1974 **E**; ditch at Tomich Farm N of Beauly, 1976 **E**; ditch under the railway bridge just W of Bunchrew, 1976 **E**; shingle of the River Truim at Drumochter.

M. luteus L.   *Blood-drop-emlets*

Recent study of *Mimulus* by R. H. Roberts and A. Silverside has proven *M. luteus* L. to be absent or very rare in the area. Only those records recently determined can be accepted as *M. luteus* with certainty. In *M. luteus sensu stricto* the calyx and pedicels are glabrous. Pl 17.

M On an island in the River Lossie, 1849 *Mr Morrison* and in the Mulben Burn, 1861 (Watson); Waterford near Forres, 1898 *JK* **FRS**; banks of the River Spey at Grantown, 1954; shingle of the River Spey at Craigellachie, Orton and Fochabers, 1973.

N Nairn and Cawdor, 1900 (Lobban).

I Culcabock Dam and in Glen Urquhart, 1885 (Aitken); ditch in a lane near the estuary below Beauly, 1947 *NDS* **K**; shingle of River Truim at Dalwhinnie, 1954 det. R.H.R.; shingle of the River Spey at Newtonmore, 1956 det. R.H.R.; flush in field below Netherwood at Newtonmore, 1969 *S. Haywood*; ditch by Mealfuar-vonie, 1971; rubbish tip at Longman Point, 1971 **E** det. A.S.; flush in Strathdearn not far from Tomatin, 1974; by the burn at Bogroy,

1974 **E** det. A.S.; ditch near Tomich Farm near Beauly and in a ditch under the railway bridge just W of Bunchrew, 1976 **E** det. A.S.

M. cupreus Dombrain× **guttatus**× **luteus**
M. tigrinus hort.; *M. tigrinoides hort.. M. cupreus* and *M. luteus* are completely interfertile and the progeny show a wide range of corolla markings in the form of blotches of various sizes and shapes. The name '*tigrinus*' covers the whole range of the corolla patterns in the hybrids. These hybrids sometimes cross with *M. guttatus* to produce sterile hybrids with larger, showier corollas and of course hardier plants for garden use (fide R. H. Roberts 15.10.1975). Pl 17.
I By a small burn running into Loch Ness at Brackla, 1975 **E** det R.H.R.

M. moschatus Dougl. ex Lindl. *Musk*
Introduced. Formerly cultivated in gardens for its musky smell. Rare on river shingle and by burns.
M In several places around Fochabers and one in Speymouth, spreading fast, *G. Birnie* (Burgess); a few plants on the SW corner of the Buckie Loch in the Culbin Forest, 1952 **E**; in a flush in the Hills of Cromdale, far from habitation, 1954; bog near Boat of Garten, by the pond at Sanquhar House, Forres and at Carrbridge, 1954; shingle of the River Spey at Blacksboat and at Craigellachie, 1973 **E**; shingle of the River Findhorn at Darnaway, 1965 **CGE**; runnel in a lane near Hazelbank, Kellas, 1968 **E**.
N Shingle of the River Nairn at Holme Rose and Nairn, 1954.
I By the Moniack Burn at Kirkhill, 1954; ditches near Lewiston kirkyard, and by Knockie Lodge; abundant by a moorland burn at Erchless in Strathglass, 1975.

Erinus L.
E. alpinus L. *Fairy Foxglove*
Introduced. Naturalised on walls.
I On the old bridge at Whitebridge, 1925; wall at Invermoriston, 1950.

Digitalis L.
D. purpurea L. *Dead Men's Bells, Foxglove*
Native. Abundant on waste ground and moorland tracks throughout the district.
M Knock of Alves (with white flowers) *G. Taylor*, Sluie, *J. G. Innes* (Gordon); Sanquhar, 1856 *JK* **FRS**; Cluny Hill, Forres, *J. G. Innes* **ABD**; W bank of the River Findhorn near Forres, 1953 *A. Melderis* **BM**; on dunes in the Culbin Forest and throughout the county, 1954.
N (Stables, *Cat.*); Ardclach (Thomson); wood at Lethen, 1961 **ABD**, and throughout the county.
I Inverness, 1831, Dalwhinnie, 1833 (Watson, *Loc. Cat.*); canal banks, Inverness, ? date ? *coll.* **INV**; Horse Shoe by Loch Ness, 1904 *G. T. West* **STA**; abundant throughout the district, 1954.

Veronica L.
V. beccabunga L.   *Brooklime*
Native. Very common in the lowlands, less so in the highlands. By
burns, in ditches and loch margins. Map 190.
M Innes House, Urquhart, 1794 *A. Cooper* **ELN**; very common
(Gordon); Manachy, 1871 *JK* **FRS**; St John's Mead, Darnaway,
1953 *A. Melderis* **BM**; ditch in the Culbin Forest and throughout
the county, 1954.
N (Stables, *Cat.*); not frequent at Ardclach (Thomson); ditches
at Holme Rose, Cawdor, Geddes Reservoir, Cran Loch, near
Ordbreck, pool at Kilravock and in a burn by Rait Castle Farm,
1954; boggy margin of the Minister's Loch, Nairn, 1961 **ABD**.
I Inverness, 1832 (Watson, *Loc. Cat.*); Englishton, 1880 ? *coll.*
INV; a small depauperate form creeping on sand, Urquhart Bay,
1904 *G. T. West* **E,STA**; shingle of the River Enrick at Milton,
1947; ditches at Fort Augustus, Tomatin, Strathglass and in a
small burn at Kingussie, etc., 1954

V. anagallis-aquatica L.   *Water Speedwell*
Native. Rare on loch margins and in ditches in the lowlands absent
from the highlands.
M Knockando (Murray); ditches at Duffus, Spynie and Mosswynd
at Elgin (Gordon); Myreside, Elgin, 1874 ? *coll.* **ELN**; parishes of
Alves and on the Leen at Garmouth (Burgess); rare by the Mosset
Burn at Forres and by Dyke pond, 1954; margin of the lake at
Gordonstoun, 1956 **E**; bog at Lochinvar near Elgin, 1960 **ABD**;
locally plentiful in a ditch to the W of Duffus Castle, 1976 **E**.
N (Stables, *Cat.*); Nairn, 1857 *JK* **FRS**!.
I Kingsmill dam at Inverness, 1839 *WAS* **E**.

V. scutellata L.   *Marsh Speedwell*
    Var. **scutellata**
Native. Abundant in acid bogs and moorland drains. Flowers very
pale blue or white. Fig 27e, Map 191.
M In Moray, *Rev. Cowie* (Murray); at Order Pot, Elgin and Roseisle
(Gordon); bog at Manachy near Forres, 1871 *JK* **FRS**; Buckie
Loch Culbin, 1930 *D. Patton* **GL**; and in 1966 *MMcCW* **E**; bog at
millbuies, 1972 **E**; on the old curling pond at Grantown, 1971 **E**;
moorland slope at Muchrach, 1972 **E**; pool on the Lossiemouth golf
course and throughout the county.
N Nairn, *A. Falconer* (Gordon); bogs surrounding Ord Loch, moor
at Lochindorb, by Dulsie Bridge and at Glenferness, 1954; margin
of Loch of the Clans, 1960 **ABD,K**.
I Dalwhinnie, 1833 (Watson, *Loc. Cat.*); by the Asylum Reservoir
at Inverness and in a marsh at Muckovie quarry (Aitken); Kincraig,
1904 *C. B. Neilson* **GL**; throughout the district, 1954; bog on the
moor at Garva Bridge, 1972 **E**; backwater of the River Beauly
below Black Bridge at Kilmorack, 1972 **E**; bog E of Boat of Garten,
1973 **BM**.

Var. **villosa** Schumach
A variant with deep blue flowers and densely hairy stems.
M Pool N of Loch Vaa near Aviemore, 1956 *C. C. Townsend* **K**;
abundant on the stony verge of Loch Vaa, 1973 **E,BM**.
I Inshriach, 1961 *P. H. Davis* **E**; bogs to the W of Insh village,
1975; backwater of the River Calder near Banchor Mains at
Newtonmore, 1975 *J. D. Buchanan*.

V. officinalis L.   *Common Speedwell*
Native. Very common on dry soils, on moors and in open woods
Flowers lilac. Fig 27f.
M Innes House, Urquhart, 1794 *A. Cooper* **ELN**; very common
(Gordon); Altyre woods, 1898 *JK* **FRS**; Culbin Sands (very
glandular) *ESM* **CGE**; throughout the county, 1954; wood at Dyke,
1967 **CGE**; dunes by the Buckie Loch, Culbin Forest, 1968 **E**.
N (Stables, *Cat.*); Ardclach, 1889 *R. Thomson* **NRN**; bank on the
moors at Drynachan and throughout the county, 1954; open wood
at Househill near Nairn, 1962 **ABD**.
I Inverness, 1832 (Watson, *Loc. Cat.*); Drumreach Farm at
Kirkhill, ? date ? *coll.* **INV**; by a burn on Cairn Gorm (glabrous
leaved) (Marshall, 1887); Glen Feshie, 1950 *C. D. Pigott*; throughout
the district, 1954.

V. montana L.   *Wood Speedwell*
Native. Frequent in deciduous woods in the western districts and
in the valleys of the rivers Spey and Findhorn often becoming
established on shingle. Fig 27d, Map 192.
M Hollybank near Gordon Castle, Fochabers, *R. Bremner* (Murray)
and *G. Don* **E,BM,K**; Newmill at Alves, *G. Wilson*, Darnaway
woods, herb. *Brodie*, Dalvey, 1843, Birdsyards woods, 1848, Cothall,
1845 (Gordon, *annot.*); in Greshop Wood, Forres, by the River Spey
at Rothes, Knockando and Fochabers, 1954; woods at Brodie
Castle, 1962 **K,CGE**; on shingle by the River Spey at Orton, 1973 **E**.
N (Stables, *Cat.*); Ardclach, not common (*New Stat. Acc. Scot.*,
1842); woods at Kilravock Castle, 1954.
I Islands at Inverness (Murray) and 1839 *WAS* **E!**; wood at the
mouth of the River Coiltie at Lewiston, 1947; policies of Auchnagairn at Kirkhill, 1954; abundant in the wood by the Falls of
Divach, mouth of the River Tarff at Fort Augustus, by the Breakachy
Burn in Strathglass and in Reelig Glen, 1970; by the Falls of Foyers,
1971 **E**; in a damp shady corner of a field near Struy Bridge in
Strathglass, 1973 *M. Barron*; by the Bunchrew Burn, 1974 **E**;
alder wood by the banks of the River Beauly below Lovat Bridge,
Beauly.

V. chamaedrys L.   *Germander Speedwell*
Native. Abundant on grassy banks, in woods and open places on
moors. Fig 27a.
M Innes House, Urquhart, 1794 *A. Cooper* **ELN**; very common
(Gordon); Forres, 1871 *JK* **FRS**; Spey shingle, Garmouth, 1953
A. Melderis **BM**; Culbin Forest and throughout the county, 1954.

N (Stables, *Cat.*); Ardclach, 1882 *R. Thomson* **NRN**; grassy bank at Drynachan and throughout the county, 1954.
I Inverness, 1832 Dalwhinnie, 1833 (Watson, *Loc. Cat.*); Ness Islands, Inverness, 1854 *HC* **INV**; throughout the district, 1954; cliff ledge at 840 m, Sgùrr na Lapaich in Glen Strathfarrar, 1972.

V. fruticans Jacq.   *Rock Speedwell*
  V. saxatilis Scop.
Native. Rare on mountain rocks.
I In Coire Garbhlach and Creag na Gaibhre in Glen Feshie, 1967 *D. A. Ratcliffe et al.*; cliff on Druim nam Bò and several places in Glen Feshie, 1971 *R. McBeath*.

V. repens Clarion ex DC.
M Introduced on the lawn at Knockomie House near Forres, 1975 **E,BM,JEL**.

V. alpina L.   *Alpine Speedwell*
Native. Frequent on wet mountain ledges and grassy slopes on the higher mountains.
I Mountains of Badenoch (Lightfoot); Cairn Gorm, 1839 *J. Johnstone* **K**; and 1874 *A. Ley* **ABD**; Braeriach, 1877 (Boyd); Coir an t-Sneachda, 1893 *AS* **E,K,OXF**; Cairn Gorm, 1898 *WASh* **K**; between Loch Einich and Coire Dhondail, 1949 *P. S. Green* **E**; grassy flush near the summit of Ben Alder, 1956; Coire Garbhlach, 1957 *E. C. Wallace*; flush below cliffs on Sgùrr na Lapaich in Glen Strathfarrar, grassy slopes in Coire nan Laogh N of Newtonmore, on Creag Liath at Gaick and on Bynack More and in all the corries of the Cairngorms.

V. serpyllifolia L.   *Thyme-leaved Speedwell*
  Subsp. **serpyllifolia**
Native. Abundant on cultivated and waste land, moorland tracks and short grassland. Fig 27c.
M Innes House, Urquhart, 1794 *A. Cooper* **ELN**; very common (Gordon); Forres, 1900 *JK* **FRS**; throughout the county, 1954; path at Brodie Castle, 1963 **BM,CGE**.
N (Stables, *Cat.*); Ardclach and Nairn, 1900 (Lobban); moorland track at Drynachan and throughout the county, 1954; garden weed at Kinsteary House, Auldearn, 1964 **ABD**.
I Dalwhinnie, 1833 (Watson, *Loc. Cat.*); Kincraig, 1891 *AS* **E**; Inverness, 1940 *AJ* **ABD**; gravel track at Gaick Lodge and throughout the district, 1954.

  Subsp. **humifusa** (Dickson) Syme
Native. Frequent in flushes and by mountain burns often washed down and becoming established on river shingle. Differs from subsp. *serpyllifolia* in the broader, almost orbicular and entire leaves, a larger pale blue flower and the inflorescence and capsule with glandular hairs. Fig 27b.
M River shingle at Darnaway, 1964 **CGE**.

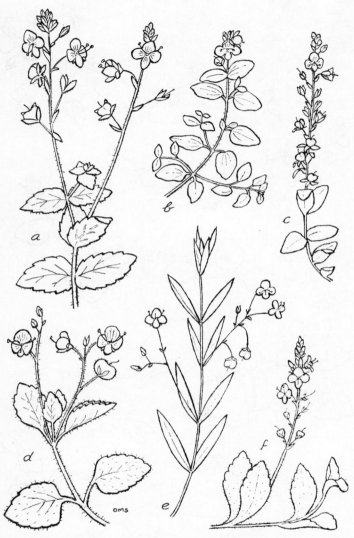

Fig 27   a, **Veronica chamaedrys**; b, **V. serpyllifolia** subsp. **humifusa**;
c, **V. serpyllifolia** subsp. **serpyllifolia**; d, **V. montana**; e, **V. scutellata**;
f, **V. officinalis**.

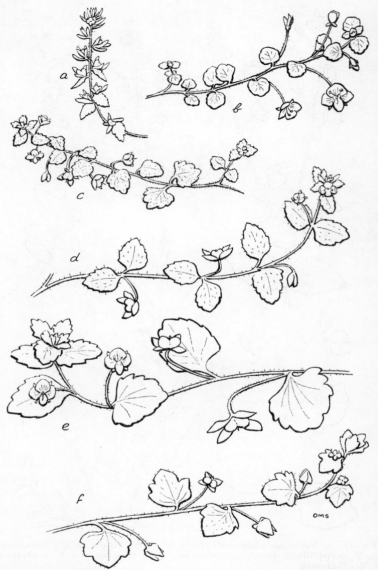

Fig 28   a, **Veronica arvensis**; b, **V. filiformis**; c, **V. polita**; d, **V. agrestis**; e, **V. persica**; f, **V. sublobata**.

I Hills in Glen Feshie, 1831 (Gordon); Braeriach, 1882 *J. Groves* **BM**; Cairn Gorm (Druce, 1888); Coire Dhubh in Glen Shirra, 1916 *ESM* **BM**; Fort Augustus, 1928 *A. N. Macdonald* **K**; Coire Chùirn, Drumochter, 1946 *R. Mackechnie* **GL,K**; Tom a' Chòinich in Glen Affric, 1947 *C. C. Townsend* **K**; Ben Alder, by the river at Loch Killin, by the burn at Gaick, by the Allt Fionndairnich in Strathdearn, common on Sgùrr na Muice in Glen Strathfarrar and on the mountains to the W of Garva Bridge etc.; banks of a small burn by Loch Conagleann at Dunmaglass, 1972 **E**.

V. arvensis L. *Wall Speedwell*
Native. Abundant on cultivated and waste land and on dry rocks and dunes. Fig 28a.
M Very common (Gordon); in the Manse garden at Forres, 1868 *JK* **FRS**; wall on a hill ½ mle S of Forres, 1953 *A. Melderis* **BM**; throughout the county, 1954; railway siding at Manachy distillery near Forres, 1964 **CGE**.
N (Stables, *Cat.*); weed at Ardclach, 1888 *R. Thomson* **NRN**; on a track at Drynachan and throughout the county, 1954.
I Inverness, 1832, Dalwhinnie, 1833 (Watson, *Loc. Cat.*); Kincraig, 1891 *AS* **E**; garden at Ben Alder Lodge and throughout the district, 1954; dry bank in a lane at Farley 1972 **E**.

V. hederifolia L. sensu lato *Ivy-leaved Speedwell*
Two taxa are now recognised in the British Isles. They are all probably referable to **V. sublobata**.

V. hederifolia L.
Native. Common on cultivated ground in the lowlands, rare in the highlands. Middle lobe of the leaf wider than long; flowers pale blue with a white centre; style 0·7–1 mm long.
M Very common (Gordon); near Forres (Druce, 1897); Forres, 1898 *JK* **FRS**; throughout the county, 1954; arable field at Coltfield, 1973 **BM**; dry bank under pines by the Moss of Meft, 1975 **E,BM**.
N (Stables, *Cat.*); in fields and on rubbish at Ardclach, 1889 *R. Thomson* **NRN**; garden weed at Holme Rose, Cawdor Castle, Glenferness House, Geddes and Blackhills Farm near Auldearn, 1954; weed at Kinsteary House, Auldearn, 1964 **ABD**; weed at Lochloy House, 1967 **BM**.
I (Watson, *Top. Bot.*); at Kirkhill, Fort Augustus, Inverness, Culloden House and in the garden at Ben Alder Lodge, 1954; wall top at Castle Stuart, 1970 **E**; garden weed at Tomatin House, newly-sown lawn at Kingussie and in the gardens at Farr and Dochfour and in a cornfield by Loch Flemington etc.; Kincraig, 1975 *J. Clark*.

V. sublobata M. Fischer
V. hederifolia subsp. *lucorum* (Klett & Richter) Hastl
Middle lobe of the leaf longer than wide; flowers pale lilac; style less than 0·5 mm long. Fig 28f.

M Garden weed at Rose Cottage, Dyke, 1966 **E,CGE** det. M. Fischer.

V. persica Poir.   *Buxbaum's Speedwell*
Introduced. A very common annual of cultivated land. Fig 28e, Map 193.
M Pinefield Nursery, 1840 *J. G. Innes* (Gordon, *annot.*); Dr Innes's garden, Forres, 1868 *JK* **FRS**; a garden weed throughout the county, 1954.
N Cawdor and Nairn, 1900 (Lobban); garden at Drynachan Lodge and throughout the county, 1954; weed at Lochloy House, 1967 **BM**.
I Drumreach Farm at Kirkhill, *c* 1880 ? *coll.* **INV**; common in gardens throughout the lowland areas, 1954; garden weed at Ben Alder Lodge, Fort Augustus and Kingussie, in an oat field at Kincraig, 1956; gardens at Newtonmore, in Glen Urquhart and at Pityoulish, Invermoriston and Dunmaglass in the highland areas.

V. polita Fr.   *Grey Field-speedwell*
Native. Rare in cultivated ground. Fig 28c.
M Casual in the parishes of Dyke and Elgin (Burgess); garden weed at Greshop House, Forres, 1952.
N Garden weed at Kinsteary House, Auldearn, 1964 **E,ABD**; flower beds at Cawdor Castle, 1976 **E**.
I Asylum grounds at Inverness (Aitken); garden at Viewhill, Kirkhill, 1954.

V. agrestis L.   *Green Field-speedwell*
Native. Common in cultivated ground. Fig 28d, Map 194.
M Innes House, Urquhart, 1794 *A. Cooper* **ELN**; Elginshire, 1837 *G. Gordon* **E**; the Manse garden at Forres, 1868 *JK* **FRS**; garden weed at Dyke and throughout the county, 1954.
N (Stables, *Cat.*); Ardclach (Thomson); garden weed at Drynachan and Dulsie Bridge and throughout the county, 1954; arable field at Geddes, 1961 **ABD**.
I Inverness, 1832 (Watson, *Loc. Cat.*); Drumreach Farm at Kirkhill *c* 1880 ? *coll.* **INV**; frequent in the lowlands reaching Ben Alder Lodge, Newtonmore, Pityoulish and Flichity House in the highlands. Appears to be absent from the west of the district.

V. filiformis Sm.   *Slender Speedwell*
Introduced and first recorded as an escape in the British Isles in 1927. Now abundant on lawns and in parks and often on river shingle. A very slender creeping plant with large blue flowers that bloom in spring. Fig 28b, Map 195.
M Abundant by the County Library in the Cooper Park at Elgin, 1948; grassy bank at Rothes and river shingle at Fochabers, 1954; lawns at Milton Brodie, Cardnach Wood Cottage at Knockando, lawn at Inverugie House, Burgie House and river shingle at Delfur House, all 1967 **BM**; bank at Dunphail, garden at Newton House and grass verge by Cromdale Kirk; shingle of the River Spey at

Orton and Fochabers and abundant on the lawn at Orton Mains, 1973.
N   Grass field at Kilravock Castle, 1966 **ABD,BM,CGE**; lawn at Lochloy House, 1967 **E,BM**; grassland at Cawdor and in Auldearn kirkyard.
I   Waste grassland at Beauly, 1947; Kirkhill, 1954; Kingussie, 1955 *A. C. Alnutt (Atlas)*!; lawn at Inchnacardoch, 1956; bank by the River Ness at Inverness, 1961 **BM**; lawn at Viewhill, Kirkhill, 1967 **ABD,BM,CGE**; Kilmorack kirkyard, 1972 **E**; also at Moy, Culloden, Tomatin, Knockie and Newtonmore etc.

Pedicularis L.

P. palustris L.   *Red Rattle*
Native. Common in bogs. Map 196.
M   Very common (Gordon); Loch Spynie (Druce, 1897); bog at the W end of the Buckie Loch, Culbin Forest, 1925 and 1968 **E,BM**; Winter Lochs near Binsness, 1953 *A. Melderis* **BM**; flush on the hill at Easter Limekilns on Dava Moor and throughout the county, 1954.
N   (Stables, *Cat.*); Littlemill, 1888 *R. Thomson* **NRN**; near Lochloy, 1901 *JK* **FRS**; in bogs throughout the county, 1954; margin of the Minister's Loch on the Nairn Dunbar golf course, 1962 **ABD**.
I   Dalwhinnie, 1833 (Watson, *Loc. Cat.*); bogs by Loch Uanagan and at Inchnacardoch Bay, 1904 *G. T. West* **E,STA**; throughout the district, 1954.

P. sylvatica L.   *Lousewort*
Native. Abundant on dry moorland. Map 197.
M   Very common (Gordon); Forres, 1902 *JK* **FRS**; in the Culbin Forest and throughout the county, 1954.
N   (Stables, *Cat.*); Ardclach (Thomson); bog on the moor at Drynachan and throughout the county, 1954; among heather on the Nairn Dunbar golf course, 1962 **ABD**.
I   Inverness, 1832, Dalwhinnie, 1833 (Watson, *Loc. Cat.*); Wester Lovat, c 1880 ? *coll.* **INV**; Nairnside, 1940 *AJ* **ABD**; moors in Glen Affric, 1947 and throughout the district.

Rhinanthus L.

In this account the leaves on the main stem which lie between the uppermost branches and the lowest flowers of the terminal spike are known as intercalary leaves. Specimens in CGE have been determined by P. D. Sell.

R. angustifolius C. C. Gmelin subsp.   *Greater Yellow Rattle*
    grandiflorus (Wallr.) D. A. Webb
    R. serotinus (Schonh.) Oborny subsp. *apterus* (Fr.) Hyl.
        R. major auct.
Native, but doubtfully so in north Scotland. Formerly found as a casual in cornfields and pastures. Not seen in recent years. Differs from all variants of *R. minor* by the long (*c.* 2 mm) teeth of the upper lip of the corolla which are twice as long as broad and in the corolla

tube being curved upwards. In *R. minor* the teeth are short and the corolla tube straight.

M Frequent but doubtfully native (Gordon); Moray, *WAS & H. C. Watson* **CGE**; Garmouth (as *R. major* Ehrh. var. *aptera* Fr.), 1898 *ESM* **E,CGE**; casual in pastures in the parishes of Dyke, Kinloss, Urquhart and Speymouth (Burgess).
N Corn and grass fields in Nairnshire, 1834 *WAS* **E,ABD,CL,CGE**; Merrytown Hill, Nairn, 1838 *WAS* **E,GL**; Nairn, 1874 *A. Ley* **CGE**; Cawdor wood, 1885 *R. Thomson* **NRN** conf. P. D. Sell.
I By Leachfield Loch near Fort George, 1844 *W. Gourlie* **GL**; between Cawdor and Culloden, 1855 *HC* **INV**; near Fort George (with wingless fruit), 1898 *ESM* and *WASh* in herb. *M. L. Wedgewood* **MBH**.

R. minor L.   *Yellow Rattle*
Native. Common in grassland and alpine meadows, on dunes and on roadside verges. The old records without herbarium specimens may refer to any of the subspecies. Map 198.

Subsp. **minor**
Occasional. Usually with short flowerless branches; internodes $\pm$ equal, leaves $2-4 \times 5-7$ mm, oblong and parallel sided for most of their length, intercalary leaves o; calyx hairy only on the margin; corolla pale yellow.
M Innes House, Urquhart, 1794 *A. Cooper* **ELN**; very common (Gordon); behind Mundole, 1871 *JK* **FRS**; fixed dunes at Findhorn, 1964 **CGE**; meadow by the bog at Drakemyres, 1972 **E,ABD**; track to the River Spey at Fochabers and at Orton, 1972 **E**.
N (Stables, *Cat.*); Ardclach (Thomson); dry bank by the road between Clephanton crossroads and Loch Flemington, 1962 **CGE**.
I Dalwhinnie, 1833 (Watson, *Loc. Cat.*); by the Holm Burn, Inverness, 1854 *HC* **INV**; pastures by the Spey at Laggan Bridge, 1916 *ESM* **CGE**; meadow by Spey Dam, Laggan, 1972 **E**; roadside verges at Braiton of Leys near Inverness, at Lundie in Glen Moriston and above Inverfarigaig on the Foyers road, 1975; meadow by Loch Ness at Lewiston; on shingle of the River Spey at Newtonmore, 1975 **E**; verge to the track on the N side of Loch Ericht near Dalwhinnie.

Subsp. **stenophyllus** (Schur) O. Schwarz
This is the common *Rhinanthus* of the district. Intermediates with *R. minor* are frequent. Abundant on short grassy banks, on moorland track verges and on dunes. Usually with long arcuate-ascending flowering branches and shorter flowering branches; leaves of the main stem $1 \cdot 5 - 4 \cdot 5$ cm $\times$ $2-5$ mm; intercalary leaves $1-2$ pairs; calyx hairy only on the margin; corolla dingy-yellow. Map 199.
M Near Boat of Garten, 1888 Druce **OXF**; by the Buckie Loch in the Culbin Forest, on the W bank of the River Findhorn near Forres and at St John's Mead at Darnaway, 1953 all *A. Melderis* **BM**; dunes at Lossiemouth, 1963 **CGE**; grass verge of moor at Carrbridge, 1972 **E**; by the River Spey at Orton; on a dry gravel bank at Garmouth, 1973 **E** and throughout the county.

N Grass verge on the shores of Loch Flemington, 1962 **ABD,CGE**; raised beach at Maviston near Nairn, 1968 **E**; moorland track at Achavraat, 1973 **E**.
I Near Fort George, 1898 *ESM* **CGE**; by the River Calder at Newtonmore, 1911 *ESM & WASh* **E,GL,STA,CGE**; Dalwhinnie, 1911 *ESM* **CGE**; throughout the district 1954; grass verge to the road at Abriachan, 1972 **E**; banks of the River Beauly at Groam of Annat, 1972 **E**; track in Affric Forest, 1971 **ABD**; meadow by Affric Lodge, 1971 **BM**.

Subsp **monticola** (Sterneck) O. Schwarz
R. spadiceus Wilmott
Frequent in the highlands on short grassy places on moors. Usually with many short or moderate flowerless branches from near the base, leaves of main stem 1–2 cm × 2–4 mm linear-lanceolate, more or less tapering from near the base; intercalary leaves 1–2 pairs; calyx hairy only on the margin; corolla dull yellow becoming treacle-brown throughout. Map 200.
M Culbin Sands, 1953 *UKD* **UKD**; grassy moor W of Carrbridge, 1972 **E**.
I Between the Hotel and the Post Office at Dalwhinnie, 1911 *ESM* **E,CGE**; Glen Feshie, 1933 *A. E. Ellis* **BM**; among short heather in Glen Mazeran, 1968 **E**; on the lower slopes of Sgùrr na Lapaich in Glen Affric, 1971 **ABD**; hillside in Glen Markie, 1972 **CGE**; roadside verges between Newtonmore and Creag Dhubh and by Loch Killin, 1972 **E,BM**; moor at Garva Bridge, 1972 **ABD**; plentiful in Strathglass and Glens Cannich and Strathfarrar.

Subsp. **lintonii** (Wilmott) Sell
R. vachelliae Wilmott
Rare on grass slopes below mountain cliffs. Leaves tapering from the base; calyx hairy all over; corolla yellow or orange-yellow.
I Near the NE end of Loch Mullardoch, 1947 *E. Vachell* **BM**! (*Watsonia* **6**: 300 (1967)); grass slope below the cliffs on Sgùrr na Lapaich in Glen Strathfarrar, at 820 m, 1972 **E**; flushed grassland below Beinn an t-Socaich, Glen Affric at 330 m, 1974 *R. W. M. Corner et al.*

Subsp. **borealis** (Sterneck) Sell
A mountain plant of rock ledges and occasionally washed down and becoming established on river banks. Easily distinguished by its parallel-sided leaves; calyx hairy all over and pale yellow corolla.
I Cliffs at 810 m above Loch Dubh, 5 miles N of Cluny Castle, Laggan, 1911 *ESM* **E,CGE**; Geal Chàrn in Glen Markie, 1916 *ESM* **E,CGE**; southern spur of Mam Sodhail, Glen Affric, 1947 *NDS* **BM** and on many mountains in that area; Leacainn, Cairngorms, 1958 *C. West* **E,CGE**; cliffs on Sgùrr na Lapaich in Glen Strathfarrar, at 800 m 1971 **E,BM**; ledge in the gully of the Allt Fionndairnich in Strathdearn, at 500 m 1973; established under alders on the banks of the River Spey above Newtonmore, 1973 **E**; cliffs on A' Chaoruinn Mòr near Garva Bridge, in Coire Odhar

of Glen Einich and by the Allt Lorgaidh in Glen Feshie, 1973;
rocks above Allt Beithe Min and in the gorge of the Allt na Faing
in Glen Affric, 1974 *R. W. M. Corner et al.*

Melampyrum L.

M. pratense L. subsp. **pratense**   *Common Cow-wheat*
    Var. **pratense**
Native. Common in birch woods, on river banks in shade and on
open moorland. Very variable in leaf shape, toothing on the bracts
and flower colour. Plants on very acid moors have flowers often
blotched with red or purple. Map 201.
M Very common (Gordon); moors at Lochindorb, Conicavel,
Dorback, Dava Moor and the Hills of Cromdale, 1954; birch wood
E of Duthil, 1972; under pines on the roadside W of Carrbridge,
1972 **E**.
N (Stables, *Cat.*); near the Hermitage at Cawdor, 1844 *A. Croall*
(Watson); Ardclach (Thomson); woods at Glenferness House,
1973.
I Dalwhinnie, 1833 (Watson, *Loc. Cat.*); Kinrara, 1843 *WAS*
(Watson); woods at Ness Castle, Inverness (Aitken); Glen Feshie,
1950 *C. D. Pigott*; Coire an Lochain of Cairn Gorm, 1953 *A. Melderis*
BM; throughout the district, 1954; banks of the River Moriston in
Glen Moriston, 1971 **E**.

Var. **hians** Druce
In woods chiefly in the highlands. Flowers deep yellow. Map 202.
M Dunphail to Forres (Druce, 1888); parks of the River Findhorn
at Forres, 1889 *W. F. Miller* **E**; Aviemore, 1894 *AS* **E**; woods above
the Divie Burn at Dunphail, 1897 *ESM* **CGE**; Clunyhill at Forres,
1900 *JK* **FRS**; Bridge of Brown, 1905 *ESM* **E**; pine woods at
Grantown, 1936 *JEL* **E**; in all the highland areas, 1954.
N Cawdor woods, 1832 *WAS* **ABD** and 1839 herb. *Goodsir* **E**;
woods at Holme Rose, Coulmony, Geddes and Glenferness, 1954;
abundant in a small wood (now felled) at Cothill Farm near Cran
Loch, 1960 **ABD**; under pines N of Cran Loch in the Culbin
Forest, 1961 **BM,K,CGE**; birch wood by Loch Belivat, 1973.
I Kilmorack (Druce, *J. Bot.* **28**: 39 (1890)); Larig Pass,
Cairngorms, at 680 m, *AS* **E,GL**; birch wood at Farley, pine woods
at Creag Dhubh near Laggan, at Loch Garten and at Breakachy
in Strathglass, birch woods by the Falls of Divach and by Loch Ness
at Aldourie, river bank at Foyers, 1972.

Var. **purpureum** C. J. Hartm.
The colour form with dark red corolla and flame-coloured lip.
M Bridge of Brown (as var. *montanum* Johnst.) 1905 (Marshall,
J. Bot. **44**: 159 (1906)).
I Strathaffric (Ball, 1851); Glen Einich and Glen Feshie, at 670 m,
1909 *J. A. Wheldon & A. Wilson* (*J. Bot.* **53**: 177 (1915)); Larig;
Pass (as var. montanum) 1892 *AS* **E**; moor in Glen Banchor and at
Drumochter, 1954.

M. sylvaticum L. *Wood Cow-wheat*
Native. Rare in mountain ravines. Flowers smaller than in *M. pratense*, pale dull yellow with lower lip deflexed and mouth wide open, tube usually shorter than the calyx.
M Woods at Muchrach, Strathspey, 1834 *J. G. Innes* **ABD** and in 1835 **E**.
N Cawdor woods (Aitken).
I Lynleish in Strathspey, *J. G. Innes* (Gordon); Kinrara, 1843 *WAS* (Gordon, *annot.*) Glen Doe, 1904 *G. T. West* **STA**; rocky ravine of Gleann na Ciche in Glen Affric at 500 m, 1960 *H. Milne-Redhead*; in Coire Ghàidheil, in a gorge on Sgùrr nan Ceathreamhnan, on Tigh Mòr na Seilge at 600 m (hundreds of plants), in Coire an Athair and lower Coire 'n t-Siosalaich, all in the Forest of Kilmorack at the head of Glen Affric, 1974 *R. W. M. Corner et al.*

Euphrasia L.

All cited specimens recorded by *MMcCW* have been determined by P. F. Yeo. Species are difficult to determine as numerous hybrids and hybrid swarms occur. Unbranched individuals can be found in any colony even in those of species normally branched.

E. micrantha Reichb. *Heath Eyebright*
  E. gracile (Fr.) Drej.
Native. Abundant on dry moors particularly among *Calluna*, and occasionally in bogs. Plants slender, usually branched, leaves frequently purplish; flowers very small, usually purple; corolla 4–6 mm, the lower lip longer than the upper with three narrow emarginate segments, the middle one the longest. Fig 29b.
M Abundant on heaths on Dava Moor, 1898 *ESM* **GL,CGE**; Culbin, 1913 *D. Patton* **GL** and 1961 *MMcCW* **K**; moor by Slochd N of Carrbridge (with white flowers) 1973 **E**; in a bog at Clachbain near Duthil, 1973 **E**; dry moorland bank by Loch Vaa, 1973 **ABD,BM**; moor by Loch Mòr near Duthil, 1974 **BM**.
N Coast near Fort George, Nairn, 1898 *ESM* **CGE**; moorland by Lochloy, 1952 **CGE**; by Loch of the Clans, 1960 **ABD,K**; moor at Clunas Reservoir, 1973 **E**; Nairn Dunbar golf course, 1974 **BM**.
I Grave at Culloden (as *E. officinalis*) 1841 *WAS* **GL**; Foyers, 1881 *F. Townsend* **CGE**; Glen More Forest, 1898 *ESM* **E**; Kincraig and Kingussie, 1903 *J. W. H. Trail* **ABD**; moors at Glen Mazeran, 1968 **CGE**; shingle bank by the River Spey at Newtonmore, 1973 **E**; in short turf below the cliffs of Creag Dhubh near Laggan, 1973 **ABD**; moor by Loch Killin, 1972 **BM**; near Glen Affric Lodge, 1971 **K**, and throughout the district.

E. scottica Wettst.
Native. Frequent in bogs. Differs from *E. micrantha* in having flexuous stems, light green leaves that are usually purple beneath, white flowers and the lower lip of the corolla with three subequal lobes, scarcely exceeding the upper lip. Fig 29a, Map 203.

M Has been observed by *F. Townsend* on damp poor soils where vegetation is scanty (Burgess); bog by the River Dulnain at Muchrach, 1972 **BM**; dry moor N of Creag na h-Iolaire above Aviemore, and on a dry bank by Loch Vaa, 1973 **E**; flush in the gorge of Huntly's Cave on Dava Moor, 1974 **ABD**.

N Flush on the raised beach by Maviston Farm, Culbin Forest, 1968 **CGE**; wet grassy bank at Clunas Reservoir, 1973 **E**.

I Moor near Clava (Trail, *ASNH* **8**: 97 (1899)); Kingussie, 1899 *ESM* **CGE**; Kingussie, by Loch Einich and on Cairn Gorm, 1902 *J. W. H. Trail* **ABD**; Strathmarkie near Laggan, 1916 *ESM* **CGE**; on An Tudair Beag in Glen Affric (as *E. frigida*), 1947 *E. F. Warburg* **OXF**; bog on the moors at Corrimony, Glen Urquhart, 1964 **CGE**; on Sgùrr Na Lapaich, at 915 m 1972 **E**; wet bank by the road bridge at Glendoe, 1972 **ABD**; bog near Garva Bridge, 1972 **BM**; flushes in Glen Banchor, 1973 **E**; bog at Gaick Lodge, 1973 **BM**.

E. frigida Pugsl.

Native. Frequent on wet rock ledges and grassy slopes of mountains over 600 m. Leaves rather broad and thick, green; cauline internodes very long, the upper usually the longest, the upper floral internodes very short. Corolla moderate, 5–8 mm, white with a lilac upper lip; lower lip longer than the upper with emarginate lobes, the medium longer and narrower than lateral.

I Coire an t-Sneachda of Cairn Gorm (as var. *laxa*) *ESM* **CGE**; Coire an Lochain of Cairn Gorm (as var. *laxa*) *R. M. Adam* (*Trans. BSE* **32**: 248 (1935)); Coire an t-Sneachda, 1955 **CGE**; Sgòran Dubh, 1956 *P. D. Sell* **CGE**; Loch Einich, 1956 *UKD* **E**; Coire nan Laogh of Glen Banchor (as var. *laxa*), 1956 *A. C. Crundwell & E. F. Warburg* **GL**; Gleann nam Fiadh, Glen Affric, *E. F. Warburg* **OXF**; slopes below cliffs of Sgùrr na Lapaich in Glen Strathfarrar, 1971 **E,ABD**; burn gully at Gaick, 1973 **E**; Coire Leachavie, Mullach Fraoch-choire, Sgùrr an Fhurail in Glen Affric, 1974 *R. W. M. Corner et al.*; cliff ledge on Sgùrr na Muice at Monar and grass slope on Meall A' Chaoruinn Mòr near Garva Bridge, 1975.

E. frigida × scottica

I Coire na Geurdain of Mullach Fraoch-choire in Glen Affric, 1974 *R. W. M. Corner et al.* ('probably this' P. F. Yeo).

E. foulaensis Townsend ex Wettst.

Native. Frequent in grassy places on the margins of salt-marshes by the sea. A simple, sometimes branched, very compact plant with rather broad, thick, glabrous leaves. Corolla small, 4–6 mm, white or sometimes pale violet, the lower lip rather longer than the upper with emarginate almost equal lobes. Capsule greatly exceeding the calyx-teeth. Fig 29d.

M Salt-marsh between Kincorth and Binsness on the Findhorn estuary, 1953 **CGE**, also collected there by *A. Melderis* **BM** and *UKD* **UKD**; common on the grass verge of the salt-marshes in the Culbin Forest, 1968 **CGE**.

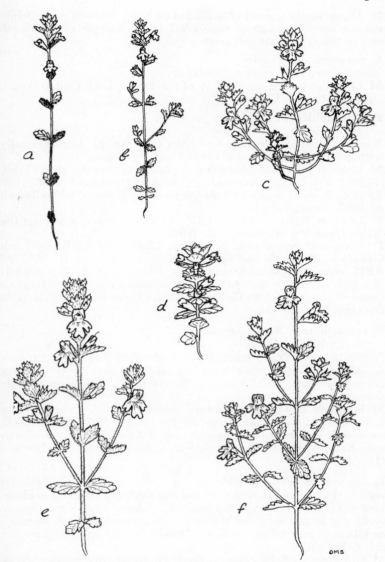

Fig 29   a, **Euphrasia scottica**; b, **E. micrantha**; c, **E. confusa**;
d, **E. foulaensis**; e, **E. borealis**; f, **E. nemorosa**.

N Damp sandy ground near the sea at Nairn (flowers small, white capsule exceeding calyx) 1898 *ESM* **CGE**; margin of salt-marsh east of Nairn Dunbar golf course, 1969 **CGE**.

E. rotundifolia Pugsley
Native. Rare on maritime grassland.
M On a flat sand dune slack W of the Buckie Loch Culbin Forest, 1954 A. Melderis **BM**.

E. ostenfeldii (Pugsley) Yeo
    E. curta auct.
Native. On short moorland turf often near the sea. Leaves usually with dense eglandular hairs. Corolla small, 4–6 mm.
M Coast of Burghead (*E. curta* var. *glabrescens* Wettst.) 1897 (Trail, *ASNH* **8**; 179 (1899)); Aviemore (*E. curta* var. *glabrescens* Wettst.) 1898 *ESM* **E**; in short grassy turf on the moor by Beum a' Chlaidheimh on Dava Moor, 1973 **E**.
N Coast at Nairn *ESM & WASh* (Lobban); grassy dunes on the Nairn Dunbar golf course, 1961 **ABD**.
I Glenmore Forest near Aviemore, 1898 *ESM* **CGE**; by Loch Gynack at Kingussie and in the Cairngorms, 1902 *J. W. H. Trail* **ABD**; river shingle bank near Loch Affric, 1947 *M. S. Campbell*; in a grassy gully on the north face of Mealfuarvonie, 1971 **E**; grass slopes of Creag Liath at Gaick and in the Coire nan Laogh in Glen Banchor, 1973 **E**.

E. tetraquetra (Breb.) Arrondeau
    E. occidentalis Wettst.
Native. Rare on grassy margins of salt-marshes. Resembles a robust *E. foulaensis*, but is more branched, and the floral leaves form a dense four-angled spike. Corolla white or lilac 5–7 mm, the capsule shorter than the calyx.
M Findhorn Bay (as *E. foulaensis*) 1898 *ESM* **CGE**; shingle by the River Spey at Garmouth, 1953; by the estuary at Findhorn (as *E. occidentalis* var. *calvescens* Pugsl.), 1953 *UKD* **E,UKD**; salt-marsh near Binsness, Culbin Forest, 1956 *A. C. Crundwell & E. F. Warburg* **GL**.

E. nemorosa (Pers.) Wallr.   *Eyebright*
Native. Occasional in river pastures and short grassland on roadside verges. Plant branched, leaves typically glabrous (but often strongly glandular in this area and difficult to distinguish from *E. borealis*), corolla 5–8 mm white to lilac, capsule usually shorter than the calyx. Fig 29f.
M Grass verge of the drive to Blairs House at Altyre, 1954; bank near Rothes distillery (finely pilose), 1956 **CGE**; embankment by the River Findhorn at Waterford near Forres, 1974 **E**.
N Nairn Dunbar golf course, 1968 **CGE**; on moorland by the Allt Breac at Drynachan (eglandular, with abnormally narrow leaves and fine teeth), 1973 **E**.
I Roadside verge near a large pool, Glen Moriston (glandular),

1971 **E**; grass verge of moor by Loch Tarff near Fort Augustus (glandular), 1972 **CGE**; banks of the River Beauly near Groam of Annat (strongly glandular) 1972 **BM**; banks of the River Spey W of Newtonmore (densely covered with very short glandular hairs) 1973 **E**; meadow by the River Spey at Newtonmore (narrow leaves, eglandular) 1973 **E**; Suidhe Hill at Kincraig, 1975.

E. confusa Pugsley
Native. Common in short grassland on moors. Stem much branched near the base giving the plant a tufted appearance, leaves often bronze in colour usually with sparse eglandular and short glandular hairs, corolla moderate in size, white to purple, lower lip longer than the upper and with emarginate lobes, the middle one the longest. Fig 29c, Map 204.

M By the Muckle Burn near Moy House, Forres, 1957 **CGE**; grass field near Hamlets S of Lossiemouth (remarkable form with long glandular hairs), 1964 **CGE**; dunes at Findhorn, 1964 **CGE**; in short turf by the lime kiln at Fae near Dorback, 1972; grass slope at Muchrach, 1972 **K**; in a bog at Dalrachney Beg near Carrbridge, 1972 **BM**; grass bank in a burn gully at Auchness near Dallas (glandular), 1973 **E**; by a small burn at Slochd (some glands), 1973 **E**; pasture at Clachbain near Duthil (with some glands), 1973 **E**; grassy slope on the moor by Easter Limekilns on Dava Moor (eglandular), 1973 **E,ABD**; moor at Kellas (eglandular), 1973 **E**; turf by the sea at Spey Bay (densely glandular), 1974 **E**; also at Roseisle, on dunes at Findhorn and at Craigellachie.
N Meadow by the farm buildings at Drynachan, 1970 **CGE**; stony ground by the Allt Breac W of Drynachan, 1973 **E**; by the Leonach Burn at Dulsie Bridge (with some glands), 1974 **ABD**; steep bank in the gorge of the River Findhorn above Dulsie Bridge, 1974 **BM**.
I Urquhart Castle, 1936 *E. O. Callen* **E**; near Lochan Dubh, Cannich, 1947 *E. F. Warburg* **OXF**; in short turf on the moors at Corrimony, Glen Urquhart, 1964 **CGE**; grass verge in the pine woods in the Affric Forest in Glen Affric, 1971 **E**; moor at Glendoe, 1972; short grass at Coignafearn, 1970 **E**; grass slope below the cliffs of Tom Bailgeann (a broad-leaved form), 1973 **E**; moors in Glen Mazeran, 1973 **CGE**; mountain slope in Glen Markie, 1972 **BM**; moor at Abriachan, 1972 **ABD**; grass slope below the cliffs at Loch Killin, 1972 **BM,K**; common in short grass in Glen Feshie, in Strathfarrar, at Moy Hall, Dochfour House and Alvie moor etc.

E. confusa × micrantha
I Slopes of Sgùrr na Lapaich, 1971 **E**.

E. confusa × scottica
M In short turf by a lime kiln on Dava Moor, 1973 **E**.
I Moorland bog at Spirean Beag near Laggan, 1973 **E**; bog in upper Glen Feshie, 1974 **E**.

E. arctica Lange ex Rostrup
   Subsp. **borealis** (Townsend) Yeo
     E. brevipila auct.
Native. Abundant on grassy banks, moorland verges and in meadows. Very variable. The former records for *E. brevipila* Burnat & Gremli belong here. A tall plant often 20 cm high and with 0–5 pairs of branches; leaves with short, and occasionally long glandular hairs; corolla large, 6–9 mm, white or lilac the lower lip longer than the upper with broad emarginate lobes, the middle one the longest; capsule equalling the calyx. Fig 29e, Map 205.
M Morayshire, 1838 *J. Cruickshank* **E**; Dunphail, 1897 *ESM* **E**; grassy plateau 4 miles E of Nairn and at Dunphail and Garmouth, 1898 all *ESM* **CGE**; grass verge by Blairs House, Forres, 1952 **CGE**; grass bank by the River Spey near the old Bridge of Spey at Grantown, 1971 **ABD**; boggy field at Drakemyres, 1972 **E**; short grassland by the lime kiln at Fae near Dorback, 1972 **BM,K**; short turf W of Carrbridge, 1972 **BM**, and throughout the county.
N Ardclach, 1884 *R. Thomson* **NRN**; Nairn, 1898 *ESM* **GL**; pasture by Lochloy, 1952 **CGE**; Cran Loch, 1962 **ABD**; by the moorland track at Loch of the Clans, 1960 **ABD,K**; in short grass by Clunas Reservoir, 1973 **E**.
I Kingussie, 1898 *ESM* **E**; near the Manse of Insh, 1902 *J. W. H. Trail* **ABD**; Loch an Eilein 1932 leg. *K. D. Little* **E**; Culloden, 1934 *E. O. Callen* **E**; pasture by the Moniack Burn at Kirkhill, 1961 **K**; grass track by Glen Shirra Lodge at Garva Bridge, 1972 **BM**; by the lime quarry at Suidhe, Kincraig, in a meadow by Loch Meiklie, grassland in Glen Feshie and at Gaick Lodge, 1973 **E**, etc.

   Var. **notata** (Towns.)
M Near Aviemore, 1908 *J. A. Wheldon & A. Wilson*, herb. *ESM* **CGE**; by the curling pond at Grantown, 1950; verge to the road by Smallburn distillery N of Rothes, 1954 **CGE**; by the A59 at Duiar near Advie, 1965 *P. D. Sell* **CGE**; by the A96 at Leitch's Wood at Fochabers, 1956 *P. D. Sell* **CGE**.
I Between Kingussie and Drumguish, 1956 *P. D. Sell* **CGE**.

E. arctica × micrantha
M Fixed dunes in the Culbin Forest, 1968 **CGE**.
N Dunes on the Nairn Dunbar golf course, 1961 **ABD** and 1968 **CGE**.
I Near the Youth Hostel at Buntail near Drumnadrochit, 1936 *E. O. Callen* **E** det. Pugsley.

E. arctica × rotundifolia
N Nairn (as *E. latifolia* Pursh) 1898 *ESM* **E**, det. E. O. Callen.

E. arctica × scottica
M Bog on the moor W of Carrbridge, 1972 **E**.
I Rothiemurchus (as *E. brevipila × scottica*) 1935 *G. Taylor* herb. *M. L. Wedgewood* **MBH**.

Odontites Ludw.

O. verna (Bellardi) Dumort. subsp. **verna** *Red Eyebright, Red Bartsia*

Native. Common in farmyards and on tracks and waste places.
Map 206.

M Very common (Gordon); Lingieston, 1856 *JK* **FRS**; tracks at Kinloss and at Findhorn, 1925; throughout the county, 1954.
N Moyness, 1832 *WAS* **ABD**; Nairn, 1931 *G. Taylor* herb. *M. L. Wedgewood* **MBH**; in the farmyard at Holme Rose, at Cawdor, Geddes and Nairn, 1954; abundant in the stony verge of Loch Flemington; boggy field at Ordbreck, 1974.
I Inverness, 1832 and Dalwhinnie, 1833 (Watson, *Loc. Cat.*); shores of Loch Uanagan at Fort Augustus, 1904 *G. T. West* **STA**; near Torvain, 1940 *AJ* **ABD**; waste ground at Ardersier, at Struy in Strathglass, near Lewiston and at Culloden, 1954; verge of Loch Flemington, 1972 **E**; track at Garva Bridge, stubble field at Lochend and in a boggy field at Essich, etc.

Parentucellia Viv.

P. viscosa (L.) Caruel *Yellow Bartsia*
Introduced on the lawns at Gordonstoun where it remained for two years, 1950 *R. Richter.*

Lathraea L.

L. squamaria L. *Toothwort*
Native. Rare. A parasitic herb without chlorophyll found on the roots of various trees and shrubs, chiefly on *Alnus* or *Corylus*.
I In the parish of Kiltarlity (J. Fraser, *New Stat. Acc. Scot.* 1845). Not seen in recent years but it was refound near Dingwall, v/c 106 in 1968 and it is possible the plant could be refound at Kiltarlity.

Mazus radicans Cheeseman
A dwarf, far-creeping prostrate plant with brownish leaves and large white flowers blotched purple. Introduced on the lawn at Blackhills House, Lhanbryde. Recorded in 1966 by *Mrs Ronald Christie*. A plant was transferred to the lawn at Rose Cottage, Dyke, 1968 **E**.

Alonsoa warscewiczii Regel
Annual with pale brick-red flowers. Introduced with seeds at Hillhead Farm near Forres in 1964, at Dyke in 1969 **E** and at Lethen House, Nairn, 1976 *D. Brodie* **E**!.

LENTIBULARIACEAE
Pinguicula L.

P. lusitanica L. *Pale Butterwort*
Native. Frequent in moorland bogs in the west of the district. Flowers pale lilac.
I Glen Strathfarrar, 1836 *WAS* **E**!; plentiful in the bogs below Sgùrr na Lapaich in Glen Strathfarrar, 1947; on Sgùrr na Muice and to the W of Creag Fhir-eoin at Monar; in a small burn at the SW corner of Aigas Loch in Strathglass, 1976 **E**.

P. vulgaris L.   *Sheep Rot, Common Butterwort*
Native. Abundant in moorland bogs, by burns and on wet rocks
and ledges. Flowers violet.
M Very common (Gordon); Califer Hill near Forres, 1871 *JK*
FRS; dune slacks near the Buckie Loch, Gulbin Forest, 1925;
Culbin, 1953 *A. Melderis* **BM** (but now dried out from there);
throughout the county, 1954.
N (Stables, *Cat.*); ditch near Lochloy, 1867 *JK* **FRS**; Ardclach
(Thomson); bog by the Minister's Loch on the Nairn Dunbar
golf course, 1962 **ABD**; flush on the raised beach at Maviston, 1968
E, and throughout the county.
I Inverness, 1832 and Dalwhinnie, 1833 (Watson, *Loc. Cat.*);
Badenoch, 1838 ? *coll.* **E**; Newton Hill ? date and *coll.* **INV**; Coire
Dhubh near Laggan, 1916 *ESM* **E**; Fort Augustus, 1904 *G. T. West*
STA; Coire an Lochain of Cairn Gorm, 1953 *A. Melderis* **BM**;
throughout the district, 1954.

Utricularia L.
U. vulgaris L. agg.   *Greater Bladderwort*
Native. Floating herb with no roots; leaves divided into filiform
segments, green with numerous bladders; flowers bright yellow.
The old records for *U. vulgaris* may all belong to *U. australis*
R. Brown. They have not been critically examined.
M Loch of Spynie, 1831 *WAS* **E**; in shallow water in the Buckie
Loch, Culbin, 1898 *ESM* **E,CGE**; Lochnellan near Lochindorb,
1959 *D. A. Ratcliffe*.
N Moss of Litie, near Auldearn, 1833 *WAS* **E**; Lochloy ? date
and *coll.* **ELN**; in the Mosses of Inshoch and Litie, *W. Brichan*
(Gordon).
I Loch Alvie near Kincraig, 1891 *AS* **E**; Lochs Ness, Uanagan,
Ruthven and Tarff, 1904 (West, *Proc. RSE* **25**: 967 (1905)); canal
below Insh village, 1975 **E**, det. P. Taylor.

U. australis R.Br.
    U. neglecta Lehm.
Native.
M Plentiful in shallow peaty water near Spynie Loch, 1898 *ESM*
BM,CGE OXF; in shallow water in the Buckie Loch, Culbin, 1898
ESM **BM,CGE,OXF**.
N Pool by the track to the sea below Lochloy House in the Culbin
Forest, 1960 **E,K**, det. P. Taylor; Loch Kirkaldy near Glenferness
(not in flower but possibly this), 1971 **E**, det. P. Taylor.
I Loch Garten, 1947 *J. Walton* **GL**; Loch Mallachie, 1973 **E**, det.
P. Taylor

U. intermedia Hayne   *Intermediate Bladderwort*
Native. Shallow peaty pools and loch margins. More slender than
the last and having leaves of two kinds, one bearing green leaves
without, or very few, bladders, the other with colourless leaves
bearing many bladders. Seldom flowers.

M Elginshire, *Miss Bell & H. Watson* **CGE**; Gordon Castle, 1830
W. J. Hooker **GL**; Grantown, 1872 *JK* **FRS**; Boat of Garten (Druce,
1888); Edinkillie, Elgin, *P. Leslie* and Cromdale, *J. Keith* (Burgess);
Lochnellan near Lochindorb and pool by Beum a'Chlaidheimh,
1959 *D. A. Ratcliffe*; abundant in Loch Mòr near Duthil, 1973
E,ABD,BM; in a small burn S of Lochan Dubh near Lochindorb,
1976 **E**.
N Peaty pool at Littlemill, 1973 *R. Richter* **E,BM** det. P. Taylor.
I Bog near Loch Ericht Hotel, Dalwhinnie (Boyd); lochs
Mallachie, Garten and Morlich, *G. C. Druce* **OXF**; Loch Alvie near
Kincraig, 1891 **E**; lochs N of Invermoriston, Loch a'Mhuilinn
near Portclair and Lochan Creag Dhubh at Fort Augustus, 1904
G. T. West **STA**; pools at Loch nan Geadas (Pealling); Loch na
Bà Ruaidhe above Milton, 1947; Loch Laide, 1954; Loch Meiklie,
1973 **E**; peaty pools near Loch Gorm above Glen Urquhart, 1973
M. Barron & J. Winham; lochan at Upper Gartally near Milton,
1975.

U. minor L.   *Lesser Bladderwort*
Native. Frequent in shallow peat bogs. Both kinds of leaves have
bladders, the plant flowers freely and the corolla is bright yellow.
For notes on *U. bremii* cf (*Phyt.* 1 :2, 259 (1843)).
M Loch Spynie, 1832 *WAS* **E** and in herb. *W. Brand* **E** (as *U. bremii*).
bremii).
N Moss of Inshoch, 1833 *WAS* **GL**, collected there in 1898, *ESM*
BM,CGE and by *P. M. Playfair*, **E**; Moss of Litie, *J. B. Brichan*
(Watson).
I Dalwhinnie (Balfour, 1868); in a small loch on Beinn a'Bha'ach
Ard, in flower, *J. Farquharson & T. Fraser* (*Trans. BSE* **9**: 474
(1868)); glen 3 miles N of Truim (Boyd); small bog at Eskdale,
1973 **E**; Loch Beannacheran in Strathglass and in Loch an Eilein;
small lochan near Loch Gorm, 1873 *M. Barron & J. Winham*; loch
at Upper Gartally above Milton, 1975.

LABIATAE
Mentha L.
A taxonomically difficult genus due to much hybridisation and the
occurrence of the cultivated forms of *M. spicata* and its hybrids. Some
of these which have spread vegetatively and become locally prominent have in the past received formal scientific names. The
following account is based on that of R. M. Harley in Tutin, T. G.
et al., eds., *Flora Europaea* 3: 183–186 (1972) and also see the account
by the same author in Stace, C.A., ed. *Hybridisation and the Flora of
the British Isles*: 383–390 (1975). Following the International Code
of Botanical Nomenclature all 'varieties' of hybrids must now be
designated as 'nothomorphs' and this has been done as necessary.
Recent records have been determined by R. M. Harley who has
also provided some of the notes.

Key to Mentha

1. Calyx hairy in throat with distinctly unequal
   teeth. Corolla tube gibbous.   **pulegium**
   Calyx glabrous in throat with more or less
   equal teeth. Corolla tube straight.   2
2. Bracts like the leaves. Inflorescence usually
   terminated by leaves.   3
   Bracts small and inconspicuous and unlike the
   leaves except sometimes below. Flowers in
   terminal heads or spikes.   6
3. Plant hairy, green, usually fertile. Calyx broadly
   campanulate with broadly triangular teeth.   **arvensis**
   Plant glabrous or hairy, often red-tinged,
   usually sterile. Calyx narrowly campanulate to
   tubular; teeth narrow.   4
4. Calyx campanulate 2–3·5 mm with teeth
   rarely more than 1 mm. Plant usually more or
   less glabrous.   × **gentilis**
   Calyx 3·5–4 mm tubular, or if shorter the teeth
   1–1·5 mm and the plant distinctly hairy.   5
5. Plant red-tinged, subglabrous. Upper bracts
   suborbicular, cuspidate.   × **smithiana**
   Plant usually green, at least not often strikingly
   red-tinged, hairy. Upper bracts ovate to ovate-
   lanceolate, not cuspidate.   × **verticillata**
6. Leaves distinctly petiolate. Flowers in a head
   or oblong spike 12–20 mm in diameter.   7
   Leaves sessile or nearly so. Flowers in a spike
   5–15 mm in diameter.   8
7. Flowers in an oblong spike, leaves usually
   lanceolate. Sterile.   × **piperita**
   Flowers in a head, sometimes with one to three
   whorls below, leaves usually ovate. Fertile.   **aquatica**
8. Leaves hairy beneath, not more than 45 mm
   long, oblong to suborbicular, obtuse, strongly
   rugose, serrate with teeth bent down towards
   leaf undersurface and therefore appearing
   crenate. Corolla off-white or pink.   **suaveolens**
   Leaves glabrous or hairy beneath, usually more
   than 45 mm long, frequently lanceolate or
   sometimes ovate to suborbicular, rugose or
   smooth, acute, serrate with spreading teeth.   **spicata** or
   Corolla lilac, white or pink.   × **villosa**

N.B. Many plants have flowers with included anthers and this has
often been considered evidence of hybridity. While hybrids usually
have such anthers, these can on rare occasions be exerted. Similarly
most mint species have a fairly high proportion of plants which are
female and also have included anthers, so this character must be
treated with caution.

M. pulegium L. *Penny-royal*
M Certainly introduced, rare in a field near the Church of Birnie, 1834 (Gordon).

M. arvensis L. *Corn Mint*
Native. Occasional in arable fields but more common in damp places by pools and burns. Map 207.
M Aviemore, 1837 *W. Brand* E; very common (Gordon); margin of Sanquhar House pond at Forres, 1953 **ABD**; banks of the River Spey at Aviemore, 1955 *J. Milne* E; on the W shore of Loch na Bo at Lhanbryde, 1972 E; stubble field E of Carrbridge and in a grass field at Kellas, 1973.
N Common at Ardclach, 1886 *R. Thomson* **NRN**; Lochloy (as var. *praecox* (Sole) Smith), 1904 *G. T. West* **E,STA**; fields at Holme Rose, by Loch Flemington and on the Carse of Delnies, 1954; root field by Cran Loch, 1960 **K**.
I Kincraig, 1891 *AS* E; shores of Loch Uanagan at Fort Augustus (as var. *agrestis* (Sole) Fraser), 1904 *G. T. West* **E,STA**; near Dunain Park, Inverness, 1940 *AJ* **ABD**; edge of a pool at Milton, 1947; rough fields at Farr and in Strathglass, 1956; edge of a pool by the River Glass at Cannich, 1975 E; cornfield at Mains of Bunachton near Dunlichity, 1976 E; rough field at Culburnie, Kiltarlity, 1976.

M. aquatica × arvensis = M. × verticillata L.
Marsh Whorled Mint
Very variable. More robust than *M. arvensis*, with the flowers in axillary whorls sometimes crowded towards the top of the flowering stem. The stamens are normally included. Native. Frequent on loch margins and pools by rivers. Map 208.
M Parishes of Dyke, Elgin, Urquhart, Birnie and Knockando (Burgess); banks of the Muckle Burn near Moy, 1952 **K, RAG & RMH**; banks of the River Lossie at Elgin, 1954 **ABD**; near Findhorn, 1954; verge of a bog at Lochinvar near Elgin, 1960 **K**; backwater of the River Spey at Garmouth, 1964 **CGE**; at the W end of the Buckie Loch in the Culbin Forest, 1967 **ABD**; grassy marshland at the NE end of Loch Spynie (with unusual toothed leaves), 1972 E; single of the River Spey at Kingston, 1972 **E,ABD,BM**; banks of the River Spey at Orton, 1973 **E,BM,CGE**; stoney shores of Loch Puladdern at Aviemore, 1973 E; backwater of the Spey at Dandaleith, 1974 **CGE**.

The following varieties were determined by Rex A. Graham. Var. ± **acutifolia** (Sm.) Fraser from the burn above Sanquhar pond at Forres, 1953 **ABD,K** and var. **paludosa** (Sole) Druce from the Buckie Loch, Culbin in 1954.
N Cran Loch, 1904 *G. T. West* E; banks of the River Nairn at Holme Rose, 1955 **ABD**; boggy moor near Lochloy, 1953 **K**; ditch by the lochan at Househill, Nairn, 1974 E.
I Near Kincraig, 1891 *AS* E; by Loch Insh at Kincraig, 1955 *J. Milne* E; in alder-willow wood near Kylechorky Lodge, Glen

Urquhart, 1959 *V. S. Summerhayes* **K**; ditch at Farr, 1973; marsh below Croftcarnach N of Kingussie, 1973 **E**.

M. arvensis × **spicata** = **M.** × **gentilis** L.
A variable hybrid rather similar to the preceding, but usually subglabrous, often red-tinged and with often more acute leaves. The calyx tube and pedicel are usually glabrous and the calyx teeth ciliate. Corolla deep lilac; stamens included. Native and introduced. Occasional on river banks.
M Speymouth, Moray, 1929 *K. D. Little* **K**; banks of the River Lossie at Hillhead, 1950 **ABD,RAG & RMH**; backwater of the River Spey at Kingston, 1972 **E,BM**; margin of the lochan below Crofts of Buinach near Kellas, 1973 **E**; by the River Lossie at Calcots, 1973 *P. Cudmore* **CGE**; by the Spey near Baxter's Factory at Fochabers, 1973 **E**.
I Margin of Raigmore pond Inverness, 1975 **E**.

M. aquatica × **arvensis** × **spicata** = **M.** × **smithiana**
R. A. Graham
A robust plant similar to *M. gentilis* but differing in the longer more tubular calyx, and the suborbicular, cuspidate upper bracts. Introduced.
M Burn below Milton Duff and the burn at Linkwood (as *M. rubra* Sm.) (Gordon); Mill of Birnie (Watson).
I Loch Uanagan, Fort Augustus (as *sativa* var. *rubra* Huds.), 1904 *G. T. West* **E,STA**; Urquhart Bay, 1904 *G. T. West* **E,STA**.

M. aquatica L.   *Water Mint*
Native. Common in ditches and bogs. Inflorescence consisting of a terminal head of several whorls with usually a few axillary whorls below. Pedicels and calyx hairy; stamens normally exerted. Map 209.
M (Watson, *Top. Bot.*); ditch by the Manse at Dyke, 1865 *JK* **FRS**; parishes of Knockando Forres, Duffus, Bellie and Speymouth (Burgess); margin of Loch Spynie and the Buckie Loch, 1930; by the River Spey at Knockando, 1973.
N Ardclach and Nairn, 1900 (Lobban); banks of the River Nairn at Holme Rose, by Geddes Reservoir, margin of Loch Flemington and bogs on the Carse of Delnies etc., 1954; margin of Loch Kirkaldy, pool near Kinsteary, by the burn at Lethen and pool at Littlemill, 1972.
I Margin of Loch Flemington, 1940 *AJ* **ABD**; by the River Enrick at Milton, 1947; burn on the golf course at Inverness, by the Spey at Newtonmore, in Strathglass and Glen Urquhart etc.

M. aquatica × **spicata** = **M.** × **piperita** L.   *Peppermint*
Introduced and completely naturalised. Frequent in ditches and wet places by rivers and burns. Cultivated as the source of peppermint and occasionally escaping. Stems purple-tinged. Inflorescence a congested terminal spike, pedicels and calyx-tube glabrous (in nm *piperita*); corolla reddish-lilac; stamens included.

Nm **piperita**

M Sanquhar House pond at Forres, 1861 *JK* **FRS**; lochan between Avielochan and Aviemore, 1955 *J. Milne* **E**; parishes of Dyke, Forres, Speymouth and Knockando (Burgess); by the River Lossie at Hillhead, at Fochabers, Elgin, Grantown Dunphail and Dava, 1954; shingle of the River Spey at Garmouth, 1964 **CGE**; on the W shore of Loch na Bo, 1972 **E**; banks of the River Spey at Orton, 1973 **E,BM**; by a small bridge over the Broad Burn N of Auchinroath near Rothes, 1973 **ABD**; by the Black Burn bridge near Pluscarden Priory, 1973 *R. Richter* **BM**; banks of the River Spey at Fochabers, 1973 **E**.

N By the River Nairn at Holme Rose, 1955; ditch between Geddes and Howford Bridge near Nairn, 1967 **E,ABD**; banks of Geddes reservoir, 1972 **E**.

I Banks of the River Ness, Laggan Pool etc. (Aitken); burn side at Tomich, 1932 *Rex A. Graham* **RAG** & **RMH**; ditch near Balcarse Farm Kirkhill, 1975 **E**; pool by the River Glass at Eskadale, 1975 **E**; by the burn on the Inverness golf course, 1976.

Nm **hirsuta** Fraser
Forma *hirsuta* (Fraser) R. A. Graham

This form, first recognised by Fraser, appears to be a mutation from the glabrous cultivated peppermint which is frequently naturalized and with which it often grows. It seems worth recognising because it is very constant in its characters and often causes confusion. There may occasionally be other hairy peppermints which arise spontaneously by hybridisation between *M. spicata* and *M. aquatica* and these usually look different, cf. Fraser *BEC Report*, 1926.

M By a small burn running into Sanquhar pond at Forres, 1952 **ABD**; by the Howie Burn at Rafford, 1955 **ABD**; on the W shore of Loch na Bo, Lhanbryde, 1972 **E**; ditch by the old railway line below Auchinroath near Rothes, 1972 **E**; in the Mosset Burn opposite Christie's Nursery at Forres, 1973 **E**; by a small burn at Stonefield E of Pluscarden, 1975 **BM**.

M. suaveolens Ehrh.   *Apple-scented Mint*
   M. rotundifolia auct.

Doubtfully native. Leaves ovate-oblong to suborbicular, strongly rugose, hairy above, white-tomentose beneath; inflorescence forming a congested spike; calyx campanulate hairy; corolla pinkish or dingy white.

M Boat of Garten (Druce, 1888); escape in the parishes of Forres and Speymouth (Burgess); farmyard at Coltfield near Alves, 1941? *coll.* **STA**.

I (Trail, *ASNH* 172 (1906)).

M. spicata L.   *Spearmint*
   M. viridis (L.) L.; *M. longifolia* auct. non (L.) Huds; *M. sylvestris* auct. non L.; *M. niliaca* auct. non Juss. et Jacq.; *M. cordifolia* auct. non Opiz; *M. scotica* R. A. Graham.

Introduced. This species includes both the common glabrous mint of gardens, often escaping, and hairy forms which have been incorrectly named *M. longifolia* (L.) Huds though this latter species does not occur in Britain. The plant known as *M. scotica* R. A. Graham (*M. niliaca* var. *sapida* auct.) is a striking and easily recognised variant of *M. spicata* with grey tomentose leaves.

M Near an old wall below Blackhills (as *M. viridis*), 1835 (Gordon); waste ground at Grantown and Boat of Garten, river bank at Elgin, 1954; banks of the River Spey at Orton, 1973 **E**; by the Spey N of Fochabers, 1973 **E,BM** etc.

Plants falling into '*M. scotica* R. A. Graham' have been recorded from Moray as follows: by the Muckle Burn *c.* 2 miles from Forres (as *M. niliaca* var. *sapida*), 1898 *ESM* **BM,K,CGE, OXF,MBH**; by the Muckle Burn near Whitebridge, Moy House, 1952 **ABD,K** and in 1953 **ABD**, also collected there by *A. Melderis* **BM**; by the pond at Sanquhar House, Forres, rubbish tip at Forres and in an old quarry near the harbour at Lossiemouth, 1974.

N Howford marsh and at Lochloy, 1888 *R. Thomson* **NRN**; Ardclach, 1900 (Lobban).

I Near Kincraig, 1891 *AS* **E**; river bank at Struy, under the arch of the bridge (hairy form) 1933 *R. A. Graham* **RAG** & **RMH**; waste land at Ardersier, 1952; roadside verge near Creagdhubh Lodge, Laggan, 1973; near a cottage at Alvie House and by the station at Newtonmore, 1974; backwater of the River Spey at the N end of the golf course at Newtonmore, 1975 **E**.

M. spicata × **suaveolens** = **M.** × **villosa** Hudson
    M. niliaca auct. non Juss et Jacq.; *M.* × *cordifolia* Opiz
An extremely variable hybrid which may be glabrous or hairy. Formerly the glabrous plants were referred to *M.* × *cordifolia* and the hairy to *M.* × *niliaca*. Most forms are intermediate in leaf shape and rugosity between *M. spicata* and *M. suaveolens*, but there are some broad-leaved rugose forms of *M. spicata* which are best distinguished from the hybrid by their fertility, as they usually produce seed. The hybrid is a sterile triploid. *M.* × *villosa* occurs in a number of forms of which all are apparently garden escapes. The most striking form is nm *alopecuroides* (Hull) Briq. which is often mistaken for *M. suaveolens*. It can be distinguished by its more robust habit, its sweet odour like *M. spicata*, the spreading teeth on its broadly ovate or orbicular leaves, and its robust spikes of pink flowers.

M Quarry near Lossiemouth harbour, 1954; by the River Lossie at Elgin, 1960 **ABD,K**.

N Near a cottage at Auldearn station, 1967 **E,ABD**.

I Roadside bank N of Invermoriston, by Loch Ness, 1973 **E**; rubbish tip in the salt-marsh at Castle Stuart, Petty, 1975 **E**.

Nm **alopecuroides** (Hull) Brig. pro var.
    M. alopecuroides Hull; *M. velutina* Lej.
M Garden escape at Kingston (as *M. alopecuroides* Hull) 1898

ESM **E,CGE**; quarry at Lossiemouth, 1970 **E,K**; bank outside cottages near Mains of Alanbuie near Keith, 1974 **BM**.

Nm **webberi** (Fraser)
    M. × *nilaca* Juss et Jacq. var. *webberi* Fraser
This interesting plant may well prove to be a form of *M.* × *villosa* but a garden escape which seemed to be this, was collected in West Ross and on examination turned out to be diploid. This suggests that it is a hybrid *M. longifolia* × *suaveolens*. Further study is needed (Harley, *pers. com.*, 1975).
M Waste ground near the Haugh at Elgin, 1953 **ABD,K**; in the quarry near the harbour at Lossiemouth, 1956 **ABD**.
I   Roadside verge on moor where garden rubbish had been tipped, N of Coulan near Inverness, 1970 **E,K**.

M. longifolia × **spicata** = **M.** × **villosa-nervata** auct. ? an Opiz
Occurs occasionally as a garden escape, it is usually a hairy plant with narrow sharply serrate leaves but has not yet been recorded for v.c. 95 or 96.

Lycopus L.
L. europaeus L.   *Gipsywort*
Native. Occasional on loch margins and in grassy bogs in the lowlands.
M Buckie Loch, Culbin Forest, scarce, 1898 *ESM* **CGE** and 1968 *MMcCW* **E**; verge of Sanquhar pond at Forres, margins of Loch Spynie, Gilston lochs, Lochinvar near Elgin, ditch at Kincorth and backwater of the River Spey at Kingston, 1954; plentiful on the edge of Millbuies loch, 1972 **E**; among stones at the artificial pond by the Mosset Burn in Forres.
N Bogside of Boath near Auldearn, 1832 *W. Brand* **E**; Cawdor, 1834 *WAS* **E**; Balloan at Cawdor, *WAS* (Gordon); margin of Loch Flemington, 1954.
I   Wet copse near Beaufort and canal banks at Muirtown, Inverness (J. Don, 1898); in a small pool by the River Beauly near Groam of Annat, 1930 *M. Cameron*; ditch by the salt-marshes at Muirtown Basin, Inverness, 1956; in the old Ice Pond at Lovat Bridge near Beauly, 1976 **E,BM**.

Origanum L.
O. vulgare L.   *Marjoram*
Native. Rare on dry base rich rocks and banks and washed down rivers where it becomes established on shingle and river banks.
M Elginshire, 1837 *G. Gordon* **E**; Pluscarden (Gordon); parishes of Duffus, Elgin, Speymouth and Knockando (Burgess); banks of the River Spey at Knockando, 1965 *G. Shepherd*; banks of the Spey between Inverallan and Grantown, 1972 **E**; a single plant on shingle consolidation at Rothes, 1973 *A. J. Souter*.
N Nairn and Cawdor (Lobban).
I   Banks of the River Ness, 1855 *HC* **INV**; Inverness, 1859 *JK* **FRS**.

Thymus L.

T. praecox Opiz subsp. **arcticus** (E. Durand) Jalas   *Thyme*
    T. drucei Ronn.
Native. Abundant on dunes, dry moorland and among scree and rocks.
M Very common (as *T. serpyllum*) (Gordon); near Glassgreen (with white flowers), 1837 *J. G. Innes* (Watson); Waterford, 1900 *JK* **FRS**; Slochd near Carrbridge, 1940 *AJ* **ABD**; banks of the River Findhorn near Forres, 1953 *A. Melderis* **BM**; abundant throughout the county 1954.
N (Stables, *Cat.*); Cawdor, 1870 *R. Thomson* **NRN**; moor at Loch of the Clans, 1960 **ABD**; river shingle at Drynachan and throughout the county.
I Dalwhinnie, 1833 (Watson, *Loc. Cat.*) and 1885 herb. *Backhouse* **E**; Kincraig and Rothiemurchus, 1891 *AS* **E**; Invermoriston, 1904 *G. T. West* **STA**; moorland track by Loch Conagleann at Dunmaglass, 1972 **E** and throughout the district.

Acinos Mill.

A. arvensis (Lam.) Dandy   *Basil Thyme*
Native. Very rare among rocks in short base rich grassland. Formerly found in fields.
M Fields at Kirkhill in Duffus, *WAS* **E,ABD**; field at Coltfield at Alves, *G. Wilson* **E,ABD**; Oldmills, *P. Cruickshank* (Gordon); field SW of Springfield at Elgin and roadside at Barefleet hills, 1862 (Watson); Muirtown of Dalvey (with white flowers) *J. B. Brichan & J. G. Innes* (Gordon, *annot.*); Brodie ? date *and coll.* **ELN**; in a small rocky depression on the Elgin golf course, 1970 **E**.
N In a field west of Nairn on the south side of the road to Fort George, *Brodie* **E**.
I (Druce, *Com. Fl.*). This may refer to the Nairn record.

Clinopodium L.

C. vulgare L.   *Wild Basil*
Native. Rare on grassy banks on base rich soil and occasional on river shingle where it has probably been introduced with distillery refuse.
M Banks of the River Spey at Rothes, 1837 *G. Gordon* **E,ABD**; haughs at Rothes and Grantown, *J. Fraser* (Gordon); by the Findhorn below the Heronry at Darnaway (Aitken); Speyside near Kinchurdy (Druce, 1888); shingle of the River Spey at Garmouth, 1953; distillery yard at Carron, 1957 **E,K**; river bank near the iron bridge at Broom of Moy 1967 **E**; bank of the river at Grantown, 1971 **E**; shingle of the River Spey at Craigellachie, 1975.
I Bank by the road approaching Ferrybrae from Kirkhill, 1954 **E**; a few plants below the cliffs of Creag Dhubh near Laggan, 1974; railway embankment at Tomachrochar S of Nethybridge, 1975 **E**.

Salvia L.

S. reflexa Hornem.
Introduced with bird seed or foreign grain. Flowers pale blue.
M Rubbish tip at Elgin, 1958 **E**.

Prunella L.

P. vulgaris L.   *Selfheal*
Native. Abundant in short grass, waste land and river shingle.
Pl 17.
M Very common (Gordon); Greshop near Forres, 1898 *JK* **FRS**;
W bank of the River Findhorn on shingle, near Forres, 1953
A. Melderis **BM**; throughout the county, 1954.
N (Stables, *Cat.*); pastures, common at Ardclach, 1887 *R. Thomson*
NRN; Nairn, 1900 *A. Lobban* **E**; throughout the county, 1954.
I Inverness, 1832 and Dalwhinnie, 1833 (Watson, *Loc. Cat.*);
moor 4 miles SE of Inverness (with white flowers), 1937 *J. Walton*
GL; Bunchrew, 1940 *AJ* **ABD**; throughout the district, 1954.

Stachys L.

S. arvensis (L.) L.   *Field Woundwort*
Native. Rare in cultivated ground.
M Frequent, Crofts at Elgin etc. (Gordon); Rafford, 1861 *JK*
FRS; Darnaway, 1946 and Buinach, 1953 *R. Richter*; garden weed
at Brodie Castle, Dallas Lodge and Fochabers 1954; rare at Moy
House, 1955; cottage garden at Rafford, 1968 **E,BM**.
N (Stables, *Cat.*); Ardclach (Thomson); garden weed at Lethen
and Geddes, 1954.
I Inverness, 1832 (Watson, *Loc. Cat.*); garden weed at Kirkhill,
1960; rubbish tip on Longman Point, 1970 **E**; waste ground at
Milton near Allanfearn, 1973 *M. Barron*; abundant in a root field
at Farley, 1975 **E**; stubble field at Lochend, 1975.

S. palustris L.   *Marsh Woundwort*
Native. Common in bogs and ditches and by rivers and burns,
occasionally in arable fields. Very variable. Leaves lanceolate the
lower with very short petioles, the upper sessile; corolla pinkish-
purple. The plant is supposed to be odourless but it often smells
slightly. Map 210.
M Frequent (Gordon); Berrigeith near Forres, 1850 *JK* **FRS**; bog
by Loch Spynie, Culbin Forest, Boat of Garten and throughout the
county 1954; ditch near Tomcork at Dallas, 1975 **E**; banks of the
old railway line near Delriach near Cromdale, 1975 **E**.
N (Stables, *Cat.*); Ardclach (Thomson); by the River Findhorn at
Drynachan, by the River Nairn at Holme Rose, at Cawdor and in
bogs by Cran Loch, Lochloy, Loch of the Clans and Loch Flemington;
boggy field by the Burn of Blarandualt at Ordbreck, 1975 **E**.
I Alvie near Aviemore, 1830 ? *coll.* **GL**; Inverness, 1832 (Watson,
Loc. Cat.); Kessock, 1854 *HC* **INV**; Culloden Moor station, 1940
AJ **ABD**; bogs around Loch Meiklie in Glen Urquhart, at Fort
Augustus, Drumnadrochit, Ardersier, Kincraig etc., 1954; dry
edge of a field at Daviot, 1972 **E**; ditches at Durris near Dores and at
Kindrummond, 1976 **E**.

S. palustris × sylvatica = S. × ambigua Sm.
Very variable in leaf shape, length of the petioles and odour.

M Meadow by the old Spey Bridge at Grantown, 1975 **E,BM**; bog by the Black Burn at Pluscarden, 1975 **E**.

N Moyness, *WAS* (Gordon); in tall vegetation by Cran Loch, by the Burn of Blarandualt near Ordbreck and in a ditch near Righoul School, all in 1975 **E,BM**.

I (Watson, *Top. Bot.*) (probably Stables record for Nairn); Strathglass, *J. Milne* (*Critical Atlas*); bogs at the NW end of Loch Meiklie, 1973 **E,BM**; grass field by the railway line below Balblair near Beauly, 1964 **E**; dry roadside verge at Dulnain Bridge, marsh by Newtonmore golf course, ditch at Cannich, by the pond at Culloden House, wood by Loch Ness at Lewiston, near Gorthleck and car park at Inshriach Nursery, 1975 **E**.

S. sylvatica L.   *Hedge Woundwort*
Native. Common in shady waste places. Foetid when bruised; corolla claret-coloured. Map 211.

M Frequent (Gordon); Greshop Wood near Forres, 1900 *JK* **FRS**; Darnaway, 1953 *A. Melderis* **BM**; throughout the county, 1954.

N (Stables, *Cat.*); river bed near Firhall, Nairn, 1885 *R. Thomson* **NRN**; throughout the county, 1954; woods at Lethen House, 1961 **ABD,K**.

I Ness Islands, Inverness, 1854 *HC* **INV**; Bunchrew, 1940 *AJ* **ABD**; waste ground at Cannich 1947; common in the lowland areas, 1954; roadside verge at Pityoulish, ravine at Inverfarigaig, in boulders at the foot of the cliffs of Creag Dhubh near Laggan, a single clump by the River Feshie in Glen Feshie, on scree below Creag nan Clag etc.

Ballota L.

B. nigra L. subsp. **foetida** Hayek   *Black Horehound*
Introduced in this area as a medicinal plant.

M Foot of the Churchyard wall, Dyke, 1840 *J. G. Innes* and at Deanshaugh near Elgin, 1858 *George Christie* (Gordon, *annot.*); Dyke, 1901 *JK* **FRS** and 1963 *MMcCW* **CGE**; roadside between Fochabers and Spey Bay, 1952.

N Nairn and by the Muckle Burn at Ardclach, 1900 (Lobban).

I Fort Augustus, ? date and *coll.* **ELN**.

Lamiastrum Heister ex Fabr.

L. galeobdolon (L.) Ehrend. & Polatschek   *Yellow Archangel*
   Galeobdolon luteum Huds.
Introduced in north Scotland.

M Shaded situation opposite Milne's Institute at Fochabers, *G. Birnie* (Burgess).

I Holm near Inverness (G. A. Lang 1905).

Lamium L.

L amplexicaule L.   *Henbit*
Native. Abundant in cultivated ground in the lowlands, less so in the highlands. Teeth of the calyx shorter than the tube, connivant in fruit. Fig 30a, Map 212.

M Innes House, Urquhart, 1794 *A. Cooper* **ELN**; very common (Gordon); Greshop near Forres, 1867 *JK* **FRS**; Aviemore, 1919 *M. L. Wedgewood* **MBH**; throughout the county, 1954; field at Barley Mill Farm, Brodie, 1964 **E,ABD**.
N Cawdor, 1832 *WAS* **ABD**; Piperhill, 1887 *R. Thomson* **NRN**; garden weed at Drynachan Lodge and throughout the county, 1954.
I (Watson, *Top. Bot.*); Kincraig, 1891 *AS* **E**; gardens at Kirkhill, Inverness, Dochfour House, Culloden House etc. in the lowlands; hotel gardens at Kingussie, and Invermoriston; cottage garden at Croachy in Strathnairn and at Baillid near Newtonmore, 1975.

L. moluccellifolium Fr. *Northern Dead-nettle*
Native. Frequent in cultivated ground up to 1954, now seldom seen. Differs from *L. amplexicaule* in being more robust, the calyx teeth longer than the tube, spreading in fruit. Fig 30b.
M Forres, *J. G. Innes* (Watson); Boat of Garten and at Forres (as *L. intermedium*) (Druce, 1888); Forres, 1887 *G. C. Druce* **OXF**; garden weed at Newton House, Elgin, at Logie House, at Castle Grant, Grantown, Findhorn tip, and in a carrot field at Covesea etc., 1954; single plant at the Aviemore sewage works, 1973.
N Nairn and Ardclach, 1900 (Lobban); garden at Drynachan Lodge, at Holme Rose and Geddes House, 1954.
I Garden weed at Milton, 1947; field at Kirkhill, 1954; Fort Augustus and Ardersier, 1956; Inverness Infirmary garden, 1971 **E**; farmyard at Banchor Mains, Newtonmore, 1975.

L. hybridum Vill. *Cut-leaved Dead-nettle*
Native. Rare in cultivated ground. Fig 30c.
M Bulletloan, Forres, (as *L. incisum*) 1830 (Gordon); Sanquhar House, 1956 **ABD**; rubbish tip, Forres, 1968 **E**; common in the Bowling Green garden at Grantown, 1971 **E**; by the Forestry offices at Kintessack, 1974.
I Lochside near Loch Flemington, Gollanfield, 1956; garden at Dochfour House and in a cottage garden at Aldourie Pier, 1975.

L. purpureum L. *Red Dead-nettle*
Native. Abundant in cultivated ground and waste places throughout the district.
M Innes House, Urquhart, 1794 *A. Cooper* **ELN**; very common (Gordon); Findhorn (Druce, 1897); Forres, 1898 *JK* **FRS**; throughout the county, 1954.
N (Stables, *Cat.*); anywhere, all the year, Ardclach 1886 *R. Thomson*, **NRN**; garden weed at Drynachan and throughout the county, 1954; garden at Geddes House, 1966 **ABD,CGE**.
I Inverness, 1832 (Watson, *Loc. Cat.*); Reelig gardens at Kirkhill, ? date and *coll.* **INV**; throughout the district, 1954; Badenoch Hotel garden at Newtonmore, 1975 *J. Clark*.

L. album L. *White Dead-nettle*
Doubtfully native in north Scotland. Occasional on waste ground, railway embankments and by buildings. Map 213.

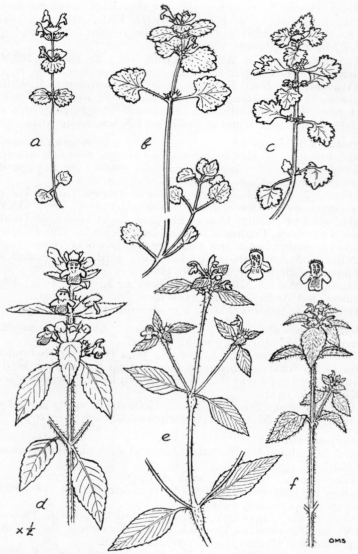

Fig 30 a, **Lamium amplexicaule**; b, **L. moluccellifolium**; c, **L. hybridum**
d, **Galeopsis speciosa**; e, **G. tetrahit**; f, **G. bifida**.

M Dunfermaline's garden, 1831, Burghead, 1841 *J. G. Innes*, Brodie House, 1842 (Gordon, *annot.*); Dunphail station, 1898 *JK* **FRS**; railway yard at Elgin, 1956; lane between Knockomie and Whiterow near Forres, at Findhorn below buildings, railway line at Blacksboat and by the Mill at Knockando, 1954; below the wall behind Brodie Castle, 1967 **CGE**; waste ground at Lossiemouth, 1967 **E,ABD,BM**; by the burn at Fochabers and at Longmorn and Newton House near Elgin, 1975; roadside near Springfield E of Forres, 1976 **BM**.

N (Stables, *Cat.*); Penick Castle *J. B. Brichan* (Gordon, *annot.*); waste land at Househill and on the railway embankment just W of Nairn, 1974.

I Not common, Inverness (J. Don, 1898); farmyard and lane at Achnagairn, Kirkhill, 1954; near Cluny, Laggan, 1956; bank in the road by Balnakyle Farm, Inverness, 1971 **E**; where rubbish was tipped in a field at Banchor Mains, Newtonmore, 1973 **BM**; raised beach at Ardersier and top of the bank above the River Calder at Baillid, Newtonmore, 1975.

L. maculatum L.   *Spotted Dead-nettle*
Garden outcast.
M Wood just S of Rafford Manse, 1967 **E,ABD**; bank opposite cottages at Delfur, Boat o' Brig, 1967 **ABD**; policies of Dalvey House near Forres (forma *immaculatum*), 1956 **CGE**, and 1960 **E,ABD,K**.
I Verge opposite house at Wellbank near Beauly, 1977 **E**.

Leonurus L.
L. cardiaca L.   *Motherwort*
Introduced.
M Knockando, *C. Watt* (Burgess); banks of the River Lossie below The Beild, Elgin 1960 *J. Harrison*.
N Doubtfully native, rare by the ruins of Penick Castle, *J. B. Brichan* (Gordon).

Galeopsis L.
G. angustifolia Ehrh. ex Hoffm.   *Narrow-leaved Hemp-nettle*
     G. ladanum auct.
Native. Rare. Formerly found in wheat fields.
M Fields near Milton Brodie, 1833 *G. Wilson* **E,ABD**; rare at Alves (Gordon); near Rafford, 1841 *J. G. Innes* and Elgin, 1867 (Watson); near Forres, 1842 *J. Fraser* **E**.
I Chapeltown, Croy, *J. Fraser & J. G. Innes* (Gordon, *annot.*).

G. tetrahit L.   *Dai Nettle, Common Hemp-nettle*
Native. Abundant in cultivated land and waste places. Middle lobe of the lower lip of the corolla entire, the network of dark markings never reaching the margin. Fig 30e, Map 214.
M Very common (Gordon); cornfields at Forres, 1856 *JK* **FRS**; field near Moy House, 1953 *A. Melderis* **BM**; root field at Kintessack, 1960 **K**, and throughout the county.

N (Stables, *Cat.*); Ardclach (Thomson); farmyard at Geddes, 1961 **ABD**, and throughout the county.
I Inverness, 1832 (Watson, *Loc. Cat*); Millstream and Bught Islands (Aitken); Bunchrew, 1940 *AJ* **ABD**; at Kincraig, Newtonmore, Fort Augustus and Glen Moriston etc., 1954; arable field at Groam of Annat near Beauly, 1972 **E**.

G. bifida Boenn.
Native. Under-recorded. Differs from *G. tetrahit* in having more glandular hairs on the nodes and the middle lobe of the lower lip of the corolla deeply emarginate and the network of markings reaching the margin, or often the whole lip dark. Fig 30f.
M Aviemore, 1920 *M. L. Wedgewood* **MBH**; waste ground at Sanquhar House, Forres, 1960 **K**; arable field near Brodie station, 1963 **E**; by the old railway line at Knockando, 1972 **E**; rubbish tip near Hillhead, 1973 **E**; waste ground by the Black Burn at Pluscarden, 1975.
N Root fields at Dulsie Bridge, Ordbreck and on the rubbish tip at Newton of Park, 1975.
I Waste land on Longman Point, 1954; field at Alturlie Point, shingle beach at Ardersier cottage garden at Aldourie, root fields at Farley, Mid Craggie and Dores; field at Dunain, 1970 *M. Barron*; root field at Tomich Farm near Beauly, 1975 **E**.

G. speciosa Mill.   *Large-flowered Hemp-nettle*
Native. Rare in cultivated land but common in vice-countires 93 and 94. Differs from the two preceding species in being larger and having large pale yellow flowers with a violet lower lip to the corolla. Fig 30d.
M Drumin (Gordon); parishes of Drainie, Forres and Speymouth (Burgess); field near Kinloss Abbey, 1925; root fields at Knockando, Rothes and Lochinvar, waste ground at Longmorn and Elgin, 1954; rubbish tip at Grantown, 1971 **E**.
N Root field at Cawdor, 1954.
I (Watson, *Top. Bot.*); waste ground at Inverness and Beauly, 1955 *J. M. Whyte*; newly sown verge at Kirkhill, 1971 **E**; abundant in a root field at Tomich Farm near Beauly 1976 **K**.

Nepeta L.
N. cataria L.   *Cat-mint*
Introduced at Ardclach, 1888 (Thomson).

N. faassenii Bergmans ex Stearn
Established on a grass bank in a field at Banchor Mains, Newtonmore from garden refuse tipped in the field, 1975 *J. D. Buchanan* **E**!.

Glechoma L.
G. hederacea L.   *Grundavy, Ground Ivy*
Native. Common in damp woods and on river banks, occasionally in dry places among shingle by the sea. Rare in the highlands. Map 215.

M Very common (Gordon); Forres, 1899 *JK* **FRS**; throughout the county, 1954; woods at Brodie Castle, 1963 **CGE**; among stones by the Muckle Burn at Earlsmill, 1967 **CGE**.
N (Stables, *Cat.*); on the banks of the River Findhorn at Ardclach, 1882 *R. Thomson* **NRN**; in woods at Cawdor, Dulsie Bridge, Glenferness and Holme Rose, 1954; grass bank at Drynachan Lodge and among stones by the River Nairn at Little Kildrummie.
I Near Inverness, 1854 *HC* **INV**; in woods at Kirkhill, Daviot, Culloden, Kiltarlity and Bunchrew etc. in the lowlands; among stones on the raised beach at Ardersier and on shingle on the margin of Loch Flemington, 1954; in a small burn at Flichity House and at Alvie Kirk, 1975.

Marrubium L.
M. vulgare L.   *White Horehound*
Doubtfully native in north Scotland. Rare on waste ground usually near buildings.
M Kingston near Garmouth and at Kinloss Abbey (Gordon); Kinloss, 1837 *G. Gordon* **E**, and collected there by *J. G. Innes* **ABD** and in 1857 *JK* **FRS**; parishes of Dyke, Forres, Kinloss, Urquhart and Speymouth (Burgess); dunes near cottages at Findhorn, 1925 and in 1953 *UKD* **UKD**.
N Introduced at Ardclach (Thomson).

Scutellaria L.
S. galericulata L.   *Skullcap*
Native. Rare on river banks, bogs and loch margins but locally common in Strathglass and by Loch Ness.
M Rare by the side of the River Spey a little above Rothes (Gordon); abundant in the old alder wood on the verge of the salt-marsh W of the Buckie Loch, Culbin Forest, 1973 *K. Allan et al.* **E**!.
I Kingsmill near Inverness, *WAS & G. Gordon* (Gordon); behind the Manse at Rothiemurchus, 1876 *JK* **FRS**; ditch near Beaufort (J. Don, 1898); shores of Loch Ness at Inverfarigaig and Foyers, 1904 *G. T. West* **E,STA**; shores of Loch Ness at Dores, 1971 **E**; abundant among stones the whole length of Loch Ness, in Strathglass, bog on the margin of Loch Meiklie in Glen Urquhart, margin of the small lochan in the pine wood at Inshriach, shores of Loch Alvie and in a small burn outlet at Loch Bunachton.

Teucrium L.
T. scorodonia L.   *Wood Sage*
Native. Common on dry moorland banks, on sand dunes and among rocks and boulders on hillsides. Map 216.
M Very common (Gordon); shingle of the River Findhorn at Broom of Moy, 1856 *JK* **FRS**; the Leen at Garmouth, 1953 *A. Melderis* **BM**; on dunes in the Culbin Forest and throughout the county, 1954.
N (Stables, *Cat.*); among rocks by the River Findhorn at Ardclach, 1884 *R. Thomson* **NRN**; dry moorland bank at Drynachan and throughout the county, 1954.

I Inverness, 1832 (Watson, *Loc. Cat.*); banks of the Ness at Inverness, 1854 *HC* **INV**; by Loch Dochfour, 1940 *AJ* **ABD**; bank at Ben Alder Lodge and throughout the district, 1954.

Ajuga L.

A. reptans L. *Bugle*

Native. Common in damp woods in grass pastures and on scree and by burns on mountains up to 610 m. Map 217.

M Innes House, Urquhart, 1794 *A. Cooper* **ELN**; very common (Gordon); Dunphail, 1898 *JK* **FRS**; with white flowers in the woods at Darnaway and Dyke (Burgess); Cothall, 1953 *A. Melderis* **BM**; throughout the county, 1954.

N Cawdor woods, 1833 *WAS* **E**; Ardclach, 1891 *R. Thomson* **NRN**; wet bank at Drynachan, 1954 **ABD**; in shade by the River Findhorn at Glenferness, 1961 **K,CGE** and throughout the county.

I Inverness, 1832 (Watson, *Loc. Cat.*); Dulnain woods, Ness Islands and Culloden etc. (Aitken); Rothiemurchus, 1891 *AS* **E**; Strathglass, 1957 *J. Milne* **E**; on scree below the cliffs of Creag Dhubh near Laggan, by the burn in Coire Liath at Gaick and well distributed through the district.

A. pyramidalis L. *Pyramidal Bugle*

Native. Rare in short moorland turf.

I Strath Errick, Inverness-shire (Gordon); six plants on Creag Dhearg, Invermoriston, 1961 *D. A. Ratcliffe*.

Sideritis montana L. occurred occasionally on river shingle at Aberlour and Rothes from a distillery above (Trail, 1904).

PLANTAGINACEAE
Plantago L.

P. major L. *Warba Blades, Greater Plantain*

Native. Abundant in open habitats, by roads and tracks and in cultivated ground.

M Very common (Gordon); Boat of Garten, 1888 ? *coll.* **GL**; Waterford, 1898 *JK* **FRS**; tracks in the Culbin Forest and throughout the county, 1954.

N Cawdor, 1832 *WAS* **ABD**; everywhere in Ardclach, 1884 *R. Thomson* **NRN**; moorland road at Drynachan and throughout the county, 1954.

I Inverness, 1832 and Pitmain, 1833 (Watson, *Loc. Cat.*); Dalwhinnie and Tomatin, 1903 (Moyle Rogers); Lentran, 1940 *AJ* **ABD**; throughout the district, 1954. A hybrid with a cultivated species occurs on the lawn at Ballindarroch, Scaniport, 1960 **E,CGE**.

P. media L. *Hoary Plantain*

Introduced in north Scotland. Rare in lawns.

M Green in front of Russel Place, Forres, 1902 *JK* **FRS**; introduced with lawn seed in the parishes of Dyke, Edinkillie, Forres, Speymouth, Urquhart and Knockando (Burgess); lawn at Milton Brodie, 1960.

N Inland at Nairn, 1900 (Lobban); in the lawn at the High Church at Nairn, 1950.
I Shingle spit at the outflow of the River Tarff into Loch Ness, at Fort Augustus, 1904 *G. T. West*; grassland at Inverness, 1968 *A. Langton*; in the Infirmary lawn at Inverness, 1973 **E**.

P. lanceolata L. *Rib Grass, Carle Doddies, Sodgers,*
Ribwort Plantain
Native. Abundant in grassland throughout the district.
M Innes House, Urquhart, 1874 *A. Cooper* **ELN**; very common (Gordon); common at Forres, 1900 *JK* **FRS**; ½ mile S of Forres station, 1953 *A. Melderis* **BM**; on fixed dunes in the Culbin Forest and throughout the county, 1954; Aviemore, 1964 *J. M. Stirling* **E**.
N (Stables, *Cat.*); Ardclach (Thomson); grass bank at Drynachan and throughout the county, 1954; in short turf at Househill, Nairn, 1962 **ABD**.
I Inverness, 1832 and Dalwhinnie, 1833 (Watson, *Loc. Cat.*); throughout the district reaching 915 m in the Cairngorms, 1954.

Var. **anthroviridis** W. Wats.
M Dunes at Findhorn, 1953 *NDS* **BM** and *A. Melderis* **BM**.

P. maritima L. *Sea Plantain*
Native. Common on salt-marsh verges, in short turf and grassy cliff ledges by the sea, and on moorland tracks and roadside verges inland Map 218.
M Seaside, also in several inland places as at the banks of the River Lossie at Mayne, 1824 (Gordon); Greshop, 1856 *JK* **FRS**; salt-marsh Culbin, 1925 and 1953 *A. Melderis* **BM**; in all the coastal areas, 1954; dry bank by Loch Vaa near Aviemore, 1973.
N (Stables, *Cat.*); frequent in intermediate localities between the seashore and the tops of the higher hills, have been observed in Nairnshire (Gordon); by the Cairn of Achagour and by Loch Belivat (*Trans. NA of Lit. & SS* (1887)); in all the coastal areas and on a track by the Loch of the Clans, 1954.
I Inverness, 1832 and shore of Loch Ericht at Dalwhinnie (Watson, *Loc. Cat*); Strath Affric (Ball, 1851); roadsides in Glen Affric (as *P. serpentina*) (Buchanan White, *J. Bot.* **8**:129 (1870)); Phopachy by Beauly Firth, *c.* 1880 ? *coll.* **INV**; common in tracks in glens Affric and Cannich, 1947; shingle island in the River Glass in Strathglass, roadside verges in Glen Moriston, at Fort Augustus, near Mealfuarvonie, at Guisachan, by Loch Meiklie, on rocks by the boat house at Knockie on Loch Ness; grass bank by the lochan near Nuide E of Newtonmore and in a bog at Essich, etc.

P. coronopus L. *Buck's-horn Plantain*
Native. Common on dry sandy places and in rock crevices by the sea, occasionally inland.
M Frequent at Burghead, Covesea and Lossiemouth and side of the Oakwood near Scroggiemill, 1823 (Gordon); Forres, 1900 *JK*

FRS; fixed dune NE of the Buckie Loch, Culbin and in all the coastal areas, 1954.
N (Stables, *Cat.*); beside the bothy at the end of the golf course at Nairn, 1887 *R. Thomson* **NRN**; fixed dunes on the E of Nairn, 1962 **ABD**.
I Fort George, 1844 *A. Croall* (Watson, *Top. Bot.*); shingle of the River Enrick at Drumnadrochit, 1947; car park at Kessock Ferry, by the salt-marsh at the Carse of Ardersier and track on the raised beach at Ardersier, 1954; edge of the salt-marsh at Castle Stuart, 1976 **E**.

P. afra L.
   P. psyllium L.
Introduced with foreign grain and bird seed.
N From 'Swoop' at Wild Goose Cottage, Nairn, 1969 **E**.
I From distillery sweepings on the rubbish tip at Longman Point, Inverness, 1970 **E,BM,CGE,JEL**.

Littorella Berg.
L. uniflora (L.) Aschers.   *Shore-weed*
Native. Abundant on stoney margins of lochs and pools on acid soil, and often forming an extensive turf in shallow water. Fig 31a, Map 219.
M Very common (Gordon); Blairs Loch near Forres, 1867 *JK* **FRS**; east loch in Culbin Forest, 1923 *D. Patton* **GL**; stoney margin of Lochindorb, 1930; Winter lochs at Binsness, Culbin, 1953 *A. Melderis* **BM**; margins of lochs Puladdern at Aviemore, Millbuies, Loch na Bo, Romach and Hatton near Kinloss, pools by Beum a' Chlaidheimh and in a depression on the shingle ridges at Kingston, 1954.
N (Stables, *Cat.*); Loch Belivat, 1887 *R. Thomson* **NRN**; margins of Loch Flemington, Loch of Boath, Loch of the Clans, Lochloy and Cran Loch, 1854; pool on the Carse of Delnies, 1956.
I Very common (Gordon); Loch Ericht (Balfour, 1868); lochs N of Invermoriston, Loch Ashie, Loch Tarff and many other lochs, 1904 *G. T. West* **E,STA**; near Torvain, 1940 *AJ* **ABD**; in all the shallow edged lochs in the district, 1954; boggy depressions on Suidhe at Kincraig and by Loch Glanaidh above Drumnadrochit, and in a pool at 900 m on Sgùrr na Lapaich in Glen Strathfarrar.

CAMPANULACEAE
Wahlenbergia Schrad. ex Roth
W. hederacea (L.) Rchb.   *Ivy-leaved Bellflower*
Introduced.
M Lawn weed at Logie House near Forres and planted from there to the lawn at Rose Cottage, Dyke, 1965 **E,BM**.
Campanula L.
C. latifolia L.   *Giant Bellflower*
Native. Occasional in woods and by rivers in the lowlands; rare, and perhaps introduced in the highlands.

M Rare (Gordon); escape in the parish of Edinkillie (Burgess); by the Muckle Burn near Barley Mill at Brodie, by the track to Newton of Darnaway, by the River Spey at Blacksboat, Rothes, Fochabers and Garmouth, ditch at Kinloss and near Longmorn, 1954; under alders by the Burnie Path, Dyke, 1968.
N Woods at Cawdor and Holme Rose, 1954; wood at Geddes, 1961 **ABD**.
I By the Grotaig Burn, Glen Urquhart (Aitken); Reelig Glen (J. Don, 1898); by the River Enrick at Drumnadrochit, 1947; in the policies of Achnagairn at Kirkhill and Farr House, by the River Spey at Inverdruie and at Ness Islands, Inverness, 1954; roadside verge at Cluny Castle near Laggan, 1956.

C. rapunculoides L. *Creeping Bellflower*
Introduced. Frequent on railway embankments and waste ground.
M Knockomie near Forres, 1865 *JK* **FRS**; railway line near Elgin (Druce, 1897); Grantown (as *C. rapunculus*) 1915 *D. Patton* **GL**; weed in a garden at Fochabers, 1919 ? *coll.* **GL**; Tarras at Alves, 1941 ? *coll.* **STA**; garden weed at Newton House, Brodie Castle, Dyke, Findhorn and Boat of Garten, 1954; railway embankments at Knockando and Dunphail, 1961.
N Garden weed at Geddes House, 1954.
I Lower Leachkin, 1940 *AJ* **ABD**; roadside at Milton, 1947; garden weed at Farr House and Creagdhubh Lodge near Laggan.

C. lactiflora Bieb.
Introduced.
M Garden outcast in the wood behind the walled garden at Altyre House, Forres, 1966 **E**.
I On the stony verge of a backwater of the River Glass on the small island near Eskadale in Strathglass, 1972 **E**; shingle of the River Farrar below Struy bridge and garden outcast on the moor by the Manse at Tomatin, 1975.

C. persicifolia L. *Peach-leaved Bellflower*
Introduced.
M River bank at Darnaway, 1964 **E** and outside the walled garden on waste ground.

C. glomerata L. *Clustered Bellflower*
Native, but introduced in north Scotland.
M Garden escape at Gordon Castle (Burgess).

C. rotundifolia L. *Witches' Thimbles, Scottish Bluebell,*
Harebell
Native. Abundant on dry banks, moors, dunes and mountain ledges.
M Very common (with white flowers at Romach and Kintrae) (Gordon); Forres, 1900 *JK* **FRS**; wet cliffs between Lossiemouth and Hopeman, 1898 *ESM* **CGE**; dunes in the Culbin Forest and throughout the county, 1954.
N (Stables, *Cat.*); Ardclach, occasionally with white flowers

(Thomson); moors at Drynachan etc., 1954; dry bank at Lethen, 1961 **ABD**.
I Inverness, 1832 Dalwhinnie and Pitmain, 1833 (Watson, *Loc. Cat.*); Kincraig and Rothiemurchus, 1891 *AS* **E**; by a stream in the Coire an Lochain (as var. *latifolia*) *R. M. Adam* (*Trans. BSE* **32**: 246 (1935)); Nairnside, 1940 *AJ* **ABD**; cliff ledge at 830 m, on Sgùrr na Lapaich in Glen Strathfarrar, 1972 and throughout the district.

C. rapunculus L.   *Rampion Bellflower*
Introduced. Garden escape.
M (A. Bennett, *Scot. Nat.* 1888); escape at Knockando (Burgess); Elgin, 1974 *R. Richter* **E**.

Jasione L.
J. montana L.   *Sheep's-bit*
Native. Rare on sandy moorland banks. Fig 31b.
M Raised beach between Lossiemouth and Garmouth 1835 *G. Gordon* **E,ABD** and *WAS* **ABD**; also collected there in 1897 *JK* **FRS**; 1898 *ESM* **E**; 1924 ? *coll.* **GL** and 1953 *A. Melderis* **BM**. Owing to excavations made into the shingle banks by the nearby brick works it is now extremely rare and only five plants were seen in 1975; a few plants on the golf course dunes at Lossiemouth, 1966 *Alan Souter Jnr.*

LOBELIACEAE
Lobelia L.
L. dortmanna L.   *Water Lobelia*
Native. Common in shallow acid water on loch margins. Map 220.
M Abundant in one or two places, Lochindorb and Loch Puladdern, 1825 (Gordon); small lochan at Boat of Garten and in Loch Vaa N of Aviemore, 1954.
N Lochloy and Cran Loch, *Brodie* (Gordon); Lochloy, 1857 *JK* **FRS**; Ord Loch, Ardclach, 1900 (Lobban); in lochs Flemington and Belivat, 1954.
I Strath Affric (Ball, 1851); lochs Garten, Mallachie, Pityoulish, Eilein and Gynack, (Druce, 1888); Coiltry Loch, 1904 *G. T. West* **E**; Loch Alvie, 1904 *G. B. Neilson* **GL**; very common in Strathspey, Strathglass and the smaller glens throughout the district, 1954.

Pratia Gaudich
P. angulata Hook. fil.
M Planted on the lawn at Rose Cottage, Dyke from a garden in Edinburgh, 1966 **E**.

RUBIACEAE
Sherardia L.
S. arvensis L.   *Field Madder*
Native. Formerly frequent in cornfields, now only seen occasionally. Fig 31c.
M Frequent in cultivated ground, 1824 (Gordon); Forres, 1898 *JK* **FRS**; near Elgin (as var. *hirsuta* Baguet) (Druce, 1897); stubble

field at Kellas, waste ground at Elgin and at Lossiemouth, 1954;
arable field and garden at Kinloss, 1973 *R. M. Suddaby*; in newly-
sown grass by the Health Centre at Forres, 1975 *K. Christie*.
N Cawdor, 1832 *WAS* **ABD**; Ardclach, 1884 *R. Thomson* **NRN**;
stubble field at Holme Rose, 1954.
I Southside road, Inverness, ? date, ? *coll.* **INV**; fields about
Inverness and at Englishton (J. Don, 1898); Fort Augustus, 1904
G. T. West **STA**; Inverness, 1940 *AJ* **ABD**; disturbed ground by the
dam by Loch Mullardoch in Glen Cannich, 1962 *U.K. Duncan et al.*

Asperula L.
A. arvensis L.
Introduced with foreign grain and bird seed.
M Frequent (Trail, 1904).
N From 'Swoop' fed to birds, Nairn, 1968 **E,ABD**.

Cruciata Mill.
C. laevipes Opiz   *Crosswort*
 Galium cruciata (L.) Scop.
Introduced in north Scotland. Rare on railway embankments and
occasionally on waste ground.
M Findhorn near Forres, *Miss Robertson* and at Relugas, *WAS*
(Murray); waste ground by the Lochloy road near Brodie, 1950
(gone with road widening); waste ground at Rothes and on a grass
verge by Carron station, 1955; single plant by the Glenbeg road at
Grantown, 1976.
I Waysides at Kilmorack and Inshes (J. Don, 1898); lane near the
railway line at Bruichnain at Bunchrew, 1954 **E,BM**; railway
embankment by Dunachton near Kincraig, 1972 *D. Hayes*.

Crucianella L.
C. angustifolia L.
Introduced with bird seed.
N From 'Swoop' fed to birds, 1969 **E**.

Galium L.
G. odoratum (L.) Scop.   *Woodruff*
 Asperula odorata L.
Native. Frequent in deciduous woods and in ravines and among
rocks on base-rich soil. Very fragrant. Map 221.
M Frequent in the Oakwood near Scroggiemill, on the banks of the
Divie and at Dunphail (Gordon); Alves, *G. Wilson* **ABD**; Dunphail,
1898 *JK* **FRS**; St John's Mead at Darnaway, 1953 *A. Melderis*
BM; common in shady places by the River Spey, 1954; in a small
wood at Burgie House, 1967 **CGE**; among rocks by the Allt
Iomadaidh near Dorback, 1974.
N (Stables, *Cat.*); Cawdor woods (Gordon); Ardclach by the
Findhorn 1886 *R. Thomson* **NRN**; among rocks in wet places by the
River Findhorn at Coulmony, Dulsie Bridge and Glenferness, 1954.
I (Watson, *Top. Bot.*); Abriachan, Holm burn and Bunchrew
woods (J. Don, 1898); Glendoe, 1904 *G. T. West* **STA**; Bunchrew,

Fig 31   a, **Littorella uniflora**; b, **Jasione montana**; c, **Sherardia arvensis**; d, **Galium saxatile**; e, **G. boreale**; f, **Adoxa moschatellina**.

1940 *AJ* **ABD**; wood at Achnagairn, Kirkhill, 1950 and in Reelig Glen; by Loch Ness at Fort Augustus, at Inverfarigaig, Invermoriston, in a small gorge in Strathglass etc., 1954; cliffs of Loch Killin; among boulders below the cliffs of Creag Dhubh near Laggan, 1972 **E**; in a small wood by the River Enrick at Milton and among rocks and boulders in the Ryvoan Pass etc.

G. boreale L. *Northern Bedstraw*

Native. Common on rocks and banks by rivers, on scree and grassy mountain slopes and often on river shingle. Leaves 4 in a whorl; flowers white; fruit rough, with hooked bristles. Fig 31e. Map 222.

M Banks of the Spey at Grantown, 1837 *WAS* **ABD**; Elginshire, 1837 *G. Gordon* **E**; Aviemore, 1891 *AS* **E,GL** and 1902 *JK* **FRS**; golf course at Garmouth, 1953 *A. Melderis* **BM**; common on the banks of the rivers Spey and Findhorn, 1954; single patch on moorland verge N of Duthil; roadside bank at Grantown 1968 **BM,CGE**; very small and stunted plants on the dry slopes of Beum a' Chlaidheimh on Dava Moor, 1973.

N (Stables, *Cat.*); by Lord Cawdor's bridge at Banchor, the only habitat now known in Nairnshire, 1885 *R. Thomson* **NRN**; beech wood at Cawdor Castle, banks of the River Findhorn at Coulmony, Dulsie Bridge and Glenferness, 1954.

I Dalwhinnie, 1833 (Watson, *Loc. Cat.*); near Foyers, 1837 *WAS* **E**; Ness Islands, 1854 *HC* **INV** and in 1858 *JK* **FRS**; shores of Loch Ness, 1904 *G. T. West* **E,STA**; Tromie Bridge, 1904 *G. B. Neilson* **GL**; Glen More (a large branching form with acuminate leaves) (Druce, 1888) (this form is also to be found on the railway embankment 2 miles N of Kingussie); bank of a small burn at Gaick, cliffs above Loch Killin, summit of Ben Alder and throughout the district, 1954.

G. mollugo L. *Hedge Bedstraw*
Subsp. **mollugo**

Doubtfully native in north Scotland. Panicle with spreading branches, leaves 6–8 in a whorl obovate-oblanceolate, cuspidate; flowers 3 mm diam., fruit 1 mm.

M Rare by the side of a dyke leading from the cornyard to the turnpike road at Ardgay, 1826, at Pittendreich and Pluscarden, *J. Fraser* (Gordon); Culbin Sands, 1911 *P. Ewing* **GL**; roadside bank S of Hillhead Farm near Forres, 1974 **E,BM**; roadside by the A95 N of Boat of Garten, 1975 **E**.

N Near Boath and Kinsteary at Auldearn, *J. B. Brichan* (Watson); by the path near Delnies school, 1960.

I Belladrum, 1833 *WAS* (Gordon); railway embankment at Allanfearn, 1974 **E,BM**; grass bank near Aldourie Castle, 1975.

G. album Miller subsp. **album**
G. mollugo subsp. *erectum* Syme

Very similar to subsp. *mollugo* but the panicle has ascending branches, which remain so in fruit, the leaves are narrower, oblanceolate, mucronate, and the flowers and fruit larger.

M Dalvey (as G. *mollugo*) *J. G. Innes* (Gordon) and in 1954 *MMcW*
CGE; grass bank by Lochinvar near Elgin, 1960 **E,ABD**; bank by
the Aviemore Centre, 1961 **E,BM**; lane to Darklass Farm near
Dyke, at Blacksboat, Grantown, Rothes and Garmouth; waste
ground by the bridge over the River Dulnain at Balnaan, grass
verge at Sanquhar pond and grass verge at Paddockhaugh; roadside
near Burnside S of Elgin, 1975 **BM**; roadside near Birnie, 1975 **BM**.
N Roadside bank near Geddes House on the road to Righoul,
1960 **ABD,CGE**; roadside near Cothill Farm E of Nairn, 1960.
I Near Dunain, 1940 *AJ* **ABD**; railway embankment at Bunchrew
and by the Kirkhill Kirk, 1954; grass bank by the Moniack Burn at
Achnagairn, 1961 **K,CGE**; by the track to Clunes at Kirkhill,
1961 **BM**; hedgerow at Auchgourish, 1973 **E,BM**.

G. mollugo × verum = G. × pomeranicum Retz.
M Roadside verge near Grantown, 1950; bank by the Aviemore
Centre (with both parents), 1961 **E,BM,CGE**; roadside at Paddock-
haugh, 1973 **E**.
N Roadside bank (with both parents) near Geddes House on the
road to Righoul, 1960 **E,ABD,CGE**; W of Cothill Farm, 1970 **E**.
I Foyers, 1942 *UKD* **E**; by the Moniack Burn at Kirkhill, 1954.

G. verum L.   *Lady's Bedstraw*
Native. Abundant on dry banks, wall tops, dunes and on moorland
tracks. Flowers yellow.
M Very common (Gordon); Knockomie near Forres, 1871 *JK*
FRS; at Findhorn and on the W bank of the river, on shingle, 1953
A. Melderis **BM**; bank at Aviemore, 1961 **E**; dunes in the Culbin
Forest and throughout the county, 1954.
N (Stables, *Cat.*); common in Ardclach, 1883 *R. Thomson* **NRN**;
moorland track at Drynachan and throughout the county, 1954;
roadside near Geddes, 1960 **ABD**.
I Inverness, 1832, Dalwhinnie and Pitmain, 1833; Tomnahurich,
Inverness, 1854 *HC* **INV**; Nairnside, 1940 *AJ* **ABD**; moorland track
in upper Glen Feshie, in a gully in the Ryvoan Pass and throughout
the district.

G. saxatile L.   *Heath Bedstraw*
    G. harcynicum Weigel; *G. hercynicum* auct.
Native. Abundant on moors and grassland on acid soil. Flowers
white. Fig 31d.
M Very common (Gordon); Aviemore, 1892 *AS* **E**; Forres, 1900
JK **FRS**; dunes in the Culbin Forest and throughout the county,
1954.
N (Stables, *Cat.*); moors at Ardclach, 1884 *R. Thomson* **NRN**;
moors at Drynachan and throughout the county, 1954; birch wood
at Geddes, 1961 **ABD,K**; grass bank near Auldearn, 1962 **CGE**.
I Dalwhinnie, 1833 (Watson, *Loc. Cat.*); Tomatin, 1903 (Moyle
Rogers); Feshie Bridge, 1904 *G. B. Neilson* **GL**; Coire an Lochain of
Cairn Gorm, 1953 *A. Melderis* **BM**; throughout the district 1954.

G. sterneri Ehrend. *Limestone Bedstraw*
Native. Frequent on base-rich grassy slopes and scree, often becoming established on river shingle. Somewhat resembles a very slender *G. saxatile* but the leaves are linear with backward-directed prickles on the margins, the flowers are cream-coloured and the whole plant of a yellowish-green colour.
M Shingle of the River Spey at Blacksboat, 1954 **CGE**; at Fochabers 1973 **BM**; and at Aviemore and Craigellachie, 1973 **E**; grass slope on the N face of Creag nan Gabhar at Aviemore, 1974 **E**.
I Larig Ghru, *Groves* **BM**; shingle of spate bed of the River Feshie and on rocks in gullies (as *G. pumilum* Murr.), 1952 *R. Mackechnie &
E. C. Wallace* **CGE**; slopes of Mealfuarvonie, 1955 *A. A. Slack* and 1971 *MMcCW* **E**; grassy slope in Glen Markie, 1972 **E**; base-rich scree in the burn gully at Gaick Lodge, 1973 *R. McBeath & MMcCW* **E**.

G. palustre L. subsp. **palustre** *Common Marsh Bedstraw*
Native. Common in ditches, by burns and in grassy bogs, and among stones on loch margins. Map 223.
M Very common (Gordon); Waterford near Forres, 1898 *JK* **FRS**; Loch Spynie (Druce, 1897); grassy bog by the Buckie Loch, Culbin 1968 **E**; trailing on stones at the margin of Lochindorb and throughout the county, 1954.
N (Stables, *Cat.*); common at Ardclach, 1887 *R. Thomson* **NRN**; Nairn (as var. *lanceolatum* Presl) 1911 *P. Ewing* **GL**; margin of Geddes Reservoir, 1960 **ABD**; bog on the raised beach at Maviston, 1968 **E**, and throughout the county.
I Pitmain, 1833 (Watson, *Loc. Cat.*); Bught and Culloden (Aitken); Loch Insh, 1891 *AS* **E**; Fort Augustus, 1904 *G. T. West* **E,STA**; throughout the district, 1954.

G. uliginosum L. *Fen Bedstraw*
Native. Rare in grassy, tussocky moorland bogs. Resembles small types of *G. palustre*, but is lighter green in colour, has ascending more rigid stems and mucronate leaves 6–8 in a whorl, the margins with backwardly directed prickles making the plant very rough to the touch.
M Abundant in one or two places (Gordon); parishes of Dyke, Speymouth and Urquhart, apt to be overlooked (Burgess); flush by a burn on the Haughs of Cromdale, 1971 **E**; grassy bog, on tussocks at Drakemyres, 1971 *A. J. Souter* **E**; flushes at Fae near Dorback, shingle at Bridge of Brown and bog on the moor by Loch an t-Sithein, 1973; grassy flushes on the hill behind Easter Limekilns on Dava Moor, 1973 **E**.
N Flush on the raised beach at Maviston, 1968 **E**.
I Wet places on the Carse, at Muckovie and Culloden (J. Don, 1898); bog on Suide hill, Kincraig, 1956; grassy bog in the Ryvoan Pass, 1971; by a small burn above Alvie House, 1974 **E**.

G. tricornutum Dandy *Corn Cleavers*
Introduced with foreign grain.

M Plentiful along the Spey (Trail, 1904).
N From 'Swoop' at Nairn, 1968 **E**.

G. aparine L.   *Bleedy Tongues, Sticky Willie, Goosegrass*
Native. Abundant in waste places, hedgerows and among shingle by
the sea.
M Very common (Gordon); Forres, 1898 *JK* **FRS**: Dava, 1954
BM; throughout the county, 1954; very fine-leaved dwarf plant
probably introduced with grain for the nearby distillery occur on
cinders at Knockando station, 1966 **E**.
N (Stables, *Cat.*); common at Ardclach, 1886 *R. Thomson* **NRN**;
cultivated ground at Drynachan and throughout the county, 1954;
garden weed at Nairn, 1969 **E**.
I Inverness, 1833 (Watson, *Loc. Cat.*) and 1940 *AJ* **ABD**; among
boulders below the cliffs of Creag Dhubh near Laggan, 1972;
throughout the lowlands, 1954.

CAPRIFOLIACEAE
Sambucus L.
S. ebulus L.   *Danewort, Dwarf Elder*
Doubtfully native.
I Rare at Ruthven Barracks at Kingussie and at Culloden, 1830
(Gordon); bank in a field behind Kincardine kirkyard, 1875 *JK* **FRS**
and 1893 *AS* **E**; planted at Ruthven Barracks, 1916 *ESM* **E** (it is
still to be found there); plentiful in a pine wood N of Coylumbridge,
1962; single plant by the River Calder at Newtonmore and a single
plant on the verge of a pine wood near Feshie Bridge.

S. nigra L.   *Bourtree, Elder*
Introduced in north Scotland. Common in bushy places and open
woods.
M Certainly introduced, frequent (Gordon); a few scattered trees
in the Culbin Forest and throughout the county, 1954.
N (Stables, *Cat*); Ardclach (Thomson); sandy coast 2 miles E of
Nairn, 1898 *ESM* **CGE**; by the River Nairn at Holme Rose, waste
land at Cantraydoune, Cawdor and Lethen etc., 1954.
I Islands at Inverness (Aitken); common in the lowlands, and
scattered over the highland areas, (often planted near old crofts and
cottages as a protection from witches), becoming bird-sown on river
banks etc.

   Var. **laciniata** L.   *Cut-leaved Elder*
A variant with deeply dissected leaves.
M Parishes of Alves, Lhanbryde and Speymouth (Gordon);
appeared in the garden of Greyfriars House, Elgin, 1972 *G. Yool*!.
I Many trees by the railway bridge at the approach to Wester
Lovat Farm, Kirkhill, 1956; several trees at a lay-by near Aigas in
Strathglass, 1972 **E**.

   Var. **viridis** Weston   *Green-berried Elder*
A variant with green fruits.

I Roadside by the Moray Firth E of Inverness, 1972 *A. Higginbottom*; many trees on the hillside above Drumnadrochit, 1976 *R. Melville & MMcCW* **E,ABD**; one or two trees by the railway line just E of the level crossing at the road junction of the A9–A96 at Inverness.

S. racemosa L. *Red-berried Elder*
Introduced. Locally abundant in open woods and on banks. Flowers greenish; fruit red.
M Abundant on a roadside bank N of Rothes Glen Hotel, 1930; Boat of Garten, Longmorn, and by the River Spey at Blacksboat and Fochabers, 1954; pine wood at Dallas, 1963 **CGE**; edge of a wood at Ardgye near Elgin, 1974.
N Drynachan, 1955; roadside near Barevan at Cawdor, 1966 **ABD**.
I Waste ground at Beauly, 1947; by the A82 at the road junction to Blackfold by Loch Ness, 1950; waste ground at Nairnside, in Reelig Glen, roadsides at Kincraig, Ruthven Barracks, Strathdearn, Lochend and between Kirkhill and Beauly; among boulders below the base-rich cliff above Inverfarigaig, by the burn at Midlairgs and by Culloden and Flichity Houses, etc.

S. sieboldiana Blaume ex Schwerin occurs in a shrubbery at Culduthel, Inverness and was doubtless planted.

Viburnum L.

V. tinus L.
Garden escape on the Kinloss road near Sueno's Stone, Forres. Well-known shrub by the name of *Laurustinus*.

V. lantana L. *Wayfaring Tree*
Planted as hedges. Fruit at first red turning to black.
M Policies of Darnaway Castle, 1950; hedge at Balnaferry, 1968 *R. M. Suddaby*.
I Planted in a hedge behind Bogroy Inn near Kirkhill, 1970 *M. Cameron*!.

V. opulus L. *Guelder Rose*
Native. Occasional in woods and among rocks on river banks.
M Frequent at Darnaway and Birdsyards woods, *J. G. Innes* (Gordon); Sanquhar House, 1857 *JK* **FRS**; in a mixed wood at Whitemire, felled wood at Binsness in the Culbin Forest, among rocks by a small burn in the Hills of Cromdale and near Elgin, 1954; roadside S of Altyre near Forres, 1966; near the Golden Gates at Darnaway Castle, 1975.
N Cawdor, *WAS* (Gordon); near Boath House, Auldearn, *J. B. Brichan* (Gordon, *annot.*); banks of the River Nairn (Aitken)!; near Cran Loch, 1960 **ABD**.
I Banks of the Ness and other streams near Inverness, *G. Anderson* (Hooker, *Fl. Scot.*); Inverness, 1832 (Watson, *Loc. Cat.*); Badenoch (Gordon); Kingussie, 1840 *A. Rutherford* **E**; among rocks by the river in Glen Moriston, 1971 **E**; cliff by the road at Eskadale, wood

at Inchnacardoch, by the River Enrick at Milton, wood by Loch
Ness, Fort Augustus, in several places in Strathglass and banks of the
river near the Ferry at Beauly.

Symphoricarpos Duham.

S. albus (L.) S. F. Blake var. **laevigatus** (Fernald) S. F. Blake
   S. rivularis Suksdorf   *Snowberry*
Introduced. Planted by cottages etc. and forming large patches by
roadsides and in shrubberies. Well-known for its white globose
fruits.

M Policies of Brodie Castle and Moy House etc., in a pinewood
near the town at Grantown, by the curling pond at Carrbridge, by
the river at Nethybridge and in a hedge near Milltown airfield etc.,
Aviemore, 1970 *P. H. Davis* **E**.

N Wood at Kinsteary, Auldearn by the River Nairn at Nairn and
on the bank at Lethen Mill, etc.

I Near Beauly, 1947; at Tomatin, Alvie and Kincraig,
Kilmorack, Cluny Castle near Laggan, at Dores, Inverdruie and
Daviot etc.

Linnaea L.

L. borealis L.   *Linnaea, Twin-flower*
Native. Frequent in pine woods but often overlooked unless in
flower for it trails through the *Vaccinium* and *Calluna* and the small
roundish leaves become obscured. Pl 15.

M Knock of Alves, 1829 *G. Gordon* **E,GL,BM**, and in 1880 *JK*
FRS!; Gordon Castle woods at Fochabers, 1830 ? *coll.* **BM**!; wood
between Lhanbryde and Urquhart, 1841 *Mr Martin* (Gordon,
annot.); pine wood at Castle Grant, Grantown, 1895 *C. Bailey* **OXF**;
by Lochindorb Lodge, 1934 *R. M. Adam* **E**!; planted near The Dell
at Nethybridge, 1948; pine wood between the town and the River
Spey at Grantown and in several other places nearby, 1954; wood
at Darnaway, 1968 *R. M. Suddaby*.

N Cawdor woods, 1830 *WAS* pine wood at Kilravock Castle, 1846
A. Croall; Dulsie wood at Daltra, 1890 *R. Thomson & Mr Moir of
Bombay* **NRN**.

I Below Drummond House near Inverness (Gordon); Glenmore
burn to Snow corrie, Cairn Gorm, 1877 (Boyd); Loch en Eilein,
1884 *JK* **FRS**; Rothiemurchus, 1892 *AS* **E**; Loch a Eilein 1892 *AS*
BM; hill above Inshriach Nursery, 1973 *D. Tennant*; Affric
Forest in Glen Affric, 1976 *G. Mackie*.

Lonicera L.

L. xylosteum L.   *Fly Honeysuckle*
Introduced in north Scotland. Flowers in pairs.
I Planted in the policies of Achnagairn House at Kirkhill, 1963
BM,CGE.

L. periclymenum L.   *Honeysuckle*
Native. Common in woods and among boulders and rocks. Map 224.
M Innes House, Urquhart, 1794 *A. Cooper* **ELN**; frequent (Gordon);

Sanquhar, 1856 *JK* **FRS**; Culbin Forest and throughout the county, 1954.
N (Stables, *Cat.*); common at Ardclach (Thomson); rare in a gully on the moors at Drynachan but common elsewhere, 1954.
I Inverness, 1832 (Watson, *Loc. Cat*); Ness Islands, 1854 *HC* **INV**; Bunchrew, 1940 *AJ* **ABD**; cliffs in Glen Feshie, among rocks in the Ryvoan Pass, boulders at Ben Alder Lodge and below the cliffs at Creag Dhubh near Laggan and throughout the district, 1954.

Leycesteria Wall.
L. formosa Wall.
Garden escape. Policies of Brodie Castle, 1930; escape near Fochabers, 1974 *A. H. A. Scott.*

ADOXACEAE
Adoxa L.
A. moschatellina L.   *Moschatel, Townhall Clock*
Native. Frequent in deciduous woods in the lowlands, rare in the highlands. A fragile plant with ternate leaves and about 5 green flowers in a terminal head smelling strongly of musk. Flowers in March–April. Fig 31f. Map 225.
M Newmill of Alves, *G. Wilson*, Birdsyards at Forres, *Miss Robertson*, Altyre and Dunphail, 1824 (Gordon); Sanquhar, 1880 *JK* **FRS**; wood by Green Gates, Brodie, 1925; banks of the River Spey at Pitcroy, 1967 **E**; under beech trees at the edge of the playing field at Fochabers, 1974 **BM**; by the burn at Boat o' Brig and in woods at Relugas, Knockando and Dava etc.
N Lethen, 1839 *WAS* **E**; in an alder wood by Howford Bridge, 1963 **ABD**; under hazel by the river at Nairn.
I By the Holm Burn near Inverness (Aitken); woods at Achnagairn, Kirkhill and under hazel by the railway bridge at Tomatin, 1954; steep bank on the hill above Milton, 1972 **E**; under oak trees near Divach and on top of a bank by the railway at Clunes, Kirkhill, 1972 **E**; in shade by the Bunchrew Burn, at Inverfarigaig, bank at Kyllachy in Strathdearn and by the River Beauly below Lovat Bridge.

VALERIANACEAE
Valerianella Mill.
V. locusta (L.) Betcke   *Lamb's Lettuce, Corn Salad*
Native. Rare on dunes and wall tops.
M Parishes of Bellie and Birnie (Burgess); dunes by the Findhorn estuary near Kinloss, 1920; garden weed at Darnaway and Dyke, 1970 **E**; top of the kirk wall at Dyke, 1974.
N Cawdor, *WAS* (Gordon).
I (Druce, *Com. Fl.*), possibly the Cawdor record.

V. eriocarpa Desv.   *Hairy-fruited Corn Salad*
Introduced. Garden weed at Greshop House, Forres, 1953.

V. dentata (L.) Poll.   *Narrow-fruited Corn Salad*
Introduced in north Scotland. Cornfields.
M Coltfield at Alves, 1834 *G. Wilson* **E**; cornfields at Innes and
Drainie (Gordon).

Valeriana L.

V. officinalis L.   *Valerian*
Native. Abundant in meadows, by rivers and burns and margins of
woods in wet places, also on dry sand dunes. Map 226.
M Very common (Gordon); Greshop, 1898 *JK* **FRS**; beside
Byers Burn at Fochabers, 1923 ? *coll.* **GL**; on dry dunes on the Bar
and in bogs by the Buckie Loch in the Culbin Forest, and throughout
the county, 1954; Dunphail, 1960 ? *coll.* **E**.
N (Stables, *Cat.*); by the River Findhorn at Ardclach, 1888
R. Thomson **NRN**; abundant by Lochloy and Cran Loch and by the
river at Nairn etc., 1954; dry dunes on the Old Bar in Culbin, 1968
E,CGE; bog by the Minister's loch on the Nairn Dunbar golf
course, 1960 **ABD**.
I Inverness, 1832 Dalwhinnie and Pitmain, 1833 (Watson, *Loc.
Cat.*); Ness Islands (Aitken); Kilmorack, 1940 *AJ* **ABD**; among
rocks and boulders in the Ryvoan Pass and below cliffs at Creag
Dhubh near Laggan and in wet places throughout the district, 1954.
An old record (as var. *montana*) 1793 (some doubt, Druce, 1888).

V. pyrenaica L.   *Pyrenean Valerian*
Introduced. Garden escape in ditches and by burns.
M Altyre Burn probably escaped from the garden, *J. G. Innes*
(Gordon, *annot.*)!; Altyre, 1857 *JK* **FRS**; abundant by Mosset
Burn above Sanquhar pond, Forres, 1963 **K**; ditch by the drive to
Darnaway near Berryley, 1962 **CGE**.
N By the Muckle Burn at Ardclach, 1900 (Lobban).
I Waste ground at Milton 1947; garden escape by the burn at
Boleskine House by Loch Ness, 1971 **E**.

DIPSACACEAE
Dipsacus L.

D. fullonum L.
Subsp. **fullonum**   *Wild Teasel*
Introduced in north Scotland. Receptacular bracts ending in a long
straight spine.
M Certainly introduced, rare near Glassgreen, 1835 (Gordon); by
the burn at Fochabers, 1961 *A. J. Souter* **E**.
I Canal banks opposite Holm near Inverness (Aitken); waste
ground Longman Point, 1970.

Subsp. **sativus** (L.) Thell.   *Fuller's Teasel*
This is the teasel used for raising the nap on various kinds of cloth.
Receptacular bracts ending in a stiff recurved spine.
I Two plants on the tip at Longman Point, Inverness, 1970 **E**.

Knautia L.

K. arvensis (L.) Coult. *Field Scabious*
Doubtfully native in north Scotland. Rare on railway embankments, yards and roadside banks.
M Rare, doubtfully native, Clashland at Urquhart, 1828, Bow Bridge, *J. Lawson* (Gordon); near the schoolhouse at Duffus ? *coll.*
ELN; Greshop by the railway, 1898 *JK* **FRS** and 1966 *MMcCW*
E,CGE; banks at Lesmurdie near Elgin at Grantown and Blacksboat, 1954; roadside bank ½ mile E of Hopeman, 1972 **E**.
N Lethen, *J. B. Brichan* (Gordon); abundant on the railway embankment by the bridge over the A96 just E of Nairn, 1975 **E**.
I Inverness, 1832 (Watson, *Loc. Cat.*); by the burn E of Parks of Inshes near Inverness (Aitken); Englishton (J. Don, 1898); among cinders in the station yard at Inverness, 1930 (still there); east end of Loch Meiklie, 1958 *E. W. Groves* **EWG**.

Succisa Haller

S. pratensis Moench *Devil's-bit Scabious*
Native. Abundant on wet and dry moors, in meadows, pastures and damp woods. Flowers usually purple, occasionally pink or white.
M Very common (Gordon); Cothall (with white flowers), 1843 *J. G. Innes* (Gordon, *annot.*); Aviemore, 1893 *AS* **E**; Fochabers, 1901 ? *coll.* **ABD**; Culbin Forest and throughout the county, 1954.
N Cawdor woods, 1832 *WAS* **ABD** and 1855 *HC* **INV**; Ardclach, 1891 *R. Thomson* **NRN**; moors at Drynachan and throughout the county, 1954.
I Dalwhinnie and Pitmain, 1833 (Watson, *Loc. Cat.*); Falls of Foyers, *W. Brand* **E**; Englishton Muir and Culcabock etc. (Aitken); near Torvaine, 1940 *AJ* **ABD**; throughout the district, 1954.

Cephalaria elata (Hornemann) Schrader occurs as a garden relic in waste ground at Cromdale, Moray.

COMPOSITAE
Helianthus L.

H. tuberosus L. *Jerusalem Artichoke*
Introduced.
M Shingle of the River Spey at Rothes (Trail, 1904).

H. annuus L. *Sunflower*
Introduced with foreign grain and bird seed. Frequent on rubbish tips at Forres and Elgin and Inverness.

Bidens L.

B. cernua L. *Nodding Bur-marigold*
Native. Margins of ponds and by burns. Not seen in recent years.
M Near Elgin, *G. Gordon* **ABD**; abundant in one or two places, pond near the Manse at Urquhart, the Leen at Garmouth and at Pittendreich (Gordon); near Forres, *J. G. Innes* **FRS**; Urquhart, ? *coll.* **ELN**.
I Bogroy (J. Don, 1898).

B. tripartita L.   *Trifid Bur-marigold*
Native. Margins of ponds and by burns. Not seen in recent years.
M Elgin, 1829 (Watson, *Top. Bot.*); beside the watercourse of the
dam of Mill of Grange, 1868 *JK* **FRS**; Kinloss, *A. MacGregor* and
Rothes by the Spey (Burgess).

Cosmos Cav.
C. bipinnatus Cav.
Introduced with foreign grain.
M Tips at Elgin and Cloddach, 1954.
I Longman tip, 1971 **E**.

Galinsoga Ruiz & Pav.
G. parviflora Cav.   *Gallant Soldier*
Introduced. Rare in cultivated ground.
M Garden weed, Elgin, introduced with plants transferred from
Hythe, Kent and apparently well established, 1972 *A. H. A. Scott* **E**;
garden weed at Kinloss, 1976 *E. M. Legge*!.

Guizotia Cass.
G. abyssinica (L. fil.) Cass.
Introduced with foreign grain. Frequent on rubbish tips but seldom
flowers in north Scotland.
M Forres and Elgin rubbish tips, appears yearly, 1950.
I The Longman tip, flowering, 1970 **E,ABD,CGE**.

Senecio L.
S. jacobaea L.   *Stinking Willie, Ragwort*
Supposedly introduced to this area and named by the Highlanders
after the battle of Culloden as *Stinking Willie* in contempt of William
Duke of Cumberland who was supposed to have introduced the
seed among his English oats (Thomson). An abundant weed of
grassland, dunes and waste land.
M Very common (Gordon); Forres, 1856 *JK* **FRS**; shingle on the W
bank of the River Findhorn near Forres, 1953 *A. Melderis* **BM**;
throughout the county, 1954; dunes in the Culbin Forest, 1968 **E**.
N (Stables, *Cat.*); Ardclach (Thomson); moorland track at
Drynachan and throughout the county 1954; pastures on the margin
of Loch Flemington, 1975 **E,BM**.
I Inverness, 1832 Pitmain, 1833 (Watson, *Loc. Cat.*); Bunchrew
Burn, ? date ? *coll.* **INV**; a form with very narrow ligules occurs at
Beauly (Druce, 1890); abundant in the lowlands, 1954; rare on a
track in upper Glen Feshie, waste ground at Ben Alder Lodge and
below a moorland cliff, where animals had sheltered on the moor by
Loch Monar.

S. aquaticus Hill   *Marsh Ragwort*
Native. Common in grassy bogs, by burns and in ditches in the
lowlands, less so in the highlands. Map 227.
M Frequent (Gordon); plentiful in the marshes at Loch Spynie,
Lochinvar and Sanquhar pond at Forres, and in backwaters of the

River Spey at Boat of Garten, Rothes and Fochabers etc.; bog by
the Black Burn at Pluscarden, 1968 **CGE**; margin of Lochindorb
Castle, 1974.
N (Stables, *Cat.*); Nairn, 1857 *JK* **FRS**; by the Loch of the Clans
etc. (Aitken); bogs by Geddes Reservoir, Culbin Forest by Cran
Loch and Lochloy and at Dulsie Bridge etc., 1954.
I (Watson, *Top. Bot.*); Carse and Inshes (J. Don, 1898); Cherry
Island in Loch Ness, by Loch Uanagan and at Coiltry Loch, 1904
G. T. West **E**; common by the River Beauly and throughout
Strathglass but scarce in the upper regions of Strathspey; margin of
Loch Pityoulish, 1954; salt-marsh at Castle Stuart and bogs about
Loch Meiklie etc.

S. aquaticus × jacobaea = S. × ostenfeldii Druce
M Among the ruins of Lochindorb Castle, 1974 **E**.
I Meadow by Corrimony school in Glen Urquhart, 1975 **E,BM**.

S. squalidus L. *Oxford Ragwort*
Introduced. Railway sidings and distillery yards.
M Station yard at Forres, 1956 **E,BM**; sidings by the Knockando
Distillery, 1966 **E,ABD,BM**.

S. sylvaticus L. *Heath Groundsel*
Native. Common in open places on sandy acid soil, on dry banks on
the margins of pine woods and on dunes and river shingle. Map
228.
M Very common (Gordon); Clunyhill at Forres, 1856 *JK* **FRS**;
Boat of Garten, 1888 ? *coll.* **GL**; Culbin, 1923 *D. Patton* **GL**!;
Newmill at Elgin, 1959 *E. S. Harrison* **E**; Grantown (as var.
auriculatus Meyer) 1872 ? *coll.* **E**; on shingle by the River Findhorn
near Forres, 1953 *A. Melderis* **BM**; throughout the county, 1954;
under pines near Binsness, Culbin Forest, 1968 **E**.
N (Stables, *Cat.*); Ardclach (Thomson); Nairn, 1899 *G. B. Neilson*
GL; dry moorland bank at Drynachan, under pines at Fornighty
Ford and gravel bank at Dulsie Bridge etc., 1954.
I Inverness, 1832 Pitmain, 1833 (Watson, *Loc. Cat.*); Englishton
Burn (Aitken); Kincraig, 1891 *AS* **E**; Kingussie, 1893 *J. Buchanan*
GL; Torvaine, 1940 *AJ* **ABD**; dry banks at Inverness, Pityoulish,
Fort Augustus etc., 1954; river shingle at Newtonmore, Tomatin
and Kincraig etc.

S. viscosus L. *Sticky Groundsel*
Introduced in north Scotland. Waste land and river shingle and now
frequent among cinders on railway embankments and station yards.
Map 229.
M Among cinders on the railway line at Grantown, 1953; station
yards at Rothes, Blacksboat and Boat of Garten, 1954; siding at
Mosstowie, 1960 **ABD,K**; gravel heap by the road at Dyke, 1963
CGE; shingle of the River Spey and Forres station, 1964 **CGE**;
station yard at Knockando, 1966 **E**; shingle of the Spey at Fochabers,
sandy bank near Elgin tip and rubbish heap by the saw-mill W of

Carrbridge, 1975; tall plants of over 60 cm on waste ground at Cloddach S of Elgin, 1976 **E,BM**.
I   Kingsmill road Inverness, *c.* 1880 ? *coll.* **INV**; railway line at Clunes near Kirkhill, 1956; siding at Inverness, 1961 **E,K**; waste ground at Ardersier, railway sidings at Kingussie, Broomhill, Tomatin, Moy and Daviot; waste land at Longman Point, 1971 **E**; soil heap at Midlairgs near the quarry, 1970; shingle of the River Spey at Newtonmore, 1971 *J. D. Buchanan*!.

S. vulgaris L.   *Grunny Swally, Groundsel*
Native. Abundant in cultivated ground, waste places and sand dunes.
M Very common (Gordon); Forres everywhere, 1857 *JK* **FRS**; throughout the county, 1954; railway line at Pitcroy, 1967 **CGE**; dunes in the Culbin Forest, 1968 **E,BM**.
N (Stables, *Cat.*); Ardclach (Thomson); garden at Drynachan Lodge, and throughout the county, 1954; dunes on the Nairn Dunbar golf course, 1962 **ABD**.
I Inverness, 1832 (Watson, *Loc. Cat.*); Craig Phadrig, *c.* 1880 ? *coll.* **INV**; Inverness (as var. *erectus* Trow.), 1931 *M. L. Wedgewood* **MBH**; Inverness, 1940 *AJ* **ABD**; garden at Dalwhinnie and throughout the district, 1954.

   Var. **hibernicus** Syme
Introduced. Rare on cultivated ground and railway sidings.
M Garden weed at Elgin, 1969 **E**.
I   Distillery and station yards at Tomatin, 1952; railway sidings at Culloden, 1976 **E**.

S. tanguticus Maxim.
Garden relic or outcast.
M Rubbish tip at Grantown, well established, 1971 **E**.
N Garden outcast near the Doocote at Auldearn, 1975.
I Garden relic in the policies of Clunes at Kirkhill, 1954.

S. fluviatilis Wallr.   *Broad-leaved Ragwort*
Introduced. Rare by ditches and river banks and on waste ground.
M A large patch by a ditch opposite the old curling pond at Grantown, 1935, collected 1963 **E,ABD,BM**.
N Certainly introduced, rare at Castle of Inshoch, *WAS & G. Gordon* (Gordon).
I Waste land near the Cumberland Stone on Culloden Moor, 1954.

S. burchellii DC.
Introduced with foreign grain in distillery refuse on the rubbish tip at Longman Point at Inverness, 1970 **E**.

Doronicum L.
D. pardalianches L.   *Leopard's-bane*
Introduced. Garden escape well established on river banks and in waste places.

M Certainly introduced, Fochabers (Gordon)!; Altyre, 1840 *J. G. Innes* (Gordon, *annot.*); woods at Grantown, Brodie Castle, Cromdale and Nethybridge and on the banks of the river the whole length of the Spey, 1954; banks of the River Findhorn at Darnaway, 1963 **CGE**.
N Lethen, *J. B. Brichan* (Gordon).
I Wood by the River Enrick at Milton, 1947; wood at Kirkhill, by the Bunchrew Burn, at Alvie and Fort Augustus, 1954; roadside verge at Newtonmore and bank at Skye of Curr.

D. plantagineum L. *Plantain Leopard's-bane*
Introduced. Garden escape.
N Lethen, 1829 *WAS* **E**; garden outcast by East Milton, 1967 **E,ABD,CGE**; roadside at Newton of Budgate, 1972 **E**.
I Roadside verge just E of the old railway bridge at Ardersier, on the Nairn road, 1971. **E**.

Tussilago L.
T. farfara L. *The Son before the Father, Coltsfoot*
Native. Abundant on waste ground, on banks and among stones by rivers and by mountain burns up to 900 m.
M Birnie, 1836 *G. Gordon* **E**; very common (Gordon); waste land throughout the county, and dunes in the Culbin Forest, 1954; among stones by the Muckle Burn at Dyke, 1964 **E**.
N (Stables, *Cat.*); a good bed by the Findhorn at Levrattich near Ardclach, 1882 *R. Thomson* **NRN**; throughout the county, 1954; river shingle at Howford Bridge, 1964 **ABD**; by the Muckle Burn at Fortnighty Ford, 1972 **E**.
I Inverness, 1832 Dalwhinnie, 1833 (Watson, *Loc. Cat.*); throughout the district, 1954; waste land at Foyers, 1964 **CGE**; gully on the cliffs of Tom Bailgeann near Torness, 1972 **E**.

Petasites Mill.
P. hybridus (L.) Gaertn., Mey. & Scherb. *Butterbur*
Native. Frequent by ditches, river banks and railway embankments in the lowlands, not yet recorded for the highlands. Pl 17, Map 230.
M Abundant in one or two places, Alves, *G. Wilson* and Pluscarden and Altyre, *J. G. Innes* (Gordon); Waterford, 1840 *J. G. Innes* (Gordon, *annot.*); Dyke, 1901 *JK* **FRS**; by the Broad Burn at Smallburn, by the Kinloss burn, at Longmorn and on Haugh Island in the River Spey at Orton, 1954; banks of the Muckle Burn at Dyke, 1964 **CGE**.
N (Stables, *Cat.*); Nairnshire, *J. B. Brichan* (Gordon, *annot.*); Auldearn and Cawdor, 1900 (Lobban).
I (Watson, *Top. Bot.*); banks of the River Ness and at Lentran (Aitken); Achnagairn, 1902 (Pollock); railway embankment from Bunchrew to Clunes, 1954.

P. albus (L.) Gaertn. *White Butterbur*
Introduced. Frequent in waste places on wood margins and banks of ditches and burns. Map 231.

M Escape at Brodie (Marshall, 1899); Altyre near Forres, 1901 *JK* **FRS**; burn at Boghead Farm, Fochabers (as forma *laciniatus* Sourek), 1922 *G. B. Neilson* **GL**; waste land at Garmouth, 1953; steep bank by the road at Orton, by Moy House garden wall, Cromdale, railway at Knockando, Dava, by the River Lossie and by Dallas Kirk, 1954 etc.; by the small burn at the Old Manse at Dyke, 1964 **CGE**; banks of the River Lossie at Kellas, 1972 **E**.

N Banks of the River Nairn at Nairn and at Holme Rose, 1954; waste ground at Geddes House, 1964; wood at Kinsteary House, Auldearn, 1964 **ABD**.

I Waste land by the road between Milton and Drumnadrochit, 1947; banks of the River Beauly near Lovat Bridge, 1947; waste land at Fort Augustus, by Loch Mhòr at Errogie, at Farr, by Dunlichity Kirk and on the Islands at Inverness, 1954; roadside bank at the road junction A9–A833 near Kiltarlity, 1972 **E**; roadside verge at Dores, wood by Loch Ness at Lewiston and by the burn at Tomich.

P. fragrans (Vill.) C. Presl   *Winter Heliotrope*
Introduced. Garden escape flowering in winter.

M Garden in Tolbooth Street, Forres, 1954 **E**; waste land at Elgin, 1955.

I By the burn at Kingussie, 1935; bank at Ardersier, 1952; on the raised beach at Ardersier, 1971 **E**; roadside verge by Loch Dochfour and in a ditch at the entrance to Culloden House.

Calendula L.
C. officinalis L.
Pot Marigold occurs as a garden outcast on all the rubbish tips in the area, sometimes escaping to waste land and river banks.

Inula L.
I. helenium L.   *Elecampane*
Introduced. Garden escape long established at Rothes, Speyside (Burgess).

Telekia Baumg.
T. speciosa (Schreb.) Baumg.
Introduced. Garden outcast often mistaken for *Inula helenium* but is more robust and the flowers a bright apricot-orange in colour.

M Ditch at Lettoch near Nethybridge, 1950; ditch by The Kennels, Brodie, 1963 **E,CGE**.

N Garden escape at Culcharry near Cawdor, 1975.

I Below Lovat Bridge near Beauly, 1914 *J. Roffey* (*Proc.* **5**: 343 (1964)), recorded there (as *Inula helenium*) by *M. Cameron* in 1930 and by *J. N. Mills* in 1970; garden escape at Drynachan Lodge (as *Inula helenium*) 1955.

Filago L.
F. vulgaris Lam.   *Common Cudweed*
    F. germanica L., non Huds.
Native. Rare in open sandy places on dunes and in cultivated fields. Fig 32f, Map 232.

M Frequent (Gordon); near the Hospital at Forres, 1900 *JK* **FRS**; Lossiemouth, 1907 *G. B. Neilson* **GL** and in 1961 *MMcCW* **K**; arable field at Lochinvar, dunes at Covesea and waste sandy ground at Elgin, 1954.

N (Stables, *Cat.*); fields at Dulsie and at the Manse Farm at Ardclach, 1871 *R. Thomson* **NRN**; dry bank by the Cawdor road near Howford Bridge, 1960 **ABD**; stubble fields at Cran Loch and at Maviston, 1968 **E,BM**.

I Occasional on banks and dry pastures at Holm, Nairnside, moors by Loch Ashie and Balcraggan (J. Don, 1898); at Croy and Dalcross, 1900 (Lobban); Loch Ness-side near Abriachan, 1940 *AJ* **ABD**; cornfield near Croy, 1954; sandy bank at Ardersier, 1956; plentiful on shingle at Whiteness Head, 1976 **E,BM**.

F. minima (Sm.) Pers.   *Small Cudweed*
Native. Abundant on dunes, sandy tracks and fixed shingle banks by rivers. Fig 32b, Map 233.
M Very common (Gordon); near the Hospital at Forres, 1900 *JK* **FRS**; Lossiemouth, 1904 *G. B. Neilson* **GL**; Culbin, 1923 *D. Patton* **GL**; the Leen at Garmouth, 1953 *A. Melderis* **BM**; throughout the county, 1954; shingle of the River Spey at Garmouth, 1971 **E**.
N (Stables, *Cat.*); Ardclach, 1887 *Dr Sclanders* **NRN**; shore at Nairn, 1916 *D. Patton* **GL**; throughout the county, 1954; track on moor by Loch of the Clans, 1960 **ABD**; abundant on the sand dunes at Nairn, 1961 **ABD**.
I Inverness, 1832, Pitmain, 1833 (Watson, *Loc. Cat.*); roadside near Struy and at Craig Phadrig (Aitken); Kincraig, 1891 *AS* **E**; sandy ground at Culloden, Ardersier, Broomhill and Kincraig etc., 1954; river shingle by the Markie Burn, by the River Spey at Newtonmore, on a small island in the River Glass at Eskadale and at Dunmaglass etc.

Gnaphalium L.
G. sylvaticum L.   *Heath Cudweed*
Native. Common on moorland and pine wood tracks, and on dunes. Fig 32d, Map 234.
M Very common (Gordon); Broom of Moy, 1856 *JK* **FRS**; Grantown, 1890 *A. Thompson* **OXF**; shingle of the River Spey at Garmouth, 1953 *A. Melderis* **BM**; Grantown, 1953 *M. C. Craig* **E**; throughout the county, 1954; dunes in the Culbin Forest, 1968 **E**.
N (Stables, *Cat.*); common at Ardclach, 1887 *R. Thomson* **NRN**; moorland track at Drynachan and throughout the county, 1954.
I Inverness, 1832 Pitmain and Dalwhinnie, 1833 (Watson, *Loc. Cat.*); near Inverness, herb. *Borrer* **K**; Islands at Inverness, *c.* 1880 *P.F.* **INV**; Asylum grounds and Culloden Moor etc. (Aitken); Feshie Bridge, 1904 *G. B. Neilson* **GL**; Dochgarroch, 1940 *AJ* **ABD**; track by Ben Alder Lodge and throughout the district, 1954.

G. norvegicum Gunn.   *Highland Cudweed*
Native. Rare on mountain ledges and scree. Pl 18.

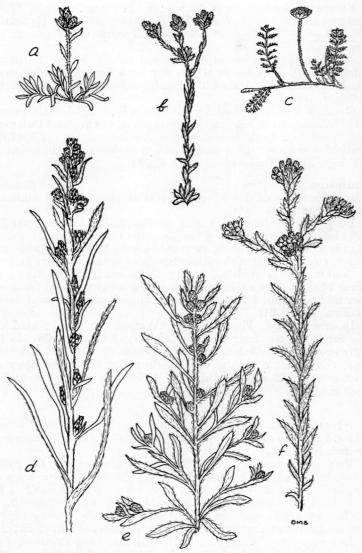

Fig 32   a, **Gnaphalium supinum**; b, **Filago minima**; c, **Cotula squalida**; d, **Gnaphalium sylvaticum**; e, **G. uliginosum**; f, **Filago vulgaris**.

I Sgùrr na Lapaich in Glen Affric, 1947 *NDS* **BM**!; Coire Leachavie of Mam Sodhail, 1947 *N. Y. Sandwith*!; cliffs of Sgùrr na Lapaich in Glen Strathfarrar, 1971 **E**.

G. supinum L. *Dwarf Cudweed*
Native. Common on bare mountain tops and grassy slopes, often in wet places, and occasionally washed down rivers where it becomes established on fixed shingle. Fig 32a.
M Parish of Knockando, *C. Watt* (Burgess).
N Càrn nan tri-tigheanan near Drynachan, 1969 *D. A. Ratcliffe*.
I Dalwhinnie, 1833 (Watson, *Loc. Cat.*); near Dalwhinnie, herb. *Borrer* **K**; Badenoch and Mamsoul (Gordon); Fraoch-choire (Ball, 1851); shores of Loch Einich, 1892 *AS* **E**; Mam Sodhail, Glen Affric, 1955 *D. M. Henderson* **E**; Coire an Lochain of Cairn Gorm, 1953 *A. Melderis* **BM**; summit of Ben Alder and all the higher hills; grass path at Coignafearn Lodge at 380 m, 1970; fixed shingle by the River Killin at 300 m, 1972.

G. uliginosum L. *Marsh Cudweed*
Native. Common in dry and wet open places on moors, farmyards and in gateways of fields. Fig 32e, Map 235.
M Frequent at Darkland and Speymouth etc. (Gordon); wet track at Knockando and dry track on the Culbin Forest etc., 1954.
N (Stables, *Cat.*); Cawdor and Nairn, 1900 (Lobban); garden path at Cawdor Castle, root field at Ordbreck and throughout the county, 1954; bog by Loch of the Clans, 1960 **ABD**.
I Inverness, 1832 (Watson, *Loc. Cat.*); Asylum grounds and Craig Phadrig (Aitken); Aldourie, 1904 *G. T. West* **STA**; wet moorland tracks at Affric Lodge and throughout the district.

Anaphalis DC.
A. margaritacea (L.) Benth. *Pearly Everlasting*
Introduced. Garden relic or escape.
M Railway embankment S of Dava station, 1953.
I Beauly, 1944 *K. N. G. Macleay*!; river bank at Clava, 1953; garden relic at Dalwhinnie station and by the river at Beauly.

Antennaria Gaertn.
A. dioica (L.) Gaertn. *Mountain Everlasting*
Native. Common on moors on base-rich soil. Male flowers usually white, female rose-pink. Map 236.
M Innes House, Urquhart, 1794 *A. Cooper* **ELN**; frequent (Gordon); Aviemore, 1871 *JK* **FRS**; banks of the River Findhorn, 1913 *D. Patton* **GL**; Culbin Sands, 1925 *G. C. Druce* **OXF**!; rocky hummocks on the moors at Duthil, Dava, Dallas etc., 1954; Hopeman golf course, 1956.
N (Stables, *Cat.*); Balachroan, 1872 *R. Thomson* **NRN**; moors at Drynachan, 1954; on rocks by the River Findhorn at Coulmony and Glenferness; moor by the Leonach Burn near Dulsie Bridge, 1960 **ABD**.
I Inverness, 1832 Dalwhinnie and Pitmain, 1833 (Watson, *Loc.*

Cat.); Strathspey, 1839 *W. B. Jnr.* herb. *Borrer* **K**; N of Dalnaspidal, 1891 *AS* **E**; Tromie moor, 1893 *J. Buchanan* **GL**; Leys, Daviot and Englishton (J. Don, 1898); Ashie moor, 1940 *AJ* **ABD**; scattered throughout the district, 1954; on a short grassy moor above Newtonmore, 1972 **E**.

Solidago L.

S. virgaurea L.   *Goldenrod*
Native. Common on moors, dry banks, margins of pine woods, rocks by rivers and mountain ledges. The mountain forms with large. flower heads have been named var. *cambrica* (Huds.) Sm. Map 237.
M Grantown, 1837 *WAS* **E**; frequent (Gordon); rocks by the river at Sluie, 1899 *JK* **FRS**; very common in the Findhorn valley and throughout the county, 1954.
N (Stables, *Cat.*); moors by the schoolhouse at Ardclach, 1880 *R. Thomson* **NRN**; on rocks by the River Findhorn at Dulsie Bridge, Coulmony and Glenferness etc., 1954.
I Dalwhinnie, 1833 (Watson, *Loc. Cat.*); Glen Einich (as var. *cambrica*) (Druce, 1888); Braeriach (as var. *cambrica*) *AS* **E**; by Loch Dochfour, 1940 *AJ* **ABD**; throughout the district, 1954.

S. canadensis L.   *Canadian Goldenrod*
Garden escape.
M Quarry near Lossiemouth harbour, 1967 **E,CGE**; outside cottages at Dallas, by a small burn at Kingston and by Sheriffston Kirk.
I Escape near Lurgmore by Loch Ness, 1960.

S. gigantea Ait. var. leiophylla Fern.
Garden escape in a quarry at Lossiemouth, 1967 **E,CGE**.

Aster L.

A. tripolium L.   *Sea Aster*
Native. Common on salt-marshes.
M Rare in Findhorn Bay, *G. Wilson* (Gordon) and 1856 *JK* **FRS**; salt-marsh NW of the Buckie Loch, Culbin, 1961 **E,ABD**; plentiful in Findhorn Bay and at Garmouth.
N (Stables, *Cat.*); by the Druim E of Nairn, 1937 *J. Walton* **GL**; salt-marshes at Carse of Delnies, Nairn and between the Old Bar and the Culbin Forest, 1954.
I By the Beauly Firth, 1850 *P. H. MacGillivray* **ABD**; Beauly (as var. *discoideus* Rchb.) (*J. Bot.* **28**: 39 (1890)); Bunchrew, 1940 *AJ* **ABD**; salt-marshes at Castle Stuart, Inverness and the length of the Firth to Beauly.

A. novi-belgii L.   *Michaelmas Daisy*
Garden escape. Common on waste land and on river banks.
M Shingle of the River Spey at Kingston, 1964 **E**; garden escape by the River Spey at Rothes (as *A. longifolius* Lam.) 1973 *A. J. Sauter*; waste land by the Old Spey Bridge at Grantown and on the margins of Loch Puladdern at Aviemore etc.
N Railway embankment at Nairn, 1970.

I Waste ground at Farr, 1954; railway embankments at Broomhill and Gollanfield, waste ground by the Longman tip, by the lagoons at Inverness, banks of the River Glass and in marshes at Beauly etc.

A. laevis L.
Garden escape on the railway embankment ½ mile E of Nairn, 1975 **E**.

A. lanceolatus× **novi-belgii**=**A.**×**salignus** Willd.
Garden escape by a small burn running into the River Spey between Kingussie and the river, 1975 **E**.

Erigeron L.
E. borealis (Vierh.) Simmons   *Alpine Fleabane*
Native. Very rare on mountain rock ledges.
I On all the Cairngorms (Trail, *Cairn Gorm Club Journal*, 1895); Bynach in Abernethy at 700 m, *R. Mackechnie* **BM**.

Bellis L.
B. perennis L.   *Gowans, Daisy*
Native. Abundant in short grassland and often in mountain flushes.
M Innes House, Urquhart, 1794 *A. Cooper* **ELN**; very common (Gordon); Forres, 1856 *JK* **FRS**; Cothall, 1953 *A. Melderis* **BM**; throughout the county, 1954.
N (Stables, *Cat.*); Ardclach (Thomson); throughout the county, 1954; river shingle at Glenferness, 1973 **E**.
I Inverness, 1832, Dalwhinnie and Pitmain, 1833 (Watson *Loc. Cat.*); Craig Phadrig, *c.* 1880 ? *coll.* **INV**; flush on Sgùrr na Lapaich in Glen Strathfarrar at 900 m and throughout the district.

Eupatorium L.
E. cannabinum L.   *Hemp Agrimony*
Native. Rare by burns and in ditches.
M At Cothall on the River Findhorn (Gordon); Broom of Moy, 1856 *JK* **FRS**; wet meadow at Darnaway opposite Cothall, 1953; river shingle at Seafield, 1975 *K. Christie*.
N Bogside near Boath, Auldearn, *J. B. Brichan* (Gordon, *annot.*); ditch by the cemetery at Nairn, 1960 **ABD**; several plants on the margin of Lochloy, 1973 *R. M. Suddaby*!.
I (Watson, *Top. Bot.* and Druce *Com. Fl.*). Both these records may refer to Brichan's record for Auldearn.

Anthemis L.
A. tinctoria L.   *Yellow Chamomile*
Introduced. Rare on waste ground and river shingle.
M Garden relic at the entrance to Brodie Castle, 1948; river shingle at Kingston, 1953.
N Waste ground by the River Nairn at Nairn, 1960 **E,ABD**.

A. cotula L.   *Stinking Chamomile*
Doubtfully native in north Scotland. Introduced with foreign grain and bird seed.

M Distillery yard at Carron, 1955 **E,ABD**; waste ground with other grain aliens in Greshop wood near Forres, 1957 **K**; railway siding at Knockando, 1966 **E,BM**; Elgin rubbish tip, and roadside verge at Lhanbryde, 1973.
N From 'Swoop' at Nairn, 1969 **ABD**; rubbish tip at Newton of Park, 1975 **E**.
I Waste ground by the bridge over the Moniack Burn between Bogroy and Kirkhill, 1972 **E**.

A. arvensis L.   *Corn Chamomile*
Native. Frequent in cultivated ground and waste places. Map 238.
M Doubtfully native, frequent in corn and grass fields about Elgin (Gordon); rubbish at Waterford, 1874 *JK* **FRS**; Garmouth (Marshall, 1889); fields at Dyke, Forres, Knockando, Kellas and Elgin etc., 1954; cornfield at Calcots, 1967 **E,BM**; waste ground by a roadside in the town of Grantown, 1974 **E**; newly-sown grass lawn by the Health Centre at Forres, 1974 *K. Christie.*
N Fields E of Nairn, 1833 *WAS* **E,GL**; Ardclach (Thomson); pasture at Maviston, 1930; field W of Gollanfield, 1955; cornfield by Cran Loch, 1961 **E,ABD**; waste ground at Nairn, 1972 **E,BM**.
I In small quantity at Fort George (Marshall, 1889); sandy field and roadside verge at Ardersier, 1952; Gollanfield, 1956 **ABD**; verge of a track above the Falls of Divach near Lewiston, 1975 **E**; waste ground by the sea, Longman Point, Inverness, 1976 **E**.

A. altissima L.
    A. cota L.
Introduced with carrot seed in a field by Brodie pond, 1963 **E,ABD,CGE,JEL**.

Chamaemelum Mill.
C. nobile (L.) All.   *Chamomile*
    Anthemis nobilis L.
Introduced at Ardclach, Nairn (Thomson).

Achillea L.
A. millefolium L.   *Milfoil, Yarrow*
Native. Abundant in short grass, on banks, roadside verges and sand dunes.
M Innes House, Urquhart, 1794 *A. Cooper* **ELN**; very common (Gordon); Forres, 1889 *JK* **FRS**; dunes in the Culbin Forest and throughout the county, 1954.
N (Stables, *Cat.*); roadsides at Ardclach, 1888 *R. Thomson* **NRN**; moorland track at Drynachan and throughout the county, 1954.
I Inverness, 1832 Dalwhinnie and Pitmain, 1833 (Watson, *Loc. Cat.*); banks of the Canal and the Ness, *c.* 1880 *P.F.* **INV**; at Ben Alder Lodge and throughout the district, 1954.

A. ptarmica L.   *Sneezewort*
Native. Common in bogs and grassy places on moors and river banks. Map 239.

M Frequent (Gordon); Greshop, 1900 *JK* **FRS**; Buckie Loch Culbin and throughout the county, 1954.
N (Stables, *Cat.*); Ardclach, 1881 *R. Thomson* **NRN**; river banks at Drynachan and throughout the county, 1954.
I Inverness, 1832 Dalwhinnie and Pitmain, 1833 (Watson, *Loc. Cat.*); banks of the River Ness, *c.* 1880 *P.F.* **INV**; Lentran, 1940 *AJ* **ABD**; gully in the Ryvoan Pass and throughout the district, 1954.

Matricaria L.

M. maritima L.   *Sea Scentless Mayweed*
  Tripleurospermum maritimum (L.) Koch
    Subsp. **maritima**
Native. Common on sand dunes and shingle beaches.
M Frequent (Gordon); Moray, *JK* **FRS**; shingle beaches at Covesea, Kingston and Lossiemouth, 1953; sand dunes in the Culbin Forest, 1968 **E**.
N Between Findhorn and Nairn (Ewing); shingle beaches at Nairn and Carse of Delnies, 1954.
I Abundant on dunes and shingle at Whiteness Head and Ardersier, 1954; edge of the salt-marshes at Beauly; by the sea at Longman Point and at Alturlie.

    Subsp. **inodora** (C. Koch) Soó   *Scentless Mayweed*
Native. Abundant in cultivated ground and waste places.
M Very common (Gordon); Forres, 1898 *JK* **FRS**; arable fields at Dyke and throughout the county, 1954.
N (Stables, *Cat.*); Ardclach (Thomson); root fields at Cawdor, Nairn, Auldearn and Ferness etc., 1954.
I Inverness, 1832 Dalwhinnie, 1833 (Watson, *Loc. Cat.*); near Englishton (Aitken); Bunchrew, 1940 *AJ* **ABD**; Parks of Inshes, *c.* 1880 **INV**; throughout the district where there is cultivated land, 1954.

Chamomilla S.F. Gray

C. recutita (L.) Rauschest   *Wild Chamomile, Scented Mayweed*
  Matricaria recutita L.
Introduced in north Scotland. Rare on tips and waste ground.
M Railway stations at Brodie and Grantown, 1954; rubbish tip at Forres and river shingle at Fochabers; Elgin tip, 1975 **E**.
N Rubbish tip at Newton of Park, 1975.
I Single plant by the roadside at Croy, 1961 **CGE**; allotment at Inverness, 1961 **E**; roadside verge at Brackla by Loch Ness, 1975 **BM**; waste ground by the Longman tip at Inverness, 1975 **E,BM**.

C. suaveolens (Pursh) Rydb.   *Pineapple Weed*
  Matricaria matricarioides auct.
Introduced. Abundant in waste places, farmyards, gateways and by tracks. Map 240.
M Supposedly introduced with wheat. Native of NW America spreading rapidly in waste places (Burgess); farmyard at Kinloss and throughout the county, 1954.

N Track near buildings at Drynachan Lodge and throughout the county, 1954.
I Railway station at Inverness, a plant new to this district, *James Macfarlane* (*Trans. ISS & FC* **8**: 42 (1912–18)); yard at Ben Alder Lodge and throughout the district, 1954.

Anacyclus L.
A. radiatus Loisel.
Introduced with foreign grain.
M One plant at Rothes (Trail, 1904).

Chrysanthemum L.
C. segetum L.   *Gule, Corn Marigold*
Doubtfully native. Abundant in cultivated ground in the lowlands, less so in the highlands. 'The Gule, the Gordon and the Hoodie Craw, are the three warst things that Moray ever saw' (*Cge. Count. Geog.* **20** (1915)). Map 241.
M Very common (Gordon); Forres, 1898 *JK* **FRS**; root fields at Boat of Garten, Grantown, Knockando, Dyke and throughout the county, 1954; Culbin, 1964 *P. How* **E**; shingle of the River Spey at Fochabers, 1971 **E,CGE**.
N (Stables, *Cat.*); cornfields at Ardclach, 1892 *R. Thomson* **NRN**; root fields at Ferness, Cawdor and Auldearn etc.
I Pitmain, 1833 (Watson, *Loc. Cat.*); Culloden, *c.* 1880 ? *coll.* **INV**; by Loch a' Chlachain, 1904 *G. T. West* **STA**; Dochgarroch, 1940 *AJ* **ABD**; arable field at Cluny Castle Laggan, farm yard at Tomich, garden weed at Pityoulish etc., 1954.

Leucanthemum Mill.
L. vulgare Lam.   *Horse Gowans, Ox-eye Daisy*
    Chrysanthemum leucanthemum L.
Native. Abundant on grass banks, rough meadows and fixed dunes on rich soil. Map 242.
M Frequent (Gordon); common in the uplands, 1900 *JK* **FRS**; dunes at Findhorn, 1925; grass banks by the River Spey at Blacksboat, Boat of Garten and Knockando etc., 1954; moorland slope at Easter Limekilns on Dava Moor and throughout the county.
N (Stables, *Cat*); Ardclach (Thomson); meadow at Drynachan and throughout the county, 1954.
I Inverness, 1832 Dalwhinnie and Pitmain, 1833 (Watson, *Loc. Cat.*); Tomatin, 1904 (Moyle Rogers); railway embankments from Drumochter to Inverness, meadows in glens Feshie, Moriston, Affric and Urquhart etc., 1954 and throughout the district.

L. maximum (Ramond) DC.
    Chrysanthemum maximum Ramond
Garden escape near the E end of Loch Meiklie, on the old tip at Kingussie and by the road, where garden refuse had been tipped on Drumashie Moor.

Tanacetum L.

T. parthenium (L.) Schultz Bip. *Feverfew*
   Chrysanthemum parthenium (L.) Bernh.
Introduced. A garden escape on waste ground and tips.
M Doubtfully native, abundant in one or two places (Gordon); walls of Elgin Cathedral (Druce, 1897); Forres, 1898 *JK* **FRS**; bank in Relugas village, waste land at Knockando, Carrbridge, Kellas and on a track in the Culbin Forest etc., 1954.
N (Stables, *Cat.*); Ardclach, 1900 (Lobban); river shingle at Nairn and Coulmony; wall in Cawdor kirkyard, waste ground at Holme Rose etc., 1954.
I (Watson, *Top. Bot.*); Beaufort (J. Don, 1898); Fort Augustus, 1904 *G. T. West* **STA**!; waste ground by Loch Mhòr at Errogie, at Dulnain Bridge, at Kingussie, Bunchrew and Drumnadrochit etc.

T. vulgare L. *Tansy*
   Chrysanthemum vulgare (L.) Bernh.
Native. Common by burns and in waste places and often by old cottages where it was probably planted for its medicinal properties. Map 243.
M Doubtfully native, frequent (Gordon); by the Hospital at Forres, 1900 *JK* **FRS**; Grantown, 1948 *I. C. Hedge* **E**; roadside bank by Blervie Mains Farm near Rafford, 1925; by the Burnie Path at Dyke, dunes at Findhorn, railway embankments at Slochd and Kinveachy, quarry at Lossiemouth etc., 1954.
N (Stables, *Cat.*); Piperhill near Cawdor, 1887 *R. Thomson* **NRN**; waste land at Holme Rose, Geddes, Auldearn and near Clunas etc., 1954.
I Inverness, 1832 (Watson, *Loc. Cat.*); Englishton Burn and Balloan road (Aitken); railway embankments at Inverness, Ardersier and near Lentran, tip at Kingussie, bank in Insh village and by cottages at Knockie etc., 1954.

T. macrophyllum (Waldst. & Kit.) Schultz Bip.
M Garden escape by the River Spey at Boat o'Brig, 1966.

Cotula L.

C. squalida Hook. fil.
Introduced. Fig 32c.
M Lawn weed at Greshop House near Forres, 1952 and 1960 **E,ABD,K**; lawn weed at Inverugie House, Hopeman, 1967 **E,CGE**.

Artemesia L.

A. vulgaris L. *Muggart Kail, Mugwort*
Native. Frequent on waste ground and roadside verges. Map 244.
M Very common (Gordon); Forres, 1898 *JK* **FRS**; roadside by Brodie station, 1925, at Boat of Garten, Knockando, Forres and Elgin etc., 1954; roadside bank at Lochinvar, 1960; Knockando station, 1967 **ABD**; gravel works at Cloddach quarry, 1976 etc.

N (Stables, *Cat.*); roadsides at Ardclach, 1884 *R. Thomson* **NRN**; Nairn and Auldearn, 1900 (Lobban).
I Dalwhinnie and Pitmain, 1833 (Watson, *Loc. Cat.*); parks of Inshes and at Clava, not uncommon (Aitken); Dalwhinnie (as var. *coartata* Horcell), 1887 *A. E. Lomax* **GL**; Culloden (J. Don, 1898) abundant on the old railway embankment at Muirtown, Inverness, 1935; roadside at Tomich and at Kincraig and Alvie, 1954; station yard at Dalwhinnie, 1975.

A. absinthium L.   *Wormwood*
Introduced. Formerly grown for its medicinal properties.
M Parishes of Dyke and Urquhart (Burgess); Hopeman and Burghead in several places, 1950 *R. Richter*.
I Rare on waste land near Kessock Ferry, 1954 (it is plentiful at North Kessock in East Ross).

Echinops L.
E. bannaticus Rochel ex Schrader   *Globe Thistle*
Garden outcast. Phyllary tips erect in bud, dark blue.
M Rubbish tip at Elgin, 1958; bank opposite Dalvey Cottage near Dyke, 1962 **E**.
N Railway embankment just E or Nairn, 1975 **E**.

E. exaltatus Schrader
Garden outcast. Phyllary tips recurved in bud, pale grey-blue.
I Outcast on Drumashie Moor, 1975 **E**.

Carlina L.
C. vulgaris L.   *Carline Thistle*
Native. Very rare on calcareous grassland. Pl 18.
M North side of the limestone quarry at Cothall, 1844 *T. Edmondston & J. G. Innes* (Gordon, *annot.*)!; Lossiemouth, 1864 *Mr Hassley* (Watson).
N Nairn and Cawdor, 1900 (Lobban).

Arctium L.
A. nemorosum Lejeune   *Burrs, Intermediate Burdock*
Native. Very common in waste places in the lowlands, less so in the highlands. Map 245.
M Very common (as *A. lappa*) (Gordon); Cothall, 1856 *JK* **FRS**; Forres and Boat of Garten (as *A. minus*) (Druce, 1888); track in the Culbin Forest and throughout the county, 1954.
N (Stables, *Cat.* as *A. lappa*); Ardclach (Thomson); Holme Rose, Cawdor, Geddes, Dulsie Bridge, Glenferness and Auldearn, 1954.
I Inverness, 1832 (Watson, *Loc. Cat.*); near Culloden Moor, 1850 *P. H. MacGillivray* **ABD**; dyke at Englishton House (as *A. lappa*) (Aitken); Kingussie (as *A. minus*) (Druce, 1888); Beauly (as *A. intermedium*) (*J. Bot.* **28**: 42 (1890)); in the old lime quarry on Suidhe at Kincraig, among boulders below the cliffs of Creag Dhubh near Laggan, roadside verge in Glen Moriston, at Farr, Fort Augustus etc., 1954.

A. minus Bernh. *Lesser Burdock*
Introduced in north Scotland.
M Distillery tip at Smallburn near Rothes, 1956 **ABD**.

Carduus L.

C. tenuiflorus L. *Slender Thistle*
Native. Waste ground usually near the sea.
M Findhorn, 1844 *J. G. Innes* (Gordon, *annot.*); Findhorn, 1856
JK **FRS**; parishes of Duffus and Kinloss (Burgess); dunes at
Findhorn, 1964 **E,CGE**; rubbish tip at Forres, 1970 **E**; Spynie Palace.
I Waste ground at Longman Point, Inverness, 1956 and 1976 **E**.

C. argentatus L.
Introduced with foreign grain from a distillery, Inverness, 1961
E,BM.

C. nutans L. *Musk Thistle*
Introduced in north Scotland. Roadside verges and borders of
fields.
M Glassgreen, 1836 *G. Gordon* **E**; doubtfully native, abundant in
one or two places, by the gas house, Elgin and near Glassgreen
(Gordon); appears with every rotation in the centre of a field at
Stynie Farm, Speymouth, *G. Birnie* (Burgess); roadside verge at the
road junction near Glassgreen (probably Gordon's locality) and
where it remained until the road was widened in 1968, **CGE**;
twelve plants on the bank of the Muckle Burn at Whitebridge, 1960
E,ABD; grass verge of a cornfield at Rosehaugh Farm near Elgin,
1966 *R. P. Petrie* **E,BM,CGE**; newly-sown roadside verge near
Dunphail and near Birnie, 1973; dunes at Findhorn, 1973 *K.
Christie*.

C. acanthoides L. *Welted Thistle*
    C. crispus auct.
Introduced in north Scotland. Rare on waste ground.
M Doubtfully native, abundant in one or two places, at Elgin
Cathedral, Spynie Castle and Duffus churchyard (Gordon);
Waterford near Forres, 1871 *JK* **FRS**; Culbin Sands and Rafford,
A. MacGregor (Burgess); chicken run at Moy House, 1950; waste
ground near the tip at Lossiemouth, 1964 **E,CGE**; single plant at
Spynie Kirk, 1976.
I (Druce, *Com. Fl.*); by the track at Alvie Kirk, 1975 *R. McBeath*
E!.

Cirsium Mill.

C. vulgare (Savi) Ten. *Scots Thistle, Spear Thistle*
Native. Abundant in fields, by roadsides, on waste ground and
grassy mountain slopes up to 900 m. This is the plant known locally
as the Scots Thistle. Pl 18.
M Very common (Gordon); frequent at Forres, 1856 *JK* **FRS**; by
tracks and on dunes in the Culbin Forest and throughout the
county, 1954.

N (Stables, *Cat.*); Ardclach, the true *Scotch Thistle* (Thomson), he writes: 'Dunbar in his poem entitled "The Thrisell and the Rois" written in 1503 in honour of the marriage of James IV with Margaret Tudor, mentions it as the badge of Scotland, and Hamilton of Bargowe especially states that the thistle was the "Monarch's Choice". For fully a century after this we find the flower heads impressed upon the Scotch coins all represented with very little change of figure' (*Trans. NA of Lit. & SS*, **2**: pt 1 (1893)–The Rarer Flora of Ardclach) (The plant known by botanists as Scotch Thistle is *Onopordum acanthium*); roadside verge at Drynachan and throughout the county, 1954.
I Inverness, 1832, Dalwhinnie and Pitmain, 1833 (Watson, *Loc. Cat.*); Abriachan, 1940 *AJ* **ABD**; waste ground at Ben Alder Lodge and throughout the district, 1954.

C. palustre (L.) Scop.   *Marsh Thistle*
Native. Abundant in bogs and wet meadows.
M Very common (Gordon); Forres, 1871 *JK* **FRS**; the Leen at Garmouth, 1953 *A. Melderis* **BM**; damp grassland by the Buckie Loch, Culbin Forest and throughout the county, 1954.
N (Stables, *Cat.*); Ardclach (Thomson); river bank at Drynachan and throughout the county, 1954.
I Inverness, 1832 Dalwhinnie, 1833 (Watson, *Loc. Cat.*); by the Bunchrew Burn, 1854 *HC* **INV**; near Loch Ness, 1940 *AJ* **ABD**; bogs on the moors at Gaick, Glen Affric and throughout the district, 1954.

C. arvense (L.) Scop.   *Creeping Thistle*
    Var. **arvense**
Native. Abundant in fields and waste places.
M Very common (Gordon); frequent, 1856 *JK* **FRS**; dunes on the Bar, Culbin and throughout the county, 1954.
N (Stables, *Cat.*); Howford, 1887 *R. Thomson* **NRN**; river bank at Drynachan and throughout the county, 1954.
I Inverness, 1832 Pitmain, 1833 (Watson, *Loc. Cat.*); by the Bunchrew Burn, 1854 *HC* **INV**; Dalwhinnie and Tomatin, 1903 (Moyle Rogers); Lower Leachkin near Inverness, 1940 *AJ* **ABD**; in short turf by Gaick Lodge, track in upper Glen Feshie and throughout the district, 1954.

    Var. **setosum** C. A. Mey.
A variant with broad, flat, hardly lobed and weakly spiny leaves, green and glabrous beneath.
M Railway station yard at Carron, 1956 **ABD**.
I Roadside bank near the bridge over the River Calder on the Laggan road at Newtonmore, 1973 **E**.

    Var. **incanum** (Fisch.) Ledeb.
A variant with leaves somewhat similar to the above but narrower and densely cottony beneath.
M Introduced with foreign barley at Carron station, *W. G. Craib* (Burgess).

C. oleraceum (L.) Scop. *Cabbage Thistle*
Introduced. Flowers dingy yellowish-white.
M Verge to the drive at Blackhills House, Lhanbryde, 1964 **CGE**.

C. helenioides (L.) Hill *Melancholy Thistle*
    C. heterophyllum (L.) Hill
Native. Common on river banks and in wet meadows. Pl 18, Map 246.
M Frequent in the upper districts (Gordon); Inverallan near Grantown, 1895 *C. Bailey* **OXF**; Waterford near Forres, 1898 *JK* **FRS**; Darnaway, 1953 *A. Melderis* **BM**; very common on the banks of the rivers Spey and Findhorn etc., 1954.
N (Stables, *Cat.*); Dulsie, 1887 *R. Thomson* **NRN**; throughout the county, 1954; grassy bog by Geddes Reservoir, 1961 **ABD,K**.
I Dalwhinnie, 1833 (Watson, *Loc. Cat.*); above Clachnaharry at Inverness, 1853 *HC* **INV**; Pass of Inverfarigaig, 1940 *AJ* **ABD**; Englishton Muir and by the River Nairn (Aitken); abundant by Spey Dam, Laggan and on railway embankments from Drumochter to Culloden etc., 1954.

C. helenioides × palustre = C. × wankelii Reichardt
I Swampy ground by a small burn near Guisachan House, Tomich, 1935 *R. Knowling* **BM,OXF**; roadside verge N of Dalwhinnie, 1960, *B. M. C. Morgan & MMcCW* **E,CGE**.

C. dissectum (L.) Hill *Meadow Thistle*
Recorded from the Culbin Sands by *D. Patton & E. J. A. Stewart* but with no voucher specimen.

Silybum Adans.

S. marianum (L.) Gaertn. *Milk Thistle*
Introduced. Easily recognised by the leaves that are shining green variegated with white along the veins above.
M Certainly introduced at Elgin and at Castlehill, Forres *G. Wilson* (Gordon); garden weed at Moy House, 1955.
N Garden escape on waste ground at Auldearn, 1954.

Onopordum L.

O. acanthium L. *Cotton Thistle, Scotch Thistle*
Introduced. Waste ground.
M Waste ground at Forres station, 1957; rubbish tip at Rothes, 1969.
N Howford near Nairn, 1887 *R. Thomson* **NRN**; garden weed at Auldearn, 1954; waste ground away from habitation, near Broomhill, Geddes, 1972 *R. K. Smith*!.

Saussurea DC.

S. alpina (L.) DC. *Alpine Saw-wort*
Native. Frequent on mountain ledges over 600 m, occasionally washed down becoming established on gravel overspills of burns. Map 247.
I Clunie mountains SW of the Inn (R. Graham, *Phyt. NS*, **1**: 527

(1843)); Glen Einich, 1887 *G. C. Druce* **OXF**; Pass of Larig on the
Rothiemurchus side of the watershed, 1882 *Dr Mackie* **FRS**; N of
Invermoriston, 1904 *G. T. West* **STA**; Coire Chùirn at Drumochter,
Glen Markie and Geal Chàrn, 1957 *D. A. Ratcliffe*; gravel overspill
of burn below Coire an Lochain, 1953; summit of Ben Alder, 1956;
cliffs of Sgùrr nan Conbhairean, Glen Affric, 1962 *H. Milne-
Redhead*; mountain ledges on Bynack More, Coire a'Bhèin at Garva
Bridge and in Glen Banchor etc.; ledge on Sgùrr na Lapaich in
Glen Strathfarrar, 1972 **E**.

Centaurea L.

C. scabiosa L.   *Greater Knapweed*
Introduced in this area. Rare on roadside verges and railway
embankments.
M Doubtfully native, roadside between Balnageith and Forres,
1831 (Gordon), and 1860 *JK* **FRS**; Limekiln and Mainwood, 1829,
Mr Martin and between Fleurs and Bilboahall at Elgin, 1853
(Watson); single plant near Brodie station (Burgess); near Forres,
Martin Barry **ABD**; waste ground at Grantown, 1954; single plant
(now gone with road widening) on the grass verge of the A96 ¾ mile
E of Tearie cross-road, 1961 **E,ABD,CGE**.
N (Stables, *Cat.*); Mains of Courage, 1843 *J. Tolmie* (Watson).
I Hillhead of Ardersier, 1843 *J. Tolmie* (Watson); top of the old
railway line at Muirton Basin, Inverness, 1950 and 1961 **E,CGE**.

C. montana L.   *Perennial Cornflower*
Garden escape.
M Roadside opposite Dalvey Cottage, Dyke, 1930; rubbish tip at
Rothes, 1954.
I Roadside verge, away from habitation, at Connage near Arder-
sier, 1972 **E**.

C. cyanus L.   *Blue Bonnet, Cornflower*
Native. Formerly very common, like all cornfield weeds, but now
rare except at Kintessack and Moy where occasionally the
cultivated fields are blue with it. Map 248.
M Very common (Gordon); Aviemore, 1892 *AS* **E**; cornfield at
Lossiemouth, 1907 *G. B. Neilson* **GL**; near Forres, 1940 *AJ* **ABD**;
Elgin, 1941 ? *coll.* **STA**; field near Moy House, Kintessack, 1953
A. Melderis **BM**; waste land at Rothes, 1966 **E,ABD,CGE**; cornfield
between East and West Manbean, cornfields at Dallas, Lossiemouth
and at Rosehaugh Farm near Elgin; root field near Duffus Castle,
1976; abundant in a root field near Kincorth House, 1976, seed
was collected there by *R. Melville and MMcCW* for the Seed Bank at
Kew.
N (Stables, *Cat.*); formerly abundant in a cornfield at Broombank
Farm, Auldearn, 1925 and 1961 **ABD,K**.
I Inverness, 1832 (Watson, *Loc. Cat.*); cornfield near Ardersier,
1952; garden outcast on the rubbish tip at Longman Point,
Inverness, 1970.

C. nigra L. subsp. **nigra**   *Hardheads, Lesser Knapweed*
    C. obscura Jord.
Native. Abundant in grassland and on roadside verges.
M Very common (Gordon); Waterford, 1856 *JK* **FRS**; Elgin,
1941 ? *coll.* **STA**; roadside verge by Dalvey Smithy and throughout
the county, 1954.
N (Stables, *Cat.*); Ardclach (Thomson); rough ground by Loch of
the Clans and throughout the county, 1954.
I Inverness, 1832, Dalwhinnie and Pitmain, 1833 (Watson, *Loc.*
Cat.); Parks of Inshes near Inverness, 1854 *HC* **INV**; Bunchrew,
1940 *AJ* **ABD**; grass bank at Ben Alder Lodge and throughout the
district, 1954.

C. calcitrapa L.   *Red Star-thistle*
Introduced.
M West side of Dykeside Farm at Birnie, 1854 *J. Weir* (Watson, cf.
George Dickie's annot. copy in the University Library at Aberdeen).

C. solstitialis L.   *Yellow Star-thistle*
Introduced.
M Park behind school at Lossiemouth, 1860 *J. Grant* (Watson, cf.
George Dickie's annot. copy in the University Library at Aberdeen).

C. diluta Aiton
Introduced with foreign grain and bird seed.
M With bird seed aliens on the Elgin rubbish tip, 1956 **ABD**; from
distillery refuse on the tip at Rothes, 1957 **K**.
N From 'Swoop' at Wild Goose Cottage, Nairn, 1969 **E,ABD,CGE**.
I From distillery refuse on the tip at Longman Point, Inverness,
E,CGE.

C. melitensis L.   *Maltese Star-thistle*
Introduced with foreign grain.
I In an allotment near a distillery at Inverness, 1961 **E,K,CGE**.

C. salmantica L.
Introduced with bird seed.
N From 'Swoop' at Wild Goose Cottage, Nairn, 1969 **E**.

Serratula L.
S. tinctoria L.   *Saw-wort*
N 'I saw a plant collected by Miss McDonald at Nairn, how far
native I cannot say' (Druce, *J. Bot.* **28**: 42 (1890)).

Cichorium L.
C. intybus L.   *Chicory*
Introduced in north Scotland. Occasional in old pastures, grass
banks, sand dunes and waste land.
M Cornfield at Stankhouse, Birnie, 1835 *G. Gordon* **E**; doubtfully
native, rare, cornfield at Gordonstoun, 1830, Lossiemouth, 1831,
Alves by *G. Wilson* and Birnie, 1835 (Gordon); Bogtown and
Invererne, 1840 *J. G. Innes* (Gordon, *annot.*); Forres station (Aitken);

grass bank at Brodie station, field at Kincorth, and at Garmouth,
1954; Elgin tip, 1973 **E**; introduced with grass seed on the artificial
dunes at Findhorn.
N (Stables, *Cat.*); Nairn and Cawdor, 1900 (Lobban); grass fields
at Holme Rose, Cawdor and Geddes, 1954.
I (Watson, *Top. Bot.*); Lower Leachkin near Inverness, 1940
AJ **ABD**; waste ground at Longman Point and Ardersier, 1954;
roadside verge at Allanfearn and on top of a wall at Lentran;
waste ground by Longman tip, Inverness, 1971 **E**.

Lapsana L.
L. communis L.   *Nipplewort*
Native. Abundant on wood margins and waste ground. Map 249.
M Very common (Gordon); Grantown, 1890 *R. Ferrier* **E**; Forres,
1900 *JK* **FRS**; Aviemore, 1948 *E. M. Burnett* **ABD**; ½ mile S of
Forres station, 1953 *A. Melderis* **BM**; Culbin Forest and throughout
the county 1954.
N (Stables, *Cat.*); Ardclach (Thomson); waste ground at Dulsie
Bridge and throughout the county, 1954.
I Inverness, 1832 (Watson, *Loc. Cat.*); Bunchrew, 1940 *AJ* **ABD**;
by Loch Meiklie, *E. W. Groves* **EWG**; Glen Affric Lodge, 1947;
waste ground at Newtonmore, only occasional in the highlands but
throughout the lowlands.

Arnoseris Gaertn.
A. minima (L.) Schweigg. & Koerte   *Lamb's Succory*
    A. pusilla Gaertn.
Introduced in north Scotland. Cultivated land on sandy soil. Not
seen in recent years.
M At Viewfield, Urquhart and Easterton at Birnie (as *Lapsana
pusilla*) (Gordon); near Coxton, 1838 *P. Cruickshank* **E**; Birnie, 1857
G. Gordon **E**; Easterton, 1897 *P. Cruickshank* **E**.
I Culloden, *Dr McNab* and at Aigas in Strathglass 1836 (Gordon).

Hypochoeris L.
H. radicata L.   *Cat's Ear*
Native. Abundant on roadside banks, in short grassland, and on
moorland tracks and sand dunes.
M Very common (Gordon); Clunyhill at Forres, 1856 *JK* **FRS**;
Culbin, Findhorn and ½ mile S of Forres, 1953 *A. Melderis* **BM**;
moorland track at Aviemore and throughout the county 1954; sand
dunes in the Culbin Forest, 1972 **E**.
N (Stables, *Cat*); Ardclach and Nairn, 1900 (Lobban); dry bank at
Drynachan and throughout the county, 1954.
I Inverness, 1832 Dalwhinnie, 1833 (Watson, *Loc. Cat.*); canal
banks, Inverness, *c.* 1880 *P.F.* **INV**; Leachkin and near Croy
(Aitken); throughout the district, 1954.

H. glabra L.   *Smooth Cat's Ear*
Native. Rare in sandy fields and dunes.
M Near Elgin, 1833 *WAS* **E**; Morayshire, 1833 *W. Brand* **ABD**;

frequent (Gordon); Culbin Sands, 1942 *UKD* **UKD**!; Lossiemouth, 1953 *A. Melderis* **BM**; sandy field at Rosehaugh Farm near Elgin, 1956 **ABD**; waste ground at Greshop House, dunes at Covesea and on shingle by the River Spey near Fochabers; sandy field at the Warren near Brodie, 1971 **E**.

N Cawdor, 1833 *WAS* **GL** and 1837 **E**; sandy bank S of Loch of the Clans and on a shingle bank by Loch Flemington, 1954; dunes by Cran Loch, 1960 **ABD**; open place under beech trees at Househill, Nairn, 1974.

I Fort George, ex herb. *Balfour* **E**; Flemington, Inverness, 1841 *C. Babington* **E** and herb. *Borrer* **K**; pasture land at Little Croy (Aitken); sandy waste ground by Loch Flemington, 1960.

Leontodon L.

L. autumnalis L.   *Autumn Hawkbit*
Native. Abundant in dry open places on moors, tracks, dunes, river shingle and grass banks. The mountain race, subsp. *pratensis* (Koch) Archangeli has black woolly heads and has often been recorded in error for *L. hispidus* L.

Subsp. **autumnalis**
M Very common (Gordon); Forres, 1859 *JK* **FRS**; Lossiemouth, 1958 *Knox* **E**; dunes in the Culbin Forest, shingle of the River Spey at Garmouth and throughout the county.
N Moss of Litie near Auldearn, 1832 *WAS* **ABD**; meadows and pastures at Ardclach, 1891 *R. Thomson* **NRN**; moorland track at Drynachan and throughout the county, 1954; on rocks by the River Findhorn at Coulmony, 1961 **ABD,K**.
I Inverness, 1832 Dalwhinnie, 1833 (Watson, *Loc. Cat.*); banks of Loch Ness (Aitken); by the Spey at Rothiemurchus, 1889 *JK* **FRS**; Glen Einich, 1891 *AS* **E**; throughout the district, 1954.

Subsp. **pratensis** (Koch) Archangeli
Frequent in mountain districts on cliff ledges and river shingle. Under-recorded.
M Shingle of the River Spey at Garmouth, 1953 *A. Melderis* **BM**.
I Cairngorm hills, (as var. *taraxaci*) 1831 (Gordon); Fraoch-choire, Cannich, (Ball, 1851); Rothiemurchus, 1884 *JK* **FRS**; cliff ledge in the Coire an Lochain of Cairn Gorm, 1953; cliff ledges on Sgùrr na Lapaich in Glen Strathfarrar and in Glen Affric, on scree in Coire Laogh of Glen Banchor, grass banks in Gaick Forest, Glen Feshie and The Fara near Dalwhinnie etc.

L. hispidus L.   *Rough Hawkbit*
Native but introduced in this area.
M Boat of Garten, 1888 *AS* **GL**.

L. taraxacoides (Vill.) Mérat   *Lesser Hawkbit*
    L. leysseri Beck
Native but introduced in this area. Rare on disturbed ground and sand dunes.
M Waste ground by Elgin rubbish tip, 1975 **E**.

N Fixed dunes between the Nairn Dunbar golf course and the sea, 1972 *C. A. Stace, J. W. Grimes & J. H. Fremlin* **E**!
I Disturbed ground on the moor by the dam at Loch Cluanie, Glen Moriston, 1968 **E,CGE**; disturbed ground by the dam at Loch Mullardoch in Glen Cannich, 1971 *A. Currie et al.*!; in similar situations in two different localities in Glen Moriston, 1974; on the hill track to Loch Toll a' Mhuic in Glen Strathfarrar, 1975.

Picris L.

P. echioides L.   *Bristly Oxtongue*
Introduced.
M Introduced with grain in the parishes of Urquhart and Speymouth (Burgess); in a carrot field at Brodie, 1963 **E**.

Tragopogon L.

T. pratensis L. subsp. **minor** (Mill.) Wahlenb.   *Goat's-beard*
Introduced in this area. Rare on waste ground and railway embankments and station yards.
M Rare near Whirlig Gates at Elgin, 1826 (Gordon); station yard and by the railway at Garmouth, 1956 **ABD**; goods yard at Elgin station.
N Nairn, 1900 (Lobban); embankment near the station at Nairn, 1961 **E,ABD**
I Railway embankment near Clunes, Kirkhill, 1952 **E**; station yard at Kingussie, 1955 **E,CGE**; frequent on waste land on Longman Point, Inverness; embankment at Culloden 1962 **E,K** and in 1976 **BM**.

T. porrifolius L.   *Salsify*
Introduced. Garden outcast on dunes in Findhorn village, 1960 *M. E. Milward et al.* **ABD,K**!.

Mycelis Cass.

M. muralis (L.) Dumort.   *Wall Lettuce*
   Lactuca muralis (L.) Gaertn.
Native. Occasional on walls and rocks.
M Main's garden and offices at Elgin and at Birnie, 1844 (Gordon, *annot.*); walls at Moy House near Forres, 1925 and 1960 **ABD,K**; walls of Darnaway Castle near the Golden Gates, 1969 **E,BM,CGE**
N By the river Nairn at Holme Rose, 1974; planted at Skene Farm near Nairn from the cliffs at Aigas in Strathglass.
I Holly hedge at Springfield, Leys Castle and Aultnaskiach at Inverness (Aitken); Drummond, Inverness, A940 *AJ* **ABD**; abundant on the conglomerate cliffs by the dam at roadside N of Crask of Aigas, 1948; on a bank and on top of a wall near Cannich in Strathglass, 1968; abundant by the tracks in the Affric Forest in Glen Affric, 1971 **E,CGE**; wall by the lodge of Ness Castle near Inverness, 1971 **E**.

Sonchus L.

S. arvensis L. var. **arvensis**   *Perennial Sow-thistle*
Native. Common on cultivated land and waste places and among

stones on the drift-line of lochs and by the sea. Rare in the highlands where it is confined to railway yards. Map 250.

M Frequent (Gordon); cornfields near Forres, 1856 *JK* **FRS**; verge of the salt-marsh, among stones, Culbin Forest and throughout the county, 1954.

N (Stables, *Cat.*); opposite Druim House near Nairn, 1937 *J. Walton* **GL**; garden pest at Geddes House, at Holme Rose and Cawdor and on the drift-line by the sea E of Nairn, 1954.

I (Watson, *Top. Bot.*); Lentran, 1940 *AJ* **ABD**; waste land at Kirkhill, Inverness, Culloden, Inverdruie and among stones on the raised beach at Ardersier, 1954; on the old station platform at Moy, in a lay-by by Loch Ness and waste ground near Daviot village, 1975.

Var. **glabrescens** Hall

M Rubbish tip at Elgin, 1959 **K,JEL**.

S. oleraceus L.   *Sow-thistle*
Native. Abundant in cultivated ground and waste places. Stem-leaves with pointed auricles; achenes rugose.

M Very common (Gordon); Castlehill at Forres, 1850 *JK* **FRS**; track in the Culbin Forest and throughout the county, 1954.

N (Stables, *Cat.*); not common at Ardclach (Thomson); garden weed at Cawdor Castle, Geddes House and Glenferness etc., 1954.

I Inverness, 1832 (Watson, *Loc.Cat.*); Millarton (Aitken); abundant in the lowlands but rare in the highlands, 1954; waste ground at Whitebridge, on the old tip at Kingussie and garden weed at Newtonmore, 1975.

S. asper (L.) Hill   *Prickly Sow-thistle*
Native. Abundant in cultivated ground and in waste places. Stem-leaves with rounded auricles; achenes smooth.

M Castlehill, Forres, 1850 *JK* **FRS**; Culbin, 1953 *A. Melderis* **BM**; waste ground at Dyke and throughout the county, 1954.

N Ardclach, Cawdor and Nairn, 1900 (Lobban); below a wall at Dulsie Bridge and throughout the county, 1954.

I (Watson, *Top Bot.*); Bunchrew, 1940 *AJ* **ABD**; garden weed at Kirkhill, Inverness, Inverfarigaig and Kingussie etc., 1954; verge of a forest track in Glen Affric, in Glen Strathfarrar, Invermoriston and Fort Augustus etc.

Cicerbita Wallr.

C. macrophylla (Willd.) Wallr.   *Blue Sow-thistle*
    subsp. **uralensis** (Rouy) P. D. Sell
Introduced. A garden escape or outcast. Flowers bluish-lilac.

M Policies of Sanquhar House, Forres, 1956; in the wood behind the old walled garden at Altyre, 1966 **E**; by the burn at Fochabers, in a field below Pitcroy House, Strathspey and by the road to Glenlivet distillery at Cromdale.

N Ruins of a distillery near Nairn, 1942 *UKD* **UKD**; Riverside at Nairn, 1953 *J. B. Simpson* **Nat. Con. E**; garden outcast at Geddes

House, 1961 **ABD**; roadside verge near Courage and on the W of
Auldearn, 1974.
I Wood between Milton and Drumnadrochit, 1947; garden
outcast at Tomatin House, 1952.

C. plumieri (L.) Kirschl
I Garden escape at Tighnabruich near Invermoriston, 1950 **E**.

C. bourgaei (Boiss) Beauv.
M Formerly abundant outside the walled garden at Brodie Castle,
now exterminated with weed killer, 1967 **E,CGE**.

Hieracium L.

In the following account the extensive notes and key have been
compiled by P. D. Sell and C. West.

The area covered by this Flora is one of the finest for Hawkweeds
in the whole of the British Isles, and eighty-eight species have been
recorded. The mountains are particularly rich in species of the Series
Alpina and *Subalpina*, and representatives of all the main groups
occurring in the British Isles are to be found in the area as a whole.

No monograph covering all the species is at present available.
An attempt is made to give an aid to their identification by
providing a key and a reference to the best available description of
each species. When it refers to H. W. Pugsley, A Prodromus of
the British Hieracia in *Jour. Linn. Soc. London (Bot.)* **54** (1948), only
the word 'Pugsley' is given, followed by a page number and in some
cases (in brackets) by the name it appears under when it differs
from that given here. Larger plants within a colony will probably
key out better than small ones as the inflorescences and leaf teeth
are likely to be better developed. Style colour (pure yellow or
discoloured) should be noted when the plant is fresh. The small
glandular hairs on the leaves of the alpine species are best seen in
good light when the plant is fresh.

All herbarium records have been determined by P. D. Sell and
C. West unless marked with an asterisk.

Key

1. Ligules green   52. **chloranthum**
   Ligules yellow (sometimes poorly developed)   2.
2. Involucral bracts without or with few simple eglandular
   hairs   3.
   Involucral bracts with numerous to dense eglandular hairs.   23.
3. Stem leaves 0–1, rarely 2   4.
   Stem leaves more than 2, often more than 6   12.
4. Inflorescence often subumbellate; involucral bracts with
   few glandular hairs   60. **aggregatum**
   Inflorescence never subumbellate; involucral bracts with
   numerous or dense glandular hairs   5.
5. Involucral bracts with numerous to dense stellate hairs
   particularly on the margins
                          59. **piligerum**
   Involucral bracts without or with few stellate hairs   6.
6. Involucre 12–15 (–17) mm long
                          14. **atraticeps**
   Involucre up to 12 mm long   7.
7. Involucral bracts mostly obtuse   8.
   Involucral bracts all or mostly acute   9.
8. Petioles usually long; peduncles long and erect;
   involucral bracts 12–14 mm
                          30. **gracilifolium**
   Petioles usually short; peduncles short and sometimes
   curved; involucral bracts 10–12 mm
                          55. **uistense**
9. Leaves usually marbled; heads rather broad on straight
   peduncles; involucral bracts porrect in bud
                          29. **clovense**
   Leaves not marbled, rarely spotted; heads narrow, often
   on curved peduncles; involucral bracts incumbent in bud   10.
10. Basal leaves few, mostly attenuated at base; stem leaves
    (1–) 2–3 (–5); glands of involucral bracts slender
                          67. **diaphanoides**
    Basal leaves numerous, mostly truncate-based; stem
    leaves 0–1 (–2); glands on involucral bracts robust (most
    exotericum agg. will key out here)   11.
11. Styles yellow; leaves subentire to sparsely dentate
                          53. **exotericum**
    Styles discoloured; leaves deeply mammiform-dentate
                          54. **grandidens**
12. Involucral bracts without glandular hairs or with only
    few glandular hairs; often only down the median line   13.
    Involucral bracts with numerous glandular hairs, although
    they are sometimes minute   17.

13. Outer involucral bracts squarrose (curved outwards); inflorescence often more or less umbellate
    86. **umbellatum**
    All involucral bracts appressed; inflorescence never umbellate   14.

14. Peduncles with numerous, minute, rigid projections; involucral bracts without or with a few minute scattered glandular hairs   85. **maritimum**
    Peduncles without rigid projections; involucral bracts usually with a row of glandular hairs down the median line   15.

15. Styles pure yellow   84. **subumbellatiforme**
    Styles discoloured   16.

16. Leaves dark green, all less than three times as long as broad.
    78. **cambricogothicum**
    Leaves glaucous, most more than three times as long as broad   88. **salticola**

17. Involucral bracts narrow, acute 67. **diaphanoides**
    Involucral bracts broad, obtuse   18.

18. Stem leaves mostly cuneate at base, not amplexicaul
    77. **scabrisetum**
    Median and upper stem leaves mostly rounded at base, more or less amplexicaul   19.

19. Styles pure yellow   20.
    Styles discoloured   21.

20. Leaves more or less glaucous, ovate-lanceolate, denticulate or with occasional longer teeth
    79. **latobrigorum**
    Leaves yellowish green, oblong-lanceolate, usually markedly serrate-dentate   83. **reticulatum**

21. Peduncles with numerous glandular hairs; heads small and narrow; achenes pale brown
    72. **prenanthoides**
    Peduncles without glandular hairs; heads large and broad; achenes dark   22.

22. Involucral bracts with few to numerous large glandular hairs, no minute glandular hairs, sometimes with an occasional simple eglandular hair
    80. **subcrocatum**
    Involucral bracts with numerous large glandular hairs, numerous minute glandular hairs and a few simple eglandular hairs   81. **drummondii**

23. Involucral bracts with numerous glandular hairs   24.
    Involucral bracts without or with few glandular hairs   65.

24. Involucral bracts obtuse   25.
    At least the inner involucral bracts acute   41.

25. Stem leaves 5 or more   26.
    Stem leaves less than 4   32.

26. Styles discoloured   27.
    Styles pure yellow   29.
27. Glandular hairs on involucral bracts minute
                    87. perpropinquum
    Glandular hairs on involucral bracts larger and obvious   28.
28. Leaves not more than three times as long as broad
                    71. dewari
    Some leaves at least four times as long as broad
                    82. strictiforme
29. Lower surface of leaves with numerous stellate hairs
                    74. uiginskyense
    Lower surface of leaves with few or no stellate hairs   30.
30. Leaves entire or remotely denticulate, often marbled
                    73. sparsifolium
    Leaves serrate-dentate, never marbled but rarely spotted   31.
31. Leaves 6–10 (–20), mostly elliptic-lanceolate, not amplexicaul, rapidly decreasing in size upwards
                    76. stewartii
    Leaves 15–30, mostly oblong-lanceolate, more or less amplexicaul, gradually decreasing in size upwards
                    83. reticulatum
32. Involucral bracts with numerous stellate hairs, especially on the margin   33.
    Involucral bracts without or with few stellate hairs mostly at the base   35.
33. At least some leaves sharply laciniate-dentate
                    20. cuspidens
    Leaves subentire to more or less dentate, never laciniate-dentate   34.
34. Most leaves with a more or less acute apex; involucral bracts with stellate hairs more or less evenly distributed and glandular hairs obscured by the more obvious simple eglandular hairs   **50. caledonicum**
    Most leaves with a broad rounded apex; involucral bracts with dense stellate hairs on the margins and at the apex and with gladular hairs obvious and numerous
                    59. piligerum
35. Upper surface of leaves with numerous hairs   36.
    Upper surface of leaves glabrous or with few hairs   38.
36. Leaves subentire to shallowly dentate, without cusped teeth; involucre 9–11 mm
                    32. dasythrix
    Leaves with some narrowly mammiform cusped teeth; involucre 11–15 mm   37.
37. Ligules with numerous short hairs at apex; capitula rounded at base   **18. molybdochroum**
    Ligules glabrous or with few hairs; capitula narrowed at base   **26. isabellae**

38. Leaves bluish-green, ovate-lanceolate; ligules glabrous
 28. **melanochlorocephalum**
 Leaves bright green, usually narrow; ligules often with hairs at apex
 39.

39. Simple eglandular hairs on involucral bracts pale or only dark at very base; styles yellow
 46. **nitidum**
 Simple eglandular hairs on involucral bracts dark or dark for a good proportion of their length; styles usually discoloured, rarely yellow
 40.

40. Leaves narrowly elliptic, usually with unequal, narrowly mammiform, cusped teeth 21. **anfractiforme**
 Leaves usually broader, lanceolate-oblong, usually only shallowly dentate 27. **vennicontium**

41. Underside of basal leaves with numerous to dense stellate hairs
 42.
 Underside of basal leaves without or with few stellate hairs
 46.

42. Involucral bracts with few stellate hairs   43.
 Involucral bracts with numerous stellate hairs at least on the margins
 44.

43. Base of basal leaves cuneate or attenuate; ligules glabrous or with an occasional hair at apex
 35. **pseudanglicoides**
 Base of basal leaves truncate, rounded or abruptly contracted; ligules with numerous hairs at apex
 36. **shoolbredii**

44. Styles discoloured; capitula narrowed at base
 24. **hyparcticoides**
 Styles pure yellow; capitula rounded at base   45.

45. Leaves mostly acute, flaccid, with long, narrowly mammiform teeth in lower half, with subrigid hairs on margin
 41. **stenopholidium**
 Leaves mostly obtuse, thick, often subentire, rarely with broad, mammiform teeth at base, with thick rigid hairs on margin 45. **lasiophyllum**

46. Leaves mostly abruptly contracted or more or less truncate at base
 47.
 Leaves mostly cuneate or attenuate at base   51.

47. Involucral bracts with numerous stellate hairs especially on the margins 43. **saxorum**
 Involucral bracts with sparse stellate hairs   48.

48. Involucre 9–11 mm, very narrow   49.
 Involucre 11–15 mm, much broader   50.

49. Leaves mostly narrowly elliptic or oblong-lanceolate and more or less obtuse, sometimes spotted but not marbled
    56. **duriceps**
    Leaves mostly ovate, more or less acute, usually marbled brownish-purple   57. **scotostictum**

50. Ligules markedly hairy at apex
    31. **laetificum**
    Ligules glabrous or with a few hairs at apex
    58. **pictorum**

51. Ligules glabrous   52.
    Ligules with hairs at apex   56.

52. Leaves narrow, very glaucous; cauline leaves spreading
    49. **argenteum**
    Leaves broader, green or slightly glaucous; cauline leaves
    not spreading   53.

53. Leaves with thick rigid hairs on the margin and sometimes
    on the upper surface   44. **schmidtii**
    Leaves with softer, more slender hairs   54.

54. Leaves usually spotted   48. **sommerfeltii**
    Leaves not spotted   55.

55. Stem leaves 0–1   61. **subtenue**
    Stem leaves 2–4 (–8)   69. **vulgatum**

56. Upper surface of leaves without or with few hairs   57.
    Upper surface of leaves with numerous hairs   59.

57. Glandular hairs on peduncles and on involucral bracts
    small and obscure   23. **glandulidens**
    Glandular hairs at apex of peduncle and on involucral
    bracts medium and obvious   58.

58. Leaves broader, cauline not amplexicaul; involucral
    bracts with shorter less dense simple eglandular hairs
    31. **laetificum**
    Leaves mostly narrowly elliptic, cauline more or less
    amplexicaul; involucral bracts with dense long dusky
    simple eglandular hairs   34. **pseudanglicum**

59. Leaves without minute glandular hairs   60.
    Leaves with minute glandular hairs   61.

60. Leaves with thick rigid simple eglandular hairs on
    margins and upper surface 44. **schmidtii**
    Leaves with soft simple eglandular hairs
    64. **dipteroides**

61. Capitula usually numerous, small, narrowed at base
    22. **centripetale**
    Capitula few, large, broad at base   62.

62. Glandular hairs of involucral bracts and peduncles small
    and hidden among the larger simple eglandular hairs
    16. **cremnanthes**
    Glandular hairs of involucral bracts and peduncles larger
    and obvious   63.

63. Involucre 12–14 mm; leaves shallowly dentate
      17. senescens
    Involucre 15–18 mm; leaves usually strongly and
    irregularly dentate   64.
64. Ligules strongly hairy on back and at apex; styles
    discoloured   **5. memorabile**
    Ligules sparingly hairy mostly at apex; styles usually
    yellow   **13. hanburyi**
65. Minute yellowish glandular hairs present on leaves;
    alpine plants with solitary or few capitula and blackish
    involucres   66.
    Leaves without minute yellowish glandular hairs   85.
66. Involucral bracts with numerous stellate hairs especially
    on the margins   **25. callistophyllum**
    Involucral bracts with few or no stellate hairs   67.
67. Ligules with few rarely numerous simple hairs at apex   68.
    Ligules with numerous to dense simple hairs at apex and
    on the back   73.
68. Involucral bracts mostly obtuse
      19. marshallii
    Involucral bracts mostly acute   69.
69. Leaves glabrous or nearly so on upper surface
      7. globosiflorum
    Leaves with numerous hairs on upper surface   70.
70. Leaves long-lanceolate or lingulate, entire or with a few
    teeth   **15. lingulatum**
    Leaves shorter and broader, dentate often markedly so   71.
71. Involucre 12–15 mm   **16. cremnanthes**
    Involucre 15–18 mm   72.
72. Leaves not spotted, narrowly oblanceolate or lanceolate
      8. larigense
    Leaves usually spotted, ovate or ovate-lanceolate
      12. pseudocurvatum
73. Lower surface of leaves with numerous stellate hairs
      6. grovesii
    Lower surface of leaves without or with few stellate hairs 74.
74. Leaves entire or denticulate   75.
    Leaves dentate, often markedly so   79.
75. Styles discoloured   **3. tenuifrons**
    Styles yellow   76.
76. Leaves mostly obtuse at apex   77.
    Leaves mostly acute at apex   78.
77. Inner involucral bracts very long and parallel sided;
    involucre with dense, long, shaggy hairs particularly at
    base   **1. holosericeum**
    Inner involucral bracts linear-lanceolate; involucral
    bracts less shaggy   **2. alpinum**

78. Stem with long shaggy hairs
4. eximium
Stem with shorter less numerous hairs
10. macrocarpum

79. Styles discoloured   80.
Styles yellow   82.

80. Leaves broadly elliptic or ovate
11. calenduliflorum
Leaves narrowly elliptic or lanceolate   81.

81. Hairs of involucral bracts long and shaggy
4. eximium
Hairs of involucral bracts short and less dense
5. memorabile

82. Leaves mostly acute   83.
Leaves mostly obtuse   84.

83. Stem with long shaggy hairs
4. eximium
Stem with shorter less numerous hairs
10. macrocarpum

84. Stem leaves large   **2. alpinum** var. **insigne**
Stem without leaves or only with bracts
9. graniticola

85. Involucral bracts without or with few stellate hairs   86.
Involucral bracts with numerous stellate hairs at least on margins   91.

86. Cauline leaves 5–16   **75. gothicoides**
Cauline leaves less than 5   87.

87. Involucral bracts acute   88.
Involucral bracts obtuse   89.

88. Cauline leaves semi-amplexicaul
40. anglicum
Cauline leaves not semi-amplexicaul
51. orimeles

89. First flowering capitulum often almost sessile; involucral bracts few and broad   **66. orcadense**
First flowering capitulum usually on long peduncle; involucral bracts more numerous and narrower   90.

90. Leaves ovate with a truncate base
63. lintonianum
Leaves narrower usually with a cuneate base
62. subhirtum

91. Leaves with numerous to dense stellate hairs beneath   92.
Leaves without or with few stellate hairs beneath   94.

92. Leaves narrowed at base, cauline (1–) 2–5(–8)
69. vulgatum
Leaves rounded, abruptly contracted or subtruncate at base, cauline 0–2   93.

93. Styles discoloured; cauline leaves semi-amplexicaul
    37. **flocculosum**
    Styles yellow; cauline leaf not amplexicaul
    47. **jovimontis**

94. Stem leaves semi-amplexicaul   95.
    Stem leaves absent or if present not semi-amplexicaul   97.

95. Leaves spotted and blotched
    33. **petrocharis**
    Leaves not spotted   96.

96. Leaves bluish-green, abruptly contracted or subtruncate
    at base   38. **ampliatum**
    Leaves yellowish-green, cuneate-based
    39. **langwellense**

97. Leaves subtruncate or abruptly contracted at base   98.
    Leaves cuneate or attenuate at base   100.

98. Leaves markedly and irregularly dentate, stem leaves
    often 2   65. **caesiomurorum**
    Leaves subentire or denticulate for most of leaf, with a
    few teeth near the base, stem leaf 0 or 1   99.

99. Leaves spotted and blotched
    33. **petrocharis**
    Leaves not spotted or blotched
    42. **dicella**

100. Leaves spotted and blotched, glabrous or nearly so on
     upper surface   48. **sommerfeltii**
     Leaves not spotted or blotched, more or less hairy   101.

101. Involucral bracts with numerous medium simple eglandular
     hairs; leaves green   69. **vulgatum**
     Involucral bracts with dense long simple eglandular
     hairs; leaves bluish-green   102.

102. Leaves ovate or broadly elliptic
     68. **rubiginosum**
     Leaves lanceolate or oblong-lanceolate
     70. **cravoniense**

Series **Alpina** (Fries) Sell & West

Dwarf, rather shaggy plants with 1 (rarely 2–3) large blackish heads.
Leaves mostly basal, if present on stem small and bractlike, with few
to numerous, minute, yellowish glandular hairs most obvious on the
margins. Ligules with hairs at apex and often on the back. These
are native plants of alpine rock ledges and bare stony and grassy
slopes, and screes on granite, mica-schist and slate, mostly at an
altitude of over 650 m (2,000 ft). They do not usually flower until
July and August.

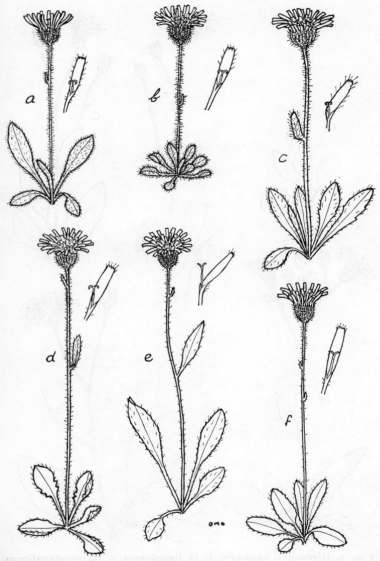

Fig 33   a, **Hieracium alpinum**; b, **H. holosericeum**; c, **H. eximium**; d, **H. graniticola**; e, **H. tenuifrons**; f, **H. globosiflorum**.

Fig 34   a, **Hieracium hanburyi**; b, **H. lingulatum**; c, **H. pseudanglicum**;
d, **H. centripetale**; e, **H. anglicum**.

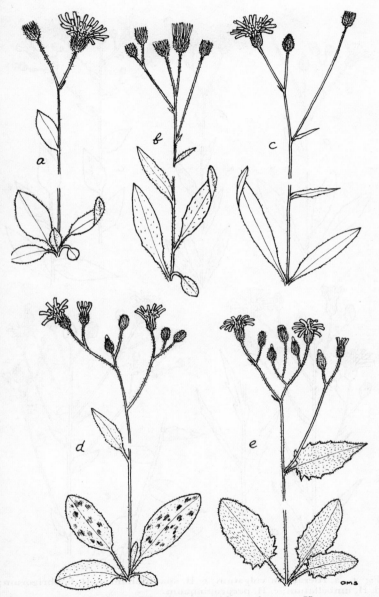

383

Fig 35   a, **Hieracium shoolbredii**; b, **H. chloranthum**; c, **H. argenteum**;
d, **H. duriceps**; e, **H. grandidens**.

Fig 36 a, **Hieracium vulgatum**; b, **H. sparsifolium**; c, **H. latobrigorum**; d, **H. umbellatum**; e, **H. perpropinquum**.

1. **H. holosericeum** Backh.   Pugsley, 40
Fig 33b.
I   South side of Glen Einich (Druce, 1888); summit of Ben Chluaig, 1891 *WASh* **CGE**; Cairn Gorm, 1898 *ESM* **CGE**; Tom a'Choinich, Glen Affric, 1947 *NDS* **BM**; A'Phòcaid, Glen Einich, 1966 *RJ & BAM* **CGE**; cliff ledge, Sgùrr la Lapaich, Glen Affric, 1,000 m, 1971 **E**; Sgùrr na Lapaich, Glen Strathfarrar, 1971 **E**; Càrn a' Chulinn in Glen Doe, 1976 *OMS & MMcCW* **E**.

2. **H. alpinum** L.   Pugsley, 39   *Alpine Hawkweed*
Fig 33a.
I   Rare. Craighue in Badenoch, 1831 (Gordon); Fraoch-choire, Toll Cregach, Cannich (Ball, 1851); rocks NE of Dalwhinnie, 1867 *A. Coore* (Balfour, 1868); Cairn Gorm, 1874 *A. Ley* **CGE**; Glen Einich and Coire an t-Sneachda, 1877 *G. C. Druce* **OXF**; Cairn Gorm, 1898 *ESM & WASh* **CGE**; Cairn Gorm, 1898 *E. F. & W. R Linton* **E**; summit of Càrn Glas, Glen Affric, 1947 *C. C. Townsend* **CGE**; bare plateau in the Coire an Lochain, 1950; Glen Einich, 1951 *J. E. Raven* **CGE**; Sgòran Dubh and Glen Einich, 1952 *C. West* **CGE**; Glen Einich, 1965 *RJ & BAM* **CGE**; Coire Garbhlach, Glen Feshie, 1966 *RJ & BAM* **CGE**; Sgùrr na Lapaid, Glen Strathfarrar, 1972 *OMS* **E**; summit of Meall Dubh S of Glen Moriston and The Fara near Dalwhinnie, 1975 **E**.

var. **insigne** Bab.   Pugsley, 40
This is almost certainly a good species.
I   Coire Garbhlach, 1966 *RJ & BAM* **CGE**.

3. **H. tenuifrons** Sell & West   Pugsley, 46 (as *H. gracilentum*)
        H. gracilentum Backh., non (Fries) Backh.
Fig 33e.
I   Coire Garbhlach, 1850 *G. A. W. Arnott* **E**; Glen Einich, 1887 *G. C. Druce* **CGE**; Sgòran Dubh, Cairngorms, 1891 *AS* **E**; Coire Chùirn at 833 m, 1911 *ESM* **E**; Sgòran Dubh Mòr, 1965 *RJ & BAM* **CGE**; Creag an Leth-choin, A' Phòcaid and Coire Garbhlach, 1966 *RJ & BAM* **CGE**; gully of burn on Bynack More, 1975 *OMS & MMcCW* **E**.

4. **H. eximium** Backh.   Pugsley, 42
Fig 33c.
Variants with discoloured styles (var. *eximium*) and pure yellow styles (var. *tenellum* Backh.) occur.
I   Braeriach, 1849 ? **E**; Beinn a' Bha'ach Ard, 1850 *J. Ball* **E**; Glen Einich, 1887 *G. C. Druce* **OXF**; Glen Einich, 1892 *AS* **E**; Coire an t-Sneachda, 1893 *AS* **E**; Cairn Gorm (Marshall, 1899); Coire an t-Sneachda, 1949 *P. S. Green* **E**; north-facing cliffs of Sgùrr na Lapaich, Glen Strathfarrar at 1,000 m, 1971 and in 1972 **E, CGE**; gully on Bynack More, 1975 *OMS & MMcCW* **E**; summit of Càrn a' Chuilinn, Glen Doe, 1976 *OMS & MMcCW* **E**.

5. **H. memorabile** Sell & West   *Watsonia* **6**: 304 (1967)
    I   Glen Einich, 1951 *J. E. Raven* **CGE**; Sgòran Dubh, Glen Einic
    1951 *J. W. Cardew & C. West* **CGE**; in two places Glen Einich, 19⟨
    RJ & BAM **CGE**; Coire Garbhlach, 1966 *RJ & BAM* **CGE**.

6. **H. grovesii** Pugsley
    I   Glen Einich, 1952 *J. W. Cardew & C. West* **CGE**; upper coi
    of Glen Einich, 1961 *RJ & BM* **CGE**; Sgòran Dubh Mòr, 19⟨
    RJ & BAM **CGE**; Creag an Leth-choin, and north of Fuara
    Diotach, Glen Einich, 1966 *RJ & BAM* **CGE**.

7. **H. globosiflorum** Pugsley   Pugsley, 46
    H. globosum Backh., non Desf.
    Fig 33f.
    I   Larig Pass, 1886 *H. & J. Groves* **CGE**; Glen Einich, 18⟨
    G. C. Druce **CGE**; Braeriach, 1892 *AS* **E**; Tom a' Chonich, 194
    NDS **BM**; Sgùrr na Lapaich, Glen Affric, 1950 *R. C. L. Burges* **CGE**
    Glen Einich, 1951 *J. W. Cardew & C. West* **CGE**; Glen Einich, 19⟨
    C. West **CGE**; A' Phòcaid, 1965 *RJ & BAM* **CGE**; Creag a
    Leth-choin, 1966 *RJ & BAM* **CGE**; gully on Bynack More, 197
    OMS & MMcCW **E**.

8. **H. larigense** (Pugsley) Sell & West   Pugsley, 47 (as *H. globos*
    florum var. *larigense*)
    I   Between Ben Macdui and Cairn Gorm, 1847 ? *coll.* ex her⟨
    Murchiston **E**; Creag an Leth-Choin, 1952 *J. W. Cardew & C We*
    CGE, and in 1966 *RJ & BAM* **CGE**.

9. **H. graniticola** W. R. Linton   Pugsley, 45
    Fig 33d.
    I   Cairn Gorm, 1898 *ESM* **CGE**; gully on Bynack More, 197
    OMS & MMcCW **E**.

10. **H. macrocarpum** Pugsley   Pugsley, 44
    I   A' Phòcaid, Glen Einich, 1950 *C. D. Pigott* **CGE**, and in 196
    RJ & BAM **CGE**; Coire Garbhlach, north of Fuaran Doitach an
    Glen Einich, 1966 *RJ & BAM* **CGE**; Sgòran Dubh in Glen Einich
    1967 *J. N. Mills & C. West* **CGE**.

11. **H. calenduliflorum** Backh.   Pugsley, 43
    I   Rocks NE of Dalwhinnie, 1867 *A. Coore* (Balfour, 1868); sout⟨
    side of Coire Garbhlach, Glen Feshie, 1966 *RJ & BAM* **CGE**.

12. **H. pseudocurvatum** (Zahn) Pugsley   Pugsley, 49
    I   Strath Affric, 1850 *J. Ball* **E**; Sgòran Dubh Mòr, 1956 *P. D. Se*
    CGE; Glen Einich and Sgòran Dubh Mòr, 1965 *RJ & BAM* **CGE**
    Creag an Leth-choin and Coire Garbhlach, 1966 *RJ & BAM* **CGE**
    A' Chailleach, Monadhliath, 1966 *A. McG. Stirling* **CGE**.

13. **H. hanburyi** Pugsley   Pugsley, 51
    Fig 34 a.
    I   Sgòran Dubh, Cairngorms, 1891 *AS* **E**; Cairn Gorm, 1898 *ESM*
    CGE; Larig Ghru, Cairngorms, 1905 *E. Edgar Evans* **E**; Gea⟨

Chàrn, Glen Markie, 1916 *ESM* **CGE**; Tom a' Choinich, 1947 *NDS* **BM**; Glen Einich, 1951 *J. W. Cardew & C. West* **CGE**; Sgòran Dubh Mòr, 1956 *P. D. Sell* **CGE**; Glen Einich in three places, 1965 *RJ & BAM* **CGE**; near summit of Sgùrr nan Conbhairen, 1968 *A. McG. Stirling* **CGE**; cliff ledge, Sgùrr na Lapaich, Glen Strathfarrar at 830 m, 1971 **E,BM**; ledge on Coire nan Laogh, Glen Banchor at 830 m, 1973 **E**; summit of Càrn a' Chuilinn, Glen Doe, 1976 *OMS & MMcCW* **E**.

14. **H. atraticeps** (Pugsley) Sell & West   Pugsley, 51 (as *H. hanburyi* var. *atraticeps*)
  I Glen Einich, 1952 *J. W. Cardew & C. West* **CGE**; Creag an Leth-Choin, 1966 *RJ & BAM* **CGE**; cliff ledge above Loch Toll a' Mhuic, Monar, 1975 *OMS & MMcCW* **E**.

Series **Subalpina** (Pugsley) Sell & West
Short to medium-sized plants with few medium-sized to large heads on straight peduncles. Leaves usually with at least a few minute glandular hairs especially on the margin, the basal few, those of the stem 0 to 4. Involucral bracts blackish-green usually with dark simple eglandular and/or glandular hairs. Native plants of rock-ledges and rocky stream-sides usually above 500 m (1,500 ft). They flower from mid-June to the beginning of August. They can be distinguished from species of the Series *Bifida* and of the Series *Vulgata* by their darker heads on straight peduncles and minute glandular hairs on their leaves.

15. **H. lingulatum** Backh. ex Hooker & Arnott   Pugsley, 55
Fig 34b.
  I South side of Glen Einich (Druce, 1888); Braeriach, 1892 *AS* **E**; Allt a' Choire Chàis, 1911 *ESM* **CGE**; cliffs, 830 m above Loch Dubh five miles north of Cluny Castle, Laggan, 1911 *ESM* **CGE**; near Aviemore, 1951 *C. West* **CGE**; Sgòran Dubh, Glen Einich, *J. W. Cardew & C. West* **CGE**; Glen Einich, 1965 *RJ & BAM* **CGE**; Creag an Leth-choin and Coire Garbhlach, 1966 *RJ & BAM* **CGE**; north-facing cliff ledge, Sgurr na Lapaich, Glen Strathfarrar, 830 m, 1971 **E,BM,CGE**; gully of burn, Gaick Lodge, 1973 **E**; mountain ledges in Coire nan Laogh, Glen Banchor, 1973 **ABD**; cliff ledge Sgùrr na Muice, 1975 *OMS & MMcCW* **E**; cliffs of Creag Dhubh S of Loch Ericht, 1976 *OMS & MMcCW*; cliffs by Lochan na Doire-uaine, Strath Mashie, 1976 *OMS & MMcCW* **E**.

16. **H. cremnanthes** (F. J. Hanb.) Pugsley   Pugsley, 57
  I South end of Ben Alder below Bealach Breabog, 1911 *ESM* **E,CGE**; cliffs of A' Chailleach, Monadhliath, 1966 *A. McG. Stirling* **CGE**.

17. **H. senescens** Backh.   Pugsley, 58
  I South side of Glen Einich, 1887 (Druce, 1888); rocks of the Piper's Burn, Glen Markie, 1916 *ESM* **CGE**; Glen Einich, 1951

J. E. Raven **CGE**; near the summit of Sgùrr nan Conbhairean, 196€
A. McG. Stirling **CGE**.

18.   **H. molybdochroum** (Dahlst.) Omang   Pugsley, 55
      I   ? locality, 1880 *G. A. W. Arnott* **E**; Sgòran Dubh in Glen Einich
      1965, *RJ & BAM* **CGE**.

19.   **H. marshallii** E. F. Linton   Pugsley, 56
      I   ? locality, 1880 *G. A. W. Arnott* **E**; Glen Markie, 1916 *ESM*
      CGE.

20.   **H. cuspidens** Sell & West   *Watsonia* **6**: 306 (1967)
      H. dissimile var. *majus* Pugsley
      I   Glenmore Lodge, 1953 *NDS* **BM**; cliffs above Lochan na
      Doire-uaine, Strath Mashie, 1976 *OMS & MMcCW* **CGE**.

21.   **H. anfractiforme** E. S. Marshall   Pugsley, 175
      I   Glen Feshie, 1966, *RJ & BAM* **CGE**; basic cliffs by the Allt
      Lorgaidh, upper Glen Feshie, 1974 **E**; cliffs on Sgùrr na Muice
      Monar, 1975 *OMS & MMcCW* **E**; rocks on The Fara near
      Dalwhinnie, 1975 **E**.

22.   **H. centripetale** F. J. Hanb.   Pugsley, 63
      Fig 34 d.
      I   Allt a' Choire Chàis, 666 m, 1911 *ESM* **CGE**; stream-side rocks
      Allt Coire Dhuibh, Laggan, 1916 *ESM* **CGE**; Piper's Burn, 500 m
      1916 *ESM* **CGE**; Glen Shira, 1916 *ESM* **CGE**; south side of Coire
      Garbhlach, 1966 *RJ & BAM* **CGE**; gorge of small burn, Lurgmore
      333 m, 1975 **E**; low cliff by Loch Monar on Creag an Fhir-eòin, 197£
      E; burn gully Clach Criche, Glen Moriston, 1975 **E**.

23.   **H. glandulidens** Sell & West   *Watsonia* **6**: 306 (1967)
      I   Bank of the Allt Fraoch-choire, Glen Strathfarrar, 1971 *A. Curri*
      AC; rock face, Loch Killin, 1972 **E,CGE**; gorge of Markie Burn
      1972 **E**; gully of burn, Gaick Lodge, 1973 **CGE**.

24.   **H. hyparcticoides** Pugsley   Pugsley, 64
      I   Stream below Coire Chomhlain, Ben Alder, Allt a' Choire
      Chàis near Dalwhinnie and Allt Beul an Sporain west of Drumochter
      at 666 m, all 1911 *ESM* **CGE**.

25.   **H. callistophyllum** F. J. Hanb.   Pugsley, 66
      I   Allt a' Choire Chàis near Dalwhinnie, 666 m, 1911 *ESM* **CGE**;
      4–5 miles north of Cluny Castle, 1911 *ESM* **CGE**; granite cliff,
      Sgòran Dubh, 1952 *C. West* **CGE**; banks of the Allt Toll a' Mhuic,
      Monar, at 666 m, 1975 *OMS & MMcCW* **E**.

26.   **H. isabellae** E. S. Marshall   Pugsley, 65
      I   Allt Beul an Sporain near Dalwhinnie above 660 m, 1911 *ESM*
      & WASh **E,CGE**; below Coire Chomhlain, Ben Alder, 1911 *ESM* **E**;
      Allt an Lochain Dubh, 4½ miles north of Cluny Castle, 1911 *ESM &*
      WASh **CGE**; Allt a' Choire Dhuibh, Glen Shirra, 1916 *ESM* **CGE**.

27. **H. vennicontium** Pugsley   Pugsley, 174
  I   Four miles north of Cluny Castle, Laggan, 1911 *ESM* **CGE**;
Sgòr Gaoith, Glen Einich, 1951 *C. West* **CGE**; rocks by the River
Coiltie at Balmacaan near Lewiston, 1972 **E**; rock face, Loch
Killin, 1972 **E,BM**; north-facing cliffs of Sgùrr na Lapaich, Glen
Strathfarrar at 900 m, 1972 **ABD**.

28. **H. melanochloricephalum** Pugsley   Pugsley, 173
  I   Sgòran Dubh, Glen Einich, 1952 *C. West* **CGE**; head of Glen
Einich and A' Phòcaid, 1966 *RJ & BAM* **CGE**; rocks in gully of
burn, Gaick Lodge, 1973 **CGE**.

29. **H. clovense** E. F. Linton   Pugsley, 117
  I   Sgòran Dubh, Glen Einich, 1952 *C. West* **CGE**; and in 1965
RJ & BAM **CGE**; banks of burn, Gaick Lodge, 1973 **CGE**.

30. **H. gracilifolium** (F. J. Hanb.) Pugsley   Pugsley, 60
  I   Allt a' Choire Chàis, near Dalwhinnie, 1911 *ESM* **CGE**; north-
facing cliffs of Sgùrr na Lapaich, Glen Strathfarrar, 915 m, 1972 **E**.

31. **H. laetificum** Sell & West   *Watsonia* **6**: 307 (1967) (as *H.*
                                    nigrisquamum)
  H. nigrisquamum Sell & West, non Hyl.
  I   North-facing cliffs of Sgùrr na Lapaich, Glen Strathfarrar,
1,000 m, 1972 **E,CGE**.

32. **H. dasythrix** (E. F. Linton) Pugsley   Pugsley, 66
  I   *An Leth-chreag, Glen Strathfarrar, 1971 *J. N. Mills* **JNM**.

33. **H. petrocharis** (E. F. Linton) W. R. Linton   Pugsley, 76
  M Rocks below Randolph's Leap, Relugas, 1970 **CGE**.
  I   Glen Feshie, 1956 *UKD* **CGE**; gully of the Allt Gharbh Chaig,
Gaick, 1974, *OMS* **E**; rocks of a small cliff south of Laggan Bridge,
1974 *OMS* **E**.

34. **H. pseudanglicum** Pugsley   Pugsley, 59
Fig 34c.
  I   Allt Gharbh Ghaig, Gaick, 1953 *C. D. Pigott* **CGE**; Creag an
Lochan, 1959 ? *coll.* **E**; A' Phòcaid, 1965, Coire Garbhlach, 1966
RJ & BAM **CGE**; north-facing cliffs of Sgùrr na Lapaich, Glen
Strathfarrar at 833 m, 1972 **E**; rock face, Loch Killin at 416 m, 1972
E; Allt Gharbh Ghaig, Gaick, 1974 *OMS* **E**; Allt Lorgaidh, upper
Glen Feshie, 1974 **E**.

35. **H. pseudanglicoides** Raven, Sell & West   *Watsonia*, **6**: 308
   (1967)
  I   Cliffs of A'Chailleach, Monadhliaths, 1966 *A. McG. Stirling* **CGE**;
south of Glen Feshie Lodge, 1969 *A. McG. Stirling* **CGE**; basic cliff
by the Allt Lorgaidh, upper Glen Feshie, 1974 **CGE**.

Series **Cerinthoidea** (Fries) Sell & West
Normally robust, softly hairy plants with few large heads on long,
straight peduncles. Leaves without glandular hairs, more or less

glaucous, basal few, those of the stem 1–6 (–8), usually more or le
amplexicaul. Involucres with dense simple eglandular hairs, brac
incumbent in bud. Ligules with hairs at apex. Styles discoloure
Margins of receptacle pits fimbriate-dentate. The species of th
group are native on cliff ledges and by rocky streams in uplan
regions, and flower from late May until September.

36. **H. shoolbredii** E. S. Marshall   Pugsley, 79
Fig 35a.
M Gorge of the Findhorn river, Darnaway, opposite Sluie, 196
CGE; earth mounds and cliffs of the old lime quarry at Cotha
near Forres, 1963 **CGE**; rocks of the Findhorn River opposit
Coulmony House, 1972 **E**.
I By the Allt Beul an Sporain, Dalwhinnie, 1911 *ESM & WAS*
E,CGE; Coire Chùirn, 1911 *ESM* **CGE**; Creag Dhubh south (
Loch Ericht above 666 m, 1911 *ESM* **E**; Coire Garbhlach, 195
M. C. F. Proctor & K. M. Goodway **CGE**; River Tromie, Tromi
Bridge, Kingussie, 1965 *J. M. Lock* **CGE**; Allt Coire Chùir
and Coire Garbhlach, 1966 *RJ & BAM* **CGE**; shingle of river
Milton, 1971 **E**; cliff ledge on the east side of Mealfuarvonie
Drumnadrochit, 1971 **ABD,CGE**; gorge of burn above Falls (
Divach, Lewiston, 1972 **E**; conglomerate cliffs below Tom Bailgeann
1972 **BM,K**; rocks by river at Spey Dam, Laggan, 1972 **K**; gully b
the Allt Gharbh Ghaig, Gaick, 1974 *OMS* **E**; basic cliff by the All
Lorgaidh, upper Glen Feshie, 1974 **ABD,BM**; ravine A' Chaoirnich
Gaick, 1974 **E**; shingle River Spey, Newtonmore, 1975; burn gull
Drumochter, 1975 **E**; cliffs by Lochan na Doire-uaine, Strat
Mashie, 1976 *OMS & MMcCW* **E**; cliffs on Creag Dhubh S o
Loch Ericht, 1976 *OMS & MMcCW* **BM**.

37. **H. flocculosum** Backh. ex Bab.   Pugsley, 77
N Ardclach by the Findhorn, 1888 *R. Thomson* **NRN**.
I Glen Einich, 1951 *C. West* **CGE**; south side of Coire Garbhlach
1966 *RJ & BAM* **CGE**; lower part of Allt Fraoch-choire, Cannicl
at 500 m, 1971 *A. Currie* **AC**.

38. **H. ampliatum** (W. R. Linton) A. Ley   Pugsley, 74
I Rocks in burn gully, Gaick Lodge, 1973 **E,CGE,BM**; basi
rocks by the track one mile north of Creag Bheag, Glen Feshie, 1974
E,BM; gully by the Allt Gharbh Ghaig, Gaick, 1974 *OMS* **E**
cliffs of Creag Dhubh, Loch Ericht, 1976 *OMS & MMcCW* **E**
cliff, Loch na Doire-uaine, 1976 *OMS & MMcCW* **E**; by the All
an t-Sluie near Dalwhinnie, 1976 *OMS & MMcCW* **E**.

39. **H. langwellense** F. J. Hanb.   Pugsley, 76
I Creag an Leth-choin, 1966 *RJ & BAM* **CGE**.

40. **H. anglicum** Fries   Pugsley, 69
Fig 34e.
I South side of Glen Einich (Druce, 1888); Creag Dhubh, Loch
Ericht, near Dalwhinnie above 610 m, 1911 *ESM & WASh* **E,CGE**
steep gully, Creag Bheag, Glen Feshie, 1951 *E. C. Wallace* **ECW!**;

Leacainn, Dalwhinnie, 1952 *C. West* **CGE**; Glen Feshie, 1956
UKD & C. C. Townsend **CGE**; east-facing cliffs of Mealfuarvonie,
Drumnadrochit, 1971 **E**; north-facing cliffs of Sgùrr na Lapaich,
Glen Strathfarrar, 1972 **E**; gully of burn, Gaick Lodge, 1973 **E**;
basic cliff by the track, Glen Feshie, 1974 **ABD**; gully of the Allt
a' Chaoirnich, Gaick, 1974 **E**; basic cliffs by the Allt Lorgaidh,
upper Glen Feshie, 1974 **K**; gully by the Allt Gharbh Ghaig, Gaick,
1974 *OMS* **E**; south of Laggan, 1974 *OMS* **E**; cliffs of Creag Dhubh
S of Loch Ericht, 1976 *OMS & MMcCW* **E**.

Series **Pallida** Pugsley ex Sell & West

Robust or slender plants with few medium to large heads on long,
straight peduncles. Leaves with long, rigid hairs particularly on
the margin, more or less glaucous, basal numerous, those of the
stem 0–2, not amplexicaul. Involucres less strongly hairy than
species of the Series *Cerinthoidea* and bracts porrect in bud. Ligules
usually glabrous. Styles usually yellow. Receptacle pits subulate-
dentate. Native species, usually on basic cliffs and flowering early
in late May and June.

41. **H. stenopholidium** (Dahlst.) Omang   *Skr. Vid.-Akad. Oslo* **3**: 36
    (1938)
    H. lasiophyllum Koch var. *euryodon* F. J. Hanb. pro parte
    I   By the river Feshie near Creag Bheag, 1951 *E. C. Wallace* **ECW**;
    north side of Coire Garbhlach, 1953 *M. C. F. Proctor & K. M.*
    Goodway **CGE**; 1966 *RJ & BAM* **CGE**.

42. **H. dicella** Sell & West   *BEC Rep.* **10**: 475 (1934) (*as H.*
    furcelliferum)
    I   Rock face by the road bridge below the Falls of Foyers, 1971
    CGE.

43. **H. saxorum** (F. J. Hanb.) Sell & West   Pugsley, 112 (as *H.*
    hypocharoides var. *saxorum*)
    M Shingle of River Findhorn north of Greshop Wood, Forres, 1956
    ABD; among rocks by the River Findhorn opposite Coulmony,
    1976 **E**.
    N River side, Ardclach, 1888 *R. Thomson* **NRN**.
    I   Scree by loch, Glen Einich, 1952 *C. West* **CGE**; shingle of River
    Enrick, Milton, 1971 **E,BM**; among rocks on steep face above
    Loch Conagleann, Dunmaglass, 1972 **E**; gully of small burn in the
    Ryvoan Pass, 1974 **E**; conglomerate cliffs of Creag nan Clag, 1976 **E**.

44. **H. schmidtii** Tausch   Pugsley, 87
    M Banks of the Findhorn, 1888 *G. C. Druce* (*ASNH* **6**: 123 (1893)).
    I   Laggan Bridge, 1916 *ESM* **CGE**; rocks near Kilmorack,
    Strathglass, 1962.

45. **H. lasiophyllum** Koch   Pugsley, 84
    I   North-facing cliff above Coire Garbhlach, 1953 *M. C. F.*
    Proctor & K. M. Goodway **CGE**.

46.   **H. nitidum** Backh.   Pugsley, 89
   I   Glen Einich, 1952 *J. A. Cardew & C. West* **CGE**

47.   **H. jovimontis** (Zahn) Roffey   Pugsley, 110
   M Cliffs in the gorge of Huntly's cave, Dava, 1973 **E**.
   I   By the Allts Coire Chùirn and Chàis near Dalwhinnie, 191
   ESM **CGE**; south side of Coire Garbhlach, 1966, *RJ & BAM* **CGE**
   cliffs and gullies of upper Glen Feshie, 1974 **E,BM**.

48.   **H. sommerfeltii** Lindeb.   Pugsley, 91
   I   Stony shore of Loch Einich at 533 m, 1892 *AS* **E**; Allt a' Choire
   Chàis, near Dalwhinnie, 1911 *ESM* **CGE**; Sgòran Dubh Mòr
   Glen Einich, 1965 *RJ & BAM* **CGE**.

Series **Oreadea** (Fries) Sell & West
Like Series *Pallida* but basal leaves fewer and stem leaves more
numerous. The species included here are not very characteristic
unless well-grown, and tend to be intermediate between extremes
of the two series.

49.   **H. argentum** Fries   Pugsley, 92
   Fig 35 c.
   Easily recognised by its narrow intensely glaucous leaves.
   M Bridge of Brown, 1905 *ESM* **CGE**; rocks to the north of the
   old Spey Bridge, Craigellachie, 1963 **CGE**; rocks by the river near
   the old bridge at Carrbridge, 1964 *C. West* **CGE**; Aviemore, 1970
   P. H. Davis **E**.
   N Bank by the road 100 yards east of Drynachan Lodge, Clunas,
   1964 **CGE**.
   I   Glen Einich, 1891 *AS* **E**; by a stream on the west side of Ben
   Alder, 1911 *ESM & WASh* **E,CGE**; plentiful on shingle of Spey,
   Laggan Bridge, 1916 *ESM* **E,CGE**; Newtonmore, 1955 *C. West*
   CGE; shingle of River Enrick, Milton, 1971 **E,BM**; dry fixed
   shingle overspill of River Spey, Newtonmore, 1973 **E**; gravel slope
   of track near Glen Feshie Lodge, 1974 **E**; cliff on S face of The Fara,
   burn shingle N of Cannich, shingle of River Farrar at Struy, burn
   gully Criche, Glen Moriston, all 1975 **E**; burn gully near Bridge of
   Oich, 1976 *OMS & MMcCW* **E**.

50.   **H. caledonicum** F. J. Hanb.   Pugsley, 105 (as *H. rubicundum*)
   M Gully of small burn on moor, Auchness, Dallas, 1973 **E**.
   I   Roadside bank below Foyers village, near Loch Ness, 1971 **E**;
   island in River Glass, near Eskadale, Strathglass, 1973 **E,BM**;
   banks of burn in gully, Gaick Lodge, 1973 **E**.

51.   **H. orimeles** F. J. Hanb. ex W. R. Linton   Pugsley, 102
   I   Sgòran Dubh Mòr, Glen Einich, 1965 *RJ & BAM* **CGE**.

52.   **H. chloranthum** Pugsley   Pugsley, 100
   Easily recognised by its green or greenish-yellow ligules. Fig 35b.
   M *Railway embankment, Aviemore, 1950 det. J. E. Lousley;
   Randolph's Leap, Reluges, 1964 *D. McClintock* **CGE**!; moorland

bank, Dalrachney Beg, Carrbridge, 1972; moorland hummock by the road east of Carrbridge, 1973 **E,ABD**; by River Lossie on the Moss of Bednawinny, Dallas, 1976 **E**.

N Shade of wood by River Findhorn, Glenferness House, 1961 **ABD,K**; by the river at Coulmony, 1961 **ABD,K**; cliff by River Findhorn, Drynachan Lodge, 1970 **CGE**; moorland by the Allt Breac, Drynachan, 1973 **E**.

I Sgùrr na Lapaich, Glen Affric, 1950 *R. C. L. Burges* **CGE**; Coylumbridge, 1955 *NDS* **BM**; by the Allt Mòr, Kingussie and rocks at Tromie Bridge, 1956 *P. D. Sell* **CGE**; bank by road, Falls of Foyers, 1971 **E**; rock face, Loch Killin, 1972 **E**; island in River Glass near Eskadale, Strathglass, 1973 **E**; rocks below monument, Creag Dhubh, Laggan, 1973 **ABD**; shingle of River Enrick, Milton, 1973 **BM**; railway embankment N of Kingussie, 1973 **E**; fixed shingle of River Spey, Newtonmore, 1973; Gaick, 1974 *OMS* **E**.

Series **Bifida** (Pugsley) Sell & West

Usually rather slender shortly hairy plants with few to numerous, small to medium-sized heads on arcuate peduncles. Basal leaves numerous, stem leaves 0–1 (–2), without glandular hairs. Involucral bracts paler than in Series *Subalpina*, with numerous simple eglandular and/or glandular hairs. The great degree of tolerance as regards habitat shown by species of this group allows them to be readily introduced into new areas, thus making it difficult to be sure whether the plants are native or not. In this account it is assumed the species are native unless otherwise stated. Some plants of this group flower throughout the year, but they show their most characteristic features between the end of May and mid-July.

53. **H. exotericum** Jord. ex Bor.   Boreau, *Fl. Centre Fr.* ed. 3, **2**: 417 (1857)

Probably introduced.

I In shade by the Big Burn at Glen Cottage by Ness Castle, Inverness, 1971 **CGE**. The above specimen is referable to *H. exotericum* in the strict sense. Those listed below belong to allied species as yet without a name. Pugsley, 142 covers the aggregate.

M Grass bank at Sanquhar House, Forres, 1961 **ABD,K,CGE**; waste land Brodie Castle, 1962 **K,CGE**; railway embankment near Spynie Palace, 1973 **E**.

N Courtyard at Cawdor Castle, 1974 **E**.

54. **H. grandidens** Dahlst.   *Symb. Bot. Upsal.* **7** (1): 182 (1943)

Probably always introduced. Fig 35e.

M Cluny Hill, Forres, 1955 *NDS* **BM**; grassy bank by Sanquhar House, Forres, 1956 **ABD**; policies of Brodie Castle by the old garden wall, 1962 **ABD,CGE**; wall near Grantown railway station, 1963 **CGE**.

I   Beauly–Kiltarlity road junction, 1962 *A. G. Kenneth* **CGE**; i
shade, Glen Cottage, Inverness, 1971 **E**; roadside bank belov
Foyers village by Loch Ness, 1971 **BM,CGE**; under trees and o
walls at Dochfour House, 1975 **E,CGE**.

55.   **H. uistense** (Pugsley) Sell & West   Pugsley, 117 (as *H. clovens*
var. *uistense*)
I   *Allt na Faing, Glen Affric, 1971 *J. N. Mills* **JNM**.

56.   **H. duriceps** F. J. Hanb.   Pugsley, 138 (as *H. killinense*)
Fig 35d.
M   Cliff on the N side of the old Spey Bridge at Craigellachie, 196
CGE; woodland path by the River Findhorn a few hundred yard
upstream on the right bank, Randolph's Leap, Relugas, 196
E,CGE; river bank opposite Coulmony, 1972 **ABD**.
N   Cliff by the River Findhorn, Drynachan, 1970 **CGE**.
I   Glen Feshie, 1956 *UKD* **CGE**; rock face, Loch Killin at 418 m
1972 **BM**; conglomerate cliffs below the Plodda Falls, Guisachan
1972 **E**; wall on steep hill by gorge, Breakachy, near Kilmorack
1973 **BM,K**; rocks by the burn, Gaick, 1974 *OMS* **E**; wood a
outlet of River Tarff at Loch Ness, Fort Augustus, 1975 **E**; path t
Loch Ness at Knockie, 1975 **E**, cliffs by Lochan na Doire-uaine
Strath Mashie, 1976 *OMS & MMcCW* **E,BM**.

57.   **H. scotostictum** Hyl.   *Symb. Bot. Upsal.* **7** (1): 127 (1943)
A species introduced into S England which has slowly sprea
northwards on railway banks and roadsides. This is the first recor
for Scotland.
I   Rocks by Loch Conagleann, Dunmaglass, 1972 **CGE**. In plent
on the garden wall at the Lodge from whence it has probabl
escaped.

58.   **H. pictorum** E. F. Linton   Pugsley, 149
I   Allt Beul an Sporain, Allt a' Choire Chàis, Allt an Lochai
Dubh 4 miles north of Cluny Castle near Laggan, and streamle
tributary of Allt an t-Sluie near Dalwhinnie, all in 1911 *ESM* **CGE**
conglomerate cliff of Tom Bailgeann by Loch Ceo Glais, 1972 **E**
rock face at Loch Killin at 418 m, 1972 **E,ABD**; conglomerat
cliff below the Plodda Falls, Guisachan near Tomich, 1972 **ABD**
BM; gully of burn by Gaick Lodge at 888 m, 1973 **ABD,K**.

59.   **H. piligerum** (Pugsley) Sell & West   Pugsley, 156 (include
H. variicolor and var. *piligerum*)
M   Rocks by the River Findhorn, opposite Coulmony, 1972
E,CGE; rocky gorge of the Allt Iomadaidh, Fae, 1974 **CGE**.
I   Allt Beul an Sporain, Allt Coire Bhathaich near Dalwhinnie
Coire Chàis and streamlet tributary of the Allt an t-Sluie, all in 191
ESM **CGE**; burn west of the valley near Drumochter Lodge, 191
WASh **CGE**; cliffs at A' Chailleach, 1966 *A. McG. Stirling* **CGE**

roadside cliff 3 miles south of Urquhart Castle, 1970 **CGE**; north-facing cliff ledge, Sgùrr na Lapaich, Glen Strathfarrar, 1971 **E,BM**; rocks by river, Tromie Bridge, 1973 **E**; burn gully at Drumochter, 1975 **E**.

50. **H. aggregatum** Backh.   Pugsley, 131
    I Sgòran Dubh, 1890 *H. Groves* **CGE**; Glen Einich, 1893 *AS* **E**.

51. **H. subtenue** (W. R. Linton) Roffey   Pugsley, 130
    I By the Piper's Burn, Glen Markie, 1916 *ESM* **CGE**; head of Glen Einich, 1965 *RJ & BAM* **CGE**; Coire Garbhlach, 1966 *RJ & BAM* **CGE**; north-facing cliff, Sgùrr na Lapaich, Glen Strathfarrar at 666 m, 1971 **E,BM**; east coire of Sgùrr nan Conbhairean, Glen Affric, 1973 *R. W. M. Corner*.

52. **H. subhirtum** (F. J. Hanb.) Pugsley   Pugsley, 160
    M Dunphail, 1886 *ESM* **CGE**.
    N Small ravine in a birch wood at Dulsie Bridge, 1963 **CGE**.
    I Allt Coire Bhathaidh near Dalwhinnie, 1911 *ESM* **CGE**; Coire Chomhlain, Ben Alder, 1911 *ESM* **CGE**; banks of the Moniack Burn, Achnagairn, Kirkhill, 1963 **CGE**; bank below Falls of Foyers, 1971 **E**; by the Allt Fraoch-choire, 1971 *A. Currie* **AC**; banks of burn gully, Gaick Lodge, 1973 **E,BM**; dry heathy slope by the Allt na Ciste, Cairn Gorm at 666 m, 1974 **E**; cliff at Creag Dhubh, Laggan, 1974 **E**; gully on Bynack More, rocks by Broulin Lodge in Glen Strathfarrar, 1975 *OMS & MMcCW* **E**; cliffs of Creag Dhubh S of Loch Ericht, 1976 *OMS & MMcCW* **E**.

53. **H. lintonianum** Druce   *J. Bot.* **49**: 355 (1911) (as *H. orithales*)
    I Allt Beul an Sporain, 1911 *ESM* **CGE**; Glen Feshie, 1953 *M. C. F. Proctor & K. M. Goodway* **CGE**; crags south of Glen Feshie Lodge, 1969 *A. McG. Stirling* **CGE**.

    ### Series **Vulgata** (Fries) Sell & West
    Like Series *Bifida* but with fewer basal leaves and more numerous stem leaves. All species are native and are found on grassy banks, streamsides, walls and rocky places. They usually flower slightly later than the Series *Bifida*, in late June and July.

54. **H. dipteroides** Dahlst.   Pugsley, 157
    N Pine wood west of Delnies, 1968 *A. Neumann & MMcCW* **CGE**.
    I Allt an Lochain Dubh, 4–5 miles north of Cluny Castle, 1911 *ESM* **CGE**; rocks Creag Dhubh, Laggan, 1973 **CGE**.

55. **H. caesiomurorum** Lindeb.   Pugsley, 176
    M Bridge of Brown, 1905 *ESM* **CGE**; 1969 *J. Mills & C. West* **CGE**; banks of the River Findhorn, Logie House, 1956 **ABD**; limestone cliffs at Cothall, Forres 1963 **CGE**; rock outcrop by the River Lossie, Kellas, 1973 **E**.
    I By the burn at Kingussie, 1898 *ESM* **CGE**; Allt Beul an Sporain, 1911 *ESM* **CGE**; shingle of the Markie Burn just above the confluence of Spey near Crathie, Laggan, 1916 *ESM* **E,CGE**; near

Aviemore, 1952 *C. West* **CGE**; Ord Ban, near Aviemore, 195
M. C. F. Proctor & K. M. Goodway **CGE**; Coylumbridge, 1955 *NL*
BM; Glen Feshie, 1956 *UKD* **CGE**; Tromie Bridge, 1956 *P. D. Sc*
CGE; cliff by Loch Ness, 3 miles south of Urquhart Castle, 197
CGE; gully of small burn by Gaick Lodge, 1973 **E,BM**.

66. **H. orcadense** W. R. Linton   Pugsley, 166 (as *H. euprepes*)
     H. euprepes F. J. Hanb., non Peter
     M Bridge of Brown, 1905 *ESM* **E**.
     I Allt Coire Chùirn, 1911 *ESM* **E,CGE**; above the Allt :
     Coire Dhuibh, Glen Shirra, 1916 *ESM* **E,CGE**; rocks by Rive
     Moriston at the Falls, Glen Moriston, 1971 **E**; east-facing cliffs (
     Mealfuarvonie, 1971 **BM**; shingle of a small burn by a bridge
     few yards from the Ceannocroc Bridge, Glen Moriston, 1974 **E**; clif
     of Creag Dhubh, S of Loch Ericht, 1971 **E**.

67. **H. diaphanoides** Lindeb.   Pugsley, 207 (as *H. praesigne*)
     M By the path a few hundred yards upstream on the right ban
     River Findhorn, Randolph's Leap, 1967 **CGE**; old railway line nea
     Rafford, 1976 **E**.
     N Gorge of River Findhorn ½ mile upstream from Dulsie Bridge
     1974 **E**.
     I The Markie Burn, Crathie, Laggan, 1916 *ESM* **E,CGE**; Lagga
     Bridge, 1955 *P. M. Garnett & C. West* **CGE**; north-facing cliffs (
     Sgùrr na Lapaich, Glen Strathfarrar at 833 m, 1972 **E**.

68. **H. rubiginosum** F. J. Hanb.   Pugsley, 181
     M Rocks in the ravine of the Allt Iomadaidh, Fae, 1974 **E**.
     N Gorge of the Allt Breac, Drynachan, 1974 **E**.
     I Banks of the Elrick Burn at Coignafearn, Strathdearn, 197
     CGE; roadside bank by Foyers village, 1971 **BM**; by the Coilti
     River, Balmacaan, Lewiston, 1972 **E**; basic cliff, Glen Feshie, 197
     E,K; basic scree by the Allt Lorgaidh in upper Glen Feshie, 197
     E,ABD,BM.

69. **H. vulgatum** Fries   Pugsley, 194
     Abundant on walls, dry grassy slopes and rocky places. Th
     commonest Hawkweed of northern Britain found on all types c
     soil.
     Fig 36a.
     M Logie (Druce, 1888); ravine of the Divie, Dunphail, 1897 *ESM*
     CGE; Grantown, 1950 *J. W. Cardew & C. West* **CGE**; among pines
     Culbin, 1953 **E**; beech wood Cluny Hill, Forres, 1954 **K**; riversid
     rocks by the old bridge at Carrbridge, 1964 *C. West* **CGE**; Cromdal
     railway station, 1968 **CGE**; gravel quarry, Loch Mòr, Duthil, 197
     ABD and throughout the county.
     N Cawdor Woods, 1833 *WAS* herb. *Borrer* **K**; Ardclach, 1887, *R*
     Thomson **NRN**; moorland bank, Clunas, 1963 **CGE**, and throughou
     the county.

I   Glen Einich, 1891 *AS* **E**; river shingle and by the Markie Burn, 1916 *ESM* **E,CGE**; Dochfour, 1940 *AJ* **ABD**; small gorge of burn, Corrimony, 1964 **CGE**; on rocks by River Affric near the bridge at entrance to Affric Forest, Glen Affric, 1971 **E**; east-facing cliffs of Mealfuarvonie, Drumnadrochit, 1971 **E**; shingle of River Enrick at Milton, 1971 **E**; rocks by a small burn, Sgùrr na Lapaich, Glen Affric, 1971 **E**; rock face, Loch Killin, 1972 **E**; beech wood, Creag Liath, Gaick at 500 m, 1973 **E**; fixed shingle by River Spey, Newtonmore, 1973 **BM**; island in River Glass, Eskadale, Strathglass, 1973 **E**, and throughout the district.

o.   **H. cravoniense** (F. J. Hanb.) Roffey   Pugsley, 180
M   Pine wood near the River Spey, Grantown, 1963 **CGE**; verge of path a few hundred yards upstream on the right bank of the River Findhorn at Randolph's Leap, 1967 **CGE**; walls of Lochindorb Castle, 1974 **E**.
N   By the Muckle Burn, Lethen, 1963 **ABD,K**; rock outcrop by the suspension bridge, Banchor, 1967 **CGE**.
I   Loch an Laghair, Glen Affric, 1947 *E. Milne-Redhead* **CGE**; roadside bank near Kirkhill, 1962 **CGE**; under trees by the burn, Kiltarlity, 1972 **E**; low cliff on Creag an Fhir-eòin, Loch Monar, 1975 **E**.

Series **Alpestria** (Fries) Sell & West
The only species of this group occurring in the region covered by the Flora is atypical of the Series. It is a native plant of rocky streamsides.

1.   **H. dewari** Syme   *Watsonia* **6**: 102 (1965)
I   Easterness, ? *coll* (*ASNH* **9**: 161 (1900)); by River Spey, Kingussie, ? *coll.* **BM** (*Watsonia* **6**: 102 (1965)).

Series **Prenanthoidea** (Koch) Sell & West
Tall robust plants with numerous small heads on relatively short peduncles. Basal leaves absent or withering early, stem leaves 12–30, semi-amplexicaul. Involucral bracts with dense glandular hairs. The single species is a native of grassy banks, rocky places and streamsides, and flowers in July and August.

2.   **H. prenanthoides** Vill.   Pugsley, 231
M   *Banks of the River Findhorn, 1837 *G. Gordon* **E**.
I   Badenoch, 1838 *A. Rutherford* **E**.

Series **Tridentata** (Fries) Sell & West
Slender to robust plants with few to numerous medium to large heads. Basal leaves usually absent, stem leaves (4–) 8–25 (–numerous), not amplexicaul, the lower often petiolate. Involucral bracts more or less obtuse, variously hairy. Margin of receptacle pits dentate. Native species of streamsides, grassy banks and rocky places. Usually flowering in July and August.

73. **H. sparsifolium** Lindeb.   Pugsley, 253 (as *H. stictophyllum*)
An easily recognisable species with narrow, usually spotted leave
Fig 36b.
I   Glen Urquhart, 1839 *J. Ball* **E**; grassy slopes between Lagga
Bridge and Cluny Castle, 1916 *ESM* **CGE**; bank by pine woo
between Cluny Castle and Laggan (probably Marshall's localit)
1974 **E**; cliffs by a lay-by Loch Ness near Lurgmore, 1975 **E,BM**
*grassy bank at Aldourie Castle, 1975.

74. **H. uiginskyense** Pugsley   Pugsley, 245
M In open birch wood by the Dulnain River, Muchrach Lodg
Dulnain Bridge, 1972 **E**; roadside bank, Ballifurth, Grantowr
1974 **E**.
I   Rothiemurchus Forest, 1892 *AS* **E**; river bank, Crathie, *ESM* **E**
between Kingussie and Kincraig, 1959 *P. D. Sell, NDS & C. We*
CGE.

75. **H. gothicoides** Pugsley   Pugsley, 244
M Near Bridge of Brown, 1905 *ESM* **E,CGE**.
I   Kincraig, 1891 *AS* **E**; in several places by the Spey, Crathi
1916 *ESM* **E,CGE**; grassy bank of main road, Kingussie, 195
J. W. Cardew & C. West **CGE**; between Newtonmore and Kinguss
and Drumguish, 1956, *P. D. Sell* **CGE**; Allt Mòr, Kingussi
1967 *C. West* **CGE**; on dry fixed shingle overspill of River Spe)
Newtonmore, 1973 **E**; grass hollow on moor *c.* 1 mile west
Newtonmore on the Laggan road, 1973 **E,CGE**; roadside ban
on the outskirts of Newtonmore on the Laggan road, 1973 **BM**.

76. **H. stewartii** (F. J. Hanb.) Roffey   Pugsley, 247 (as
                                                      H. backhousianum
N Dry bank by the road, Dulsie Bridge, 1967 **CGE**.
I   Kingussie, 1953 *C. West* **CGE**; near Laggan Bridge, 195
C. West **CGE**; fixed river shingle near golf course, Newtonmor
1973 **CGE**.

77. **H. scabrisetum** (Zahn) Roffey   Pugsley, 249 (as
                                                      H. pseudacrifolium
I   Between Kincraig and Kingussie, 1959 *P. D. Sell, NDS &*
C. West **CGE**.

78. **H. cambricogothicum** Pugsley   Pugsley, 238
The only natural habitat in which this species has been recorde(
M Right bank of the River Findhorn, Greshop Wood, Forres, 195
K (see *Crit. Atlas* 124).

Series **Foliosa** (Fries) Sell & West
Robust plants with a large panicle of medium to large head
Leaves all cauline, numerous, crowded, more or less amplexica)
and reticulately veined beneath, with more or less thickened margir
Involucral bracts broad, mostly obtuse, usually sparsely hair)
Receptacle pits with dentate or rarely fimbriate-dentate margin

Native species of grassy and rocky places flowering in July and August.

79. **H. latobrigorum** (Zahn) Roffey   Pugsley, 272
Fig 36c.
M Bank by Findhorn River, Broom of Moy, Forres, 1954 **ABD,K**; railway embankment between Knockando station and the distillery, 1956 **E**; banks of the Spey, Grantown, 1966 **CGE**.
I Grassy bank on the Laggan road, Newtonmore, 1966 **CGE**; bank by Loch Ness, Drumnadrochit, 1964 **CGE**; railway embankment, Croftcarnach, north of Kingussie, 1973 **E**; meadow at east end of Loch Meiklie in Glen Urquhart, 1973 **E**; steep bank by road, 3 miles south of Urquhart Castle, 1974 **E**; by Spey at Laggan Bridge, 1974 **E**; wood at mouth of River Tarff at its entrance to Loch Ness, Fort Augustus, 1975 **E,BM**; wall near old kirkyard at Lewiston, 1975 **E,BM**; shingle bank by River Glass, Cannich, 1975 **E**.

80. **H. subcrocatum** (E. F. Linton) Roffey   Pugsley, 277
M Shingle of River Findhorn at Darnaway opposite Sluie, 1964 **CGE**; railway embankment near Edinkillie Kirk, 1973 **E**.
N Railway embankment at Auldearn, 1956 **ABD**.
I Kingussie, 1887, *G. C. Druce* (*J. Bot.* **26**: 17 (1888)).

81. **H. drummondii** Pugsley   Pugsley, 274
M Banks of the River Spey, upstream on left bank, Boat of Garten, 1973 **E**.

82. **H. strictiforme** (Zahn) Roffey   Pugsley, 281
M Near Dunphail, 1897 *ESM* **CGE**; dry bank, Randolph's Leap, 1956 **ABD**; banks of Spey, Grantown, 1953 **K**; rocks of River Dulnain under the road bridge, Carrbridge, 1963 **CGE**; bank, Muchrach Lodge, 1972 **E**; banks of the Spey by Boat of Garten station and below Craigellachie, 1973 **E**; island in the River Spey, Aviemore, 1974 **BM**.
I Falls of Foyers, 1846 ex herb. *Mrs Osborne* **CGE**; Beauly River, 1892 *ESM* **CGE**; Kingussie, 1898 *ESM* **CGE**; Feshie Bridge, 1955 *C. West* **CGE**; between Kincraig and Kingussie, 1959 *P. D. Sell, NDS & C. West* **CGE**; railway embankment north of Kingussie, 1973 **E**.

83. **H. reticulatum** Lindeb.   Pugsley, 279
M Kellas, 1836 *G. Gordon*, **E**; Bridge of Brown, 1905 *ESM* **CGE**; riverside rocks, Bridge of Brown, 1967 *J. N. Mills & C. West* **CGE**.
N Ardclach, banks of the Findhorn, 1888 *R. Thomson*, **NRN**; by the river above Nairn, 1898 *ESM* **E,CGE**; small island in the River Nairn, Holme Rose, Croy, 1955 **ABD**.
I Banks of the river at Foyers, 1881 *F. Townsend* **CGE**; River Spey below Crathie, 1916 *ESM* **CGE**; conglomerate cliffs, Aigas, Strathglass, 1971 **E**; on Garva Bridge, 1972 **BM**; in shade of burn, Loch Meiklie, Glen Urquhart, 1973 **E**; bank at Polmaily, 1973 **K**; wall by Spey, Laggan, 1973 **E**; bank by Loch Ness at Lurgmore,

outlet of river at Inverfarigaig, 1975 **E**; embankment by Rive
Beauly between Tomich Farm and Tarradale, 1976 **E,BM**.

84.   **H. subumbellatiforme** (Zahn) Roffey   Pugsley, 284 (as *H*
      pseudamplidentatum)
      M Banks of the Spey, Aviemore, 1893 *AS* **CGE**; Dunphail, 189
      ESM **CGE**; banks of the Findhorn, Greshop Wood, 1967 **CGE**
      waste ground near Cromdale station, 1967 *C. West* **CGE**; ¾ mil
      upstream on right bank, Randolph's Leap, 1967 **CGE**; banks o
      Spey by the old Spey Bridge at Grantown, 1971 **E,CGE**; roadsid
      near the entrance to Muchrach Lodge, 1972 **E**.
      I By the Allt Mòr, Kingussie, 1898 *ESM & WASh* **E,CGE**; b
      the Spey, scarce between Newtonmore and Kingussie, 1911 *ESM* **E**
      east end of Loch Meiklie, Glen Urquhart, 1958 *E. W. Groves* **EWG**

85.   **H. maritimum** (F. J. Hanb.) F. J. Hanb.   Pugsley, 289
      I Rothiemurchus Forest, 1893 *AS* **E**; grassy slopes betwee:
      Laggan Bridge and Cluny Castle, 1916 *ESM* **E**; Kingussie, 195
      C. West **CGE**.

Series **Umbellata** (Fries) Sell & West
Robust plants with few to numerous heads in elongate panicles
the upper part often more or less umbellate. Leaves all caulin
15–50 (–numerous), not amplexicaul, margins revolute. Involucra
bracts broad obtuse, glabrous or nearly so, the outer squarrose wit
recurved apices. Margins of receptacle pits dentate or fimbriate
dentate. The only species of the group is native on dunes an
grassy banks particularly by streams. It flowers from July t
September.

86.   **H. umbellatum** L.   Pugsley, 292
      Fig 36d.
      M Logie, and Kinchurdy, Boat of Garten (Druce, 1888); bank
      of the Spey, Aviemore, 1893 *AS* **E**; Bridge of Brown, 1905 *ESM* **E**
      sand dunes on the 'back shore', Findhorn, 1953 **ABD**; under beecl
      trees by the loch in the policies of Gordon Castle, Fochabers, 197'
      E,ABD; among short Calluna-Festuca moorland at Wester Galavie
      east of Carrbridge, 1974 **E**.

Series **Sabauda** (Fries) Sell & West
Robust plants with numerous heads in a large, often elongat
panicle. Leaves all cauline, usually numerous, often aggregate
below, margin not or only slightly thickened. Involucral bract
broad, more or less obtuse, glabrous or hairy. Margins of receptacl
pits distinctly fimbriate-dentate. They occur on river banks, woo
margins and roadsides. Both species recorded for this Flora ma
have been introduced into the area. They flower from late July t
September.

7.   **H. perpropinquum** (Zahn) Druce   Pugsley, 302
     Fig 36e.
     M Moray (A. Bennett in *Scot. Nat.* 1886 cites *Gordon*, this may be
     Rev. G. Birnie's record from Speymouth (Burgess)); near Relugas
     House (Marshall, 1897); pine wood in the Culbin Forest, 1953
     CGE,E; Greshop Wood, Forres, 1954 **E,ABD**; lane by station,
     Grantown, 1963 **CGE**; Forres station yard, 1966 **CGE**; wall to the
     east of Dallas Lodge, 1966 **CGE**; old Spey Bridge, Grantown, 1971
     BM.
     N River bank Howford Bridge, Nairn, 1960 **ABD,CGE**; river bank
     Holme Rose, 1955 **ABD**.
     I Cliff by road N of Drumnadrochit, 1970 **CGE**; drive to Ness
     Castle, Inverness, 1971 **E,BM**; railway embankment at Moy
     station, 1975.

8.   **H. salticola** (Sudre) Sell & West   Pugsley, 308 (as
     H. sublactucaceum)
     M By River Spey at Garmouth, 1956 **ABD**.

Pilosella Hill

In the following account the extensive notes have been compiled
by P. D. Sell and C. West. All the British species of this genus can
be distinguished from those of *Hieracium* by their stoloniferous habit
and by their achenes being not more than 2·5 mm long. All
Herbarium specimens have been determined by P. D. Sell and
C. West unless marked with an asterisk.

P. officinarum C. H. & F. W. Schultz   *Mouse-ear Hawkweed*
     Hieracium pilosella L.
Common native of grassy and rocky places and on dunes.
Characterised by each flowering stem having a single head of
yellow flowers with a red stripe on their outer face. Fig 37a.

     Subsp. **micradenia** (Naegeli & Peter) Sell & West
     Subsp. *concinnata* (F. J. Hanb.) Sell & West
Perhaps the most common subspecies in the British Isles as a whole.
Involucral bracts with dense short glandular hairs usually not more
than 0·5 mm long, without or with an occasional simple eglandular
hair. Well distributed throughout the area. Fig 37b.
M Bridge of Brown, 1905 *ESM* **CGE**; Cromdale station, 1959
P. D. Sell & C. West **CGE**; river shingle below Kellas House, 1968
CGE; banks of River Dulnain at Muchrach, 1972 **E**; grass bank
by the graveyard by Loch Vaa, Aviemore, 1973 **E,BM**.
N Gravel spill at the mouth of the Allt Breac, Drynachan, 1970
ABD; grass verge by the ford, Achravaat, 1973 **E**; grassy bank in
the gorge of the Findhorn, Dulsie Bridge, 1974 **E**.
I Laggan, 1961 *P. M. Garnett & C. West* **CGE**; roadside bank by
Loch Ness, 3 miles south of Urquhart Castle, 1970 **CGE**; grassy
bank below Mealfuarvonie, 1971 **E**; sandstone bank by gorge at

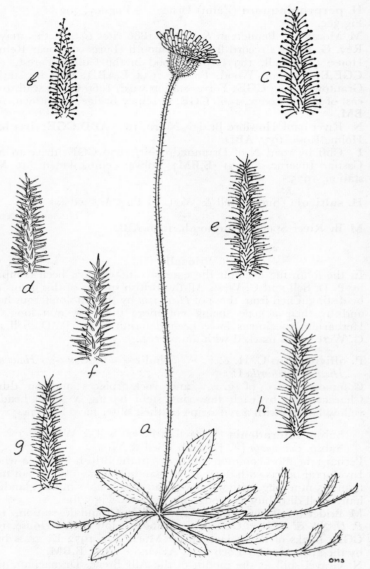

Fig 37 a, **Pilosella officinarum agg.**; b, subsp. **micradenia**; c, subsp. **euronota**; d, subsp. **officinarum**; e, subsp. **trichosoma**; f, subsp. **tricholepia**; g, subsp. **melanops**; h, subsp. **trichoscapa**.

top of Falls of Divach, Lewiston, 1972 **E**; grass by limestone quarry at Suidhe, Kincraig, 1973 **E**; grassy bank, Loch Meiklie, Glen Urquhart, 1973 **E**; overflow scree by burn, Gaick Lodge, 1973 **E**; shingle of River Spey, Newtonmore, 1973 **E**; island in River Glass near Eskadale, 1973 **E**; track in pine wood, Glen Feshie Lodge, 1974 **E**; grassy bank by a small burn at Coignashie, Strathdearn, 1974 **E**; pine wood near Dalcross, bank at Torness, by Loch Beannacheran in Glen Strathfarrar, track to Loch Toll a' Mhuic, grass track in pine wood at Tromie Bridge, and roadside verge by Hilton loch at Guisachan, all 1975 in **E**.

Subsp. **euronota** (Naegeli & Peter) Sell & West comb. nov.
   Hieracium pilosella subsp. *euronotum* Naegli & Peter, *Hier. Mittel-Eur.* 1: 155 (1885)
Involucral bracts with dense glandular hairs up to 1 mm long, stronger and darker than in subsp. *micradenia*. Fig 37c.
I Bank by Loch Ness near Inverfarigaig, 1971 **CGE**; bank on east side of Loch Beinn a' Mheadhoin in Glen Affric, 1971 **E**; rocks by river at Spey Dam, 1972 **E**; grass bank by Glen Feshie Lodge and by the Allt Lorgaidh in upper Glen Feshie, 1974 **E**; canal bank at Fort Augustus, 1975 **CGE**.

Subsp. **officinarum**
Involucral bracts with numerous pale simple eglandular hairs and few to numerous small glandular hairs. A rather rare subspecies in northern Britain where it is usually not typical. Fig 37d.
M Buckie Loch, Culbin, 1953 *NDS* **BM**; grassy bank near Cromdale station, 1967; car park at Elgin Golf Club, 1970 **CGE**; grass mound at Avielochan near Aviemore, 1973 **E**.
N Howford, 1889 *R. Thomson* **NRN**; sandy bank, south of Drynachan Lodge, 1970 **E**.
I Bank on Shenachie road, 1975 **CGE**.

Subsp. **trichosoma** (Peter) Sell & West comb. nov.
   Hieracium pilosella subsp. *trichosoma* Peter in *Bot. Jahrb.* 5: 254 (1884)
   Subsp *nigrescens* (Fries) Sell & West
Involucral bracts with numerous dark, simple eglandular hairs and numerous short glandular hairs. The most common subspecies of northern Britain. Fig 37e.
M Bank at Cothall, 1963 **CGE**; quarry at Lossiemouth, 1967 **CGE**; roadside bank on the Hill of Mulundy near Dallas, 1968 **CGE**; roadside verge Dava Moor, 1975 **BM**.
N Railway yard by Auldearn station, 1972 **E,CGE**.
I Roadside bank, Kingussie, 1959 *C. West* **CGE**; shingle of River Enrick at Milton, 1971 **E**; rough ground by pool in Glen Moriston, 1971 **CGE**; near the Lodge in Glen Affric and bank by farm below Mealfuarvonie, by the Monar dam in Glen Strathfarrar, 1971; gravel lay-by at Creag Dhubh near Laggan **E**; roadside south of Ness Castle, **E**, dry bank in Glen Markie, turf by Garva Bridge, field below crags at Ballmishag, Laggan, roadside bank near main

road on Abriachan hill, **ABD**, sandstone cliffs below Plodda Falls near Guisachan, **E**, roadside bank by the waterfall at Glendoe, **CGE**, at Dunmaglass, **CGE**, bank by lay-by at Crask of Aigas, **BM**, path by the Beauly river at Groam, **E,BM**, near Beaufort Castle in 1972 **CGE**; shingle of River Calder at Newtonmore, **K**, bank by burn on moor at Dalballoch in Glen Banchor, roadside bank by bridge over burn at Culbirnie west of Kiltarlity, **K**, verge of track on moor at Creag na h-Iolaire near Aviemore and track in moorland near road at Loch na Ba Ruaidhe, 1973 **K**; grass verge of moor in the Ryvoan Pass, 1975 **BM**; lawn at Ballindarroch, Scaniport, **BM**, flood bank of the Spey at Broomhill, **CGE**, Lundie near Glen Moriston, **E**, bank by the tennis court at Dochfour House, **BM**, edge of the south drive at Moy Hall, **E**, bank by the burn at Balnespick Lodge near Tomatin, **ABD**, Fasnakyle near Cannich in Strathglass, **ABD**, verge of track by Loch Ericht near Dalwhinnie **E**, bank near Inverdruie, **E**, shingle of burn in Glen Strathfarrar, 1975 **BM**, Ruisure near Beauly, 1975 **CGE**, and Strathdearn near Coignafearn, 1975 **CGE**.

Subsp. **tricholepia** (Naegeli & Peter) Sell & West
Involucral bracts with numerous pale simple eglandular hairs, without or with inconspicuous glandular hairs. The plants of northern Britain tend to be less pure than those of the south and to approach subsp. *melanops*. Fig 37f.
M Culbin, 1953 *E. J. Gibbons* **CGE**; near Hazelbank on Pluscarden road, 1968 **CGE**.
N Culbin forest near Maviston, 1968 **CGE**.
I Kingussie and Tromie Bridge, 1956 *P. D. Sell* **CGE**; dry edge of field, Auchnagairn, Kirkhill, 1972 **CGE**; steep slope on scree at Creag Dhearg by Lochend, 1975 **E**.

Subsp. **melanops** (Peter) Sell & West comb. nov.
Hieracium pilosella subsp. *melanops* Peter in *Bot. Jahrb.* 5: 254 (1884)
Involucral bracts with moderately dense, dark simple eglandular hairs up to 2 mm long, without or with inconspicuous glandular hairs. Fig 37g.
M Grassland by Craigellachie Nature Reserve, Aviemore, 1967 *C. West* **CGE**; roadside bank at Mundole, 1972 **CGE**; banks of River Dulnain at Muchrach, 1972 **E**; bank by bridge over River Dulnain at Balnaan, 1973 **E**; grass mound on moorland at Easter Limekilns, Dava, 1973 **E**; verge of track near railway bridge at Slochd N of Carrbridge, 1973 **E**.
I Bank by Pityoulish Kirk, 1971 **E**; bank of the Allt Coire Fhàr, Geal-chàrn, Drumochter, 1972 **E**.

Subsp. **trichoscapa** (Naegeli & Peter) Sell & West
Involucral bracts with dense dark simple eglandular hairs up to 5 mm long, without or with inconspicuous glandular hairs. Fig 37h.
M Railway station yard among cinders, Knockando, 1970 **CGE**.
I Invermoriston, 1975 **E**.

P. caespitosa (Dumort.) Sell & West subsp. **colliniforme** (Naegeli & Peter) Sell & West
Presumably introduced in all its British localities. Seems to have spread along railways and roadsides, but is so well established in some places that it it appears to be native. Easily recognised by its compact inflorescence of small concolorous yellow flowers.
M Cluny hill, Forres, *JK* **ABD,FRS**; waste land at Forres station, 1952 **BM,CGE**; abundant on the railway embankment opposite Greshop House, 1960 **ABD,K**; fixed dunes at Culbin, 1963 **K,CGE**; waste ground by Invererne House near Forres, 1968 **E,CGE**; railway embankment at Slochd near Carrbridge, 1973 **E**; bank between Braemoray Hotel and Edinkillie Kirk at Dunphail, 1975 **E**.
N Garden path at Holme Rose, 1955 (*Proc. BSBI* **2**: 2 (1956)); Coulmony river bank, 1960; roadside bank between Kilravock and Croy, 1974 **E**.
I Station road at Tomatin, 1962 **CGE**.

P. aurantiaca (L.) C. H. & F. W. Schultz   *Fox and Cubs*
A native of Central Europe, which has escaped from gardens and spread along railways and roadsides. Easily recognised by its corymbose inflorescence of orange-brown or orange-red flowers.

Subsp. **aurantiaca**
Stolons rather short and mostly underground. Rosette-leaves 100–200 (–300) × 22–60 (–70) mm. Involucral bracts 8–11 mm.
M Coniferous wood near Grantown, 1950 *J. W. Cardew & C. West* (*Watsonia* **2**: 1 (1951)); Aviemore, 1951 *C. West* **CGE**; railway embankment at Cromdale, 1954 and 1957 **E,ABD,CGE**; yard behind the new post office at Aviemore, 1975.
I *Garden weed at Kincraig, 1890 *AS* **GL**; grassy bank at Dochfour House, 1956; path in field near main road, Achnagairn, 1972 **E**; ruins of Moy Hall and at Broomhill station, 1975; old railway at Tomachrochar near Nethybridge, 1975 **E**; railway embankment north of Kingussie, 1975 **CGE**.

Subsp. **carpathicola** (Naegeli & Peter) Sojak
      P. aurantiaca subsp. *brunneocrocea* (Pugsley) Sell & West
Stolons often long and leafy, usually above ground. Rosette-leaves 60–100 (–160) × 12–30 (–30) mm. Involucral bracts 5–8 mm.
M Dunes near Buckie Loch, Culbin, 1953 **ABD**; same locality and date *E. J. Gibbons* **CGE**, and by *A. Melderis* **BM**; railway embankment, Blacksboat, 1975 **E**; railway embankment at Knockando, 1967 **CGE**; embankment at Forres, 1966 **CGE**; pine wood near the tip, Grantown, 1971 **E**; roadside verge at Nethybridge, 1973 **BM**.
N About the policies of Holme Rose and Drynachan Lodge, 1955; railway embankment at Nairn, 1961 **ABD,CGE**.
I Railway embankment near Inverness, 1942 *UKD*, **CGE**; roadside near Foyers post office, 1971 **E**; embankment at Culloden wood, 1975.

P. aurantiaca× **officinarum** = **P.**× **stoloniflora** (Waldst. & Kit.)
    C. H. & F. W. Schultz
These hybrids show considerable morphological variation, which
suggests back-crossing occurs. Some differ from *P. officinarum* only
in having long black hairs, some have in addition reddish flowers,
some have two and some several reddish flowers on a stem, and
some differ from *P. aurantiaca* only in the larger size of the capitula.
M Railway embankment at Cromdale station, with single reddish
flower, 1956 **E,ABD**, and in 1957 **E,BM,CGE**.

Crepis L.

Former records for *C. vesicaria* L. and *C. setosa* Haller fil. are errors.

C. mollis (Jacq.) Aschers.   *Northern Hawk's-beard*
Native. Rare by burns in mountain areas. Not seen in recent years
and as there are no voucher specimens the records must be treated
with some doubt.
M (Druce, *Com. Fl.*); between Findhorn and Nairn (Ewing).

C. capillaris (L.) Wallr.   *Smooth Hawk's-beard*
    C. virens L.
Native. Abundant on dry banks, walls and waste places. The
common plant in this area is the robust variant *glandulosa* Druce
with hairs on the lower part of the stem and the involucre densely
covered with blackish-green glandular hairs. It has not been
separated here. Map 251.
M Very common (Gordon); Forres, 1898 *JK* **FRS**; railway bank
near Elgin (as *C. setosa* Haller fil.) 1896 (Druce, *ASNH* 6: 54 (1897));
Findhorn and Forres station, 1953 *A. Melderis* **BM**; dunes in the
Culbin Forest and throughout the county, 1954; waste ground at
Brodie (as var. *glandulosa*), 1962 **K**.
N (Stables, *Cat.*); Ardclach (Thomson); near Nairn, 1898 *ESM*
E,CGE; dry moorland bank at Drynachan and throughout the
county, 1954; waste ground at Holme Rose (as var. *glandulosa*),
1962 **ABD**.
I (Watson, *Top. Bot.*); Dalwhinnie and Tomatin, 1903 (Moyle
Rogers); Scaniport, 1940 *AJ* **ABD**; throughout the district, 1954;
embankment at Culloden station (as var. *glandulosa*), 1962 **K**.

C. paludosa (L.) Moench   *Marsh Hawk's-beard*
Native. Common in wet meadows, by rivers and burns and mountain
flushes up to 1,000 m. Fig 38a, Map 252.
M Frequent (Gordon); Aviemore, 1891 *AS* **E**; Waterford, 1900
JK **FRS**; abundant in the Spey and Findhorn valleys and throughout
the county, 1954; bog at Drakemyres, 1972 **E**.
N (Stables, *Cat.*); Ardclach, 1891 *R. Thomson* **NRN**; throughout
the county, 1954; river bank at Dulsie Bridge, 1960 **ABD**; bog by
the Muckle Burn at Lethen, 1961 **ABD,K**; flush on the 25 foot
raised beach at Maviston, 1968 **E**.

I Badenoch, 1838 *A. Rutherford* E; Leachkin near Inverness (Aitken); near Dores, 1940 *AJ* ABD; throughout the district, 1954.

Taraxacum Weber

The following records have been arranged according to A. J. Richards, *Watsonia*, 9: 1972 *The Taraxacum Flora of the British Isles* from which short notes have been added to assist in identification. Unless so stated the species may be regarded as native. All herbarium records have been determined by A. J. Richards.

Section Erythrosperma Dahlst.   *Lesser Dandelion*
Frequent on sand dunes, dry moorland and in short grassland. Small delicate plants with unspotted, strongly dissected leaves; heads pale yellow; exterior bracts usually corniculate, achenes of varying colour.

T. brachyglossum (Dahlst.) Dahlst.
Recognised by its spreading glaucous-purple exterior bracts, discoloured styles and reddish achenes. Fig 38b.
M Findhorn (Druce, 1897); sand dunes at Findhorn, 1969 CGE; dunes near the club house on Lossiemouth golf course, 1970 CGE; dry wall by the aerodrome at Lossiemouth, 1973 E; dunes on the E side of Lossiemouth, 1974 K; moorland path on the cliff top E of Hopeman, 1975 E.
N Dunes on the Nairn Dunbar golf course, 1962 CGE and in 1972 E.
I Shingle beach at Ardersier, 1970 CGE; grass bank by the sea at Castle Stuart and on dunes at Fort George, 1970 CGE; below a cottage wall at Ardersier, 1971 E; station platform at Dalwhinnie, 1975 BM; track by the Canal at Fort Augustus, 1975 ABD.

T. argutum Dahlst.
Leaves dark green, regularly triangular-lobate; capitula habitually closed, ligules involute.
M Dunes on the E side of Hopeman, 1971 E.

T. lacistophyllum (Dahlst.) Raunk.
Leaves with abruptly expanded bases to the leaf-lobes.
M Findhorn dunes, 1973 E; moorland path on the cliffs at Hopeman, 1975 E.
N Dunes between the sea and the Nairn Dunbar golf course, 1967 CGE and in 1972 E.
I Near the sea at Fort George, 1970 CGE.

T. rubicundum (Dahlst.) Dahlst.
Leaves glabrous, purplish, strongly dissected.
M Cliff by the old Spey Bridge at Craigellachie, 1964 CGE; dunes at Lossiemouth, 1970 CGE.
N Nairn Dunbar golf course, 1967 CGE and in 1972 E.

T. laetum (Dahlst.) Dahlst.
Exterior bracts erect, pale green with purple horns; styles yellow, achenes dark violet.

M Dunes at Lossiemouth, 1972 **E,K**; dunes at Findhorn, 1972 **E**; lawn at Kinveachy near Aviemore, 1974 **E,BM**.
I   Near the sea at Fort George, 1970 **CGE**.

T. laetiforme Dahlst.
Somewhat similar to *T. laetum*, differing in the long recurved exterior bracts and reddish achenes.
M Dunes at Lossiemouth, 1970 **CGE** (the third British record).

T. fulvum Raunk.
Exterior bracts narrow, recurved, achenes cinnamon.
M Dunes on the E side of Hopeman, 1971 **E**.
N Between the sea and the Nairn Dunbar golf course, 1967 **CGE**.

T. proximum (Dahlst.) Dahlst.
Achenes dark purple-brown.
M Dunes at Hopeman, 1967 **CGE**.

Section Obliqua Dahlst.
Rare on sandy turf near the sea. Plants small, superficially similar to certain species of section *Erythrosperma* but distinguished by the many narrow, oblong, highly dissected leaves, almost orange heads and greyish or pale brown achenes; exterior bracts usually corniculate.

T. obliquum (Fries) Dahlst.
M Sand dunes on the E of Hopeman harbour, 1967 **CGE** (new to Britain) and in 1971 **E**; dunes at Findhorn, 1970 **CGE** and in 1972 **E**.

T. platyglossum Raunk.
M Sand dunes on the E side of Hopeman harbour, 1967 **CGE** (fifth British record); dunes at Lossiemouth, 1970 **CGE**.
N Nairn Dunbar golf course, 1967 **CGE**.

Section Spectabilia Dahlst.   *Broad-leaved Marsh Dandelion*
Common on wet moorlands, grass slopes on mountains and by rivers and burns. Plants medium-sized; leaves rarely linear or highly dissected, often dark, hairy and spotted with red petiole and midrib; ligules usually striped carmine or purple; achenes straw of brown-coloured; exterior bracts spreading-erect.

T. unguilobum Dahlst.
Common in wet places. Leaf-lobes recurved, exterior bracts glaucous, pink-tipped, achenes rust-coloured.
M Rocks by the River Findhorn at Randolph's Leap, Relugas, 1967 **CGE**; track in the Culbin Forest N of Binsness, 1967 **CGE**; by a small bridge on the hill road above Cromdale, 1971 **ABD**; rocks by Daltulich Bridge, 1972 **E**; railway cutting at Clashdhu, 1972 **E**; gorge of burn by Huntly's Cave on Dava Moor, 1973 *R. Richter* **ABD**!; station yard at Boat of Garten, 1973 **ABD**; wall near Lossiemouth aerodrome, 1973 **BM**; walls of Lochindorb Castle 1974 **E**; garden path at Glenerney, 1974 **K**; Balnabreich Kirk near

Mulben, 1974 **K**; rock face at Wester Elchies, pine wood verge at Snab of Moy, near Carrbridge and at Kinveachy.
N Wall at Nairn, 1962 **CGE**; roadside verge at Cantraydoune, 1970 **CGE**; roadside near Righoul, 1973 **E**; moorland by the Leonach Burn, Dulsie, 1974 **E**.
I Borlum Bridge, Drumnadrochit, 1947 *A. J. Wilmott* **BM**; by Loch Ness, 1965 (cult.) *A. J. Richards* **OXF**; bank by Loch Ness, 1970 **CGE**; dunes at Fort George, 1970 **CGE**; moorland road near Ness Castle, Inverness, 1971 **CGE**; waste land at Torgyle Bridge, Glen Moriston, 1971 **K**; path to the Falls of Foyers, 1971 **BM**; near Balblair, Kirkhill, railway embankment between Daviot and Culloden and moorland track at Drumguish, 1971; by the Craggie Burn at Daviot, 1971 **K**; raised beach at Ardersier, 1972 **E**; roadside verge at Ceannocroc, 1972 **E**; gravel of drive at Netherwood, Newtonmore, 1972 **E**; grass bank by Loch Conagleann, Dunmaglass, 1972 **E**; cliff above Eskadale in Strathglass 1973 **ABD**; shores of Loch Flemington, 1973 **BM**; wall near Gorthleck, 1973 **K**; bank at Invermoriston, 1973 **ABD**; moorland track in upper Glen Feshie, 1974 **BM**; Alturlie Point and Pass of Inverfarigaig, 1975 **E**; station yard at Dalwhinnie, 1975 **CGE**; moorland track at Lundie, Glen Moriston 1975.

T. landmarkii Dahlst.
Leaves smooth, narrow, unspotted, with narrow lobes; heads small with erect bracts.
M Moorland pasture Cromdale, 1971 **E**; Clashdhu railway cutting, 1972 **E**; grass bank in the gorge of Huntly's Cave on Dava Moor, 1973 **E,BM**; gravel path at Essle cemetery, 1973 **E**; short grass near the south lodge at Castle Grant, 1974 **K**.
N Dunes on the Nairn Dunbar golf course, 1970 **CGE**.
I Roadside verge near Drumnadrochit, 1970 **CGE**; shingle of the River Enrick at Milton, 1971 **E**; under cliffs in upper Glen Feshie, 1974 **E,BM**; grass slope below the cliffs at Lochan na Doire-uaine, Strath Mashie, 1976 **E**.

T. faeroense (Dahlst.) Dahlst.
Very common on acid moors. Leaves dark green often spotted, sometimes not lobed but usually with 2–3 slightly recurved leaf-lobes; exterior bracts adpressed-erect; achenes 4 mm, straw-coloured. Fig 38c.
M Moray, 1928 *K. D. Little* **BM**; damp dune at the E end of the Buckie Loch in the Culbin Forest, 1964 **CGE**; railway embankment below Knockando House, 1967 **CGE**; by a small burn at Cromdale, 1971; grass bank by the old Spey Bridge at Grantown, 1972 **E**; boggy ground at Huntly's Cave, Dava, 1973 **E,BM**; banks of the River Spey at Boat of Garten, 1973 **ABD**; moorland track at Slochd, 1973 **K**; Bridge of Brown, 1973 etc.
N Bog on the moor above the gorge of the small burn at Clunas Reservoir, 1973 **E**.
I Braeriach, 1887 *G. C. Druce* **OXF**; Kincraig, 1891 *AS* **BM**; Coire

Leachavie of Mam Sodnail in Glen Affiric, 1947 *A. J. Wilmott*
BM; bog at Culloden, 1962 **CGE**; moor at Coignafearn, 1970
CGE; grass slopes of Sgùrr na Lapaich in Glen Strathfarrar, 1971;
moor at Coillenaclay near Farley, 1972 **E**; east side of the River
Spey at Boat of Garten, 1972 **E**; by Loch Conagleann, Dunmaglass,
1972 **E**; among stones in the river bed at the Plodda Falls,
Guisachan, 1972 **E**; ditch on the moor at Abriachan, 1972 **ABD**;
below roadside cliffs at Eskadale, 1973 **BM**; island in the River Glass
near Eskadale, 1973 **K**; by the burn at Gaick Lodge, flush in Coire
nan Laogh of Glen Banchor, 1973; wet moorland by the lochan near
An Leacainn at Blackfold, 1973 **ABD**; burn at Coignashie in
Strathdearn, 1974 **K** etc.

T. spectabile Dahlst.
Very common. Very similar to *T. faeroense* from which it can be
distinguished by the more numerous (4–8) and more recurved
leaf-lobes. Fig 38d.
M Rocks by the River Findhorn at Randolph's Leap, Relugas,
1968 **CGE**; by the old Spey Bridge at Grantown, 1972 **E**; bog on
Romach road, Phorp, 1972 **E**; Huntly's Cave, 1973 **ABD,BM**; near
Knockando station, 1973 **ABD**; flush at Bridge of Brown, 1973 **K**;
grass bank at Glenerney House, 1974 **K**; by the old lime kiln at Fae
near Dorback and on Carn Dearg Mòr at Aviemore, etc.
N Banks of the River Findhorn at Glenferness House, 1973
E,ABD,BM,CGE; grass bank N of Lochloy, 1974 **E**.
I (Marshall, *J. Bot.* **51**: 164 (1913)); by the Allt Coire Dubh in
Glen Shirra, 1916 *ESM* **BM**; west end of Loch Einich, 1922 *C. E.
Salmon* **BM**; gully on Sgòran Dubh Mòr, Glen Einich, 1965 *R. Jones
& B. A. Miles* **BM**; flush on the north face of Sgùrr na Lapaich,
Glen Strathfarrar, 1971 **E,BM**; slopes of Mealfuarvonie, 1971; by the
dam at Loch Loyne, 1971 **CGE**; moorland bog below Netherwood,
Newtonmore, 1972 **E**; by Loch Conagleann, 1972 **E**; Coire nan
Laogh in Glen Banchor, 1973 **E,ABD,CGE**; Sgòr Bhothain at
Gaick, at 700 m, 1973 **E**; grass slope in the Ryvoan Pass, 1974 **K**;
Coignashie in Strathdearn, 1974 **K**; Pass of Inverfarigaig, 1975 **E**;
moor at Drumochter, 1975 **CGE** etc.

T. eximium Dahlst.
Scarce, mostly in the highlands. Very similar to *T. faeroense* but
with spathulate, usually unlobed, shining dark-green leaves and
larger heads and achenes.
M Railway embankment near Knockando House, 1967 **CGE**;
rocks by the River Findhorn at Randolph's Leap, 1967 **CGE**;
Glenbeg road Grantown, 1972 **E**; grass bank by Huntly's Cave,
Dava, 1973 **E**.
N Dunes on the Nairn Dunbar golf course, 1967 **CGE**.
I Damp bank by a moorland track at Farley, 1972 **E**; moor NE of
Dalwhinnie, 1973 *R. McBeath*.

T. fructiosum A. J. Richards inedit.
I By the Elrick Burn at Coignafearn in Strathdearn, 1970 **CGE**.

Fig 38   a, **Crepis paludosa**; b, **Taraxacum brachyglossum**; c, **T. faeroense**; d, **T. spectabile**.

A. J. Richards writes 'a remarkable plant clearly related to *T. faeroense*, *T. spectabile* and *T. eximium*, but only possibly *T. spectabile* on leaf shape of these, and then aberrant. The achenes at *c.* 5·2 mm are the longest known outside Asia. Probably a new species which I name *T. fructiosum* ad inter. I must see more flowering material before publication'.

T. euryphyllum (Dahlst.) M. P. Chr.

Frequent in the north. Usually on base-rich grassland. Leaves spotted, ligules striped red-purple, exterior bracts spreading, glaucous; achenes dark straw-coloured; pollen absent.

M East side of the old Spey Bridge at Grantown, 1972 **E**; banks of the Allt Iomadaidh at Fae, Dorback, 1974 **E**.

I Grass bank by the track to the Falls of Foyers, 1971 **E**.

T. maculigerum H. Lindb. fil.

Common. Leaves spotted; exterior bracts erect, dark green; pollen absent.

M Wood near Clashdhu, Altyre, 1967 **CGE**; rocks by the River Findhorn at Randolph's Leap, 1967 **CGE**; by a small bridge on the hill road at Cromdale, 1971; dunes between the golf course and the sea at Lossiemouth, 1972 **E**; Buckie Loch in the Culbin Forest, 1972 **E**; Clashdhu railway cutting, 1972 **E**; banks of the River Spey at Pitcroy, 1973 **E**; by the Spey at Knockando station, 1973 **ABD**; verge of a track at Douneduff, Darnaway, 1973 **BM**; track at Wester Elchies, 1973 **K**; roadside W of Dallas Lodge, 1973 **ABD**; Hill of Dalnapot, 1973 **BM**; by a bridge in the drive of Castle Grant, 1974 **K**; at Balnabreich, Kinveachy, Carrbridge, Dyke and Darnaway, 1974.

N Moorland Cantraydoune, 1970 **CGE**; garden path at Holme Rose, 1974 **E**; path by the River Nairn near Cantraydoune, 1974 **K**.

I Track at Glen Feshie Lodge, 1971 **ABD**; path by the Falls of Foyers, 1971 **E**; glen road Pass of Inverfarigaig, 1971 **BM**; river shingle and on a bank of the Beauly road at Milton, 1971 **CGE**; bridge over the burn at Inchnacardoch, 1972 **E**; on rocks by Loch Killin, 1972; low rocks by the road at Eskadale, 1973 **ABD**; Glen Feshie, 1974 **K** etc.

T. praestans H. Lindb. fil.

Leaves usually unspotted, smooth with a bright purple petiole and mid-rib; exterior bracts erect, pruinose; pollen present.

M Sand dunes in the Culbin Forest, 1967 **CGE**; base-rich flush near the railway bridge in the policies of Altyre House near Forres, 1967 **CGE**; garden weed at Rose Cottage, Dyke, 1969 **CGE**; shingle of River Spey at Fochabers, 1972 **E**; track at Sanquhar House, Forres, 1974 **E**; bank opposite Sluie, Darnaway, 1974 **E**; station road Carrbridge, 1974 **K**; by the pillar box at Dyke, 1974 **BM**; on the parapet of Daltulich Bridge, 1974 **ABD**.

N Boggy field near East Milton Farm, 1967 **CGE**; bank by the river path at Holme Rose, 1974 **E,K**.

I Yard at Glen Cottage, Ness Castle, 1971 **CGE**; bank at East Craggie near Daviot, 1971 **BM**; verge of a moorland track at Bunchrew, 1971 **E**; hillside at Milton, 1972 **E**; upper Glen Feshie, 1974 **E**; gully on S side of Sgùrr Gaorsaic in Glen Affric, 1974 *R. W. M. Corner*; near the power station at Kilmorack, 1975 **K**; an unusual spotted form on river shingle at Newtonmore, 1975 **E**.

T. naevosiforme Dahlst.

Leaves spotted, lobes recurved; exterior bracts spreading not glaucous, pollen present.

M Foot of a small bridge by the road W of Pluscarden Priory, 1968 **CGE**; by the river at the old Spey Bridge, Grantown, 1972 **E**; stoney verge of the River Spey at Boat o' Brig, 1972 **E**; Clashdhu railway cutting, 1972 **E**; banks of the Spey at Knockando station, 1973 **BM**; by the Spey at Boat of Garten, 1973 **ABD**; lay-by on Dava Moor, 1974 **K**.

N Verge to the road at Holme Rose, 1967 **CGE**; dunes on the Nairn Dunbar golf course, 1970 **CGE**; dry bank by a lay-by near Lethen, 1973 **E**; pathway at Holme Rose, 1974 **K**.

I Railway embankment near Daviot, 1971 **BM**; waste ground by the Foyers Post Office, 1971 **K**; shingle of the River Enrick at Milton, 1971 **CGE**; below the parapet of Torgyle Bridge in Glen Moriston, 1972 **E**; bog at Coillenaclay near Farley, 1972 **E**; near Eskadale in Strathglass, 1973 **ABD**; roadside at Fort Augustus, 1973 **CGE**; Inshriach Nursery, 1973 *R. McBeath*; shingle of the River Truim at Drumochter, 1975 **BM**.

T. subnaevosum Richards

Rare, only found in north Britain.
Leaves pale green, spotted: plant small and delicate, exterior bracts narrow and more or less recurved; styles yellow, pollen present.

M Station yard at Carrbridge, 1974 **E**; garden path at Glenerney House, Dunphail, 1974 **E**.

I Dry grass bank by the Breakachy Burn in Strathglass, 1975 **E**; waste ground between Foyers and Inverfarigaig, 1975 **E**; roadside verge at Banchor Mains, Newtonmore, 1975 **BM**.

T. drucei Dahlst.

Rare on base-rich cliffs. Leaves spotted with rounded terminal lobe.

I Cliffs above Loch Dubh in Glen Banchor, and at Gaick, 1974 *R. McBeath*.

T. naevosum Dahlst.

Leaves spotted, large, bristly; heads very large.

M Church Avenue, Grantown, 1974 **E**.

N Shingle spit in the River Nairn ½ mile upstream from Nairn, 1975 **E**.

I Bank by the road to Auchteraw in the town of Fort Augustus, 1973 **E**; Glen Markie, 1973 *R. McBeath*.

T. laetifrons Dahlst.
Frequent in open woods, Leaves unspotted, leaf-lobes recurved, toothed.
M Walls of Lochindorb Castle, 1974 **E**.
I Wood at Glen Cottage, Ness Castle, 1971 **E,CGE**; wall W of Kirkhill, 1972 **E**; in shade below the cliffs of Creag Dhubh near Laggan, 1972 **E**; by the Breakachy Burn in Strathglass, 1975 **BM**.

T. croceum Dahlst.
An arctic plant of high mountain ledges. Leaves bright green, unspotted, narrow. Heads orange-yellow.
I Coire an t-Sneachda of Cairn Gorm at 1,100 m, 1919 *G. C. Druce* **OXF** and *M. L. Wedgewood* **MBH**; cliff ledges in Glen Markie and Glen Einich, 1973 R. McBeath: gully at 610 m by the Allt Coire Chùirn at Drumochter, 1975 **E**.

T. ceratolobum Dahlst.
Rather similar to the preceding but with more leaf-lobes.
I Rock crevice on 'The Apron' in the Coire an Lochain of Cairn Gorm, 1969 **CGE**.

T. stictophyllum Dahlst
Rare. Leaves broad, stiff and spotted. Heads small.
M Grass bank by the River Spey at Fochabers, 1972 **E**.
I Grass verge of moor by Ceannocroc Bridge, Glen Moriston, 1974 **E**.

T. caledonicum Richards
Rare. Restricted to the east highlands. Leaves narrow, dark green, unspotted; exterior bracts erect, dull, pruinose.
M In the burn gorge at Huntly's Cave, Dava Moor, 1973 *R. McBeath & R. Richter* **E** (third British record).
I Eskadale in Strathglass, 1973 **E**; gorge of a small burn on Creag Dhearg near Lurgmore, 1975 **E**; hillsides at Gaick and in Glen Banchor, *R. McBeath*.

T. nordstedtii Dahlst.
Common in wet places. Leaf-lobes with a concave angle; exterior bracts erect, pruinose.
M Sandy track in the Culbin Forest, 1967 **CGE**; below the parapet of Daltulich Bridge, 1972 **E**; waste sandy ground at Burghead, 1972 **E**; by the old Spey Bridge at Grantown, 1972 **E**; railway cutting at Clashdhu, 1972 **E**; gorge at Huntly's Cave, Dava, 1973 **BM**; roadside bank by a pine wood near the lodge of Logie House, 1973 **ABD,K**; gravel drive at Sheriffston, Lhanbryde, 1973 **CGE**; track by the River Spey at Wester Elchies, 1974 **BM**; station yard Dunphail, and by the kennels at Kinveachy, 1974; shingle of the River Spey at Broomhill, 1975 **CGE**.
N Dunes at Nairn, 1962 and moorland verge at Cantraydoune, 1970 **CGE**; marsh by the River Findhorn at Ardclach, 1973 **E**; roadside at Righoul, 1973 **ABD**; roadside verge near the Ardclach Kirk, 1973 **BM**; moor between Littlemill and Auldearn, 1973 **K**.

I   Roadside at Milton of Balnagowan near Ardersier, 1971 **E**;
shingle of the River Enrick at Milton, 1971 **BM**; on a wall W of
Kirkhill, 1972 **E,CGE**; track to the waterfall at Farley, 1972 **E**;
Creag Dhubh near Laggan, 1972 **E**; bank on the glen road near
Invermoriston, 1973 **ABD**; Gaick. 1973 *R. McBeath*; Alturlie Point,
1975; roadside verge at Inverfarigaig, 1975 **E**; grass bank at
Drumochter, 1975 **CGE**.

T. adamii Claire
Leaves dark green, unspotted with purple mid-rib; exterior bracts
adpressed, white-margined.
M Railway embankment near Knockando House, 1967 **CGE**;
Clashdhu railway embankment, 1972 **E**.
I   Grass bank in Glen Moriston, 1972 **E**; on low rocks by the road
S of Eskadale in Strathglass, 1973 **E**.

Section **Vulgaria** Dahlst.   *Common Dandelion*
Robust plants of waste places, roadsides and grassland. Leaves never
spotted; ligules usually striped grey-violet or purple but never
carmine. Stigmas usually discoloured. Pollen usually present.

T. cyanolepis Dahlst.
Very common. Exterior bracts spreading, violet-purple.
M Hill road at Cromdale, 1971 **E**; railway line at Forres, 1972 **E**;
by the old Spey Bridge, Grantown, 1972 **E**; waste ground at Boat
of Garten, 1973 **ABD,BM,CGE**; track at Nethybridge, 1973 **E**;
bank at Thomshill Distillery, 1973 **K**; bank E of Knockando village,
1973 **ABD**; by the South Lodge at Castle Grant, 1973 **BM**; grass
verge in Church Avenue, Grantown, 1974 **K**; yard at Elgin station,
1974.
N Waste ground by the Muckle Burn near Lethen House, 1973 **E**;
football field at Nairn, 1975 **E**.
I   Yard at Glen Cottage, Ness Castle, 1971 **E**; roadside verge by
Loch Pityoulish, 1971 **CGE**; roadside at Inchnacardoch, 1972 **E**;
farmyard at Charleston of Dunain, 1972 **E**; bank at Castle Stuart,
1973 **ABD**; roadside verge at Fort Augustus, 1973 **K,CGE**; Struy in
Strathglass, 1975 **ABD**; waste land near the sea at Allanfearn, 1975
BM; station yard at Dalwhinnie, 1975 **CGE**.

T. sellandii Dahlst.
Exterior bracts recurved, narrow, green or somewhat violet, purple-
tipped.
M Sand dunes at Lossiemouth, 1970 **CGE**; old railway track at
Forres (a blotched form due to chemical damage), 1972 **E**; under
Hippophae at Rose Cottage, Dyke, 1974 **E**.
I   Garden weed at Viewhill, Kirkhill, 1967 **CGE**.

T. monochroum Hagl.
Doubtfully native.
I   Below the parapet of the river bridge at Struy, 1975 **E**; roadside
verge by the Crask of Aigas Post Office, 1975 **E**.

T. ancistrolobum Dahlst.
Petioles green; leaf-lobes few, large and rounded.
N   By the mill path at Cantraydoune, 1974 **E**.

T. sublaciniosum Dahlst. et H. Lindb. fil.
A tall dark green species with slightly recurved, narrow leaf-lobes
and large dark green spreading exterior bracts.
M Garden, Rose Cottage, Dyke, 1974 **E**; waste ground Elgin
station, 1974 **E**; waste ground Carrbridge station, 1974 **E**.

T. pannucium Dahlst.
Exterior bracts large, erect.
M Garden weed at Rose Cottage, Dyke, 1969 **CGE**; car park,
Forres, 1974 **K**; Elgin station, 1974 **E**; roadside verge, Wester
Greens, Dunphail, 1974 **E**.
N By the river Nairn ¾ mile upstream from the town, 1975 **E**.

T. linguatum Dahlst. ex M.P. Chr. & Wiinst.
Distinguished by its almost entire, pale leaves and short, broad
dark green exterior bracts.
M Waste ground at Grantown and at Kinveachy, 1974 **E**.

T. alatum H. Lindb. fil.
Petioles long, green and winged; exterior bracts spreading, glaucous
and pink-tipped.
M Near bridge, Glenerney, 1972 **E**; garden at Rose Cottage, Dyke,
1972 **E**; car park, Garmouth (not typical), 1973 **E**; playing field at
Fochabers, 1975 **E**.

T. lingulatum Markl.
Exterior bracts strongly recurved, glaucous above, dark green below.
I Waste ground east of the Spey bridge at Inverdruie near
Aviemore, 1973 **E**; track in a pine wood near Culloden House,
1975 **E**; track near the old kennels at Culloden House, 1975 **BM**.

T. expallidiforme Dahlst.
Exterior bracts small, pale and recurved.
I Bank by the Big Burn at Glen Cottage, Ness Castle, 1971 **E**.

T. ekmanii Dahlst.
Petiole coloured; exterior bracts spreading-recurved, suffused with
violet, paler above.
M Garden at Rose Cottage, Dyke, 1974 **E**; by the entrance gate to
Sanquhar House, Forres, 1974 **E**.
I Inshriach Nursery, 1974 *R. McBeath*.

T. porrectidens Dahlst.
Petiole reddish-purple, unwinged, exterior bracts recurved, green,
darker below.
M Roadside bank east of Urquhart Kirk, 1973 **E**.

T. pectinatiforme H. Lindb. fil.
Doubtfully native. Unmistakable leaf shape with many long very
narrow linear processes.
M Church Avenue, Grantown, 1974 **E**.

T. aurosulum H. Lindb. fil.
Leaf-lobes recurved; exterior bracts very wide-spreading.
N Field at Cawdor Castle crossroads, 1973 **E**.

T. cordatum Palmgr.
M Path at Kinloss House, 1971 **E**; dunes at Lossiemouth, 1972 **E**;
Burghead, 1972 **E**; car park by Garmouth golf course, 1973 **CGE**;
farmyard at Mains of Allanbuie, 1974 **K**.
N Dunes, Nairn Dunbar golf course, 1972 **E**; near the house at
Holme Rose, 1974 **E**.
I Below a wall by the keeper's cottage, Beaufort Castle, 1972 **E**;
shore of Loch Flemington, 1973 **E**; side road at Fort Augustus, 1973
ABD; waste land by bridge between Bogroy and Kirkhill **CGE**.

T. longisquameum H. Lindb. fil.
Leaf-lobes few, recurved, triangular; exterior bracts narrow,
recurved or spreading.
M Car park by Old Spey Bridge, Grantown, 1972 **E**; track in
village of Nethybridge, 1973 **E**; car park by Garmouth golf club,
1973 **E,BM**; roadside bank at Tulcan, 1973 **ABD**; bank by road in a
pine wood near Logie Lodge, 1973 **CGE**; station yard, Dava, 1974
K.
N Roadside at Wester Lochend near Loch Flemington, 1973 **E**;
path by River Nairn, Cantraydoune, 1974 **E**.
I Gravel drive at Viewhill, Kirkhill, 1972 **E**; dry bank,
Kilmorack, 1972 **E**; roadside verge Glen Moriston, 1973 **E**; bank
at Invermoriston, 1973 **ABD,K,CGE**; near Kingussie, 1973, *R.
McBeath*.

T. dahlstedtii H. Lindb. fil.
Petiole narrow, vivid crimson-purple; exterior bracts green,
recurved.
M Railway embankment near Knockando House, 1967 **CGE**; sand
dunes at Findhorn and on The Bar, Culbin, 1969 **CGE**; garden at
Rose Cottage, Dyke, and dunes, Lossiemouth, 1970 **CGE**; by River
Spey, Boat of Garten, 1973 **E**.
N Dunes on the Nairn Dunbar golf course, 1970 **CGE**; moor
between Littlemill and Auldearn, 1973 **E**.
I Roadside verge near Kincraig station, 1970 **CGE**.

T. duplidens H. Lindb. fil.
Doubtfully native. The only species in the section without pollen
and with yellow styles.
M Garden weed at Rose Cottage, Dyke, 1970 **CGE**.
I Railway line between Daviot and Culloden, 1971 **E**; in the old
kirkyard at Kilmorack, 1972 **E**.

T. christiansenii Hagl.
Doubtfully native. Exterior bracts small, pale and spreading with
wide hyaline borders.

M By the Muckle Burn, Dyke, 1973, *Robert Smith* (age 13) **E** (second record for Scotland).

T. bracteatum Dahlst.

M Gorge of the Divie burn at Glenerney, 1972 **E**; station yard, Boat of Garten, 1973 **E**; station yard, Dunphail, 1974 **BM**; wall at the North Lodge, Castle Grant, 1974 **E,CGE**; grass verge of the playing field at Fochabers, 1975 **ABD**.
N In shade by the river at Nairn, 1975 **E**.
I Verge to road near Kincraig station, 1970 **CGE**; garden weed at Banchor Mains, Newtonmore, 1975 **E**.

T. hamatum Raunk.

Very common. Leaf-lobes 3–5, recurved, terminal one small, triangular; exterior bracts erect or spreading, very dark, glaucous-green outside.
M Roadside bank at Tulcan, Strathspey, 1973 **E,ABD**; bank by road in a pine wood near Logie Lodge, 1973 **E**; Elgin station, 1974 **K**; Rose Cottage, Dyke (summer form), 1974 **E**; cliff opposite Sluie, Darnaway, 1974 **CGE**; roadside verge at Bogroy, Slochd, 1974 **E**.
N Near bridge over Riereach burn at Glengeoullie, Cawdor, 1970 **CGE**; roadside bank, at Inchettle, 1973 **E**; roadside verge, Brightmony, 1973 **ABD**; path by river at Holme Rose, 1974 **K**; river bank at Nairn, 1975 **BM**.
I Braeriach (*Druce*, 1887); moorland near sea, Fort George, 1970 **CGE**; grounds of Raigmore Hospital, 1973 **E**; waste ground east of bridge, Inverdruie near Aviemore, 1973 **E**; near Kingussie, 1973, *R. McBeath*; roadside bank on the hill at Foyers, 1975 **BM**; roadside at Struy bridge, Strathglass, 1975 **K**.

T. hamatiforme Dahlst.

Leaf-lobes few, recurved; exterior bracts spreading, pale and somewhat glaucous above, dark green below.
M Between Relugas and the river bridge at Daltulich, 1972 **E**; bank at Willowbank, Knockando, 1975 **E**.
N Bridge over Riereach burn at Glengeoullie, Cawdor, 1970 **CGE**; Cameron Crescent, Nairn, 1971 **E**; roadside bank near Easter Cotterton, Lethen, 1973 **E**; waste ground near Lethen House lodge, 1973 **ABD**.
I Roadside verge between Cantraydoune and Clava, 1970 **CGE**; lay-by at Crask of Aigas and in the burn ravine at Inverfarigaig, 1975 **E**.

T. polyhamatum Øllgaard

Doubtfully native.
M Grass verge of the hill road at Wester Greens, Dunphail, 1975 **E** (new to British Isles).

T. marklundii Palmgr.

Common. Leaves often suffused purple; leaf-lobes 4–5, recurved, narrow; exterior bracts spreading purple above, dark green suffused purple beneath.

M Roadside between Grantown and Nethybridge, 1971 **E,BM**; verge to road just west of Urquhart, 1973 **E**; road east of the Kirk at Urquhart, 1973 **ABD**; wall by Lossiemouth aerodrome, 1973 **E**; roadside verge at Balnacoul wood, 1973 **CGE**; Rose Cottage garden, Dyke, 1974 **K**; bank of the burn at Fochabers, 1975 **K**; car park at Garmouth, 1975 **BM**; old railway track at Dalvey Mains near Cromdale, 1975.

N Dry bank of lay-by near Lethen Lodge, 1973 **E,ABD**; roadside bank, Wester Galcantray, 1974 **K**.

I Railway line near Daviot, 1971 **E,CGE**; bank by the old Kirk at Kilmorack, 1972 **E**; Crask of Aigas (as *T. subhamatum* M. P. Chr.), 1975 **E**.

T. latisectum H. Lindb. fil.
Leaf-lobes broadly triangular, slightly recurved.
M By a track at Wester Elchies, 1974 **E**.

T. fasciatum Dahlst.
Exterior bracts wide, imbricate, pale green.
N Below the bridge over Riereach burn at Glengeoullie, Cawdor, 1970 **CGE**.

T. raunkiaerii Wiinst.
    T. duplidentifrons Dahlst.
Very common. Leaves dull, broad-lobed, dentate; exterior bracts small and spreading.
M Railway embankment near Knockando House, 1967 **CGE**; Rose Cottage garden, Dyke, 1972 **E**; wall near Castle Grant, Grantown, 1972 **E**; lane near Church Avenue, Grantown, 1974 **K**.
N Dunes on the Nairn Dunbar golf course, 1970 **CGE**; roadside at Righoul near Geddes, 1973 **E**; roadside verge by Ardclach Kirk, 1973 **E**; path by River Nairn, Cantraydoune, 1974 **K**; garden at Holme Rose, 1974 **ABD,BM**.
I Roadside near Kincraig station, 1970 **CGE**; roadside verge near Kingussie, 1971 **E**; Inshriach Nursery, 1971 **ABD**; Raigmore hospital grounds, 1973 **E**; station yard at Dalwhinnie, 1975 **BM, CGE**; Kilmorack, 1975; Banchor Mains, Newtonmore, 1975 **E**; grass bank ½ mile S of Kingussie and by the Breakachy burn in Strathglass, 1975 **BM,K**.

T. hemipolyodon Dahlst.
Doubtfully native. Leaves large, strongly dentate; exterior bracts large and erect.
M Rose Cottage, Dyke, 1974 **E**; station yard, Carrbridge (typical), 1974 **E**.

T. ordinatum Hagendyk, van Soest and Zevenbergen
Doubtfully native.
M Church Avenue and near Castle Grant, Grantown, 1974 **E**.

T. parvuliceps H. Lindb. fil.
Doubtfully native. Plant delicate, petioles red, leaves highly dentate; heads small; exterior bracts strongly recurved pale green above. Styles discoloured, pollen absent.
M Grantown, 1972 **E**.

T. polyodon Dahlst.
Common. Distinguished from the other highly dentate species with coloured petioles by the dark interlobes and recurved exterior bracts.
M Sand dunes at Hopeman, 1967 **CGE**; dunes near golf club at Lossiemouth, 1970 **CGE**; gravel drive at Kinloss House, 1971 **ABD**; path to the Muckle Burn at Brodie, 1971 **BM**; station yard, Boat of Garten, 1973 **E**; track at Wester Elchies, 1973 **E**; Rose Cottage, Dyke, 1974 **K**; Elgin station, 1974 **BM**; car park at Forres, 1974 **E**; Church Lane, Grantown, 1974 **ABD**; below walls at Nethybridge, 1975 **K**; at Hopeman and in Garmouth car park, 1975.
N Wall at Tradespark, Nairn, 1972 **E**; by the playing fields at Nairn, 1975 **BM**.
I Railway embankment between Daviot and Culloden, 1971 **E**; waste ground near the hotel at Dalwhinnie, 1975 **BM**; roadside at Struy, Strathglass, 1975; grass bank ½ mile S of Kingussie, 1975 **E**.

T. reflexilobum H. Lindb. fil.
Leaves narrow, dark with black interlobes; leaf-lobes many, narrow and recurved.
I Waste ground by the hotel at Dalwhinnie, 1975 **E**.

T. crispifolium Lindb. fil.
Leaves dark and narrow; leaf-lobes highly crisped.
M Kirkyard at Dyke, 1970 **CGE**; path by the greenhouse at Brodie Castle, 1970 **E,CGE**; farm yard at St Mary's Well, Orton (very narrow-lobed form), 1973 **E**; garden at Rose Cottage, Dyke, 1974 **K**; garden at Pylon Cottage, Brodie, 1974 **BM**; flower border by a garage at Hopeman, 1975 **E**.
I Roadside just north of Ruthven Barracks, Kingussie, 1971 **E**; Glen Strathfarrer, 1972 **E**; roadside verge by Loch Ness near Dochfour, 1972 **E**; by the Breakachy burn in Strathglass, 1975 **ABD,K**; gateway Dochfour, 1975 **ABD**; waste land at Drumochter, 1975 **CGE**; below the bridge at Struy, Strathglass, 1975 **BM**.

MONOCOTYLEDONES
ALISMATACEAE
Baldellia Parl.

B. ranunculoides (L.) Parl.   *Lesser Water-plantain*
   Alisma ranunculoides L.
Native. Rare on loch margins and in ditches. Fig 39e.
M Frequent, the Leen at Garmouth, Boghead and Trochail (Gordon); Kintessack, 1843 *J. G. Innes* (Gordon, *annot.*); parishes of

Dyke, Urquhart and Edinkillie (Burgess); verge of Gilston Loch near Elgin, 1950.

N Loch Litie, 1832 *WAS* **E,ABD**; Nairn, *A. Falconer* (Gordon); Lochloy, 1860 *JK* **FRS**; the Minister's Loch on the Nairn Dunbar golf course, 1887 *R. Thomson* **NRN** and 1967 *MMcCW* **E**; first loch E of Nairn and at Lochloy, 1904 *G. T. West* **STA**; at the W end of Lochloy, 1973 *R. M. Suddaby.*

I Peat holes above the River Beauly at Beauly, 1892 *ESM* **E,BM**; between Wester Lovat Farm and the Ferry, 1947 *A. J. Wilmott et al.* **BM**; growing in mud with *Isoetes* and *Littorella* at a depth of 1 m, near the outlet of the Caledonian Canal at Muirtown Basin, Inverness, 1975 **E,BM**.

Luronium Raf.

L. natans (L.) Raf.   *Floating Water-plantain*
   Alisma natans L.
I Floating in Loch Meiklie, Glen Urquhart where it was probably introduced by waterfowl, 1975 **E**.

Alisma L.

A. plantago-aquatica L.   *Water-plantain*
Native. Occasional in pools, in ditches and by slow-flowing burns in the lowlands. Map 253.

M Very common (Gordon); ditch at Moycarse near Forres, *J. G. Innes* **ABD**; Greshop, 1856 *JK* **FRS**; pools at Spynie Loch, Dyke, Newton of Dalvey, and Kincorth, 1954; bogs by Gilston Lochs and in a field near Sanquhar House at Forres; backwater of the River Spey at Fochabers, 1973.

N (Stables, *Car.*); pond near Nairn (Aitken); at Kingsteps and at Geddes Reservoir, 1887 *R. Thomson* **NRN**; boggy ground on the margin of Loch Flemington, by Loch of the Clans and the Minister's Loch at Nairn, 1954; pool by the River Nairn at Kilravock Castle, 1966.

I Pond near Culloden Tower and ditch at Bunchrew (Aitken); Urquhart Bay, 1904 *G. T. West* **STA**; pond at Achnagairn, Kirkhill, margin of Loch Flemington, ditch at Muirtown Basin at Inverness and at Inchnacardoch Bay near Fort Augustus, 1954; sandy spit of land at the outflow of the River Coiltie in to Loch Ness at Urquhart Bay.

BUTOMACEAE
Butomus L.

B. umbellatus L.   *Flowering Rush*
Doubtfully native in north Scotland. Probably introduced by water-fowl.

M Pond at Brodie Castle, *Brodie* **E**; ditch near the quarry-hole at Hamlets by Spynie Loch, 1966 **E**.

N In the golf burn by the Nairn coast, 1900 (Lobban).

I In the old ice pond at Lovat Bridge, Beauly (Pollock); Raigmore pond at Inverness, 1973.

HYDROCHARITACEAE
Elodea Michx.
E. canadensis Michx.   *Water Thyme, Canadian Pondweed*
Introduced. A dark green, submerged plant with leaves 3 in a whorl, and flowers on very long filiform stalks. Very common 20 years ago, now only occasional in pools, lochs and lochans. Fig 39a, Map 254.

M Loch Spynie (Druce, 1897); Garmouth, 1900 *JK* **FRS**; ponds at Sanquhar House and Altyre, pool by the River Findhorn near Moy House, 1954; abundant in Gilston and Spynie lochs, 1960; pool by the River Spey at Boat o' Brig, 1972; Loch na Bo, 1972 **E**; backwater of the River Spey at Essle, 1976.
N Cawdor curling pond, 1900 (Lobban); Loch Flemington, 1954.
I Loch Flemington and Loch Insh, 1954; Loch a' Chlachain and pond at Culloden House, 1975.

JUNCAGINACEAE
Triglochin L.
T. palustris L.   *Marsh Arrow-grass*
Native. Abundant in moorland bogs and flushes and occasionally on grass tussocks on salt-marsh verges. Fig 39c, Map 255.
M Frequent (Gordon); Loch Spynie (Druce, 1897); Forres, 1857 *JK* **FRS**; the Leen at Garmouth, 1953 *A. Melderis* **BM**; throughout the county, 1954; Buckie Loch, Culbin Forest, 1968 **E**; bog at Drakemyres, 1972 **E**.
N Moss of Litie near Auldearn, 1832 *WAS* **ABD**; Daltra near Ardclach, 1887 *R. Thomson* **NRN**; throughout the county, 1954; moor on Nairn Dunbar golf course, 1962 **ABD**; bog N of Boghole Farm, 1967 **CGE**.
I Inverness, 1832 Dalwhinnie, 1833 (Watson, *Loc. Cat.*); lochs about Invermoriston and Fort Augustus, 1904 *G. T. West* **E,STA**; Ryvoan Pass, 1937 *J. Walton* **GL**; in moorland bogs throughout the district, 1954; margin of the salt-marsh at Muirtown Basin and in a fixed dune on the raised shingle beach at Ardersier.

T. maritima L.   *Sea Arrow-grass*
Native. Common in salt-marshes the whole length of the coast. Fig 39b.
M Abundant in one or two places, Loch of Cotts, Spynie and Kincorth (Gordon); near Findhorn Bridge, 1863 *JK* **FRS**; salt-marsh between the Bar and the Culbin Forest, 1968 **E**.
N (Stables, *Cat.*); by Druim House near Nairn, 1937 *J. Walton* **GL**; salt-marsh E of Nairn, 1960 **ABD**.
I Inverness, 1832 (Watson, *Loc. Cat.*); salt-marshes at Castle Stuart, Beauly, Carse of Delnies and in the lagoons at Muirtown Basin at Inverness, 1954.

423

Fig 39 a, **Elodea canadensis**; b, **Triglochin maritima**; c, **T. palustris**; d, **Zostera angustifolia**; e, **Baldellia ranunculoides**.

APONOGETONACEAE
Aponogeton L. fil.
A. distachyos L. fil.   *Cape Pondweed*
Introduced. Planted in the lochan at Hilton near Guisachan House,
Tomich, 1953 *R. E. Palmer* and 1964 *MMcCW* **E**.

ZOSTERACEAE
Zostera L.
Z. marina L.   *Eelgrass, Grass-wrack*
Native. On fine sand, gravel or mud in the sea down to 4 m. Only
observed washed up on the shore line, the old records may refer to
either of the other two species, which are frequent in estuaries.
M Burghead, 1836 *G. Gordon* **E**; parishes of Duffus, Kinloss and
Drainie (Burgess); washed up on the Old Bar, Culbin, 1969 **E,CGE**.
N Sea coast E of Nairn, 1833 *G. Gordon* (Watson, *Top. Bot.*);
washed up on the Old Bar, Culbin, 1969 **ABD**.
I Sands at Fort George, 1833 *WAS* (Watson); near Fort George,
1846 (? *A. Croall*) **STI**; floating in Loch Tarff where it must have
been brought in by wild fowl, 1972 **E**.

Z. angustifolia (Hornem.) Reichb.   *Narrow-leaved Eelgrass*
Native. On mud flats and in shallow water. Fig 39d.
I Sands east of Fort George, 1832 *WAS* **E** and 1833 *W. Brand*
ABD; Beauly Firth, 1850 *P. H. MacGillivray* **ABD**; mud flats at
Lentran, 1892 *ESM* **E,OXF**; Bunchrew, 1940 *AJ* **ABD**; abundant
in the Beauly Firth, 1954.

Z. noltii Hornem.   *Dwarf Eelgrass*
M Mouth of the River Findhorn near Kinloss, 1913 *D. Patton* **GL!**;
on the mud flats and in runnels between the Old Bar and the
Culbin Forest, 1953.
I (Trail, *ASNH* **8**: 172 (1899)); Beauly Firth at Inverness, 1954.

POTAMOGETONACEAE
Potamogeton L.
In the following account all herbarium specimens have been seen
by J. E. Dandy unless preceded by an asterisk.

P. natans L.   *Broad-leaved Pondweed*
Native. Common in lochs, lochans and moorland ditches.
Recognised by the floating leaves appearing jointed just below the
blade and the very long stipules. Fig 40f, Map 256.
M Frequent (Gordon); Boat of Garten, 1888 *G. C. Druce* **BM**;
lochan near Aviemore, 1894 *H. Groves* **BM**; Lochindorb, 1933 and
Gilston Lochs, 1946 *G. Taylor* **BM**; Loch Oir near Lhanbryde,
1946 *G. Taylor* **BM**; fire pool in the Culbin Forest near Snab of
Moy, 1967 **E,BM,CGE**; Loch Romach and Millbuies, 1972 **E**;
Gordonstoun Lake, 1972 **E,BM**, etc.
N *Ardclach, 1888 *R. Thomson* **NRN**; Lochloy, 1904 *G. T. West*
STA; small loch E of Nairn town, 1939 *G. Taylor* **BM**; Loch

Flemington, 1956 **ABD**; Lochloy, 1972 **E,BM**; Geddes Reservoir, 1972 **E**.
I *Near Foyers House, 1855 *HC* **INV**; Loch Knockie, 1875 *H. A. Hurst* **MANCH**; Loch Morlich, 1894 *H. Groves* **BM**; tributary of the River Spey at Kingussie, 1889 *W. F. Miller* **BM**; estuary of the River Enrick in Urquhart Bay, at Inchnacardoch, in lochs Meiklie, Killin, Càrn a'Chuilinn, Tarff and Loch nan Faoileag, 1904 *G. T. West* **STA**; Nairnside, 1940 *AJ* **ABD**; in lochs Tarff and Knockie, canal by Insh village, Culloden House pond and pools in Strathglass etc. all **BM**; in Blackfold lochs, Loch Killin, Loch Mallachie etc. all **E**.

P. polygonifolius Pourr. *Bog Pondweed*
Native. Abundant in acid bogs and lochans, usually in shallow water. Fig 40e.
M Marsh near Manachy, Forres, 1856 *JK* **FRS**; near Carrbridge, 1888 *G. C. Druce* **BM**; Castle Grant at Grantown, 1895 *C. Bailey* **MANCH**; fire pool in the Culbin Forest near Snab of Moy, 1969 **BM**; pool near Lochindorb, 1972 **E,BM**; bog at Drakemyres, 1972 **E**, etc.
N *Common in ditches at Ardclach, 1890 *R. Thomson* **NRN**; at Lochloy and in a small pool **E** of Nairn, 1939 *G. Taylor* **BM**; Geddes Reservoir, 1961 **ABD**; in Lochan Tutach, at Achavraat, in a small burn at Clunas Reservoir and at Littlemill, 1973 **E**, etc.
I Strath Affric (Ball, 1851); N end of Loch Ericht, 1841 *H. C. Watson* **K**; near Kincraig, 1891 *AS* **BM**; in lochs Uanagan, Tarff, a'Mhuilinn etc., 1904 *G. T. West* **STA**; Loch Pityoulish, 1935 *G. Taylor* **BM**; moorland bog on Mealfuarvonie, 1971 **E**; small burn near Dalballoch in Glen Banchor and at Gaick etc., 1973 **E**.

P. lucens L. *Shining Pondweed*
Native. Only recorded with certainty from ponds at Brodie Castle, *c.* 1800 *Brodie* **E**. Not seen in recent years.

P. gramineus L. *Various-leaved Pondweed*
  P. heterophyllus Schreb.
Native. Frequent in the larger lochs. Much branches at the base; submerged leaves sessile, linear-lanceolate; floating leaves (sometimes absent) very long-stalked, broadly elliptical-oblong. Fig 40g, Map 257.
M Ditches at Alves, Mosstowie and Duffus, *WAS* (Murray); Mosstowie and Duffus, 1832 *G. Wilson* (Gordon); Loch Spynie, 1896 *G. C. Druce* **OXF** and 1946 *G. Taylor*, **E,BM**; Loch na Bo and Loch Oir near Lhanbryde, 1946 *G. Taylor* **E,BM**; Loch an t-Sithein S of Lochindorb, 1973 **E**.
N Inshoch, 1809 Brodie **E**; Lochloy, 1809 Brodie **E**, in 1833 *J. B. Brichan* **E**, in 1867 *JK* **FRS**, in 1898 *ESM & WASh* and 1972 *MMcCW* **E,BM**; Cran Loch, 1904 *G. T. West* **STA** and in 1975 *MMcCW* **E**; Loch Flemington, 1946 *G. Taylor* **BM**!.
I Loch Garten, 1887 *G. C. Druce* **OXF**; lochs Morlich and pityoulish, 1894 *H. Groves* **BM**; Loch nam Faoileag (as *P. lucens*),

Fig 40   a, **Potamogeton perfoliatus**; b, **P. praelongus**; c, **P.** × **zizii**; d, **P. alpinus**; e, **P. polygonifolius**; f, **P. natans**; g, **P. gramineus.**

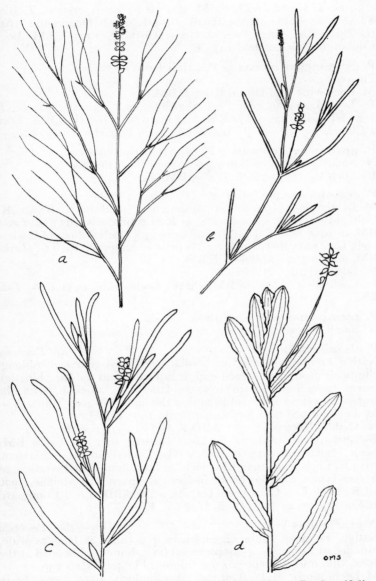

Fig 41   a, **Potamogeton pectinatus**; b, **P. berchtoldii**; c, **P. obtusifolius**; d, **P. crispus**.

1904 *G. T. West* **STA**; Loch Meiklie (as *P. lucens*), 1904 *G. T. West* **STA**; Loch Laide at Abriachan, 1954 **E**; Loch Flemington, 1946 *G. Taylor* **BM**; small loch at Upper Gortally, 1975 **E,BM**; Loch Coulan near Balnafoich, 1975 **E**.

P. gramineus×lucens = P.×zizii Koch ex Roth
Fig 40c.
M Loch Mòr E of Duthil, 1973 **E,BM**.
I Loch Uanagan near Fort Augustus (as *P. lucens*), 1904 *G. T. West* **STA**, in 1973 *UKD* **BM**, and in 1975 *MMcCW* **E,BM**; Loch a'Chlachain, 1904 *G. T. West* **STA**; Loch Bunachton 1975 **E**.

P. gramineus×natans = P.×sparganifolius Laest. ex Fries
N Small loch east of Nairn town, 1939 *G. Taylor* **BM**.
I Loch Garten, 1939 *J. Walton* **BM,GL**.

P. gramineus×perfoliatus = P.×nitens Weber
M In the River Lossie at Aldroughty, 1828 herb. *Hooker* **K**; Coltfield at Alves, 1832 *G. Wilson* **K**; Lochindorb, 1933 *G. Taylor* **BM**, in 1955 *MMcCW* **CGE**, and in 1973 **E**; loch at Boat of Garten, 1935 *G. Taylor* **BM**; Loch na Bo near Lhanbryde, 1946 *G. Taylor* **BM**, and in 1972 *MMcCW* **E,BM**.
N Lochloy, 1833 *WAS* **CGE**.
I Loch Alvie, 1891 *AS* **BM,NMW**; Loch Ashie, 1934 *E. S. Todd* **EST**.

P. gramineus×polygonifolius
N Small loch east of Nairn town, 1953 *G. Taylor* **BM**.

P. alpinus Balb.   *Red Pondweed*
Native. Frequent in acid lochs. Submerged leaves narrowly oblong-elliptical, narrowed to each end, translucent and reddish. Floating leaves (sometimes absent) obovate-lanceolate, thickish, with the transverse veins easily visible against the light. Fig 40d, Map 258.
M Loch an t-Sithein S of Lochindorb, 1973 **E,BM**.
N Geddes Reservoir, 1961 **ABD,K** and in 1972 **E,BM**.
I Loch Alvie, 1891 *AS* **E**; Loch Garten, 1942 *J. Walton* **BM**; Loch Mallachie, 1942 *J. Walton* **GL**; Loch Laide at Abriachan, 1973 **E**; Loch Insh, 1975 **BM**; lochs at Ladystone near Inverness, at Abbar Water near Lochend, in lochs Ruthven and Meiklie, 1976 all **E**; Loch Ceo Glais near Torness, 1976 **ABD**; Loch Comhnard at Corrimony, 1976 *OMS & MMcCW* **E**.

P. praelongus Wulf.   *Long-stalked Pondweed*
Native. Frequent. Submerged leaves 4–6 times as long as wide, rounded at the sessile ± amplexicaul base, blunt and hooded at the apex, thin and translucent, not shining. Fig 40b, Map 259.
M Lochindorb, *WAS* (Gordon); Lochindorb, 1834 ? *coll.* **E**, in 1903 *JK* **FRS**, in 1933 *G. Taylor* and in 1972 *MMcCW* **E**.
N In the Moss of Litie and in Lochloy, 1832 *J. B. Brichan* **K** (new to the British Flora: *Phyt.* 1: 236 (1843)); Lochloy, 1834 *WAS* **E**; Loch of Litie, 1835 *WAS* **E,BM**; loch east of Nairn, 1946 *G. Taylor* **BM**.

I Loch Garten, 1887 *G. C. Druce* **OXF**; Loch Morlich, 1894 *H. Groves* **BM**; Loch Uanagan (as *P. lucens*), 1904 *G. T. West* **STA**; Loch nam Faoileag, 1904 *G. T. West* **STA** and * 1956 *UKD* **E**; Blackfold lochs and Loch na Ba Ruaidhe near Milton, 1973 **E**; Loch Ceo Glais near Torness, 1976 **E**; Loch Ashie, 1976 **BM**; Loch Glanaidh above Drumnadrochit, 1976 **ABD**.

P. perfoliatus L. *Perfoliate Pondweed*
Native. Frequent. Submerged leaves all sessile and amplexicaul at the wide cordate base. Fig 40a, Map 260.
M Loch of Spynie, 1824 (Gordon); Loch Spynie, 1937 *G. Gordon* **E,MANCH**, and in 1972 *MMcCW* **E**; Loch na Bo, 1972 **E**; Lochindorb, 1972 **E,BM**.
N Loch near Nairn, 1894 *H. Groves* **BM**; Lochloy, 1972 **E**.
I Loch Gynack, *c.* 1847 *A. Rutherford* **BM**; Loch Ashie, 1904 *G. T. West* **STA**; Loch Garten, 1939 *J. Walton* **GL**; Loch Mallachie, 1942 *J. Walton* **BM,GL**; Loch Pityoulish, 1935 *G. Taylor* **BM**; lochs Garten and Mallachie, 1973 **E**; Loch Insh, 1975 **BM**; Loch Uanagan, 1975; Loch Ruthven, 1975 **E**; Loch Bunachton, 1975 **E,BM**; Loch Mhòr near Errogie, 1976.

P. rutilus Wolfg. *Shetland Pondweed*
Native. Very rare. The second record for the mainland of Scotland. Fig 42d.
I Loch Flemington, 1975 **E,BM**.

P. pusillus L. *Lesser Pondweed*
    P. panormitanus Biv.
Native. Often confused with *P. berchtoldii* but distinguished by the absence of lacunae along the midrib, the tubular stipules and the slender axillary turions. The old records without a voucher specimen have been omitted. Fig 42c.
M Spynie Canal, Drainie, 1939 *G. Taylor* **BM**; ponds in the Cooper Park at Elgin and in Gilston Lochs, 1946 *G. Taylor*, **E,BM!**.
N Abundant in Cran Loch, 1975 **E,BM**.
I Pools near the River Beauly at Kilmorack, 1892 *ESM* **BM,CGE**.

P. obtusifolius Mert. & Koch *Blunt-leaved Pondweed*
Native. Frequent. Submerged leaves linear narrowed to the sessile base, dark green and translucent. Fig 41c, Map 261.
M Abundant in the Altyre Burn near Chapelton, Forres and in Blairs House pond, 1953 **BM**; Loch Puladdern, 1966 *UKD* **UKD!**; in the River Lossie at Calcots, 1967 **E,BM**; Loch na Bo, 1972; Millbuies, 1972 **E,BM**; Dyke pond, 1976 **E,BM**.
N Lochloy, 1936 *A. S. Sandeman* **BM**; loch at Nairn and in Loch Flemington, 1946 *G. Taylor* **BM**; and 1967 *MMcCW* **BM**.
I Pool adjoining the River Spey below Kingussie, 1898 *ESM* **BM**; *Loch Flemington, 1930; Loch Laide, 1973 *UKD* **BM!**; lochan near Polchar E of Alvie, 1973 *R. McBeath* **E,BM**; lochan in the town of Newtonmore, 1975 *J. D. Buchanan & MMcCW* **E,BM**; Abbar Water at Lochend and in Loch Bunachton, 1975 **E,BM**.

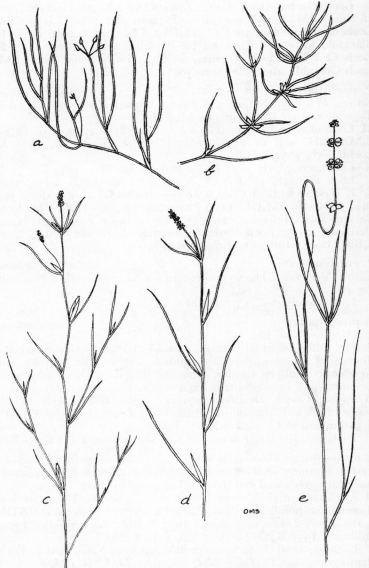

Fig 42   a, **Ruppia maritima**; b, **Zannichellia palustris**; c, **Potamogeton pusillus**; d, **P. rutilus**; e, **P. filiformis**.

P. berchtoldii Fieb. *Small Pondweed*
Native. Frequent. Submerged leaves linear, shortly mucronate, dark green, thin and translucent, the longitudinal veins always 3. Fig 41b, Map 262.

M Near Elgin, 1869 *N. E. Brown* **BM**; pool beside Greshop wood, 1867 *JK* **FRS**; by the west pier at Lochindorb, 1955 **CGE**; Millbuies loch and pond at Gordon Castle, Fochabers, 1972 **E,BM**; Loch na Bo, 1972 **E**; Loch an t-Sithein S of Lochindorb, 1973 **E,BM**.

N Pool by the River Nairn at Nairn, 1898 *ESM* **E,BM**; Lochloy, 1936 *A. S. Sandeman.* **BM**; lochs at Nairn, 1946 *G. Taylor* **BM**; Cran Loch, 1960 **ABD**; Loch Flemington, 1972 **E**.

I Loch Alvie, 1891 *AS* **BM**; Loch Morlich, 1894 *H. Groves* **BM**; Loch Laide near Abriachan, 1964 **E**; Loch Flemington, 1972 **E**; Loch Mallachie, 1973 **E**; Loch Garten, 1973 **E,BM**; Loch Uanagan by Fort Augustus, 1975.

P. crispus L. *Curled Pondweed*
Native. Frequent. Leaves undulate on the margins when mature, shining, translucent, sometimes reddish but usually dark green. Fig 41d.

M Loch Spynie *WAS* (Murray); Forres, 1902 *JK* **FRS**; Altyre Burn at Forres and River Lossie at Elgin, 1939 *G. Taylor* **BM**; Innes Canal, Urquhart and Gilston Lochs, 1946 *G. Taylor* **BM**; *Sanquhar pond, Forres, 1954; Gordonstoun lake, 1972 **E,BM**; distillery pond at Miltonduff and Gilston lochs, 1972 **E**.

N Lochloy, 1939 *G. Taylor* **BM**!.

I Loch-na-Shanisch (Aitken); Loch Ashie, 1904 *G. T. West* **STA**; Loch Garten, 1944 *R. Mackechnie* **BM**, and in 1973 *R. McBeath & MMcCW* **E**; Loch Ruthven, 1975 **E,BM**.

P. filiformis Pers. *Slender-leaved Pondweed*
Native. Rare. Leaves filiform, linear-setaceous; stipular sheath tubular below when young. Fig 42e.

M Brodie Castle, 1780 *Brodie* **E**; Rosevalley, near Burghead, *W. Wilson* **MANCH**; Loch Oir near Lhanbryde, 1946 *G. Taylor* **BM**; ditch W of Burghead, 1974 **E,BM**.

N Small loch near Nairn, 1884 *J.G. herb Groves* **BM**; Lochloy, 1936 *A. S. Sandeman* **BM**; near Nairn, 1946 *G. Taylor* **BM**; Lochloy, 1972 **E**.

I *Loch Flemington, 1968 **E**.

P. filiformis × pectinatus L. = **P. × suecicus** K. Richt.
M Innes Canal, Urquhart, 1946 *G. Taylor* **BM**; in the River Lossie just S of Calcots station, 1967 **BM**; and in 1972 **ABD**; Spynie Loch, 1972 **E,BM**.

P. pectinatus L. *Fennel-leaved Pondweed*
Native. Rare. Leaves very slender; stipular sheaths open and convolute with a whitish margin. Fig 41a.

M Loch Spynie, 1831 *WAS*, and *G. Gordon herb. Borrer* **K**; Loch of

Cotts, *WAS* (Murray); Spynie, 1867 *JK* **FRS**, in 1946 *G. Taylor* **BM** and in 1972 *MMcCW* **E**; Spynie canal, 1939 *G. Taylor* **E,BM**. **N** Nairnshire, *A. Falconer* (Gordon); Lochloy, 1973 **E**.

RUPPIACEAE
Ruppia L.
R. maritima L. *Beaked Tasselweed*
   R. rostellata Koch
Native. Rare in brackish ditches. Fig 42a
M Near the school at Kinloss, 1832 *G. Wilson* (Gordon) and *WAS* (Murray); Kinloss Bay 1834 *J. Fraser* **GL**; near the schoolhouse, Kinloss, 1865 *JK* **FRS**, and 1972 *MMcCW* **E**.
I (Trail, *ASNH* 8: 172 (1899)); salt-marsh at Muirtown Basin, 1958 *E. W. Groves* **EWG**!; brackish pool by the Beauly Firth at Balcarse Farm, Kirkhill, 1975 **E**; salt-marsh near Lentran, 1976.

ZANNICHELLIACEAE
Zannichellia L.
Z. palustris L. *Horned Pondweed*
Native. Very rare in fresh or brackish water. Fig 42b.
N Locally abundant in Cran Loch, 1960 **ABD,K** and in 1975 **E**.

LILIACEAE
Tofieldia Huds.
T. pusilla (Michx.) Pers. *Scottish Asphodel*
Native. Frequent in wet grassy places by burns and in mountain flushes usually over 300 m. Flowers greenish-white. Fig 43a, Map 263.
M Moray (A. Bennett, *Scot. Nat.* 1886).
I Dalwhinnie, 1833 (Watson, *Loc. Cat.*); Glen Feshie (Gordon); Cairn Gorm, 1870 *J. F. Duthie* **E**; Loch Ruthven and Glen Urquhart (Aitken); flush by Loch na Ba Ruaidhe above Milton, 1947; Coire an Lochain of Cairn Gorm, 1953 *A. Melderis* **BM**; in flushes on all the higher mountains, 1954; Mealfuarvonie at 600 m, 1971 **E**; Glen Convinth, 1971 *K. Christie*.

Narthecium Huds.
N. ossifragum (L.) Huds. *Bog Asphodel*
Native. Abundant in moorland bogs and wet mountain slopes. Has the reputation of being poisonous and causing the softening of the bones of animals.
M Very common (Gordon); Altyre, 1856 *JK* **FRS**; Winter lochs Culbin Forest, 1930; throughout the county, 1954.
N (Stables, *Cat.*); Ardclach (Thomson); moorland bogs at Drynachan, Dulsie Bridge and Glenferness etc., 1954.
I Inverness, 1832, Dalwhinnie, 1833 (Watson, *Loc. Cat.*); marshes round Inverness (Aitken); Newtonmore, 1914 *T. McGrouther* **GL**; Craig Dunain, 1940 *AJ* **ABD**; throughout the district, 1954; Glen Feshie, 1960 *E, Rosser* **E**.

Hemerocallis L.

H. fulva (L.) L. *Day Lily*

Garden out-cast in the wood opposite Mudhall, Dyke, 1964.

Convallaria L.

C. majalis L. *Lily-of-the-Valley*

Doubtfully native in north Scotland. Rare in woods and on river banks.

M Innes House, Urquhart, 1794 *A. Cooper* **ELN**; Darnaway woods, herb. *Brodie* and Birdsyards woods at Forres, *J. G. Innes* (Gordon); Sanquhar House, Forres, 1856 *JK* **FRS**; rare in the beech woods at Darnaway Castle (now planted with conifers), 1950.

N Abundant in one or two places, Cawdor woods *WAS* (Gordon); in a small wood (now felled) near Cothill Farm, 1920 *M. Baillie*; banks of the River Findhorn upstream from Coulmony House, 1961 **ABD**; banks of the River Nairn at Holme Rose, 1974 **E**.

I Clearly indigenous in the birch banks of Badenoch, 1827 (Gordon); perfectly wild on the birch banks of Badenoch, 1831 ? *coll.* **STA**; in the drive at Ballindoun near Kirkhill, 1950 *M. Cameron*.

Polygonatum Mill.

P. multiflorum (L.) All. *Solomon's Seal*

Introduced in north Scotland.

M Innes House, Urquhart, 1794 *A. Cooper* **ELN**; certainly introduced at Gordon Castle by the lake (Gordon); garden outcast at Cooperhill, 1964 **CGE**; in the woods at Wester Elchies and on the roadside opposite Dalvey Cottage near Dyke, 1972 **E**.

N Banks of the River Nairn at Holme Rose, 1955.

I Shingle of the River Enrick at Drumnadrochit, 1944 *M. S. Campbell*; Balblair Ness, 1950 *M. Cameron*; railway embankment near Culloden, 1954, by a small burn in Fort Augustus and woods at Cluny Castle near Laggan.

Ruscus L.

R. aculeatus L. *Butcher's Broom*

Introduced in north Scotland.

M Certainly introduced at Gordon Castle (Gordon); Altyre shrubberies, 1881 *JK* **FRS**!.

Lilium L.

L. martagon L. *Martagon Lily*

Introduced in north Scotland.

M Naturalized in the woods at Innes House, Urquhart, 1967.

L. pyrenaicum Gouan *Pyrenean Lily*

Introduced in north Scotland.

M Bank by the road near Dunphail station, 1950.

I Field at Glenmore Lodge, 1953.

A species of *Lilium* is established on the dunes of Lossiemouth golf course. It has not flowered since it was first seen in 1972.

Tulipa L.

T. sylvestris L.   *Wild Tulip*
Introduced in north Scotland.
M Formerly naturalized in the woods at Gordonstoun, 1966
E,CGE. The area is now covered with buildings.

Erythronium L.

E. dens-canis L.   *Dog's-tooth Violet*
N Garden escape, far from habitation, on the raised beach between
Druim and Lochloy in the Culbin Forest, 1966.

Gagea Salisb.

G. lutea (L.) Ker-Gawl.   *Yellow Star-of-Bethlehem*
Introduced in north Scotland. Very rare.
M Moray (A. Bennett, *Scot. Nat.*, 1886); planted in Greshop Wood,
1892 *JK* **FRS**; in several places in Greshop Wood by the path of the
river, 1964 **E,CGE**; a few plants lower down the river bank. This
area was made into a caravan site in 1974 and the plants are now
very rare.
N Rare at Blackhills near Auldearn, 1835 *WAS & J. B. Brichan*
E,ABD,GL.
I (Druce, *Com. Fl.*) Probably Stables' record for Auldearn.

Ornithogalum L.

O. umbellatum L.   *Star-of-Bethlehem*
Introduced.
M Garden escape by a track in the wood at Berryley Farm,
Darnaway, 1950; waste ground at Dunphail station and at Milton
Brodie, 1954; rough field at Hamlets near Loch Spynie, 1966 **E**;
shingle of the River Spey at Fochabers, bank at Urquhart and grass
verge to the road at Garmouth.
N Lethen, 1837 *WAS* **ABD**; grass bank by the road near
Broombank Farm, Auldearn, 1962 **ABD**; roadside verges at
Geddes, Cantraydoune and Holme Rose.
I Waste ground between Drumnadrochit and Milton, 1952;
railway embankment at Bunchrew, roadside verge $\frac{3}{4}$ mile east of
Fort Augustus, in the lane above the Kirk at Kilmorack and on a
grass verge at Cabrich.

O. nutans L.   *Drooping Star-of-Bethlehem*
Garden outcast by the burn at Brackla Distillery, 1966.

Scilla L.

S. verna Huds.   *Spring Squill*
Native. Rare in short turf on sea cliffs.
M East of Burghead, 1832 *WAS* **ABD** and *G. Gordon* **GL**!; rare at
Covesea, 1834 *Miss Johnston* and at Cummingston, *A. Falconer &*
G. Gordon (Gordon); Covesea, 1865 *JK* **FRS** and 1879 ? *coll.* **ELN**.
It can still be found in small quantity on the cliff tops between
Burghead, Hopeman and Covesea.

Endymion Dumort.

E. non-scriptus (L.) Garcke   *Wild Hyacinth, Bluebell*
   Scilla non-scripta (L.) Hoffm. & Link
Native. Common in deciduous woods, on riverbanks and under
bracken on open moorlands. Map 264.
M Innes House, Urquhart, 1794 *A. Cooper* **ELN**; Elginshire, 1837
G. Gordon **E**; Birdsyards, Forres *J. G. Innes* **ABD**; Sanquhar, 1898
JK **FRS**; by the Muckle Burn at Dyke, 1930 and throughout the
county, 1954; near Dalvey Smithy (with white flowers) 1970.
N (Stables, *Cat.*); Cawdor woods, 1888 *R. Thomson* **NRN**; woods at
Lethen, Holme Rose and Geddes etc., 1954, wood at Auldearn,
1962 **ABD**; on the raised beach by Druim House E of Nairn.
I Ness Islands, 1854 *HC* **INV**; common in woods round Inverness
(Aitken); common in the woods by Loch Ness and in Strathglass etc.,
1954; among bracken in rocky places on the moors in Strathdearn,
Mealfuarvonie and Tomich etc.

E. hispanicus (Mill.) Chouard
Garden escape.
M By the Burnie Path at Dyke, 1967 **CGE**.

Paris L.

P. quadrifolia L.   *Herb Paris*
Native. Very rare. Shady places among rocks on base-rich soil.
I Sides of Loch Ness, *A. Murray* (Hooker, *Fl. Scot.*); Ness Islands
at Inverness, 1832 *W. Brand* **GL** and 1834 **E**; Ness Islands, *G.
Anderson* (Gordon); on lower slope of Mam Suim near the entrance
to the Ryvoan Pass, Glenmore at 450 m, 1935 *K. W. Braid* **GL** and
in 1937 *J. Walton* **GL**, it is still to be found under juniper in the
Ryvoan Pass.

JUNCACEAE
Juncus L.

J. squarrosus L.   *Sprotts, Heath Rush*
Native. Abundant on moorlands in wet and dry places.
M Very common (Gordon); Moray, *G. Wilson* **ABD**; Culbin
Forest, 1953 *A. Melderis* **BM**; throughout the county, 1954.
N (Stables, *Cat.*); at Ardclach, known locally as *Moss Rush-goose
Corn* (Thomson); moors at Cawdor and throughout the county,
1954; near Dulsie Bridge, 1961 **ABD**; dunes on the Nairn Dunbar
golf course, 1961 **ABD**.
I Inverness in 1832 and Dalwhinnie and Pitmain in 1833 (Watson,
Local Cat.); Badenoch, 1838 *A. Rutherford* **E**; Culloden Moor,
W. Brand **ABD**, in 1855 *HC* **INV**, and (Aitken); Balmacaan forest,
1904 *G. T. West* **STA**; Feshie Bridge, 1904 *G. B. Neilson* **GL**;
Glenmore and Coire an Lochain of Cairn Gorm, 1953 *A. Melderis*
BM; throughout the district 1954.

J. tenuis Willd.   *Slender Rush*
Introduced. Rare on roadside verges and railway yards. Map 265.
M Path to the hill by Grantown railway station, 1963 **E,CGE**.

I   Ryvoan Pass, Cairngorms, 1904 *J. Walton* **GL**; station yard at
Tomatin, 1952; railway yard at Culloden, 1954; roadside verge
¾ mile N of Cannich, 1971 *A. Currie et al.* **E,CGE**; gateway to the hill
track opposite Erchless Castle and in a lay-by at Breakachy bridge
in Strathglass.

J. gerardii Lois.   *Saltmarsh Rush*
Native. Abundant on salt-marshes the whole length of the coast.
Very variable in size, a very dwarf form of 5 cm or less grows on the
edge of the Culbin Forest. Fig 43e.
M   Lossiemouth and Netherton of Grange (as *J. compressus*)
(Gordon); Netherton of Grange, 1856 *JK* **FRS**; near Forres, 1888
G. C. Druce **OXF**; Findhorn Bay, Culbin and Garmouth golf course,
1953 *A. Melderis* **BM**; salt-marsh at Garmouth, 1965 **E**.
N   (Stables *Cat.*); Nairn coast (as *J. compressus*) *J. B. Brichan*
(Gordon, *annot.*); salt-marsh at Kingsteps, 1961 **ABD,K**.
I   Salt-marsh at Beauly, 1947; abundant by the Beauly Firth from
Wester Lovat to Alturlie Point, 1954; in short turf at Ardersier and
on the Carse of Delnies.

J. trifidus L.   *Three-leaved Rush*
Native. Common on detritus, on rock ledges and on bare mountain
tops. Fig 43f.
I   Dalwhinnie, 1933 (Watson, *Loc. Cat.*) and herb. *Borrer* **K**;
abundant in one or two places in Glen Feshie and Mam Sodhail
(Gordon); Fraoch-choire (Ball, 1851); Dunain (Aitken); Cairn
Gorm and Braeriach, 1893 *AS* **E**; outflow of the River Tarff at
Loch Ness, 1904 *G. T. West*; Cairn Gorm, 1953 *A. Melderis* **BM**; on
all the mountain tops in the district over 900 m.
(*J. hostii* Tausch or *J. trifidus* subsp. *monanthes* (usually considered
only a native of Alps and Appenines)—J. G. Baker, (*J. Bot.* **9**: 112
(1871)) writes—'In Gay's herbarium there is a specimen of a rush
sent by Dr Greville from Braeriach which is referred by Gay to
J. hostii of Tausch and which is evidently substantially identical
with the plants given under that name in Reichb. *Exsicc.* no. 1614.
F. Schulz Herb. *Norm.* no. 52 and F. Schultz *Flora Gall.* et Germ.
Exsicc. no. 1333. Along with it Gay writes "Ab simillino J. trifida
differt culmis 1-2 foliatus non aphyllis". A full description will be
found in Koch's *Synopsis* where it is placed as a distinct species; but
although I have not yet looked specially to see, I believe that it will
be found that Scotch specimens have stem leaves developed not
infrequently'.)

J. bufonius L.   *Toad Rush*
Native. Abundant in gateways, in ditches and on moorland tracks
and loch margins. Fig 43d.
M   Frequent (Gordon); Forres, 1867 *JK* **FRS**; tracks in the Culbin
Forest and throughout the county 1954.
N   (Stables, *Cat.*); Ardclach (Thomson); throughout the county,
1954; track by the Loch of the Clans, 1960 **ABD,K**.

I Inverness, 1832 Dalwhinnie and Pitmain, 1833 (Watson, *Loc. Cat.*); Dalwhinnie, herb. *Borrer* **K**; Coiltry Loch near Fort Augustus, 1904 *G. T. West* **STA**; Loch Duntelchaig, 1940 *AJ* **ABD**; throughout the district, 1954; track by Loch na Doirb above Inverfarigaig, 1975 **E**.

Subsp. **foliosus** (Desf.) Maire & Weiller
More robust than subsp. *bufonius*, leaves wider and the perianth segments with a dark stripe on each tepal. The plant is yellowish-green in colour.
I Rare in mud at the outlet of the small burn on the now dried-up Loch na Doirb above Inverfarigaig, 1975 **E**.

J. inflexus L.   *Hard Rush*
Doubtfully native in north Scotland.
M The Leen at Garmouth, *G. Birnie* (Burgess). Not seen in recent years.

J. effusus L.   *Rashes, Soft Rush*
Native. Abundant in wet fields, moorland bogs and ditches. Fig 43b.
M Very common (Gordon); Forres, 1867 *JK* **FRS**; Dava Moor, 1940 *AJ* **ABD**; on shingle on the W bank of the River Findhorn near Forres, 1953 *A. Melderis* **BM**; Buckie Loch in the Culbin Forest and throughout the county, 1954.
N (Stables, *Cat.*); Ardclach (Thomson); Lochloy (as var. *conglomeratus*), 1904 *G. T. West* **STA**; meadows by the River Findhorn at Drynachan and throughout the county, 1954.
I Inverness, 1832, Dalwhinnie and Pitmain, 1833 (Watson, *Loc. Cat.*); Culloden, 1855 *HC* **INV**; by Loch Ness at Fort Augustus and Inchnacardoch, 1904 *G. T. West* **STA**; throughout the district, 1954; bog by a pool N of Kingussie, 1973 **E**.

J. conglomeratus L.   *Compact Rush*
In *Watsonia* 7: 169 (1969) the name for this rush is given as *J. subuliflorus* Drejer. In a personal communication (1974) J. E. Dandy informed me that this was an error. Fig 43c.
Native. Abundant on acid soil. Often confused with *J. effusus* var. *compactus* but readily distinguished by the prominent ridges on the stem and the sheathing base of the bract being widely expanded.
M Very common (Gordon); moors at Boat of Garten, in the Culbin Forest and throughout the county, 1954.
N (Stables, *Cat.*); in wet places, common at Ardclach 1889 *R. Thomson* **NRN**; Lochloy, 1904 *G. T. West* **STA**; moors at Drynachan and throughout the county, 1954.
I Dalwhinnie and Pitmain, 1833 (Watson, *Loc. Cat.*); banks of the River Nairn (Aitken) Culloden, 1855 *HC* **INV**; moors in upper Glen Feshie, at Gaick and throughout the district, 1954.

J. filiformis L.   *Thread Rush*
Native. Very rare on loch margins. Inflorescence low down on the stem. Fig 43g.

15

438

Fig 43 a, **Tofieldia pusilla**; b, **Juncus effusus**; c, **J. conglomeratus**; d, **J. bufonius**; e, **J. gerardii**; f, **J. trifidus**; g, **J. filiformis**; h, **J. balticus**.

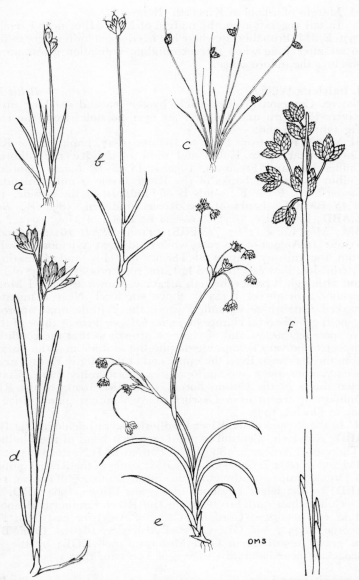

Fig 44 a, **Juncus biglumis**; b, **J. triglumis**; c, **Scirpus setacea**; d, **Juncus castaneus**; e, **Luzula arcuata**; f, **Scirpus tabernaemontani**.

M Margin of a pool at Kincorth (Burgess).
I In tall vegetation on the margin of Loch Mhòr near Farraline, 1976 **E,BM**. Probably elsewhere but overlooked for it is very difficult to see after June when the surrounding vegetation grows up and obscures the inflorescence.

J. balticus Willd.   *Baltic Rush*
Native. Common in dune slacks by the sea and reaching up the Rivers Findhorn and Dulnain for over 20 miles from the coast. Fig 43h, Map 266.
M Moray Firth from Spey to Delnies, 1827, banks of the River Lossie as far up as Kellas and those of the River Findhorn to Cuilleachan and Freeburn, 1832 *WAS & G. Gordon* (Gordon); Culbin Sands and banks of the River Lossie 7 miles inland, *G. Gordon* herb. *Borrer* **K**; sands E of Findhorn at the seaside, 1832 *WAS* **BM,GL**; banks of the River Findhorn, 1832 *W. Brand* **E,ABD**; Elginshire, 1837 *G. Gordon* **E**; Laich of Moy, 1932 ? *coll.* **BM**; Moy Carse, 1867 *JK* **FRS**; (Trail, *ASNH* **16**: 251 (1907)) writes: 'In August of this year, while residing at Aviemore, I walked from the railway station at Daviot by Moy and Tomatin to Carrbridge. Between the two last, the road crosses a range of hills, and although it passes through a ravine, known as Slochd Mòr, it reaches a height of 1,327 ft above sea level. Near a milestone marked Carrbridge 6½ miles where the altitude must approach 1,300 ft grow several clumps of *Juncus balticus*. Two or three of these are some feet across and of vigorous growth so that the conditions appear to be vary favourable despite the altitude and the distance from the sea-coast. Even the upper end of the Moray Firth is nearly twenty miles away and the open sea considerably more distant'; fixed dunes in the Culbin Forest, 1950 **BM**; banks of the River Dulnain upstream from Carrbridge, very robust plants, 1972 **E**; bog at Slochd, 1972.
N In the streams of the River Findhorn and at Lochloy, 1832 *WAS* **ABD**; Ardclach, plentiful at the Levrattich bend of the Findhorn (Thomson); Ardclach, 1898 *P. M. Playfair* **BM**; Nairn, 1874 *A. Ley* **BM** and in 1884 *J. Groves* **GL,STI,BM**; banks of the River Findhorn at Drynachan, 1954; dunes by the Nairn Dunbar golf course, 1960 **ABD**; by the River Findhorn at Ferness Bridge, 1964 **ABD**.
I Cuilleachan and Freeburn on the River Findhorn at 290 m, *WAS & G. Gordon* (Gordon); Carse of Ardersier east from Fort George, 1833 ? *coll.* **GL**; Carse of Ardersier, 1846 *A. Croall* **BM**; on grassy moorland in Glen Mazeran, 1968 **CGE**; by the Allt Bruachaig near Balvraid Lodge E of Tomatin, 1975.

J. maritimus Lam.   *Sea Rush*
Native. Rare on salt-marsh margins and runnels.
M Kinloss Bay, 1837 *G. Gordon* **E!**; near Findhorn, 1832 *WAS* **GL**; Findhorn Bay, 1898 *ESM* **CGE**; a few plants in a hollow by the salt-marsh W of Buckie Loch in the Culbin Forest, 1954.

N Rare at Lochloy, *J. B. Brichan* (Gordon); E of Nairn (Marshall, 1899)!.

J. subnodulosus Schrank   *Blunt-flowered Rush*
  J. obtusiflorus Ehrh. ex. Hoffm.
Native. Rare or wrongly identified. There is no voucher specimen for either record.
M Culbin Sands, 1923 *D. Patton & E. J. A. Stewart*.
N At one spot on the River Findhorn at the Levrattich bend (Thomson).

J. acutiflorus Ehrh. ex Hoffm.   *Sharp-flowered Rush*
  J. sylvaticus auct.
Native. Common in wet meadows and grassy moors forming large dense patches. Map 267.
M Frequent (Gordon); Craigellachie Reserve at Aviemore, at Rothes, near Dunphail and at Lochinvar near Elgin, 1954 etc. but more common in the highland areas than the lowland.
N Loch Belivat, 1888 *R. Thomson* **NRN**; grass haughland by the River Findhorn at Drynachan, 1954.
I Dalwhinnie and Pitmain, 1833 (Watson, *Loc. Cat.*); near Aldourie Pier and on the shores of Loch Ness, 1904 *G. T. West* **STA**; abundant in the highlands but only occasional in the lowlands, 1954; by Loch Pityoulish, 1958 *B. Roberts* **E**.

J. articulatus L.   *Jointed Rush*
Native. Abundant in ditches, by lochs and burns on acid soil.
M Frequent (Gordon); Waterford, 1867 *JK* **FRS**; Aviemore, 1893 *AS* **E**; Grantown, 1903 *T. A. Sprague* **K**; Garmouth golf course, 1953 *A. Melderis* **BM**; Buckie Loch, Culbin Forest, 1954 and throughout the county.
N Loch Belivat, 1888 *R. Thomson* **NRN**; moorland ditch at Drynachan and throughout the county, 1954; shingle of the river at Nairn, 1954 **K**.
I Culloden, 1855 *HC* **INV**; Inverness (Aitken); by lochs Uanagan, Meiklie, nan Eun and by Aldourie Pier, 1904 *G. T. West* **STA**; Loch Flemington, 1940 *AJ* **ABD**; throughout the district, 1954; flush W of Affric Lodge, 1971 **E**.

J. alpinoarticulatus Chaix   *Alpine Rush*
  J. alpinus Vill.
Native. Very rare in mountain flushes.
  I Glen Einich, 1887 *G. C. Druce* **OXF**.

J. bulbosus L.   *Bulbous Rush*
Native. Stamens 3. Rare and confused with *J. kochii* F. W. Schultz. The following records have not been critically examined.
M Near Forres, 1867 *JK* **ABD**; Brodie, 1869 *JK* **FRS**; Winter lochs, Culbin, 1953 *A. Melderis* **BM**; Avielochan, 1955 *J. Milne* **E**.
N Ardclach, 1888 *R. Thomson* **NRN** (conf. S. M. Walters).
I Dalwhinnie and Pitmain, 1833 (Watson, *Loc. Cat.*); Culloden, 1855 *HC* **INV**, and (Aitken); Loch an Eilein, 1955 *J. Milne* **E**.

J. kochii F. W. Schultz
Native. Abundant in cart tracks, moorlands and stoney margins of lochs and lochans on acid soil. Stamens 6, capsule brown, sharply trigonous. Very variable. Var. *fluitans* Fr. occurs in deep water often with many spreading filiform red stem branches, non flowering, and is often mistaken for a pondweed.
M Near Brodie House, from *J. Brodie of Brodie*, *G. Don* **BM,K**; winter lochs, Culbin 1954 and throughout the county; moorland flush on the hill of Mulundy, 1968 **BM**.
N By a pool on the moor at Drynachan, by the Minister's Loch at Nairn, at Dulsie Bridge and Glenferness etc., 1954; bog below the raised beach at Maviston, 1968 **E,CGE**.
I Loch Alvie, 1891 *AS* **E,GL**; Feshie Bridge, 1904 *G. B. Neilson* **GL**; Loch nam Faoileag and Coiltry Loch, 1904 *G. T. West* **STA**; abundant throughout the district, 1954.

J. castaneus Sm.　　　　　　　　　　　　　　　　　　　*Chestnut Rush*
Native. Rare in mountain flushes. Fig 44d.
I (Trail, *ASNH*, **15**: 231 (1906)); grassy flush below north-facing cliffs of Sgùrr na Lapaich in Glen Strathfarrar, 1971 **E**; E coire of Sgùrr nan Conbhairean, Glen Affric, 1973 *R. W. M. Corner et al.*

J. biglumis L.　　　　　　　　　　　　　　　　　*Two-flowered Rush*
Native. Very rare on wet rock ledges on high mountains. The two flowers are exceeded by the lowest bract. Fig 44a.
I Cairngorms, *G. & D. Don* (Hooker, *Fl. Scot.*); five plants on a wet cliff ledge, Sgùrr na Lapaich in Glen Strathfarrar, 1971.

J. triglumis L.　　　　　　　　　　　　　　　*Three-flowered Rush*
Native. Frequent in flushes and bogs in mountain areas on acid soil. The three flowers are at one level, exceeding the lowest bract. Fig 44b, Map 268.
I Rare in Glen Feshie, 1831 (Gordon); Dalwhinnie, 1833 (Watson, *Loc. Cat.*); NE of Dalwhinnie, herb. *Borrer* **K**; rare in Rothiemurchus, 1842 *J. Fraser* **GL**; in a fir wood 4 miles from Inverness on the Edinburgh road (Aitken); Braeriach and Glen Einich, 1892 *AS* **E,GL**; in flushes in all the higher mountains, 1954; flush by the Elrick Burn at Coignafearn in Strathdearn, 1970 **E**.

Luzula DC.

L. pilosa (L.) Willd.　　　　　　　　　　　　　　*Hairy Woodrush*
Native. Common in woods, on moorland banks and in burn gullies. Map 269.
M Frequent (Gordon); Burgie, 1871 *JK* **FRS**; Aviemore, 1953 *P. H. Davis* **E**; throughout the county, 1954; bank by the River Findhorn in shade, Cothall, 1963 **CGE**; Craigellachie Reserve Aviemore, 1971 *A. Angus* **STA**.
N Cawdor woods, 1832 *WAS* **ABD**; Ardclach (Thomson); moorland bank at Drynachan and throughout the county, 1954; moor by Loch of the Clans, 1960 **ABD**.

I Dalwhinnie, 1833 (Watson, *Loc. Cat.*); Dunain and Leys near Inverness (Aitken); Feshie Bridge, 1904 *G. B. Neilson* **GL**; Loch an Eilein, 1953 *P. H. Davis* **E**; throughout the district, 1954; wood by Ness Castle, Inverness, 1971 **E**.

L. sylvatica (Huds.) Gaudin   *Great Woodrush*
Native. Abundant in woods, dominating large areas to the detriment of young trees, also among rocks by burns and on mountain cliff ledges.
M Frequent (Gordon); Greshop Wood, 1871 *JK* **FRS**; occasional in the Culbin Forest but common elsewhere in the county; Greshop Wood, 1953 *A. Melderis* **BM**.
N Cawdor woods, 1832 *WAS* **ABD**; Ardclach (Thomson); rare on cliffs on the moors at Drynachan but common in the valleys of the Rivers Findhorn and Nairn, 1954; wood at Dulsie Bridge, 1961 **ABD**.
I Dalwhinnie, 1833 (Watson, *Loc. Cat.*); Ness Islands and the banks of the river, Inverness (Aitken); rock ledge on Ben Alder, gully of a burn in the Ryvoan Pass and woods at Abriachan, Glen Shirra, Glen Strathfarrar etc., 1954.

L. luzuloides (Lam.) Dandy & Wilmott   *White Woodrush*
Introduced. In woods of the larger houses, usually associated with *Poa chaixii* Vill. Flowers pinkish-brown.
M Policies of Newton House near Elgin, 1930; Rothes, 1954; in the woods by the lodge at the Golden Gates of Darnaway Castle, 1954 **E**; policies of Cathay House near Forres, 1970 *K. Christie*; shrubbery at Brodie Castle, 1974 **E**.
N Under trees by the walled garden at Holme Rose, 1955 **E,ABD**; wood at Geddes House, 1961 **E,ABD,K**; also at Lethen House and Drynachan Lodge.
I Kincraig, 1892 *AS* **E**; woods at Achnagairn and Reelig House at Kirkhill, 1954; also at Flichity House, Aldourie Castle, Moy Hall and Inverbrough Lodge near Tomatin; on a dry bank of cottages S of Tomatin, 1975.

L. spicata (L.) DC.   *Spiked Woodrush*
Native. Common on scree and open moorland on the higher hills, occasionally washed down and becoming established on river shingle. Flowers in a dense spike-like inflorescence. Pl 19, Map 270.
M Speymouth, *G. Birnie* (Burgess); shingle of the River Spey at Kingston, 1948.
I Dalwhinnie, 1833 (Watson, *Loc. Cat.*); abundant in one or two places in Badenoch and Mam Sodhail (Gordon); Fraoch-choire (Ball, 1851); Braeriach, 1847 ? *coll.* **E**; Sgòran Dhubh at 1,200 m 1891 *AS* **E**; and in 1904 *G. B. Neilson* **GL**; Coire Chùirn at Drumochter, 1911 *ESM* **E,BM**; on all the higher hills, 1954; on scree in the Coire an Lochan of Cairn Gorm, 1953 **CGE**.

L. arcuata Sw.   *Curved Woodrush*
Native. Locally frequent on scree and stoney summits of mountains

usually over 610 m. Inflorescence in 2- to 5-flowered clusters, the
outer on recurved stalks. Fig 44e.

I Cairn Gorm, 1822 *W. J. Hooker* **OXF**; and herb. *Borrer* **K**;
Cairngorm mountains, 1836 *G. A. Walker Arnott* **STA**; rare on Mam
Sodhail, 1836 *WAS* and Cairn Gorm and Braeriach (Gordon);
west coire of Cairn Gorm, 1893 *AS* **BM**; east-facing coire between
Mam Sodhail and Carn Eige and on the summit, 1947 *E. F.
Warburg & S. M. Walters*; abundant in Coire na Ciste of Cairn
Gorm at 800 m, 1975.

L. campestris (L.) DC.   *Field Woodrush*
Native. Abundant in short turf and grassy banks.
M Very common (Gordon); grassy dunes in the Culbin Forest, in
Dyke kirkyard and throughout the county, 1954.
N (Stables *Cat.*); Ardclach (Thomson); grass bank by the River
Findhorn at Drynachan and throughout the county, 1954.
I Inverness, 1832, Dalwhinnie and Pitmain, 1833 (Watson, *Loc.
Cat.*); common round Inverness (Aitken); lochan Creag Dhubh and
on the island in Loch Tarff, 1904 *G. T. West* **STA**; throughout the
district but absent from the higher hills, 1954.

L. multiflora (Retz.) Lejeune   *Heath Woodrush*
Native. Abundant on dry open places in woods and moors.
M Edinkillie and Manachy, 1871 *JK* **FRS**; Brodie and Garmouth
(Marshall, 1899); shingle of the River Spey at Garmouth, 1950 **E**;
pine woods in the Culbin Forest and throughout the county,
1954.
N Ardclach (Thomson); moorland track at Drynachan and
throughout the county, 1954; moors at Achavraat and at Dulsie
Bridge, 1961 **ABD**.
I Bunchrew Burn, 1855 *HC* **INV**; near Leys Castle, 1940 *AJ* **ABD**;
Loch an Eilein and Glen Einich, 1953 *P. H. Davis* **E**; Glenmore,
1953 *A. Melderis* **BM**; track verges at Ben Alder Lodge, in Glen
Feshie and throughout the district, 1954.

Var. **congesta** (Thuill.) Koch
A common variant with flowers in a rounded head. Distribution not
fully investigated.
M Common at Forres, 1871 ? *coll.* **ELN**; Moray (Druce, 1888); by
the Buckie Loch Culbin Forest, 1953 *A. Melderis* **BM**; river shingle
at Garmouth, 1973 **E**; dunes on the Culbin Forest, 1973 **E**; by the
River Dulnain and at Balnabreich near Mulben.
N Ardclach (Thomson); track in the birch wood by Loch Belivat
and by the River Findhorn at Ardclach, 1973.
I Bunchrew (Aitken); Coire an Lochain of Cairn Gorm, 1953
A. Melderis **BM**; banks of the River Spey at Boat of Garten, in the
Coire nan Laogh in Glen Banchor, in Glen Mazeran, in the
Ryvoan Pass, at Skye of Curr, Glendoe, Ballindarroch, Moy Hall,
moors at Ruthven near Tomatin and at Fasnakyle in Strathglass
etc.

AMARYLLIDACEAE
Allium L.

A. oleraceum L.   *Field Garlic*
Doubtfully native in north Scotland.
M Parish of Forres, also in gardens (Druce, 1889); side of the road
to Forres, 1944 ? *coll.* **STA**

A. scorodoprasum L.   *Sand Leek*
Introduced in north Scotland.
M Darnaway woods, 1903 *JK* **FRS**.

A. carinatum L.   *Keeled Garlic*
Introduced. Locally plentiful between Moy House and Broom of
Moy and occasionally on roadside and river banks. Flowers pinkish-
purple the stamens conspicuously exserted; reproduced by many
small bulbils on the inflorescence.
M Near Moy, 1913 *D. Patton* **GL**; plentiful by the roadside near
Broom of Moy and on the banks of the Muckle Burn at Moy, 1930
and 1960 **E,ABD,K**; by the Muckle Burn at Dyke, 1963 *Eric
Kleintges* **CGE**; roadside verge between Muirtown and Mudhall
Farm, Dyke, 1973 *W. G. Munro*!.
N Banks of the River Nairn upstream from the town, 1948.
I Bank on top of a wall at Lentran, 1930; Culcabock, 1970
M. Barron; bank between Moniack and Knockbain, 1975 *M.
Cuthbert.*

A. schoenoprasum L.   *Chives*
Garden escape.
M Left bank of the River Lossie below Dun Cow's Leap, *c.* 2 miles
downstream from Kellas House, 1844 (Gordon, *annot.*); on shingle
of the River Spey at Kingston, 1953; single clump in grass field near
Hamlets by Loch Spynie, 1966 **E**.
I (Trail, *ASNH* **15**: 231 (1906)).

A. paradoxum (Bieb.) G. Don   *Few-flowered Leek*
Introduced. Locally abundant in woods and by burns. Flower few,
white, with many bulbils.
M Garden weed in Tolbooth Street Forres, 1953; by the Muckle
Burn at Dyke and near Moy House, 1964 **E**.
N Abundant in an ash wood behind Kinsteary House and on the
roadside by Blackhills Farm near Auldearn; roadside verge near
the school at Earlseat and in the woods at Lethen House, 1950.
I Abundant on a bank in Godman's Brae, Inverness, 1963
M. Cameron!; Culcabock, 1970 *M. Barron.*

A. ursinum L.   *Ramsons*
Native. Occasional in woods and burn ravines. Map 271.
M In the wood by the burn at Burgie Lodge, 1920 and in 1967
E,CGE; side of a flush by the River Spey S of Fochabers and at
Rothes, 1954; near Sanquhar pond, Forres, 1969 *R. M. Suddaby*;
in a beech and oak plantation at Elgin, 1969 *R. M. Suddaby.*

N Nairn and Auldearn, 1900 (Lobban).
I Inverness, 1832 (Watson, *Loc. Cat.*); rare on the Islands, Inverness (Gordon); Ness Islands, 1854 *HC* **INV**; Abriachan burn (Aitken); Falls of Divach, Glen Urquhart, 1901 *JK* **FRS**; Moniack Burn, Reelig, ? date ? *coll.* **INV**; wood by Loch Ness at Fort Augustus, at Inverfarigaig, at Dores and Brackla by Loch Ness, 1954; wood near the waterfall at Glendoe, ravine of the burn at Abriachan hill, wood at Milton; by the Breakachy Burn in Strathglass; ravine near the boat House at Knockie Lodge by Loch Ness and in a wood by the outlet of the River Coiltie at Lewiston.

Leucojum L.
L. aestivum L.   *Summer Snowflake*
Garden escape or planted.
I A single plant in a crevice of a huge rock in Levishie Forest in Glen Moriston, 1930 *M. Cameron*.

Galanthus L.
G. nivalis L.   *Snowdrop*
Doubtfully native. Planted in the policies of many of the larger houses and becoming naturalized and escaping to river banks and roadside verges.
M Introduced, rare at Gordon Castle, Fochabers (Gordon); by the Dyke pond and on the banks of the Muckle Burn, 1971 **E,CGE** etc.
N Lethen, 1833 *WAS* **ABD** and (Gordon); woods at Cawdor Castle, 1930; on the raised beach below Druim House E of Nairn, 1966.
I Policies of Achnagairn House and in Reelig Glen at Kirkhill, 1952; by the Allt Mòr at Kingussie, 1975 etc.

Narcissus L.
N. pseudo-narcissus L.   *Daffodil*
Garden escapes, not the true Wild Daffodil.
M Certainly introduced at Gordon Castle and Castle Grant (Gordon); garden outcasts at Dyke, Bellie and Cromdale (Burgess); several cultivated variants in the small wood near Brodie station, 1954, and on several roadside verges in the county.
N Almost naturalized at Ardclach (Thomson); introduced at Cawdor and Nairn (Lobban).

N. majalis Curt.
   N. poeticus auct.
Garden escape by the Muckle Burn at Dyke in Moray, in a bog near Kingussie in Easterness and in several other places.

IRIDACEAE
Sisyrinchium L.
S. bermudiana L.   *Blue-eyed-grass*
   S. angustifolium Mill.
M Garden escape at The Muiry, Forres, 1968 *Peter Gordon*.

Iris L.
I. foetidissima L.   *Stinking Iris*
Garden escape. Flowers dingy purple, seeds bright orange.
M Near Brodie pond (Burgess); roadside opposite Dalvey Cottage,
Dyke, 1948.

I. pseudacorus L.   *Yellow Flag, Yellow Iris*
Native. Common by lochs and in ditches and bogs in the lowlands,
less so in the highlands. Map 272.
M Common (Gordon); Buckie Loch, Culbin, 1899 *JK* **FRS**;
Spynie and Gilston lochs, bogs at Boat of Garten, Kellas and
Lochinvar etc., 1954.
N (Stables, *Cat*); rare and fast disappearing from Ardclach
(Thomson); bog by the River Findhorn at Drynachan, at Loch
Flemington, Geddes Reservoir, Cran Loch and Littlemill etc., 1954.
I (Watson, *Top. Bot.*); ditches round Inverness (Aitken); Cherry
Island Loch Ness, 1904 *G. T. West*; near Moniack bridge, 1940 *AJ*
ABD; by the River Moriston at Dundreggan, abundant in
Strathglass, meadow at Glendoe, bogs surrounding Loch Meiklie, by
Loch Coulan etc., 1954.

I. cf chrysographes Dykes
Introduced. A single clump on the edge of Loch Moy near Moy
Hall, 1975 **E**. Known there for many years.

Crocosmia Planch.
C.× crocosmifolia (Lemoine) N. E. Brown   *Montbretia*
Garden escape. Occasional on roadside banks and by ditches near
habitations.
M Ditch near Fochabers, 1954; by the small burn at Conicavel,
1962; roadside verge at Kingston, 1975.
I Banks of the River Ness at Inverness, 1955; ditch on the Braes of
Kilmorack, at Gorthleck, bank opposite cottages at Leachkin and
on the banks of the River Beauly by the ferry.

ORCHIDACEAE
Cephalanthera Rich.
C. longifolia (L.) Fritsch   *Long-leaved Helleborine*
 C. ensifolia (Schmidt) Rich.
Native. Rare in woods. Pl 19.
I Drumnadrochit, 1901 *JK* **FRS**; near Castle Urquhart, 1954
M. S. Campbell!; among bracken and heather in an open birch
wood near Lochend, 1961 *D. A. Ratcliffe*.

Epipactis Sw.
E. helleborine (L.) Crantz   *Broad-leaved Helleborine*
 E. latifolia (L.) All.
M Greshop Wood near Forres, 1856 *JK* **FRS**; Urquhart, *A.*
MacGregor (Burgess).

Listera R. Br.

L. ovata (L.) R. Br.   *Common Twayblade*
Native. Rare in woods on base-rich soil and occasionally on open
moorland. Fig 45a.
M Opposite Cothall, 1839 *J. G. Innes* (Gordon, *annot.*)!; Grantown,
1888 *R. Thomson* **NRN**; Cothall lime quarry, 1913 *D. Patton* **GL**
and 1953 *A. Melderis* **BM**; wood at Gordonstoun, 1936 *R. Richter*;
in shade of alders by the Buckie Loch in the Culbin Forest, 1932;
on open moorland (now planted with conifers) on the Hill of
Mulundy, 1969 *G. Shepherd*; a single plant in an open moorland
flush at Johnstripe, 1973 *R. Richter*; in a small wood by the old kirk
at Balnabreich near Mulben and on the railway embankment
W of Rafford; in a small wood near Duffus, 1973 **E.**
N Rare at Cawdor, *WAS* (Gordon); meadow by the walled garden
at Glenferness House, 1961 **ABD.**
I Moniack Burn, 1854 *HC*, **INV**!; Cairngorms, 1935 *J. Walton*
GL!; in a small wood at the outlet to the River Tarff to Loch Ness
at Fort Augustus, 1954; in the beech wood at Achnagairn House,
Kirkhill, 1954; many plants in a hollow on the open moor between
the road and Tomachrochar S of Nethybridge, 1975.

L. cordata (L.) R. Br.   *Lesser Twayblade*
Native. A delicate small orchid with inconspicuous brownish
flowers. Common in pine woods, under heather on open moorland
and in sphagnum bogs. Fig 45b, Map 273.
M Near Brodie House, *Brodie*, **E**; Elginshire, 1837 *G. Gordon* **E**;
near Grantown, 1837 *WAS* **E**; Grantown, 1859 *JK* **FRS**; Culbin
Forest, 1948 **E,BM,CGE** and 1953 *A. Melderis* **BM**; throughout the
county, 1954.
N Hardmuir, 1833 *WAS* **GL**; Ferness wood at Ardclach, 1896
R. Thomson **NRN**; pine wood by Lochloy, 1960 **ABD**; moors at
Drynachan and pine woods at Loch Belivat.
I Strathaffric, 1850 *J. Ball* **E**; Dunain and Leys near Inverness
(Aitken); Rothiemurchus, 1893 *AS* **E**; moors at Dunmaglass,
Drumochter and throughout the district.

Neottia Ludw.

N. nidus-avis (L.) Rich.   *Bird's-nest Orchid*
Native. Rare in deciduous woods. Fig 45f.
M Sanquhar woods at Forres, 1857 *JK* **FRS**; Darnaway Forest
(Burgess); 12 flowering plants by the path to the river near the
Heronry at Darnaway, 1957 **E.**
N Cawdor woods, 1845 *Miss Grant* (Gordon, *annot.*).
I Moniack Burn at Reelig *c.* 1880 ? *coll.* **INV**; woods at Foyers
and Moniack (Aitken); near Dores, 1940 *AJ* **ABD**; in shade of
hazel by Lochness N of Drumnadrochit, 1950; bank of the Grotaig
Burn, 1953 *A. A. Slack*; beech wood at Achnagairn, Kirkhill,
1963 **CGE**; S side of Loch Ness, 1970 *E. M. Davidson* **BM**; wood
at Dunain, 1970 *M. Barron*; wood at the side of the ravine below
the Falls of Divach near Lewiston.

Goodyera R. Br.

G. repens (L.) R. Br.   *Creeping Lady's-Tresses*
Native. Abundant in all the old pine woods. Fig 45d, Map 274.
M Wood at Brodie House, 1794 *J. Mackay & Brodie* **E,BM** and
herb. *G. Don* **BM,K**; Gordon Castle, 1832 *J. Macnab & J. Hoy* **BM**;
near Elgin, 1836 *W. Robertson* **BM**; Grantown, 1850 herb. *Borrer* **K**;
Altyre near Forres, 1871 *JK* **FRS**; from near sea level at Kinloss
and throughout the Province in the older pine woods (Gordon);
Grantown, 1891 *AS* **E,OXF**; abundant under pines in the Culbin
Forest and throughout the county, 1954.
N Cawdor woods, 1833 *WAS* **E,GL,OXF** and herb. *Borrer* **K**;
Dulsie woods and at Glenferness (Thomson); pine wood by the
Carse of Delnies, on the Hill of Urchany, by Loch Belivat and
Coulmony etc., 1954; pine wood at Glenferness, 1961 **ABD**.
I In a wood opposite to Moy Hall, on the south side of the road
to Inverness, *Dr Hope*. The first authentic record (Lightfoot, **1**
(1792)); between Rothiemurchus and Glenmore, 1829 *G. A. Walker
Arnott* **BM**; Rothiemurchus, 1893 *AS* **E,BM**; near Fort George,
1932 *G. Taylor* **BM**; pine wood at Croy, 1940 *AJ* **ABD**; in all the
old pine woods throughout the district, 1954.

Hammarbya Kuntze

H. paludosa (L.) Kuntze   *Bog Orchid*
  Malaxis paludosa (L.) Sw.
Native. Rare on loch margins, usually in sphagnum. Fig 45e.
M Blacksboat, 1874 *Dr Bannerman* **E**; near Lettoch Farm on
Dorback Burn, Abernethy, 1881 *Dr McTier* **FRS**; west side of the
River Spey at Boat of Garten, 1881 *JK* and *G. C. Druce* **OXF**; near
Loch Romach (Burgess).
N Formerly common near the end of the golf course at Nairn but
the horse-mower seems to have well-nigh extinguished it, 1900
(Lobban).
I Rare near Loch Affric, 1836 *WAS* (Gordon); margin of Seafield
Loch E of Fort George, 1846 *A. Croall* **E**; small loch on Beinn a'
Bha'ach Ard, 1867 *J. Farquharson & T. Fraser* (*Trans. BSE* **9**: 474
(1868))!; bog at the foot of Creag Dhubh near Laggan (Boyd); in
two places in Strathspey, one on the E side of the Spey, *JK*; by
Loch nan Geadas (Pealling); peaty bog in the Monadhliaths, 1907
W. Edgar-Evans **E**; Struy, 1932 *Rex A. Graham*; Loch Morlich near
Sluggan Pass road, 1934 *J. Walton* **GL**; Glen Einich, 1948 *J. & I.
Ounstead*; in sphagnum at the W end of Loch a' Chlachain, 1961
M. Cuthbert & J. Roberts; shores of Loch Morlich, there until 1968,
now feared lost owing to the trampling of tourists.

Corallorhiza Chatel

C. trifida Chatel   *Coral-root Orchid*
Native. Frequent in damp dune slacks, or beneath willows, and in
dry places in pine woods. A colourless orchid with no green leaves
and roots resembling coral. Fig 45g, Pl 19.
M Culbin Sands, 1842 (Gordon, *annot.*); Culbin Sands, 1910

A. *MacGregor* and in four places in Dyke, three of which are in the Culbin Forest (Burgess); collected in the Culbin Forest in 1923 *D. Patton* **GL**; 1941 ? *coll.* **STA**; 1964 *P. H. Davis* **E** and 1953 *A. Melderis* **BM**; Spynie and Loch of the Cotts, 1957 *R. Richter*; sandy bank near the Winter lochs, Culbin Forest, 1963 **CGE**; pool by the River Findhorn N of Forres, under birch in a damp wood at Darklass Farm, Dyke, in a pine wood at Carrbridge and under willows on a shingle bank at Kingston; margin of Blairs Loch, 1973 *K. Christie.*

N Culbin, near Nairn, 1930 *G. Taylor* **BM**; under willows at Lochloy, 1954.

I In the wood behind Eskadale House towards the ferry, 1840 *Mr Lawrence* (Gordon, *annot.*); near Coylumbridge, 1916 *J. Roffey & H. W. Pugsley* **BM**; near Blackpark, Aviemore, 1922 *C. E. Salmon* **BM**; Ladystone near Inverness, 1970 *M. Barron et al.*; Leachkin, 1970 *M. Barron*; wood above Torgomack near Farley, 1973 *M. Barron*; wood above Reelig House, 1973 *M. R. Fraser*; abundant in boggy moorland under juniper and willows on Englishton Muir, *M. Cameron & MMcCW.*

A peloric form of this orchid was found in quantity at the Buckie Loch, Culbin Forest, in 1965 (McCallum Webster, *Proc. BSBI*, 1967). The following extract is of interest:

'On 25th May, 1965, a peloric form of *Corallorhiza trifida* was found in long grass at the eastern end of the area known as the Buckie Loch, Culbin State Forest, Moray (v.c. 95). Associated species observed were *Iris pseudacorus*, *Agrostis tenuis* and *Potentilla palustris*, with *Salix aurita c.* 30 feet away. Further search to the west revealed many more abnormal specimens of varying degrees of "perfection", these were in wet open ground with *Salix repens*, *Potentilla anserina*, *Veronica scutella*, *Eleocharis palustris*, etc.

'A coloured transparency of the plant was sent to Mr V. S. Summerhayes, who commented: "I do not remember seeing or hearing about this type of abnormality in *Corallorhiza* but I am not surprised, as peloria (the replacement of the outer petals by labella or vice-versa) has been recorded from many different genera of orchids. I have actually seen examples in *Ophrys insectifera* and *O. apifera*, and in *Gymnadenia conopsea*, and variants have been recorded in other British Orchids." In a later letter, Mr Summerhayes writes: "I visited Kew last week and looked up the literature on *Corallorhiza* to see if there were any references to peloriate flowers. To my great surprise I found a reference to Greville, *Flora Ediniensis*, p. 187 (1824), which reads as follows, on looking up the original.

'"I possess a highly curious monstrosity of the plant from Raveling-toll [west of Edinburgh, but now extinct there]. In all the flowers of one individual the two outer [should be inner I think] of the connivent segments of the perianth are converted into lips, as large as the true lip and beautifully spotted; the

3 remaining segments appear between them as a triphyllous calyx."

'This seems to agree very well with your plant but it is the only record I could find in the standard European monograph. Evidently such peloriate forms are scarce and have so far been recorded only from Scotland!'

Coeloglossum Hartm.

C. viride (L.) Hartm. *Frog Orchid*
Native. Rare in short base-rich grassland in the lowlands but frequent on cliff ledges in the highlands. Fig 45i, Map 275.
M Frequent at Aviemore (Gordon); parishes of Dyke and Duffus (Burgess); golf course at Boat of Garten, 1954; rocky outcrop by the sea at Burghead, 1953 and 1960 and grass bank on the moors of Edinkillie, *R. Richter*; plentiful in short turf by the old lime kiln at Fae, Dorback, 1972 **E**; grass slope in the gorge of Huntly's Cave on Dava Moor, 1973 *J. B. Dougan*!.
N Dry pasture at Assich (Aitken); on the golf course at Nairn, 1900 (Lobban).
I Ardersier Church, 1840. Painting by *S. Bland* **E**; NE of Dalwhinnie, *A. T. Coore* (Balfour 1868); low ground at Kingussie, 1877 (Boyd); Glen Einich (Druce, 1888); Kincraig, 1893 *AS* **E**; Blackpark near Aviemore, 1907 *W. Edgar-Evans* **E**; Cairngorms, 1935 *J. Walton* **GL**; Newtonmore golf course, 1914 *T. McGrouther* **GL**!; grass slopes below Ruthven Barracks at Kingussie, 1955 *C. Murdoch*; mountain ledges on Ben Alder, in Coire Chùirn, Coire and Lochain of Cairn Gorm, grass bank by the burn at Gaick Lodge, ledge in the burn at Coignashie in Strathdearn and grass track by Loch Tarff etc.; short grass in a field on the moor above Newtonmore, 1970 *S. Haywood*.

Gymnadenia R. Br.

G. conopsea (L.) R. Br. subsp. **conopsea** *Fragrant Orchid*
Native. Very common in short base-rich grassland. Very fragrant. Fig 45h, Pl 20, Map 276.
M Very common (Gordon); Grantown, 1891 *JK* **FRS**; rare in the Culbin Forest and in the last remaining corner of the old Moss of Dyke at Darklass Farm but common everywhere on the grassy moors etc.
N (Stables, *Cat.*); Ardclach (Thomson); meadow at the E end of Geddes Reservoir, by Loch of the Clans, at Dulsie Bridge, Glenferness and by the Minister's Loch on the Nairn Dunbar golf course, 1954; moor by the ford at Achavraat, 1960; on grass hummocks on the moor at Balmore near Clunas, 1963 **ABD**; in short turf by the River Nairn at Little Kildrummie.
I Pitmain, 1833 (Watson, *Loc. Cat.*); Craig Phadrig near Inverness, 1854 *HC* **INV**; low ground at Kingussie (Boyd); at Cantraybruaich, Leachkin, Dalcross and Clava etc. (Aitken); Fort Augustus, 1904 *G. T. West* **E**; abundant throughout the district, 1954.

Pseudorchis Séguier
P. albida (L.) Á. & D. Löve   *Small White-orchid*
Leucorchis albida (L.) E. Mey.
Native. Frequent in short moorland turf. Fig 45c, Pl 20, Map 277.

M Frequent in the upper districts (Gordon); Grantown, 1859 and by the brae leading to Romach, 1871 *JK* **FRS**; Boat of Garten, 1889 *G. C. Druce* **OXF**; parishes of Dyke, Edinkillie, Speymouth, Rafford and Urquhart (Burgess); moors at Kellas and Dava, 1954; bank by road at Grantown, 1974.
N Cawdor, 1833 *WAS* **E**; Ardclach (Thomson); grass hummock on the moor at Balmore near Clunas, 1963 **ABD**; in rough grassland by the view-point at Dulsie Bridge, 1963.
I Near the Toll House at Dalwhinnie and at Pitmain, 1833 (Watson, *Loc. Cat.*); low ground at Kingussie (Boyd); Kincraig, 1891 *AS* **E,OXF**; in Glen Urquhart and at Dunain (Aitken); Forst Augustus, 1904 *G. T. West* **STA**; Newtonmore golf course, 1914 *T. McGrouther* **GL**; NW of Affric Lodge and N of Loch Mullardoch, 1947 *R. D. Graham & C. Wickham*!; rare in deep heather by the Allt Greag an Lèth-choin, Glenmore, 1953; in Glen Strathfarrar and at Farr, 1954; in Glen Feshie and by Loch an Eilein, 1959 *C. Murdoch*; plentiful on the grassy lower slopes of Sgùrr na Lapaich, 1971 **E**; moor at Garva Bridge, by Loch Meiklie and on An Leacainn at Blackfold, 1973; on grassy hummocks in a birch wood below Netherwood at Newtonmore, 1973 *S. Haywood*.

Platanthera Rich.
P. chlorantha (Cust.) Reichb.   *Greater Butterfly-orchid*
Native. Rare on short base-rich grassland in the lowlands, locally abundant in the highlands especially in Strathspey. Flowers greenish-white, the pollinia divergent downwards. Fig 45k, Pl 20, Map 278.

M Near Cothall, 1871 *JK* **FRS**; at Mundole on the River Findhorn (Burgess); formerly frequent on the Califer Hill and by the Buckie Loch in the Culbin Forest, but feared lost there with forestry drainage and planting.
N Dulsie, 1903 *JK* **FRS**; Nairn, 1911 *P. Ewing* **GL**; abundant in a field at the W end of Geddes Reservoir at Burnside, 1954, not seen recently owing to heavy grazing; grass slope in a field near Ferness village and on the moor at Achavraat, 1954.
I Arniston, 1855 *HC* **INV**; low ground at Kingussie (Boyd); Glen Urquhart (Aitken); lower slopes of Mealfuarvonie, 1954; grass hillside at Drumnadrochit, 1947; bog in a grass field at Essich, banks of the River Beauly at Groam of Annat, roadside verge at Femlock near Kiltarlity, bogs at Insh village and at Inverdruie, abundant by the banks of the River Spey at Laggan, Newtonmore and Boat of Garten; bog below Netherwood, Newtonmore 1972 *S. Haywood*; small bog at Millstone near Culloden, 1975 *A. Williams*!.

Platanthera chlorantha × ½
Greater Butterfly-orchid

P. bifolia × ⅔
Lesser Butterfly-orchid

Pseudorchis albida × ⅔
Small white-orchid

Dactylorhiza maculatum
subsp. ericetorum × ⅔
Heath Spotted-orchid

D. purpurella × ½
Northern Marsh-orchid
PLATE 20

Gymnadenia conopsea × 1½
Fragrant Orchid   J. Whitcomb

P. bifolia (L.) Rich. *Lesser Butterfly-orchid*
Native. Frequent in similar situations to the former and closely resembling it but smaller in all its parts, the inflorescence denser and the pollinia parallel. Fig 45j, Map 279.

M Very common (Gordon); Aviemore, 1850 *J. Ball* **E**; Spynie, 1876 *G. C. Druce* **OXF**; in long grass by the Buckie Loch in the Culbin Forest and at the Leen, Garmouth, 1953 *A. Melderis* **BM**; pasture at Grantown, Boat of Garten, Kellas, Longmorn and near Elgin; moors at Lochindorb and on the Califer Hill near Forres, 1954; in short grass to the W of Buckie Loch, Culbin, 1968 **E**.

N (Stables, *Cat.*); Ardclach, not common, 1887 *R. Thomson* **NRN**; E of Nairn (Marshall, 1899); Nairn, 1911 *P. Ewing* **GL**; pasture at the west end of Geddes Reservoir, 1950; edge of a moorland bog at Achavraat, by Cran Loch and on a grassy roadside verge on Urchany Hill.

I Pitmain, 1833 (Watson, *Loc. Cat.*); Dochfour, 1854 *HC* **INV**; Loch an Eilein, 1865 *JK* **FRS**; by the Moniack Burn, *c.* 1880 ? *coll.* **INV**; low ground at Kingussie (Boyd); Balblair (Druce, 1888); golf course at Newtonmore, 1914 *T. McGrouther* **GL**!; moors by Loch Affric Lodge; pasture near Loch Loyne, grass slope at Suidhe, Kincraig, by the track to Loch Toll a' Mhuic in Glen Strathfarrar etc.

Orchis L.

O. mascula (L.) L. *Early-purple Orchid*
Native. Rare on base-rich grassland and grassy sea cliffs in the lowlands, more frequent on mountain ledges in the highlands. Fig 45l.

M Abundant in one or two places, between the Manse of St Andrew's and Foresterseat, at the W end of Aldroughty farm, and at Alves, *G. Wilson* (Gordon); Cothall, 1903 *JK* **FRS**; grassy roadside verge on the edge of a small wood to the E of Burgie distillery; on the golf course at Boat of Garten, by the River Spey at Fochabers and on the shingle island at Aviemore.

N (Stables, *Cat.*); Lethen, 1837 *J. B. Brichan* (Gordon, *annot.*); Ardclach (Thomson); grassy roadside verge W of Cantraydoune, 1954.

I Islands at Inverness (Gordon); grassy hillside, Milton, 1947, by a small burn at Farr, rocky slope at Fort Augustus and on Mealfuarvonie, 1956, a single plant, growing on a fallen tree trunk by the Falls of Foyers, rare in a hazel wood by the burn at Finglack, burn gully at Lurgmore, cliffs in Glen Feshie and at Gaick and a few plants on a ledge near Eskadale and on the raised beach at Ardersier; plentiful on grass tussocks in a bog in Dirr Wood near Dores, 1972 **E**; conglomerate cliff ledges on Creag nan Clag, 1975; rock ledge above the Allt Gharbh Chaig at Gaick, 1974 *OMS*.

Dactylorhiza Nevski

D. fuchsii (Druce) Soó subsp. **fuschii** *Common Spotted-orchid*
Orchis fuchsii Druce; *Dactylorchis fuchsii* (Druce) Vermeul.

454

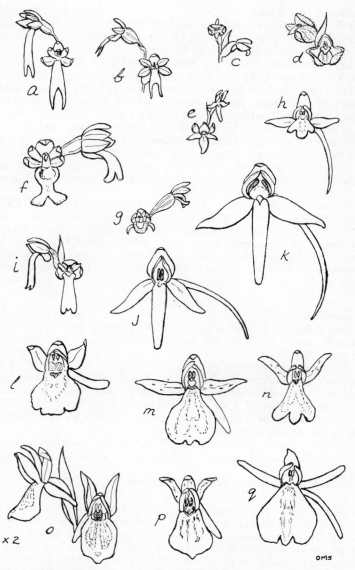

Fig 45 a, **Listera ovata**; b, **Listera cordata**; c, **Pseudorchis albida**; d, **Goodyera repens**; e, **Hammarbya paludosa**; f, **Neottia nidus-avis**; g, **Corallorhiza trifida**; h, Gymnadenia conopsea; i, Coeloglossum viride; j, **Platanthera bifolia**; k, **P. chlorantha**; l, **Orchis mascula**; m, **Dactylorhiza maculata** subsp. **ericetorum**; n, **D. fuchsii**; o, **D. incarnata**; p, **D. purpurella**; q, **Gymnadenia × D. maculata** subsp. **ericitorum**.

Native. Rare in woods, bogs and burn ravines on base-rich soil. Usually a taller more slender plant than the heath spotted-orchid and with the middle lobe of the labellum narrower and longer than the laterals. Fig 45n.

M A few plants at the west end of the Buckie Loch, Culbin Forest, 1930; rare on the Leen at Garmouth, 1953; in a wet flush by the woodland path upstream from Randolph's Leap at Relugas, 1960; Dunphail, 1961 *R. Richter.*

N Grassy flush on the raised beach below Maviston, 1954; on a small island in the Allt Dearg at Cawdor Castle, 1954 **K**; bogs by the Loch of the Clans and by Loch Belivat.

I Grassy slope at Drumnadrochit, 1947; plentiful in an alder wood by the Moniack Burn at Kirkhill, 1954 and 1972 **E**; ledge by the Glendoe waterfall, two plants on the hill road to Breakachy, meadow by the Bunchrew Burn, flush near Deanie Lodge in Glen Strathfarrar, banks of the River Spey at Boat of Garten and in a bog below Netherwood at Newtonmore.

D. fuchsii × purpurella

I In an alder wood by the Moniack Burn, Kirkhill, with both parents, 1972 **E,K**.

D. maculata (L.) Soó subsp. ericetorum (E. F. Linton)
Hunt & Summerhayes   *Heath Spotted-orchid*
Orchis maculata L.; *Dactylorchis maculata* (L.) Vermeul.

Native. Abundant in moorland bogs on acid soil. Very variable in the markings on the labellum, the central lobe is smaller and shorter than the very broad lateral lobes. Fig 45m, Pl 20.

M Very common (Gordon); Culbin Forest, Dava Moor, Boat of Garten and throughout the county, 1954.

N (Stables, *Cat.*); Ardclach (Thomson); moors at Drynachan and throughout the county, 1954; moorland bog below Maviston in the Culbin Forest, 1968 **E**.

I Dalwhinnie and Pitmain, 1833 (Watson, *Loc. Cat.*); Craig Phadrig, 1854 *HC* **INV**; Westhill S of Inverness, 1937 *J. Walton* **GL**; Glen Einich, 1949 *P. S. Green* **E**; Coire an Lochain of Cairn Gorm, 1953 *A. Melderis* **BM**; throughout the district, 1954.

D. maculata × purpurella

Frequent. Under-recorded. Very large plants can occur.

M The Leen at Garmouth, bog in a field at Auchinroath N of Rothes, grassy moor at Kellas and on the island of Lochindorb Castle, 1973.

N Boggy moor at Achavraat, 1960 **K**.

I Bog below Netherwood at Newtonmore, 1971; in a birch-oakwood at Boblainy, bog by the River Nairn at Nairnside, bog to the S of Broomhill station and bogs at Braiton of Leys near Inverness etc.

D. incarnata (L.) Soó   *Early Marsh-orchid*
Orchis strictifolia Opiz
    Subsp. **incarnata**
Native. Rare in grassy bogs and dune slacks on base-rich soil.
Flowers flesh-pink, the labellum strongly reflexed. Fig 45o.
Very common (as *O. latifolia* (Gordon); near Hopeman (Marshall,
1899); Boat of Garten (Druce. 1888); Loch Spynie (as *O. latifolia*)
(Druce, 1897); Cothall (as *O. latifolia*) 1858 *JK* **FRS**; parishes of
Dyke, Cromdale and Urquhart (Burgess); at the W end of the
Buckie Loch in the Culbin Forest, 1950; on tussocks in Dyke Moss
behind Darklass Farm, Dyke, 1953, grassy bog on Dava Moor at
Wester limekilns, 1974.
N (Stables, *Cat.*); Ardclach (as *O. latifolia*) (Thomson); between
Findhorn and Nairn (as *O. latifolia*), 1911 *P. Ewing* **GL**; bog by the
old ford at Achavraat and by Cran Loch, 1954.
I Rare in bogs on the N side of Loch Coulan near Balnafoich on
Drummossie Muir, 1976.

    Subsp. **pulchella** (Druce) Soó
Native. Rare, or over-looked, in base-rich flushes and bogs. Flowers
magenta-red, labellum reflexed. Usually flowers 2 weeks later than
the other orchids.
I Base-rich bog on the hillside by Lundie in Glen Moriston,
1975 **K**; bog by the road S of Inverdruie; flush at Farletter near
Insh, 1975 **E**.

D. incarnata × **maculata**
Flowers pale flesh-pink.
M Several plants, with both parents in a boggy field on the east side
of Wester Limekilns, Dava Moor, 1974 **E**.

D. purpurella (T. & T. A. Steph.) Soó   *Northern Marsh-orchid*
Orchis purpurella T. & T. A. Steph.
Native. Very common in base-rich bogs and river banks. Flowers
deep purple, labellum usually not reflexed. The old records for
O. latifolia in Gordon and Burgess as 'very common' must surely
belong here and not to *D. incarnata*. Fig 45p, Pl 20, Map 28o.
M Grass verge to the drive at Brodie Castle, 1930; common by the
River Spey at Boat of Garten, in the Buckie Loch, Culbin and
Dyke Moss, 1950; meadow at St John's Mead, Darnaway and on
the Leen at Garmouth, 1953 *A. Melderis* **BM**; bogs by Loch Spynie,
on the island of Lochindorb Castle, at Kellas and in slacks on
Lossiemouth golf course etc., 1954.
N Margin of the Minister's Loch at Nairn, 1930; by the River
Nairn and at Holme Rose and throughout the county, 1954;
moor near Lethen, 1961 **ABD**; on the raised beach at Maviston,
1968 **E**.
I Ness, 1854 *HC* **INV**; grassy bog near Moniack Castle, 1925;
common in Strathglass, by Loch Flemington, dune slack at Ardersier
and in the Spey valley, 1954, etc.

D. maculata×**Gymnadenia conopsea** (L.) R. Br.
N Bog on the moor by the old ford at Achavraat, 1960 **K.**
I By the track to Loch Toll a' Mhuic at Monar, 1975.

ARACEAE
Lysichiton Schott
L. americanus Hultén & St John *Skunk Cabbage*
Introduced. A large leaved arum-like plant with yellow spathes.
M Escaped from the ornamental pond at Kellas House to the
alder wood by the River Lossie, 1967 **E.**

Arum L.
A. maculatum L. *Cuckoo Pint, Lords and Ladies*
Introduced in north Scotland. Usually in woods and by burns in
the policies of the larger houses.
M Doubtfully native, rare at the Castle of Old Duffus (Gordon)
and in 1879 ? *coll.* **ELN**; believed planted in the parishes of Elgin,
Drainie and Bellie (Burgess); wood at Newton House and at Moy
House, at Rothes, Longmorn and Sheriffston House at Lhanbryde,
1954; woods at Innes House, Urquhart, 1967 **E,CGE.**
N Nairn basin, inland, 1900 (Lobban); in a wood opposite the
Home Farm at Cawdor, 1954.
I A large patch with very large unspotted leaves occurs in a small
damp wood near the railway line at Clunes House, Kirkhill, 1950
M. Cameron and in 1954 *MMcCW* **CGE**; wood at Achnagairn
House, Kirkhill and planted on a bank at Glen Cottage by Ness
Castle Inverness, 1971 **E.**

LEMNACEAE
Lemna L.
L. minor L. *Common Duckweed*
Native. Frequent in still water and ditches in the lowlands but less
so in the highlands.
M Frequent but not observed to flower (Gordon); parishes of
Dyke, Lhanbryd, Bellie, Speymouth and Knockando (Burgess);
pond by the Muckle Burn at Dyke, waterhole in the Culbin Forest,
ditch in the Hills of Cromdale and at Blacksboat etc., 1954; ditch
on the moor at Kellas, 1973 **E.**
N Near Druim House, Nairn, 1886 *R. Thomson* **NRN**; Ardclach
(Thomson); pool by the River Nairn at Kilravock, in Geddes
Reservoir, Loch of the Clans, Kinsteary and Cran Loch, 1954.
I Inverness, 1832 (Watson, *Loc. Cat.*); near Kirkhill, 1940 *AJ*
ABD; Loch Flemington, 1954; Kingussie, 1955 *C. D. Pigott (Atlas)*;
ditch by Balbeg Farm at Bunloit, pond at Achnagairn House,
Kirkhill, ditch at Boblainy, in Raigmore pond at Inverness, pond
at Culloden House, muddy pool in a field on the hill road to
Divach, small burn at Foyers and in the canal below Insh village,
1975.

SPARGANIACEAE

Sparganium L.

S. erectum L. var. **microcarpum**   *Branched Bur-reed*
    (Neuman) C. Cook

Native. Frequent in the lowland areas on pond margins and in ditches. Perianth segments black-tipped. The old records for *S. ramosum* Huds. belong here. Map 281.

M Frequent (as *S. ramosum*) (Gordon); Sanquhar pond, Forres, 1856 *JK* **FRS**; loch near Forres (Druce, 1888); Loch Spynie (Druce, 1897); Boat of Garten, 1888 ? *coll.* **GL**; ditch in the Culbin Forest, at Brodie pond, in the burn at Pittendreich etc., 1954; pond at Dipple, 1967 *R. P. Petrie* **E**; by the Woollen Mill at Dallas, 1976 *James Stewart*.

N (Stables, *Cat.*); Ardclach (Thomson); near Nairn (Marshall, 1899); margin of Geddes Reservoir, Cran Loch and in a ditch at Boghole, 1954; ditch on the Carse of Delnies, 1960 **ABD**; by the Minister's Loch at Nairn, 1960 *K* and in 1967 **E,BM**; at Loch Kirkaldy, Loch of the Clans and inburns at Ordbreck and by Rait Castle Farm.

I (Watson, *Top. Bot.*); by the River Spey at Inverdruie (Druce, 1888); pond at Essich (Aitken); Urquhart Bay, 1904 *G. T. West*; Feshie Bridge, 1904 *G. B. Neilson* **GL**; Drumnadrochit, 1940 *AJ* **ABD**; ditch at Dalcross, 1956; pool by the River Calder at Newtonmore, canal below Insh village, lochan at Ness Castle near Inverness, pool in the quarry at Moniack near Kirkhill, ditches at Kingussie, junction of the Moy Burn with the River Findhorn near Tomatin and banks of the river at Beauly; Raigmore pond, 1975 **E**.

S. emersum Rehm.   *Unbranched Bur-reed*
    S. simplex pro parte

Native. Rare in similar situation to the former. Stem simple, perianth segments not black-tipped.

M Frequent, with *S. erectum* on the Lossiemouth road (Gordon); Moss of Chapelton, *J. G. Innes* (Gordon, *annot.*); dried up swamp by the railway station at Garmouth, 1898 *ESM* **E,BM**; Aviemore (Marshall, 1899); by the River Lossie at Calcots, 1967 **CGE**; in the burn at Pittendreich S of Elgin and in a ditch on the Leen at Garmouth.

N Nairn, 1857 *JK* **FRS**; in a ditch on the north side of the Minister's Loch on the Nairn Dunbar golf course, 1971 **E**.

I Leachfield loch near Fort George, 1846 *A. Croall* (Watson); Loch Flemington, 1940 *AJ* **ABD**; canal below Insh village, 1975 **E**.

S. angustifolium Michx.   *Floating Bur-reed*
    S. affine Schnizl.

Native. Rare in the lowlands, common in the highlands. Floating in pools and lochans in acid water. The leaf-like bract of the lowest female flower is twice as long as the inflorescence. Male flowers 2. Map 282.

M Abundant in one or two places, at Aldroughty, Romach, Burgie (as *S. natans*) and Duffus, *G. Wilson* (Gordon); Loch Romach, 1850 *JK* **FRS**; Avielochan (Druce, 1888); moorland pool between Dava and Lochindorb (as var. *microcephalum* Neuman, a form with small female heads and subsolitary male heads) 1898 *ESM* **BM,CGE**; parishes of Dyke, Forres, Elgin, Rafford, Duffus and Cromdale (Burgess); pool on the moor W of Lochindorb, 1954 **E**; common in Loch Vaa N of Aviemore.
N Near Budgate of Cawdor, *WAS* (Gordon); in the Moss of Litie and at Inshoch, *J. B. Brichan* (Watson); peaty pool on the Hill of Urchany, 1974 **E**.
I Pond within the Asylum grounds, Inverness (Aitken); in Loch Uanagan, Loch Killin, Loch Meiklie, Coiltry Loch and Cherry Island in Loch Ness, 1904 *G. T. West* **STA**; near Moy Loch, 1940 *AJ* **ABD**; Loch Beannacheran in Glen Strathfarrar, in Loch Dochfour and at Tomich, 1956; backwater of the River Spey at Newtonmore, in Loch Toll a'Mhuic and Loch Garbh-choire at Ryvoan etc., in a small pool by the dam at Loch Mullardoch, 1971 **E**.

S. minimum Wallr.   *Least Bur-reed*
Native. Frequent on loch margins in shallow water and in dried up muddy pools. Leaf-like bract of the lowest flower barely exceeding the inflorescence. Male flower 1. Map 283.
M Loch on Bognach (? Buinach) Hill near Kellas, *P. Leslie*, and in the parishes of Dyke, Edinkillie and Knockando (Burgess); pool at Grantown, 1954; pool in a pine wood at Miltonduff, 1955 *R. Richter*; Loch Romach, 1972; muddy pool by Loch Vaa, 1973 **E**.
I Loch Mallachie, 1887 *G. C. Druce* **OXF**; Glenmore, 1889 *G. C. Druce* **OXF**; Abernethy, 1893 *JK* **FRS**; at the W end of Loch Meiklie, Lochan Coire Doe, loch N of Invermoriston and Lochan na Stàirne SE of Fort Augustus, 1904 *G. T. West* **STA**; Loch Lapaich Glen Affric, runnel on Mealfuarvonie, 1971; Coillenaclay and Glen Urquhart, 1973 *M. Barron*; dried up pool near Crelevan in Strathglass, 1975 **E**; pool by the River Spey S of Laggan Bridge, 1975 **E**; pool by the road at Balcraggan and in a pool by the Mains of Bunachton, 1976.

TYPHACEAE
Typha L.
T. latifolia L.   *Bull's Eggs, Reedmace, Bulrush*
Native. In swamps, loch margins and slow-flowing burns. Occasionally planted.
M Loch Spynie, 1833 *Brodie & G. Gordon* **E**; a plant or two at Waulkmill, Alves, *G. Wilson* (Gordon); Buckie Loch, Culbin (Marshall, 1899); Brodie pond, 1925; Dubh lochan in the Darnaway Forest, Sanquhar pond, Forres, lochan at Brokentore near Kellas, at Lochinvar, and in a pool on the Lossiemouth golf course, 1954.
N Moss of Inshoch near Auldearn, 1898 *ESM* **CGE**; Lochloy and

Cran Loch, 1904 *G. T. West* **E,STA**; bog below Maviston in the
Culbin Forest, 1950; by the Minister's Loch on the Nairn Dunbar
golf course, abundant at Loch of the Clans and rare at Loch
Flemington.
I (Watson, *Top. Bot.*); pond at Clunes, Kirkhill, 1950 *M. Cameron*;
lagoons at Inverness, Loch Flemington, by a small lochan at
Inverdruie, 1954; abundant in the marshes by the River Beauly
between Tomich Farm and Beauly; planted in the marshes N of
Kingussie; several patches by Loch Coulan at Balnafoich.

CYPERACEAE
Eriophorum L.

E. angustifolium Honck.   *Common Cottongrass*
Native. Abundant in pools margins of lochs and lochans on all the
moors.
M Innes House, Urquhart, 1794 *A. Cooper* **ELN**; very common
(Gordon); Califer Hill, 1860 *JK* **FRS**; Culbin Sands at Kincorth,
1937 *J. Walton* **GL**; near Grantown, 1876 *J.F.* (? *James Fraser*) **E**;
Findhorn Bay and Winter lochs Culbin, 1953 *A. Melderis* **BM**;
throughout the county, 1954.
N (Stables, *Cat.*); turfy moors at Ardclach, 1882 *R. Thomson* **NRN**;
throughout the county, 1954; moor at Glenferness, 1960 **ABD**;
marshy field near Boghole, 1967 **E,CGE**.
I Dalwhinnie and Pitmain, 1833 (Watson, *Loc. Cat.*); Badenoch,
1838 *A. Rutherford* **E**; Glen Strathfarrar and Glen Affric, 1850
J. Ball **E**; Lentran, *c.* 1880 ? *coll.* **INV**; Inverness (Aitken); by
Coiltry Loch etc., 1904 *G. T. West* **STA**; throughout the district,
1954; Sgùrr na Lapaich at 915 m and in a bog by Loch Cluanie in
Glen Moriston, 1972 **E**.

E. latifolium Hoppe   *Broad-leaved Cottongrass*
Native. Frequent in base-rich flushes and moorland bogs. Easily
recognised by its olive green colour, stiff stems and flat wide leaves
Map 284.
I In a bog in Culloden Wood, *Brodie* **E**; bog in the Allt an t-Sluie
near Dalwhinnie, 1911 *ESM* **E,BM**!; flush by Loch na Ba Ruaidhe
N of Milton, 1947; bog near the bothy on the moor of Corrimony,
1964; flushes on the moor S of Convinth, 1968; on the NE slopes
of Mealfuarvonie, 1971 **E,BM**; plentiful in Dirr Wood near Dores,
1972 **E**; flushes by Blackfold lochs, 1973 **E**; bogs by Loch Coulan
and Loch Bunachton, in a small bog near Loch nan Bonnagh near.
Farley, on the hill above Aberchalder House at Bridge of Oich,
and abundant on the moors on Drumossie Muir.

E. vaginatum L.   *Hare's-tail Cottongrass*
Native. Abundant on moors throughout the area.
M Glen of Rothes, Kellas and Romach etc. (Gordon); Edinkillie,
1899 *JK* **FRS**; bog in the Culbin Forest and throughout the county,
1954; near Blervie House, 1962 **E**; near Pitcroy, 1968 **CGE**.
N (Stables, *Cat.*); Nairn, 1887 *JK* **FRS**; near the schoolhouse

Ardclach, 1882 *R. Thomson* **NRN**; moors at Drynachan and throughout the county, 1954; moor by Cran Loch, 1960 **ABD**.
I Dalwhinnie, 1833 (Watson, *Loc. Cat.*); Badenoch, 1838 *A. Rutherford* **E**; near Inverness (Aitken); Lochan Creag Coire Doe, 1904 *G. T. West* **E,STA**; Coire an Lochain of Cairn Gorm, 1953 *A. Melderis* **BM**; throughout the district, 1954.

Scirpus L.
S. cespitosus L. subsp. **germanicus** (Palla)   *Deergrass*
     Broddesson
     Trichophorum cespitosum subsp. *germanicus* (Palla) Hegi
Native. Locally dominant in wet peaty moorland. Fig 46a.
M Common (Gordon); Culbin Forest and on all the moors throughout the county, 1954.
N (Stables, *Cat.*); moors at Ardclach, 1888 *R. Thomson* **NRN**; moors at Drynachan and throughout the county, 1954; moor at Achavraat, 1960 **ABD**.
I Dalwhinnie and Pitmain, 1833 (Watson, *Loc. Cat.*); moist heathy places about Inverness (Aitken); Carn Eilrig, 1894 *AS* **E**; Coire an Lochain of Cairn Gorm, 1953 *A. Melderis* **BM**; throughout the district, 1954.

S. maritimus L.   *Sea Club-rush*
Native. Occasional in runnels in salt-marshes.
M Parishes of Dyke and Urquhart (Burgess); plentiful by the Findhorn estuary, 1955; salt-marsh pool W of the Buckie Loch in the Culbin Forest, 1961 **ABD,K**.
N Ditch near the salt-marsh at Kingsteps, Nairn, 1960 **ABD,K**; pool by Cran Loch and on the Carse of Delnies, 1954.
I Beauly Firth near Inverness, 1930; salt-marsh at Castle Stuart, in runnels to the seat Bunchrew, by the lagoons at Muirtown Basin at Inverness, a few plants in the mud of the firth at Lentran and plentiful by the River Beauly near Tomich Farm, Beauly.

S. sylvaticus L.   *Wood Club-rush*
Native. Rare, not seen in recent years.
M In the Altyre woods, Forres, 1901 *C. Bailey* herb. *ESM* **CGE**; Speymouth near a clump of *S. lacustris*, only a few plants which have disappeared, *G. Birnie* (Burgess).

S. lacustris L.   *Common Club-rush*
Native. Rare on loch margins and muddy pools. Stems green, not glaucous, stigma 3.
M Loch Spynie but not common, *WAS* (Murray); Moy Carse, *J. G. Innes* (Gordon); Loch Spynie, 1842 ? *coll.* **E**; muddy pool by Lochinvar and in the pond at Dallas Lodge, 1954.
I Lakes in Badenoch (Gordon); by lochs Uanagan, Meiklie and nam Faoileag, 1904 *G. T. West* **STA**; Loch Insh and Loch Beag at Alvie 1956; by a small lochan at Inverdruie, at Upper Gartally, Loch Bunachton and Loch Coulan.

S. tabernaemontanae C. C. Gmel.   *Grey Club-rush*
Native. Common in brackish pools and runnels by the sea. Stems
glaucous-green, stigmas 2. Fig 44f.
M Marsh near the Old Bar of Findhorn, 1843 *J. B. Brichan &
J. G. Innes* (Gordon, *annot.*); Findhorn Bay, 1860 *JK* **FRS**; between
Forres and the Culbin Sands, 1889 *G. C. Druce* **OXF**; Loch Spynie
(Druce, 1897); Laich of Moy, 1953 *B. Welch* **BM**; bog by Lochinvar,
1961 **K**.
N *A. Ley* ms, 1873 (Watson, *Top. Bot.*); salt-marsh E of Nairn,
1896 *ESM* **E,CGE**; by Cran Loch, 1960 **ABD**.
I Rare in the lagoons at Muirtown Basin, Inverness, 1975 and
in pools by the Beauly Firth near Tomich Farm, Beauly.

S. setaceus L.   *Bristle Club-rush*
   Isolepis setacea (L.) R. Br.
Native. Common in bare wet, gravelly places in ditches on moorland
tracks by lochs. Fig 44c, Map 285.
M Frequent (Gordon); Seapark, Kinloss, 1857 *JK* **FRS**; Boat of
Garten, 1887 *G. C. Druce* **OXF**; Culbin Sands, 1911 *P. Ewing* **GL**;
by the Muckle Burn near Moy House, 1953 *A. Melderis* **BM**; in
gravel by the drive at Brodie Castle, 1962 **BM,K**; and throughout
the county.
N (Stables, *Cat.*); moors at Ardclach, 1888 *R. Thomson* **NRN**;
Culbin Sands towards Nairn, 1913 *D. Patton* **GL**; throughout the
county, 1954; Loch of the Clans, 1960 **ABD,K**.
I Pitmain, 1833 (Watson, *Loc. Cat.*); ditch at Fort Augustus,
wet track on Newtonmore golf course, gravel verge of Loch Garten,
ditch in Glendoe, in Glen Affric etc., 1954; and well distributed
throughout the district.

S. cernuus Vahl.   *Slender Club-rush*
Recorded from between Findhorn and Nairn (Ewing). No voucher
specimen.

S. fluitans L.   *Floating Club-rush*
   Eleogiton fluitans (L.) Link
Native. Rare in the lowlands, frequent in the highlands. An
inconspicuous, light green floating plant of peaty pools and ditches.
Fig 46b, Map 286.
M Abundant in one or two places, in pools between Brodie and the
sea and in a Moss W of Aldroughty (Gordon); near Elgin, ? date
herb. *N. Tyacke* **OXF**; peaty bog hole at Lochindorb, ditch at
Relugas and in a pool at Conicavel, 1954.
N (Stables, *Cat.*); Cawdor, *W. Brand* **E**; Nairn, 1911 *P. Ewing* **GL**;
Ardclach (Thomson); pools by Loch Allan and by the Minister's
Loch, 1954; shallow water at Loch of the Clans, 1960 **ABD**; boggy
pool near Banchor, 1963 **ABD**.
I Leachfield near Ardersier, 1846 *A. Croall* (Watson); Cherry
Island Loch Ness, Loch Ashie, Loch Tarff, Coiltry Loch and at the
NE end of Loch Ness, 1904 *G. T. West*; Loch an Eilein, 1950

R. R. Roberts **ABD**; pool near Fasnakyle at Cannich, 1956; in Loch Meiklie, Loch Mallachie, Loch Laide and Loch Gynack etc., muddy pools at Crelevan in Strathglass, and by Coulan; backwater of the River Spey at Newtonmore, 1973 **E**.

Eleocharis R. Br.

E. acicularis (L.) Roem. & Schult.   *Needle Spike-rush*
I   (Druce, *Com. Fl.*). No voucher specimen.

E. quinqueflora (F. X. Hartmann) Schwarz   *Few-flowered*
  E. pauciflora (Lightf.) Link   *Spike-rush*
Native. Abundant in moorland bogs and flushes. The lowest glume at least half as long as the spikelet. Fig 46c.
M Stotfield, Rosevalley near Burghead and Culbin (Gordon); Dunphail (Druce, 1888); in ruts on the fixed dunes W of Buckie Loch in the Culbin Forest, 1950; Findhorn Bay, 1953 *A. Melderis* **BM**; throughout the county, 1954; bogs at Drakemyres, Fae and Muchrach, 1972 **E**.
N Sands E of Nairn, 1833 *WAS* **STA**; Ardclach, not common, 1888 *R. Thomson* **NRN**; throughout the county, 1954; wet moorland by the Minister's Loch at Nairn, 1960 **ABD**; flush by the Geddes Reservoir, 1961 **ABD**; bog below Maviston in the Culbin Forest, 1968 **E**; very large plants on the muddy margin of Lochloy, 1973 **E**.
I   (Watson, *Top. Bot.*); Kincraig, 1891 *AS* **E**; by Loch Morlich, 1944 *R. Mackechnie* **GL**; in flushes throughout the district, 1954; bog by Tromie Bridge, 1973 **E**.

E. multicaulis (Sm.) Sm.   *Many-stalked Spike-rush*
Native. Rare in bogs on acid soil. The lowest glume less than quarter length of the spikelet.
M Loch Spynie, *P. Cruickshank* and marshy ground from Auldearn to Dyke, *J. B. Brichan* (Gordon); Winter lochs, Culbin, 1953 *A. Melderis* **BM**!.
N Lochloy, 1838 *WAS* **E** and 1867 *JK* **FRS**; bog at Achavraat, 1960 **ABD**; bog by the pool at Keppernach near Achavraat, 1973 **E,BM**.
I   Carse of Ardersier, 1846 *A. Croall* (Watson); bogs below Sgùrr na Lapaich in Glen Strathfarrar and at Fort Augustus, 1954; by a roadside lochan between Inverton and Knappach E of Newtonmore, 1975 **E**.

E. palustris (L.) Roem. & Schult. subsp **vulgaris**   *Common*
  S. M. Walters   *Spike-rush*
Native. Common on loch margins, banks of rivers and in salt-marshes. Very variable. A dwarf well marked variety occurs on the salt-marsh verge of the Culbin Forest and needs further study. Fig 46e, Map 287.
M Common (Gordon); near Kinloss school, 1857 *JK* **FRS**; Loch Spynie (Druce, 1897); Culbin Sands (well marked variety needing study) 1953 *NDS* **BM**!; throughout the county, 1954; verge of Loch Vaa, 1973 **ABD**; at Avielochan, Longmorn, pool at the Snab,

bog E of Carrbridge and at Loch Ban near Boat of Garten, 1973
E.
N (Stables, *Cat.*); Nairn, *J. B. Brichan* **GL**; Loch Belivat, 1888
R. Thomson **NRN**; Geddes Reservoir and throughout the county,
1954; Loch Flemington, 1956 **ABD**; ditch near the sea at Kingsteps
near Nairn, 1960 **ABD**; bog below Maviston in the Culbin Forest,
1968 **E**; pool at the road junction near Littlemill, 1973 **BM**; muddy
verge of Ord Loch, 1973 **E.**
I (Watson, *Top. Bot.*); Beauly, 1889 *G. C. Druce* **OXF**; Loch
Uanagan and Coiltry Loch near Fort Augustus, 1904 *G. T. West*
STA; Rothiemurchus, 1949 *P. S. Green* **E**; shores of Loch Crunachdan,
Glen Shirra, 1972 **E**; in deep water by the River Spey at
Newtonmore, 1973 **ABD,CGE**; shores of Loch Morlich, 1973
E,BM etc.

E. palustris × uniglumis
I Bog at the W end of Loch Meiklie, 1973 **E.**

E. uniglumis (Link) Schult.   *Slender Spike-rush*
Native. Frequent in bogs on the coastal belt. Lowest glume encircles
the base of the spikelet giving it a bent appearance. Fig 46d.
M Findhorn, 1896 *G. C. Druce* **OXF**; by the golf course at
Garmouth, 1953 *A. Melderis* **BM**; salt-marsh W of the Buckie Loch,
Culbin Forest, 1968 **E,CGE**.
N Salt-marsh E of Nairn, 1898 *ESM* **CGE** and *WASh* herb. *M. L.
Wedgewood* **MBH**! and in 1961 *MMcCW* **ABD,K**; Carse of Delnies,
1954.
I Near Blackness Castle, 1855 *A. How & HC* **INV**; Easterness
(*ASNH* **9**: 161 (1900)) salt-marsh at Beauly, 1930; between Wester
Lovat and the ferry near Beauly, 1947 *S. M. Walters*; bog E of the
tow path at Clachnaharry, Inverness, 1954; salt-marsh near
Tomich Farm near Beauly, 1976 **E.**

Blysmus Panz.
B. rufus (Huds.) Link   *Narrow Blysmus, Salt-marsh Flat-sedge*
Native. Locally abundant in the salt-marshes from Findhorn to
Inverness. Fig 46f.
M Culbin, 1830, by Loch Spynie and between Findhorn and
Burghead (Gordon); sands west of Seapark, Kinloss, 1832 *G. Gordon*
GL; Loch Spynie, 1832 *WAS* **E,GL**; Kinloss, 1857 *JK* **FRS**;
Findhorn and Loch Spynie (as var. *bifolius*), 1886 *G. C. Druce* **OXF**;
salt-marsh W of Buckie Loch in the Culbin Forest, 1953 *A. Melderis*
BM and 1968 *MMcCW* **E,K,CGE**.
N Delnies, 1833 *WAS* and Nairn, *A. Falconer* (Gordon); salt-
marsh 4½ miles E of Nairn (as var. *bifolius* Wallroth) in pools with
the type, *ESM* **CGE**; sea coast at Nairn, 1870 *J. F. Duthie* **STI**;
near Nairn, 1884 *J. Groves* **E,GL,OXF**; salt-marsh at Kingsteps,
1960 **ABD** and on the Carse of Delnies.
I Salt-marsh below Castle Stuart, 1954 **E**; abundant by Longman
Point Inverness now mostly covered with rubbish and earth, and
in the salt-marsh at Lentran Point, 1954.

Fig 46 a, **Scirpus cespitosa**; b, **S. fluitans**; c, Eleocharis quinqueflora; d, **E. uniglumis**; e, **E. palustris**; f, **Blysmus rufus**; g, **Schoenus nigricans**.

Schoenus L.

S. nigricans L.   *Black Bog-rush*
Native. Frequent in base-rich flushes and bogs and occasionally on salt-marsh verges. Fig 46g, Map 288.
M Marshes near the sea near Brodie House, ? *coll.* **E** and *G. Don* **BM**!; in the Moss of Coxton, the Leen at Garmouth, in an alder plantation near the Manse of Duffus and on the Hill of Monachty, *G. Wilson* (Gordon); the Leen Garmouth, 1953 *A. Melderis* **BM**!; rare in the Moss of Dyke at Darklass Farm, Dyke, 1954.
N Moss of Litie, *WAS* (Gordon)!; wood W of Feddan, 1902 *JK* **FRS**; Nairn, 1911 *P. Ewing* **GL**; fixed dunes near the sea on the Nairn Dunbar golf course, 1960 **ABD**; bog below Maviston in the Culbin Forest, 1968 **E**.
I Flushes by Loch na Ba Ruaidhe above Milton, 1947; abundant in bogs in Glen Convinth 1954; at Fort Augustus and at Upper Gartally 1956; in flushes by Loch Bunachton and in bogs on Drumrossie Muir near Essich; plentiful in a bog by Loch Coulan, 1976, **E,BM**.

Rhynchospora Vahl

R. alba (L.) Vahl   *White Beak-sedge*
Native. Rare in acid bogs.
M Dyke Moss between Brodie House and the sea, 1802 *Brodie* **E**, herb. *G. Don* **BM,K**.
I (Druce, *Com. Fl.*); bogs below Sgùrr na Lapaich and to the west of Loch Monar, 1947.

Carex L.

C. laevigata Sm.   *Smooth-stalked Sedge*
Native. Frequent in damp woods and grassy moorland banks and by rivers. Map 289.
M Inverallan near Grantown, 1895 *C. Bailey* **OXF**; Greshop Wood, Forres, 1950 *A. C. G. Gough & M. Hunter* and in 1954 *MMcCW* **E**; by the River Spey near Advie bridge; ditch in the woods by the River Findhorn, opposite Sluie, at Darnaway, 1963 **CGE**; at Douneduff, 1967 **E,BM**; open wood near Pluscarden, 1968 **E**; by the River Lossie at Kellas, 1973 **E**, and in a field at Auchinroath near Rothes.
N By Loch Belivat (as *C. binervis*) 1885 *R. Thomson* **NRN**; birch woods at Geddes and Dulsie Bridge, 1960 **ABD**; roadside bank at Carnoch, 1963 **BM,CGE**; banks of the River Findhorn at Ardclach, 1973 **E**, and at Coulmony.
I Fraoch-choire (Ball, 1851) near Kilmorack, 1892 *ESM* **CGE**; in shade by the River Enrick at Milton, 1956; meadow by the Moniack Burn at Kirkhill, 1972 **E**; by the Bunchrew Burn, 1974 **E**; bank by the Breakachy road in Strathglass and abundant on the grassy hillside between Lurgmore and Creag Dhearg by Loch Ness; a single plant in an unusual habitat under a boulder at 800 m in the Coire Laogh of Cairn Gorm, 1975 **BM**.

C. distans L. *Distant Sedge*
Native. Rare in short turf by the sea.
M Near the coast in Speymouth (Burgess); in short turf by the salt-marsh W of the Buckie Loch in the Culbin Forest, 1950 and in 1968 **E**; estuary of the river at Findhorn 1925.
N Lochloy and Nairn, 1900 (Lobban); verge of the salt-marsh at Kingsteps E of Nairn, 1950 and 1961 **ABD**.
I Rare at the edge of the salt-marsh below Castle Stuart, 1976 **E**.

C. hostiana DC. *Tawny Sedge*
Native. Common in base-rich bogs and moorland flushes. Map 290.
M Moray, *G. Gordon* **E**; at Linkwood and Duffus (Gordon); Califer Hill near Forres, 1857 *JK* **FRS**; Winter lochs in the Culbin Forest and throughout the county, 1954; bog at Drakemyres, 1972 **E**.
N (Stables, *Cat.*); flushes at Drynachan, on the Nairn Dunbar golf course, bog at Littlemill etc., 1954; bank by Geddes Reservoir, 1961 **ABD**; bog below Maviston in the Culbin Forest, 1968 **E**.
I Dalwhinnie, 1833 (Watson, *Loc. Cat.*); Glen Urquhart, 1850 *J. Ball* **E**; Asylum grounds at Inverness (Aitken); Kingussie, 1887 *G. C. Druce* **OXF**; Alvie, 1887 *JK* **FRS**; throughout the district, 1954; bog below Insh village, 1975 **E**.

C. hostiana × lepidocarpa subsp. **scotica**
I Bog by a small burn between the lime quarry on Suidhe Hill and Kincraig 1973 **E**; by the Big Burn near the lochans on the moor above Essich, 1976 **E**.

C. hostiana × serotina
M Near Aviemore in great plenty with the parents, 1898 *ESM* **E**.
I Loch Alvie (Marshall, 1899).

C. binervis Sm. *Green-ribbed Sedge*
Native. Abundant on dry moorland.
M Very common (Gordon); near Forres, 1860 *JK* **FRS**; Culbin Forest, 1930 and throughout the county, 1954.
N (Stables, *Cat.*); Ardclach (Thomson); moors at Drynachan and throughout the county, 1954; bank of a ditch at Dulsie Bridge, 1960 **ABD**.
I Inverness, 1832, Dalwhinnie, 1833 (Watson, *Loc. Cat.*); Fort Augustus, 1850 *J. Ball* **E**; Glen Einich and Coire an t-Sneachda of Cairn Gorm, 1887 *G. C. Druce* **OXF**; Craig Dunain, 1940 *AJ* **ABD**; Glenmore, 1933 *A. Melderis* **BM**; throughout the district, 1954.

C. lepidocarpa Tausch *Long-stalked Yellow Sedge*
    Subsp. **lepidocarpa**
Native. Rare in base-rich bogs and loch margins. Stems flexuous; fruit arcuate-deflexed; female glumes pale yellow-brown, hyaline, caducous. Fig 47a.
M Findhorn, 1888 *G. C. Druce* **OXF**; margin of the lochan just N of Boat of Garten, 1954.

N About 2½ miles above the town of Nairn, 1898 *ESM & WASh*
BM; Nairn, 1911 *P. Ewing* **GL**; boggy ground on the verge of the
Minister's Loch at Nairn, 1961 **E,ABD,K**; in mud on the S shore of
Lochloy, 1973 **E**.
I Backwater of the River Calder at Newtonmore, 1973 **E**; bog
in a grass field at Essich, 1975 **E**; flushes at Loch Bunachton, at
Loch Coulan and on the moors by the Big Burn on Drumossie Muir.

Subsp. **scotica** E. W. Davies
The common subsp. in the area. Frequent on base-rich moorland
bogs and flushes. Stems stiff; fruit arcuate-deflexed; female glumes
dark chestnut-brown, usually persistent.
M Winter lochs in the Culbin Forest, 1925; flushes in the Hills of
Cromdale and at Grantown, 1954; bog at Drakemyres, 1972 **E**;
flushes on the moors at Fae, on Dava Moor and in the Craigellachie
Reserve at Aviemore, 1973; bog on the moor at Johnstripe by
Dunphail.
N Bogs at Holme Rose and by Cran Loch, 1954; bog E of Loch
Litie, 1957 **K**; by the Minister's Loch at Nairn, 1963 and in the
bog below Maviston in the Culbin Forest, 1968 **E**; moor above
Clunas Dam, 1973 **E**.
I Badenoch, 1838 *A. Rutherford* **E**; Loch an Eilein (as subsp.
lepidocarpa) 1887 *G. C. Druce* **OXF**; Kincraig, 1891 *AS* **E**; Carn
Eige, Glen nam Fiadh, 1947 *S. M. Walters*; bog on Sgùrr na Lapaich
in Glen Strathfarrar, 1971 **E**; by Loch Duntelchaig, 1972 **E**; bogs
by Blackfold lochs, in Coire nan Laogh, Glen Banchor and in a
flush at Gaick, 1973 **E** etc.

C. lepidocarpa subsp. **lepidocarpa×serotina = C.×schatzii**
Kneuck.
N Locally plentiful on the muddy south shore of Lochloy, with
both parents, 1973 **E,BM**.

C. demissa Hornem.   *Common Yellow Sedge*
Abundant in damp and dry places on moors and dunes. Fig 47b.
M Frequent (as *C. flava*) (Gordon); Forres, 1860 *JK* **FRS**;
Aviemore, 1891 *AS* **E**; dune slack W of the Buckie Loch in the
Culbin Forest and on moors throughout the county, 1954.
N (Stables, *Cat.*); common at Ardclach, 1888 *R. Thomson* **NRN**;
moors at Drynachan and throughout the county, 1954.
I Inverness, 1832 Dalwhinnie and Pitmain, 1833 (Watson, *Loc.
Cat.*); Asylum grounds Inverness (Aitken); Loch Morlich, 1887
Loch an Eilein and Glenmore, *G. C. Druce* **OXF**; at 800 m on
Sgùrr na Lapaich, 1971 **E**, and throughout the district.

C. demissa×hostiana
M Near Aviemore, 1898 *ESM* **OXF**; bog at Easter Limekilns on
Dava Moor, 1973 **E**.
I Loch Alvie, 1898 *WASh* **BM**; at the SE corner of Loch
Duntelchaig, 1971 **BM**; bog below Insh village, 1975 **E**.

C. scandinavica E. W. Davies
Native. Frequent on dune slacks and muddy margins of lochans.
Salt-marsh W of the Buckie Loch, Culbin, 1953 **CGE**; gravel path
by Loch Romach, 1956.
N Margin of Cran Loch, 1950 and on sandy tracks near the sea
at Kingsteps near Nairn, 1954.
I Dune slacks on the raised shingle beach at Ardersier, 1956.

C. serotina Mérat   *Small-fruited Yellow Sedge*
    C. oederi auct., non Retz.
Native. Occasional on muddy loch margins and in dune slacks.
Fig 47c.
M Buckie Loch, Culbin and at Garmouth (Marshall, 1899); by
Loch Ban at Boat of Garten, 1922 *C. E. Salmon* **BM**; round a small
pool W of Binsness, Culbin, 1935 *A. Stewart Sandeman* **BM**; Buckie
Loch, Culbin, 1953 **K**; shores of Loch Vaa N of Aviemore, 1973 **E**.
N Sands near Nairn, 1874 *A. Ley* **BM**; wet sandy ground near the
coast 1 mile E of Nairn (as *C. oederi* Retz. var. *cyperoides*), 1898 *ESM*
BM,CGE; sandy shore of Cran Loch, 1904 *G. T. West* **STA**; shores
of Loch Flemington, 1960; bog below Maviston in the Culbin
Forest, 1968 **E,BM**; sandy shores of Lochloy, 1973 **E,BM**.
I Loch Garten, 1887 *G. C. Druce* **OXF**; Loch Alvie (Marshall,
1899); stony margin of Loch na Ba Ruaidhe above Milton, 1947;
stony margin of Loch Flemington, 1954; Loch Garten, 1973 **E**;
shores of Loch Ashie, by a small loch at Upper Gartally and flush
on the moor by the Big Burn at Essich; margin of Loch Ceo Glais,
1976 **E,BM**.

C. extensa Gooden.   *Long-bracted Sedge*
Native. Frequent in short turf on salt-marsh margins.
M Forming a close turf at Kinloss, 1844 *T. Edmondston* (Gordon,
annot.); Findhorn and Culbin Sands (as var. *pumila*) 1896 *G. C. Druce*
OXF; salt-marsh W of Buckie Loch, Culbin, 1953 *A. Melderis* **BM**
and 1968 *MMcCW* **E**; salt-marsh at Kingston, 1953.
N Salt-marsh E of Nairn, 1898 *ESM* **E,BM,CGE**; and 1960
MMcCW **ABD,K**.
I (Druce, *Com. Fl.*); salt-marsh Castle Stuart, 1954; Inverness,
1955 *J. M. Whyte*.

C. sylvatica Huds.   *Wood Sedge*
Native. Frequent in woods, shady places by rivers and occasionally
on mountain ledges. Map 291.
M Rare on the banks of the River Findhorn, 1837 (Gordon);
banks of the Divie, 1860 *JK* **FRS**; Greshop Wood near Forres,
1930; in woods at Newton House near Elgin, at Rothes and at
Fochabers and by the River Findhorn at Relugas, 1954; by the
River Findhorn opposite Coulmony House, 1972 **E**.
N Cawdor woods, 1833 *WAS* **GL**; Ardclach (Thomson); woods
at Holme Rose, Cawdor Castle and Coulmony House, 1954; wood
near the house at Geddes, 1961 **E,ABD,CGE**; banks of the River

Findhorn at Ardclach, 1972; under beech trees at Kilravock Castle, 1974 **E**.

I In shade by the River Enrick at Milton, 1947; in a small wood at the outlet to the River Tarff at Fort Augustus and in Reelig Glen, 1954; policies of Achnagairn House, Kirkhill, 1967 **BM,K,CGE**; among boulders below the cliffs of Creag Dhubh near Laggan, 1972 **E**; cliff ledge at Loch Killin, 1972 **E**; meadow at Kirkhill, 1972 **E**; Aultnaskiach Dell, Inverness, 1975 *M. Barron*.

C. capillaris L.   *Hair Sedge*
Native. Rare on base-rich mountain slopes. Pl 22.
I Coire Chùirn near Drumochter (Marshall, *J. Bot.* **51**: 164 (1913)); basic flush in Coire Coulavie of Mam Sodhail in Glen Affric at 915 m, 1961 *R. W. M. Corner*; Creag na Caillich in Glen Feshie, 1972 *R. McBeath*!; locally plentiful by the Allt Lorgaidh in upper Glen Feshie, 1974.

C. pseudocyperus L.   *Cyperus Sedge*
M Pond in front of Birdsyards (Sanquhar) House at Forres, *W. Brand* (Graham, *Trans. BSE* **1**: 26 (1840)) collected there in 1841 *WAS* **GL**; in 1857 *JK* **FRS**; in 1868 *J. Roy* **ABD**; in 1953 *MMcCW* **E** and 1961 **E,K,CGE**.

C. rostrata Stokes   *Bottle Sedge*
Native. Abundant on loch margins in pools and moorland bogs and ditches throughout the area. Leaves glaucous-green.
M Very common (Gordon); pond near Forres, 1841 *WAS* **E**; Greshop, 1857 *JK* **FRS**; Loch Spynie (Druce, 1897); the Leen at Garmouth, 1953 *A. Melderis* **BM**; Buckie Loch in the Culbin Forest, 1954 and throughout the county.
N (Stables, *Cat.*); Levrattich mill dam at Ardclach, 1888 *R. Thomson* **NRN**; Cran Loch, 1904 *G. T. West* **E**; Nairn (as var. *brunnescens* Anders.), 1911 *P. Ewing* **GL**; in deep water at Geddes Reservoir, 1960 **ABD**, and throughout the county.
I Dalwhinnie, 1833 (Watson, *Loc. Cat.*); Strathaffric, 1850 *J. Ball* **E**; by the River Spey near Kingussie (as var. *brunnescens*), 1898 *ESM* **E,OXF**; Inverness, 1871 ? *coll.* **ELN**; Loch Flemington, 1940 *AJ* **ABD**; Coire an Lochain of Cairn Gorm, 1953 *A. Melderis* **BM**; abundant throughout the district, 1954.

C. rostrata×vesicaria = **C.×involuta** (Bab.) Syme
M (Watson, *Top. Bot.*); with both parents by the old curling pond at the south side of Grantown, 1963 **BM,CGE** and 1971 **E**.
I (Trail, *ASNH* **15**: 233 (1906)); marshy mouth of the River Coiltie, Drumnadrochit, with both parents, 1947 *A. J. Wilmott et al.*; lochan near Inshriach Nursery, 1975 *OMS & MMcCW* **E**.

C. vesicaria L.   *Bladder Sedge*
Native. Frequent in bogs and river banks preferring humus-rich soil. Leaves yellowish-green. Often with *C. rostrata*. Map 292.
M Aviemore, 1870 *JK* **FRS**; Boat of Garten, 1888 ? *coll.* **GL**; margin of Blairs Loch 1950; by the old curling pond at Grantown,

on Dava Moor, by the River Spey at Blacksboat and Rothes, 1954; pool in the field by Newton of Dalvey Farm, 1956 **ABD**; bog behind the school at Boat of Garten, ditch on the island in the River Spey at Aviemore and in backwaters at Broomhill station.
N Near Auldearn, 1833 *WAS* **GL**; dam at Millhill, 1838 *WAS* **E,OXF**; margin of Loch Flemington, 1954.
I Near Ruthven Barracks, 1837 *G. Gordon* **E**!; Upper Islands at Inverness, 1871 ? *coll*. **ELN**; common in the Spey marshes from Kingussie to Kincraig, 1898 *ESM* **E**; bogs around Loch Meiklie, Loch Flemington, by Loch Ness at Lewiston and in most of the bogs in Strathglass, 1954; in Glen Strathfarrar, by the River Spey at Newtonmore and at Fort Augustus etc.; banks of the River Beauly at Kilmorack, 1972 **E**; by a pool E of Dundreggan in Glen Moriston, 1971 **E**.

C. saxatilis L. *Russet Sedge*
Native. Occasional in base-rich mountain flushes over 800 m. Fig 47d.
I Dalwhinnie, 1833 (Watson, *Loc. Cat.*); eastwards of Dalwhinnie, 1841 herb. *Borrer* **K**; Ben Alder, 1875 *J. T. Carrington* **BM**; Glen Einich, 1887 *G. E. Druce* **OXF**; bog above Loch Einich, 1899 *J. Groves* **BM**; south end of Glen Einich, 1902 *J. W. H. Trail* **FRS**; SW side of Coire Coulavie in Glen Affric, 1947 *E. F. Warburg & S. M. Walters*; SE slopes of Càrn nan Gobhar, Cannich, 1958 *D. A. Ratcliffe*; flush at 820 m on Sgùrr na Lapaich in Glen Strathfarrar, 1971 **E,CGE**.

C. riparia Curt. *Greater Pond-sedge*
Introduced in north Scotland.
M (Druce, *Com. Fl.*); Speymouth, *G. Birnie* (Burgess); planted in the ornamental pond at Kellas House, 1966 **E,BM,CGE**.
I (Druce, *Com. Fl.*). No voucher specimen.

C. acutiformis Ehrh. *Lesser Pond-sedge*
C. paludosa Gooden.
Native. Bogs on base-rich soil.
M Rare near Calcots, *P. Cruickshank* (Gordon); Speymouth, *G. Birnie* (Burgess); a large patch in a bog by the railway line SE of Calcots station, 1967 **E,ABD,BM,CGE**.
I Compensation pond, Raigmore, 1855 *HC* **INV**.

C. pendula Huds. *Pendulous Sedge*
Introduced in north Scotland.
M On a sandstone rock near the Heronry at Darnaway, 1834 *J. G. Innes* **E**; banks of the Findhorn at Logie, *J. G. Innes* (Gordon); Greshop Wood, 1877 *JK* **FRS**; Dalvey, near Forres, 1875 ? *coll*. **ELN**; woods by Kincorth House, 1950 and banks of the River Findhorn below Sluie.

C. pallescens L. *Pale Sedge*
Native. Common in damp grassy places by burns, on banks and in ravines, usually in shade. Fig 47e, Map 293.

M Frequent (Gordon); birch wood bank by Grantown station, 1950; wood at Kincorth House and in a ravine on the Hills of Cromdale, 1954; in a damp field at Auchinroath near Rothes, by a burn at Slochd and by the River Spey at Aviemore and Boat of Garten.

N Bank at Drynachan and by the River Findhorn at Dulsie Bridge and Coulmony, 1954; wood at Lethen, 1961 **ABD,K**; by the Leonach Burn and the River Findhorn at Ardclach.

I Inverness, 1832, Dalwhinnie, 1833 (Watson, *Loc. Cat.*); Kingussie, 1840 *A. Rutherford* **E**; Strathaffric, 1850 *J. Ball* **E**; Beauly, 1875 *? coll*. **ELN**; Kincraig, 1891 *AS* **E**; throughout the district, 1954; in shade of birch trees below the cliffs of Creag Dhubh near Laggan, 1972 **E**; wood by the Moniack Burn at Kirkhill, 1972 **E**.

C. panicea L.   *Carnation Grass*
Native. Abundant in wet grassy places and moorland bogs. Fig 47g.
M Gathered in Rape Park at Brodie House, 1804 *Brodie* **E**; very common (Gordon); Califer Hill near Forres, 1957 *JK* **FRS**; the Leen at Garmouth, 1953 *A. Melderis* **BM**; in the Buckie Loch Culbin Forest and throughout the county, 1954.

N (Stables, *Cat.*); frequent in wet pastures at Ardclach, 1889 *R. Thomson* **NRN**; Nairn (as forma *gracilis* Lange), 1911 *P. Ewing* **GL**; moorland bog at Drynachan and throughout the county, 1954; bog by the River Findhorn at Dulsie Bridge, 1960 **ABD**.

I Inverness, 1832, Dalwhinnie and Pitmain, 1833 (Watson, *Loc. Cat.*); Rarehig Bog, 1855 *HC* **INV**; Culloden Moor, 1871 *? coll*. **ELN**; Kingussie, 1887 *G. C. Druce* **OXF**; marshes between Kingussie and Kincraig (as var. *tumidula* Laest.) (Marshall, 1899); Sgòran, Glen Einich, 1891 *AS* **E**; throughout the district, 1954.

C. vaginata Tausch   *Sheathed Sedge*
Native. Occasional in damp grassy places by burns and on rock ledges in the higher mountains. Pl 21.
M (Druce, *Com. Fl.*).
I Mountains of Cairn Gorm, 1802 *Brodie & G. Don* **E**; rocks on the south side of Loch Ericht, 1867 *P. N. Fraser & A. Craig Christie* (Balfour, 1868); Glen Einich, 1887 and Braeriach, 1888 *G. C. Druce* **OXF**; Coire Chùirn and Loch Dubh in Glen Banchor (Marshall, 1913); SW of Coire Coulavie in Glen Affric, 1947 *E. F. Warburg*; flush near the summit of Ben Alder, 1956, by a small burn on Sgùrr na Lapaich in Glen Strathfarrar, in the Coire nan Laogh in Glen Banchor, cliff ledges on Sgùrr na Muice and in grassy flushes on Bynack More etc.

C. limosa L.   *Bog-sedge*
Native. In wet peaty places on loch margins and moorland bogs. Rare in the lowlands more frequent in the highlands. Fig 47f. Map 294.
M In sphagnum on the margin of the small lochan at Lochnellan on Dava Moor, 1953 **CGE** and 1968 **E**.
N Bog near the road junction near Loch Belivat, 1973 **E**.

Fig 47   a, **Carex lepidocarpa**; b, **C. demissa**; c, **C. serotina**; d, **C. saxatilis**;
e, **C. pallescens**; f, **C. limosa**; g, **C. panicea**; h, **C. flacca**; i, **C. pilulifera**.

Fig 48 a, **Carex remota**; b, **C. curta**; c, **C. lachenalii**; d, **C. ovalis**; e, **C. echinata**; f, **C. rupestris**; g, **C. pauciflora**; h, **C. dioica**; i, **C. pulicaris**.

I Small loch on Beinn a' Bha'ach Ard, *J. Farquharson & T. Fraser* (*Trans BSE* **9**: 474 (1868)); Loch Mallachie in Strathspey (undated) *JK* **FRS**!; Auchindoun (? Cawdor) (Aitken); Loch Mallachie, 1887 *G. C. Druce* **BM,OXF**!; swamp round Loch nan Geadas (Pealling)!; bog S of Loch Affric, 1947 *A. J. Wilmott* **BM**!; bog at Drumnadrochit, and at Loch na Ba Ruaidhe above Milton, 1947; pool verge in sphagnum in the Affric Forest, Glen Affric, 1971 **E**; abundant in a dried-up pool on the moor near Loch Glanaidh, above Drumnadrochit, 1976, with *Tofieldia* and *Littorella*.

C. magellanica Lam.   *Tall Bog-sedge*
    C. paupercula Michx.
Native. In similar situations to the preceding species and often growing with it and from which it differs in having smooth stems, the lower bracts exceeding the inflorescence and fewer flowers (up to 10) on the female spikes.
I Bog S of Loch Affric, 1947 *M. S. Campbell* **BM**, and *E. F. Warburg* **OXF**!; by several pools in the Affric Forest near Affric Lodge, 1971 **E**.

C. rariflora (Wahlenb.) Sm.   *Mountain Bog-sedge*
Native. Locally plentiful in wet grassy places at burn sources on mountain tops. Pl 21.
I West of Loch Einich, 1842 *J. Fraser* **GL**; Braeriach, 1877 (Boyd); bogs near the head of the Allt Coire Chùirn and the Allt a' Choire Chàis near Dalwhinnie, 1911 *ESM & WASh* **E,BM**; Càrn Bàn Mòr above Loch Einich, 1956 *E. C. Wallace*, **ABD,GL, ECW**; Mòine Mhòr, 1961 *D. A. Ratcliffe*; grassy bog at the source of a small burn on Coire Liath at Gaick, 1973; Càrn Dearg in Glen Banchor, 1973 *R. McBeath*.

C. flacca Schreb.   *Glaucous Sedge*
Native. Common in flushes, on grassy river banks, on cliff ledges on mountains and by the sea on base-rich soil; also in short turf on the margins of salt-marshes. Fig 47h.
M Manachy near Forres, 1871 *JK* **FRS**; wet sea cliffs between Lossiemouth and Hopeman, 1898 *ESM* **CGE**; Buckie Loch in the Culbin Forest, 1953 *A. Melderis* **BM**; throughout the county, 1954; margin of the salt-marsh between the Old Bar and the Culbin Forest, 1969 **E**; flush in a field at Muchrach, 1972 **E**.
N Cawdor woods, 1837 *WAS* **E**; Ardclach (Thomson); Cran Loch (as var. *stictocarpa*) 1904 *G. T. West* **STA**; flush on the moor at Drynachan and throughout the county, 1954; grassy banks of the River Findhorn at Dulsie Bridge, 1960 **ABD**.
I Rothiemurchus Forest, 1892 *AS* **E**; marshy places in Glen Doe (as var. *metaliana*) 1904 *G. T. West* **STA**; Cairn Gorm, 1935 *J. Walton* **GL**; throughout the district, 1954; cliff-face ledges on Sgùrr na Lapaich in Glen Strathfarrar and on Tom Bailgeann etc.

C. hirta L.   *Hairy Sedge*
Native. Very rare in damp grassy woods and on river banks.

M Spynie, 1837 *G. Gordon* **E**; rare near Viewfield, 1837 *Mr Grey* (Gordon); parishes of Dyke, Elgin and Knockando (Burgess); grass verge to the field between Dyke village and the Muckle Burn, 1963 **CGE**; boggy ground at the top of the river gorge at Logie House, 1966.
N West of Mills of Nairn, 1833 *WAS* **GL**; Nairn, 1857 *JK* **FRS**.
I In the meadow and the woods by the Moniack Burn at Achnagairn House, Kirkhill, 1962 **E,ABD,BM**.

C. lasiocarpa Ehrh.   *Slender Sedge*
    C. filiformis auct.
Native. Frequent in shallow water on the margins of lochs and in moorland pools. Map. 295.
M Bog near Aviemore, *W. Borrer* (Hooker, *Fl. Scot.*); Aviemore (Gordon); margin of lochan on Dava Moor, 1930 and 1963 **CGE**; pool at Boat of Garten, ditch at Lochindorb and near Relugas, 1954; margin of Loch Mòr at Duthil, 1973.
N Achagour, W of Ardclach, 1888 *R. Thomson* **NRN**.
I Beside the water-lily lochan behind the Manse, Rothiemurchus, 1883 *JK* **FRS**!; Loch Alvie, 1891 *AS* **E**; Strath Mashie near Laggan, 1916 *ESM* **E**; Balmacaan Forest, 1904 *G. T. West* **STA**; N side of Loch Pollan Buidhe S of Loch Affric, 1947 *E. F. Warburg*; bog by Loch Carrie in Glen Cannich, 1947; margin of Loch na Ba Ruaidhe at Milton, at Fort Augustus, by Loch Laide, Loch Meiklie and Loch Insh at Kincraig, 1956; peaty pool in the Affric Forest in Glen Affric, 1971 **E**; edge of the loch in *Myrica* scrub at Lochan na Curra, 1972 *M. Barron*; by Loch an Eilein, Loch Geadas and in a ditch by the A9 at Drumochter etc.

C. pilulifera L.   *Pill-headed Sedge*
Native. Abundant on dry acid moorland hummocks and in short grassy places in birch woods. Fig 47i.
M Frequent (Gordon); Tarras moor, 1860 *JK* **FRS**; Aviemore, 1893 *AS* **E**; throughout the county, 1954; track verge in the Culbin Forest, 1956; grassy verge of the drive at Brodie Castle, 1962 **ABD,K**.
N (Stables, *Cat.*); heaths and wet places near the schoolhouse Ardclach, 1889 *R. Thomson* **NRN**; moor at Drynachan and throughout the county, 1954; birch wood near Loch na Litie E of Auldearn, 1968 **CGE**.
I Dalwhinnie and Pitmain, 1833 (Watson, *Loc. Cat.*); Ben Alder, 1867 *A. Craig Christie* **E**; Strathaffric, 1889 *T. A. Cartwright* **OXF**; Coire an Lochain of Cairn Gorm, 1953 *A. Melderis* **BM**; throughout the district, 1954.

C. caryophyllea Latourr.   *Spring Sedge*
    C. praecox auct.
Native. Frequent in short base-rich grassland. Map 296.
M Bank in Dyke Moss, *Brodie* **E**; banks of the River Findhorn at Logie, *J. G. Innes* (Gordon); birch wood by Grantown station, 1930

and 1963 **CGE**; grass slope on the moors in the Hills of Cromdale, 1957 **ABD**; grass mound by the old lime kiln at Fae and on the banks of the River Spey at Grantown, 1972 **E**; common on Boat of Garten golf course and in short grass by Loch Mòr at Duthil and at Slochd near Carrbridge.

N Lawn at Cawdor Castle, 1842 *WAS* **E**!; grass bank at Drynachan, at Dulsie Bridge and in Cawdor kirkyard, 1954.

I Common on heaths (Aitken); Kincraig, 1891 *AS* **E**; Rothiemurchus, 1892 *AS* **E**; grass bank at Drumochter, 1956; at Abriachan and Drumnadrochit, 1957; in short grass on burn overspills in Glen Feshie and at Gaick, in Glen Mazeran, near Laggan, on the Carse of Ardersier, grass slopes below Ruthven Barracks and by the River Spey at Kingussie, on a small base-rich hummock in the moors at Alvie, etc.; grassy moor above Netherwood at Newtonmore, 1972 **E**.

C. buxbaumii Wahlenb.   *Club Sedge*
Native. Margin of a loch near Milton, 1954 *D. S. Kettle* **BM**; and 1956 *MMcCW* **E**. Pl 19.

C. atrata L.   *Black Alpine-Sedge*
Native. Occasional on rock ledges of the higher mountains. Pl 22.
I Braeriach, 1877 (Boyd); Larig Pass, 1886 *H. &. J. Groves* **BM**; Glen Einich, 1887 *G. C. Druce* **OXF**; Glen Cannich, 1887 *G. C. Druce* **BM**!; Coire Chùirn at Drumochter, at 800 m, 1911 *ESM* **E,BM**!; Glen Feshie, 1950 *C. D. Pigott* **CGE**!; Allt Bhrodainn at Gaick, 1949 *N. Y. Sandwith*!; cliffs on Sgùrr na Lapaich in Glen Strathfarrar, 1971.

C. elata All.   *Tufted-sedge*
Native. Very rare, perhaps misidentified.
M Brodie House (as *C. stricta*) *Brodie* herb. (Gordon). No voucher specimen.
I Inchnacardoch Bay, Loch Ness, 1904 *G. T. West* **STA**.

C. aquatilis Wahlenb.   *Water Sedge*
Native. Abundant on river banks in Strathspey, rare elsewhere.
M By the River Spey at Kinchurdy, 1887 *G. C. Druce* **OXF**; pool near the mouth of the Spey at Garmouth, 1898 *ESM* **BM**; banks of the River Spey at Cromdale, 1955 **K**; rare in a ditch in an alder-birch wood W of the Buckie Loch in the Culbin Forest, 1961 **K**; Grantown and Aviemore, 1972 **E**; bog in Monahondie Moss near Archiestown, 1975 *R. Richter* **E**.
N Findhorn basin on the south side of Wade's Road, 1900 (Lobban); the Minister's Loch, Nairn 1954.
I Banks of the Spey near Manse of Inver, 1860 *JK* **FRS**; by the Spey at Kincraig (as var. *elatior* Bab. = var. *watsonii* Syme fide *ESM*) 1891 *AS* **E,GL**; below Kingussie (as var. *typica* Kuk.) 1898 *ESM* **E,BM**; Inchnacardoch Bay (with very hard trunks), Loch Ness, NW of Loch Uanagan, by Loch Ashie, Loch Ruthven and Loch Tarff, 1904 *G. T. West*; Loch Meiklie, 1904 *G. T. West* **STA**;

Loch a' Choire and Loch nan Geadas (Pealling); banks of the
River Findhorn at Tomatin and by Loch Morlich, 1953; W end
of Loch Affric, 1962 *UKD et al.*

C. aquatilis × nigra
I   By the Allt a' Choire Chàis near Dalwhinnie, 1960 *R. W. David*
BM.

C. aquatilis × recta = C. × grantii A. Bennett
M Greshop Wood near Forres, *Mr Grant* (Burgess) 'This plant was
discovered in sands at Wick by Mr Grant, late editor of the *Elgin
Courant* and given to Dr Keith who planted it in Greshop Wood,
Forres where it is still flourishing. Mr Arthur Bennett reported the
plant as *C. aquatilis × salina = C. × grantii'*. Not seen in recent years
and probably now gone with the drying up of the pools that were
formerly to be found in the wood.

C. recta Boott   *Estuarine Sedge*
    C. kattegatensis Fr. ex Krecz.
Native. River banks in tidal water.
I   Shores of the Beauly Firth, *G. C. Druce* (A. Bennett, *Trans. BSE*
18: 254 (1891)); Beauly river, 1892 *ESM* **GL!**.

C. nigra (L.) Reichard   *Common Sedge*
Native. Abundant in bogs, by rivers and in grassy places on acid
and base-rich soil. Very variable.
M Very common (Gordon); near Old Duffus Castle, 1833 *WAS*
GL; Greshop, 1871 *JK* **FRS**; Buckie Loch in the Culbin Forest,
1953 *A. Melderis* **BM**; banks of the River Findhorn, 1956 **CGE**;
bog in the Hills of Cromdale, 1956 **K**; throughout the county 1954.
N (Stables, *Cat.*); moorland marshes at Ardclach, 1888 *R. Thomson*
NRN; moor at Drynachan and throughout the county, 1954; bog
by Loch of the Clans, 1957 **K**.
I   Dalwhinnie and Pitmain, 1833 (Watson, *Loc. Cat.*); Glen
Urquhart and Strathaffric, 1850 *J. Ball* **E**; Culloden Moor, 1871
? *coll.* **ELN**; Kingussie, 1887 *G. C. Druce* **OXF**; shores of Loch
Pityoulish, 1892 *AS* **E**; common round Inverness (Aitken);
Glenmore, 1953 *A. Melderis* **BM**; throughout the district, 1954; edge
of a small moorland pool E of Dundreggan in Glen Moriston, 1971
E,BM.

C. bigelowii Torr. ex Schwein   *Stiff Sedge*
    C. rigida Good.
Native. In bare stony places on mountains usually over 600 m.
M Summit of Càrn Eachie in the Hill of Cromdale, 1950; Càrn
Dearg Mòr above Aviemore 1974 **E**.
I   Dalwhinnie, 1833 (Watson, *Loc. Cat.*); mountains near
Dalwhinnie, 1841 herb. *H. C. Watson* **K**; Strathaffric, 1850 *J. Ball* **E**;
Ben Alder, 1867 *A. Craig Christie* **E,ABD** Sgòran in Glen Einich,
1891 *AS* **E**; Coire an Lochain of Cairn Gorm, 1953 *A. Melderis* **BM**;
throughout the district on all the higher mountains.

C. bigelowii × nigra = C. × decolorans Wimm.
I Bog by the Allt Coire Chùirn at 823 m, *ESM & WASh* (*J. Bot.*
51: 164 (1913)).

C. paniculata L.   *Greater Tussock-sedge*
Native. Frequent in base-rich bogs in the lowlands, rare in acid
bogs in the Highlands. Map 297.
M Frequent (Gordon); Birdsyards (Sanquhar) House, Forres, 1839
W. Brand (*Trans. BSE* **1**: 26 (1840)) and in 1866 *JK* **FRS** and in
1961 *MMcCW* **E,BM,K**; salt-marsh verge W of the Buckie Loch in
the Culbin Forest, 1953 **CGE**; plentiful in the marshes at the Leen,
Garmouth, 1953 *A. Melderis* **BM**; a few plants at Lochindorb,
pond at Kincorth and in Brodie Pond, 1954; bog opposite Croft
of Ryeriggs, 1972 *A. J. Souter.*
N Rare by the burn at the village of Ferness, 1891 *R. Thomson*
NRN; S side of Lochloy, 1953; plentiful at Loch Litie, 1963
ABD,CGE; by the lochan at Househill near Nairn, 1967.
I Rothiemurchus, 1892 *AS* **E**; on the E shore of Loch Morlich,
1964 **CGE**; bog to the N of Loch Conagleann at Dunmaglass, 1972;
a few tussocks by Loch Mallachie and in a ditch on the edge of the
moor above Inverfarigaig on the Foyers road.

C. paniculata × remota = C. × boenninghausiana Weihe
M With both parents by the ditch at the edge of the alder wood
W of Buckie Loch in the Culbin Forest, 1961 **E,ABD,BM**.

C. diandra Schrank   *Lesser Tussock-sedge*
    C. teretiuscula Good.
Native. Rare in peaty meadows and bogs.
M Abundant in one or two places, W of Aldroughty Farm, 1830,
Springfield, 1837 *P. Cruickshank*, and Rape Park at Brodie House,
herb. *Brodie* (Gordon); Urquhart, *G. Birnie* (Burgess); pond in
front of Birdsyards (Sanquhar) House, *W. Brand* (*Trans. BSE* **1**: 26
(1840)); W end of the Buckie Loch in the Culbin Forest, 1953
CGE; and in 1968 **E,K**; bog at Lochinvar near Elgin, 1957 **K**;
rare in the Winter lochs near Binsness and by the small lochan at
Lochnellan on Dava Moor, 1954.
N (Stables, *Cat.*); edge of the Minister's Loch on the Nairn Dunbar
golf course, 1950 **CGE** and in 1960 **ABD**; Cran Loch, 1960 **ABD**,
and on both shores of Lochloy.
I (Druce, *Com. Fl.*) (probably Stables' record for Nairn); bog
surrounding Loch na Ba Ruaidhe above Milton, 1947; N side of
two small lochans at Essich, 1972 *M. Barron*!; bog in a rough
moorland pasture opposite Coulan Wood near Balnafoich, 1975
E,BM.

C. otrubae Podp.   *False Fox-sedge*
Native. Rare in marshes near the sea.
M At the Old Castle of Duffus, *G. Gordon* **E**; abundant in one
or two places (Gordon); Drainie, 1865 *JK* **FRS**; roadside at

Lossiemouth, 1875 ? *coll.* **ELN**; a few plants by the ditch in the alder wood W of Buckie Loch in the Culbin Forest, 1953 **CGE**.
I Bog on the edge of the lagoons at Clachnaharry, Inverness, 1954 **E**.

C. otrubae× remota = C.× pseudoaxillaris K. Richt.

M Gordon Castle, Banffshire, 1842 *WAS* **GL** (this could be in v/c 94 Banff, but without exact locality given it is difficult to judge without refinding the plant. The v/c boundary runs just east of the loch at Gordon Castle).

C. disticha Huds.   *Brown Sedge*
Native. Very rare, marshes and river banks.
M Banks of the River Spey at Grantown, 1953 **E**.

C. arenaria L.   *Sand Sedge*
Native. Abundant on sand dunes and occasionally inland where it is probably introduced.
M Abundant on the sea shore and near the west gate of Lesmurdie Cottage, Elgin (Gordon); Findhorn, 1857 *JK* **FRS**; Culbin Sands, 1953 *A. Melderis* **BM** and *MMcCW* **CGE**; by the track in the Darnaway woods near Berryley Farm and on dunes from Speybay to Nairn.
N (Stables, *Cat.*); on the shore at Nairn, 1889 *R. Thomson* **NRN**; by Lochloy, 1904 *G. T. West* **STA**; dunes by Cran Loch, 1960 **ABD**.
I (Watson, *Top. Bot.*); shingle beach at Ardersier and on the dunes at Fort George, 1930; Carse of Delnies to Whiteness Head and on Longman Point at Inverness, 1954.

C. maritima Gunn.   *Curved Sedge*
       C. incurva Lightf.
Native. Rare in sandy mud at the outlets of burns and estuaries by the sea. Not seen in recent years.
M Elginshire (about 15 feet long) 1802 *G. Gordon* **E**; in great quantity around the Mill of Outlet between Findhorn and Burghead, 1809 *W. Borrer, W. J. Hooker & Brodie* **E**; also found by Brodie 'among the peat holes in the small sand hills on this side going from Brodie House to the Old Bar, near the shore where the ground had been overflowed in winter'; with *Juncus balticus* on a grassy flat near the mouth of the River Spey (as var. *erecta* Lange), *G. Gordon* **K**; sands of Rosevalley, *WAS* herb. *H. C. Watson* **BM**. There are many specimens of this sedge collected from the same localities at later dates in the herbaria at **GL,BM** and **OXF**.
N Coast E of Nairn, 1833 *WAS* **GL,BM** and *W. Brand* **ABD**; sandy coast 1 mile E of Nairn, 1898 *ESM* **E,BM**.

C. spicata Huds. and **C. muricata** L. both recorded from Loch Spynie by Gordon, (Watson, *Top. Bot.*) and Burgess, must be regarded as doubtful in the absence of voucher specimens.

C. echinata Murr.   *Star Sedge*
Native. Abundant on boggy moorland. Fig 48e.

M Frequent (Gordon); Loch of Blairs near Forres, 1871 *JK* **FRS**; Altyre, 1877 *G. C. Druce* **OXF**; St John's Mead Darnaway, 1953 *A. Melderis* **BM**; moorland bog on the Culbin Forest, 1954 and throughout the county, 1954.
N (Stables, *Cat.*); common in boggy meadows at Ardclach, 1889 *R. Thomson* **NRN**; Nairn, 1911 *P. Ewing* **GL**; moors at Drynachan and throughout the county, 1954.
I Dalwhinnie and Pitmain, 1833 (Watson, *Loc. Cat.*); Cairn Gorm, 1863 *J. P. Bisset* **BM**; Asylum grounds Inverness (Aitken); near Fort Augustus, 1904 *G. T. West* **STA**; throughout the district, 1954; Loch an Eilein, 1955 *J. Milne* **E**.

C. remota L. Remote Sedge
Native. Frequent in damp shady places. Fig 48a, Map 298.
M Culreach near Gordon Castle, 1842 *WAS* **E**; frequent (Gordon) Greshop, 1871 *JK* **FRS**; Fochabers, 1894 ? *coll.* **ABD**; among alders in the Culbin Forest, at Dunphail, Conicavel and Garmouth, 1954; in a flush by the River Findhorn at Relugas and in the woods near the river in Darnaway Forest; in shade by the pond at Sanquhar House, Forres, 1961 **K**.
N Polneach, Nairnshire, 1833 *WAS* **GL**; by the Geddes Reservoir, 1954; in the gorge of the River Findhorn at Deeside Bridge, 1974 **E**.
I Ness Islands, Inverness, 1855 *HC* **INV** and 1871 ? *coll.* **ELN**; wood by Loch Ness at Fort Augustus and at Inverfarigaig, 1956; plentiful in the woods at Achnagairn House at Kirkhill, 1972 **E**; by the waterfall at Glendoe, 1972 **E**; on a shingle island in the River Enrick at Milton, by the Bunchrew Burn, by the well at Breakachy school, plentiful in runnels to Loch Ness S of Dores, by the Falls of Divach and at Muirtown lagoons at Inverness; Glassburn in Strathglass, 1973 *M. Barron*.

C. curta Gooden. White Sedge
C. canescens auct.
Native. Common in grassy moorland bogs. Fig 48b, Map 299.
M Frequent (Gordon); Manachy near Forres, 1871 *JK* **FRS**; Loch Spynie and Altyre (Druce, 1896); Buckie Loch, Culbin and Loch Spynie (Marshall, 1899); Winter Lochs in the Culbin Forest and throughout the county, 1954.
N (Stables, *Cat.*); moors at Ardclach, 1887 *R. Thomson* **NRN**; by the Minister's Loch, Nairn, at Achavraat and by Loch of the Clans, 1954; bog at Dulsie Bridge and at Cran Loch, 1960 **ABD**.
I Badenoch, 1838 *A. Rutherford* **E**; Aultnacaber, Glenmore (Druce, 1888); Glen Einich (as var. *alpicola* Wahl.) (Druce, 1888); Loch Alvie and Loch Morlich, 1891 *AS* **E**; by the Allt a' Choire Chàis near Dalwhinnie, 1911 *ESM* **E** and *WASh* **OXF**; Allt Coire Chùirn near Dalwhinnie (as var. *fallax* F. Kuntz) 1911 *ESM* **E,GL**; Strath Mashie (as var. *robustior* Blytt) 1916 *ESM* **E**; throughout the district, 1954; Sgùrr na Lapaich at 823 m, 1972 **E**.

C. ovalis Gooden.   *Oval Sedge*
Native. Common in grassy places on moors and in short turf on roadside verges. Fig 48d, Map 300.
M Frequent (Gordon); Culbin Sands, 1911 *P. Ewing* **GL**; throughout the county, 1954; by Loch nan Carraigean, N of Aviemore, 1955 *J. Milne* **E**; tall plants in a bog to the W of Dava station, 1956 **K**; grass verge of the drive at Brodie Castle, 1962 **BM,K,CGE**.
N (Stables, *Cat.*); near Daltra, Ardclach, 1889 *R. Thomson* **NRM**; track verge on the moors at Drynachan and throughout the county, 1954.
I Dalwhinnie and Pitmain, 1833 (Watson, *Loc. Cat.*); Kingussie, 1838 *A. Rutherford* **E**; Beauly, 1870 ? *coll.* **ELN**; Kincraig, 1891 *AS* **E**; by the Spey between Newtonmore and Kingussie, 1911 *ESM* **E**; Moy (as var. *argyrolochin* Koch) 1919 *M. L. Wedgewood & G. C. Druce* **OXF,MBH**; throughout the district, 1954.

C. lachenalii Schkuhr   *Hare's-foot Sedge*
   C. *lagopina* Wahlenb.
Native. Rare on cliff ledges in the Cairngorm mountains. Fig 48c.
I Coire an t-Sneachda of Cairn Gorm, 1886 *G. C. Druce* **BM,OXF**!, in 1887 **BM** and 1888 **E**; Bynack, Abernethy at 600 m *R. Meinertzhagen* **BM**; ledges on 'The Apron' in the Coire an Lochain of Cairn Gorm, 1955 **K**.

C. rupestris All.   *Rock Sedge*
Native. Rare in crevices of base-rich rocks. Fig 48f.
I Cairngorms 1842 (Gordon, *addenda*); calcareous schist in Coire Garbhlach in Glen Feshie, 1966 (*Trans. BSE* **40**: 337 (1967)); dry rocks in Coire Chùirn, 1970 *R. McBeath*.

C. pauciflora Lightf.   *Few-flowered Sedge*
Native. Common in wet moorland bogs. Fig 48g. Map 301.
M Aviemore, *WAS* and Dyke Moss, *J. B. Brichan* (Gordon); bogs at Grantown, Dava Moor, Kellas and Fae etc., 1954; bog at Dallas, 1954 **K**; by a small lochan near Lochindorb, 1963 **CGE**; boggy flush by the road W of Carrbridge, 1971 **E**.
N Highland Boath. *WAS* (Gordon); sphagnum bog W of Lochindorb, 1963 **BM,CGE**; moor at Achavraat and at Glengeouillie at Cawdor.
I Craiganain, Inverness-shire, 1822 (Gordon); Badenoch, 1838 *A. Rutherford* **E**; marsh at Belladrum near Kiltarlity (Aitken); Glenmore, 1887 *G. C. Druce* **OXF**; Lochan Creag Dhubh, Fort Augustus, 1904 *G. T. West* **STA**; Strath Mashie, 1916 *ESM* **E**; throughout the district, 1954; bog in Glen Mazeran, 1968 **CGE**; bogs in Glen Moriston and by Blackfold lochs, 1973 **E**.

C. pulicaris L.   *Flea Sedge*
Native. Common on base-rich grassland, moorland flushes and mountain cliff ledges. Fig 48i, Map 302.
M At the W end of the great pond at Brodie (in plenty where the tall specimens in fruit were gathered, they also grow in water),

1804 *Brodie* **E**; very common (Gordon); Relugas, 1850 *JK* **FRS**; on grass tussocks in the Buckie Loch, Culbin and throughout the county, 1954; bog at Drakemyres, 1972 **E**.
N (Stables, *Cat.*); wet moors at Ardclach, 1888 *R. Thomson* **NRN**; by the Minister's Loch on the Nairn Dunbar golf course, 1950; at Drynachan etc., 1954; Dulsie Bridge, 1960 **ABD**; bog below Maviston Farm in the Culbin Forest, 1968 **E**.
I Dalwhinnie, 1833 (Watson, *Loc. Cat.*); Drumochter, 1841 *H. C. Watson* **OXF**; Glen Einich, 1887 *G. C. Druce* **OXF**; Asylum grounds Inverness (Aitken); Rothiemurchus, 1892 *AS* **E**; throughout the district, 1954; cliff ledge at 823 m; Sgùrr na Lapaich in Glen Strathfarrar **E**.

C. dioica L. *Dioecious Sedge*
Native. Common in base-rich flushes on moors and mountain slopes. Fig 48h, Map 303.
M Very common (Gordon); Boat of Garten (with rather deflexed fruit) (Druce, 1888); Dyke Moss, Carrbridge, Dava Moor, Hills of Cromdale and Dunphail etc., 1954; bog at Drakemyres, 1972 **E**.
N Coarsach Park, Cawdor, 1832 *WAS* **GL**; Cawdor, 1833 *WAS* **E**; moors at Drynachan, west of Lochindorb, by Cran Loch and at Littlemill, 1954; flush by the Minister's Loch on the Nairn Dunbar golf course (as forma *isogyna* (Angstr. ex Fries) Kük. a male flower with a few female spikelets), 1960 **ABD** and 1963 **CGE**.
I Dalwhinnie, 1833 (Watson, *Loc. Cat.*); Dalwhinnie, 1841 *H. C. Watson* **OXF**; Glen Urquhart, 1850 *J. Ball* **E**; Creag Dhubh N of Invermoriston, 1904 *G. T. West* **STA**; well distributed through the district, 1954; very large-headed female flowers on tussocks by the Cluanie Dam in Glen Moriston, 1971 **E**; bog at Hughtown, 1972 **E**.

GRAMINEAE
Phragmites Adans.
P. australis (Cav.) Trin. ex Steud. *Streeds, Star Reeds,*
  P. communis Trin. *Common Reed*
Native. Locally abundant in swamps, ditches and loch margins, occasionally on dry acid moorland. Map 304.
M Loch Spynie, 1837 *G. Gordon* **ABD**!; abundant in one or two places, at Spynie and Kinloss, *J. G. Innes* (Gordon); near Forres, 1857 *JK* **FRS**; Aviemore, 1891 *AS* **E**; ditches near Duffus Castle, at Lochinvar near Elgin and by pools at Kingston, 1954; Buckie Loch in the Culbin Forest, 1956 **CGE**; margins of Loch Mòr at Duthil and by Loch an t-Sithein S of Lochindorb.
N Moss of Little E of Auldearn, *WAS* (Gordon)!; margins of Loch Flemington and Cran Loch, 1954; by the Minister's Loch on the Nairn Dunbar golf course, 1960 **K**.
I Loch Alvie (Murray)!; lochs Garten and Pityoulish (Druce, 1888); Loch a' Chlachain, Dunlichity (Pealling); Loch Meiklie, 1904 *G. T. West* **STA**; abundant in the Spey marshes N of Kingussie and at Beauly, 1954; ditch N of Gollanfield station, at Knockie,

Loch Bunachton, Fort Augustus, Loch Bran and in Strathglass etc.; dry moorland in Glen Einich, 1973; with variegated leaves at Raigmore Pond near Inverness.

Molinia Schrank

M. caerulea (L.) Moench   *Purple Moor-grass*
Native. Locally abundant on wet moorland throughout the district.
M Abundant at Elgin, Edinkillie and Grantown woods, Aviemore, 1887 *G. C. Druce* **OXF**; Culbin Forest and throughout the county, 1954.
N (Stables, *Cat.*); occasional on dry moors at Ardclach, 1885 *R. Thomson* **NRN**; moors at Dulsie Bridge, Drynachan and Cawdor and throughout the county, 1954.
I Inverness, 1832 Dalwhinnie, 1833 (Watson, *Loc. Cat.*); Cairn Gorm, 1832 *WAS* **GL**; Culloden, 1855 *HC* **INV**; Glen Urquhart and N of Fort Augustus, 1904 *G. T. West* **STA**; Drumashie Moor, 1940 *AJ* **ABD**; throughout the district, 1954; common in wet places, at Inchnacardoch and 1 mile S of Dalwhinnie, 1970 *C. E. Hubbard*.

Sieglingia Bernh.

S. decumbens (L.) Bernh.   *Heath-grass*
    Triodia decumbens (L.) Beauv.
Native. Abundant on dry acid moorland verges and occasional on base-rich grassland. Fig 49a.
M At Urquhart, Roseisle, Rothes and Dunphail (Gordon); moors above Forres, 1871 *JK* **FRS**; in damp grass hollows in the Culbin Forest and throughout the county, 1954; dry moorland bank at Knockando station, 1966 **E,ABD,CGE**.
N Cawdor woods, 1832 *WAS* **ABD**; Ardclach (Thomson); moors at Drynachan, Glenferness, Coulmony and Dulsie Bridge, etc., 1954.
I Inverness, 1832 (Watson, *Loc. Cat.*); Ardersier, 1846 *A. Croall* (Watson); throughout the district, 1954; Inchnacardoch, 1970 *C. E. Hubbard*.

Stipa L.

S. variabilis Hughes
M Introduced with wool 'shoddy' at Greshop House near Forres, 1958 **K**.

Glyceria R. Br.

G. fluitans (L.) R. Br.   *Floating Sweet-grass*
Native. Common on loch margins, in ditches, bogs and slow-flowing burns. Lemmas 6–7·5 mm long, blunt or slightly pointed; anthers 2–3 mm long; panicle narrow after flowering.
M Innes House, Urquhart, 1794 *A. Cooper* **ELN**; frequent (Gordon); Waterford near Forres, 1857 *JK* **FRS**; west bank of the River Findhorn, on shingle, 1953 *A. Melderis* **BM**; throughout the county, 1954; muddy shores of Loch Romach, 1955 **CGE**; ditch at Lochinvar near Elgin, 1957 **K**; by Loch Vaa N of Aviemore, 1961 *P. H. Davis* **E**; bog between Carrbridge and Dulnain Bridge, 1973 **E**.

arex rariflora × 1
Mountain Bog-sedge

M. C. F. Proctor

Carex vaginata × 1 *M. C. F. Proctor*
Sheathed Sedge

PLATE 21

Carex capillaris × 2   *M. C. F. Proctor*
Hair Sedge

Carex atrata × 1   *M. C. F. Pro*
Black Alpine-sedge

PLATE 22

Poa glauca × 1
Glaucous Meadow-grass

M. C. F. Proctor

Poa flexuosa × ⅔
Wavy Meadow-grass

M. C. F. Proctor

PLATE 23

Alopecurus alpinus × 1
Alpine Foxtail

M. C. F. Proctor

Phleum alpinum × 1
Alpine Cat's Tail

M. C. F. Proct

PLATE 24

N (Stables, *Cat.*); common in ditches at Ardclach, 1891 *R. Thomson*
NRN; throughout the county, 1954; margin of Loch of the Clans,
1960 **ABD,K**; pool at Dulsie Bridge, 1960 **K**; Geddes Reservoir, 1960
ABD; bog in a field at Dulsie Bridge, 1961 **ABD**; ditch at Nairn,
1961 **ABD,K**; boggy field N of Boghole Farm near Brodie, 1967
E,ABD,CGE.
I Dalwhinnie and Pitmain, 1833 (Watson, *Loc. Cat.*); lochs
Uanagan and Killin, 1904 *G. T. West* **STA**; by lochans in Glen
Moriston, Strathglass, Strathspey and throughout the district, 1954;
verge of a small lochan by the Mullardoch Dam in Glen Cannich
(as forma *triticea*) 1971 **E.**

G. fluitans×**plicata** = **G.**×**pedicellata** Townsend
N Muddy margin at the west end of Geddes Reservoir, 1961 **K.**

G. plicata Fr.   *Plicate Sweet-grass*
Native. Rare in ditches and on pond margins. Spikelets breaking up
beneath each lemma at maturity; lemmas 3–5·5 mm long; anthers
1 mm.
N Ardclach (Thomson); ditch by the cemetery at Nairn, 1960
ABD,K; margin of Geddes Reservoir, 1960 **ABD** and in 1961 **E,K**;
mouth of a small burn at Nairn, 1961 **ABD,BM.**
I Kilmorack (*forma*), 1892 *ESM* **BM**; farm pond at Essich, 1975 **E**;
in a small burn near Viewhill, Ardersier, 1976 **E,BM,K.**

G. declinata Bréb.   *Small Sweet-grass*
Native. Frequent in ditches, margins of moorland bogs and in wet
cart tracks. Stems usually ascending from a curved or bent base;
leaves greyish-green or tinged purple; lemmas 3-toothed or 3-lobed
at the tip, 4–5 mm long; anthers very small, 0·8–1 mm long. Fig 49b,
Map 305.
M Glen Brown (Marshall, *J. Bot.* 44: 154 (1906)); by the pond at
Sanquhar House, Forres, 1954; bog at Lochinvar near Elgin, 1957
BM, and in 1961 **E,ABD,K**; boggy field at Innes House, Urquhart,
1967 **E,CGE**; bogs at Auchness near Dallas, at Lettoch near
Nethybridge and in a dried out lochan at Boat of Garten, 1972;
ditch by the River Lossie at Kellas, 1973 **E**; cart track in a birch
wood at Muchrach, 1973 **E**; runnel at Lyntellach near Bridge of
Brown, 1973 **ABD**; ditch by the Glenbeg road at Grantown.
N Bogs at Drynachan, Coulmony and Dulsie Bridge, 1954;
muddy margin of Geddes Reservoir, 1960 **E,ABD**; pool by the
River Nairn at Kilravock Castle, 1974 **E.**
I (Trail, *ASNH* **16**: 228 (1907)); near Cannich Hotel, Strathglass,
1947 *T. G. Tutin* **BM!**; backwater of the River Findhorn at Tomatin,
1952; moorland bogs at Farr and at the foot of Sgùrr na Lapaich
in Glen Strathfarrar, 1956; by a small burn in Glen Mazeran, 1968;
margin of a small lochan at Mullardoch Dam, 1971 **E**; runnel above
Alvie House, 1974 **E**; also at Dalwhinnie, Laggan Bridge, Abriachan
and Breakachy in Strathglass, etc.; in the farm pond at Essich
(a luxuriant plant with a longer panicle (35 cm) than had been
seen previously, CEH), 1975 **K.**

G. maxima (Hartm.) Holmberg   *Reed Sweet-grass*
     G. aquatica (L.) Wahlberg., non J. & C. Presl
Native. Very rare in rivers, lochs and ditches in deep water.
Recorded from Cherry Island in Loch Ness (G. T. West, *Proc.
Royal Society of Edinburgh* **25**: 967 (1905)) (but without a voucher
specimen).

Festuca L.

F. pratensis Huds.   *Meadow Fescue*
     F. elatior L. pro parte
Native, but perhaps introduced in this area. Rare in pastures and
on railway embankments. Auricles of the basal leaves hairless.
M Province of Moray, *WAS* (Murray); by the Bridge of Findhorn,
1857 and at Waterford near Forres, 1871 *JK* **FRS**; railway bank
near Forres (Druce, 1897); parishes of Dyke, Forres, Speymouth
and Knockando (not mentioned by Gordon), a recent introduction
sown along with grass seed (Burgess); distillery yard at Carron,
1953; pasture at St John's Mead, Darnaway, 1956 **BM,K**; roadside
verges at Dunphail, Kinloss and Covesea; pasture at Duffus Castle,
1972.
I Grassy roadside verge at Milton, Glen Urquhart, 1947, pasture
at Kincraig and at Ruthven Barracks, Kingussie, by the River Spey
at Newtonmore and by the River Beauly at Groam of Annat;
waste land on the Longman Point at Inverness, 1971 **E**; pasture by
the Moniack Burn at Kirkhill, 1972 **E**; bare ground by the Cluanie
Dam in Glen Moriston, 1974 **E**; rough grassland by old buildings
near Easter Dalziel, Petty, 1975 **E,K**.

F. arundinacea Schreb.   *Tall Fescue*
Native. Frequent in the lowlands, less so in the highlands. On
grassy banks and roadside verges. Auricles of the basal leaves fringed
with very short hairs. Map 306.
M In Moray rare, *WAS* (Murray); frequent at Ardivot Bank, at
Sheriffmill Bridge, on the haugh of Rothes and at Moy near
Kincorth (Gordon); Findhorn-side, 1888 *G. C. Druce* **OXF**;
Garmouth and Spynie (Marshall, 1899); on dunes near the mouth
of the River Findhorn, 1953 *A. Melderis* **BM**; shingle of the River
Findhorn, Darnaway, 1956 **BM**; Greshop wood, 1962 **E**; at
Lochinvar, banks of the Spey at Fochabers, dry grassy bank at
Milton Hill, policies of Brodie Castle, by the golf course at
Garmouth and at Spynie Kirk etc.; roadside verge between Banrach
Farm and Dyke village, 1968 **E,K,CGE**.
N Auldearn, 1835 *WAS & J. B. Brichan* **E**; by the river at Nairn
(Marshall, 1899); grass bank by the cemetery at Nairn, 1961
ABD,K.
I (Watson, *Top. Bot.*); banks of the River Ness and at Culcabock
Dam (Aitken); banks of the River Ness at Inverness, 1956 **BM**;
rough field by the Moniack Burn at Kirkhill, 1961 **K**; bank of a
ditch at Allanfearn, pasture by Loch Ness at Aldourie Castle, edge
of a salt-marsh W of Bunchrew; roadside near Ardersier, several

clumps on the Longman Point tip at Inverness, by the burn on the Inverness golf course and roadside bank near Croy.

Var. **strictior** (Hack.) K. Richt.
A variant with leaf-blades shorter and narrower (3–5 mm) than var. *arundinacea*, rather rigid. Panicles narrower and shorter (up to 15 cm), mostly erect, contracted after flowering.
M Banks of the River Spey at Dandaleith, 1974 **E**.
I Newly-sown verges at the Cluanie Dam in Glen Moriston, 1974 **E**.

F. arundinacea×**gigantea** = **F.**×**gigas** Holmberg
M In Greshop wood, near Forres (with both parents), 1953 *B. M. C. Morgan* herb. *A. Melderis* **BM** and 1956 *MMcCW* **E**.
I Plentiful (with both parents) in a pasture by the Moniack Burn at Achnagairn House, Kirkhill, 1961 **E,ABD,BM,K,CGE**.

F. gigantea (L.) Vill. *Giant Fescue*
Bromus giganteus L.
Native. Occasional in woods and shady places by rivers and lochs. Leaves bright green, awns very long 10–18 mm.
M Dunphail 1827, at Kinloss and by the Spey at Fochabers (Gordon); Greshop wood, 1856 *JK* **FRS** and 1953 *MMcCW* **K**; river bank at Darnaway, 1964 **K**; common in the oak wood W of Elgin, 1973.
N Banks of the River Nairn at Nairn, 1960 **K**; by the River Findhorn at Ferness Bridge, 1964 **ABD**.
I The islands at Inverness, 1833 *WAS* **GL**; banks near Inverness (as var. *triflora* Koch) 1943 *UKD* **K**; wood at Invermoriston, and in Reelig Glen, 1954; by the Breakachy Burn in Strathglass, 1956; lane at Achnagairn, Kirkhill, 1961 **K**; in Dirr Wood near Dores, gorge of a burn at Kilmorack, wood by Loch Ness at Lewiston, by the Falls of Divach and in a hazel wood at Milton, 1975; shores of Loch Ness at Brackla, 1975 **K**.

F. altissima All. *Wood Fescue*
F. sylvatica Vill. non Huds.
Native. Occasional in shady burn ravines and woods.
M Burn of Denaira, Darnaway, *Brodie* **E**; Dunphail, 1827 and Darnaway woods (Gordon); Dunphail (as var. *calamaria*) 1888 *JK* **FRS**; ravine of the Divie at Dunphail, 1897 *ESM* **BM,CGE** and in 1956 *MMcCW* **E**, and 1963 **K,CGE**; burn ravine at Darnaway, 1960 **ABD,K**.
N Cawdor Castle woods, 1834 *WAS* **E** and at later dates at **GL,BM** and herb. *Borrer* **K**; gorge of the Reireach Burn at Cawdor, 1960 **ABD,K**.
I Near Fort Augustus, *G. Don* (Hooker, *Fl. Scot.*); on rocks near old lime kiln in Reelig Glen, Kirkhill, 1964; ravine in Glen Tarff near Fort Augustus, 1973 *V. Gordon & E. Young*; ravine of the River Coiltie near Divach, 1973; in a small wood at the outlet of the River Tarff to Loch Ness, 1975 **E,BM**.

F. heterophylla Lam.   *Various-leaved Fescue*
Doubtfully native in north Scotland. Rare in woods. Leaves of
two kinds, the basal fine, 0·3–0·5 mm wide, infolded, those of the
culm flat, 2–4 mm wide; lemma with awn up to 6 mm long.
I Grounds of Moy Hall (probably introduced with rhodo-
dendrons), 1919 *G. C. Druce* **OXF**, and in 1975 *MMcCW* **E,K**.

F. rubra L.
Tentative key to subspecies of *Festuca rubra* L. by C. E. Hubbard.
  1. All the leaf-blades setaceous (bristle-like), usually plicate, up to
     1·3 mm wide in side-view; anthers mostly 2–3 mm long:
       2. Plants without rhizomes, densely tufted; outer basal leaf-
          sheaths glabrous or minutely pubescent; upper glume 3·5–6 mm
          long; lower lemmas 5–6·5 mm long, glabrous or rarely hairy
          (cultivated or waste ground)   .   .   .   subsp. *commutata*
       2. Plants with short to long rhizomes, loosely to densely tufted, or
          with scattered shoots and culms:
            3. Plants compactly tufted or mat-forming, with comparatively
               short very slender rhizomes; culms 10–48 cm high; panicles
               3–8 cm long:
                 4. Plants mat-forming; outer basal leaf-sheaths usually
                    glabrous; leaf-blades 0·5–0·7 mm wide in side-view, dark
                    green; lower lemmas 6–7·5 mm long, rarely less (–5 mm),
                    always glabrous (salt-marshes)   .   .   subsp. *litoralis*
                 4. Plants forming loose to dense tufts; outer basal leaf-sheaths
                    mostly retrorsely pubescent; leaf-blades 0·6–1·3 mm wide
                    in side-view, becoming whitish-pruinose, or green; lower
                    lemmas 5–5·5 mm long, rarely more (–6 mm), glabrous,
                    or pubescent (rocky coasts and cliffs)   .   subsp. *pruinosa*
            3. Plants loosely to closely tufted, or with scattered vegetative
               shoots and culms, with relatively long to very long rhizomes:
                 5. Lemmas (lower) awnless or mucronate, or the upper with
                    awns 0·5–1·5 mm long; culms mostly scattered, 12–30 cm
                    high; panicles dense, narrow, 3–5 cm long; upper glume
                    3–4·5 mm long, pubescent in the upper part; lemmas
                    4–5·5 mm long, densely short villous to hirsute, or ?
                    occasionally glabrous (high altitudes on mountains).
                       .   .   .   .   subsp. *arctica*
                 5. Lemmas usually awned, the awns 0·5–3 mm long; glumes
                    glabrous; culms –85 cm high:
                      6. Lemmas 4·5–5·5 (rarely –6) mm long, glabrous, or
                         shortly pubescent; upper glume 3–4·5 mm long; basal
                         outer leaf-sheaths densely retrorsely pubescent
                           .   .   .   .   subsp. *rubra*
                      6. Lemmas 6–8 mm (rarely 9 mm) long, usually densely
                         hairy with spreading hairs, rarely glabrous; upper glume
                         4·5–7·5 mm long:
                           7. Plant with long branching rhizomes, on sand-dunes;
                              basal outer leaf-sheaths usually glabrous; panicles
                              7–16 cm long   .   .   .   .   subsp. *arenaria*

7. Plant with slender relatively short rhizomes, at high
elevations on mountains; basal outer leaf-sheaths
pubescent; panicles 4·5–6·5 cm long
. . . . subsp. [ex Cairngorms]
1. All the leaf-blades flat, or only the basal blades plicate and
setaceous and the culm-blades flat; anthers 3–4 mm long; culms up
to 110 cm high, relatively stout; basal leaf-sheaths densely
pubescent; lemmas 6–8 mm long, usually glabrous; panicles
12–22 cm long, up to 15 cm wide:
8. Basal leaf-blades plicate and setaceous, 0·5–1 mm wide in side-
view; culm-blades flat, 1·5–4 mm wide, 7–11-ribbed above, the
ribs pubescent . . . . subsp. *megastachys*
8. All leaf-blades flat, 2–5 mm wide, up to 13-ribbed above, the
ribs pubescent . . . . . subsp. *multiflora*

Subsp. **rubra**   *Red Fescue*
Native. Abundant in short grassland, on dunes, moors and mountain
slopes very variable. Unless stated the following records have not
been critically examined.
M Frequent (Gordon); common at Forres, 1871 *JK* **FRS**; St John's
Mead at Darnaway, 1953 *B. Welch* **BM**; throughout the county,
1954; dry bank in a quarry at Lossiemouth, 1956 **K** det. CEH;
waste land on Forres tip, 1957 **E,K** det. CEH; river shingle at
Darnaway, 1960 **ABD,K,CGE** det. CEH; yard at Manachy
distillery near Forres, 1967 **K** det. CEH.
N (Stables, *Cat.*); Ardclach (Thomson); on moors at Drynachan,
Dulsie Bridge etc. 1954; river bank at Glenferness, 1960 **ABD,BM,K**
det. CEH; Cawdor Castle 1960 **ABD,BM,K** det. CEH.
I Dalwhinnie, 1833 (Watson, *Loc. Cat.*); abundant throughout
the district, 1954; mountain slope in Coire nan Laogh of Glen
Banchor and meadow at Gaick Lodge, 1973 **E** det. CEH; grass
verge to road at Ceannocroc Bridge in Glen Moriston, 1974 **BM**
det. CEH; abundant in grassy places 1 mile S of Dalwhinnie and
at Inchnacardoch, 1970 *C. E. Hubbard*.

Var. **barbata** (Schrank) K. Richt.
M Salt-marsh, Culbin (not typical) 1975 **K** det. CEH.
I Grass verge at Tullochgrue Rothiemurchus, 1975 **K** det. CEH.

Var. **glaucescens** (Heget. & Heer) Nyman
Common on salt-marsh turf; leaves and panicles bluish-green;
spikelets hairy.
M Findhorn and the Leen at Garmouth, 1953 *A. Melderis* **BM**.

Subsp. **arenaria** (Osb.) Syme   *Sand Fescue*
Var. **arenaria** (Osb.) Fries
Native. Frequent on sand dunes. Leaves bluish-white; spikelets
densely hairy.
M Seaside variety (Burgess); short turf near the Buckie Loch in
the Culbin Forest and at Findhorn, 1953 *A. Melderis* **BM**!

Subsp. **commutata** Gaud.   *Chewings Fescue*
I Raised shingle beach at Ardesier, 1975 **K**; roadside verge near
Lundie, Glen Moriston 1975 **E,K**; shingle of the River Farrar at
Struy, 1975 **E** all det. CEH.

Subsp. **litoralis** (G. W. F. Meyer) Auquier
Forms a close green turf above the salt-marsh level of *Puccinellia
maritima* and has short rhizomes and culms and narrow erect
panicles.
M Culbin Sands and on shingle by the River Spey at Garmouth,
1953 *A. Melderis* **BM**; dunes at Findhorn, 1953 **BM**.
I Salt-marsh W of Bunchrew on the Beauly Firth, 1975 **K** det.
CEH.

Subsp. **pruinosa** (Hack.) Piper
Native. Salt-marsh turf; leaves and panicles bluish-green or green;
spikelets hairless.
M Findhorn estuary, 1953 *NDS & MMcCW* **BM**.

Subsp. **arctica** (Hack.) Govar.
        Var. **mutica** Hartm.
I Wet grassy flush by a small burn running into Loch Toll
a'Mhuic, Monar, 1975 **K**; grass slope on The Fara near Dalwhinnie,
1975 **K**, both det. CEH. Mountain ledge at 915 m on Sgùrr na
Lapaich in Glen Strathfarrar (aristate form) 1972 **E**, det. CEH.

Subsp. **megastachys** Gaud.
I Grass bank at Torness and on waste ground by the old road
bridge near Erchless Castle in Strathglass and roadside verge at
Tullochgrue, 1975, all **K**; waste ground on Longman Point
Inverness, by Loch Mòr at Errogie, pasture at Moy Hall, 1975
all **E**; grass bank at Balvraid near Tomatin, 1975 **BM**, all det.
CEH.

Subsp. **multiflora** (Hoffm.) Jirasek ex Dostal
        F. rubra var. *planifolia* (Trautr.) Hack.
M Margin of grassland, Darnaway, 1953 *A. Melderis* **BM** and
MMcCW **E**; pasture at St John's Mead Darnaway, 1956 **ABD,K**
det. CEH.

F. ovina L.   *Sheep's Fescue*
Native. Abundant on dry grassy moorland and on dunes. Lemma
with a fine short awn.
M Frequent (Gordon); Forres, 1857 *JK* **FRS**; Aviemore, 1892
AS **E**; dunes at Findhorn, 1953 *NDS* **BM**; moor at Fochabers, 1950
CGE; dunes in the Culbin Forest and throughout the county, 1954.
N (Stables, *Cat.*); common in upland pastures at Ardclach, 1891
R. Thomson **NRN**; moors at Drynachan, Cawdor and Glenferness
etc., 1954; moor at Dulsie Bridge, 1960 **ABD,K** det CEH; dunes
on the Nairn Dunbar golf course, 1961 **ABD,K** det. CEH.
I Dalwhinnie and Pitmain, 1833 (Watson, *Loc. Cat.*); Sgòran
Dubh Mòr, 1891 *AS* **E**; Rothiemurchus, 1949 *P. S. Green* **E**; river

terraces in Glen Feshie, 1950 *C. D. Pigott*; throughout the district, 1954.

F. tenuifolia Sibth.   *Fine-leaved Sheep's Fescue*
Native. Common in dry places in pine and birch woods, on wall tops and open moorland, especially in the west of the district. Lemma finely pointed, very rarely awned. Map 307.
M Elgin, 1883 (as *F. ovina*) ? *coll.* **ELN**; Findhorn, 1953 *NDS* **BM**; St John's Mead at Darnaway, 1953 *A. Melderis* **BM**; pine wood in Culbin and Roseisle Forests, railway embankment at Knockando and banks at Smallburn and Aviemore, etc., 1954; on tussocks in a bog at Drakemyres and on a moorland slope at Fae near Dorback, 1972 **E** det. CEH.
N Near Millhill, Nairn, 1832 *J. B. Brichan* **ABD**; Cawdor woods, 1943 *UKD* **UKD** det. CEH; bank at Ferness Bridge, 1955; birch wood at Coulmony, in woods at Lochloy House and at Kilravock Castle etc.
I Beauly, 1889 *T. A. Cartwright* **OXF**; moors near Fort Augustus and Milton, 1952; pine woods at Guischan, Scaniport, Broomhill, Tullochgrue and Tomatin etc., 1954; plentiful in birch woods in Strathglass and moors at Boat of Garten, Dunlichity and in Strathdearn; moor at Loch Killin, 1972 **E** det. CEH; birch wood by Creag Dhubh near Laggan, 1973 **E**, det. CEH; 1 mile S of Dalwhinnie, 1970 *C. E. Hubbard*.

F. vivipara (L.) Sm.   *Viviparous Fescue*
Native. Common on moorland tracks and river shingle. Spikelets always proliferating vegetatively (viviparous). Fig 49c.
M Castle of Spynie (Gordon); near Forres, 1876 *JK* **FRS**; banks of the River Findhorn at Boom of Moy and in Darnaway Forest, 1953 *A. Melderis* **BM**; shingle of the River Findhorn opposite Sluie, 1956 **K**; on rocks by the river at Relugas, 1963 **CGE**; Grantown, 1963 *A. Cgekalowska* **E**; on river shingle from Aviemore to Garmouth; stoney verge of Lochindorb Castle, 1974.
N (Stables, *Cat.*); occasional at Ardclach, 1888 *R. Thomson* **NRN**; moors at Drynachan, Coulmony, Dulsie Bridge and Glenferness, 1954; banks of the River Findhorn at Ardclach, 1973 **E**.
I Summit of Cairn Gorm (Balfour, 1868) shores of Loch Ness at Fort Augustus, 1904 *G. T. West* **E,STA**; Newtonmore, 1927 *T. Wise* **GL**; low down by a burn on Mam Sodhail and at Affric Lodge, 1947 *A. J. Wilmott* **BM**; Am Beanaidh in Rothiemurchus, 1949 *P. S. Green* **E**; on moorland tracks in Glen Feshie, at Gaick and throughout the district, 1954; 1 mile S of Dalwhinnie, 1970 *C. E. Hubbard*.

Festuca × Lolium = × **Festulolium** Aschers. & Graebn.
M Gordon Castle, Fochabers *c.* 1800 *J. Hoy* herb. *Brodie* **E**.

Lolium L.

L. perenne L.   *Perennial Rye-grass*
Native. Abundant on roadside verges and waste places. Planted as a crop.

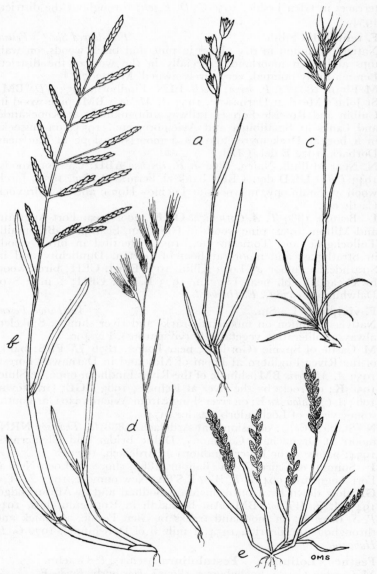

Fig 49 a, **Sieglingia decumbens**; b, **Glyceria declinata**; c, **Festuca vivipara**; d, **Vulpia bromoides**; e, **Catapodium marinum**.

Var. **perenne**
M Innes House, Urquhart, 1794 *A. Cooper* **ELN**; very common
(Gordon); Kinloss (as var. *cristatum*), 1860 *JK* **FRS**; ½ mile S of
Forres and on shingle of the River Spey at Garmouth, 1953
A. Melderis **BM**; track verge in the Culbin Forest and throughout
the county, 1954; Lossiemouth, 1963 *V. Mackinnon* **E**.
N (Stables, *Cat.*); Ardclach (a congested form), 1899 *R. Thomson*
NRN; moorland tracks at Drynachan, Cawdor and Glenferness
etc., 1954; roadside verge at Nairn, 1960 **K**.
I Inverness, 1832 and Pitmain, 1833 (Watson, *Loc. Cat.*); Dores,
1940 *AJ* **ABD**; throughout the district, 1954; along roadsides
1 mile S of Dalwhinnie and tracks at Inchnacardoch, 1970 *C. E.
Hubbard*.

Var. **tenue** (L.) Syme
Small compact tufts, with narrower leaf-blades and more slender
spikes with fewer-flowered spikelets than in var. *perenne*.
I The Islands, Inverness (as forma *tenue*) 1943 *UKD* **UKD** det.
CEH.

L. multiflorum Lam. *Italian Rye-grass*
   L. italicum A. Braun
Introduced. Cultivated as a crop and naturalized on waste ground.
M Near Forres (Druce, 1897); fields at Kinloss and Findhorn, 1925;
½ mile S of Forres station, 1953 *A. Melderis* **BM**; by a track in the
Culbin Forest and on waste ground throughout the county, 1954.
N Sown as a crop at Drynachan Lodge, at Holme Rose and at
Cawdor, 1954; roadside near Nairn, 1961 **E,K,CGE**.
I Windhill near Beauly, 1947 (*BSBI Year Book* **39** (1949));
waste ground at Milton, 1947; Kingussie station yard, newly-
sown roadside verge in Glen Moriston, at Kirkhill, Inverness,
Inverfarigaig, Fort Augustus and Inverdruie etc., 1954.

L. multiflorum × **temulentum**
I On the tip at Longman Point, Inverness, 1970 **E** det. **CEH**.

L. temulentum L. *Darnel*
Introduced with foreign grain.
M Gordon Castle, 1811 *J. Hoy* herb. *Brodie* **E**; near Forres, *G. Don*
herb. *Brodie* **E**; cornfield near Elgin, 1833 *G. Gordon* **ABD**; Springfield
near Elgin, 1825 *WAS* and Birnie, 1835 (Gordon); Garmouth,
G. Gordon **GL**; Birnie near Elgin, 1837 *G. Gordon* **E,ABD**; shingle of
the River Spey at Rothes, 1899 *J. W. H. Trail*; rubbish tip at
Elgin, 1956 **ABD**, and 1957 **K,CGE**; tip at Forres, 1956 **BM** and
1960 **K**; Elgin, 1961 **E,K,CGE**.
N Auldearn, *J. B. Brichan* (Gordon); from 'Swoop', Wild Goose
Cottage, Nairn, 1969 **E,ABD**.
I Rubbish tip on Longman Point, Inverness, 1970 **E**.

L. rigidum Gaud. × **temulentum** L.
I Rubbish tip on Longman Point, Inverness, 1971 **E,K,CGE** det.
CEH.

Vulpia C. C. Gmel.
V. membranacea (L.) Dumort.   *Dune Fescue*
Festuca uniglumis Ait.
Introduced in north Scotland.
I   Farm yard at Affric Lodge, 1971 **E** det. CEH.

V. bromoides (L.) Gray   *Barren Fescue, Squirreltail Fescue*
Native. Frequent on sand dunes, on roadside verges and tracks and
river shingle. Fig 49d, Map. 308.
M Ditch banks opposite the old Kennels at Brodie House, 1811
Brodie **E**; Morayshire, 1837 *G. Gordon* **ABD**; roadside on the way to
Invererne near Forres, 1880 *JK* **FRS**; by the Muckle Burn at Moy
and on Spey shingle at Garmouth, 1953 *A. Melderis* **BM**; dunes at
Lossiemouth and in the Culbin Forest, 1954; station yard at Forres,
1957 **ABD,K**; track at Garmouth, 1957 **CGE**; Elgin, 1961 **K**;
railway line at Brodie, 1963 **BM,CGE**; at Knockando and
Lossiemouth, 1963 **CGE**; shingle of the River Spey at Fochabers,
1971 **E**.
N (Stables, *Cat.*); not common at Ardclach, 1887 *R. Thomson* **NRN**;
roadside verge at Littlemill, 1954; Nairn, 1961 **ABD**; shingle of the
River Nairn at Little Kildrummie, 1974 **E**.
I By Loch Ness, 1839 *J. Ball* **E**; waste ground by Loch Loyne
and at Ceannocroc Bridge in Glen Moriston, 1950; waste ground
at Inverness, Invermoriston, Glen Strathfarrar, Milton and Daviot,
1956; roadside verge at Whitebridge, 1961 **E,K**; railway embankment
at Culloden, 1976 **BM,CGE**; gravel drive at Kirkhill, 1976 **BM**.

V. myuros (L.) C. C. Gmel.   *Rat's-tail Fescue*
Introduced in north Scotland. Rare on railway tracks and in
distillery and station yards.
M Introduced with wool 'shoddy' at Greshop House near Forres,
1958 **ABD,CGE**; railway line at Forres, 1961 **E,ABD,BM,K**; on
sandy soil by the Forestry Office at Kintessack, 1968 **E**.
N Railway line at Nairn station, 1961.
I   Introduced with foreign grain in a distillery yard at Inverness,
1961 **E,BM,K,CGE**; railway siding at Inverness, 1962 **ABD,CGE**.

Puccinellia Parl.
P. maritima (Huds.) Parl.   *Sea Poa, Common Salt-marsh-grass*
Glyceria maritima (Huds.) Wahlberg
Native. Common in salt-marshes and muddy estuaries.
M Moray, not rare, *WAS* (Murray); abundant in one or two
places (Gordon); Findhorn Bay, 1857 *JK* **FRS**; Culbin and Findhorn
Bay, 1953 *A. Melderis* **BM**; dunes NW of Buckie Loch in the
Culbin Forest, 1968 **E,BM,K,CGE**; salt-marshes at Kingston and
Lossiemouth.
N Salt-marsh E of Nairn, in the Culbin Forest and on the Carse
of Delnies, 1954; salt-marsh verge E of the Nairn Dunbar golf course,
1961 **ABD,K**.
I   East shore of Fort George, 1846 *A. Croall* (Watson)!; salt-

marshes from Beauly to Inverness, 1954; Carse of Ardersier, 1956; in short turf by the sea at Allanfearn.

Catapodium Link
C. marinum (L.) C. E. Hubbard   *Darnel Poa, Sea Fern-grass*
   Poa loliacea Huds.
Native. Very rare in sand dunes. Fig 49e.
M Stoney dunes at Lossiemouth, 1966 *A. J. Souter* **E**!; sandy grass bank by the lighthouse at Burghead, 1966 *R. Richter*!.

Poa L.
P. annua L.   *Annual Meadow-grass*
Native. Abundant in cultivated ground, waste places, moorland tracks and in grassy places on mountains and under rocks and cliffs where animals have sheltered, reaching 1,200 m in the Cairngorms.
M Abundant (Gordon); Forres, 1860 *JK* **FRS**; on shingle of the River Findhorn near Forres and ½ mile S of the railway station, 1953 *A. Melderis* **BM**; track in the Culbin Forest and throughout the county, 1954.
N (Stables, *Cat.*); Ardclach (Thomson); throughout the county, 1954; river shingle at Nairn, 1960 **ABD**.
I Inverness, 1832 Dalwhinnie and Pitmain, 1833 (Watson, *Loc. Cat.*); Tomatin (Moyle Rogers); near the summit of Cairn Gorm and throughout the district, 1954; Inchnacardoch, 1970 *C. E. Hubbard*.

P. alpina L.   *Alpine Meadow-grass*
Native. Occasional on scree and rock ledges in the higher mountains. Usually viviparous.
M Brought down the River Spey to Speymouth, *G. Birnie* (Burgess).
I Cairn Gorm, *J. B. Brichan* (Gordon, *annot.*)!; Fraoch-choire, Cannich (Ball, 1851); Glen Einich and Coire an t-Sneachda of Cairn Gorm, 1884 *H. & J. Groves* **BM**!; Coire an Lochain of Cairn Gorm, 1887 *G. C. Druce* **OXF**!; in a rock gully in upper Glen Feshie (the non viviparous form), 1951 *E. C. Wallace* (*Watsonia* **2**: 355 (1953)); Coire Garbhlach in Glen Feshie, 1957 *E. C. Wallace*; damp alpine slopes of Sgùrr nan Conbhairean at 800 m, *H. Milne-Redhead* (*Atlas*); scree in a burn gully on Sgùrr na Lapaich in Glen Strathfarrar at 820 m, 1971; rock ledge on Bynack More, 1975.

P. flexuosa Sm.   *Wavy Meadow-grass*
Native. Very rare on cliff ledges and rocks in the higher mountains. Pl 23.
I Ledge in a chimney on the buttress above the lochan in the Coire an Lochain of Cairn Gorm, 1953 **E,BM**.

P. nemoralis L.   *Wood Meadow-grass*
Native. Common in woods, on walls and on base-rich mountain ledges. Very variable. Map 309.

Var. **nemoralis**
M Rare (Gordon); Darnaway woods opposite Logie, 1839 *J. G.*

Innes (Gordon, *annot.*); rocks at Sluie, 1871 *JK* **FRS**; near Forres (Druce, 1897); ½ mile S of Forres station, 1953 *A. Melderis* **BM**; Culbin Forest, woods at Brodie, Dunphail, Grantown and Boat of Garten etc., 1954; shingle of the River Spey near Fochabers, 1956 **BM**; rubbish tip at Elgin and shingle of the River Findhorn at Darnaway, 1964 **E,CGE**; by the old bridge at Carrbridge, 1972.
N Rocks by the River Findhorn at Drynachan, in woods at Holme Rose, Dulsie Bridge and by Cran Loch etc., 1954; wood at Glenferness, and Cawdor, 1960 **K**; by the River Nairn, 1961 **ABD,K**; cliff in the gorge of the Allt Breac at Drynachan, 1973 **E**; wall at Kilravock Castle, 1974 **E**.
I Speyside, Easterness, 1887 *G. C. Druce* **OXF**; cliffs above Loch Dubh, 5 miles N of Cluny Castle, 1911 *ESM* **E,BM**; wall top at Polchar S of Aviemore, 1955 *J. R. Laundon* **BM**; wall tops at Bught, Inverness, at Newtonmore, near Loch Meiklie and at Kincraig etc., 1954; cliff ledges on Sgùrr na Lapaich in Glen Strathfarrar (var.) at 800 m, 1972 **E**; cliffs at Loch Killin and on rocks by the Falls of Divach, 1972 **E**; meadow by Gaick Lodge (var.) 1973 **E**; wall by Loch Meiklie in Glen Urquhart (var.) 1973 **ABD**; walls of the lime quarry on Suidhe Hill, Kincraig (var.), 1973 **BM**; on shingle of a small island near Eskadale in Strathglass (var.), 1973 **E**; shingle of the River Spey at Newtonmore (var.), 1973 **E**; cliffs in Glen Feshie, 1973 **E**; rock face on Creag Dhubh near Laggan, 1973 **BM**; in woods at Lewiston, Knockie Lodge, Fort Augustus and Inverness etc.

Var. **coarctata** Gaud.
Culms stiff, mostly erect, panicles contracted and rather dense after flowering; spikelets 3–5 flowered. Mountain ledges.
I Rock face in a burn gully in Glen Feshie, and on rocks on Creag Bheag, 1974 **E**, det. CEH.

Var. **parnelli** Hook. & Arnott
Differs from *P. nemoralis* in having more erect closer panicles; slightly longer ligules and upper sheath longer than the blade, but the differences are trivial.
M Speyside at Kinchurdy (Druce, 1888); ravine of the Divie on shaded rocks near Dunphail, 1897 *ESM* **BM,CGE**; cliff ledge by the River Findhorn a short way upstream from Randolph's Leap at Relugas, 1967 **ABD,BM,K,CGE**.
I Speyside, Easterness, *G. C. Druce* **OXF** (probably the record from Kinchurdy, a place Druce often mistook as being in v/c 96).

P. glauca Vahl   *Glaucous Meadow-grass*
Native. Rare on mountain rock ledges. Stems stiff, glaucous. Pl 23.
I Coire Garblach in Glen Feshie and in the Coire an t-Sneachda of Cairn Gorm, 1950 *C. D. Pigott* **CGE**; cliffs in Glen Einich, 1972 *D. Tennant.*

P. balfourii Parnell
Resembles *P. glauca* but stems less stiff, more flaccid leaf-blades,

loose-inclined or nearly nodding panicle, narrower glumes (lower lanceolate, upper ovate) and less hairy lemmas. The ligules are longer than in *P. nemoralis*.
I Glen Einich, 1888 *G. C. Druce* **OXF**.

P. compressa L.   *Flattened Meadow-grass*
Doubtfully native in north Scotland. Stems strongly compressed.
M Darnaway, 1839 *J. G. Innes* (Watson); parishes of Dyke and Speymouth, on dry refuse and walls (Burgess); dry bank by the railway siding at Knockando station, 1954 **E** and 1956 **ABD**; tip at Elgin, 1964 **E,K,CGE**.
I East shore at Fort George, 1846 *A. Croall* (Watson).

P. pratensis L.   *Smooth Meadow-grass*
Native in river pastures and grassy banks, frequently sown as a crop and escaping to roadside verges, not nearly as common as *P. subcaerulea*. Culms smooth; lower panicle branches 4–5 together. Ligule short, 1 mm or less, truncate; lower glume not acuminate.
M Very common (Gordon); Forres, 1857 *JK* **FRS**; grass verge near Kinloss Kirk, 1925; pasture at Darnaway, 1957 **CGE**; by the River Findhorn at St John's Mead, Darnaway, 1960 **E,K**; by the Muckle Burn at Dyke, 1963 **BM,CGE**; grass bank at Sanquhar House, Forres and on a roadside verge at Balnabriech near Mulben.
N (Stables, *Cat.*); Ardclach, 1888 *R. Thomson* **NRN**!; pasture by the River Findhorn at Glenferness, 1960 **K**
I Dalwhinnie and Pitmain, 1933 (Watson, *Loc. Cat.*); Beauly, 1919 *G. C. Druce* **OXF**; pastures at Fort Augustus, Invermoriston, Milton, Inverness and in Strathglass etc., 1956; meadow by Loch an t-Seilich at Gaick Lodge, 1973 **E**.

P. subcaerulea Sm.   *Spreading Meadow-grass*
   P. irrigata Lindm.
Native. Abundant on sand dunes, river shingle and by moorland roads and tracks. Rhizomes extensive; lower panicle branches usually 2, sometimes up to 5; ligule up to 2 mm, membranous; lower glume acuminate.
M Findhorn and Forres (Druce, 1897); dunes at Findhorn, 1953 **K**; near Binsness and in the Culbin Forest, 1953 *A. Melderis* **BM**; fixed dunes W of the Buckie Loch in the Culbin Forest, 1955 **E**; station yard at Knockando and throughout the county, 1954; verge to a track by Loch Mòr near Duthil, 1973 **E**; shingle of the River Spey at Garmouth 1973 **ABD**; roadside verge near Nethybridge, 1973 **BM**.
N Abundant on the sand dunes at Nairn, 1925 and in 1960 **ABD,K**; moorland track at Drynachan and throughout the county, 1954; roadside verge at Glenferness, 1961 **ABD,K**.
I Cairn Gorm, *G. C. Druce*; track verge at Beauly, 1947; throughout the district 1954; shingle of a burn in Glen Banchor and by a track at Gaick Lodge, 1973 **E**.

P. trivialis L.   *Rough Meadow-grass*
Native. Abundant in meadows, waste places and wet moorland
tracks. Ligule long (up to 8 mm) acute. Panicle branches scabrid.
M Common (Gordon); Waterford, 1871 *JK* **FRS**; waste land at
Kinloss and Findhorn, 1925; by the Muckle Burn at Moy, and on
Spey shingle at Garmouth, 1953 *A. Melderis* **BM**; track in the
Culbin Forest and throughout the county, 1954.
N (Stables, *Cat.*); Ardclach, 1894 *R. Thomson* **NRN**; waste ground
at Drynachan and throughout the county, 1954; among stones by
the river Nairn, 1960 **ABD**.
I Dalwhinnie and Pitmain, 1833 (Watson, *Loc. Cat.*); waste ground
at Newtonmore, in Glen Affric at Inverness and throughout the
district, 1954; tracks at Inchnacardoch, 1970 *C. E. Hubbard*; meadow
by Loch an t-Seilich by Gaick Lodge, 1973 **E**.

P. chaixii Vill.   *Broad-leaved Meadow-grass*
Introduced. Occasional in woods in the policies of the larger houses.
A stout, tufted bright-green perennial with very wide leaves. 5–10
mm wide.
M Policies of Newton House near Elgin, 1952 **E,K**; Innes House,
Urquhart, 1967 **E,CGE**; quarry at Newton Toll, 1961 **ABD,CGE**;
woods at Darnaway Castle near the Golden Gates, 1963 **E,BM**;
Brodie Castle, 1961 **ABD,BM,K**; also at Altyre House, Wester
Elchies and Leuchars near Lhanbryde.
N Policies of Coulmony House, 1960 **ABD**; wood at Geddes
House, 1962; beyond the kitchen garden at Acharaidh at Nairn,
1965 *D. McClintock*.
I Moy Hall, Easterness, 1919 *G. C. Druce* **OXF**!; Achnagairn
House, Kirkhill, 1967 **ABD,K**; at Aldourie Castle, Dochfour House,
Reelig House and Inverbrough Lodge near Tomatin.

Catabrosa Beauv.

C. aquatica (L.) Beauv.   *Whorl-grass*
Native. Rare in muddy margins of ponds and ditches.
M In a burn that falls into the sea between Burghead and Findhorn,
1828, ditches on the E side of Birkenhill and on the Leen at
Garmouth (Gordon); Lingieston near Forres and in Dyke Moss, 1843
J. G. Innes (Gordon, *annot.*); a dwarf prostrate form was noticed near
Lossiemouth, the var. *littoralis* Parn or very near it (Marshall, 1899);
sandy spit of land in the Sanquhar House pond at Forres, 1961
E,ABD,BM,K,CGE; abundant in a drain running to the sea 1 mile
W of the Buckie Loch in the Culbin Forest (? var. *grandiflora* Haek.)
det. CEH; small burn running from the distillery to Burgie school,
1967 **E,ABD,BM,K,CGE**.
N The record from Inshoch by *J. G. Innes* is an error (Watson).
I Ditches on the Carse of Ardersier, 1846 *A. Croall* (Watson); at
the outlet of a small burn at Ardersier, 1975 **E**; abundant in the
same burn but further inland near the farm of Viewhill, 1976
E,ABD,BM,K.

Fig 50　a, **Briza media**; b, **Bromus mollis**; c, **Koeleria cristata**; d, **Trisetum flavescens**; e, **Aira praecox**; f, **A. caryophyllea**; g, **Anthoxanthum odoratum**; h, **Nardus stricta**.

Dactylis L.
D. glomerata L.   *Cock's-foot*
Native. Abundant in meadows, waste places and roadside verges.
M Banks of the River Findhorn, 1837 *G. Gordon* **ABD**; Waterford,
1871 *JK* **FRS**; ½ mile S of Forres station, 1953 *A. Melderis* **BM**;
dunes in the Culbin Forest and throughout the county, 1954.
N (Stables, *Cat.*); near the schoolhouse at Ardclach, 1888 *R. Thomson* **NRN**; grass verge to a track at Drynachan and throughout the county, 1954; meadow at Nairn, 1960 **ABD,K**.
I Inverness, 1832 (Watson, *Loc. Cat.*); Dalwhinnie and Tomatin,
1903 (Moyle Rogers); Scaniport, 1940 *AJ* **ABD**; roadside verge at
Drumochter and throughout the district, 1956.

Var. **collina** Schlechtendal
M Sand dunes at Hopeman (as var. *abbreviata* Drej.), 1943 *UKD*
UKD,K.

Cynosurus L.
C. cristatus L.   *Crested Dog's-tail*
Native. Abundant in short turf, on roadside verges and moorland
tracks.
M Lossie at the influx of Tayoch etc. (occasionally viviparous)
(Gordon) Forres, 1871 *JK* **FRS**; ½ mile S of Forres station, 1953
A. Melderis **BM**; Culbin Forest and throughout the county, 1954;
Aviemore, 1964 *M. Cowan et al.* **E.**
N (Stables, *Cat.*); Ardclach (Thomson); short turf by a path on
the moors at Drynachan and throughout the county, 1954; roadside
verge at Glenferness, 1960 **K**; by the river at Nairn, 1961 **ABD**.
I Inverness, 1832 Pitmain, 1833 (Watson, *Loc. Cat.*); Dalwhinnie,
1903 (Moyle Rogers) short grassland at Ben Alder Lodge and
throughout the district, 1956; Inchnacardoch, 1970 *C. E. Hubbard*.

C. echinatus L.   *Rough Dog's-tail*
Introduced.
I Roadside near Coylumbridge, 1944 *R. Mackechnie*.

Briza L.
B. media L.   *Datheron Ducks, Shak and Trammel, Quaking-grass*
Native. Occasional on base-rich grassland. Fig 50a.
M Locally frequent. Spynie Loch (Gordon); banks of the River
Findhorn, 1837 *G. Gordon* **ABD**; parishes of Dyke, Elgin, Rothes,
Lhanbryd, Urquhart, Bellie, Speymouth and Knockando (Burgess);
Hills of Cromdale, by Gilston Lochs and at Relugas, 1954; Boat
of Garten golf course, 1956; pasture by the River Spey at Fochabers,
1957 **CGE**; on grass tussocks in the bog at Drakemyres, and by the
old lime kiln at Fae near Dorback, 1972 **E**; on shingle by the river
at Bridge of Brown, 1973 **BM**.
N (Stables, *Cat.*).
I Kincraig, 1891 *AS* **E**; towpath at Muirtown, Inverness, 1930
M. Cameron; locally plentiful in short grass in Glen Feshie and at
Gaick Lodge; rare on Suidhe Hill at Kincraig, 1973.

Melica L.

M. uniflora Retz.   *Wood Melick*
Native. Occasional in shady burn ravines, on rocks and in woods.
Sheaths pubescent; panicle spreading.
M Darnaway woods, 1839 *J. G. Innes* (Gordon, *annot.*)! and 1960
MMcCW **ABD**.
N Rare on the river bank at Nairn, 1957 **E**.
I About the Dream at Kilmorack, *WAS* (Murray); rocks in
Reelig Glen, 1925; ravine at Inverfarigaig and in a small wood
at Fort Augustus, 1954; shaded rocks on the banks of the Grotaig
Burn, 4 miles SW of Strone by Loch Ness, 1954 *A. A. Slack &
A. McG. Stirling*; at Inchnacardoch, Tomich, Dores and Strathglass,
1956; top of the path at the Falls of Foyers, 1971 **E,CGE**;
conglomerate cliffs of Tom Bailgeann near Torness, on rocks by the
boat house at Knockie Lodge.

M. nutans L.   *Mountain Melick*
Native. Frequent on rock ledges in mountain ravines and by rivers
on base-rich soil. Sheaths glabrous; spikelets secund, larger and
fewer than the preceding species. Map 310.
M Dunphail (Gordon); Darnaway, 1839 *J. G. Innes* (Gordon,
annot.); by the River Findhorn at Logie, 1860 *JK* **FRS**; Dunphail,
1886 *ESM* **CGE**; rocky bank by the River Spey at Wester Elchies,
1954; ravine of a small burn in the Darnaway Forest, 1960 **ABD**;
by the path on the right bank of the River Findhorn at Daltulich
Bridge, 1972.
N Cawdor woods, 1832 *WAS* **GL**!; by the burn of Cawdor, 1864
R. Thomson **NRN**; by the River Findhorn at the foot of the Bell
Tower at Ardclach, 1888 (Thomson) and in 1973 *MMcCW* **E**;
rock crevice by the River Findhorn at Banchor, 1951 **E**, and in
several places lower down the river.
I ? locality, 1838 *A. Rutherford* (Watson, *Top. Bot.*); Dream near
Beauly, 1875 ? *coll.* **ELN**; Strathglass, 1832 *J. Walton* **GL**; ravine
by the River Cannich, and on a cliff below the Plodda Falls at
Guisachan, 1947; Allt Coire Bhlair in Glen Feshie, 1951 *E. C.
Wallace* **BM**; Fort Augustus, Inchnacardoch and in the burn
ravine at Inverfarigaig, 1956; Creag nan Eun, Invermoriston, 1961
D. A. Ratcliffe; burn at Corrimony, 1964; banks of the River Enrick
at Milton, and among rocks by the river in Glen Moriston, 1971 **E**;
on a scree in Glen Feshie; rocks by the burn at Lochan an Tairt
above Glen Urquhart, 1973 *M. Barron*; burn gully at Lurgmore and
in the river gorge at Tromie Bridge, 1975.

Sesleria Scop.

S. albicans Kit. ex Schult.   *Blue Moor-grass*
   S. caerulea (L.) Ard. subsp. *calcarea* (Celak.) Hegi
Native. Rare on micaceous schists on mountains.
I (Druce, *Com. Fl.*). ? error. There is no trace of this record.

17

Bromus L.

B. erectus Huds.   *Upright Brome*
    Zerna erecta (Huds.) S. F. Gray
Introduced in north Scotland.
I   Casual at Inverness, 1871 ? *coll.* **ELN**.

B. ramosus Huds.   *Hairy Brome*
    Zerna ramosa (Huds.) Lindm.
Native. Frequent in shady places on wood margins and by
rivers. Absent from the highlands. A tall grass up to 100 cm, with
puberulent stems and large purplish-glaucous panicle with nodding
spikelets and awned lemmas. Map 311.
M Dunphail, Darnaway and Pluscarden woods (Gordon); Greshop
wood and Cothall near Forres, 1856 *JK* **FRS!**; St John's Mead,
Darnaway, 1953 *A. Melderis* **BM**; edge of the wood by the Muckle
Burn at Brodie, wood at Dallas, at Rothes, Longmorn and
Fochabers, 1954; banks of the River Findhorn opposite Coulmony,
1972.
N Ardclach and Nairn, 1900 (Lobban); wood by the River Nairn
at Holme Rose, and at Cawdor Castle, 1954.
I Banks of the Moniack Burn near Kirkhill, 1925; by the River
Enrick at Milton, 1947; woods at Foyers, Divach, Invermoriston
and near Glassburn in Strathglass, 1954; by Loch Ness S of Dores,
1971 **E**; woods at Glendoe, Fort Augustus, Lewiston, policies of
Ness Castle at Inverness and in the ravine of the burn at
Inverfarigaig; by the Big Burn at Essich, 1975.

B. inermis Leyss.   *Hungarian Brome*
Introduced with foreign grain and grass seed. Awnless.
M Washed down the River Findhorn in a sack of foreign grain
from a distillery above and established in Greshop wood near Forres,
1956 **E,K,CGE**; on a newly-sown roadside near Roseisle, 1971 **E**.

B. sterilis L.   *Barren Brome*
    Anisantha sterilis (L.) Nevski
Native. Frequent on sandy banks and roadside verges on wall tops
and waste ground in the lowlands. Absent from the highlands.
Map 312.
M On an old wall in College Street, Elgin, 1824 and at Pluscarden
(Gordon); Lossiemouth, 1867 *J. Roy* (Watson); Forres, *J. B.
Brichan* (Gordon, *annot.*); Waterford, 1871 *JK* **FRS**; roadside by
Moy House, and on sandy ground in Findhorn village, 1925;
on the old wall at Greengates near Brodie, dunes at Hopeman,
roadside bank to the E of Westfield near Elgin etc., 1954, bank at
Dallas, 1970 **BM**.
N Not indigenous *WAS* (Murray); wall top at Cawdor Castle, 1960
ABD,K; sandy bank at Nairn, 1961 **E,ABD,K**; bank by the gates
to Kinsteary House at Auldearn, 1974.
I Roadside verge at Allanfearn, 1955; waste ground on Longman
Point, Inverness, 1956; gravel drive of Viewhill at Kirkhill, 1972 **E**;
raised beach at Ardersier, 1972.

B. madritensis L. *Compact Brome*
Introduced with foreign grain.
M From wool 'shoddy' at Greshop House 1958 **K** and (var.
ciliatus Guss.) **CGE**; railway siding at Knockando distillery, 1966 **E**;
garden weed at Dyke (probably introduced with plants collected
from wool 'shoddy'), 1972 **E**, det. CEH.

B. diandrus Roth *Great Brome*
   B. gussonii Parl.
Introduced with foreign grain.
M From distillery sweepings on Rothes tip, 1955 **ABD** and 1957
BM; railway siding at Blacksboat, 1956 **CGE**; siding at Mosstowie
station near Elgin, 1956; railway siding at Knockando, 1966
E,CGE.
I Distillery yard at Inverness, 1961 **E,ABD,BM,K,CGE**; distillery
at Tomatin, 1961 **K**; railway siding at Inverness, 1962 **CGE**.

B. mollis L. *Geese Grass, Soft Brome*
   B. hordeaceous auct; *Serrafalcus mollis* (L.) Parl.
Native. Common in waste places, on dunes and in meadows in the
lowlands. Rare in the highlands where it is confined to railway
station yards and sidings, or as an escape from sown meadows.
Spikelets 15–20 mm; lemmas pubescent, rarely glabrous, 8·5–11 mm
long; grain shorter than the palea. Fig 50b.
M Very common (Gordon); Forres, 1871 *JK* **FRS**; Lossiemouth,
1907 *C. B. Neilson* **GL**; dunes at Findhorn, 1925; waste ground
Dunphail, Cromdale and Blacksboat etc., 1954; track in the Culbin
Forest, 1956 **CGE**; garden weed at Greshop House, Forres, 1958 **K**.
N (Stables, *Cat.*); Cawdor, 1855 *HC* **INV**; hayfields at Ardclach,
1888 *R. Thomson* **NRN**; farmyard at Geddes House, 1954; waste
ground by the harbour at Nairn, 1960 **K**; dry bank 1½ miles E of
Geddes, 1976 **E**.
I Inverness, 1832 Pitmain, 1833 (Watson, *Loc. Cat.*); raised
shingle beach at Ardersier, 1952; roadside verge by Ceannocroc
Bridge in Glen Moriston, at Milton, Kirkhill and Fort Augustus,
1954; waste ground by the Torvaine golf course at Inverness, 1972 **E**;
sown in a hayfield at Boblainy, 1973 **E**; grass slope at Ruthven
Barracks, 1973 **E**; roadside bank at Loch Flemington, Culloden
station yard and on newly-sown roadside verges at Inverfarigaig,
Inchnacardoch and Fort Augustus; garden path at Dochfour
House, 1975.

B. thominii Hardouin *Lesser Soft Brome*
   B. hordeaceous subsp. *thominii* (Hard.) Hyl.
Native. Rare on sandy ground and waste places near the sea.
Spikelets 10–15 mm, glabrous.
M Sand dunes at Hopeman, 1943 *UKD* **UKD** (*Bot. Soc. & Ex.
Club Report* **13**: 321 (1946–47)); bank at Rothes and sand dunes at
Lossiemouth, 1954, det. A. Melderis; dunes at Lossiemouth 1956
BM.

N Sea bank at Nairn, 1957 **BM,K**; Carse of Delnies and on a sandy bank by Loch of the Clans.

B. lepidus Holmberg   *Slender Soft Brome*
Doubtfully native in north Scotland. Frequent on waste ground and sandy banks in the lowlands. Rare on newly-sown roadside verges and meadows in the highlands. Spikelets 10–15 mm, glabrous; lemma 4·5–5·5 mm long; grain exceeding the palea.
M Near the Royal Hotel at Forres, 1953 *A. Melderis* **BM**!; rubbish tip at Rothes, 1954 **BM**; sandy ground at Lossiemouth, 1956 **K**; at Garmouth, Culbin, Fochabers, near Cromdale and in the station yard at Blacksboat etc., 1954; meadow at Innes House, Urquhart, 1967 **CGE**; garden path at Brodie Castle, 1974.
N Roadside verge at Cantraydoune and farm yard at Geddes House, 1954; meadow by Loch of the Clans, 1960 **ABD**.
I Hayfield at Allanfearn, 1943 *UKD* **UKD** det. CEH; station yard at Beauly, 1947 *NDS* **BM**; meadow near the mouth of the River Coiltie at Drumnadrochit, 1947 *A. J. Wilmott* **BM**; wall top near Milton, 1956 **BM**; hayfield at Ceannocroc, Glen Moriston, 1957 **BM**; waste ground at Kirkhill, 1967 **K**; newly-sown roadside verge by Loch Ness S of Dores, 1971 **E**.

Var. **micromollis** (Krösche) C. E. Hubbard
A variant with hairy spikelets.
I Newly-sown roadside verge by Loch Ness S of Dores, 1971 **E** det. CEH.

B. lepidus × mollis = **B. × pseudothominii** P. Smith
B. thominii auct. mult.
Spikelets glabrous; lemmas 7–8 mm long; grain equalling the palea.
M Newly-sown grass by the Forres Health Centre, 1975 **E,K**; dunes by the old railway S of Lossiemouth, 1976 **E**, det. CEH.
N Roadside near Washingwells Farm, Auldearn, 1972 **E** det. CEH; bank E of Geddes, 1976 **E**.
I Roadside by the Dog Falls, Glen Affric (as *B. thominii* Hard.) 1962 *P. F. Hunt* **K**; waste ground at Dochfour House, Dochgarroch, 1975 **E**, det. CEH.

Var. **hirsutus** (Holmb.) ? P. Smith.
Lemmas hairy.
M Roadside bank by the railway bridge at Dalvey Smithy W of Forres, 1975 **K**; dry bank near Miltonhill E of Forres, 1976 **E**, det. CEH.
I Farm yard at Affric Lodge, 1971 **E**; sown meadow at Boblainy, 1973 **E,BM**; bank by Ruthven Barracks, Kingussie, newly-sown verge at Inverfarigaig and at Fort Augustus, all 1975 **E**; dunes at Fort George, 1976 **E**; grass verge to road by Loch Flemington, 1976 **BM**. All det. CEH.

B. racemosus L.   *Smooth Brome*
Rare. Waste land. Inflorescence a raceme, narrow, erect; lemmas 6·5–8 mm long, glabrous.

I Sandy bank at the road junction of the A96/B9039 near Castle Stuart, 1976 **E,BM**.

B. commutatus Schrad. *Meadow Brome*
Introduced in north Scotland. Panicle nodding; lemma *c.* 9 mm long.
M Roadside on way to Waterford near Forres, 1898 *JK* **FRS**; introduced with grass seed in the parishes of Dyke and Speymouth (Burgess).
I Cornfield at Dalwhinnie, 1833 (Watson, *Loc. Cat.*); waste ground at Allanfearn, 1943 *UKD* **UKD**.

B. arvensis L. *Field Brome*
Introduced.
M Gordon Castle, *c.* 1800 *J. Hoy* herb. *Brodie* **E**; Waterford near Forres, 1871 *JK* **FRS**.

B. secalinus L. *Rye Brome*
Introduced.
M Spynie and Drainie, scarcely a native, disliked by cattle, *WAS* (Murray); cornfields at Spynie, 1828, at Drainie, Birnie and Rafford, *J. G. Innes* (Gordon); Forres, 1839 *J. G. Innes* (Gordon, *annot.*); near Elgin, 1830 ? *coll.* **E**; Birnie near Elgin, ? date, herb. *Maclagan* **E**; Rafford, 1834 *J. G. Innes* **E**; Drainie, 1865 *JK* **FRS**; Pilmuir near Forres, 1878 *JK* **FRS**.
N Hill of Penick, 1833 *J. B. Brichan & WAS* **E,ABD**.

B. willdenowii Kunth *Fescue Brome*
B. unioloides Kunth; *Ceratochloa unioloides* (Willd.) Beauv.
Introduced with foreign grain. Rare on rubbish tips. Tall grass up to 70 cm; spikelets strongly compressed.
M Rubbish tip at Elgin, 1972 **E**; tip at Grantown, 1973.
I From distillery refuse on the rubbish tip on Longman Point at Inverness, 1971 **E**.

B. alopecuroides Poir.
Introduced with wool 'shoddy' at Greshop House, Forres, 1958 **K**.

B. lanceolatus Roth var. **lanuginosus** (Poir.) Dinsmore
Introduced.
M Rubbish tip at Grantown, 1973 **E**, det. CEH.

B. rubens L.
Introduced with foreign grain.
M Railway siding Knockando distillery, 1966 **E**, det. *A. Melderis*.

Brachypodium Beauv.
B. sylvaticum (Huds.) Beauv. *False Brome*
Native. Frequent on rocky banks and in deciduous woods on base-rich soil. Map 313.
M By the River Findhorn below Logie (Gordon; wood opposite Cothall, 1839 *J. G. Innes* (Gordon, *annot.*); Greshop Wood, 1856 *JK* **FRS**!; sandy track verges in the Culbin Forest, 1950; on shingle on the W bank of the River Findhorn near Forres, 1953 *A. Melderis*

BM; woods at Brodie Castle, Grantown, Hills of Cromdale and on the banks of the rivers Spey and Findhorn, 1954.
N Cawdor woods, *WAS* (Gordon); frequent at Ardclach, 1888 *R. Thomson* **NRN**; banks of the River Nairn at Holme Rose and by the River Findhorn at Coulmony, Dulsie Bridge, Glenferness and Ardclach, 1954; by the Muckle Burn at Lethen House, 1961 **ABD**.
I Inverness, 1832 (Watson, *Loc. Cat.*); islands at Inverness, *WAS* (Gordon); Dores, 1940 *AJ* **ABD**; wood by the River Enrick at Milton, 1947; at Inverdruie, Foyers, Reelig Glen, Farr and Invermoriston etc., 1954; among boulders below the cliffs of Creag Dhubh near Laggan and plentiful by Loch Ness and in Strathglass, 1972.

Agropyron Gaertn.

A. caninum (L.) Beauv.   *Bearded Couch*
Frequent on rock ledges, in woods and on river banks and shingle. Glumes persistent when fruit is shed; awn longer than the lemma. Map 314.

Var. **caninum**
M Frequent on the Findhorn above Sluie, 1831 (Gordon)!; Greshop wood, 1867 *JK* **FRS**!; near Forres (Druce, 1897); plentiful on shingle of the River Spey at Garmouth, 1950 **E**; St John's Mead at Darnaway, 1953 *A. Melderis* **BM**; common on river shingle from Aviemore to the sea, 1954.
N On a small island in the River Nairn near Holme Rose, 1955 **ABD**; river bank at Nairn, 1960 **E,K**.
I Islands at Inverness, 1833 *WAS* **GL**!; verge of a small wood by the River Enrick at Milton, 1947; at Invermoriston, Kincraig, by Loch Meiklie in Glen Urquhart, among boulders below the cliffs of Creag Dhubh near Laggan etc., 1956; lane at Kirkhill, 1961 **K,CGE**; ravine at Inverfarigaig, 1971 **E**; cliff ledge by Loch Killin, 1972 **E**; plentiful on both shores of Loch Ness and in Strathglass, in a burn gully at Tomich and banks of the River Spey at Inverdruie etc.

Var. **glaucum** Lange.
Shady places. Leaves, culms and spikes bluish-green.
I Windhill near Beauly, 1947 *NDS* **BM**.

A. repens (L.) Beauv.   *Quickens, Common Couch*
Native. Abundant in cultivated and waste ground and forming thick swards among stones on the tide-line. Glumes falling with the fruit, lemma acute or awned.
M Very common (Gordon); Waterford, 1871 *JK* **FRS**; margin of the Findhorn estuary at Kinloss, 1925; Findhorn Bay, 1953 *JK* **FRS**; margin of the Findhorn estuary at Kinloss, 1925; Findhorn Bay, 1953 *A. Melderis* **BM**; throughout the county 1954; by the Spey at Aviemore, 1955 *J. Milne* **E**; field at Mosstowie, 1956 **K**; dunes at Hopeman, 1967 **K**.

N (Stables, *Cat.*); Ardclach (Thomson); near Drynachan Lodge throughout the county, 1954; among stones on the tide-line at Nairn, 1960 **K**.

I (Watson, *Top. Bot.*); throughout the district wherever cultivated land or railway embankments occur, 1954 and reaching Ben Alder Lodge and Gaick Lodge.

Var. **aristatum** Baumg.
Var. *leersianum* Gray
M (Trail, *ASNH* **16**: 229 (1907)).
I A purple-spiked form (as var. *barbatum* (Duval–Jouve)) near Kingussie (Druce, 1888); grass verge to a cornfield near Cluny Castle, Laggan, 1972 **E**; verge of salt-marsh W of Bunchrew, 1976 **E**.

A. junceiforme (A. & D. Löve) A. & D. Löve   *Sand Couch*
Native. Common in all the sand dune areas.
M Frequent at Stotfield, Covesea, Burghead and Culbin (Gordon); coast near Forres (as var. *acutum* DC) *G. C. Druce* (Burgess); Findhorn, 1857 *JK* **FRS**; Culbin and Findhorn, 1953 *A. Melderis* **BM**; dunes in the Culbin Forest, 1960 **K**; east beach at Lossiemouth, 1963 *V. Mackinnon* **E**.
N Nairn, *WAS* (Gordon); common on the sandy shore at Nairn, 1889 *R. Thomson* **NRN**; among shingle on the Carse of Delnies, 1954.
I Near Fort George, 1846 *A. Croall* (Watson); on the raised beach at Ardersier, 1954.

Secale L.
S. cereale L.   *Rye*
Introduced as a crop, or with distillery refuse, escaping to waste ground.
M Railway siding Blacksboat, 1956 **ABD**; distillery tip Carron, 1957 **E**; tips at Elgin and Forres, 1961 **E**; Lossiemouth (as *Hordeum jubatum*) 1963 *V. Mackinnon* **E**.
I Relic from a crop, Nairnside, 1974.

Triticum L.
T. aestivum L.   *Wheat*
Introduced as a crop or with distillery refuse. Waste ground.
M Roadside verge near Burghead, distillery tips at Knockando and Carron.
I Longman Point, 1970 **E,K**; tip by Muirtown Basin, Inverness, 1958 *E. W. Groves* **EWG**.

Elymus L.
E. arenarius L.   *Lyme Grass*
Native. Common on sand dunes.
M On the Agaty rock at Stotfield village, 1828 and on the shore at Burghead, *Mr Fraser* (Gordon); Findhorn, 1856 *JK* **FRS**; Findhorn, 1953 *A. Melderis* **BM**; shingle bank by the sea at Garmouth and plentiful on the Old Bar, Culbin, 1954.

508 KOELERIA

N Nairn, 1884 *J. Groves* **GL**; dunes at Kingsteps near Nairn, 1962
ABD.
I Dunes at Fort George, 1925; inland by the A9 near Daviot
Kirk, 1964, probably introduced with sand from the coast.

Hordeum L.

H. murinum L. *Wall Barley*
Native. Waste sandy ground chiefly near the sea.
M Alves churchyard and Spynie, *WAS* (Murray); certainly
introduced, rare on walls about Greyfriars at Elgin, 1824,
churchyards at Drainie, Duffus, Alves and Forres and at Spynie
Castle. The localities of this plant mark its introduction into the
province at a very early date (Gordon); Elginshire, *W. J. Hooker*
herb. *Trail* **ABD,K**; Elginshire, 1837 *G. Gordon* **E**; Forres, 1856 *JK*
FRS; introduced with wool 'shoddy' (forma) at Greshop House
near Forres, 1958 **K**; Forres station yard, 1960 **ABD,K**; sandy ground
at Lossiemouth, 1961 **E,K,CGE.**
N About houses at Nairn harbour, 1954.
I Ardersier, 1943 *UKD* **K**!; by the sea at Longman Point,
Inverness, 1971 **E,BM.**

H. distichon L. *Two-row Barley*
Introduced. Railway sidings and waste ground.
M Station yard at Blacksboat, 1956 **ABD**; garden weed at Greshop
House, Forres, 1957 **E**; railway siding by the distillery at Knockando,
1966 **E.**
N Introduced with 'Swoop' at Wild Goose Cottage, Nairn, 1969.
I Railway siding at Inverness, 1962 **E**; rubbish tip on Longman
Point, Inverness, 1970.

H. vulgare L. *Six-row Barley, Bere, Barley*
Relic of cultivation.
M Roadside verge near Burghead, 1972 **E.** Also on tips and in root
fields.
N Waste ground at Newton of Park, 1970.
I Rubbish tip on Longman Point, Inverness, 1961; newly-sown
grass verge by Loch Ness S of Dores, 1970.

The following appeared in wool 'shoddy' at Greshop House near
Forres in 1958. **H. glaucum** Steud. **K,CGE**; **H. geniculatum** All.
K,CGE; **H. flexuosum** Steud. **K,** and **H. pusillum** Nutt. **K.**

Koeleria Pers.

K. cristata (L.) Pers. *Crested Hair Grass*
Native. Frequent on short base-rich turf on moorland banks and
sea cliffs. Fig 50c, Map 315.
M Burghead and mouth of the Forres Burn, *J. G. Innes* (Gordon);
near Forres, 1835 *J. G. Innes* **E**; Strathspey, 1861 *JK* **FRS**; Boat
of Garten, 1887 *G. C. Druce* **OXF**; wall top on the road from
Speybridge to Nethybridge, 1895 *C. Bailey* **OXF**; bank of the River

Findhorn in Greshop Wood, at Darnaway and Findhorn Bay, 1953
A. Melderis **BM**; sandy ground in Greshop wood, 1953 **E**; dunes
at Findhorn, 1964 **K,CGE**; in short turf on a roadside bank between
Grantown and Cromdale, 1971 **E**; golf courses at Lossiemouth and
Hopeman, dry bank in the village of Aviemore, etc.
N (Stables, *Cat.*); frequent in pastures at Ardclach, 1888 *R.*
Thomson **NRN**; in short turf at Holme Rose, Cawdor, Loch of the
Clans and at the 'Doocot' at Auldearn, 1954; dunes on the Nairn
Dunbar golf course, 1960 **ABD**; grass bank near the cemetery at
Nairn, 1961 **ABD,K**; grass bank by Cran Loch and at Drynachan.
I Short distance from Dalwhinnie, 1833 (Watson, *Loc. Cat.*); near
Fort George, 1845 *A. Croall* **STI**; Kincraig, 1891 *AS* **E,OXF**; dry
heath at Laggan, 1922 ? *coll.* **GL**; banks at Milton, Inverness,
Culloden and Newtonmore, etc., 1954; common in glens Mazeran,
Kyllachy and Urquhart; on Cromalt Mount at Ardersier, 1974 **E**;
common on grass banks in Strathglass, 1973 **BM**.

Trisetum Pers.

T. flavescens (L.) Beauv.   *Yellow Oat Grass*
    Avena flavescens L.
Native. Occasional on dry base-rich grassland. Spikelets shining,
yellowish, awn geniculate from above the middle of the lemma.
Fig 50d, Map 316.
M Certainly introduced, rare in the lawns at Westerton House,
Pluscarden (Gordon); near the Manse at Dyke, 1858 *JK* **FRS**; at
Innes House, Urquhart and in three stations in Dyke where it has
maintained its site on top of a dyke between the village and Brodie
station for over eighty years (Burgess); grass bank at Sanquhar
House, Forres and in a meadow at Darnaway Castle, 1954.
N Bank at Holme Rose, 1954; roadside near Ferness Bridge, 1955;
meadow at Kilravock Castle, 1974 **E,BM**.
I At the Manse of Rothiemurchus, 1883 *JK* **FRS**; by the tow-path
of the Caledonian Canal at Bught, Inverness, and at Kirkhill, 1954;
in short turf at the edge of Newtonmore golf course, 1973 **E,BM**;
on the roadside verge near the bridge over the River Calder on the
Laggan road W of Newtonmore, 1973 **E**; roadside verge at
Ardersier, 1974 **E**; by the drive at Dochfour House and roadside
verge near Reelig House, 1975.

Avena L.

A. fatua L. var. **fatua**   *Wild Oat*
Introduced. Formerly frequent on waste ground and in cornfields,
now abundant and a troublesome weed in many crops.
M Doubtfully native, frequent (Gordon); Forres, 1857 *JK* **FRS**;
shingle of the River Spey at Garmouth and in station yards at
Blacksboat and Grantown, 1954; rubbish tip at Forres, 1957 **CGE**;
from wool 'shoddy' at Greshop House, Forres, 1958 **CGE**; waste
ground at Elgin, 1961 **E,BM,K**; farm yard at Kirkhill near Elgin,
1967 **E**; distillery yard at Carron, 1968 **K,CGE**; now abundant in
root and barley fields throughout the county.

N Blackhills, Nairn, 1833 *WAS* **E**; barley field at Howford Bridge, 1972; root field on the Lochloy road near Nairn and in a barley field at Kilravock, 1974.

I Waste ground by Loch Morlich, 1953; waste ground at Kingussie, 1954; fields at Dores, Daviot, Gollanfield and all the lowland areas; distillery yard at Inverness, 1961 **K**; garden weed at Pityoulish, root field at Foxhole in Glen Convinth, on the raised beach at Ardersier and among stones by the burn at Corrigarth etc.

Var. **pilosa** Syme

A variant with only a few hairs on the back of the lemma in the region of the awn; the lemmas turn grey (not reddish-brown as in *A. fatua*) at maturity.

M Distillery yard at Carron, 1968 **K,CGE**, det. CEH.

I Waste ground on Longman Point, Inverness, 1975 **E,K** det. CEH.

Var. **glabrata** Peterm.

A variant with hairless lemmas, except for a short basal tuft, becoming yellowish when ripe.

M Rubbish tip at Rothes from distillery refuse, 1958 **E**, det. CEH.

Var. **villis** (Wallr.) Hausskn.
Var. *hybrida* Aschers

A variant with lemmas glabrous on the back and with a basal tuft of hairs 1–2 mm long.

N Root field at Holme Rose, 1955 **ABD**, det. CEH.

A. strigosa Schreb.   *Bristle Oat*

Introduced. Occasional in cornfields and waste ground. Formerly much cultivated in the Hebrides. Differs from *A. fatua* in the yellow-green colour, the panicle being secund, the lemma having two straight scabrid bristles and the glumes being purple-tinged at the base.

M Cornfields at Birnie, 1840 (Gordon, *annot.*); casual at Brodie, Forres, Garmouth and Elgin (Marshall, 1899); rubbish tip at Forres, 1956 **ABD**; tip at Elgin, 1958 **E**.

N (Stables, *Cat.*); common in cornfields at Ardclach, 1888 *R. Thomson* **NRN**; plentiful in oatfields about Nairn, 1898 *ESM* **E,CGE**.

I Roadside verge at Culloden, 1955 **E**; tow-path banks at Muirtown Basin, Inverness, 1958 *E. W. Groves* **EWG**; rubbish tip on Longman Point, Inverness, 1970 **E**; raised beach at Ardersier, 1975; plentiful in an oat field at Mains of Bunachton near Dunlichity, 1976 **E,K**.

A. sativa L.   *Cultivated Oat*

Relic of cultivation in root fields and waste ground. Under-recorded.

M Distillery yard at Carron, 1968 **K**.
N Daless near Drynachan, 1965 *D. McClintock*.
I Station yard at Beauly, 1947; waste ground on Longman Point and root field at Flichity.

Helictotrichon Bess.
H. pratense (L.) Pilg.   *Meadow Oat Grass*
Avena pratensis L.
Native. Frequent on base-rich grassy places on moors and river banks. Leaves glaucous-green, stiff; sheaths glabrous; lemma biaristate at the tip; awns from above the middle of the lemma. Map 317.
M Stotfield, west of Burghead but rare in Moray, *WAS* (Murray); locally common (Gordon); Greshop near Forres, 1866 *JK* **FRS**; Darnaway, by the river at St John's Mead, 1953 *A. Melderis* **BM**; banks of the River Findhorn at Broom of Moy, 1955 **K**; grassy banks by the River Spey at Grantown and Rothes, and Knockando, 1954; banks of the River Findhorn at Relugas, 1967 **BM,K,CGE**; grass slope at Castle Roy N of Nethybridge, 1974.
N Ardclach, 1889 *R. Thomson* **NRN**; river bank at Glenferness, 1954; by the River Findhorn at Coulmony, 1960 **ABD** and 1961 **ABD,K**; Ardclach, 1973 **E**; Drynachan, 1973.
I West side of Loch Ericht near Dalwhinnie, 1833 *G. Gordon* (Watson, *Top. Bot.*); by the River Enrick at Milton, 1947; plentiful by the River Spey from Spey Dam to Kincraig, moorland slopes on Mealfuarvonie, Eskadale, Coignashie, Knockie and in Glen Feshie; grassy roadside verges at Coylumbridge, at Inchully, moor above Creag Dhubh near Laggan and on a cliff ledge at Monar, hillside at Gaick Lodge, 1972 **BM**.

H. pubescens (Huds.) Pilg.   *Downy Oat Grass*
Avena pubescens Huds.
Native. In similar situations to the former but rare in the lowlands. Leaves soft, green; lower sheaths pubescent; lemma with four points at the tip. Map 318.
M Knock of Alves, *WAS* (Murray); Findhorn at Logie (Gordon); dry bank at Lynemore N of Grantown, bank near Cromdale and by the River Spey at Rothes, 1954.
N On the 'Doocot' hill at Auldearn, *WAS* (Murray); meadow at Kilravock Castle, 1954; short grassy verge W of Cantraydoune, 1962 **ABD,BM,K,CGE**.
I Craghue in Badenoch (Gordon); bank at Milton, 1947; moorland slopes of Creag na h-Iolaire above Aviemore, at Shenachie near Tomatin and at Kincraig, 1954; abundant on the NE slopes of Mealfuarvonie, 1971 **E,BM,CGE**; grass verge by the Torvaine golf course at Inverness, and at Spey Dam near Laggan, and banks of the River Beauly at Groam of Annat, 1972 **E**; common in Badenoch and on the banks of the River Spey between Boat of Garten and Broomhill and in several places in Strathglass; hillside on Cromalt Mound at Ardersier, 1974 **E**.

Arrhenatherum Beauv.
A. elatius (L.) Beauv. ex J. & C. Presl   *Swine's Arnuts,*
   Subsp. **bulbosum** (Willd.) Hylander   *Knot Grass, False Oat*
   A. tuberosum (Gilib.) F. W. Schultz
Native. Abundant on grassy roadside verges, on waste land, in crops
and on river shingle. Base of stems with strongly swollen tubers.
M Very common (Gordon); cornfields frequent, 1856 *JK* **FRS**;
St John's Mead at Darnaway, 1953 *A. Melderis* **BM**; on dune
margins in the Culbin Forest and throughout the county, 1954.
N (Stables, *Cat.*); Ardclach (Thomson); roadside verge at
Drynachan and throughout the county, 1954; waste ground at Nairn,
1960 **K**; bank at Geddes, 1961 **ABD,K,CGE**.
I Near Inverness, 1855 *HC*, **INV**; waste ground at Fort Augustus,
at Inverfarigaig, Tomatin, and Strathglass etc., 1954; newly-sown
roadside verge at Dalwhinnie and in a meadow at Gaick Lodge,
1973.

Holcus L.
H. lanatus L.   *Pluff Grass, Yorkshire Fog*
Native. Abundant in meadows and on waste ground. Nodes not
bearded; awns included within the glumes.
M Very common (Gordon); Forres, 1871 *JK* **FRS**; ½ mile S of
Forres station, 1953 *A. Melderis* **BM**; on dunes in the Culbin Forest
and throughout the county, 1954.
N (Stables, *Cat.*); common at Ardclach, 1891 *R. Thomson* **NRN**;
meadow at Drynachan and throughout the county, 1954; waste
land at Nairn, 1960 **K**, and 1961 **ABD,K**.
I Inverness, 1832, Dalwhinnie, 1833 (Watson, *Loc. Cat.*); Tomatin,
1903 (Moyle Rogers) Dores, 1940 *AJ* **ABD**; abundant throughout
the district, 1954; common at Inchnacardoch, and 1 mile S of
Dalwhinnie, 1970 *C. E. Hubbard*; burn gully at Gaick Lodge, 1973.

H. mollis L.   *Soft Grass*
Native. Common in woods and on wet moorland and river banks.
Stem nodes hairy; awns long, exerted.
M Oakwood near Elgin herb. *Brodie* (Gordon); Greshop, 1857
JK **FRS**; Forres, 1891 *G. C. Druce* **OXF**; very common under pines
in the Culbin Forest and throughout the county 1954.
N Cawdor woods, *WAS* (Gordon); common by the schoolhouse at
Ardclach, 1888 *R. Thomson* **NRN**; wooded bank at Drynachan and
throughout the county, 1954; wood at Lethen, 1960 **ABD,K**; by
the river at Nairn, 1960 **K**.
I Badenoch (Gordon); Dalwhinnie, 1903 (Moyle Rogers)
Speyside, *G. C. Druce* **OXF**; in wet places on moors at Alvie, Gaick
and throughout the district, 1954.

Deschampsia Beauv.
D. caespitosa (L.) Beauv.   *Tufted Hair Grass*
Native. Abundant on roadside verges, in bogs, on moors and
mountain slopes up to 1,050 m in the Cairngorms. A very variable

species for which numerous varieties have been described. Awns
arising from near the base of the lemma.
M Frequent (Gordon); Waterford, 1871 *JK* **FRS**; in shade by
Loch Spynie (as var. *argentea* Gray) *ESM* **CGE**; Brodie (as var.
argentea Gray) *G. C. Druce* **OXF**; banks of the River Findhorn, on
heath, 1954 *A. Melderis* **BM**; in the Culbin Forest and throughout
the county, 1954; Greshop wood near Forres, 1956 **K.**
N (Stables, *Cat.*); very robust in moist shady places at Ardclach,
1894 *R. Thomson* **NRN**; moors at Drynachan and throughout the
county, 1954; near Nairn, 1960 **K.**
I Inverness, 1832 Dalwhinnie and Pitmain, 1833 (Watson, *Loc.
Cat.*); Ness Islands at Inverness, 1855 *HC* **INV**; Coire an t-Sneachda
of Cairn Gorm, 1887 *G. C. Druce* **OXF**; a common form in the
coiries above 610 m (as var. *brevifolia* (Parn.), in the Coire an
t-Sneachda, in Glen Einich (as var. *pseudo-alpina* Syme, viviparous
and dark coloured) and Cairngorms (as var. altissima Lamk.)
(Druce, 1888); Coire an Lochain and Coire Leacainn (as var.
alpina Gaud.) 1887 *G. C. Druce* **OXF**; Coire an Lochain (viviparous
form) 1953 *A. Melderis* **BM**; throughout the district, 1954; in
depressions at Dalwhinnie and at Inchnacardoch, 1970 *C. E.
Hubbard.*

Var. **parviflora** (Thuill.) Coss. & Germ.
A variant differing from the type in the narrower, less rough leaf-
blades (up to 2·5 mm wide) and smaller spikelets (2·5–3·5 mm long).
I Glen Urquhart, 1947 *NDS* **BM**; by the Beauly river E of Beauly,
1947 *A. J. Wilmot* **BM.**

D. alpina (L.) Roem. & Schult.   *Alpine Hair Grass*
Native. On rocks and scree in the higher mountains. Inflorescence
usually viviparous; awn from above the middle of the lemma.
M Brought down the River Spey to Speymouth, *G. Birnie* (Burgess).
I Cairngorms, *W. J. Hooker* **CGE!**; Braeriach, 1869 *JK* **FRS**;
Cairn Gorm, 1887 and Braeriach 1888 *G. C. Druce* **OXF**; Carn Eige
in Glen Affric, 1958 *D. A. Ratcliffe*; cliff ledge at 1,100 m in the Coire
an Lochain of Cairn Gorm, 1955 **K,CGE**; scree by a small burn on
Bynack More, 1975.

D. flexuosa (L.) Trin.   *Wavy Hair Grass*
Native. Abundant on moors and in open woods on acid ground.
M Frequent (Gordon); Califer near Forres, 1871 *JK* **FRS**; pine
woods in the Culbin Forest and throughout the county, 1954.
N (Stables, *Cat.*); common on dry heaths, 1894 *R. Thomson* **NRN**;
moors at Drynachan and throughout the county, 1954; moor at
Geddes, 1954 **ABD,BM**; Glenferness, 1960 **K**; near Nairn, 1971
K,CGE.
I Dalwhinnie and Pitmain, 1833 (Watson, *Loc. Cat.*); Inches near
Inverness, 1871 ? *coll.* **ELN**; Braeriach (as var. *montana* Huds.) 1888
G. C. Druce **OXF**; Coire an t-Sneachda and by a waterfall in Glen
Einich (as var. *montana* Huds.), 1887 *G. C. Druce* **OXF**. In true
montana the panicle is closed in fruit and the glumes rich purple

(Druce, 1888); Sgòran Dubh, Glen Einich, 1904 *G. B. Neilson* **GL**; Coiltry near Fort Augustus, 1904 *G. T. West* **STA**; throughout the district, 1954; Inchnacardoch, 1970 *C. E. Hubbard.*

D. setacea (Huds.) Hack.　　　　　　　　　　　*Bog Hair Grass*
Native. Occasional in acid moorland bogs and on stony loch margins. Differs from *D. flexuosa* in the very fine leaves, the upper sheaths smooth not rough, the acute ligules of 4–5 mm, the more distant spikelets and the panicle branches being covered with asperites giving them a dull appearance. Map 319.
M Blairs Loch at Altyre near Forres, and by Loch Dallas at Kinchurdy, 1887 *G. C. Druce* **OXF**; Culbin ½ mile NW of Kincorth (as *D. flexuosa*), 1937 *J. Walton* **GL**!; loch near Aviemore golf course, 1919 *G. C. Druce* **OXF** and *M. L. Wedgewood* **STA**; Winter lochs in the Culbin Forest, 1953 *NDS, JEL & A. Melderis* **BM** and in 1954 *MMcCW* **ABD,CGE**; wet muddy depression on the shingle beach W of Kingston, 1975.
N Wet heaths 3 miles E of Nairn, 1898 *ESM & WASh* **BM,K**; moorland bog by the ford at Achavraat, 1960 **ABD**.
I Loch an Eilein, Mr Groves (Druce, 1888); Loch Pityoulish and Loch Gamha, 1887 *G. C. Druce* **OXF**; Rothiemurchus, 1888 *JK* **FRS**; stoney margin of Loch Morlich, 1954 **E**; in a small bog near Spey Bridge at Boat of Garten, 1973 **E,BM**; at the S end of a lochan near Nuidhe, in bogs at Tromie Bridge; muddy shores of Loch Ceo Glais near Torness, 1976 **E**.

Aira L.
A. praecox L.　　　　　　　　　　　　　　　*Early Hair Grass*
Native. Very common on dry open moorland, track verges, sand dunes, rock pockets. Fig 50e.
M Frequent (Gordon); Clunyhill at Forres, 1856 *JK* **FRS**; at Findhorn, Garmouth and dunes at Binsness in the Culbin Forest, 1953 *A. Melderis* **BM**; among rocks at Lossiemouth, 1955; sandy track in Greshop Wood, 1960 **ABD**, and throughout the county.
N (Stables, *Cat.*); not common at Ardclach, 1891 *R. Thomson* **NRN**; open dry places on the moors at Drynachan and throughout the county, 1954; track at Dulsie Bridge, 1960 **ABD,K**; path at the cemetery at Nairn, 1961 **ABD**.
I Dalwhinnie and Pitmain, 1833 (Watson, *Loc. Cat.*); Kingussie, 1887 *G. C. Druce* **OXF**; 4 miles SE of Inverness, 1937 *J. Walton* **GL**; abundant throughout the district.

A. caryophyllea L.　　　　　*Mouse Grass, Silver Hair Grass*
Native. Common in dry gravelly places, among cinders in railway yards and in shallow soil round rocks and on wall tops.

Subsp. **caryophyllea.** Fig 50f. Map 320.
M Frequent (Gordon); Forres, 1881 *JK* **FRS**; shingle of the River Spey at Garmouth, 1953 *A. Melderis* **BM**; throughout the county, 1954; dunes at Lossiemouth, 1955 **CGE**; gravel bank in Greshop Wood, 1960 **ABD**; dry bank at Grantown, 1962 **CGE**; railway

track at Brodie station, 1963 **CGE**; yard at Manachy distillery
near Forres, 1967 **K**.
N (Stables, *Cat.*); common at Ardclach, 1899 *R. Thomson* **NRN**;
throughout the county 1954; dunes on the Nairn Dunbar golf course,
1960 **ABD,K,CGE**.
I Inverness, 1832 Dalwhinnie and Pitmain, 1833 (Watson, *Loc.
Cat.*); dry banks in Glen Strathfarrar, Invermoriston, Tomatin and
Inverfarigaig etc., 1954; shingle of the River Spey at Newtonmore,
1972 *J. D. Buchanan*; railway embankment at Culloden station,
1976 **E,BM**.

Subsp. **multiculmis** (Dumort.) Aschers.&Graebn. ex Hegi
Introduced. Occasional in similar situations to the preceding. Some
of the records for *A. caryophyllea* may belong here, and from which
it differs in being usually taller (up to 60 cm) with many stems,
ascending panicle branches, the spikelets crowded at their ends.
I Gravel path of the old Kirk at Ardersier, 1973 **K**; paths at
Ballindarroch **E,K** and at Dochfour House, 1975 **E**; top of a wall
by Loch Ness near Lurgmore, in Culloden wood and roadside
verge on the hill road at Inverfarigaig, 1975 **E**; shingle bar at
Whiteness Head, 1976 **E**; gravel drive at Viewhill, Kirkhill, 1975
E,K; at Cullochy Lock S of Fort Augustus **E**; railway track at
Broomhill, 1976 **BM**.

Corynephorus Beauv.
C. canescens (L.) Beauv.   *Grey Hair Grass*
Introduced in north Scotland. Locally plentiful on moorland dunes
near the sea. In view of the proposed development of the area at
Lossiemouth where the plant occurs, three plants were transferred
to the shingle west of Kingston and to the dunes on the golf course
at Lossiemouth in 1974 by the author and R. Richter.
M A few plants among Canary Grass and other aliens at Kingston,
and on the W side of Lossiemouth, 1900 *G. Birnie* (Burgess); dunes
on the E side of Lossiemouth, 1932 *E. H. Chater* **BM**; abundant at
Lossiemouth, 1953 *A. Melderis* **BM** and in 1955 *MMcCW*
E,ABD,K,CGE.

Ammophila Host
A. arenaria (L.) Link   *Bents, Marram Grass*
Native. Common on sand dunes the whole length of the coast.
Planted inland.
M All along the sandy places of the shore and a few plants grow
on the N side of the turnpike in the middle of the Oakwood near
Elgin (probably an old resting place of the fish women, where the
plant had fallen from their creels) (Gordon); seaside near Forres,
1857 *JK* **FRS**; Culbin, 1953 *A. Melderis* **BM**; plentiful from
Lossiemouth to Culbin 1954.
N (Stables, *Cat.*); between Findhorn and Nairn (Ewing)!;
dunes from Carse of Delnies to the Culbin Forest, 1954.
I Planted at the N end of Loch Morlich by a member of the
Forestry Commission who had taken the plants from Culbin

between 1930 and 1940; seen here by *MMcCW* in 1952 and collected
from there in 1956 **CGE**; salt-marsh verges from Lentran to
Ardersier, 1954; dunes at Fort George and at Whiteness Head, 1956.

Calamagrostis Adans.

C. epigejos (L.) Roth   *Wood Smallreed*
Native. Rare by burns and ditches.
I Ditch at the NE corner of Loch Morlich, 1956 **E**; burn near
Loch Pityoulish, 1971 *R. McBeath*.

C. canescens (Weber) Roth   *Purple Smallreed*
Native.
I Ness Islands, Inverness (as *C. lanceolata*) (Aitken).

Agrostis L.

A. canina L.   *Brown Bent*
The following *MMcCW* herbarium records have been determined
by C. E. Hubbard.

Subsp. **canina**.
Native. Frequent but over-looked. On wet moorland bogs and in
ditches. Stolons present; panicle little contracted in fruit. Awns
present. The form without an awn (forma *mutica*) is common in
the lowlands.
M Edge of a moorland bog E of Carrbridge and by Loch Mòr near
Duthil (forma *mutica*), 1973 **E**; moorland bog by Beum a'
Chlaidheimh on Dava Moor, 1973 **ABD**; ditch at Easter Limekilns,
Dava Moor (forma *mutica*) 1973 **BM**; bog E of Nethybridge (forma
mutica) 1973 **K**; ditch by the Black Burn at Pluscarden, 1976.
N Pool on the moor by the Allt Breac at Drynachan, 1973 **E**;
sandy shore of Lochloy, 1973 **E**; moorland pool at Achavraat
(forma *mutica*), 1973 **ABD**; ditch at Ordbreck, 1975.
I Kilmorack Falls near Beauly (forma *mutica*), 1889 *G. C. Druce*
OXF; Inshriach, 1961 *P. H. Davis* **E**; ditch in Glen Cannich, 1947;
by Loch Gynack at Kingussie, at Kincraig, by loch Flemington
and ditch at Foyers etc., 1954; bog to the NW of Loch Meiklie
in Glen Urquhart, and in a small bog at Boat of Garten, 1973 **E**;
roadside bog at Eskdale 1973 **ABD**; Newtonmore golf course
(forma *mutica*), 1973 **K**.

Subsp. **montana** (Hartm.) Hartm.
Native. Abundant on dry moorland. Stolons absent but rhizomes
well-developed; panicle strongly contracted and spike-like after
flowering.
M Frequent (Gordon); moors throughout the county, 1954; moor
by the Dorback Burn near Dava, 1955 **E**; moor W of Nethybridge,
1976 **E**.
N Moors at Drynachan, Holme Rose, Cawdor and Kilravock
Castle, 1954.
I Dalwhinnie and Pitmain, 1833 (Watson, *Loc. Cat.*); Cairn Gorm,
1919 *G. C. Druce* **OXF**; moors throughout the district, 1954;
abundant on grassy slopes 1 mile S of Dalwhinnie and at

Inchnacardoch, 1970 *C. E. Hubbard*; moor below Sgùrr na Lapaich in Glen Strathfarrar, 1972 **E**.

A. tenuis Sibth. *Common Bent*

Native. Abundant on grassy moors, in birch and pine woods and on dunes. Panicle spreading in fruit; lemma awnless, or very occasionally a short awn from near the top. Ligule of the sterile shoots shorter than broad.

M Very common (Gordon); Forres, 1871 ? *coll*. **ELN**; Fochabers, 1923 *G. B. Neilson* **GL**; Winter Lochs in the Culbin Forest and shingle of the River Spey at Garmouth, 1953 *A. Melderis* **BM**; birch wood at Aviemore and throughout the county, 1954; Greshop Wood near Forres, 1955 **E**.

N (Stables, *Cat*.); very common at Ardclach, 1889 *R. Thomson* **NRN**; moors at Drynachan and throughout the county, 1954; grass bank at Nairn, 1960 **K**; birch wood at Glenferness, 1964 **ABD**.

I Dalwhinnie, 1833 (Watson, *Loc. Cat*.); shores of Loch Ness, 1904 *G. T. West* **STA**; Scaniport, 1940 *AJ* **ABD**; river terraces in Glen Feshie, 1950 *G. D. Pigott*; abundant throughout the district, 1954; car park at Coire na Ciste of Cairn Gorm, 1974 **E**. Dwarf forms known as *A. pumila* L. are frequently found on moorland tracks and open places. They are infected with the smut *Tilletia decipiens*.

M Findhorn (Druce, 1897); parishes of Dyke, Kinloss and at Kingston in Speymouth (Burgess); Boat of Garten, Fochabers and Culbin Forest etc., 1954.

I (Trail, *ASNH* **16**: 227 (1907)); disturbed waste ground by Cluanie Dam in Glen Moriston, track in Glen Feshie, bank at Ruthven Barracks, Kingussie, lay-by at Breakachy bridge, and on moorland tracks by Loch Toll a' Mhuic, by the Moy Burn and Sronlarig etc., 1974; old railway siding at Daviot station, 1975.

A. gigantea Roth *Black Bent*

Native. Frequent on cultivated ground, waste land and in woods. Ligule of the sterile shoots longer than broad; panicle larger than the preceding, spreading in flower and fruit. Map 321.

M Greshop Wood near Forres, 1953 **BM** and 1955 **K**; quarry at Lossiemouth, waste ground at Burghead, railway yard at Blacksboat and at Nethybridge, 1954; banks of Spey at Aviemore, 1955 *J. Milne* **E**; root field at Milton of Grange near Forres, 1973; barley field at Dyke, Invererne Farm and in the garden at Darnaway Castle.

N Root field at Auldearn, 1954; ditch by a cornfield at Foynesfield, 1974 **E**.

I Waste land at Kingussie and on Longman Point, 1954; rubbish tip on Longman Point, Inverness, 1970 **E**; verge of the car park at Drumnadrochit, roadside bank at Gollanfield, fields at Loch Meiklie, Erchless in Strathglass, Pityoulish, Daviot, Dalveallan, Dores and Dalcross; garden weed at Inverbrough Lodge near Tomatin, at Culloden House and by cottages at Allanfearn.

A. stolonifera L.   *Creeping Bent*
Native. Common in the lowlands, less so in the highlands. In grassy
salt-marshes and loch margins, on waste ground, often in farm yards.
Stolons present; panicle closed in fruit.
M Very common (Gordon); Waterford near Forres, 1871 *JK* **FRS**;
Findhorn (as var. *maritima* Meyer) (Druce, 1897); Culbin Sands
at Kincorth, 1937 *J*. *Walton* **GL**; shingle of the River Spey at
Garmouth and at the winter Lochs near Binsness, 1953 *A. Melderis*
BM; and 1955 *MMcCW* **E**; ditch at Rothes, 1956 **K**; edge of
the salt-marsh W of the Buckie Loch, Culbin, 1961 **K,CGE**; at
Grantown, Lochindorb and Elgin etc., 1954; dried up loch at
Speyslaw E of Lossiemouth, 1972 **E**.
N (Stables, *Cat.*); near the Druim, Nairn, 1937 *J*. *Walton* **GL**;
salt-marsh Culbin, ditches at Holme Rose and Cawdor, 1954;
sandy margin of Lochloy, 1973 **E**.
I Inverness by the sea, 1810 herb. *Borrer* **K**; Inverdruie (as
A. palustris Huds.), 1887 *G. C. Druce* **OXF**; Beauly river, 1892
(as var. *maritima* Meyer, det. W. R. Philipson as var. *palustris* Huds.),
1892 *ESM* **E**; Dalwhinnie and Tomatin, 1903 (Moyle Rogers);
dune slack at Ardersier, 1930; ditch at Drumnadrochit, 1947;
ditches at Fort Augustus, Strathglass, at Farr, Culloden and
Inverness, etc., 1954; backwater of the River Spey at Newtonmore,
1973 **E**.

A. stolonifera × tenuis
M Grassland under willows at the NE end of the Buckie Loch in
the Culbin Forest, 1955 **E**, det. CEH.

A. avenacea J. G. Gmel.
Introduced with wool 'shoddy' at Greshop House, Forres, 1958 **K**.

Apera Adans.
A. spica-venti (L.) Beauv.   *Loose Silky-bent*
Introduced in north Scotland, probably with foreign grain.
M Waterford near Forres, 1871 *JK* **FRS**.

Phleum L.
P. bertolonii DC.   *Small Timothy Grass*
Native. Frequent on dry banks and in short grassland. Plant up to
50 cm; spikelets 2–3 mm (excluding awn).
M In poor soil by roadsides in the parishes of Dyke, Forres,
Lhanbryd, Urquhart and Speymouth (Burgess); in short turf at
Carrbridge, Lochindorb, Dava Moor, Hills of Cromdale and in
the Culbin Forest, etc., 1954; grass bank near Wellhill, Kintessack,
1957 **K**; by the path to the old curling pond at Grantown, 1971
E,BM,CGE; grass slope below Castle Roy at Abernethy, 1974 **E**.
N Ardclach, 1888 *R. Thomson* **NRN**; pasture at Dulsie Bridge,
Loch of the Clans, Holme Rose and Glenferness, 1954; grass
verges at Achavraat and by the river at Nairn, 1960 **ABD,K**;
on the grass bank below the 'Doocot' at Auldearn, 1975.
I In short grass at Inverdruie, Kincraig, Kirkhill, Inverness and

Culloden, 1954; bank near Beaufort Castle, 1972 **E**; entrance to a field at Croft near Laggan and in a farm yard at Glendoe, 1972; Cromalt Mound at Ardersier, 1975 **K**; farm yard at Kirkton near Bunchrew, 1975 **E**.

P. pratense L. *Cat's Tail, Timothy Grass*
Native. Common in meadows where it is often sown for hay, and on waste ground. Plant larger than the preceding, up to 100 cm; spikelets 3–4 mm long (awn excluded). Map 322.
M Occasionally in fields and pastures throughout the county (Gordon); Balnaferry, 1871 *JK* **FRS**; waste ground at Boat of Garten and throughout the county, 1954; field at Darnaway, 1960 **K**; Aviemore, 1964 *J. M. Stirling* **E**.
N (Stables, *Cat.*); occasional at Ardclach, 1880 *R. Thomson* **NRN**; yard at Drynachan and throughout the county, 1954.
I Dalwhinnie (Moyle Rogers); at Fort Augustus, Tomich, Strathglass, Kirkhill and Glen Moriston etc., 1954; Inchnacardoch, 1970 *C. E. Hubbard*; sown along with other grasses on the ski-slopes of Cairn Gorm, 1972.

P. alpinum L. *Alpine Cat's Tail*
  P. commutatum Gaudin
Native. Rare on grass slopes and among rocks and scree on the higher mountains. Pl 24.
I Garvay Moor (Hooker *Fl. Scot.*); Badenoch, 1838 *A. Rutherford* **E**; Braeriach and Coire an t-Sneachda of Cairn Gorm, 1887 *G. C. Druce* **OXF**!; Glen Einich, 1889 *H. Groves* **BM**!; rocks at the head of Loch Einich, 1889 *JK* **FRS**; Coire an t-Sneachda of Cairn Gorm, 1893 *AS* **E**; top of Coire Garbhlach in Glen Feshie, 1944 *Rex A. Graham* (*Atlas*); Coire a' Bhein near Garva Bridge, *A. Marsland*!; on scree by the Allt a' Choire Dheirg of Bynack More, 1976.

P. arenarium L. *Sand Cat's Tail*
Doubtfully native in north Scotland. Sand dunes.
M Parishes of Drainie and Urquhart, *G. Birnie* (Burgess). Not seen in recent years.

Alopecurus L.

A. myosuroides Huds. *Black Foxtail*
Introduced in north Scotland usually with foreign grain from distillery refuse.
M Arable field near Moy House, 1952; distillery yard at Smallburn near Rothes, 1956; rubbish tip at Rothes, 1956 **K**; waste ground in Greshop Wood, 1957 **E,K**; distillery yard at Carron, 1968 **E,CGE**.
N Shingle of the River Nairn near Holme Rose, 1955; introduced with 'Swoop' at Wild Goose Cottage, Nairn, 1969 **E,ABD,CGE**; rubbish tip at Newton of Park, 1975.
I Distillery yard at Tomatin, 1961 **E,K,CGE**; distillery yard at Inverness, 1961 **K**; rubbish tip on Longman Point, Inverness, 1971 **E,BM**.

A. pratensis L.   *Meadow Foxtail*
Native. Very common on grassy roadside verges and in pastures.
Map 323.

M Frequent in the upper districts, Castle of Spynie, Pluscarden
Abbey and on the island of Lochindorb (Gordon); Waterford, 1871
JK **FRS**; pasture in the Culbin Forest and throughout the district,
1954; edge of a wood at Invererne House near Forres, 1967 **E,BM,K.**
N (Stables, *Cat.*); occasional at Ardclach, 1891 *R. Thomson* **NRN**;
meadows at Cawdor, Geddes, Glenferness and on the Carse of
Delnies etc., 1954; roadside verge at Cantraydoune, 1962 **ABD,K.**
I Dalwhinnie and Pitmain, 1833 (Watson, *Loc. Cat.*); Kingussie,
1896 *G. C. Druce* **OXF**; Inverness, 1940 *AJ* **ABD**; pastures at Ben
Alder Lodge, at Gaick, Garva Bridge and well distributed throughout
the district, 1954.

A. geniculatus L.   *Marsh Foxtail*
Native. Common in bogs and by pools in meadows, by slow-running
burns and occasionally in dry situations in farmyards, arable
fields and on waste land. Awns exceeding the glumes by 2–3 mm.
Map 324.
M Innes House, Urquhart, 1794 *A. Cooper* **ELN**; very common
(Gordon); Aviemore, 1892 *AS* **E**; Findhorn, 1896 *G. C. Druce* **OXF**;
Buckie Loch in the Culbin Forest and throughout the county, 1954;
bog at Lochinvar near Elgin, 1957 **K**; pond by the Muckle Burn
at Dyke, 1963 **ABD,BM,CGE**; shores of Loch Spynie, 1964 **K**; bog
at Darnaway, 1967 **E**.
N (Stables, *Cat.*); by the schoolhouse at Ardclach, 1890 *R. Thomson*
NRN; runnel on the moors at Drynachan and throughout the
county, 1954; bog at Dulsie Bridge, 1960 **ABD,K**; ditch by the
river at Nairn, 1961 **ABD,K.**
I Inverness, 1832, Pitmain, 1833 (Watson, *Loc. Cat.*); Dalwhinnie,
1840 herb. *Borrer* **K**; well distributed in bogs throughout the district,
1954; farm yard at Affric Lodge, 1947; rootfield at Daviot, dry
waste ground at Inverness and farm yard at Glendoe, etc., 1975.

A. aequalis Sobol.   *Orange Foxtail*
Native. Very rare in bogs. Similar in general appearance to the
preceding but with the awn from near the middle, straight, included
in the glumes or very slightly exceeding them.
I In a dried up pool at the edge of a grass field N of Croy, 1955
E,K,CGE. New to Scotland.

A. alpinus Sm.   *Alpine Foxtail*
Native. On wet grassy slopes on the higher mountains. Pl 24.
I Hill in Glen Feshie, 1831 (Gordon); Braeriach and Glen Einich,
1888 *G. C. Druce* **OXF**; by the side of a torrent in Glen Einich, 1893
AS **E**!; by the Allt Coire Chùirn near Dalwhinnie, 1911 *ESM*
BM,CGE!; Allt a' Choire Chàis, 1911 *ESM* **BM**; Bynack More,
Cairngorms, 1927 *R. Meinertzhagen* **BM**!; grass slope below the
cliffs of Coire an t-Sneachda of Cairn Gorm, 1953; Mòine Mhòr in

Glen Feshie, 1961 *D. A. Ratcliffe*; slopes above Loch Dubh in Glen Banchor and by a burn on the E side of Coire an t-Slugain W of Garva Bridge, 1973 *R. E. Groom* **E**; grass slopes on Meall a' Chaoruinn Mòr and in Coire a' Bhein near Garva Bridge, 1975!.

Milium L.
M. effusum L. *Wood Millet*
Native. Occasional in damp woods. Map 325.
M Elginshire, 1837 *G. Gordon* **E**; Dunphail and Oakwood near Elgin, wood at Darnaway, herb. *Brodie* (Gordon); wood at Sanquhar House, Forres, 1956 **E**; policies of Brodie Castle, 1963 **ABD,BM, CGE**; policies of Newton House and in the oak wood at the Beild near Elgin; by the River Findhorn at Relugas; in a small wood E of Burgie distillery, 1967 **CGE**; in a ravine by a small burn at Slochd N of Carrbridge, 1973 **E**; banks of the Divie Burn at Glenerney, 1974 **E**. **N** Cawdor, 1830 (Gordon); Cawdor woods, 1834 *WAS* **E**; by the river at Nairn, 1962 **ABD,CGE**.
I Bught islands, Inverness (Aitken); lane by Achnagairn Farm at Kirkhill, 1967 **K** and 1974 **E**; Reelig Glen, 1968.

Anthoxanthum L.
A. odoratum L. *Sweet Vernal Grass*
Native. Abundant in pastures, moors and mountain slopes on acid or base-rich soil. Fig 50g.
M Very common (Gordon); Waterford, 1871 *JK* **FRS**; the Leen at Garmouth, 1953 *A. Melderis* **BM**; dunes in the Culbin Forest and throughout the county, 1954; Lossiemouth, 1963 *V. Mackinnon* **E**. **N** (Stables, *Cat.*); common in pastures at Ardclach, 1889 *R. Thomson* **NRN**; moors at Drynachan and throughout the county, 1954; river bank at Glenferness, 1960 **K**; grass verge to the golf course at Nairn, 1961 **ABD,K**.
I Inverness, 1832, Dalwhinnie and Pitmain, 1833 (Watson, *Loc. Cat.*); Càrn a' Chuilinn near Glen Doe, 1904 *G. T. West* **STA**; Tomatin, 1903 (Moyle Rogers); river terraces in Glen Feshie, 1950 *C. D. Pigott*; Coire an Lochain of Cairn Gorm, 1953 *A. Melderis* **BM**; throughout the district, 1954; Inchnacardoch, 1970 *C. E. Hubbard*.

Phalaris L.
P. arundinacea L. *Reed Canary Grass*
Native. Abundant in wet places by rivers, burns, marshes and ditches. The dead leaves persist throughout the winter. Map 326. A variant (var. *picta* L.) with variegated leaves sometimes escapes from gardens.
M By the River Lossie at Deanshaugh, at Aldroughty and on the banks of the River Spey above Rothes, 1838 *G. Gordon & WAS* (Gordon); Birdsyards (Sanquhar) House, Forres, 1857 *JK* **FRS**; Buckie Loch in the Culbin Forest and throughout the county, 1954. **N** (Stables, *Cat.*); Loch Belivat and Ardclach, 1888 *R. Thomson* **NRN**; ditch at Holme Rose and throughout the county, 1954; margin of a pool at Kilravock Castle, 1975.

I Inverness, 1832 (Watson, *Loc. Cat.*); by Loch Ness at Fort
Augustus, 1850 *J. Ball* **E**; pond near Daviot (Aitken); by the River
Oich and on Cherry Island in Loch Ness, 1904 *G. T. West* **E,STA**;
common in Strathspey and Strathglass and in all the lowland areas,
but rare in the lesser glens, 1954; pasture at Dalwhinnie, 1974.

P. canariensis L.   *Canary Grass*
Introduced with foreign grain and bird seed. Occasionally sown
for pheasant food.
M Potato field at Garmouth, 1898 *ESM* **CGE**; parishes of Dyke,
Forres, Urquhart, Knockando and Speymouth (Burgess); rubbish
tip at Forres, 1950; shingle of the River Spey at Fochabers and
from bird seed at Asliesk Farm near Burgie, 1954; from distillery
refuse at Rothes, 1956 **CGE**; and in 1966 **E**; rubbish tip at Elgin,
1961 **E,ABD,K**; sown for pheasant food in the woods at Brodie
Castle, 1963.
N Nairn, 1860 *JK* **FRS**; Ardclach (Thomson); rubbish tip at
Nairn and Newton of Park, 1954.
I Inverness, 1871 ? *coll.* **ELN**; waste land at Beauly, 1957; from
distillery refuse on the rubbish tip at Longman Point, Inverness,
1970 **E**; waste land at Clava, 1973 *M. Barron*; raised beach at
Ardersier, 1975.

P. minor Retz.   *Lesser Canary Grass*
Introduced with wool 'shoddy'.
M Greshop House, Forres, 1958 **E**.

P. paradoxa L.   *Awned Canary Grass*
Introduced with wool 'shoddy'.
M Greshop House, Forres, 1958 **CGE**.

  Var. **praemorsa** Coss. & Dur.
M Greshop House, Forres, 1958 **CGE**.

Hainardia Greuter
H. cylindrica (Willd.) Greuter
  Monerma cylindrica (Willd.) Coss. et Dur.
N Introduced with 'Swoop', Wild goose Cottage, Nairn, 1969 **E**.

Parapholis C. E. Hubbard
P. incurva (L.) C. E. Hubbard   *Curved Hard-grass*
M Introduced with wool 'shoddy' at Greshop House, Forres,
1958 **K**.

Nardus L.
N. stricta L.   *Mat Grass*
Native. Abundant on moors and mountains.
M Very common (Gordon); Forres, 1857 *JK* **FRS**; Buckie Loch,
Culbin, 1953 *A. Melderis* **BM**; throughout the county, 1954.
N (Stables, *Cat.*); common on moors at Ardclach, 1887 *R. Thomson*
NRN; moors at Drynachan and throughout the county, 1954; golf
course at Nairn, 1960 **ABD**; moor by Loch of the Clans, 1960 **K**.

I Inverness, 1832 Pitmain, 1833 (Watson, *Loc. Cat.*); Badenoch, 1838 *A. Rutherford* **E**; Tomatin, 1903 (Moyle Rogers); Loch an Eilein, 1954 *A. Currie* **Nat. Con. E**; throughout the district, 1954.

Echinochloa Beauv.
E. crus-galli (L.) Beauv. *Cockspur*
Panicum crus-galli L.
Introduced with foreign grain and bird seed. Occasional on rubbish tips.
M Elgin tip, 1975 **E.**
N Newton of Park tip, 1975.
I The Longman tip at Inverness, 1970 **E,K.**

E. utilis Ohwi & Yabuno
E. frumentacea auct.
Introduced with foreign grain and bird seed. Rare on rubbish tips.
M Elgin tip, 1972 **E.**
I The Longman tip at Inverness, 1970 **E,K,CGE.**

Panicum L.
P. capillare L. *Old-witch Grass*
Introduced with bird seed.
N From 'Swoop' fed to birds at Wild Goose Cottage, Nairn, 1969 **E.**

P. miliaceum L.
Introduced with foreign grain and bird seed. Frequent on rubbish tips but rarely flowers in the north.
M Elgin tip, 1925 etc. and 1970 **E.**
N From 'Scoop' at Nairn, 1969 and on Newton of Park tip, 1975.
I From distillery refuse on the Longman tip at Inverness, 1970 **E.**

Digitaria P. C. Fabr.
D. ciliaris (Retz.) Koel.
D. sanguinalis (L.) Scop.
Introduced with bird seed.
N From 'Swoop' fed to birds at Wild Goose Cottage, Nairn, 1969 **E,ABD,BM,K,CGE** det. CEH.

Setaria Beauv.
S. viridis (L.) Beauv. *Green Bristle-grass*
Introduced.
M Speymouth, *G. Birnie* (Burgess); Forres rubbish tip, 1970 **E,K.**
N From 'Swoop' at Nairn, 1969 **E,ABD,BM,CGE.**

S. glauca (L.) Beauv. *Yellow Bristle-grass*
S. lutescens (Weigel) Hubbard
Introduced with foreign grain and bird seed.
M Elgin rubbish tip, 1975 **E.**

S. italica (L.) Beauv. *Italian Millet*
Introduced with foreign grain and bird seed. Common on rubbish tips.

M Elgin tip, 1958 E; Forres, tip 1970 **E.**
N Newton of Park tip (not flowering), 1975.
I The Longman tip at Inverness, 1970.

Sasa Makino & Shibata
S. palmata (Mitford) E. G. Camus   *Bamboo*
Planted in the policies of some of the larger houses.
M Brodie Castle (flowering), 1967 **E,BM,CG.**

ADDENDA

Lycopodiella inundata (L.) Holub
M Clashdhu, Altyre Woods near Forres, 1977 *R. Richter.*

Cardaria draba (L.) Desv.
I Waste land near Fort George, 1977 *MMcCW* **E.**

Cerastium arctium Lange
I Coire Garbhlach, Glen Feshie, 1977 *D. Tennant.*

Lathyrus aphaca L.
M Bird seed alien Elgin tip, 1977 *OMS & MMcCW* **E.**

Lens culinaris Medicus
M Bird seed alien Elgin tip, 1977 *OMS & MMcCW* **E.**

Rosa coriifolia Fr. var. **lintoni** Scheulz
Pl 8 f.
M Roadside verge by the old railway line near Lossiemouth
cemetery, 1977 **E.**

Rosa mollis Sm. var. **relicta** H. Harrison
Common throughout both vice-counties. Flowers white with pink
stripe on reverse of petals. Needs further study. Pl 9b.
M Hedge at Kintessack, 1952; hedge at Dyke etc., 1976.
N Roadside scrub at Kingsteps near Nairn, 1954; Cawdor and
Lethen etc., 1976.
I Grass bank by Borlas House, Fort Augustus, 1954; Newtonmore
golf course, roadside south of Nethybridge etc., 1977.

Sedum lydium Boiss
N Abundant on the gravel drive at Geddes House, 1954 **E.**

Oenothera erythrosepala Borbás
N Newton of Park rubbish tip, 1977 **E.**

Eryngium maritimum L.   *Sea Holly*
M Shingle bar Findhorn, 1952.

Amsinckia intermedia Fisch. & Mey
M Cloddach tip near Hillhead, Elgin, 1977 **E.**

Leontodon taraxacoides (Vill.) Mérat
I Ten plants on the newly sown verge of the A9 near Dalwhinnie, 1977 *S. & V. Heyward* !.

Hieracium hastiforme Sell & West
I Sgùrr na Lapaich, Glen Strathfarrar, 1971 *J. N. Mills* **JNM**.

Potamogeton crispus L.
I Loch Etteridge, Glen Truim, 1977 *S. Haywood & MMcCW*.

Allium vineale L.   *Crow Garlic*
I Embankment of the Caledonian Canal at Fort Augustus, 1977 *J. Clark* ! E.

Aira caryophyllea L. subsp. **multiculmis** (Dum.) Asch. & Graebn. ex Hegi
M Abundant in a forest ride near the Dubh Loch, Darnaway, 1977 **E**.

ERRATUM

p. 136. **M. siberica**—for *Spring Beauty* read *Pink Purslane*. *Spring Beauty* refers to **M. perfoliata**.

DISTRIBUTION MAPS

The following maps represent the distribution of 326 selected species on a 10-kilometre basis of the National Grid, Ordnance Survey. Species occurring *only* on the sea coast or on high mountains, have, in general, been omitted. The very common species have also been omitted.

Symbols

● = Records seen by the author
L = Records in herbaria or literature

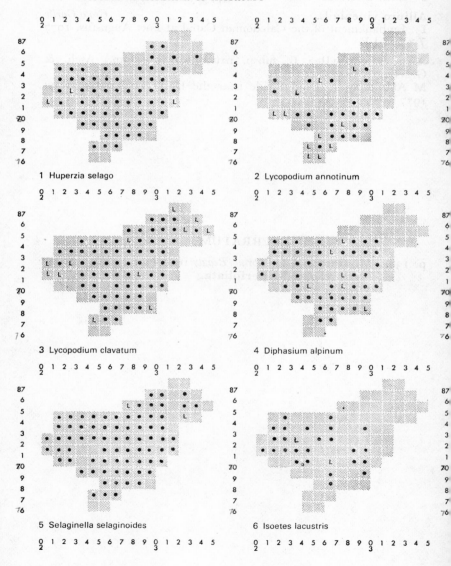

1 Huperzia selago

2 Lycopodium annotinum

3 Lycopodium clavatum

4 Diphasium alpinum

5 Selaginella selaginoides

6 Isoetes lacustris

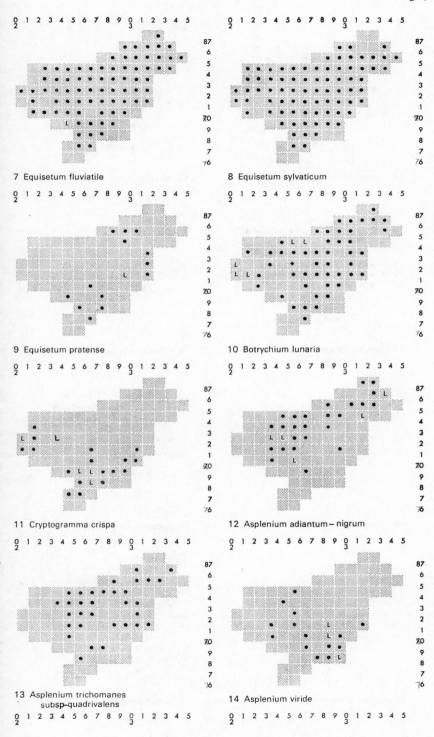

7 Equisetum fluviatile

8 Equisetum sylvaticum

9 Equisetum pratense

10 Botrychium lunaria

11 Cryptogramma crispa

12 Asplenium adiantum—nigrum

13 Asplenium trichomanes
   subsp-quadrivalens

14 Asplenium viride

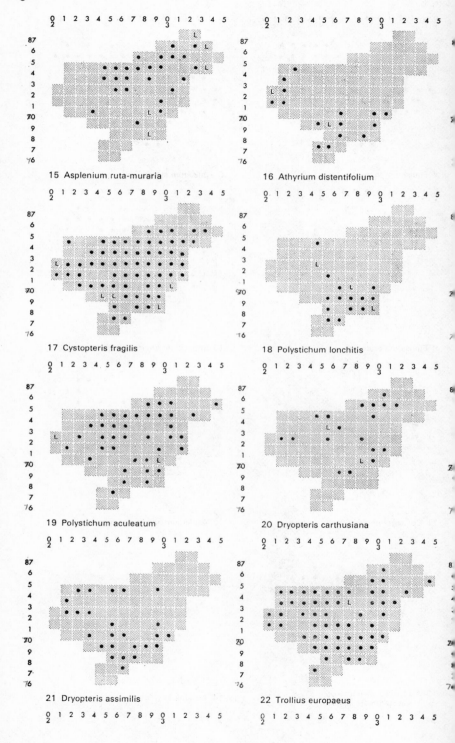

15 Asplenium ruta-muraria

16 Athyrium distentifolium

17 Cystopteris fragilis

18 Polystichum lonchitis

19 Polystichum aculeatum

20 Dryopteris carthusiana

21 Dryopteris assimilis

22 Trollius europaeus

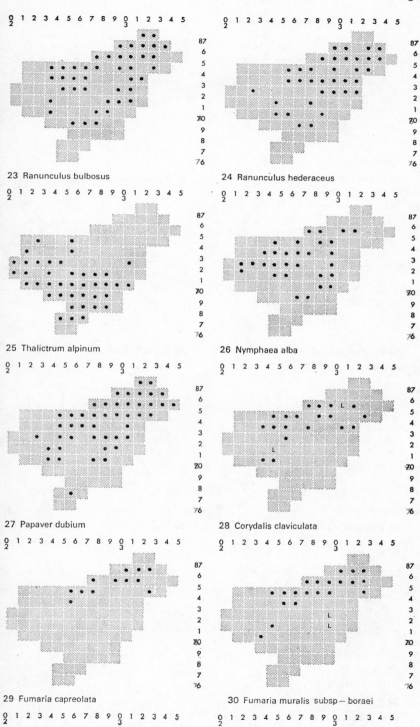

23 Ranunculus bulbosus

24 Ranunculus hederaceus

25 Thalictrum alpinum

26 Nymphaea alba

27 Papaver dubium

28 Corydalis claviculata

29 Fumaria capreolata

30 Fumaria muralis subsp – boraei

530

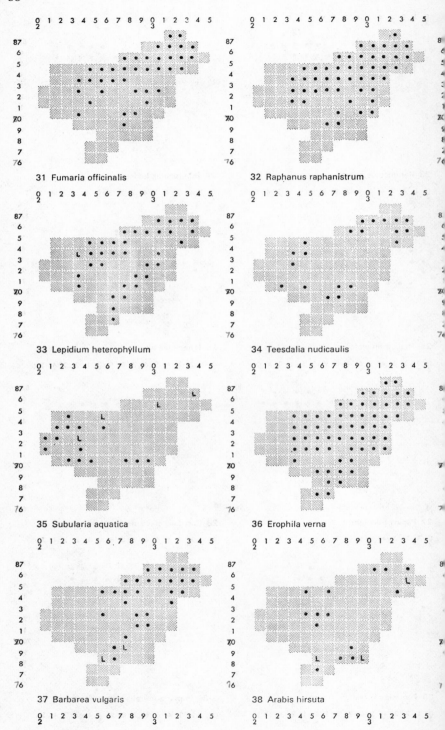

31 Fumaria officinalis

32 Raphanus raphanistrum

33 Lepidium heterophyllum

34 Teesdalia nudicaulis

35 Subularia aquatica

36 Erophila verna

37 Barbarea vulgaris

38 Arabis hirsuta

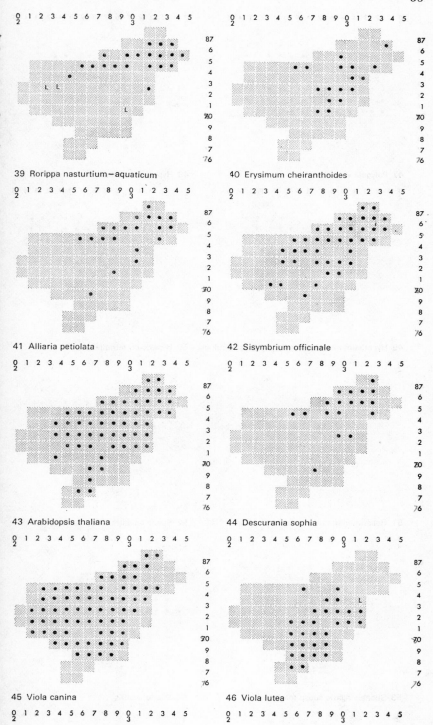

39 Rorippa nasturtium–aquaticum

40 Erysimum cheiranthoides

41 Alliaria petiolata

42 Sisymbrium officinale

43 Arabidopsis thaliana

44 Descurania sophia

45 Viola canina

46 Viola lutea

532

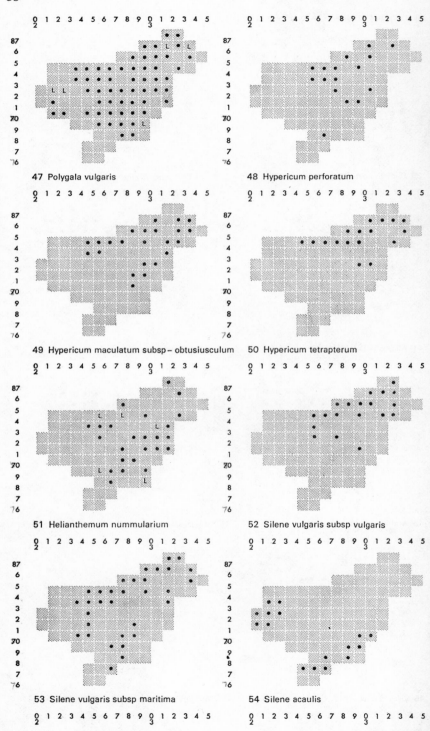

47 Polygala vulgaris

48 Hypericum perforatum

49 Hypericum maculatum subsp - obtusiusculum

50 Hypericum tetrapterum

51 Helianthemum nummularium

52 Silene vulgaris subsp vulgaris

53 Silene vulgaris subsp maritima

54 Silene acaulis

533

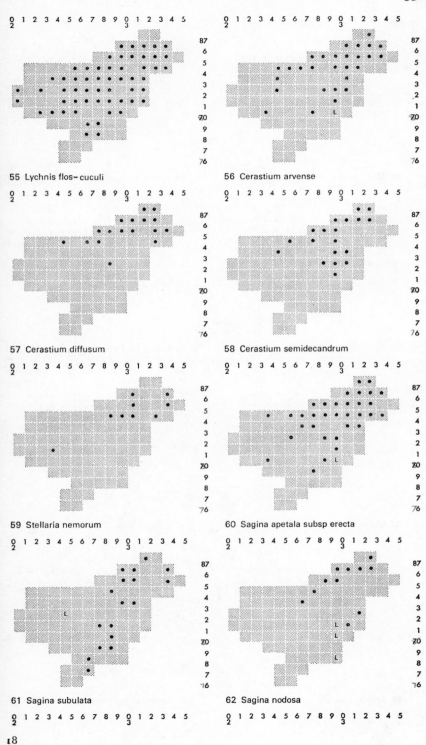

55 Lychnis flos-cuculi

56 Cerastium arvense

57 Cerastium diffusum

58 Cerastium semidecandrum

59 Stellaria nemorum

60 Sagina apetala subsp erecta

61 Sagina subulata

62 Sagina nodosa

18

534

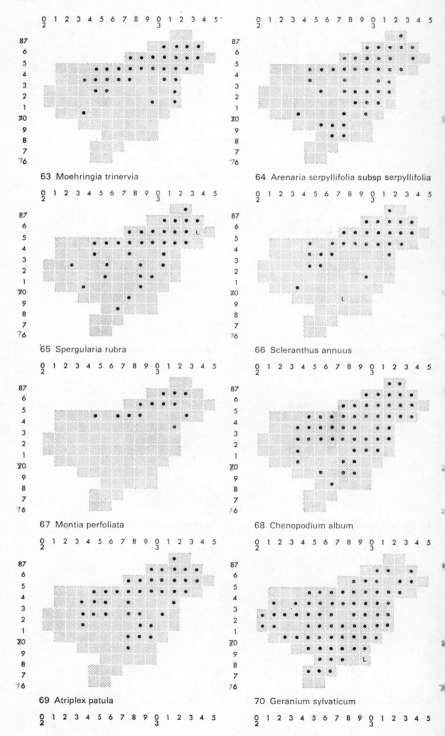

63 Moehringia trinervia

64 Arenaria serpyllifolia subsp serpyllifolia

65 Spergularia rubra

66 Scleranthus annuus

67 Montia perfoliata

68 Chenopodium album

69 Atriplex patula

70 Geranium sylvaticum

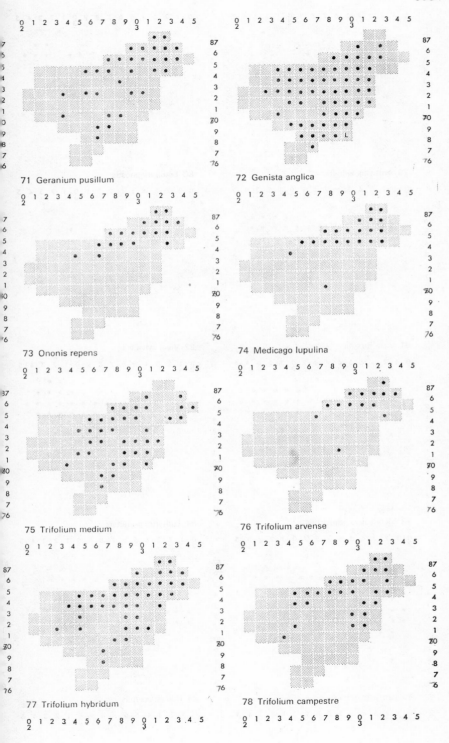

71 Geranium pusillum

72 Genista anglica

73 Ononis repens

74 Medicago lupulina

75 Trifolium medium

76 Trifolium arvense

77 Trifolium hybridum

78 Trifolium campestre

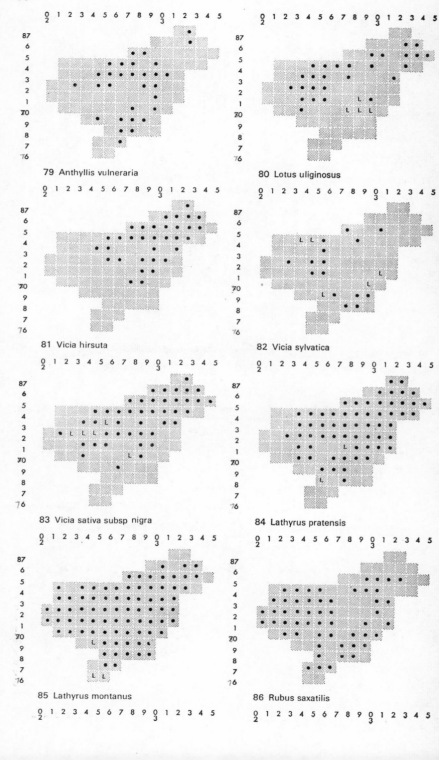

536

79 Anthyllis vulneraria

80 Lotus uliginosus

81 Vicia hirsuta

82 Vicia sylvatica

83 Vicia sativa subsp nigra

84 Lathyrus pratensis

85 Lathyrus montanus

86 Rubus saxatilis

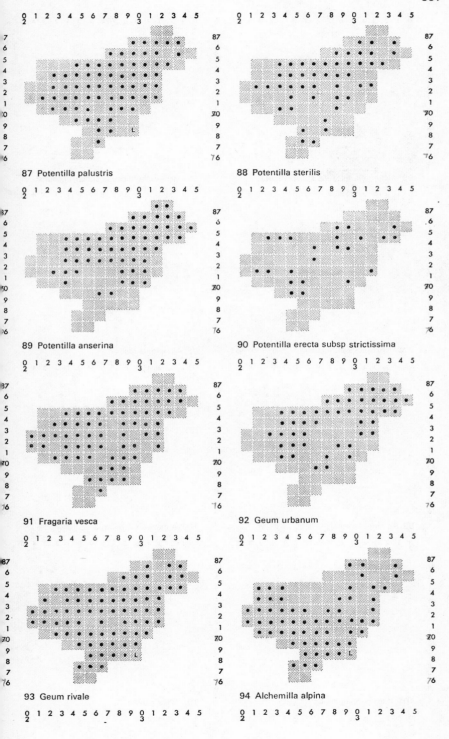

87 Potentilla palustris

88 Potentilla sterilis

89 Potentilla anserina

90 Potentilla erecta subsp strictissima

91 Fragaria vesca

92 Geum urbanum

93 Geum rivale

94 Alchemilla alpina

538

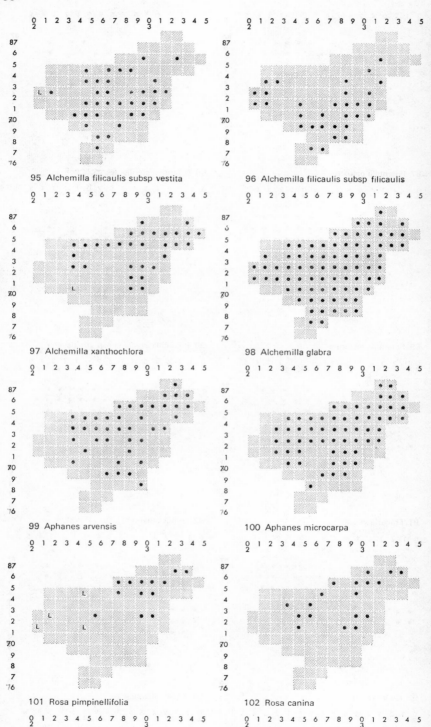

95 Alchemilla filicaulis subsp vestita

96 Alchemilla filicaulis subsp filicaulis

97 Alchemilla xanthochlora

98 Alchemilla glabra

99 Aphanes arvensis

100 Aphanes microcarpa

101 Rosa pimpinellifolia

102 Rosa canina

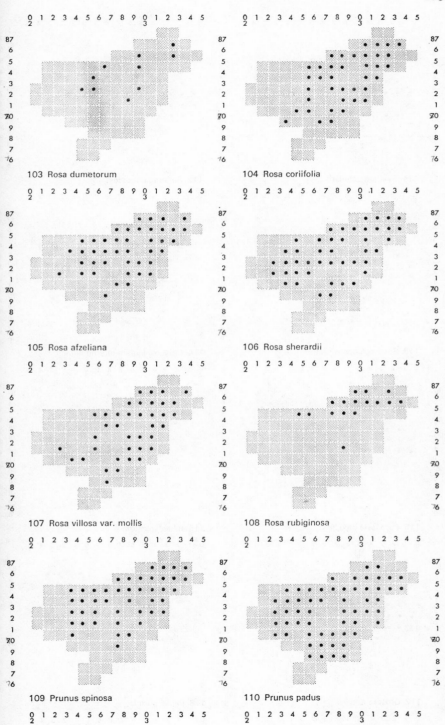

103 Rosa dumetorum

104 Rosa coriifolia

105 Rosa afzeliana

106 Rosa sherardii

107 Rosa villosa var. mollis

108 Rosa rubiginosa

109 Prunus spinosa

110 Prunus padus

540

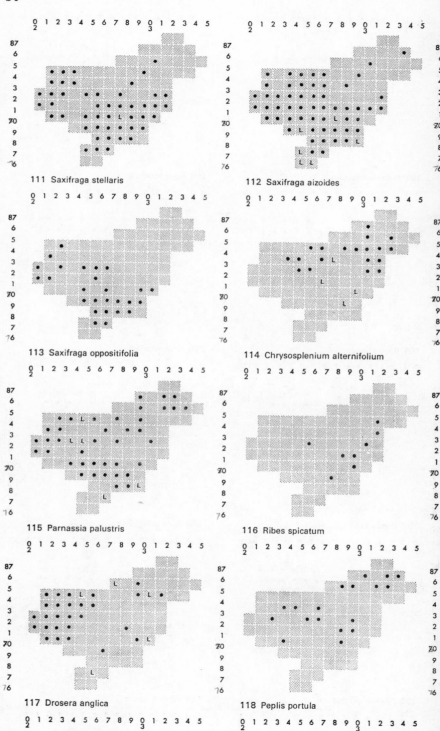

111 Saxifraga stellaris

112 Saxifraga aizoides

113 Saxifraga oppositifolia

114 Chrysosplenium alternifolium

115 Parnassia palustris

116 Ribes spicatum

117 Drosera anglica

118 Peplis portula

54 I

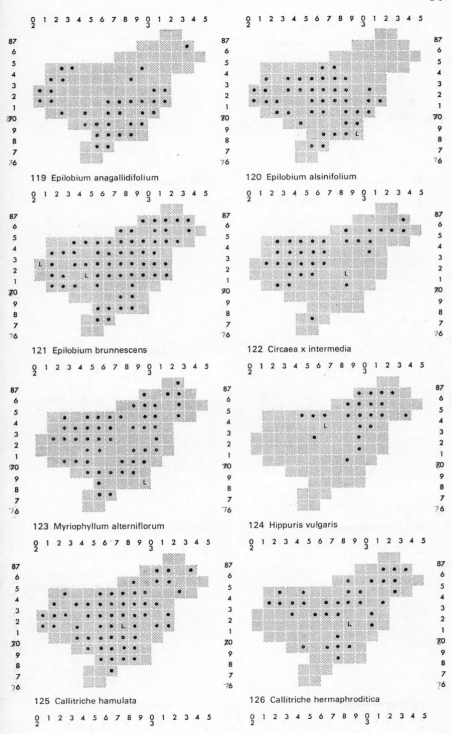

119 Epilobium anagallidifolium

120 Epilobium alsinifolium

121 Epilobium brunnescens

122 Circaea x intermedia

123 Myriophyllum alterniflorum

124 Hippuris vulgaris

125 Callitriche hamulata

126 Callitriche hermaphroditica

542

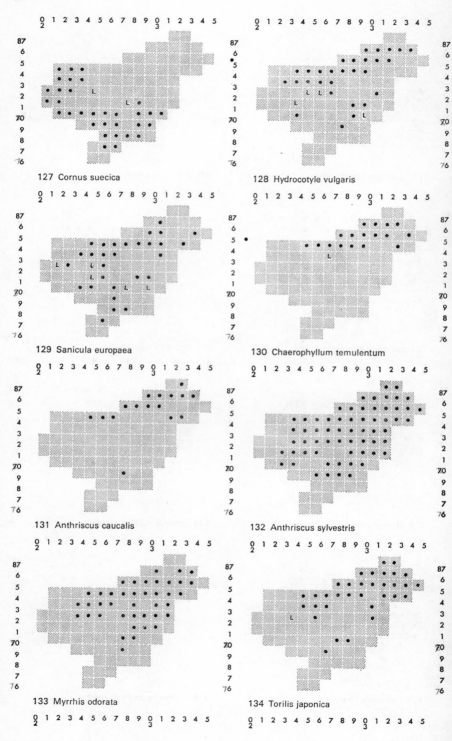

127 Cornus suecica

128 Hydrocotyle vulgaris

129 Sanicula europaea

130 Chaerophyllum temulentum

131 Anthriscus caucalis

132 Anthriscus sylvestris

133 Myrrhis odorata

134 Torilis japonica

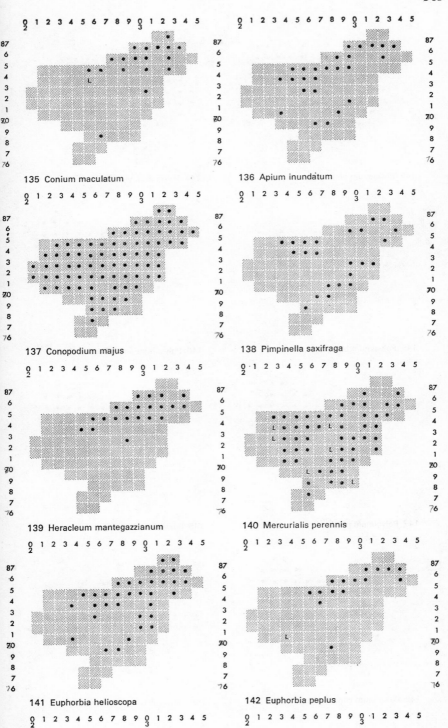

135 Conium maculatum

136 Apium inundatum

137 Conopodium majus

138 Pimpinella saxifraga

139 Heracleum mantegazzianum

140 Mercurialis perennis

141 Euphorbia helioscopa

142 Euphorbia peplus

544

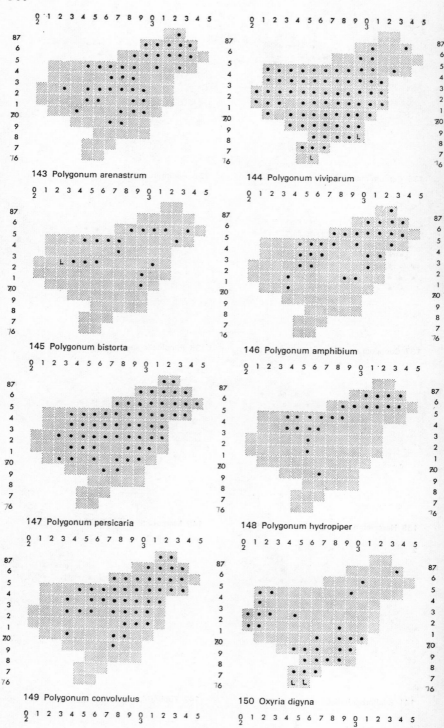

143 Polygonum arenastrum

144 Polygonum viviparum

145 Polygonum bistorta

146 Polygonum amphibium

147 Polygonum persicaria

148 Polygonum hydropiper

149 Polygonum convolvulus

150 Oxyria digyna

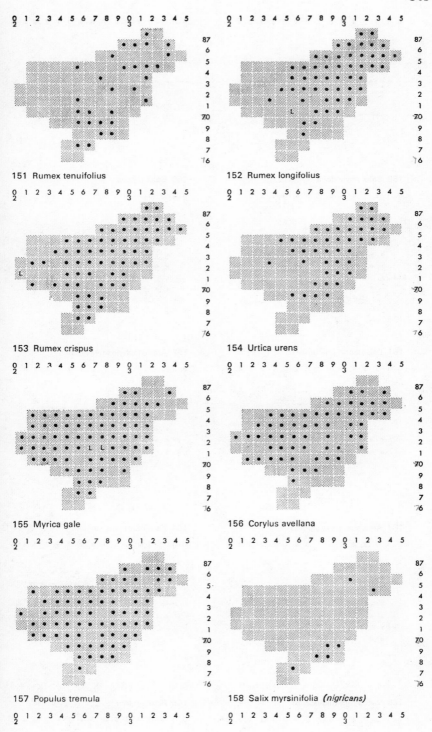

151 Rumex tenuifolius

152 Rumex longifolius

153 Rumex crispus

154 Urtica urens

155 Myrica gale

156 Corylus avellana

157 Populus tremula

158 Salix myrsinifolia *(nigricans)*

546

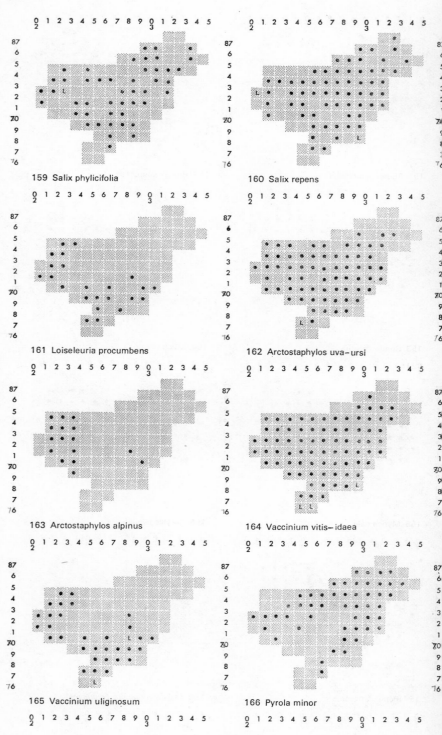

159 Salix phylicifolia

160 Salix repens

161 Loiseleuria procumbens

162 Arctostaphylos uva-ursi

163 Arctostaphylos alpinus

164 Vaccinium vitis-idaea

165 Vaccinium uliginosum

166 Pyrola minor

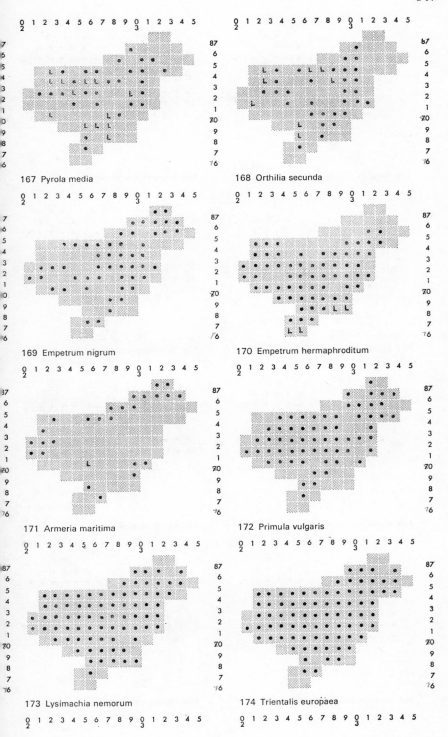

167 Pyrola media

168 Orthilia secunda

169 Empetrum nigrum

170 Empetrum hermaphroditum

171 Armeria maritima

172 Primula vulgaris

173 Lysimachia nemorum

174 Trientalis europaea

548

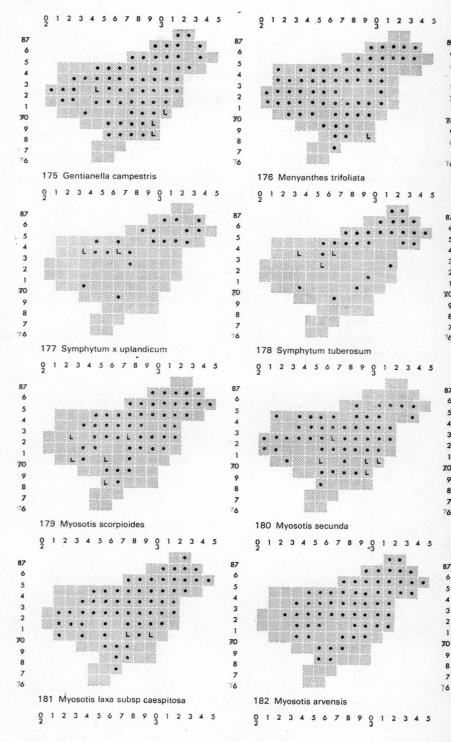

175 Gentianella campestris

176 Menyanthes trifoliata

177 Symphytum x uplandicum

178 Symphytum tuberosum

179 Myosotis scorpioides

180 Myosotis secunda

181 Myosotis laxa subsp caespitosa

182 Myosotis arvensis

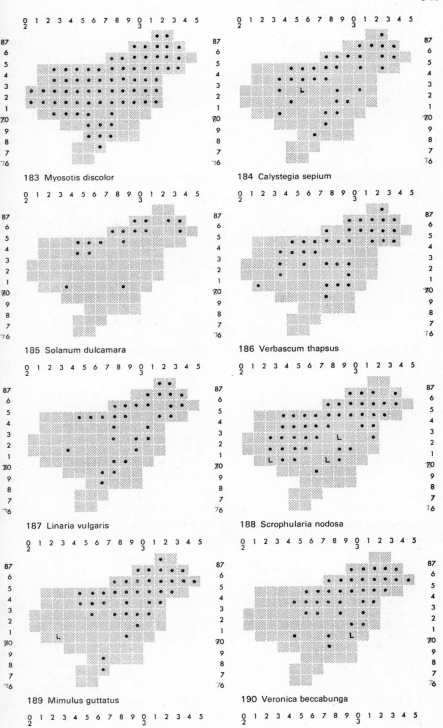

183 Myosotis discolor

184 Calystegia sepium

185 Solanum dulcamara

186 Verbascum thapsus

187 Linaria vulgaris

188 Scrophularia nodosa

189 Mimulus guttatus

190 Veronica beccabunga

550

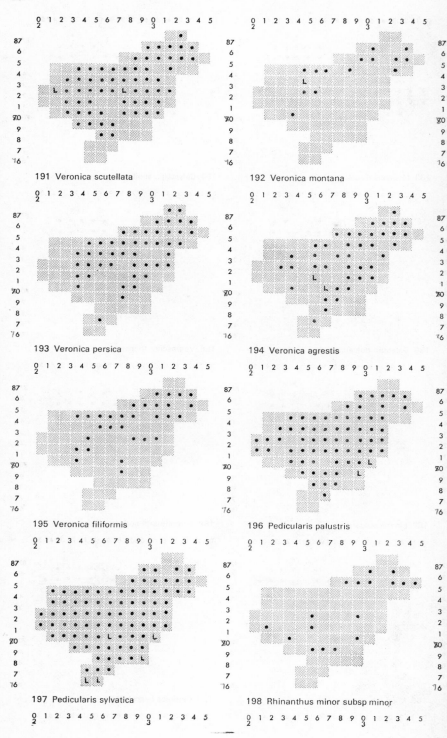

191 Veronica scutellata

192 Veronica montana

193 Veronica persica

194 Veronica agrestis

195 Veronica filiformis

196 Pedicularis palustris

197 Pedicularis sylvatica

198 Rhinanthus minor subsp minor

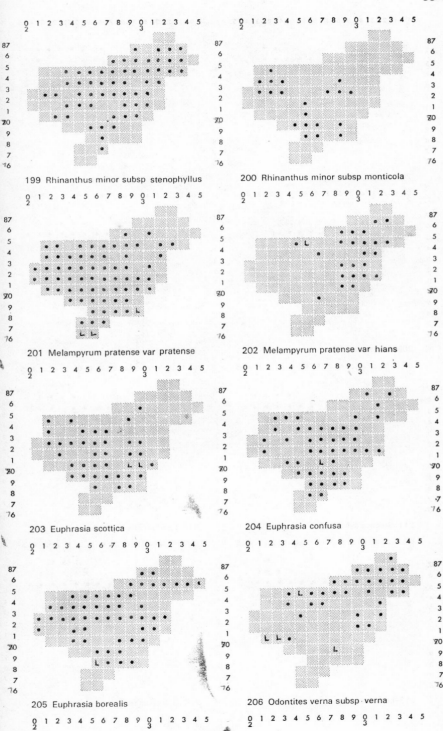

199 Rhinanthus minor subsp stenophyllus

200 Rhinanthus minor subsp monticola

201 Melampyrum pratense var pratense

202 Melampyrum pratense var hians

203 Euphrasia scottica

204 Euphrasia confusa

205 Euphrasia borealis

206 Odontites verna subsp verna

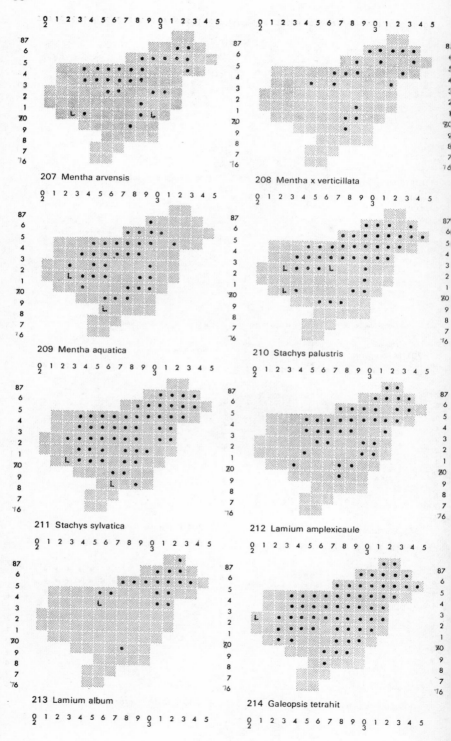

207 Mentha arvensis

208 Mentha x verticillata

209 Mentha aquatica

210 Stachys palustris

211 Stachys sylvatica

212 Lamium amplexicaule

213 Lamium album

214 Galeopsis tetrahit

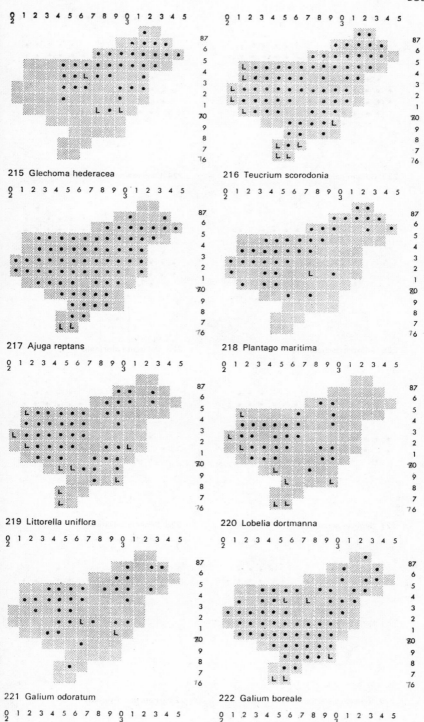

215 Glechoma hederacea

216 Teucrium scorodonia

217 Ajuga reptans

218 Plantago maritima

219 Littorella uniflora

220 Lobelia dortmanna

221 Galium odoratum

222 Galium boreale

554

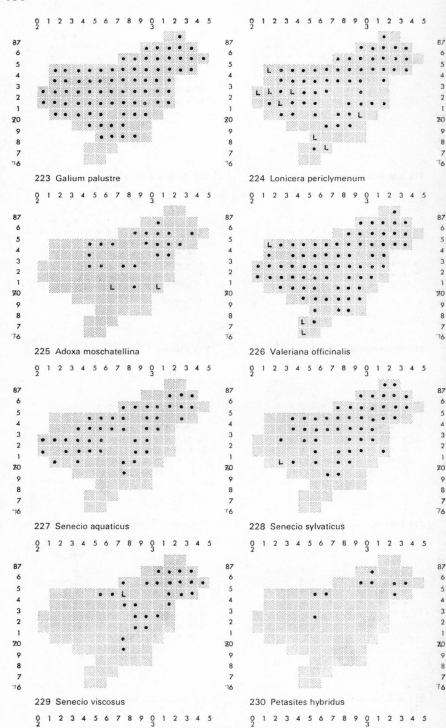

223 Galium palustre

224 Lonicera periclymenum

225 Adoxa moschatellina

226 Valeriana officinalis

227 Senecio aquaticus

228 Senecio sylvaticus

229 Senecio viscosus

230 Petasites hybridus

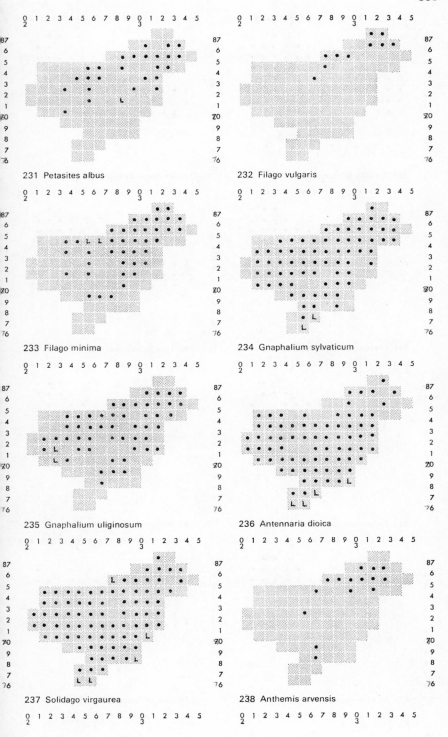

231 Petasites albus

232 Filago vulgaris

233 Filago minima

234 Gnaphalium sylvaticum

235 Gnaphalium uliginosum

236 Antennaria dioica

237 Solidago virgaurea

238 Anthemis arvensis

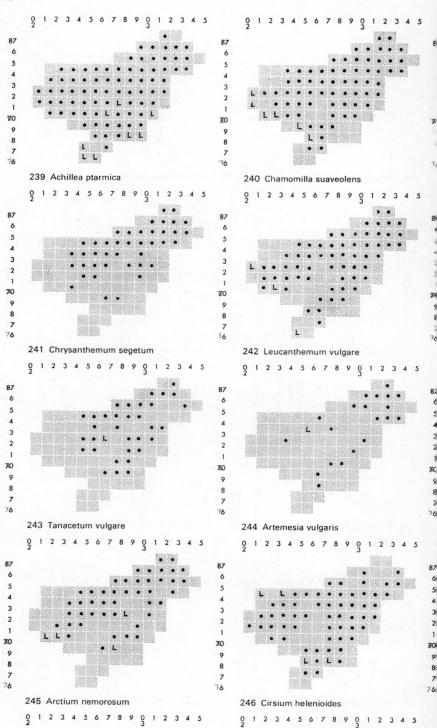

239 Achillea ptarmica

240 Chamomilla suaveolens

241 Chrysanthemum segetum

242 Leucanthemum vulgare

243 Tanacetum vulgare

244 Artemesia vulgaris

245 Arctium nemorosum

246 Cirsium helenioides

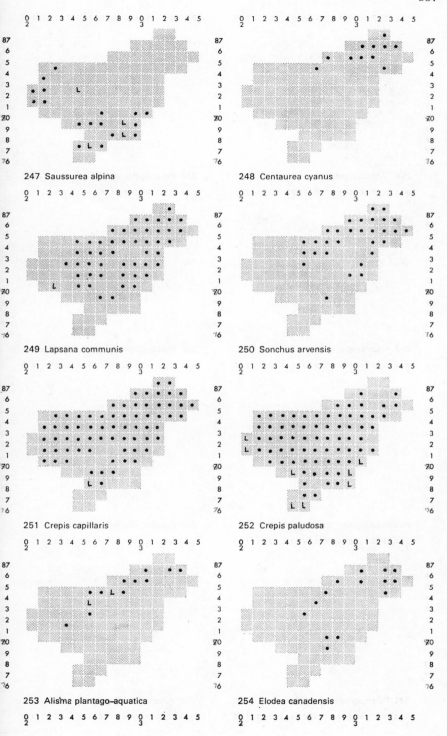

247 Saussurea alpina

248 Centaurea cyanus

249 Lapsana communis

250 Sonchus arvensis

251 Crepis capillaris

252 Crepis paludosa

253 Alisma plantago–aquatica

254 Elodea canadensis

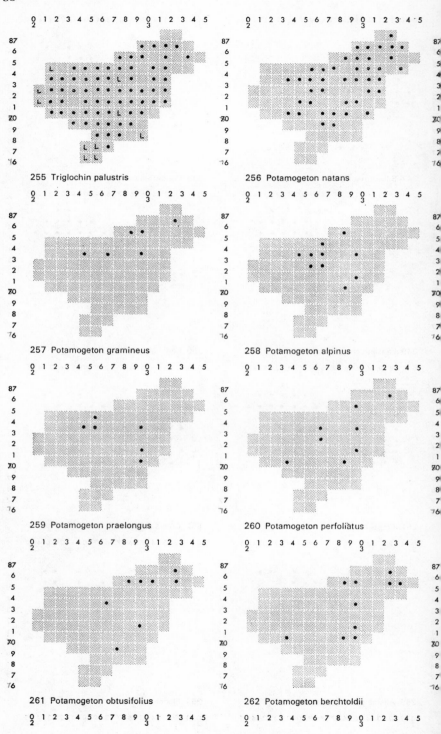

255 Triglochin palustris

256 Potamogeton natans

257 Potamogeton gramineus

258 Potamogeton alpinus

259 Potamogeton praelongus

260 Potamogeton perfoliatus

261 Potamogeton obtusifolius

262 Potamogeton berchtoldii

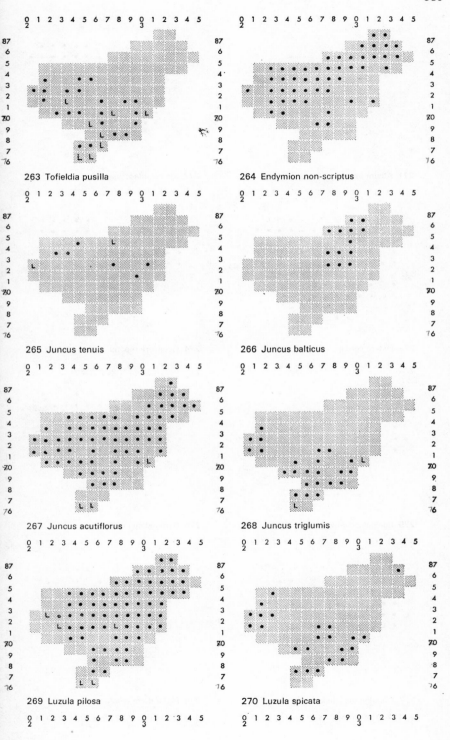

263 Tofieldia pusilla

264 Endymion non-scriptus

265 Juncus tenuis

266 Juncus balticus

267 Juncus acutiflorus

268 Juncus triglumis

269 Luzula pilosa

270 Luzula spicata

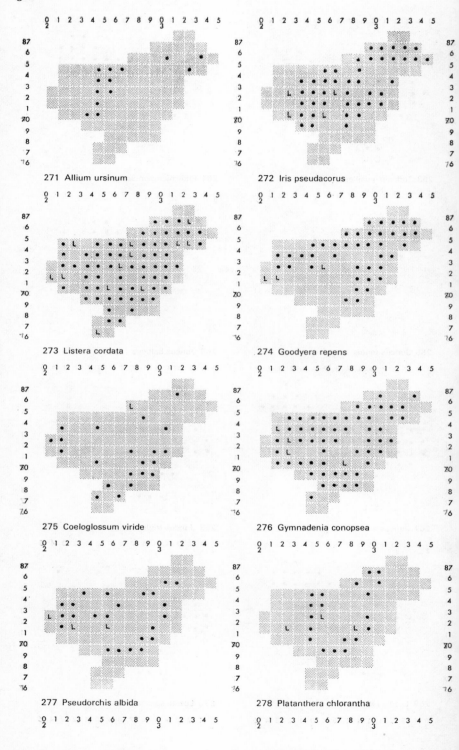

271 Allium ursinum

272 Iris pseudacorus

273 Listera cordata

274 Goodyera repens

275 Coeloglossum viride

276 Gymnadenia conopsea

277 Pseudorchis albida

278 Platanthera chlorantha

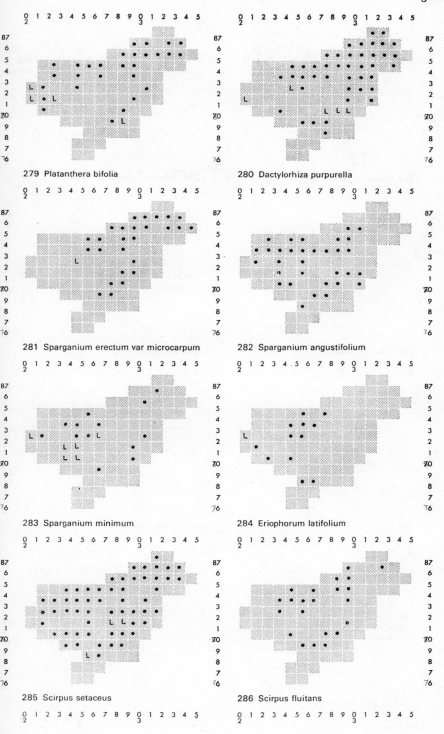

279 Platanthera bifolia

280 Dactylorhiza purpurella

281 Sparganium erectum var microcarpum

282 Sparganium angustifolium

283 Sparganium minimum

284 Eriophorum latifolium

285 Scirpus setaceus

286 Scirpus fluitans

562

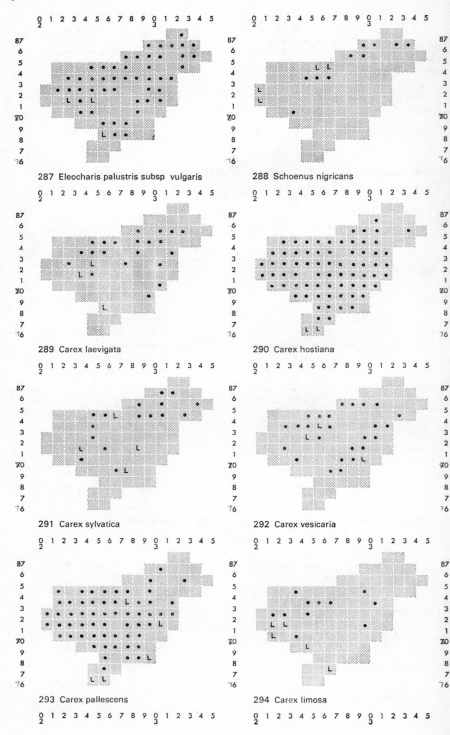

287 Eleocharis palustris subsp vulgaris

288 Schoenus nigricans

289 Carex laevigata

290 Carex hostiana

291 Carex sylvatica

292 Carex vesicaria

293 Carex pallescens

294 Carex limosa

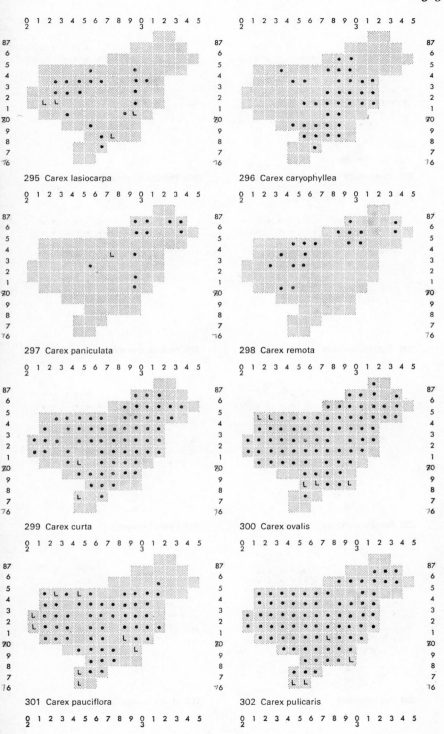

295 Carex lasiocarpa

296 Carex caryophyllea

297 Carex paniculata

298 Carex remota

299 Carex curta

300 Carex ovalis

301 Carex pauciflora

302 Carex pulicaris

564

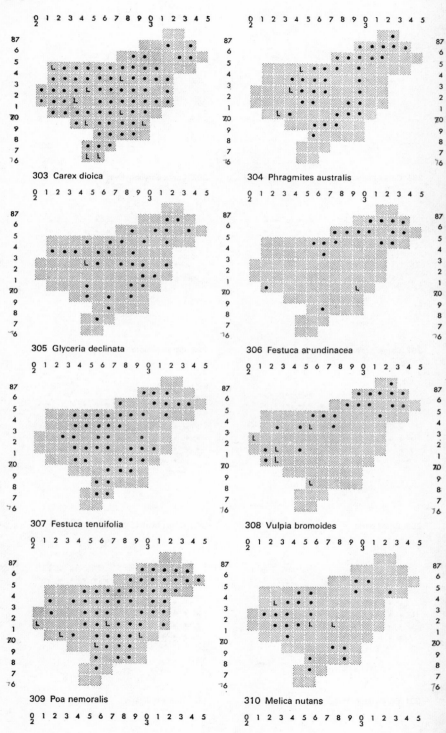

303 Carex dioica

304 Phragmites australis

305 Glyceria declinata

306 Festuca arundinacea

307 Festuca tenuifolia

308 Vulpia bromoides

309 Poa nemoralis

310 Melica nutans

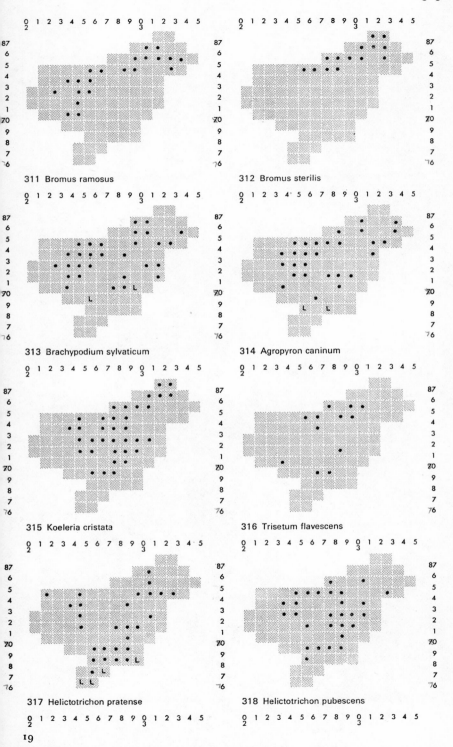

311 Bromus ramosus

312 Bromus sterilis

313 Brachypodium sylvaticum

314 Agropyron caninum

315 Koeleria cristata

316 Trisetum flavescens

317 Helictotrichon pratense

318 Helictotrichon pubescens

566

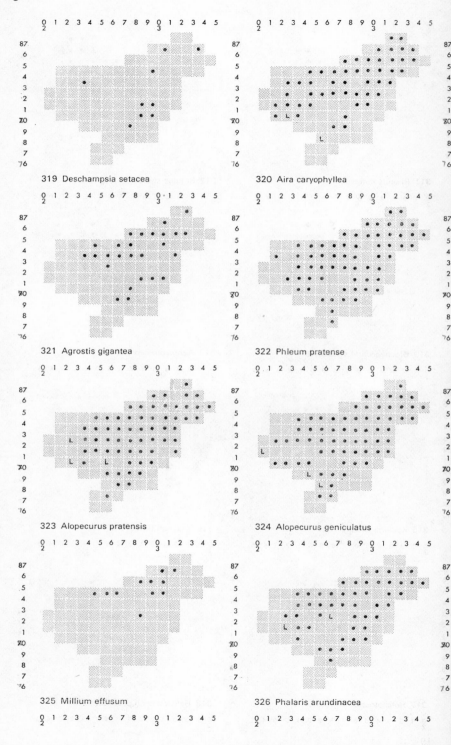

319 Deschampsia setacea

320 Aira caryophyllea

321 Agrostis gigantea

322 Phleum pratense

323 Alopecurus pratensis

324 Alopecurus geniculatus

325 Millium effusum

326 Phalaris arundinacea

BIBLIOGRAPHY

A number of important papers on Elgin and Easterness plants are referred to in Simpson (1960).

Aitken, T. Plant list, *Transactions of the Inverness Scientific and Field Club* (1875–85).

Birse, E. L. & Dry, F. T. Assessment of climatic conditions in Scotland, *Soil Survey of Scotland* (1970).

Birse, E. L. & Robertson, J. S. Natural and semi-natural vegetation of the Nairn and Cawdor district, *The Macaulay Institute for Soil Research* (1976).

Burgess, J. J. (ed.) *Flora of Moray* (1935).

Burnett, J. H. (ed.) *The Vegetation of Scotland* (1964).

Clapham, A. R., Tutin, T. G. & Warburg, E. F. *Flora of the British Isles*, edn. 2 (1962).

Druce, G. C. *The Comital Flora of the British Isles* (1932).

Gordon, G. *Collectanea for a Flora of Moray* (1839).

Murray, A. *The Northern Flora*, 1 (1836).

Perring, F. H. & Walters, S. M. (ed.), *Atlas of the British Flora* (1962),

Perring, F. H. (ed.) & Sell, P. D. *Critical Supplement to the Atlas of the British Flora* (1968).

Ross, S. M. The climate of the district of Moray in Omand, D. (ed.), *the Moray Book*, 28–48 (1976).

Simpson, N. D. *The Bibliographical Index of the British Flora* (1960).

Sinclair, Sir J. (ed.). *Statistical Accounts for Scotland* (1792–00).

Stables, W. A. Marked copy of Collectanea for a Flora of Moray, G. Gordon by W. A. Stables, no. 176, *County Catalogues*, H. C. Watson library Kew herbarium.

Watson, H. C. *Local Catalogues*, nos. 28, 43 and 44, library Kew herbarium.

Watson, H. C. *The new botanists guide*, 2 (1837).

Webster, M. McC. *Check List of the Flora of the Culbin State Forest* (1968).

APPENDIX 1

LIST OF CONTRIBUTORS OF THE FLORA

fl = flourished in the year stated

Robert M. Adam (1885–1967), R. S. Adamson (fl 1894), Thomas Aitken (fl 1885), McAllan (fl 1900), Miss J. E. Allan (fl 1838), Keith Allan, Mrs N. Allardyce, A. C. Aluntt, A. H. G. Alston (1902–58), G. Anderson (1803–78), A. Angus, E. Armitage (fl 1909), Prof. G. A. Walker Arnott (1799–1868), G. M. Ash (1900–59), A. Atkins (fl 1883).

Prof. C. C. Babington (1808–95), Backhouse (fl 1885), C. Bailey (fl 1893), Miss M. Baillie (fl 1900), Lady Maud Baillie (fl 1963), Dr Bainbridge (fl 1824), J. C. Baker (fl 1871), Prof. Sir I. B. Balfour (1855–1922), Prof. John Hutton Balfour (1808–84), John Ball (1818–89), Dr Bannerman, P. Barnes, Mrs M. Barron, Miss Bell (fl 1835), Dr R Beveridge (fl 1843), Rev. George Birnie (1860–1941), J. P. Bisset (fl 1863), Miss S. Bisset (fl 1923), Miss Sarah Bland (artist, fl 1840), R. A. Boniface, W. Borrer (1781–1862), Dr Bostock (fl 1831), W. B. Boyd (fl 1877), McBraid (fl 1840), W. Brand (1807–69), R. Bremner (fl 1800), Rev. J. B. Brichan (fl 1844), W. Brichan (fl 1832), P. A. Briggs, James Brodie of Brodie (1744–1824), Mrs D. Brodie, Miss F. Brooke, Mr Brown of Perth (nurseryman fl 1812), N. E. Brown (fl 1869), F. Browne (fl 1859), J. Buchanan (fl 1953), Mrs J. D. Buchanan, Rev. John Buchanan (fl 1893), Mrs Eleonar Bulloch, E. M. Burnett, Dr J. S. Bushman (fl 1839), Mrs. A. Berens.

Dr E. O. Callan (1912–70), Mrs E. A. Cameron, Mrs M. D. Cameron (1893–), Mrs V. Cameron, Mrs Campbell of Jura (fl 1947), Miss A. Campbell of Jura, Miss M. S. Campbell, Mrs P. Campbell, Mr Campbell (fl 1840), P. F. Cannon, J. F. Cardew (fl 1951), J. T. Carrington (fl 1875), T. A. Cartwright (fl 1885), A. Cgekalowska, E. H. Chater (fl 1932), A. Craig Christie (fl 1867), George Christie (fl 1858), Mrs Kate Christie, Mrs Ronald Christie, Mr Clark (fl 1950), Mrs Joan W. Clark, L. Clarke (fl 1934), Mr Coles, Alexander Cooper (1816–87), A. T. Coore (fl 1867), Dr R. W. M. Corner, R. H. Corstorphine (fl 1897), M. Cowan, Rev. Cowie (fl 1800), W. G. Craib (1907), M. C. Craig, E. Crapper, F. C. Crawford (fl 1902), Alexander Croall (1809–85), J. Cruickshank (fl 1830), P. Cruickshank (fl 1839), A. C. Crundwell, Mrs P. Cudmore, Andrew Currie, Mrs M. Cuthbert, John Cuthbertson.

R. W. David, Miss E. M. Davidson, Dr Peter H. Davis, Mrs E. T. Dawson, Mr Dick (fl 1839), James Dickson (1738–1822), G. Dodds (fl 1836), David Don (1799–1841), George Don (1764–1814), George Don 1798–1856, J. Don (fl 1898), J. Donald (fl 1900), Miss J. B. Dougan, Dr G. C. Druce (fl 1920), A. Duff (fl 1839), Dr U. K. Duncan, P. Dunnett, H. Du Pre (fl 1955), J. C. Duthie (fl 1870), J. W. Dyce.

Eric S. Edees, Thomas Edmondston (1805–46), A. E. Ellis, E. Edgar Evans (fl 1905), P. Ewing, Dr A. W. Exell.

P.F. (fl 1880), Alexander Falconer (fl 1839), Rev. James Farquharson (1781–1843), Rev. James Farquharson (son of preceding, fl 1868), Miss R. E. Ferrier (fl 1890), James, Earl of Fife (fl 1800), Mrs H. M. Finlay, Miss Catherine Finlayson, J. Forbes (fl 1837), Hon. Miss Fraser of Lovat (fl 1868), Rev. A. Fraser (fl 1793), James Fraser (fl 1834), Rev. James Fraser (1814–1902), Rev. John Fraser (fl 1793), M. R. Fraser, N. Fraser (fl 1863), P. N. Fraser (1830–1905), Rev. Thomas Fraser (fl 1880), John Fraser (1750–1811), John Fraser (son of the preceding, 1799–1860), John Fraser (horticulturist, 1854–1935), Prof. J. H. Fremlin.

Mr Galloway (fl 1800), J. S. Gamble (fl 1871), Rev. P. M. Garnett (fl 1960), Miss E. J. Gibbons, David Gill, Rev. J. Gillan (fl 1800), Dr K. M. Goodway, Rev. George Gordon (1801–93), Peter Gordon, Miss V. Gordon, W. Gordon (fl 1886), James A. Gossip (fl 1900), A. C. C. Gough, W. Gourlie (fl 1844), N. A. Graham, R. D. Graham (fl 1947), Rex A. H. Graham (1915–58), Prof. Robert A. Graham

(1786–1845), Lt. Col. James A Grant (1827–92), Miss Grant (fl 1835), James Grant (fl 1885), P. S. Green, A. G. Gregor (fl 1881), Greville (fl 1822), Mr Grey (fl 1837), Mrs E. R. Grigor, John Grigor (arborticulturist, 1832–81), J. W. Grimes, R. E. Groom, E. W. Groves, H. Groves (1855–1912), J. Groves (fl 1889).

Miss G. Haines, Mrs E. MacAlister Hall (fl 1929), P. M. Hall (fl 1939), W. Hall (fl 1787), Sir Thomas Hanbury (1832–1907), F. J. Hanbury (1801–93), J. A. Hankey (fl 1839), E. S. Harrison, Mrs Jean Harrison, Mrs Y. Heslop Harrison, Mr Hassley (fl 1864), David Hayes, Mrs Sally Haywood, I. C. Hedge, D. M. Henderson, Stanley Heyward, Mrs Vera Heyward, Miss A. Higginbottam, Sir W. J. Hooker (1785–1865), Dr Hope (fl 1790), G. A. Hosking, A. How (fl 1855), P. How, James Harlow Hoy (fl 1800), Dr C. E. Hubbard, Mr Hughes (fl 1838), P. F. Hunt, Mrs Mary Hunter, Miss Hutchins (fl 1808).

B. Ing, Cosmos Innes (fl 1840), Dr John G. Innes (1815–81).

Roy Jones, Miss Johnston (fl 1834), Anthony Johnston, Dr J. Johnstone (fl 1839).

I.K. (fl 1832), Rev. James Keith (1825–1905), A. G. Kenneth, Dr. D. S. Kettle, Robert Kidston (1852–1924), Rev. D. Kingston, Eric Kleintges, F. W. Knaggs.

J. Lammond, Miss A. Langton, G. V. C. Last (1875–1945), E. R. Laundon, J. R. Laundon, Miss Susan Laurenson, Mr Lawrence (fl 1840), John Lawson (fl 1830), Miss E. M. Legge, P. Leslie, Ronald, Earl of Leven and Melville (fl 189), Alexander, Earl of Leven and Melville, Rev. Augustin Ley (1842–1911), Rev. John Lightfoot (1735–1788), D. W. Lindsay, E. F. Linton (fl 1898), W. R. Linton (fl 1898), K. D. Little, Alexander Lobban (fl 1900), A. E. Lomax (fl 1883), J. E. Lousley (1907–76), F. J. Lyne (fl 1879).

H. A. McAllister, R. McBeath, D. McClintock, A. N. Macdonald (fl 1928), Miss McDonald (fl 1890), J. Macfarlane (fl 1912), Prof. W. MacGillivray (1796–1852), Dr P. H. MacGillivray (1834–95), John MacGillivray, A. MacGregor (fl 1935), T. McGrouther (1858–1941), Miss McInnes (pre. 1859), J. C. MacIver, J. Mackay (fl 1794), Miss Ruth Mackay, R. Mackechnie, Dr Mackie (fl 1882), Mrs Gertrude Mackay, Miss V. Mackinnon, Mr Maclagan, Miss McLaughlan (fl 1837), Prof. K. N. G. MacLeay, Mr Macleod (fl 1970), Dr Gilbert McNab (fl 1839), James Macnab (1810–78), W. R. Macnab (1780–1848), Dr McTier (fl 1882), Anthony McTier (fl 1891), Rev. J. G. Macvicar (1801–84), Rev. E. S. Marshall (1858–1919), Miss Ann Marsland, Mr Martin (fl 1853), Chaworth Masters, W. Mathews (fl 1870), A. H. Maude (fl 1925), 'Medicus' (fl 1888), R. Meinertghagen, Dr A. Melderis, Dr R. Melville, Beverley A. Miles (1937–70), W. F. Miller (fl 1889), Prof. J. N. Mills, J. Milne, E. Milne-Redhead, H. Milne-Redhead (1906–1974), Miss M. E. Milward, Miss J. Mitchell, Mr Moir, K. Moore, Miss B. M. C. Morgan, Mr Morrison (fl 1849), Miss C. W. Muirhead, Mrs Anne Munro, W. G. Munro, Colin Murdoch, Alexander Murray (1798–1838), Mrs J. A. Murray, Miss M. Murray.

G. B. Neilson, Alfred Neumann, J. G. Newbould, George Nicholson (1847–1908), Dr P. Nicholson (fl 1835), Mr Nicholson (fl 1918), Dr V. Norris.

M. Orran, Mrs Osborne (fl 1846), Mrs Osgood (fl 1910), I. Ounstead, J. Ounstead.

F. Palmer (fl 1903), H. C. Palmer (fl 1903), R. E. Palmer, Dr Parsons (fl 1821), Dr Donald Patton (1884–1959), R. J. Pealling (fl 1923), E. Pelham-Clinton, Roy P. Petrie, M. T. T. Phillips, Prof. C. D. Pigott, Dr P. M. Playfair (fl 1898), Mr Potts (fl 1877), J. A. Power (1810–1886), Dr N. M. Pritchard, Dr M. C. F. Proctor, H. W. Pugsley (1868–1947).

Miss A. Radford, A. G. Rait, Dr D. A. Ratcliffe, J. E. Raven, Dr A. J. Richards, Dr R. Richter, Miss L. Riddell-Webster, B. Roberts, Mrs Jenny Roberts, R. H. Roberts, Mrs Ruth Rymer Roberts (fl 1930), Miss Robertson (fl 1835), Major-General I. A. Robertson, J. Robertson (fl 1847), W. Robertson (fl 1836), Mrs Robinson (fl 1835), Rev. John Roffey (1860–1927), J. G. Roger, F. A. Rogers (fl 1903), Rev. W. Moyle Rogers (1835–1920), I. M. Roper (fl 1920), Mr Ross (fl 1930), Dr E. M. Rosser, John Roy (1826–93), Mrs B. H. S. Russell, Rev. A. Rutherford (fl 1838), W. Rutherford (fl 1840).

C. E. Salmon (1872–1930), Mrs A. Stewart Sandeman (fl 1936), Noel Y. Sandwith (1901–65), Miss Sangster (fl 1890), Dr Sclanders (fl 1887), Alastair H. A. Scott, Peter D. Sell, Dr George Shepherd, J. Shier (fl 1824), W. A. Shoolbred (fl 1889), A. J. Silverside, Dr J. B. Simpson (fl 1940), Mrs J. B. Simpson (fl 1925), Norman D. Simpson, A. A. Slack, J. Stirling Smith (fl 1889), Mrs Janet Smith, Robert Smith, Robert Smith Jnr., Alexander Somerville (1842–1907), J. E. Souster, Alan J. Souter (1916–76), Alan Souter Jnr., Mrs Marion Souter, M. L. Sprague (fl 1903), T. A. Sprague (1903), William Alexander Stables (1810–90), Dr Clive A. Stace, Edward Stewart (fl 1856), E. J. A. Stewart (fl 1913), Miss F. Stewart, James Stewart, Mrs Olga M. Stewart, Miss P. Stewart, A. McG. Stirling, Miss Janet Stirling, J. M. Stirling, Dr James Straith (1765–1815), R. M. Suddaby, V. S. Summerhayes (1897–1974).

G. Taylor (fl 1839), Sir George Taylor, David Tennant, Miss A. Thomson (fl 1877), Miss Elsie M. Thomson, Dr Robert Thomson (fl 1880), Miss E. S. Todd (1859–1949), Dr J. Tolmie (fl 1843), C. C. Townsend, Frederick Townsend (1822–1905), Prof. J. W. H. Trail (1851–1919), Dawson Turner (1775–1858), Prof. T. G. Tutin, Dr N. Tyacke.

Miss E. Vachell (1879–1948).

E. M. Wakefield (fl 1938), E. C. Wallace, Dr S. M. Walters, Prof. J. Walton (fl 1938), Dr E. F. Warburg (1908–66), H. Ward (fl 1886), H. C. Watson (1804–1881), Charles Watt (fl 1913), W. M. Watt (fl 1930), Webb (fl 1921), C. Webster, Mrs M. L. Wedgewood (1854–1953), Hon. Douglas N. Wier, Rev. J. Weir (fl 1854), Mrs B. Welch, Dr Cyril West, G. T. West (fl 1904), W. West (fl 1880), J. A. Wheldon (fl 1909), Miss Beatrice E. White (fl 1900), Dr F. Buchanan White (1842–94), Miss Jean M. Whyte, Miss C. Wickham (fl 1947), Miss Wilkinson (fl 1915), Mrs Agnes Williams, A. J. Wilmott (1888–1950), Albert Wilson (1862–1949), Rev. G. Wilson (fl 1835), P. F. Wilson, W. Wilson (fl 1834), John Winham, Thomas Wise (1854–1932), R. S. Wishart (fl 1888), G. E. Woodroffe.

G. Yool, Dr D. P. Young (1917–72), Miss E. Young, E. A. Younger.

APPENDIX 2

One kilometre national grid references of less well-known places

Abriachan	28/55.35	Corrimony	28/37.30
Achavraat	28/91.48	Corrour	28/89.11
Achnagairn	28/55.45	Cothall	38/01.54
Aigas	28/45.41	Cothill	28/95.58
Aldroughty	38/18.62	Coulan	28/68.36
Auchness	38/11.49	Coulmony	28/97.47
		Cran Loch	28/94.59
Balcarse	28/52.45	Creag Leacainn	27/65.84
Ballachraggan	38/13.49	Creag nan Clag	28/60.28
Ballindarroch	28/61.39	Cruives	28/50.43
Ballochrochin	28/84.36	Cullochy Lock	28/34.04
Balnabreich	38/34.50		
Balnageith	38/02.59	Dalaschyle	28/82.49
Beinn a' Chrasgain	27/60.98	Daltulich	28/98.48
Beum a' Chlaidheimh	28/94.30	Delriach	38/08.31
Blackfold lochs	28/58.41	Divach	28/49.27
Blacksboat	38/18.38	Dochfour	28/60.34
Boat o' Brig	38/31.51	Drakemyres	38/39.54
Boblainy	28/49.39	Drynachan	28/86.39
Borlum	28/38.08	Drumguish	27/79.99
Buinach	38/18.55	Dunlichity	28/65.32
Carn Chuilinn	28/41.03	Dunmaglass	28/59.22
Castle Stuart	28/74.49	Dyke Moss	28/97.58
Ceannocroc Br.	28/22.10		
Clach Criche	28/22.06		
Clashdhu	38/03.52	Easter Limekilns	28/98.36
Cloddach	38/20.58	Easter Muckovie	28/70.44
Clunas	28/88.46	Eskadale	28/45.39
Coignafearn	28/67.15	Essich	28/64.39

Fae	38/09.17	L. Loyne	28/19.07
Farley	28/47.45	L. na Geadas	28/59.30
Farr	28/68.31	L. nan Lann	28/44.12
Flıchity	28/67.28	L. of the Clans	28/83.53
Fraoch-choire	28/20.28	L. Uanagan	28/37.07
		Lower Derraid	38/03.32
Gaick Lodge	27/75.84	Lurgmore	28/59.37
Garva Bridge	27/52.94	Lynemore	38/03.31
Geal-chàrn Beag	28/84.14		
Glassgreen	38/22.60	Mealfuarvonie	28/45.07
Glenerney	38/01.46	Meall Dubh	28/24.07
Glen Shirra Lodge	27/54.93	Mid Lairgs	28/71.36
Greshop Wood	38/02.59	Midmorile	28/79.26
Guisachan	28/28.25	Monchondie Moss	38/22.42
		Moss of Kinudie	28/90.55
Hill of Monaughty	38/13.59	Moss of Petty	28/77.51
Hilton Cottage	28/28.25	Moy Hall	28/76.35
Holme Rose	28/80.48	Moy House	38/01.59
Huntly's Cave	38/03.32		
		Nairnside House	28/73.42
Inchnacardoch	28/37.10	Newton Toll	38/16.63
Inverfarigaig	28/52.24		
		Ordbreck	28/87.49
Kellas House	38/16.53	Ord Hill	28/89.49
Kiltarlity	28/50.41		
Kilravock Castle	28/81.49	Pilmuir	38/02.58
Kinchurdy	28/93.15	Pitmain	28/74.02
Kingston	38/33.65	Pittendreich	38/19.61
Kinveachy	28/90.18	Pityoulish	28/92.14
Kirkton	28/60.45	Plodda Falls	28/27.23
		Polmaily	28/47.30
Lingeiston	38/03.60		
Loch a' Chlachain	28/65.32	Reelig Glen	28/55.42
L. an Ordain	28/55.24	Relugas	28/99.48
L. an t' Sithein	28/97.32	Righoul	28/88.51
L. Bunachton	28/66.34		
L. Ceo Glais	28/59.29	Sanquhar pond	38/04.58
L. Coiltry	28/35.06	Shenachie	28/82.34
L. Comhnard	28/36.27	Sluie	38/00.52
L. Conagleann	28/58.21		
L. Crunachdan	27/54.93	Tarff bridge	28/38.08
L. Doirb	28/53.24	Tom Bailgeann	28/58.29
L. Glanaidh	28/52.32	Tomich	28/30.27
Lochinvar	38/17.61	Torness	28/57.26
L. Kirkaldy	28/96.41		
L. Laide	28/54.35	Wester Elchies	38/25.43
L. Litie	28/94.55	Whitebridge	28/48.15

INDEX